Organometallic Compounds of the

Lanthanides, Actinides and Early Transition Metals

Organometallic Compounds of the Lanthanides, Actinides and Early Transition Metals

Edited by

D. J. Cardin
Trinity College, Dublin

S. A. Cotton
Stanground School, Peterborough

M. Green
University of Bristol

J. A. Labinger
Atlantic Richfield Company
Chatsworth, California

LONDON NEW YORK
CHAPMAN AND HALL

First published 1985 by Chapman and Hall Ltd
11 New Fetter Lane, London EC4P 4EE
733 Third Avenue, New York NY 10017

Phototypeset in the United States of America by
Mack Printing Company, Easton, Pennsylvania 18042
Printed in Great Britain by J.W. Arrowsmith Ltd, Bristol

ISBN 0 412 26830 2

Library of Congress Cataloging in Publication Data

Main entry under title:

Organometallic compounds of the lanthanides, actinides, and early transition metals.
(Chapman and Hall chemistry sourcebooks)
Includes index.
1. Organometallic compounds—Handbooks, manuals, etc.
2. Rare earth metal compounds—Handbooks, manuals, etc.
3. Actinide elements—Handbooks, manuals, etc.
4. Transition metal compounds—Handbooks, manuals, etc.
I. Cardin, D. J., 1941- . II. Series.
QD411.'73 1985 547'.054 84-29363

ISBN 0-412-26830-2

British Library Cataloguing in Publication Data

Organometallic compounds of the lanthanides, actinides and early transition metals.—(Chapman and Hall chemistry sourcebooks)
1. Organometallic compounds 2. Rare earth metal compounds 3. Actinide elements
I. Cardin, D.J.
547'.054 QD411

ISBN 0-412-26830-2

Contents

Preface

This is one of the first volumes to be published in the series of *Chapman and Hall Chemistry Sourcebooks* which provides carefully tailored information to workers in specialized areas of chemistry. The information contained in this book is derived from the *Dictionary of Organometallic Compounds*, published in November 1984.

The wide range of metals in this particular volume is of great interest to inorganic and organic chemists alike. Organic derivatives of several of the early transition elements are involved in commercial catalytic processes, particularly important examples being the use of titanium compounds in Ziegler-Natta polymerisation and molybdenum and tungsten intermediates in alkene metathesis. Reactions involving the organolanthanides and organoactinides have as yet been less well explored, but there is burgeoning interest in their synthesis. The large variety of compounds in this particular compendium will ensure its appeal to a wide readership.

The databank on the properties of organometallic compounds, which is represented in its current form by the *Dictionary of Organometallic Compounds* and its subset publications such as this volume, will be kept continuously up-to-date. Supplements to the main *Dictionary* will appear annually and revised editions of this *Sourcebook* will be published from time to time as demands permits.

D.J. Cardin
S.A. Cotton
M. Green
J.A. Labinger

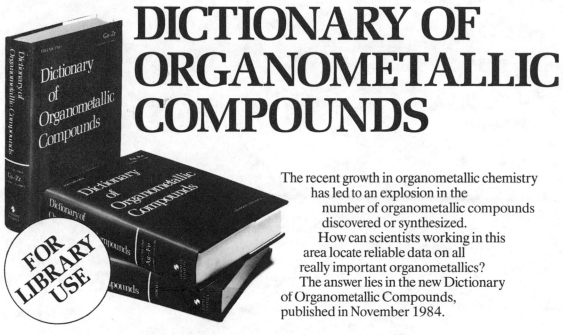

Introduction

1. Using the Sourcebook

The *Sourcebook* is divided into element sections: within each section the arrangement of entries is in order of molecular formula according to the Hill convention (i.e. C, then H, then other elements in alphabetical sequence of element symbol; where no carbon is present, the elements including H are ordered strictly alphabetically).

Every entry is numbered to assist ready location and the entry number consists of a metal element symbol followed by a five-digit number.

Indexes

There are three printed indexes: a name index which lists every compound name or synonym in alphabetical order; a molecular formula index which lists all molecular formulae, including those of derivatives, in Hill convention order; and a CAS registry number index listing all CAS numbers included in the *Sourcebook* in serial order. All indexes refer to the entry number. In the name index an entry number which follows immediately upon an index term means that the term itself is used as the entry name but an entry number which is preceded by the word 'see' means that the term is a synonym to an entry name. In all three indexes an entry number which is preceded by the word 'in' refers the reader to a specified stereoisomer or derivative which is to be found embedded within the particular entry.

In addition to the three printed indexes, each element section (except where the section is very short) is preceded by a graphical structure index allowing the rapid visual location of compounds of interest. The structure index reproduces all structure diagrams present in that element section in reduced size and printed in entry number order.

The following paragraphs summarize important considerations in compiling the information in this *Sourcebook*. For more detailed information, see the Introduction to the *Dictionary of Organometallic Compounds* from which this *Sourcebook* derives.

2. Compound Selection

In compiling this *Sourcebook* the aim has been to include from the primary literature up to mid 1983:

(1) Compounds representative of all important structural types (typically, the parent member of each series, where known, together with a selection of its homologues).

(2) Any compound with an established use, such as in catalysis, as a synthetic reagent or starting material.

(3) Other compounds of particular chemical, structural, biological or historical interest, especially those thought to exhibit unusual bonding characteristics.

Some compounds which are not considered sufficiently important to justify separate entries of their own have been included as derivatives in the entries of other compounds. These may include for example:

(1) Organic derivatives in the classical sense.

(2) Donor-acceptor complexes.

(3) The various salts of an anion or cation. In nearly every case, the entry for an ionic substance refers to the naked anion or cation, and the molecular formula, molecular weight and CAS registry number given for the main entry are those of the ion, in agreement with current CAS practice. Salts of the ion with various counterions are then treated as derivatives and the molecular formulae of all of these are given.

(4) Oligomeric compounds. Where a compound is known in several states of molecular aggregation these are all included in the one entry, which usually refers to the monomer. Compounds which are known only in dimeric form are entered as such, but the hypothetical monomers are included as derivatives to ensure that the names and molecular formulae of the monomeric forms occur in the indexes.

All names and molecular formulae recorded for derivatives occur in the Name and Molecular Formula Indexes respectively.

3. Chemical Names and Synonyms

The naming of organometallic compounds is frequently problematic and so in selecting the range of alternative names to present for each compound or derivative, editorial policy has been to report names which are found in the literature, including *Chemical Abstracts*, and not to attempt to impose a system of nomenclature. The editorial generation of new names has therefore been kept to a minimum required by consistency. Most names given in the *Sourcebook* are those given in the original paper(s) and in *Chemical Abstracts*.

Names corresponding to those used by CAS during

the 8th, 9th, and 10th collective index periods (1967-71, 1972-6 and 1977-81 respectively) are labelled with the suffixes 8CI, 9CI and 10CI respectively.

4. Toxicity and Hazard Information

Toxicity and hazard information is highlighted by the sign ▷ which also appears in the indexes.

All organometallic compounds should be treated as if they have dangerous properties.

The information contained in the *Sourcebook* has been compiled from sources believed to be reliable. No warranty, guarantee or representation is made by the Publisher as to the correctness or sufficiency of any information herein, and the Publisher assumes no responsibility in connection therewith.

The specific information in this publication on the hazardous and toxic properties of certain compounds is intended to alert the reader to possible dangers associated with the use of those compounds. The absence of such information should not, however, be taken as an indication of safety in use or misuse.

5. Bibliographic References

The selection of references is made with the aim of facilitating entry into the literature for the user who wishes to locate more detailed information about a particular compound. Reference contents are frequently indicated using mnemonic suffixes. In general recent references are preferred to older ones, and the number of references quoted does not necessarily indicate the relative importance of a compound.

Journal abbreviations generally follow the practice of *Chemical Abstracts Service Source Index* (CASSI). In patent references, no distinction is made between patent applications and granted patents.

6. Sources of Further Information

The following books and review series provide more information about various aspects of organometallic chemistry. Lists of reviews specific to organic compounds of particular metals may be found in the introductory sections of the metals concerned.

General

Comprehensive Organometallic Chemistry, Wilkinson, G. *et al.* Eds, Pergamon, Oxford, 1982. This book represents the most complete and up to date review of the whole subject. In addition to sections for each element there are chapters on the use of organometallics in organic synthesis and catalysis.

Comprehensive Inorganic Chemistry, Trotman-Dickenson, A.F. *et al.* Eds, Pergamon, Oxford, 1973. Contains information about organometallics as well as discussions of oxidation states, coordination chemistry and analysis of the metals.

Gmelins Handbuch der Anorganischen Chemie, 8th Edn, Springer-Verlag, Berlin. Some volumes of Gmelin covering organometallic compounds have been updated relatively recently and can therefore be consulted for comprehensive data on some types of organometallics. Some Gmelin element sections, however, are many years out of date.

Houben-Weyls Methoden der Organischen Chemie, 4th Edn, Band XIII, *Metallorganische Verbindungen*, Thieme-Verlag, Stuttgart.

The Chemistry of the Carbon-Metal Bond, Hartley, F.R. and Patai, S. Eds, Wiley, New York, 1982-. Contains sections on the synthesis, analysis and thermochemistry of various classes of organometallic compounds.

Transition-Metal Complexes of Phosphorus, Arsenic and Antimony Ligands, McAuliffe, C.A. Ed., Macmillan, London, 1973.

Methods of Elemento-Organic Chemistry, Kocheshkov, K.A. Ed., North Holland, Amsterdam, 1967.

MTP International Review of Science: Inorganic Chemistry, Series 2, Emeléus, H.J. Ed., Butterworths, London; University Park Press, Baltimore, 1974-5.

Advances in Organometallic Chemistry, Academic Press, 1964-.

Annual Surveys of Organometallic Chemistry, Elsevier, 1964-7.

Organometallic Chemistry Reviews, Elsevier, 1966-7.

Organometallic Chemistry Reviews, Section A: Subject Reviews 1968-72.

Organometallic Chemistry Reviews, Section B: Annual Surveys 1968-74.

Journal of Organometallic Chemistry: This incorporates reviews and surveys after the discontinuation of the two series of *Organometallic Chemistry Reviews*.

Organometallic Chemistry, 1972-, (Specialist Periodical Reports), RSC.

Coordination Chemistry Reviews, Elsevier, 1966-.

Progress in Inorganic Chemistry, Interscience, 1959-.

Advances in Inorganic Chemistry and Radiochemistry, Academic Press, 1959-.

Analysis

Scott's Standard Methods of Chemical Analysis, Furman, N.H. Ed., 6th Edn, Van Nostrand, New York, 1962.

Crompton, T.R., *Chemical Analysis of Organometallic Compounds*, Academic Press, London, 1973.

Spectroscopy

Nuclear Magnetic Resonance Spectroscopy of Nuclei Other than the Proton, Axenrod, T. and Webb, G.A. Eds, Wiley, London, 1974.

NMR and the Periodic Table, Harris, R.K. and Mann, B.E. Eds, Academic Press, London, 1978.

^{13}C NMR Data for Organometallic Compounds, Mann, B.E. and Taylor, B.F. Eds, Academic Press, London, 1981.

Spectroscopic Properties of Inorganic and Organometallic Compounds, 1968-, (Specialist Periodical Reports), RSC.

Handling

Shriver, D.F., *The Manipulation of Air-Sensitive Compounds*, McGraw-Hill, 1969.

Organometallic Syntheses, Academic Press, New York, 1965, Vol. 1.

Am Americium

S. A. Cotton

Américium (Fr.), Americium (Ger.), Americio (Sp., Ital.), Америций (Amieritsii) (Russ.), アメリシウム (Japan.)

Atomic Number. 95

Atomic Weight. 243 (stablest isotope)

Electronic Configuration. [Rn] $5f^7\,7s^2$

Oxidation State. +3

Coordination Number. See under Uranium

Colour. Reported Am(III) compounds are flesh-coloured.

Availability. Americium compounds are not generally available. The most suitable material for synthesis is $AmCl_3$. The two most common isotopes are ^{241}Am ($t_{1/2}$ = 458y) and the longer lived ^{243}Am ($t_{1/2} = 8.8 \times 10^3$y).

Handling. Extreme care is necessary in the handling of americium due to its acute toxicity.

Toxicity. Americium is an extremely toxic α- and γ-emitter.

Spectroscopy. Optical spectra have been used to study americium compounds.

Analysis. X-ray techniques are used to demonstrate that compounds are isomorphous with analogues of the stabler actinides.

References. See under Uranium

$C_{15}H_{15}Am$ **Am-00001**

Tris(η^5-2,4-cyclopentadien-1-yl)americium

Tricyclopentadienylamericium

[1295-18-7]

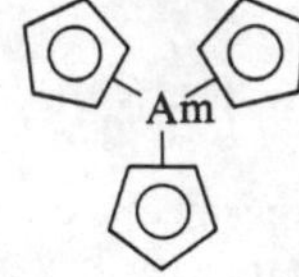

M 438.284

Rose solid. Sol. C_6H_6, THF. Mp >330°. Subl. at 160-200° *in vacuo*. Glows in the dark. Slowly dec. in air.

Baumgartner, F. *et al*, *Angew. Chem., Int. Ed. Engl.*, 1966, **5**, 134 (*synth*)

Kanellakopulos, B. *et al*, *Radiochim. Acta*, 1978, **25**, 89 (*synth*, *ir*, *uv*)

Bk Berkelium

S. A. Cotton

Berkélium (Fr.), Berkelium (Ger.), Berquelio or Berkelio (Sp.), Berchelio or Berkelio (Ital.), Берклий (Bierklii) (Russ.), バークリウム (Japan.)

Atomic Number. 97

Atomic Weight. 247 (stablest isotope, $t_{1/2} = 1.4 \times 10^3$y)

Electronic Configuration. [Rn] $5f^7\,7s^2$

Oxidation State. +3

Coordination Number. See under Uranium.

Colour. Amber.

Availability. Berkelium is not readily available. Anhydrous $BkCl_3$ formed by treatment of the oxide with HCl gas has been used as a starting material. See Laubereau, P. G. *et al.*, *Inorg. Chem.*, 1970, **9**, 109.

Handling. See under Plutonium.

Toxicity. See under Plutonium.

Analysis. See under Americium.

References. See under Uranium.

$C_{15}H_{15}Bk$ **Bk-00001**
Tris(η^5-2,4-cyclopentadien-1-yl)berkelium
Tricyclopentadienylberkelium

Bk

M 442.284
Amber solid. Subl. at 135-65° *in vacuo.*
Laubereau, P.G. *et al*, *Inorg. Chem.*, 1970, **9**, 1091 (*synth, uv*)

Ce Cerium

S. A. Cotton

Cérium (Fr.), Cer (Ger.), Cerio (Sp., Ital.), Церий (Tserii) (Russ.), セリウム (Japan.)

Atomic Number. 58

Atomic Weight. 140.12

Electronic Configuration. [Xe] $4f^1\ 5d^1\ 6s^2$

Oxidation States. +3 (+4 for certain incompletely characterised compounds).

Coordination Number. Usually 6 or greater.

Colour. The few reported complexes are generally yellow or green.

Availability. Common starting materials are CeO_2 and the more expensive anhydrous $CeCl_3$.

Handling. See under Lanthanum.

Toxicity. See under Lanthanum.

Isotopic Abundance. ^{138}Ce, 0.25%; ^{140}Ce, 88.48%; ^{142}Ce, 11.1%.

Spectroscopy. The Ce^{3+} ion has electronic configuration f^1 and its compounds would be expected to give ^1H nmr spectra with relatively small paramagnetic shifts, but no data have yet been reported for organometallics.

Analysis. See under Lanthanum.

References. See under Lanthanum.

C_8H_8CeCl **Ce-00001**

Chloro(cyclooctatetraene)cerium

Di-μ-chlorobis(η^8-1,3,5,7-cyclooctatetraene)dicerium, 9CI

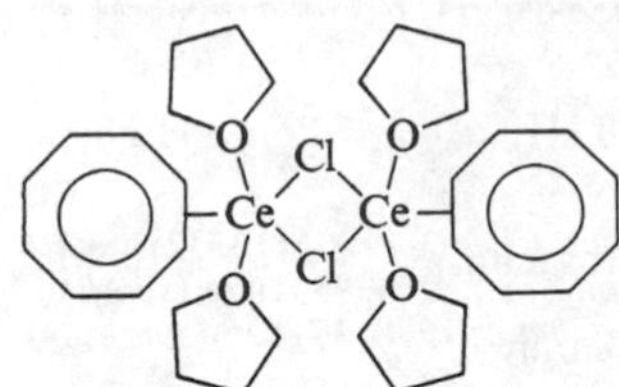

Bis-THF adduct

M 279.724

Bis-THF adduct is dimeric.

Bis-THF adduct: [12701-52-9].

$C_{32}H_{48}Ce_2Cl_2O_4$ M 847.875

Yellow-green cryst. Sol. dioxan, THF. Mp >345°. $\mu = 1.79\mu_B$. Extremely air- and moisture-sensitive.

Hodgson, K.O. *et al*, *Inorg. Chem.*, 1972, **11**, 171 (*struct*)
Hodgson, K.O. *et al*, *J. Am. Chem. Soc.*, 1973, **95**, 8650 (*synth, ir, uv, props*)

$C_{13}H_{14}CeCl$ **Ce-00002**

Chloro[1,3-propanediylbis(η^5-2,4-cyclopentadien-1-ylidene)]-cerium

[75861-70-0]

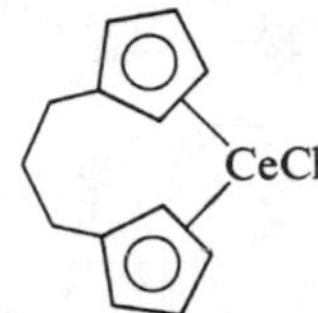

M 345.827

Probably dimeric. Sol. THF. Mp >100°. Air- and moisture-sensitive. $\mu_{eff} = 2.00\mu_B$ (306K).

John, J.N. *et al*, *J. Coord. Chem.*, 1980, **10**, 177 (*synth*)

$C_{15}H_{15}Ce$ **Ce-00003**

Tris(η^5-2,4-cyclopentadien-1-yl)cerium, 9CI

Tri-π-cyclopentadienylcerium, 8CI

[1298-53-9]

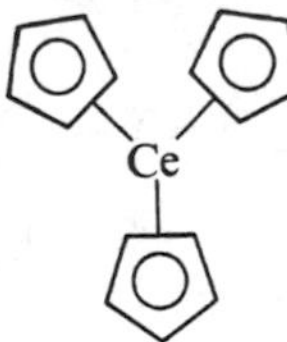

M 335.404

Orange cryst. Mp 435°. Subl. at 200-50° *in vacuo*, hydrol. by H_2O.

Cyclohexylisocyanide adduct: [37298-97-8]. Sol. C_6H_6. Mp 134°.

Wilkinson, G. *et al*, *J. Am. Chem. Soc.*, 1954, **76**, 6210 (*synth*)
Reid, A.F. *et al*, *Inorg. Chem.*, 1966, **5**, 1213 (*synth*)
Von Ammon, R. *et al*, *Ber. Bunsenges. Phys. Chem.*, 1972, **76**, 995 (*nmr, ir*)
Devyatykh, G.G. *et al*, *Dokl. Akad. Nauk. SSSR*, 1973, **208**, 1094 (*ms*)

$C_{15}H_{15}CeCl$ **Ce-00004**

Chlorotris(η^5-2,4-cyclopentadien-1-yl)cerium

Tricyclopentadienylcerium chloride

[12636-79-2]

M 370.857

Dark-brown cryst. Insol. aromatic hydrocarbons, sol. most other org. solvs., e.g. pet. ether, THF, $CHCl_3$, MeOH. Mp 176° dec.

Kalsotra, B.L. *et al*, *Isr. J. Chem.*, 1971, **9**, 569 (*synth, ir*)

$C_{16}H_{16}Ce^{\ominus}$ **Ce-00005**

Bis(η^8-1,3-5,7-cyclooctatetraene)cerate(1−), 9CI

Dicyclooctatetraenecerate(1−)

M 348.422 (ion)

K salt: [51187-43-0].

$C_{16}H_{16}CeK$ M 387.521

Pale-green solid. Mp >345°. Extremely air- and moisture-sensitive. $\mu = 1.88\mu_B$.

K salt, diglyme adduct: [37328-32-8].

$C_{22}H_{30}CeKO_3$ M 521.696

Bright-green plates. Air- and moisture-sensitive.

Mares, F. *et al*, *J. Organomet. Chem.*, 1970, **24**, C68 (*synth*)
Hodgson, K.O. *et al*, *Inorg. Chem.*, 1972, **11**, 3030 (*struct*)
Hodgson, K.O. *et al*, *J. Am. Chem. Soc.*, 1973, **95**, 8650 (*synth, ir, raman*)
Clack, D.W. *et al*, *J. Organomet. Chem.*, 1976, **122**, C28.

$C_{18}H_{14}CeCl_2$ **Ce-00006**

Dichlorobis[(1,2,3,3*a*,7*a*-η)-1*H*-inden-1-yl]cerium, 9CI

[36354-35-5]

$(C_9H_7)_2CeCl_2$

M 441.335

Red-brown cryst. Sol. THF, Me_2CO, C_6H_6, spar. sol. toluene, xylene. Mp 129-30° dec. Unaffected by water.

Kalsotra, B.L. *et al*, *Isr. J. Chem.*, 1971, **9**, 569 (*synth, ir*)

$C_{20}H_{20}Ce_2Cl_2$ **Ce-00007**

Dichlorotetrakis(η^5-2,4-cyclopentadien-1-yl)dicerium

Chlorobis(η^5-2,4-cyclopentadien-1-yl)cerium dimer

$[(C_5H_5)_2CeCl]_2$

M 611.524

$C_{21}H_{20}Ce$ **Ce-00008**

Tris(η^5-2,4-cyclopentadien-1-yl)phenylcerium, 9CI

Tricyclopentadienylphenylcerium

[39330-90-0]

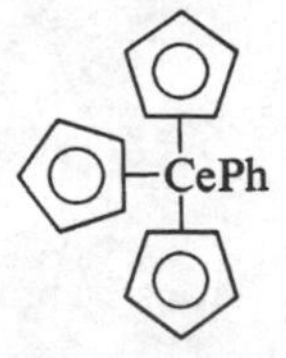

M 412.509

Red-brown cryst. Sol. pet. ether. Dec. by air.

Kalsotra, B.L. *et al*, *J. Inorg. Nucl. Chem.*, 1973, **35**, 311 (*synth*, *ir*)

$C_{24}H_{24}Ce_2$ **Ce-00009**

Tris(1,3,5,7-cyclooctatetraene)dicerium, 9CI

[60605-80-3]

$Ce_2(C_8H_8)_3$

M 592.694

Bright-green solid. Sol. THF. $\mu_{eff} = 1.86\mu_B$. Highly oxygen-sensitive. Stable to 250°.

Greco, A. *et al*, *J. Organomet. Chem.*, 1976, **113**, 321 (*synth*)
Ely, S.R. *et al*, *J. Am. Chem. Soc.*, 1976, **98**, 1624 (*synth*)
de Kock, C.W. *et al*, *Inorg. Chem.*, 1978, **17**, 625 (*synth*, *ir*)

Cf Californium

S. A. Cotton

Californium (Fr., Ger.), Californio (Sp., Ital.), Калифорний (Kalifornii) (Russ.), カリホルニウム or カリフォルニウム (Japan.)

Atomic Number. 98

Atomic Weight. 251 (stablest isotope, $t_{1/2} \sim 800y$)

Electronic Configuration. [Rn] $5f^{10}\ 7s^2$

Oxidation State. +3

Coordination Number. See under Uranium.

Colour. Red.

Availability. ^{249}Cf ($t_{1/2} = 360y$) is preferred over ^{252}Cf ($t_{1/2} = 2.65y$) although the latter is produced in larger quantities.

Handling. See under Uranium and Plutonium.

Toxicity. See under Plutonium. ^{252}Cf is a particularly strong neutron emitter.

Analysis. See under Americium.

References. See under Uranium.

$C_{15}H_{15}Cf$ **Cf-00001**

Tris(η^5-2,4-cyclopentadien-1-yl)californium

Tricyclopentadienylcalifornium

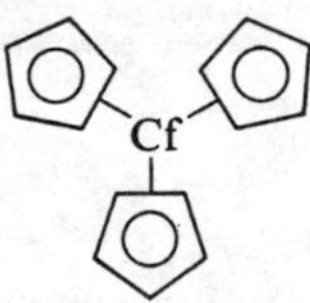

M 446.284

Ruby-red solid. Subl. at 135-320° *in vacuo*.

Laubereau, P.G. *et al*, *Inorg. Chem.*, 1970, **9**, 1091 (*synth*)

Cm Curium

S. A. Cotton

Curium (Fr., Ger.), Curio (Sp., Ital.), Кюрий (Kiurii) (Russ.), キュリウム (Japan.)

Atomic Number. 96

Atomic Weight. 247 (stablest isotope, $t_{1/2} = 1.6 \times 10^7$y)

Electronic Configuration. [Rn] $5f^7\ 6d^1\ 7s^2$

Oxidation State. +3

Coordination Number. See under Uranium.

Colour. Colourless.

Availability. ^{244}Cm ($t_{1/2} = 17.6$y), formed by neutron bombardment of plutonium, is the most easily obtained isotope, although like the other transuranium elements it is not generally available. $CmCl_3$ is probably the best starting material.

Handling. See under Uranium and Plutonium.

Toxicity. Like plutonium, curium is extremely toxic, and accumulates in the body.

Analysis. See under Americium.

References. See under Uranium.

$C_{15}H_{15}Cm$ **Cm-00001**

Tris(η^5-2,4-cyclopentadien-1-yl)curium

Tricyclopentadienylcurium

[11077-58-0]

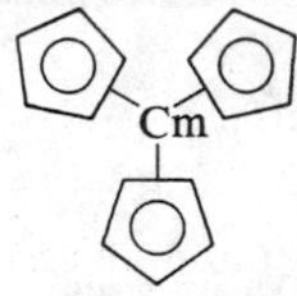

M 442.284

Colourless solid. Mp >300°. Subl. at 180° *in vacuo*. Fluoresces red in uv light.

Baumgartner, F. *et al*, *J. Organomet. Chem.*, 1970, **22**, C17 (*synth*, *ms*)

Laubereau, P.G. *et al*, *Inorg. Nucl. Chem. Lett.*, 1970, **6**, 59 (*synth*)

Cr Chromium

M. Green

Chrome (Fr.), Chrom (Ger.), Cromo (Sp., Ital.), Хром (Khrom) (Russ.), クロム (Japan.)

Atomic Number. 24

Atomic Weight. 51.996

Electronic Configuration. [Ar] $3d^5\,4s^1$

Oxidation States. −4, −2, −1, 0, +1, +2, +3, +4.

Coordination Number. The most common coordination numbers with conventional ligands are 4, 5 and 6. Tetrahedral geometry is particularly common in oxidation states −1, 0, +1.

Colour. Variable including colourless, yellow, red, green and blue.

Availability. A range of starting materials are available including $Cr(CO)_6$, anhydrous $CrCl_3$, $CrCl_3{\cdot}3THF$ and $Cr(C_6H_6)_2$. The latter is moderately expensive. $Cr(CO)_6$ is the starting material for almost all chromium-carbonyl chemistry and for many other Cr(0) species.

Handling. The group 6 and 7 carbonyls $[M(CO)_6]$ (M = Cr, Mo, W) and $[M_2(CO)_{10}]$ (M = Mn, Tc, Re) are air-stable materials and can be handled without special precautions. However, the simple anions $[M(CO)_5]^{2-}$, $[M(CO)_5]^-$, which are synthetically useful species are moisture-sensitive, and should normally be handled under a dry oxygen-free nitrogen atmosphere. The recently prepared group 6 anions $[M(CO)_4]^{4-}$ are extremely pyrophoric. In contrast the halo-substituted anions $[XM(CO)_5]^-$ (M = Cr, Mo, W) and the related group 7 species $[XM(CO)_5]$ (M = Mn, Tc, Re) are air- and moisture-stable materials. Other important starting materials vary in stability, for example, simple η^5-cyclopentadienyl complexes such as $[MoCl(CO)_3(\eta^5\text{-}C_5H_5)]$ and $[Mn(CO)_3(\eta^5\text{-}C_5H_5)]$ can be handled without special precautions, whereas potentially important complexes such as $[Mo(\eta^6\text{-}C_6H_6)_2]$, $[Cr(\eta^5\text{-}C_5H_5)_2]$ and $[Cr(\eta^3\text{-}C_3H_5)_3]$ are air-sensitive and must be handled with care. Simple molecules such as WMe_6 containing carbon to metal σ-bonds are sensitive, whereas, $[MeMn(CO)_5]$ or the rhenium analogue can be handled in air. Molecules containing metal to carbon multiple bonds such as carbene and carbyne complexes vary in stability and it is not possible to generalise. In all cases the original papers should be consulted, and in the case of air-sensitive materials suitable techniques such as those described in Shriver's book adopted.

Toxicity. Chromium compounds are not regarded as highly toxic, but the carbonyl complexes should, as for other metals, be regarded as such.

Isotopic Abundance. ^{50}Cr, 4.31%; ^{52}Cr, 83.76%; ^{53}Cr, 9.54%; ^{54}Cr, 2.38%. The radioisotope commonly used in tracer work is ^{51}Cr ($t_{1/2} \sim 27.7$d).

Spectroscopy. ^{53}Cr has $I = \frac{3}{2}$ and a low nmr detection sensitivity and there have been few reports of the use of this nucleus as an nmr probe.

Analysis. For standard methods of analysis, see *Scott's Standard Methods of Chemical Analysis*, Furman, N. H. Ed., 6th Edn, 1962.

References. General reviews are listed in the introduction to the *Sourcebook*.

Cr-00001

$[Cr(CO)_4]^{\ominus\ominus\ominus\ominus}$

Cr-00002

R = H

Cr-00003

$(Me_2Tl)_2CrO_4$

Cr-00004

$(OC)_5CrBr^{\ominus}$ (O_h)

Cr-00005

$[(OC)_5CrCl]^{\ominus}$ (O_h)

Cr-00006

$[(OC)_5CrI]^{\ominus}$ (O_h)

Cr-00007

$Cr(CO)_5^{\ominus\ominus}$

Cr-00008

$[Cr(CO)_5OH]^{\ominus}$

Cr-00009

$(OC)_5CrAsH_3$ (O_h)

Cr-00010

$(OC)_5CrNH_3$ (O_h)

Cr-00011

$Cr(CO)_5PH_3$

Cr-00012

$(OC)_5CrSbH_3$ (O_h)

Cr-00013

$(OC)_5CrNH_2NH_2$ (O_h)

Cr-00014

Cr-00015

$[(OC)_5CrCN]^{\ominus}$ (O_h)

Cr-00016

Cr-00017

$(OC)_5CrCS$ (O_h)

Cr-00018

$(OC)_5CrCSe$ (O_h)

Cr-00019

$Cr(CO)_6(O_h)$

Cr-00020

$(OC)_5CrCNBH_3^{\ominus}$ (O_h)

Cr-00021

X = Cl

Cr-00022

As Cr-00022 with
X = I

Cr-00023

$[MeCr(CO)_5]^{\ominus}$

Cr-00024

Cr-00025

Cr-00026

Li salt, dioxane complex

Cr-00027

$(OC)_5CrNCSMe$

Cr-00028

$[(OC)_5CrCOCH_3]^{\ominus}$ (O_h)

Cr-00029

Cr-00030

As Cr-00032 with
X = NS

Cr-00031

X = NO

Cr-00032

$(OC)_5Cr{=}C(NH_2)CH_3$

Cr-00033

$(OC)_5CrSMe_2$ (O_h)

Cr-00034

Cr-00035

$[Cr(CH_2Ph)(H_2O)_5]^{\oplus\oplus}$

Cr-00036

Cr-00037

Cr-00038

$[(OC)_5Cr{\equiv}CNMe_2]^{\oplus}$

Cr-00039

Cr-00040

Cr-00041

$(OC)_5Cr{=}C(OMe)CH_3$

Cr-00042

Cr-00043

$(OC)_5Cr{=}CHNMe_2$

Cr-00044

$(OC)_5Cr{=}C(CH_3)NHMe$

Cr-00045

Cr-00046

$Cr(CO)_5PMe_3$

Cr-00047

$(OC)_5CrS{=}PMe_3$ (O_h)

Cr-00048

Cr-00049

$(Et_2Tl)_2CrO_4$

Cr-00050

Cr-00051

Cr-00052

Cr-00053

Cr-00054

$(OC)_5Cr{=}C(CN)NMe_2$

Cr-00055

Cr-00056

Cr-00057

Cr-00058

Cr-00059

Cr-00060

Cr-00061

X = Br

Cr-00062

As Cr-00062 with
X = Cl

Cr-00063

Cr-00064

$[(OC)_5Cr{-}I{-}Cr(CO)_5]^{\ominus}$

Cr-00065

Cr-00066

Cr-00067

$(OC)_5Cr—H_2P—PH_2—Cr(CO)_5$

Cr-00068

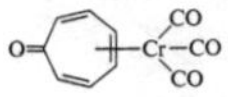

Cr-00069

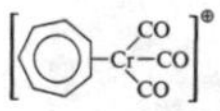

Cr-00070

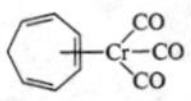

Cr-00071

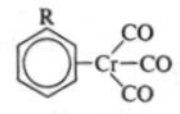

R = CH_3

Cr-00072

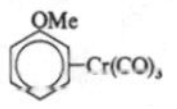

Cr-00073

$(OC)_5Cr{=}CClNEt_2$

Cr-00074

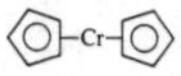

Cr-00075

$(OC)_5Cr(NCNEt_2)$

Cr-00076

$(OC)_5Cr{=}C(SEt)_2$

Cr-00077

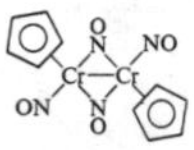

Cr-00078

$(OC)_5Cr{=}C(NMe_2)_2$

Cr-00079

$(OC)_5Cr{=}C(CH_3)OSiMe_3$

Cr-00080

As Cr-00003 with
R = Me

Cr-00081

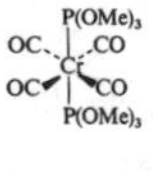

Cr-00082

$[(OC)_5Cr—SMe—Cr(CO)_5]^{\ominus}$

Cr-00083

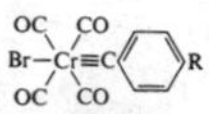

R = H

Cr-00084

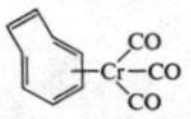

Cr-00085

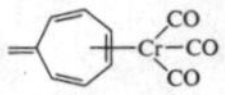

Cr-00086

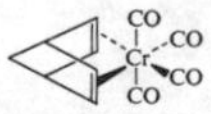

Cr-00087

As Cr-00072 with
R = $COCH_3$

Cr-00088

As Cr-00072 with
R = COOMe

Cr-00089

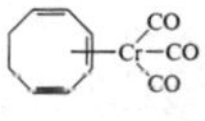

Cr-00090

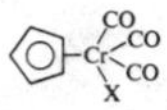

X = $-GeMe_3$

Cr-00091

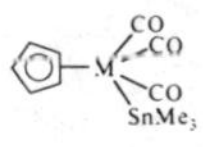

M = Cr

Cr-00092

$(OC)_5CrTe(SnMe_3)_2$ (O_h)

Cr-00093

$[(OC)_5CrCOPh]^{\ominus}$ (O_h)

Cr-00094

As Cr-00084 with
R = CH_3

Cr-00095

$(OC)_5Cr{=}CPhNH_2$

Cr-00096

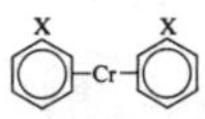

X = Cl

Cr-00097

As Cr-00097 with
X = F

Cr-00098

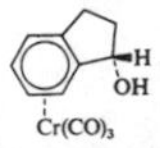

1(*S*)-(–)-*form*

Cr-00099

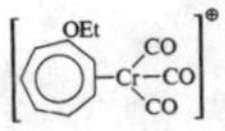

Cr-00100

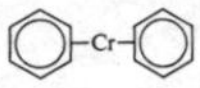

Cr-00101

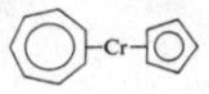

Cr-00102

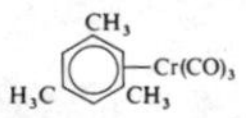

Cr-00103

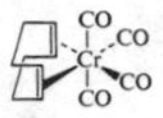

Cr-00104

As Cr-00072 with
R = $SiMe_3$

Cr-00105

As Cr-00072 with
R = $SnMe_3$

Cr-00106

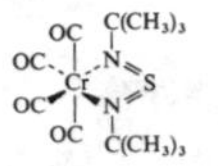

Cr-00107

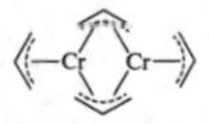

Cr-00108

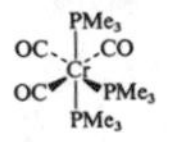

Cr-00109

$Cr[CH(CH_3)_2]_4$ (T_d)

Cr-00110

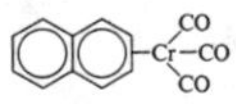

Cr-00111

$(OC)_5Cr{=}CPhOMe$

Cr-00112

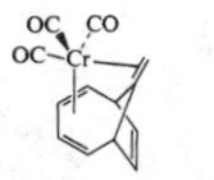

Cr-00113

$(OC)_5CrPMe_2PMe_2Fe(CO)_4$

Cr-00114

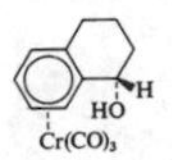

(*S*)-(+)-*form*

Cr-00115

As Cr-00072 with
R = CH_2SiMe_3

Cr-00116

As Cr-00072 with
R = CH_2SnMe_3

Cr-00117

$[MeCrCl_2(THF)_3]$

Cr-00118

$(OC)_5Cr{=}C(CH_3)N{=}CHPh$

Cr-00119

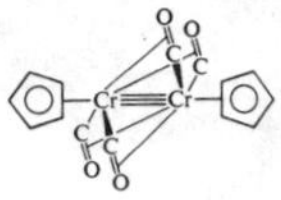

Cr-00120

As Cr-00097 with
X = CH_3

Cr-00121

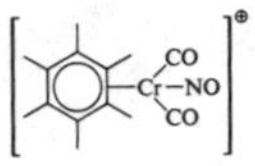

Cr-00122

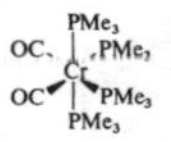

Cr-00123

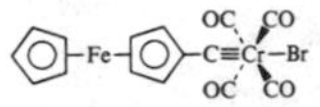

Cr-00124

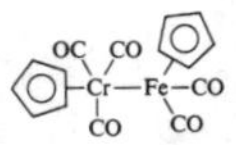

Cr-00125

$(OC)_5Cr{=}C(OMe)CH{=}CHPh$

Cr-00126

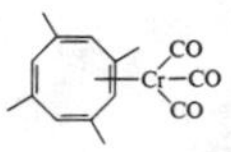

Cr-00127

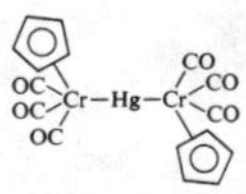

Cr-00128

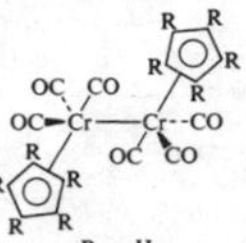

R = H

Cr-00129

$(OC)_5Cr{=}C{=}C{=}CPhNMe_2$

Cr-00130

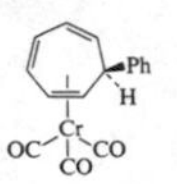

exo-form

Cr-00131

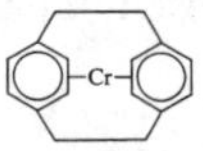

Cr-00132

$[MeCrCl_2(Py)_3]$

Cr-00133

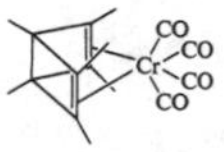

Cr-00134

As Cr-00097 with
R = CH_2CH_3

Cr-00135

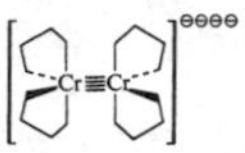

Cr-00136

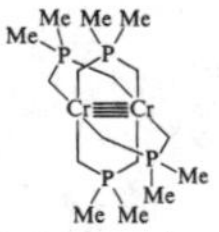

Cr-00137

$[Cr(CH_2SiMe_3)_4]$ (T_d)

Cr-00138

$Cr(CO)[P(OMe)_3]_5$

Cr-00139

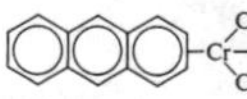

Cr-00140

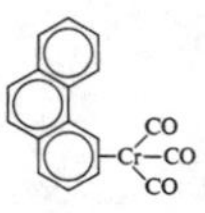

Cr-00141

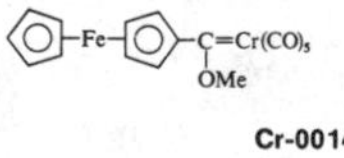

Cr-00142

$(OC)_5Cr{=}CPh_2$

Cr-00143

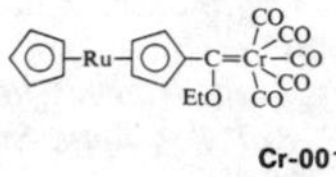

Cr-00144

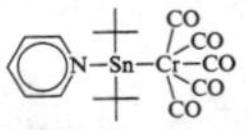

Cr-00145

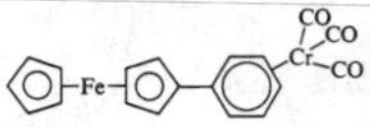

Cr-00146

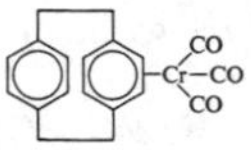

Cr-00147

$[(Me_3Si)_2CH]_2SnCr(CO)_5$

Cr-00148

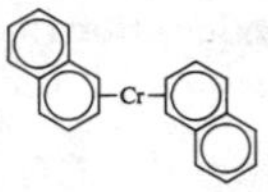

Cr-00149

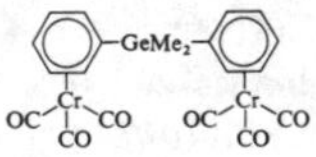

Cr-00150

$Cr[CH_2C(CH_3)_3]_4$

Cr-00151

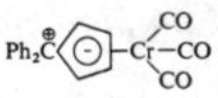

Cr-00152

$(OC)_5Cr{=}C(OEt)N{=}CPh_2$

Cr-00153

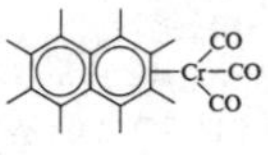

Cr-00154

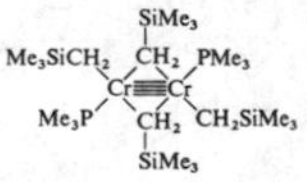

Cr-00155

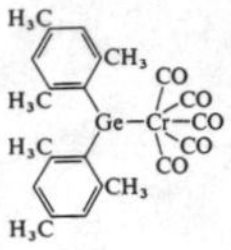

Cr-00156

$(Ph_2Tl)_2CrO_4$

Cr-00157

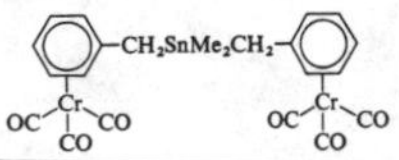

Cr-00158

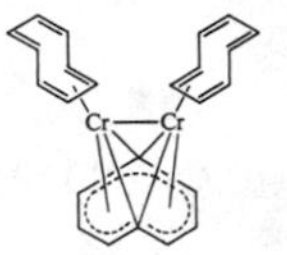

Cr-00159

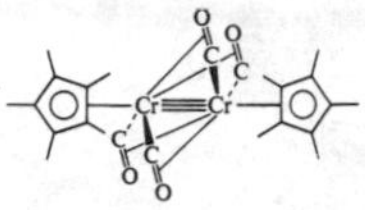

Cr-00160

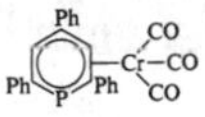

Cr-00161

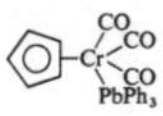

Cr-00162

$(OC)_5Cr{=}C(NMe_2)SiPh_3$

Cr-00163

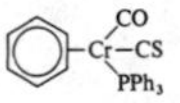

Cr-00164

As Cr-00129 with
R = CH_3

Cr-00165

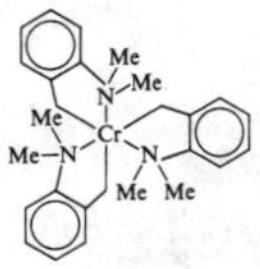

Cr-00166

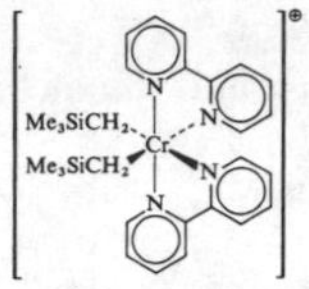

Cr-00167

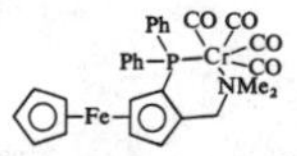

Cr-00168

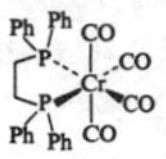

Cr-00169

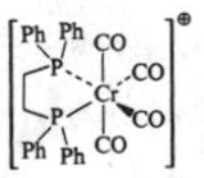

Cr-00170

$[CrPh_3(THF)_3]$

Cr-00171

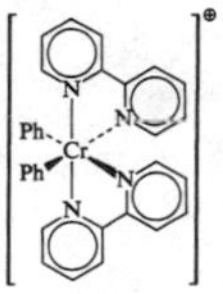

Cr-00172

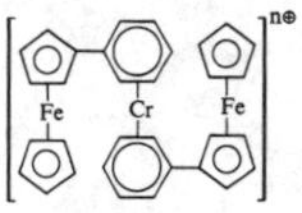

n = 0

Cr-00173

As Cr-00173 with
n = 1

Cr-00174

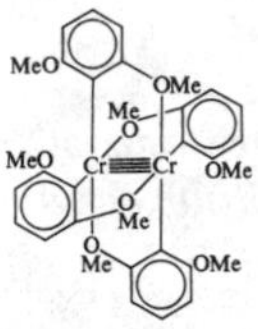

Cr-00175

$[Cr(CO)_2(PPh_3)_2(NO)I]$

Cr-00177

$Cr[CH_2CPh(CH_3)_2]_4$ (T_d)

Cr-00178

$C_3H_9CrO_3P_3$ Cr-00001

Tricarbonyltris(phosphine)chromium, 9CI

[25885-03-4]

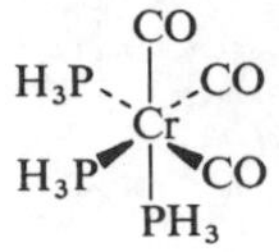

M 238.020

Mp 130° dec.

fac-form

Yellow cryst.

Fischer, E.O. *et al*, *J. Organomet. Chem.*, 1969, **18**, P26 (*synth*)
Guggenberger, L.S. *et al*, *Inorg. Chem.*, 1973, **12**, 1143.
Huttner, G. *et al*, *J. Organomet. Chem.*, 1973, **47**, 383 (*cryst struct*)

$C_4CrO_4^{\ominus\ominus\ominus\ominus}$ Cr-00002

Tetracarbonylchromate(4–)

$[Cr(CO)_4]^{\ominus\ominus\ominus\ominus}$

M 164.038 (ion)

Tetra-Na salt: [67202-62-4].
$C_4CrNa_4O_4$ M 255.997

▷Pyrophoric

Ellis, J.E. *et al*, *J. Am. Chem. Soc.*, 1983, **105**, 2296 (*synth*)

$C_4H_6CrO_4P_2$ Cr-00003

Tetracarbonylbis(phosphine)chromium, 9CI

[25885-01-2]

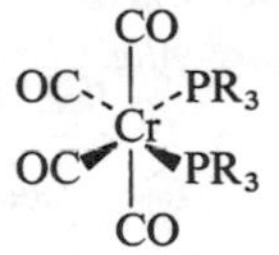

R = H

M 232.033

cis-form

Pale-yellow cryst.

Fischer, E.O. *et al*, *Chem. Ber.*, 1969, **102**, 2547 (*synth*)
Fischer, E.O. *et al*, *J. Organomet. Chem.*, 1969, **18**, P26.
Guggenberger, L.S. *et al*, *Inorg. Chem.*, 1973, **12**, 1143 (*cryst struct*)

$C_4H_{12}CrO_4Tl_2$ Cr-00004

Tetramethyl[μ-[chromato(2–)-*O*:*O'*]]dithallium

Dimethylthallium chromate

$(Me_2Tl)_2CrO_4$

M 584.898

Yellow cryst. (Me_2CO aq.). Sol. H_2O, EtOH. Mp 255°.

▷Toxic

Goddard, A.E., *J. Chem. Soc.*, 1921, 672 (*synth*)

$C_5BrCrO_5^{\ominus}$ Cr-00005

Bromopentacarbonylchromate(1–), 9CI

[14911-72-9]

$(OC)_5CrBr^{\ominus}$ (O_h)

M 271.952 (ion)

Tetramethylammonium salt: [14780-93-9].
$C_9H_{12}BrCrNO_5$ M 346.098
Orange cryst.

Abel, E.W. *et al*, *J. Chem. Soc.*, 1963, 2068 (*synth*)
Knoll, L., *J. Organomet. Chem.*, 1979, **182**, 77 (*props*)
Chatt, J. *et al*, *J. Chem. Soc., Dalton Trans.*, 1980, 2032.
Cotton, F.A. *et al*, *Inorg. Chem.*, 1981, **20**, 578 (*props*)
Nixon, J.F. *et al*, *J. Chem. Soc., Chem. Commun.*, 1981, 199 (*props*)

$C_5CrClO_5^{\ominus}$ Cr-00006

Pentacarbonylchlorochromate(1–), 9CI

[14911-56-9]

$(OC)_5CrCl^{\ominus}$ (O_h)

M 227.501 (ion)

Tetraethylammonium salt: [14780-95-1].
$C_{13}H_{20}CrClNO_5$ M 357.754
Yellow cryst. Sol. $CHCl_3$, insol. Et_2O.

Bis(triphenylphosphine)iminium salt: [65650-76-2].
$C_{41}H_{30}ClCrNO_5P_2$ M 766.088
Yellow cryst. ir ν_{CO} 2059, 1926, 1854 cm^{-1}.

Abel, E.W. *et al*, *J. Chem. Soc.*, 1963, 2068 (*synth*)
Grillone, M.D., *Transition Met. Chem.*, 1981, **6**, 93 (*synth*)
Cotton, F.A. *et al*, *J. Am. Chem. Soc.*, 1981, **103**, 398 (*props*)

$C_5CrIO_5^{\ominus}$ Cr-00007

Pentacarbonyliodochromate(1–), 10CI

[14780-98-4]

$(OC)_5CrI^{\ominus}$ (O_h)

M 318.953 (ion)

Tetraethylammonium salt:
$C_{13}H_{20}CrINO_5$ M 449.205
Orange cryst.

Methyltriphenylphosphonium salt: [76309-27-8].
$C_{24}H_{18}CrIO_5P$ M 596.277
Orange cryst.

Abel, E.W. *et al*, *J. Chem. Soc.*, 1963, 2068 (*synth*)
Organomet. Synth., 1965, **1**, 122 (*synth*)
Angelici, R.J. *et al*, *Inorg. Chem.*, 1980, **19**, 3853 (*props*)
Grim, S. *et al*, *Inorg. Chem.*, 1980, **19**, 2475 (*props*)
Magomedov, G.K. *et al*, *Zh. Obshch. Khim.*, 1980, **50**, 2623 (*props*)

$C_5CrO_5^{\ominus\ominus}$ Cr-00008

Pentacarbonylchromate(2–), 9CI

$Cr(CO)_5^{\ominus\ominus}$

M 192.048 (ion)

Trigonal bipyramidal struct.

Di-Na salt: [51233-19-3].
$C_5CrNa_2O_5$ M 238.028
Reactive anion, readily oxidised.

Di-Cs salt: [57127-95-4].
$C_5CrCs_2O_5$ M 457.859
Reactive anion.

Behrens, H. *et al*, *Z. Anorg. Allg. Chem.*, 1960, **306**, 94 (*props*)
Behrens, H. *et al*, *Chem. Ber.*, 1967, **96**, 2220 (*props*)
Behrens, H. *et al*, *Z. Naturforsch., B*, 1973, **18**, 276 (*synth*)
Ellis, J.E. *et al*, *J. Am. Chem. Soc.*, 1974, **96**, 7825 (*synth*)
Ellis, J.E. *et al*, *J. Organomet. Chem.*, 1975, **97**, 79 (*props*)
Ellis, J.E. *et al*, *Inorg. Chem.*, 1977, **16**, 1357 (*synth, props*)

$C_5HCrO_6^{\ominus}$ Cr-00009

Pentacarbonylhydroxychromate(1–)

[55000-28-7]

$[Cr(CO)_5OH]^{\ominus}$

M 209.055 (ion)

K salt, Dibenzo-18-crown-6-complex: Yellow cryst.

Cihonski, J.L. *et al*, *Inorg. Chem.*, 1975, **14**, 1717 (*synth, ir, raman*)

$C_5H_3AsCrO_5$ Cr-00010

(Arsine)pentacarbonylchromium, 8CI
Pentacarbonylarsinechromium
[29454-58-8]

$(OC)_5CrAsH_3$ (O_h)

M 269.993

Yellow cryst. Mp 69-70°.

Dobson, G.R. *et al*, *Inorg. Chim. Acta*, 1967, **1**, 287 (*synth*)
Fischer, E.O. *et al*, *Chem. Ber.*, 1970, **103**, 1815.
Fischer, E.O. *et al*, *Chem. Ber.*, 1971, **104**, 986 (*synth*)

$C_5H_3CrNO_5$ Cr-00011

Amminepentacarbonylchromium, 9CI
[15228-27-0]

$(OC)_5CrNH_3$ (O_h)

M 209.078

Yellow cryst. Mp 150°.

Behrens, H. *et al*, *Chem. Ber.*, 1964, **97**, 433 (*synth*)
Strohmeier, W. *et al*, *Chem. Ber.*, 1966, **99**, 3419 (*synth*)
Lloyd, D.R. *et al*, *J. Chem. Soc., Faraday Trans. 2*, 1974, 1418 (*pe*)
Sellmann, D. *et al*, *Angew. Chem.*, 1975, **87**, 772 (*props*)
Sellmann, D. *et al*, *J. Organomet. Chem.*, 1976, **111**, 303 (*props*)

$C_5H_3CrO_5P$ Cr-00012

Pentacarbonyl(phosphine)chromium, 9CI
[18116-53-5]

$Cr(CO)_5PH_3$

M 226.045

Pale-yellow cryst. by subl. Mp 116°.

Fischer, E.O. *et al*, *Chem. Ber.*, 1969, **102**, 2547 (*synth*)
Fischer, E.O. *et al*, *J. Organomet. Chem.*, 1969, **18**, P26.
Guggenberger, L.S. *et al*, *Inorg. Chem.*, 1973, **12**, 1143.

$C_5H_3CrO_5Sb$ Cr-00013

Pentacarbonyl(stibine)chromium, 8CI
[32356-12-0]

$(OC)_5CrSbH_3$ (O_h)

M 316.822

Orange-yellow cryst. Mp 44°.

Fischer, E.O. *et al*, *Chem. Ber.*, 1971, **104**, 986 (*synth, ir, nmr*)

$C_5H_4CrN_2O_5$ Cr-00014

Pentacarbonyl(hydrazine-*N*)chromium, 9CI
[30092-08-1]

$(OC)_5CrNH_2NH_2$ (O_h)

M 224.093

Yellow cryst. Mp 120-8°.

Sellman, D., *Z. Naturforsch., B*, 1970, **25**, 890 (*synth*)
Sellman, D. *et al*, *J. Organomet. Chem.*, 1976, **111**, 303 (*synth, props, ir, nmr*)

$C_5H_5ClCrN_2O_2$ Cr-00015

Chloro(η^5-2,4-cyclopentadien-1-yl)dinitrosylchromium, 9CI
Cyclopentadienyldinitrosylchlorochromium
[12071-51-1]

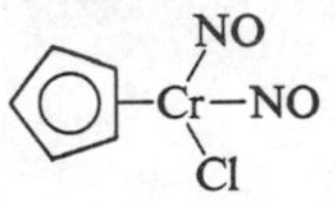

M 212.556

Golden needles by vac. subl. at ca. 70°. Dec. ca. 140° without melting.

Carter, O.L. *et al*, *J. Chem. Soc. (A)*, 1966, 1095 (*struct*)
Legzdins, P. *et al*, *Inorg. Chem.*, 1975, **14**, 1875 (*synth, ms*)
Herberhold, M. *et al*, *Isr. J. Chem.*, 1977, **15**, 206 (*synth*)
Inorg. Synth., 1978, **18**, 129 (*synth, pmr*)
Müller, J. *et al*, *J. Organomet. Chem.*, 1979, **169**, 25 (*ms*)

$C_6CrNO_5^{\ominus}$ Cr-00016

Pentacarbonyl(cyano-*C*)chromate(1−), 9CI
[14971-28-9]

$[(OC)_5CrCN]^{\ominus}$ (O_h)

M 218.066 (ion)

Na salt: [14971-41-6].
$C_6CrNNaO_5$ M 241.055
White cryst.

Bis(triphenylphosphine)iminium salt: [22043-86-3].
$C_{42}H_{30}CrN_2O_5P_2$ M 756.653
White cryst.

(THF)$_3$Cr(3+) salt: [62997-61-9].
$C_{30}H_{24}Cr_4N_3O_{18}$ M 922.513
Red-brown cryst. Sol. C_6H_6.

Behrens, H. *et al*, *Z. Anorg. Allg. Chem.*, 1960, **306**, 94 (*props*)
Seyferth, D. *et al*, *J. Am. Chem. Soc.*, 1960, **82**, 1080 (*props*)
King, R.B., *Inorg. Chem.*, 1967, **6**, 25 (*synth, props*)
Behrens, U. *et al*, *J. Organomet. Chem.*, 1977, **131**, 65 (*synth, struct*)

$C_6CrN_2O_4^{\ominus\ominus}$ Cr-00017

Tetracarbonyldicyanochromate(2−)

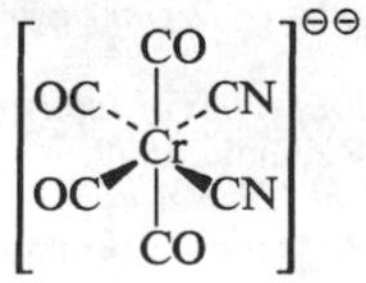

M 216.073 (ion)

cis-form

Bis(triphenylphosphine)iminium salt: [22142-87-6].
$C_{42}H_{30}CrN_3O_4P_2$ M 754.660
White, air-stable cryst.

Behrens, H. *et al*, *Z. Anorg. Allg. Chem.*, 1960, **306**, 94.
Behrens, H. *et al*, *Chem. Ber.*, 1964, **97**, 433 (*synth*)
Ruff, J.K., *Inorg. Chem.*, 1969, **8**, 86 (*synth*)

C_6CrO_5S Cr-00018

(Carbonothioyl)pentacarbonylchromium, 9CI
[50358-90-2]

$(OC)_5CrCS$ (O_h)

M 236.119

Air- and moisture-stable yellow cryst. Sol. hexane.

Angelici, R.J. *et al*, *J. Am. Chem. Soc.*, 1973, **95**, 7516 (*synth*)
Butler, I.S. *et al*, *Inorg. Chem.*, 1976, **15**, 2602 (*ir*)
Poliakoff, M. *et al*, *Inorg. Chem.*, 1976, **15**, 2022.
Lichtenberger, D.L. *et al*, *Inorg. Chem.*, 1976, **15**, 2015 (*uv*)
Butler, I.S. *et al*, *J. Organomet. Chem.*, 1977, **132**, C1 (*synth*)
Saillard, J.Y. *et al*, *Acta Crystallogr., Sect. B*, 1978, **34**, 3318 (*cryst struct*)

C_6CrO_5Se — Cr-00019

(Carbonoselenoyl)pentacarbonylchromium, 10CI

[63356-87-6]

$(OC)_5CrCSe$ (O_h)

M 283.019

Highly volatile, air-stable deep-yellow cryst.

Butler, I.S. *et al*, *J. Organomet. Chem.*, 1977, **132**, C1 (*synth*)
Butler, I.S. *et al*, *Z. Anorg. Allg. Chem.*, 1978, **446**, 17.
Butler, I.S. *et al*, *Inorg. Chim. Acta*, 1979, **32**, 113 (*pe*)
Wilkinson, J.R. *et al*, *J. Organomet. Chem.*, 1979, **179**, 159 (*nmr*)
English, A.M. *et al*, *J. Organomet. Chem.*, 1981, **205**, 177 (*synth*)
Plowman, K.R. *et al*, *Inorg. Chem.*, 1981, **20**, 2553 (*ir*)

C_6CrO_6 — Cr-00020

Chromium carbonyl, 10CI, 9CI, 8CI

Hexacarbonylchromium. Chromium hexacarbonyl

[13007-92-6]

$Cr(CO)_6(O_h)$

M 220.058

Catalyst for hydrogenation, isomerisation, water gas shift reaction and alkylation of aromatic hydrocarbons. Colourless cryst. (methylcyclohexane or by subl.). Sl. sol. CCl_4, insol. H_2O, EtOH, Et_2O. Mp 152-5°.

▷Explodes at 210°. GB5075000.

Podall, H.E. *et al*, *J. Am. Chem. Soc.*, 1961, **83**, 2057 (*synth*)
Whitaker, A. *et al*, *Acta Crystallogr.*, 1967, **23**, 977 (*cryst struct*)
Rieke, R. *et al*, *J. Org. Chem.*, 1974, **76**, C19 (*synth*)
Bodner, G.M., *Inorg. Chem.*, 1975, **14**, 2694 (*cmr*)
Jost, A. *et al*, *Acta Crystallogr., Sect. B*, 1975, **31**, 2649 (*nd*)
Michels, G.D. *et al*, *Inorg. Chem.*, 1980, **19**, 479.
Fieser, M. *et al*, *Reagents for Organic Synthesis*, Wiley, 1967-83, **8**, 110.
Sax, N.I., *Dangerous Properties of Industrial Materials*, 5th Ed., Van Nostrand-Reinhold, 1979, .
Hazards in the Chemical Laboratory, (Bretherick, L., Ed.), 3rd Ed., Royal Society of Chemistry, London, 1981, .

$C_6H_3BCrNO_5^{\ominus}$ — Cr-00021

Pentacarbonyl[(cyano-*C*)trihydroborato(1–)-*N*]chromate(1–), 9CI

$(OC)_5CrCNBH_3^{\ominus}$ (O_h)

M 231.899 (ion)

Mp >170° dec.

Tetramethylammonium salt: [52138-78-0].
$C_{10}H_{15}BCrN_2O_5$ M 306.045
Yellow cryst.

King, R.B. *et al*, *J. Organomet. Chem.*, 1974, **65**, 71 (*synth, ir, nmr*)

$C_6H_3ClCrO_4$ — Cr-00022

Tetracarbonylchloroethylidynechromium, 10CI

[62938-49-2]

$$X{-}Cr(CO)_4{\equiv}CCH_3$$

X = Cl

M 226.536

Yellow cryst. Struct. by analogy with Tetracarbonyliodoethylidynechromium, Cr-00023 .

Fischer, E.O. *et al*, *Chem. Ber.*, 1976, **109**, 1673.
Hans, J. *et al*, *J. Organomet. Chem.*, 1981, **216**, 235.

$C_6H_3CrIO_4$ — Cr-00023

Tetracarbonyliodoethylidynechromium, 9CI

[50701-14-9]

As Tetracarbonylchloroethylidynechromium, Cr-00022 with

X = I

M 317.988

Cryst.

Fischer, E.O. *et al*, *Angew. Chem., Int. Ed. Engl.*, 1973, **12**, 564 (*synth, nmr*)
Huttner, G. *et al*, *Angew. Chem., Int. Ed. Engl.*, 1974, **13**, 609 (*cryst struct*)

$C_6H_3CrO_5^{\ominus}$ — Cr-00024

Pentacarbonylmethylchromate(1–), 9CI

Methylpentacarbonylchromium(–1)

$[MeCr(CO)_5]^{\ominus}$

M 207.083 (ion)

$(Ph_3P)_2N^{\oplus}$ salt: [62197-85-7]. (T-4)*Triphenyl*(P,P,P-*triphenylphosphineimidato*-N)*phosphorus*(1+) *pentacarbonylmethylchromate*(1–), 9CI. *Bis*(*triphenylphosphine*)*iminium methylpentacarbonylchromate.*
$C_{42}H_{33}CrNO_5P_2$ M 745.670
Yellow cryst.

Ellis, J.E. *et al*, *Inorg. Chem.*, 1977, **16**, 1357.

$C_6H_8CrN_2O_2$ — Cr-00025

(η^5-2,4-Cyclopentadienyl)methyldinitrosylchromium

Methyldinitrosyl(η^5-*2,4-cyclopentadien-1-yl*)*chromium*

[53522-59-1]

$$(\eta^5\text{-}C_5H_5){-}Cr(NO)_2Me$$

M 192.137

Red cryst.

Piper, T.S. *et al*, *J. Inorg. Nucl. Chem.*, 1956, **3**, 104 (*synth*)
Wojcicki, A. *et al*, *Inorg. Chem.*, 1980, **19**, 3803 (*synth, props*)

$C_6H_{10}CrI$ Cr-00026

Iodobis(η^3-2-propenyl)chromium, 9CI

Bis(allyl)chromium iodide. Diallyliodochromium

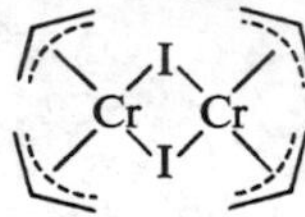

M 261.046

Dimeric.

Dimer: [12154-13-1]. *Di-μ-iodotetrakis(η^3-2-propenyl-)dichromium,* 9CI. *Tetraallyldiiododichromium.*

$C_{12}H_{20}Cr_2I_2$ M 522.091

Air-sensitive red cryst.

Wilke, E. *et al*, *Angew. Chem., Int. Ed. Engl.*, 1966, **5**, 151 (*synth, props*)

Aumann, R. *et al*, *Chem. Ber.*, 1968, **101**, 954 (*props*)
Fischer, E.O. *et al*, *J. Organomet. Chem.*, 1972, **40**, 159 (*props*)
Inorg. Synth., 1977, **17**, 95 (*synth*)
Klinger, R.J. *et al*, *Inorg. Chem.*, 1981, **20**, 34 (*synth*)

$C_6H_{18}Cr^{\ominus\ominus\ominus}$ Cr-00027

Hexamethylchromate(3−), 9CI

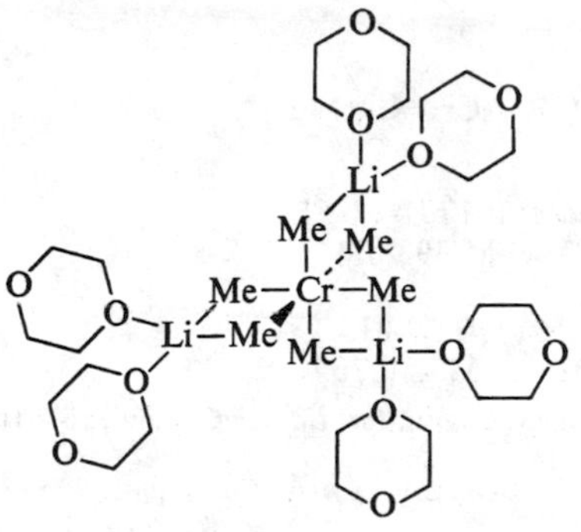

Li salt, dioxane complex

M 142.204 (ion)

Tri-Li salt: [15335-69-0].

$C_6H_{18}CrLi_3$ M 163.027

Obt. as dioxan complex.

Krausse, J. *et al*, *J. Organomet. Chem.*, 1974, **65**, 215 (*synth, cryst struct*)

$C_7H_3CrNO_5S$ Cr-00028

Pentacarbonyl(methyl thiocyanate-*N*)chromium, 10CI

[57196-01-7]

$(OC)_5CrNCSMe$

M 265.160

Yellow cryst. Mp 39-41° dec.

Quick, M.H. *et al*, *Inorg. Chem.*, 1976, **15**, 160 (*synth*)
Goddard, R. *et al*, *J. Chem. Soc., Dalton Trans.*, 1978, 1255 (*synth, cryst struct*)

$C_7H_3CrO_6^{\ominus}$ Cr-00029

Acetylpentacarbonylchromate(1−), 9CI

$[(OC)_5CrCOCH_3]^{\ominus}$ (O_h)

M 235.093 (ion)

Li salt: [38883-45-3].

$C_7H_3CrLiO_6$ M 242.034

Reactive.

Tetramethylammonium salt: [75112-13-9].

$C_{11}H_{15}CrNO_6$ M 309.239

Pale cryst. Mp 175° dec.

Fischer, E.O. *et al*, *Chem. Ber.*, 1967, **100**, 2445 (*synth*)

$C_7H_4CrO_3S$ Cr-00030

Tricarbonylthiophenechromium, 8CI

[12078-15-8]

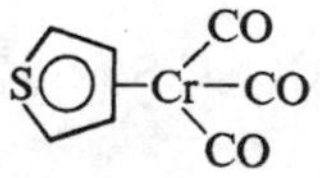

M 220.163

Red cryst.

Bailey, M.F. *et al*, *Inorg. Chem.*, 1965, **4**, 1306 (*cryst struct*)
Ofel, K. *et al*, *Chem. Ber.*, 1966, **99**, 1732.
Fischer, E.O. *et al*, *J. Organomet. Chem.*, 1967, **8**, P5 (*synth*)
Ofel, K. *et al*, *J. Organomet. Chem.*, 1971, **30**, 211 (*cryst struct*)

$C_7H_5CrNO_2S$ Cr-00031

Dicarbonyl(η^5-2,4-cyclopentadien-1-yl)thionitrosylchromium, 10CI

[66539-91-1]

As Dicarbonyl(η^5-2,4-cyclopentadien-1-yl)nitrosylchromium, Cr-00032 with

X = NS

M 219.178

Red cryst. Mp 68-9°.

Greenhough, T.J. *et al*, *Inorg. Chem.*, 1979, **18**, 3548 (*synth, cryst struct*)
Hubbard, J.L. *et al*, *Inorg. Chem.*, 1980, **19**, 1388 (*pe*)

$C_7H_5CrNO_3$ Cr-00032

Dicarbonyl(η^5-2,4-cyclopentadien-1-yl)nitrosylchromium, 10CI

Cyclopentadienyldicarbonylnitrosylchromium

[36312-04-6]

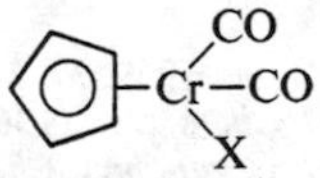

X = NO

M 203.117

Orange cryst. Mp 67-8°. Subl. at 60-80° *in vacuo*.

Fischer, E.O. *et al*, *Chem. Ber.*, 1961, **94**, 93 (*synth*)
Müller, J., *J. Organomet. Chem.*, 1970, **23**, C38 (*ms*)
Ball, D.E. *et al*, *J. Organomet. Chem.*, 1973, **55**, C24 (*props*)
Mintz, E.A. *et al*, *J. Organomet. Chem.*, 1977, **137**, 199 (*props*)
Inorg. Synth., 1978, **18**, 126 (*synth*)
Atwood, J.L. *et al*, *J. Organomet. Chem.*, 1979, **165**, 65 (*props, cryst struct*)
Herberhold, M. *et al*, *J. Organomet. Chem.*, 1980, **191**, 79 (*props*)

$C_7H_5CrNO_5$ Cr-00033

(1-Aminoethylidene)pentacarbonylchromium, 9CI

[22852-50-2]

$(OC)_5Cr{=}C(NH_2)CH_3$

M 235.116

Pale-yellow cryst. Sol. org. solvs. Mp 76-7°.

Klabunde, U. *et al*, *J. Am. Chem. Soc.*, 1967, **89**, 7141.

Moser, E. *et al*, *J. Organomet. Chem.*, 1968, **12**, P1 (*synth*)
Fischer, E.O. *et al*, *Chem. Ber.*, 1971, **104**, 1339 (*synth*)
Block, T.F. *et al*, *J. Organomet. Chem.*, 1977, **139**, 235.
Doetz, K.H. *et al*, *Chem. Ber.*, 1980, **113**, 3597 (*props*)

$C_7H_6CrO_5S$ Cr-00034

Pentacarbonyl(dimethyl sulfide)chromium
Pentacarbonyl[thiobis[methane]]chromium, 9CI
[31172-83-5]

$(OC)_5CrSMe_2$ (O_h)

M 254.177
Yellow oil. Mp 9°.

Ehrl, W. *et al*, *Chem. Ber.*, 1970, **103**, 3563 (*synth*)
Raubenheimer, H.G. *et al*, *J. Organomet. Chem.*, 1976, **112**, 145 (*synth, derivs*)
Herberhöld, M. *et al*, *J. Chem. Res. (S)*, 1977, 246 (*synth*)

$C_7H_8CrN_3O_2^{\oplus}$ Cr-00035

Acetonitrile(η^5-2,4-cyclopentadien-1-yl)dinitrosylchromium(1+)
[74924-58-6]

M 218.155 (ion)
Hexafluorophosphate: [74924-59-7].
$C_7H_8CrF_6N_3O_2P$ M 363.119
Useful starting material. Green cryst.

Wojcicki, A. *et al*, *Inorg. Chem.*, 1980, **19**, 3803 (*synth, props*)

$C_7H_{17}CrO_5^{\oplus\oplus}$ Cr-00036

Pentaaquabenzylchromium(2+)
Pentaaqua(phenylmethyl)chromium(2+)
[34788-74-4]

$[Cr(CH_2Ph)(H_2O)_5]^{\oplus\oplus}$

M 233.204 (ion)

Anet, F.A.L., *Can. J. Chem.*, 1959, **37**, 58.
Kochi, J.K. *et al*, *J. Am. Chem. Soc.*, 1961, **83**, 2017.
Sneeden, R.P.A. *et al*, *J. Organomet. Chem.*, 1966, **6**, 542.
Taube, H. *et al*, *J. Am. Chem. Soc.*, 1971, **93**, 1117.

$C_8H_2Cr_2O_8^{\ominus\ominus}$ Cr-00037

Octacarbonyl-μ-dihydrodichromate(2−)

M 330.091 (ion)
Bis(tetraethylammonium) salt: [84850-72-6].
$C_{24}H_{42}Cr_2N_2O_8$ M 590.596
Air-sensitive burgundy cryst.

Ellis, J.E. *et al*, *J. Am. Chem. Soc.*, 1983, **105**, 2296 (*synth*)

$C_8H_5CrO_3^{\ominus}$ Cr-00038

Tricarbonyl(η^5-2,4-cyclopentadien-1-yl)chromate(1−)
[48121-47-7]

M 201.122 (ion)
Has wide synthetic utility.
Na salt: [12203-12-2].
$C_8H_5CrO_3Na$ M 224.111
Air- and moisture-sensitive.

Inorg. Synth., 1963, **7**, 99 (*synth*)
Burlitch, J.M. *et al*, *J. Chem. Soc., Dalton Trans.*, 1974, 828 (*props*)
Ellis, J.E. *et al*, *J. Organomet. Chem.*, 1975, **99**, 263 (*synth*)
Hackett, P. *et al*, *J. Chem. Soc., Dalton Trans.*, 1975, 1606 (*props*)
Pedersen, S.E. *et al*, *Inorg. Chem.*, 1975, **14**, 2365 (*props*)

$C_8H_6CrNO_5^{\oplus}$ Cr-00039

Pentacarbonyl[(dimethylamino)methylidyne]chromium(1+), 9CI
[61170-98-7]

$[(OC)_5Cr{\equiv}CNMe_2]^{\oplus}$

M 248.135 (ion)
Hexafluorophosphate: [61771-01-5].
$C_8H_6CrF_6NO_5P$ M 393.099
Red cryst.
Tetrachloroborate: [61770-99-8].
$C_8H_6BCl_4CrNO_5$ M 400.757

Fischer, E.O. *et al*, *Angew. Chem., Int. Ed. Engl.*, 1976, **15**, 616 (*synth*)
Hartshorn, A.J. *et al*, *J. Chem. Soc., Chem. Commun.*, 1976, 761 (*synth, props*)
Fischer, H. *et al*, *Angew. Chem.*, 1981, **93**, 483 (*props*)

$C_8H_6CrN_2O_2$ Cr-00040

(η^6-Benzene)dicarbonyl(dinitrogen)chromium, 9CI
Dicarbonyl(dinitrogen)(η^6-benzene)chromium
[38904-62-0]

M 214.144
Somewhat unstable, disproportionates to $[(\eta^6\text{-}C_6H_6)(OC)_2CrN_2Cr(CO)_2(\eta^6\text{-}C_6H_6)]$.

Strohmeier, W. *et al*, *Chem. Ber.*, 1964, **97**, 1877 (*synth*)
Sellmann, D. *et al*, *Z. Naturforsch., B*, 1972, **27**, 465, 718 (*synth*)

$C_8H_6CrO_3$ Cr-00041

Tricarbonyl(η^5-2,4-cyclopentadien-1-yl)hydrochromium, 9CI
Tricarbonyl-π-cyclopentadienylhydrochromium,, 8CI. *Cyclopentadienyltricarbonylchromium hydride. Cyclopentadienylhydridotricarbonylchromium. Hydridotricarbonyl(π-cyclopentadienyl)chromium*
[36495-37-1]

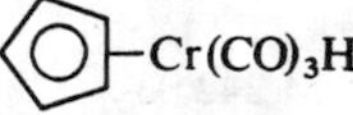

M 202.130
Yellow cryst. Mp 57-8° dec. V. air-sensitive. Dec. slowly at r.t., rapidly at Mp, evolving H_2. Readily volatile. Acidic; dissolves in alkalis → solution of $[C_5H_5Cr(CO)_3]^{\ominus}$.

Fischer, E.O. *et al*, *Z. Naturforsch., B*, 1955, **10**, 140 (*synth*)
Piper, T.S. *et al*, *J. Inorg. Nucl. Chem.*, 1956, **3**, 104 (*pmr, ir*)
Fischer, E.O. *et al*, *Chem. Ber.*, 1961, **94**, 2413 (*synth*)
Inorg. Synth., 1963, **7**, 136 (*synth*)
Hart, W.P. *et al*, *J. Am. Chem. Soc.*, 1980, **102**, 1196 (*synth, props*)
Nesmeyanov, A.N. *et al*, *J. Organomet. Chem.*, 1980, **184**, 63 (*props*)

$C_8H_6CrO_6$ Cr-00042

Pentacarbonyl(1-methoxyethylidene)chromium, 10CI

[20540-69-6]

$(OC)_5Cr{=}C(OMe)CH_3$

M 250.128

Yellow cryst. Mp 34°.

Aumann, R. *et al*, *Chem. Ber.*, 1968, **101**, 964 (*synth*)
Fischer, E.O. *et al*, *J. Organomet. Chem.*, 1971, **28**, 367 (*props*)
Connor, J.A. *et al*, *J. Chem. Soc., Dalton Trans.*, 1972, 2419 (*nmr*)
Bodner, G.M. *et al*, *Inorg. Chem.*, 1973, **12**, 1071 (*nmr*)
Inorg. Synth., 1977, **17**, 99 (*synth*)
Brown, F.J., *Prog. Inorg. Chem.*, 1980, **27**, 1 (*props*)

$C_8H_7CrNO_3$ Cr-00043

Tricarbonyl(1-methylpyrrole)chromium, 8CI

[33506-43-3]

MeN(pyrrole)–Cr(CO)3

M 217.144

Yellow cryst. Dec. ca. 110°.

Fischer, E.O. *et al*, *J. Organomet. Chem.*, 1967, **8**, P5 (*synth*)
Ofel, K. *et al*, *J. Organomet. Chem.*, 1971, **30**, 211 (*synth, props*)

$C_8H_7CrNO_5$ Cr-00044

Pentacarbonyl[(dimethylamino)methylene]chromium, 10CI

[38893-15-1]

$(OC)_5Cr{=}CHNMe_2$

M 249.143

Yellow cryst. Mp 64-6°.

Hartshorn, A.J. *et al*, *J. Chem. Soc., Dalton Trans.*, 1978, 348 (*synth*)
Hill, J.E. *et al*, *Transition Met. Chem.*, 1978, **3**, 315 (*props*)

$C_8H_7CrNO_5$ Cr-00045

Pentacarbonyl[1-(methylamino)ethylidene]chromium, 9CI

[12387-00-7]

$(OC)_5Cr{=}C(CH_3)NHMe$

M 249.143

Red cryst.

Baikie, P.E. *et al*, *J. Chem. Soc., Chem. Commun.*, 1967, 1199.
Connor, J.A. *et al*, *J. Chem. Soc. (A)*, 1969, 578 (*synth*)
Fischer, E.O. *et al*, *Chem. Ber.*, 1973, **106**, 3893 (*props*)
Lloyd, M.K. *et al*, *J. Chem. Soc., Dalton Trans.*, 1973, 1743.

$C_8H_9CrNO_2$ Cr-00046

Carbonyl(η^5-2,4-cyclopentadien-1-yl)(η^2-ethene)nitrosylchromium, 10CI

Carbonylnitrosyl(η^5-cyclopentadienyl)(η^2-ethylene)chromium

[38816-34-1]

(Cp)Cr(C2H4)(NO)(CO)

M 203.161

Orange-yellow cryst. (pentane at −80°). ir ν_{CO} 1971, ν_{NO} 1664 cm^{-1} (CH_2Cl_2). Darkens in air at 47° and dec. to black oil at 86°.

Herberhold, M. *et al*, *J. Organomet. Chem.*, 1972, **42**, 407 (*synth, props*)
Alt, H. *et al*, *J. Organomet. Chem.*, 1974, **77**, 353 (*nmr*)
Kreiter, C.G. *et al*, *Chem. Ber.*, 1981, **114**, 1845 (*nmr*)

$C_8H_9CrO_5P$ Cr-00047

Pentacarbonyl(trimethylphosphine)chromium, 8CI

[26555-09-9]

$Cr(CO)_5PMe_3$

M 268.126

Yellow low-melting solid.

Jenkins, J.H. *et al*, *Inorg. Chem.*, 1967, **6**, 2250.
Mathieu, R. *et al*, *Inorg. Chem.*, 1970, **9**, 2030.
Connor, J.A. *et al*, *J. Organomet. Chem.*, 1972, **43**, 357.

$C_8H_9CrO_5PS$ Cr-00048

Pentacarbonyl(trimethylphosphine sulfide-*S*)chromium, 9CI

[35244-74-7]

$(OC)_5CrS{=}PMe_3$ (O_h)

M 300.186

Yellow cryst. Sol. C_6H_6. Mp 87°.

Ainscough, E.W. *et al*, *J. Chem. Soc., Dalton Trans.*, 1973, 2360 (*synth, props*)
Doorman, B.M. *et al*, *Inorg. Nucl. Chem. Lett.*, 1973, **9**, 941.
Baker, E.N. *et al*, *J. Chem. Soc., Dalton Trans.*, 1973, 2205 (*cryst struct*)
Raubenheimer, H.G. *et al*, *J. Organomet. Chem.*, 1976, **112**, 145 (*ir*)

$C_8H_{12}CrN_3O_3^{\oplus}$ Cr-00049

(η^5-2,4-Cyclopentadien-1-yl)[methoxy(methylamino)methylene]dinitrosylchromium(1+)

Bisnitrosyl[methoxy(methylamino)methylene](η^5-2,4-cyclopentadien-1-yl)chromium(1+)

[74924-70-2]

[(Cp)Cr(NO)2=C(OMe)(NHMe)]⊕

M 250.197 (ion)

Hexafluorophosphate: [74924-71-3].
$C_8H_{12}CrF_6N_3O_3P$ M 395.161
Green cryst. (Me_2CO/Et_2O).

Wojcicki, A. *et al*, *Inorg. Chem.*, 1980, **19**, 3803 (*synth*)

$C_8H_{20}CrO_4Tl_2$ Cr-00050

(Chromato-*O*)bis(diethylthallium)

Diethylthallium chromate. Bis(diethylthallium) chromate

$(Et_2Tl)_2CrO_4$

M 641.006

Yellow cryst. (EtOH). Sol. H_2O, EtOH. Mp 193° dec.

▷Violently explosive; toxic

Goddard, A.E., *J. Chem. Soc.*, 1921, 672 (*synth*)

$C_9H_5ClCrHgO_3$ Cr-00051

Tricarbonyl(chloromercury)[μ-[(1-η:1,2,3,4,5,6-η)phenyl-]]chromium

[41576-43-6]

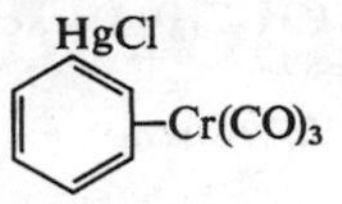

HgCl

-Cr(CO)₃

M 449.176

Light-yellow solid. Mp 139° dec.

Magomedov, G.K.I. *et al*, *J. Gen. Chem. USSR* (*Engl. Transl.*), 1972, **42**, 2443 (*synth*, *ir*)

$C_9H_5ClCrO_3$ Cr-00052

Tricarbonyl(η^6-chlorobenzene)chromium, 9CI

(η^6-Chlorobenzene)tricarbonylchromium

[12082-03-0]

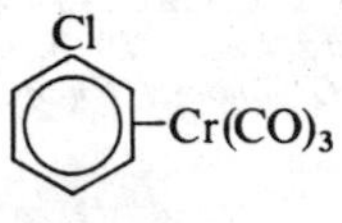

Cl

-Cr(CO)₃

M 248.586

Yellow cryst. (Et_2O/pet. ether). Mp 102-3°. Sublimes *in vacuo*.

Fischer, E.O. *et al*, *Z. Naturforsch.*, *B*, 1958, **13**, 458 (*synth*)
Nicholls, B. *et al*, *J. Chem. Soc.*, 1959, 551 (*synth*)
Fischer, R.D., *Chem. Ber.*, 1960, **93**, 165 (*ir*)
Gaivonski, P.E. *et al*, *Izv. Akad. Nauk. SSSR*, *Ser. Khim.*, 1973, **11**, 2618; *CA*, **80**, 59047 (*ms*)
Bodner, G.M. *et al*, *Inorg. Chem.*, 1974, **13**, 360 (*cmr*)
Inorg. Synth., 1978, **19**, 154 (*synth*, *ir*)

$C_9H_5CrFO_3$ Cr-00053

(η^6-Fluorobenzene)tricarbonylchromium

Tricarbonyl(η^6-fluorobenzene)chromium, 9CI

[12082-05-2]

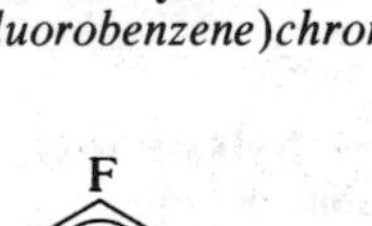

F

-Cr(CO)₃

M 232.131

Yellow cryst. (heptane). Mp 123-4°. Sublimes *in vacuo*.

Nicholls, B. *et al*, *J. Chem. Soc.*, 1959, 551 (*synth*)
Fischer, R.D., *Chem. Ber.*, 1960, **93**, 165 (*ir*)
Strohmeier, W., *Chem. Ber.*, 1961, **94**, 2490 (*synth*)
Müller, J. *et al*, *Chem. Ber.*, 1969, **102**, 3314 (*ms*)
Bodner, G.M. *et al*, *Inorg. Chem.*, 1974, **13**, 360 (*cmr*)
Inorg. Synth., 1979, **19**, 154 (*synth*)

$C_9H_5CrLiO_3$ Cr-00054

[μ(η:η^6-Phenyl)](tricarbonylchromium)lithium, 10CI

[67421-16-3]

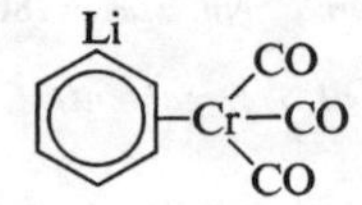

Li

CO

-Cr—CO

CO

M 220.074

Formed as soln. in THF.

Rausch, M.D. *et al*, *J. Organomet. Chem.*, 1978, **153**, 59 (*synth*)
Semmelhack, M.F. *et al*, *J. Am. Chem. Soc.*, 1979, **101**, 768 (*synth*)

$C_9H_6CrN_2O_5$ Cr-00055

Pentacarbonyl[cyano(dimethylamino)methylene]chromium, 10CI

[61771-03-7]

$(OC)_5Cr{=}C(CN)NMe_2$

M 274.153

Hartshorn, A.J. *et al*, *J. Chem. Soc.*, *Chem. Commun.*, 1976, 761 (*synth*, *ir*, *nmr*)

$C_9H_6CrO_2S$ Cr-00056

(η^6-Benzene)(carbonothioyl)dicarbonylchromium, 10CI

Carbonothioyl(dicarbonyl)(η^6-benzene)chromium

[63356-86-5]

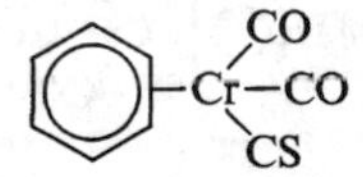

CO

-Cr—CO

CS

M 230.201

Yellow cryst. Mp 125° dec. Related complexes have been synthesised.

Butler, I.S. *et al*, *J. Organomet. Chem.*, 1977, **133**, 59 (*synth*)
Herberhold, M. *et al*, *Chem. Ber.*, 1978, **111**, 2273 (*synth*)
Inorg. Synth., 1979, **19**, 197.

$C_9H_6CrO_3$ Cr-00057

(η^6-Benzene)tricarbonylchromium, 9CI

Tricarbonylbenzenechromium. Benzenechromium tricarbonyl

[12082-08-5]

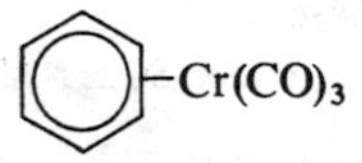

-Cr(CO)₃

M 214.141

Yellow cryst. (Et_2O/pet. ether). Mp 161.5-163°. Sublimes at 130°/2 mm.

▷GB4700000.

Gmelin Suppl., **3**, 183.
Natta, G. *et al*, *Chem. Ind.* (*Milan*), 1958, **40**, 287 (*synth*)
Corradini, P. *et al*, *J. Am. Chem. Soc.*, 1959, **81**, 2271 (*struct*)
Strohmeier, W. *et al*, *Chem. Ber.*, 1964, **97**, 1877 (*pmr*)
Inorg. Synth., 1978, **19**, 154 (*synth*, *ir*)
Federov, L.A. *et al*, *J. Organomet. Chem.*, 1979, **182**, 499 (*cmr*)

$C_9H_6CrO_4$ Cr-00058

Tricarbonyl(η^6-phenol)chromium, 9CI
(η^6-Phenol)tricarbonylchromium
[32802-03-2]

OH —Cr(CO)$_3$

M 230.140
Yellow cryst. Mp 113-5°.

Natta, G. *et al*, *Chem. Ind. (Milan)*, 1958, 1003 (*synth*)
Fischer, R.D., *Chem. Ber.*, 1960, **93**, 165 (*ir*)
Jackson, W.R. *et al*, *J. Chem. Soc.*, 1960, 469 (*synth*)
Fischer, E.O. *et al*, *Chem. Ber.*, 1961, **94**, 93 (*synth*)
Lundquist, R.T. *et al*, *J. Org. Chem.*, 1962, **27**, 1167 (*uv*)
Müller, J. *et al*, *Chem. Ber.*, 1969, **102**, 3314 (*ms*)

$C_9H_7CrNO_3$ Cr-00059

(η^6-Aniline)tricarbonylchromium
(η^6-Benzenamine)tricarbonylchromium, 9CI. Tricarbonyl(aniline)chromium
[12108-11-1]

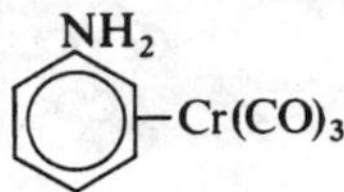

M 229.155
Yellow cryst. Mp 161-2°. Sublimes *in vacuo*.

Fischer, E.O. *et al*, *Chem. Ber.*, 1958, **91**, 2763 (*synth*)
Natta, G. *et al*, *Chem. Ind. (Milan)*, 1958, 1003 (*synth*)
Fischer, E.O. *et al*, *Chem. Ber.*, 1961, **94**, 93 (*synth*)
Brown, D. *et al*, *J. Chem. Soc.*, 1962, 3849 (*ir*)
Lundquist, R.T. *et al*, *J. Org. Chem.*, 1962, **27**, 1167 (*uv*)
Wu, A. *et al*, *J. Organomet. Chem.*, 1971, **33**, 53 (*pmr*)
Bodner, G.M. *et al*, *Inorg. Chem.*, 1974, **13**, 360 (*cmr*)

$C_9H_8CrO_3$ Cr-00060

Cyclopentadienyltricarbonylmethylchromium
Tricarbonyl(η^5-2,4-cyclopentadien-1-yl)methylchromium, 9CI
[41311-89-1]

OC—Cr—Me
OC CO

M 216.156
Yellow cryst. Thermally unstable.

Piper, T.S. *et al*, *J. Inorg. Nucl. Chem.*, 1956, **3**, 104 (*synth, ir, pmr*)
Alt, H.G., *J. Organomet. Chem.*, 1977, **124**, 167 (*synth*)
Samuel, E. *et al*, *J. Organomet. Chem.*, 1979, **172**, 309.

$C_9H_8CrO_6$ Cr-00061

Pentacarbonyl(tetrahydrofuran)chromium, 10CI
[15038-41-2]

O
OC CO
Cr
OC CO
CO

M 264.155
Yellow cryst.

Strohmeier, W. *et al*, *Chem. Ber.*, 1961, **94**, 398 (*synth*)
Schubert, U. *et al*, *J. Organomet. Chem.*, 1978, **144**, 175 (*cryst struct*)
Veith, M. *et al*, *J. Organomet. Chem.*, 1981, **216**, 377 (*props*)
Roesky, H.W. *et al*, *Inorg. Chem.*, 1981, **20**, 2910 (*props*)

$C_9H_{10}BrCrNO_4$ Cr-00062

Bromotetracarbonyl[(diethylamino)methylidyne]chromium, 10CI
[57091-06-2]

OC CO
X—Cr≡CNEt$_2$
OC CO
X = Br

M 328.082
Orange cryst. Mp 47°.

Fischer, E.O. *et al*, *Chem. Ber.*, 1978, **111**, 3542 (*synth, cryst struct*)

$C_9H_{10}ClCrNO_4$ Cr-00063

Tetracarbonylchloro[(diethylamino)methylidyne]chromium, 10CI
[59218-84-7]
As Bromotetracarbonyl[(diethylamino)methylidyne]-chromium, Cr-00062 with

X = Cl

M 283.631
Yellow cryst. Mp 52° dec. Struct. by analogy with Bromotetracarbonyl[(diethylamino)methylidyne]chromium, Cr-00062 .

Fischer, E.O. *et al*, *J. Organomet. Chem.*, 1976, **107**, C23 (*synth*)
Hartshorn, A.S. *et al*, *J. Chem. Soc., Chem. Commun.*, 1976, 761 (*synth*)
Fischer, H. *et al*, *Angew. Chem.*, 1981, **93**, 483 (*props*)

$C_9H_{15}Cr$ Cr-00064

Tris(η^3-2-propenyl)chromium, 9CI
Tris(π-allyl)chromium. Triallylchromium. Chromium triallyl
[12082-46-1]

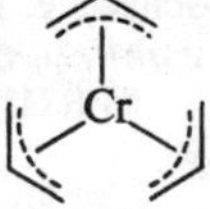

M 175.214
Monomeric in C_6H_6 and dioxan. Dark-red cryst. Mp 77-9° dec. Paramagnetic.

Wilke, G. *et al*, *Angew. Chem., Int. Ed. Engl.*, 1966, **5**, 151 (*props*)
Inorg. Synth., 1972, **13**, 73 (*synth, props*)

$C_{10}CrIO_{10}^{\ominus}$ Cr-00065

Decacarbonyl-μ-iododichromate(1−), 8CI

$[(OC)_5Cr—I—Cr(CO)_5]^{\ominus}$

M 459.005 (ion)
Bis(triphenylphosphine)iminium salt: [21637-80-9]. $C_{46}H_{30}CrINO_{10}P_2$ M 997.592
Red cryst. Mp 109-10°.

Ruff, J.K., *Inorg. Chem.*, 1968, **7**, 1821 (*synth*)
Dahl, L.F. *et al*, *J. Am. Chem. Soc.*, 1970, **92**, 7327 (*cryst struct*)

$C_{10}Cr_2O_{10}^{\ominus\ominus}$ Cr-00066

Decacarbonyldichromate(2−), 9CI
[45264-01-5]

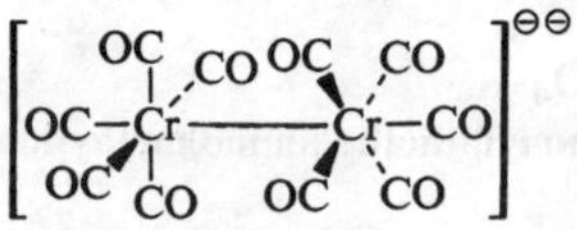

$$\left[(OC)_5Cr{-}Cr(CO)_5 \right]^{\ominus\ominus}$$

M 384.096 (ion)
Di-Na salt: [15616-67-8].
$C_{10}Cr_2Na_2O_{10}$ M 430.076
Yellow cryst.
Di-K salt: [57348-61-5].
$C_{10}Cr_2K_2O_{10}$ M 462.293
Yellow cryst.
Bis(tetraethylammonium) salt: [15682-68-5].
$C_{26}H_{40}Cr_2N_2O_{10}$ M 644.601
Yellow cryst.
Bis(bis(triphenylphosphine)iminium)salt: [49736-18-7].
$C_{82}H_{60}Cr_2N_2O_{10}P_4$ M 1461.270
Dark-yellow cryst. Sol. in CH_2Cl_2. Mp 185-9° dec.

Hayter, R.E., *J. Am. Chem. Soc.*, 1966, **88**, 4376 (*props*)
Ruff, J.K., *Inorg. Chem.*, 1967, **6**, 2080 (*props*)
Dahl, L.F. *et al*, *J. Am. Chem. Soc.*, 1970, **92**, 7312 (*cryst struct*)
Ruff, J.K. *et al*, *J. Organomet. Chem.*, 1971, **33**, C64 (*props*)
Inorg. Synth., 1974, **15**, 89 (*synth*)
Unguvenasu, C. *et al*, *J. Chem. Soc., Chem. Commun.*, 1975, 388 (*synth*)
Behrens, H. *et al*, *Z. Naturforsch., B*, 1977, **32**, 1105 (*props*)

$C_{10}HCr_2O_{10}^{\ominus}$ Cr-00067

Decacarbonyl-μ-hydrodichromate(1−), 10CI
[73740-63-3]

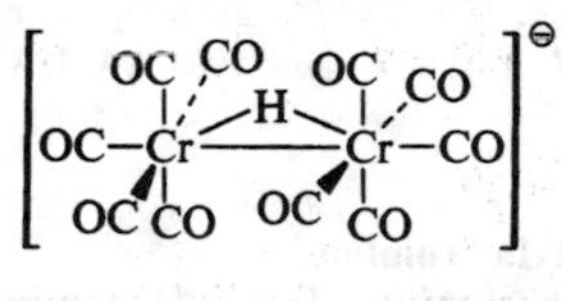

$$\left[(OC)_5Cr{-}H{-}Cr(CO)_5 \right]^{\ominus}$$

M 385.104 (ion)
Tetraethylammonium salt: [16924-36-0].
$C_{18}H_{21}Cr_2NO_{10}$ M 515.357
Yellow cryst., v. air-sensitive.
Bis(triphenylphosphine)iminium salt: [62341-83-7].
$C_{46}H_{31}CrNO_{10}P_2$ M 871.695
Yellow cryst.

Hayter, R.J., *J. Am. Chem. Soc.*, 1966, **88**, 4376 (*synth*)
Dahl, L.F. *et al*, *J. Am. Chem. Soc.*, 1970, **92**, 7312 (*cryst struct*)
Dahl, L.F. *et al*, *J. Am. Chem. Soc.*, 1977, **99**, 4497.
Cooper, C.B. *et al*, *Adv. Chem. Ser.*, 1978, **167**, 232 (*ir, raman*)

$C_{10}H_4Cr_2O_{10}P_2$ Cr-00068

Decacarbonyl[μ-(diphosphine-*P*:*P'*)]dichromium
[52589-03-4]

$$(OC)_5Cr{-}H_2P{-}PH_2{-}Cr(CO)_5$$

M 450.075
Yellow powder. Sol. THF, Me_2CO, insol. nonpolar solvs.
THF adduct: Colourless needles.

Sellman, D., *Angew. Chem., Int. Ed. Engl.*, 1973, **12**, 1020 (*synth*)

$C_{10}H_6CrO_4$ Cr-00069

Tricarbonyl[(2,3,4,5,6,7-η)-2,4,6-cycloheptatrien-1-one-]chromium, 8CI
Tricarbonyl(η⁶-tropone)chromium
[32648-85-4]

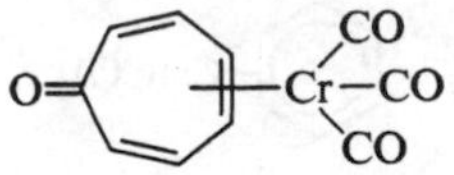

M 242.151
Orange-red cryst. Sol. Et_2O. Mp 120.5-121.5°.

Pauson, P.L. *et al*, *J. Chem. Soc. (C)*, 1970, 2315 (*synth, props*)
Barrow, M.J. *et al*, *J. Chem. Soc., Chem. Commun.*, 1971, 119 (*cryst struct*)

$C_{10}H_7CrO_3^{\oplus}$ Cr-00070

Tricarbonyl(η⁷-cycloheptatrienylium)chromium(1+), 10CI
Tricarbonyltropyliumchromium(1+)
[46238-43-1]

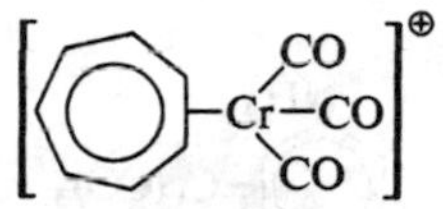

M 227.160 (ion)
Tetrafluoroborate: [12170-19-3].
$C_{10}H_7BCrF_4O_3$ M 313.963
Red cryst. or orange solid. Sol. CH_2Cl_2. Mp >300°.
Triiodide:
$C_{10}H_7CrI_3O_3$ M 607.873
Brown solid. Mp >150° dec.
Perchlorate:
$C_{10}H_7ClCrO_7$ M 326.610
Red cryst.
▷Dec. explosively >170°

Monro, J.D. *et al*, *J. Chem. Soc.*, 1961, 3475 (*props*)
Connor, J.A. *et al*, *J. Organomet. Chem.*, 1970, **24**, 441 (*synth*)
Deganello, G. *et al*, *J. Organomet. Chem.*, 1975, **97**, C46 (*props*)
Hackett, P. *et al*, *Inorg. Chim. Acta.*, 1975, **12**, L19 (*props*)
John, G.R. *et al*, *J. Organomet. Chem.*, 1976, **120**, C45 (*props*)
Olah, G. *et al*, *J. Org. Chem.*, 1976, **41**, 1694 (*pmr, cmr*)
Alton, J.G. *et al*, *Inorg. Chim. Acta.*, 1980, **41**, 245 (*props*)

$C_{10}H_8CrO_3$ Cr-00071

Tricarbonyl[(1,2,3,4,5,6-η)-1,3,5-cycloheptatriene]chromium, 9CI
[12125-72-3]

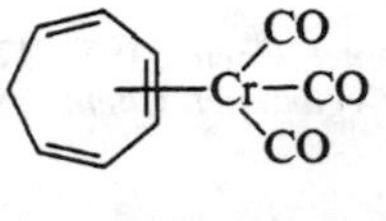

M 228.167
Orange-red cryst. (hexane). Mp 128-30°.

Abel, E.W. *et al*, *J. Chem. Soc.*, 1958, 4559 (*synth*)
Fischer, R.D., *Chem. Ber.*, 1960, **93**, 165 (*ir*)
Bennett, M.A. *et al*, *J. Chem. Soc.*, 1961, 2037 (*pmr*)
Deckelmann, E. *et al*, *Helv. Chim. Acta*, 1969, **52**, 892 (*props*)
Müller, J. *et al*, *Chem. Ber.*, 1970, **103**, 3128 (*ms*)
Mann, B.E., *J. Chem. Soc., Chem. Commun.*, 1971, 976 (*cmr*)
Pauson, P.L. *et al*, *J. Chem. Soc., Perkin Trans. 2*, 1972, 1141.
Adedeji, F.A. *et al*, *J. Organomet. Chem.*, 1975, **97**, 221.
Kreiter, C.G. *et al*, *Chem. Ber.*, 1975, **108**, 1502 (*nmr*)
Gower, M. *et al*, *Inorg. Chim. Acta*, 1979, **37**, 79 (*props*)

$C_{10}H_8CrO_3$ **Cr-00072**

Tricarbonyl[(1,2,3,4,5,6-η)methylbenzene]chromium, 9CI
Tricarbonyl(η⁶-toluene)chromium. (η⁶-Toluene)-tricarbonylchromium
[12083-24-8]

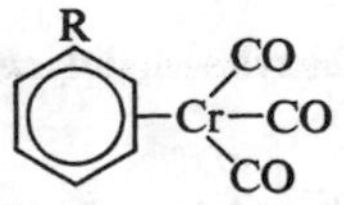

R = CH_3

M 228.167
Mp 80-1°.

Nicholls, B. *et al*, *J. Chem. Soc.*, 1959, 551 (*synth*)
Fischer, R.D., *Chem. Ber.*, 1960, **93**, 165 (*ir*)
Jackson, W.R. *et al*, *J. Chem. Soc.*, 1960, 469 (*synth*)
Lloyd, M.K. *et al*, *J. Chem. Soc., Dalton Trans.*, 1973, 1768 (*electrochem*)
Jackson, W.R. *et al*, *Aust. J. Chem.*, 1975, **28**, 1535 (*cmr*)
Koerner von Gustorf, E.A. *et al*, *Adv. Inorg. Chem. Radiochem.*, 1976, **19**, 65 (*props*)
v. Meurs, F. *et al*, *J. Organomet. Chem.*, 1976, **113**, 341 (*pmr*)
Bamford, C.H. *et al*, *J. Chem. Soc., Faraday Trans. 1*, 1977, **73**, 1406 (*props*)

$C_{10}H_8CrO_4$ **Cr-00073**

Tricarbonyl-(1,2,3,4,5-η-methoxybenzene)chromium, 9CI
Anisoletricarbonylchromium, 8CI. Tricarbonylanisolechromium
[12116-44-8]

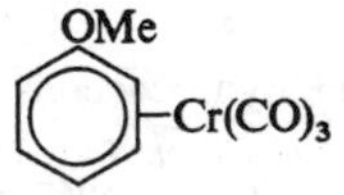

M 244.167
Catalyst for the selective hydrogenation of unsaturated fats. Yellow cryst. (C_6H_6/pet. ether or by subl.). Mp 84-5°. Air sensitive.

Mangini, A. *et al*, *Inorg. Chim. Acta*, 1968, **2**, 8 (*pmr*)
Roques, B.P. *et al*, *J. Organomet. Chem.*, 1974, **73**, 327 (*cmr*)
Semmelhack, M.F., *Ann. N.Y. Acad. Sci.*, 1977, **295**, 36 (*use*)
Inorg. Synth., 1979, **19**, 154 (*synth, props*)
Semmelhack, M.F. *et al*, *J. Org. Chem.*, 1979, **44**, 3275.

$C_{10}H_{10}ClCrNO_5$ **Cr-00074**

Pentacarbonyl[chloro(diethylamino)methylene]chromium, 10CI
[59218-83-6]

$(OC)_5Cr{=}CClNEt_2$

M 311.642
Yellow cryst. Dec. >30° to $Cl(OC)_4Cr{\equiv}CNEt_2$.

Fischer, E.O. *et al*, *J. Organomet. Chem.*, 1976, **107**, C23 (*synth*)
Huttner, G. *et al*, *J. Organomet. Chem.*, 1977, **141**, C17 (*synth, cryst struct*)
Fischer, H. *et al*, *Angew. Chem., Int. Ed. Engl.*, 1978, **17**, 842 (*props*)
Fischer, H. *et al*, *Angew. Chem., Int. Ed. Engl.*, 1981, **20**, 463 (*props*)

$C_{10}H_{10}Cr$ **Cr-00075**

Chromocene, 9CI
Di-π-cyclopentadienylchromium, 8CI. Biscyclopentadienylchromium
[1271-24-5]

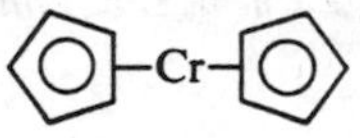

M 182.185
Red needles. Mp 173°. V. air-sensitive. Paramagnetic.

Cotton, F.A. *et al*, *Z. Naturforsch., B*, 1954, **9**, 417 (*synth*)
Fischer, E.O. *et al*, *Z. Naturforsch., B*, 1954, **9**, 503 (*synth*)
Friedman, L. *et al*, *J. Am. Chem. Soc.*, 1955, **77**, 3689 (*ms*)
Fischer, E.O., *Recl. Trav. Chim. Pays-Bas*, 1956, **75**, 629 (*struct*)
Fritz, H.P., *Chem. Ber.*, 1959, **92**, 780 (*ir*)
Koehler, F.H., *J. Organomet. Chem.*, 1976, **110**, 235 (*synth, cmr, pmr*)

$C_{10}H_{10}CrN_2O_5$ **Cr-00076**

Pentacarbonyl(diethylcyanamide-N′)chromium, 10CI
[66915-54-6]

$(OC)_5Cr(NCNEt_2)$

M 290.195
Air-sensitive, pale-yellow cryst. (Et_2O/pentane). Sol. Me_2CO, CH_2Cl_2. Mp 77°.

Schubert, U. *et al*, *Acta Crystallogr., Sect. B*, 1978, **34**, 2293 (*cryst struct*)
Fischer, E.O. *et al*, *J. Organomet. Chem.*, 1978, **149**, C40 (*synth, ir, cmr, pmr, cryst struct*)

$C_{10}H_{10}CrO_5S_2$ **Cr-00077**

[Bis(ethylthio)methylene]pentacarbonylchromium, 10CI
[67483-97-0]

$(OC)_5Cr{=}C(SEt)_2$

M 326.302
Air-stable, orange-red cryst. Mp 52°.

Lappert, M.F. *et al*, *J. Chem. Soc., Chem. Commun.*, 1978, 146 (*synth*)

$C_{10}H_{10}Cr_2N_4O_4$ **Cr-00078**

Bis(η⁵-2,4-cyclopentadien-1-yl)di-μ-nitrosyldinitrosyldichromium, 9CI
Bis(η⁵-cyclopentadienyl)trinitrosyldichromium
[36607-01-9]

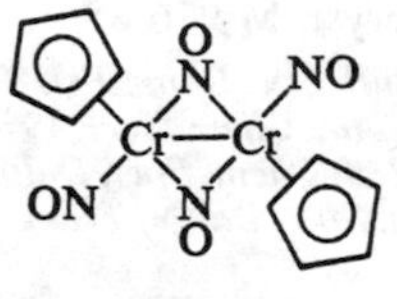

M 354.205
Purple solid. Mp 146.5-147°.

Kirchner, R.M. *et al*, *J. Am. Chem. Soc.*, 1973, **95**, 6602 (*synth, cryst struct*)
Calderon, J.L. *et al*, *J. Organomet. Chem.*, 1974, **64**, C10.
Kolthammer, B.W.S. *et al*, *Tetrahedron Lett.*, 1978, 323 (*props*)
Inorg. Synth., 1979, **19**, 208.

$C_{10}H_{12}CrN_2O_5$ **Cr-00079**

[Bis(dimethylamino)methylene]pentacarbonylchromium, 10CI
[61771-06-0]

$(OC)_5Cr{=}C(NMe_2)_2$

M 292.211

Yellow needles. Mp 102°.

Hartshorn, A.J. *et al*, *J. Chem. Soc., Chem. Commun.*, 1976, 761 (*synth, ir, nmr*)
Hartshorn, A.J. *et al*, *J. Chem. Soc., Dalton Trans.*, 1978, 348 (*synth, props*)

$C_{10}H_{12}CrO_6Si$ Cr-00080

Pentacarbonyl[1-(trimethylsiloxy)ethylidene]chromium, 8CI

[12307-27-6]

$(OC)_5Cr{=}C(CH_3)OSiMe_3$

M 308.283

Orange cryst.

Fischer, E.O. *et al*, *J. Organomet. Chem.*, 1968, **12**, P1 (*synth, ir, nmr*)

$C_{10}H_{18}CrO_4P_2$ Cr-00081

Tetracarbonylbis(trimethylphosphine)chromium, 9CI

[17548-29-7]

As Tetracarbonylbis(phosphine)chromium, Cr-00003 with

R = Me

M 316.193

cis-form

Pale-yellow cryst. Mp 71°.

Jenkins, J.H. *et al*, *Inorg. Chem.*, 1967, **6**, 2250.
Mathieu, R. *et al*, *Inorg. Chem.*, 1970, **9**, 2030.
Connor, J.A. *et al*, *J. Organomet. Chem.*, 1972, **43**, 357.

$C_{10}H_{18}CrO_{10}P_2$ Cr-00082

Tetracarbonylbis(trimethyl phosphite-*P*)chromium, 8CI

[21370-42-3]

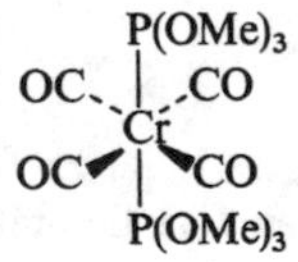

M 412.190

trans-form

Yellow air-stable cryst. Mp 76-8°.

Poilblanc, R. *et al*, *Bull. Soc. Chim. Fr.*, 1962, 1301.
Ogilvie, F.B. *et al*, *J. Am. Chem. Soc.*, 1970, **92**, 1916 (*nmr*)
Braterman, P.S. *et al*, *J. Chem. Soc., Dalton Trans.*, 1973, 1027.
Mann, B.E., *J. Chem. Soc., Dalton Trans.*, 1973, 2012.

$C_{11}H_3Cr_2O_{10}S^{\ominus}$ Cr-00083

Decarbonyl(*μ*-methanethiolato)dichromate(1−)

$[(OC)_5Cr{-}SMe{-}Cr(CO)_5]^{\ominus}$

M 431.191 (ion)

Yellow. Diamagnetic.

Tetraethylammonium salt: [24604-40-8].
$C_{19}H_{23}Cr_2NO_{10}S$ M 561.443
Sl. air-sensitive orange-yellow cryst. (Me_2CO aq.).

Bis(triphenylphosphine)iminium salt: [22017-07-8].
$C_{47}H_{33}Cr_2NO_{10}P_2S$ M 969.778
No phys. props. reported.

Behrens, H. *et al*, *Z. Anorg. Allg. Chem.*, 1969, **369**, 131 (*synth, ir, nmr*)

$C_{11}H_5BrCrO_4$ Cr-00084

Bromotetracarbonyl(phenylmethylidyne)chromium, 9CI

[50701-13-8]

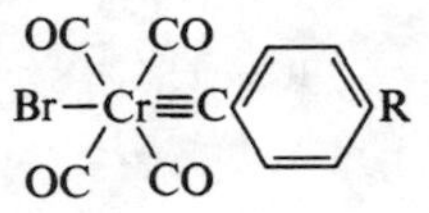

R = H

M 333.058

Yellow needles (pentane).

Fischer, E.O. *et al*, *Chem. Ber.*, 1976, **109**, 1673; 1977, **110**, 805 (*synth, ir, nmr*)
Frank, A. *et al*, *J. Organomet. Chem.*, 1978, **161**, C27 (*cryst struct*)

$C_{11}H_8CrO_3$ Cr-00085

Tricarbonyl[(1,2,3,4,5,6-*η*)-1,3,5,7-cyclooctatetraene]chromium, 10CI

[12093-03-7]

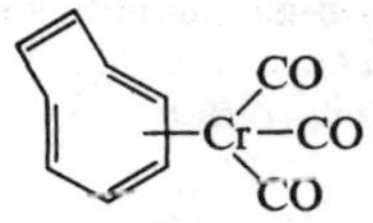

M 240.178

Red-brown cryst. Sol. hydrocarbons.

Kreiter, C.G. *et al*, *J. Am. Chem. Soc.*, 1966, **88**, 3444 (*synth*)
King, R.B., *J. Organomet. Chem.*, 1967, **8**, 139 (*synth*)
Mann, B.E. *et al*, *J. Chem. Soc., Dalton Trans.*, 1979, 1021 (*nmr*)

$C_{11}H_8CrO_3$ Cr-00086

Tricarbonyl[(1,2,3,4,5,6-*η*)-7-methylene-1,3,5-cycloheptatriene]chromium, 9CI

Tricarbonyl($η^6$-heptafulvene)chromium

[52472-03-4]

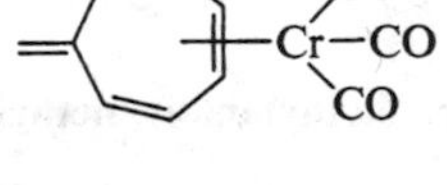

M 240.178

Unstable red oil, characterised by ir and nmr.

Howell, J.A.S. *et al*, *J. Chem. Soc., Dalton Trans.*, 1974, 293 (*synth, nmr*)

$C_{11}H_8CrO_4$ Cr-00087

[(2,3,5,6-*η*)-Bicyclo[2.2.1]hepta-2,5-diene]tetracarbonylchromium, 9CI

Tetracarbonyl(2,5-norbornadiene)chromium, 8CI. Tetracarbonylnorbornadienechromium

[12146-36-0]

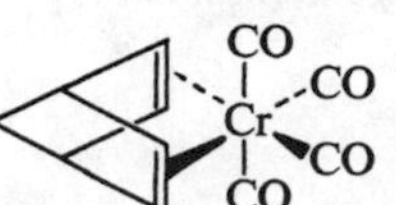

M 256.178

Orange-yellow or white cryst. Mp 92-3°.

Bennett, M.A. *et al*, *J. Chem. Soc.*, 1961, 2037 (*synth, ir, pmr*)

Dub, M. *Organometallic Compounds*, Springer-Verlag, Berlin, 2nd Ed., 1966, **1**.
Mann, B.E., *J. Chem. Soc., Dalton Trans.*, 1973, 2012 (*cmr*)
Rietvelde, D. *et al*, *J. Organomet. Chem.*, 1976, **118**, 191 (*props*)
Darensbourg, D.J. *et al*, *J. Am. Chem. Soc.*, 1977, **99**, 896 (*synth, props*)

$C_{11}H_8CrO_4$ Cr-00088

Tricarbonyl[1-(η^6-phenyl)ethanone]chromium, 9CI

Tricarbonyl(η^6-acetophenone)chromium. (Acetylbenzene)tricarbonylchromium

[12153-11-6]

As Tricarbonyl[(1,2,3,4,5,6-η)methylbenzene]chromium, Cr-00072 with

R = $COCH_3$

M 256.178

Mp 91-2°.

Nicholls, B. *et al*, *J. Chem. Soc.*, 1959, 551 (*synth*)
Ercoli, R. *et al*, *Chim. Ind. (Milan)*, 1959, **41**, 404 (*synth*)
Jackson, W.R. *et al*, *J. Chem. Soc.*, 1960, 469 (*synth*)
Dusausoy, Y., *C.R. Hebd. Seances Acad. Sci.*, 1970, **270**, 1792 (*cryst struct*)
Gubin, S.P. *et al*, *J. Organomet. Chem.*, 1970, **22**, 149 (*props*)
Lumbroso, H. *et al*, *J. Organomet. Chem.*, 1973, **61**, 249.

$C_{11}H_8CrO_5$ Cr-00089

Tricarbonyl[(1,2,3,4,5,6-η)methyl benzoate]chromium, 9CI

(Benzoic acid)tricarbonylchromium methyl ester, 8CI. *(Methyl benzoate)tricarbonylchromium*

[12125-87-0]

As Tricarbonyl[(1,2,3,4,5,6-η)methylbenzene]chromium, Cr-00072 with

R = COOMe

M 272.177

Catalyst for hydrogenation of unsatd. fats, esp. 1,3-dienes to *cis*-monoenes. Red cryst. (Et_2O/pet. ether or by subl.). Mp 95-6°.

Fischer, E.O. *et al*, *Chem. Ber.*, 1958, **91**, 2763 (*synth*)
Jackson, W.R. *et al*, *J. Chem. Soc.*, 1960, 469 (*synth*)
Fischer, R.D., *Chem. Ber.*, 1960, **93**, 165 (*ir*)
Sim, G.A. *et al*, *J. Chem. Soc. (A)*, 1967, 1619 (*cryst struct*)
Frankel, E.N. *et al*, *J. Org. Chem.*, 1969, **34**, 3936, 3960 (*use*)
Lloyd, M.K. *et al*, *J. Chem. Soc., Dalton Trans.*, 1973, 1768.
van Meurs, F. *et al*, *J. Organomet. Chem.*, 1976, **113**, 341 (*synth*)
Inorg. Synth., 1979, **19**, 154 (*synth*)

$C_{11}H_{10}CrO_3$ Cr-00090

Tricarbonyl[(1,2,3,4,5,6-η)-1,3,5-cyclooctatriene]chromium, 9CI

[55928-13-7]

Cr–CO, CO, CO

M 242.194

Red cryst.

Fischer, E.O. *et al*, *Chem. Ber.*, 1959, **92**, 2645 (*synth*)
Prout, C.K. *et al*, *J. Chem. Soc.*, 1962, 3770 (*cryst struct*)
Aumann, R. *et al*, *Tetrahedron Lett.*, 1970, 903 (*synth*)
Kreiter, C.G. *et al*, *Chem. Ber.*, 1975, **108**, 1502 (*nmr*)

$C_{11}H_{14}CrGeO_3$ Cr-00091

Tricarbonyl-π-cyclopentadienyl(trimethylgermyl)chromium, 8CI

[31884-35-2]

CO, CO, Cr, CO, X

X = $-GeMe_3$

M 318.816

Yellow cryst. Mp 86-8°. $Bp_{0.001}$ 50° subl.

Cardin, D.J. *et al*, *J. Chem. Soc. (A)*, 1970, 2594 (*synth, props*)
Keppie, S.A. *et al*, *J. Chem. Soc. (A)*, 1971, 3216 (*synth*)

$C_{11}H_{14}CrO_3Sn$ Cr-00092

Tricarbonyl-π-cyclopentadienyl(trimethylstannyl)chromium, 8CI

[31854-87-2]

M 364.916

Lemon-yellow cryst. Mp 109-10°. $Bp_{0.001}$ 80° subl.

Patil, H.R.H. *et al*, *Inorg. Chem.*, 1966, **5**, 1401 (*synth*)
Cardin, D.J. *et al*, *J. Chem. Soc. (A)*, 1970, 2594 (*synth, props*)
Keppie, S.A. *et al*, *J. Chem. Soc. (A)*, 1971, 3216 (*synth, props*)

$C_{11}H_{18}CrO_5Sn_2Te$ Cr-00093

Pentacarbonyl(hexamethyldistannatellurane)chromium, 9CI

[28480-04-8]

$(OC)_5CrTe(SnMe_3)_2$ (O_h)

M 647.236

Solid. Mp 85°.

Schumann, H. *et al*, *Chem. Ber.*, 1973, **106**, 2049 (*ir, raman, nmr*)

$C_{12}H_5CrO_6^{\ominus}$ Cr-00094

Benzoylpentacarbonylchromate(1–), 9CI

$[(OC)_5CrCOPh]^{\ominus}$ (O_h)

M 297.164 (ion)

Li salt: [39912-34-0].

$C_{12}H_5CrLiO_6$ M 304.105

Reactive.

Tetramethylammonium salt: [54817-05-9].

$C_{16}H_{17}CrNO_6$ M 371.309

Pale-yellow cryst. Mp 92°.

Fischer, E.O. *et al*, *Chem. Ber.*, 1967, **100**, 2445 (*synth*)
Aumann, R. *et al*, *Chem. Ber.*, 1968, **101**, 954 (*props*)
Fischer, E.O. *et al*, *J. Organomet. Chem.*, 1972, **46**, C15 (*props*)

Fischer, E.O. *et al*, *Chem. Ber.*, 1974, **107**, 3554 (*props*)
Fischer, E.O. *et al*, *J. Organomet. Chem.*, 1975, **91**, C23 (*props*)

$C_{12}H_7BrCrO_4$ Cr-00095

Bromotetracarbonyl[(4-methylphenyl)methylidyne]chromium, 10CI

[57987-16-3]

As Bromotetracarbonyl(phenylmethylidyne)chromium, Cr-00084 with

R = CH_3

M 347.085

Yellow powder.

Fischer, E.O. *et al*, *Monatsh. Chem.*, 1977, **108**, 759 (*synth*)
Kreissl, F.R. *et al*, *Chem. Ber.*, 1978, **111**, 3283 (*props*)

$C_{12}H_7CrNO_5$ Cr-00096

(Aminophenylmethylene)pentacarbonylchromium, 10CI

[32370-44-8]

$(OC)_5Cr{=}CPhNH_2$

M 297.187

Yellow needles. Mp 79°.

Klabunde, U. *et al*, *J. Am. Chem. Soc.*, 1967, **89**, 7141.
Moser, E. *et al*, *J. Organomet. Chem.*, 1968, **12**, P1 (*synth*)
Fischer, E.O. *et al*, *Chem. Ber.*, 1971, **104**, 1339 (*synth*)
Block, T.F. *et al*, *J. Am. Chem. Soc.*, 1977, **99**, 4321 (*pe*)
Block, T.F. *et al*, *J. Organomet. Chem.*, 1977, **139**, 235.
Doetz, K.H. *et al*, *Chem. Ber.*, 1980, **113**, 3597 (*props*)

$C_{12}H_{10}Cl_2Cr$ Cr-00097

Bis(η^6-chlorobenzene)chromium, 9CI

[42087-89-8]

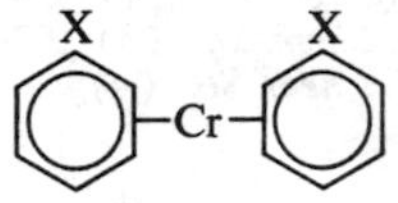

X = Cl

M 277.113

Black cryst. Mp 135° dec.

Graves, V. *et al*, *Inorg. Chem.*, 1976, **15**, 577 (*synth, props*)

$C_{12}H_{10}CrF_2$ Cr-00098

Bis(η^6-fluorobenzene)chromium, 9CI

[42087-90-1]

As Bis(η^6-chlorobenzene)chromium, Cr-00097 with

X = F

M 244.204

Black cryst. Mp 204° dec.

Skell, P.S. *et al*, *J. Am. Chem. Soc.*, 1973, **95**, 3337 (*synth, props*)
Graves, V. *et al*, *Inorg. Chem.*, 1976, **15**, 577 (*synth, props*)

$C_{12}H_{10}CrO_4$ Cr-00099

Tricarbonyl[(3a,4,5,6,7,7a-η)-2,3-dihydro-1*H*-inden-1-ol-]chromium, 9CI

Tricarbonyl-η^6-1-indanolchromium

[12088-83-4]

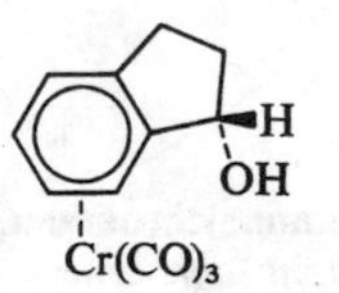

1(*S*)-(–)-*form*

M 270.205

(R)-endo-form

Yellow needles. $[\alpha]_D^{22}$ +60.9° (c, 2.07 in $CHCl_3$).

(S)-endo-form

Yellow prismatic cryst. Mp 110°. $[\alpha]_D^{22}$ –60.7° (c, 2.14 in $CHCl_3$).

(±)-endo-form

Yellow cryst. Sol. CCl_4. Mp 105°.

Monosuccinate:

$C_{16}H_{14}CrO_7$ M 370.278

Cryst. (Et_2O/pentane). Mp 123°.

Jaouen, G. *et al*, *J. Am. Chem. Soc.*, 1975, **97**, 4667 (*synth*)

$C_{12}H_{11}CrO_4^{\oplus}$ Cr-00100

Tricarbonyl(ethoxy-π-cycloheptatrienylium)chromium(1+), 8CI

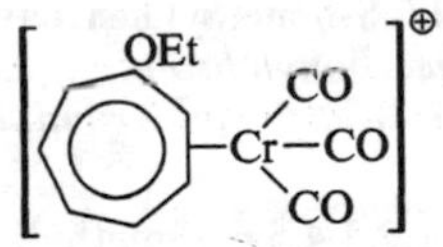

M 271.213 (ion)

Tetrafluoroborate: [32825-28-8].

$C_{12}H_{11}BCrF_4O_4$ M 358.016

Orange cryst. Mp 110°.

Pauson, P.L. *et al*, *J. Chem. Soc. (C)*, 1970, 2315 (*synth*)

$C_{12}H_{12}Cr$ Cr-00101

Bis(η^6-benzene)chromium, 9CI

Dibenzenechromium

[1271-54-1]

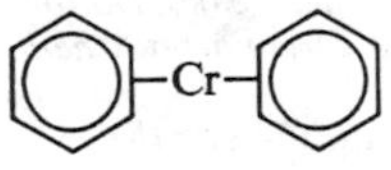

M 208.223

Polymerisation catalyst. Brown-black air-sensitive cryst. Sol. C_6H_6, sl. sol. Et_2O. Mp 284-5° (280-1°). Adopts strict D_{6h} symmetry in the solid state. Subl. at 160° *in vacuo*. Dec. on pyrolysis to metallic Cr.

▷GB5850000.

Fischer, E.O. *et al*, *Z. Anorg. Allg. Chem.*, 1961, **312**, 244 (*synth, props*)
Keulen, E. *et al*, *J. Organomet. Chem.*, 1966, **5**, 490 (*cryst struct*)
Emanuel, R.V. *et al*, *J. Chem. Soc. (A)*, 1969, 3002 (*nmr*)
Timms, P.L., *J. Chem. Soc., Chem. Commun.*, 1969, 1033 (*synth*)
Timms, P.L., *J. Chem. Educ.*, 1972, **49**, 782 (*synth*)
Connor, J.A. *et al*, *J. Chem. Soc., Faraday Trans. 1*, 1973, **69**, 1218.
Graves, V. *et al*, *Inorg. Chem.*, 1976, **15**, 577 (*cmr*)
Graves, V. *et al*, *J. Organomet. Chem.*, 1976, **120**, 397 (*pmr*)

$C_{12}H_{12}Cr$ **Cr-00102**

(η^7-Cycloheptatrienylium)(η^5-2,4-cyclopentadien-1-yl)chromium, 9CI

π-Cycloheptatrienyl-π-cyclopentadienylchromium

[12093-81-1]

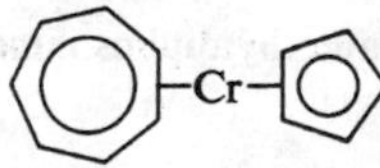

M 208.223

Blue-black cryst. Mp ca. 230° dec. Sublimes at 80-100°/0.1 mm.

King, R.B. *et al*, *Inorg. Chem.*, 1964, **3**, 785 (*synth*)
Fischer, E.O. *et al*, *Chem. Ber.*, 1966, **99**, 2905 (*synth*)
Elschenbroich, C. *et al*, *J. Am. Chem. Soc.*, 1973, **95**, 6956.
Groenenboom, C.J. *et al*, *J. Organomet. Chem.*, 1974, **76**, C4, 229 (*cmr, pmr*)
v. Oven, H.O. *et al*, *J. Organomet. Chem.*, 1974, **81**, 379 (*synth*)

$C_{12}H_{12}CrO_3$ **Cr-00103**

Tricarbonyl[(1,2,3,4,5,6-η)-1,3,5-trimethylbenzene]chromium, 9CI

Tricarbonyl(mesitylene)chromium, 8CI. (η^6-1,3,5-Trimethylbenzene)tricarbonylchromium. Mesitylenetricarbonylchromium

[12129-67-8]

M 256.221

Yellow needles (diisopropyl ether). Mp 172-4°. Sublimes *in vacuo*.

Fischer, E.O. *et al*, *Z. Naturforsch.*, *B*, 1958, **13**, 458 (*synth*)
Nicholls, B. *et al*, *J. Chem. Soc.*, 1959, 551; 1960, 469 (*synth*)
Fischer, R.D., *Chem. Ber.*, 1960, **93**, 165 (*ir*)
Strohmeier, W. *et al*, *Chem. Ber.*, 1964, **97**, 1877 (*pmr*)
Jackson, W.R. *et al*, *Aust. J. Chem.*, 1975, **28**, 1535 (*cmr*)
Inorg. Synth., 1978, **19**, 154 (*synth*)

$C_{12}H_{12}CrO_4$ **Cr-00104**

Tetracarbonyl[(1,2,5,6-η)-1,5-cyclooctadiene]chromium, 9CI

[12301-34-7]

M 272.220

White cryst.

Bennett, M.A., *Adv. Organomet. Chem.*, 1966, **4**, 353 (*synth, props*)
Rietvelde, D. *et al*, *J. Organomet. Chem.*, 1976, **118**, 191 (*synth, props*)
Darensbourg, D.J. *et al*, *J. Am. Chem. Soc.*, 1977, **99**, 896 (*synth, props*)

$C_{12}H_{14}CrO_3Si$ **Cr-00105**

[Trimethyl(η^6-phenyl)silane]tricarbonylchromium, 9CI

Tricarbonyl(η^6-trimethylsilylbenzene)chromium

[33248-13-4]

As Tricarbonyl[(1,2,3,4,5,6-η)methylbenzene]chromium, Cr-00072 with

R = $SiMe_3$

M 286.322

Mp 72-3°.

Seyferth, D. *et al*, *Inorg. Chem.*, 1963, **2**, 417 (*synth*)
Hagen, A.P. *et al*, *Inorg. Chem.*, 1976, **15**, 1512 (*synth*)
Van Meurs, F. *et al*, *J. Organomet. Chem.*, 1976, **113**, 353 (*props*)

$C_{12}H_{14}CrO_3Sn$ **Cr-00106**

Tricarbonyl(trimethylphenylstannane)chromium, 8CI

Tricarbonyl(trimethylstannylbenzene)chromium

[31868-92-5]

As Tricarbonyl[(1,2,3,4,5,6-η)methylbenzene]chromium, Cr-00072 with

R = $SnMe_3$

M 376.927

Mp 78-9°.

Seyferth, D. *et al*, *Inorg. Chem.*, 1963, **2**, 417 (*synth*)
Poeth, T.P. *et al*, *Inorg. Chem.*, 1971, **10**, 522 (*synth*)

$C_{12}H_{18}CrN_2O_4S$ **Cr-00107**

(Di-*tert*-butylsulfurdiimide-*N,N'*)tetracarbonylchromium

*[Bis(1,1-dimethylethyl)sulfurdiimide-*N,N'*]tetracarbonylchromium, 9CI*

[55257-39-1]

M 338.341

Purple cryst. Mp 111° dec.

Lindsell, W.E. *et al*, *J. Chem. Soc., Dalton Trans.*, 1975, 40 (*synth*)
Meij, R. *et al*, *J. Organomet. Chem.*, 1976, **110**, 219 (*synth*)

$C_{12}H_{20}Cr_2$ **Cr-00108**

Tetra-2-propenyldichromium, 9CI

Tetra(π-allyl)dichromium

[12295-17-9]

M 268.282

Brown-red cryst. Mp 121-9° dec.

Aoki, T. *et al*, *Bull. Chem. Soc. Jpn.*, 1969, **42**, 545 (*synth, props*)

$C_{12}H_{27}CrO_3P_3$ **Cr-00109**

Tricarbonyltris(trimethylphosphine)chromium

[30476-93-8]

M 364.261

mer-form
Yellow low-melting solid.

Jenkins, J.H. *et al*, *Inorg. Chem.*, 1967, **6**, 2250.
Mathieu, R. *et al*, *Inorg. Chem.*, 1970, **9**, 2030.
Connor, J.A. *et al*, *J. Organomet. Chem.*, 1972, **43**, 357.

$C_{12}H_{28}Cr$ Cr-00110

Tetraisopropylchromium
Tetrakis(1-methylethyl)chromium, 9CI.
Tetrakis(isopropyl)chromium
[38711-69-2]

$Cr[CH(CH_3)_2]_4$ (T_d)

M 224.349
Dark brown-red cryst. Sol. pentane.
▷Pyrophoric

Kruse, W., *J. Organomet. Chem.*, 1972, **42**, C39 (*synth*)
Müller, J. *et al*, *Angew. Chem., Int. Ed. Engl.*, 1975, **14**, 760 (*synth*)

$C_{13}H_8CrO_3$ Cr-00111

Tricarbonyl[(1,2,3,4,4*a*,8*a*-η)naphthalene]chromium, 9CI
Naphthalenetricarbonylchromium
[12110-37-1]

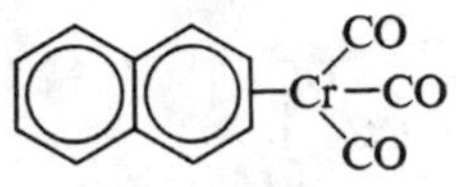

M 264.200
Red cryst.

Fischer, E.O. *et al*, *Chem. Ber.*, 1958, **91**, 2763 (*synth*)
Natta, G. *et al*, *Chem. Ind. (Milan)*, 1958, **40**, 1003 (*synth*)
Kunz, V. *et al*, *Helv. Chim. Acta*, 1967, **50**, 1052 (*cryst struct*)
Nicholas, K.M. *et al*, *Inorg. Chem.*, 1971, **10**, 1519 (*pmr, ir*)
Nazarova, E.B. *et al*, *Izv. Akad. Nauk SSSR, Ser. Khim.*, 1979, 58; *CA*, **90**, 151182 (*ir, raman*)

$C_{13}H_8CrO_6$ Cr-00112

Pentacarbonyl(methoxyphenylmethylene)chromium, 9CI
[27436-93-7]

$(OC)_5Cr{=}CPhOMe$

M 312.199
Orange-red cryst. Mp 46°.

Fischer, E.O. *et al*, *Chem. Ber.*, 1967, **100**, 2445 (*synth*)
Aumann, R. *et al*, *Chem. Ber.*, 1968, **101**, 964 (*synth*)
Bodner, G.M. *et al*, *Inorg. Chem.*, 1973, **12**, 1071 (*nmr*)
Fischer, E.O. *et al*, *Chem. Ber.*, 1976, **109**, 1868 (*props*)
Doetz, K.H. *et al*, *Chem. Ber.*, 1976, **109**, 2033 (*props*)
Brown, F.J., *Prog. Inorg. Chem.*, 1980, **27**, 1 (*props*)

$C_{13}H_{10}CrO_3$ Cr-00113

Tricarbonyl(2,3,4,5-η:9,10-η-9-methylenebicyclo[4.2.1]nona-2,4,7-triene)chromium
[70585-10-3]

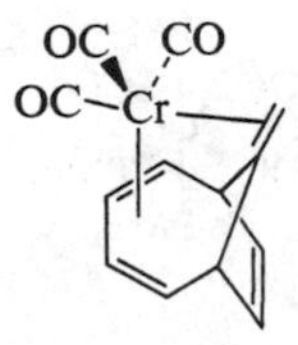

M 266.216
Red cryst. (hexane).

Jameson, E.B. *et al*, *Organometallics*, 1982, **1**, 689 (*synth, cryst struct*)

$C_{13}H_{12}CrFeP_2O_9$ Cr-00114

Tetracarbonyl[μ-(tetramethyldiphosphine)](pentacarbonylchromium)iron, 8CI
[34406-98-9]

$(OC)_5CrPMe_2PMe_2Fe(CO)_4$

M 482.023
Yellow. Mp 96°.

Vahrenkamp, H. *et al*, *Angew. Chem., Int. Ed. Engl.*, 1971, **10**, 513 (*synth*)

$C_{13}H_{12}CrO_4$ Cr-00115

Tricarbonyl[(4*a*,5,6,7,8,8*a*-η)-1,2,3,4-tetrahydro-1-naphthalenol]chromium, 9CI
Tricarbonyl-η⁶-1-tetralolchromium
[12126-74-8]

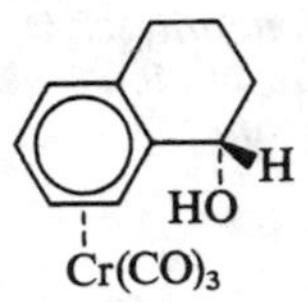

(*S*)-(+)-*form*

M 284.231

(*R*)-*endo-form*
Cryst. Mp 140°. $[\alpha]_D^{22}$ −20° (c, 2.245 in $CHCl_3$).

(*S*)-*endo-form*
Needles (Et_2O/pet. ether). Mp 138°. $[\alpha]_D^{22}$ +17° (c, 2.1 in $CHCl_3$), (85% optical purity).

(±)-*endo-form*
Cryst. (Et_2O/pet. ether). Mp 130°.
Monosuccinate:
$C_{17}H_{16}CrO_7$ M 384.305
Cryst. (Et_2O/pet. ether). Mp 143°.

Jaouen, G. *et al*, *J. Am. Chem. Soc.*, 1975, **97**, 4667 (*synth, ir, resoln, props*)

$C_{13}H_{16}CrO_3Si$ Cr-00116

Tricarbonyl(η⁶-benzyltrimethylsilyl)chromium
Tricarbonyl[trimethyl(η⁶-phenyl)methylsilane]chromium, 9CI
[53470-78-3]
As Tricarbonyl[(1,2,3,4,5,6-η)methylbenzene]chromium, Cr-00072 with

$R = CH_2SiMe_3$

M 300.349
Mp 143-4°.

Bly, R.S. *et al*, *J. Am. Chem. Soc.*, 1969, **91**, 4221 (*synth*)
Dosser, P.J. *et al*, *J. Organomet. Chem.*, 1974, **71**, 207 (*props*)

$C_{13}H_{16}CrO_3Sn$ Cr-00117

(Benzyltrimethylstannane)tricarbonylchromium, 8CI
Tricarbonyl(η⁶-benzyltrimethylstannane)chromium
[31724-95-5]
As Tricarbonyl[(1,2,3,4,5,6-η)methylbenzene]chromium, Cr-00072 with

$R = CH_2SnMe_3$

M 390.954
Mp 146-8°.

Harrison, P.G. *et al*, *J. Am. Chem. Soc.*, 1971, **93**, 5398 (*nmr*)
Poeth, T.P. *et al*, *Inorg. Chem.*, 1971, **10**, 522 (*synth, ir, raman*)

$C_{13}H_{27}Cl_2CrO_3$ Cr-00118

Dichloromethyltris(tetrahydrofuran)chromium, 9CI

[36153-92-1]

$[MeCrCl_2(THF)_3]$

M 354.257
Green cryst. μ_B = 3.85 bm.

Nishimura, K. *et al*, *J. Organomet. Chem.*, 1972, **37**, 317 (*synth*)
Yamamoto, A. *et al*, *J. Organomet. Chem.*, 1975, **102**, 57 (*synth*)

$C_{14}H_9CrNO_5$ Cr-00119

[1-(Benzylideneamino)ethylidene]pentacarbonylchromium, 8CI

[31246-02-3]

$(OC)_5Cr{=}C(CH_3)N{=}CHPh$

M 323.225
Deep-yellow needles. Mp 93°.

Fischer, E.O. *et al*, *Chem. Ber.*, 1970, **103**, 1262, 3744 (*synth, ir, nmr*)

$C_{14}H_{10}Cr_2O_4$ Cr-00120

Tetracarbonylbis(η^5-2,4-cyclopentadien-1-yl)dichromium, 10CI

[54667-87-7]

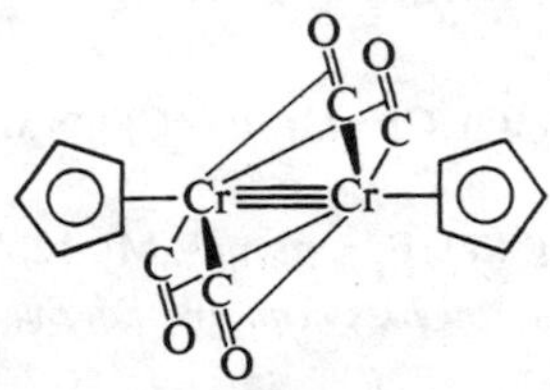

M 346.223
Four "semi-bridging" carbonyls in cryst. Cryst. (toluene/pet. ether). Mp 205-6° (sealed tube).

Hackett, P. *et al*, *J. Chem. Soc., Dalton Trans.*, 1974, 1625 (*synth, ir*)
Curtis, M.D. *et al*, *J. Organomet. Chem.*, 1978, **155**, 131 (*cryst struct*)
Jemmis, E.D. *et al*, *J. Am. Chem. Soc.*, 1980, **102**, 2576 (*struct*)

$C_{14}H_{16}Cr$ Cr-00121

Bis[(1,2,3,4,5,6-η)methylbenzene]chromium, 9CI

Bis(toluene)chromium

[12087-58-0]

As Bis(η^6-chlorobenzene)chromium, Cr-00097 with

X = CH_3

M 236.276
Green-black cryst. Mp 86-7°.

Fischer, E.O. *et al*, *Z. Anorg. Allg. Chem.*, 1956, **286**, 146 (*synth*)
Fischer, E.O. *et al*, *Z. Naturforsch., B*, 1956, **11**, 758 (*synth*)
Fischer, E.O. *et al*, *Chem. Ber.*, 1959, **92**, 938.
Skell, P.S. *et al*, *J. Am. Chem. Soc.*, 1973, **95**, 3337 (*synth, props*)

Graves, V. *et al*, *Inorg. Chem.*, 1976, **15**, 577 (*nmr*)
Graves, V. *et al*, *J. Organomet. Chem.*, 1976, **20**, 397 (*nmr*)

$C_{14}H_{18}CrNO_3^{\oplus}$ Cr-00122

Dicarbonyl[(1,2,3,4,5,6-η)hexamethylbenzene]nitrosylchromium(1+), 9CI

Dicarbonylnitrosyl(η^6-hexamethylbenzene)-chromium(1+)

CO / Cr—NO / CO ⊕

M 300.297 (ion)
Mp >170° dec.

Hexafluorophosphate: [43105-73-3].
$C_{14}H_{18}CrF_6NO_3P$ M 445.261
Yellow air-stable cryst.

Connelly, N.G. *et al*, *J. Chem. Soc., Dalton Trans.*, 1974, 2334; 1975, 2335 (*synth, ir, pmr*)

$C_{14}H_{36}CrO_2P_4$ Cr-00123

Dicarbonyltetrakis(trimethylphosphine)chromium, 8CI

[30476-94-9]

PMe_3, OC, PMe_3, Cr, OC, PMe_3, PMe_3

M 412.328

cis-form

Yellow cryst. Mp 150° dec.

Jenkins, J.H. *et al*, *Inorg. Chem.*, 1967, **6**, 2250 (*synth*)
Mathieu, R. *et al*, *Inorg. Chem.*, 1970, **9**, 2030 (*synth*)
Connor, J.A. *et al*, *J. Organomet. Chem.*, 1972, **43**, 357.

$C_{15}H_9BrCrFeO_4$ Cr-00124

Bromotetracarbonyl(ferrocenylmethylidyne)chromium, 10CI

Fe—C≡Cr—Br (OC, CO, OC, CO)

M 440.981

trans-form [68830-17-1]

Red microcryst. (MeOH). Mp >70° dec. Cl and I analogues also known.

Fischer, E.O. *et al*, *Chem. Ber.*, 1978, **111**, 3530 (*synth, struct, ir, pmr, cmr*)
Fischer, E.O. *et al*, *J. Organomet. Chem.*, 1980, **191**, 261 (*ir, pmr, cmr*)

$C_{15}H_{10}CrFeO_5$ Cr-00125

Tricarbonyl(η^5-2,4-cyclopentadien-1-yl)[dicarbonyl(η^5-2,4-cyclopentadien-1-yl)iron]chromium, 10CI
Pentacarbonylbis(η^5-2,4-cyclopentadien-1-yl)chromiumiron
[75339-31-0]

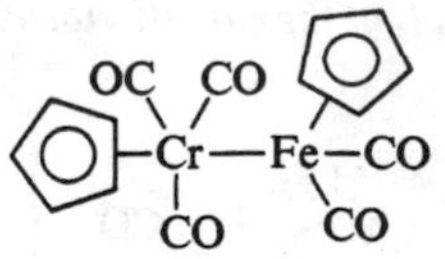

M 378.084
Green-black solid. Mp 107°.

Madach, T. *et al*, *Chem. Ber.*, 1980, **113**, 2675 (*synth, ir, pmr, uv*)

$C_{15}H_{10}CrO_6$ Cr-00126

Pentacarbonyl(1-methoxy-3-phenyl-2-propenylidene)chromium, 9CI
[54873-11-9]

$(OC)_5Cr{=}C(OMe)CH{=}CHPh$

M 338.236
Deep-red cryst. Mp 73-6°.

Casey, C.P. *et al*, *J. Organomet. Chem.*, 1974, **77**, 345 (*synth*)
Casey, C.P. *et al*, *Inorg. Chem.*, 1977, **16**, 391 (*synth, ir, nmr*)

$C_{15}H_{16}CrO_3$ Cr-00127

Tricarbonyl[(1,2,3,4,5,6-η)-1,3,5,7-tetramethyl-1,3,5,7-cyclooctatetraene]chromium, 9CI
[12302-33-9]

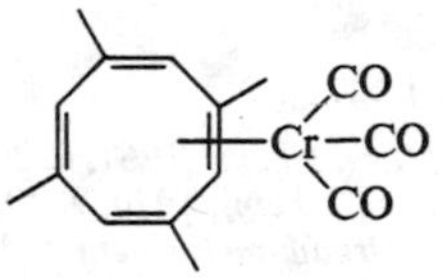

M 296.286
Red microcryst. Mp 65-9°.

Cotton, F.A. *et al*, *J. Am. Chem. Soc.*, 1968, **90**, 1438 (*synth*)
Bennett, M.J. *et al*, *J. Am. Chem. Soc.*, 1968, **90**, 903.
Cotton, F.A. *et al*, *J. Am. Chem. Soc.*, 1974, **96**, 7926 (*synth, nmr*)

$C_{16}H_{10}Cr_2HgO_6$ Cr-00128

Hexacarbonyldi-π-cyclopentadienyl-μ-mercuriodichromium, 8CI
Bis[tricarbonyl(η^5-cyclopentadienyl)chromium]mercury
[12194-11-5]

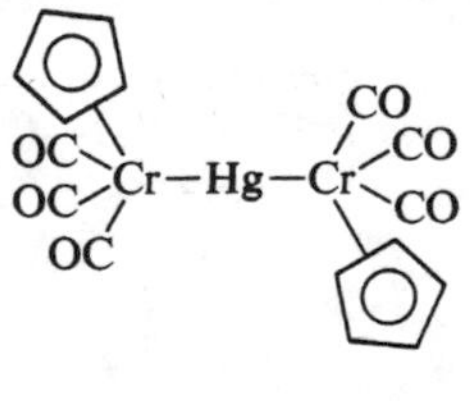

M 602.833
Yellow cryst. Mp 201-3°. $Bp_{0.1}$ 130° subl.
▷GB6225000.

Inorg. Synth., 1963, **7**, 99 (*synth*)
Dub, M., *Organometallic Compounds*, Springer-Verlag, Berlin, 2nd Ed., 1966, **1**.
Burlitch, J.M. *et al*, *Inorg. Chem.*, 1970, **9**, 563 (*props*)
Manning, A.R. *et al*, *J. Chem. Soc. (A)*, 1971, 637 (*synth, props*)

$C_{16}H_{10}Cr_2O_6$ Cr-00129

Hexacarbonylbis(η^5-2,4-cyclopentadien-1-yl)dichromium, 10CI
Bis[tricarbonyl(π-cyclopentadienyl)chromium]
[12194-12-6]

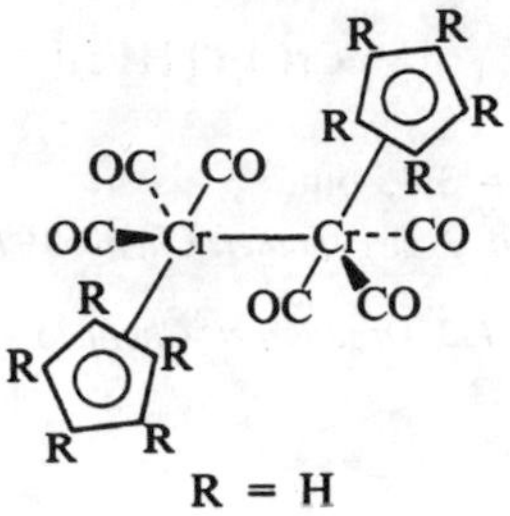

R = H

M 402.243
Deep-green cryst. $Bp_{0.1}$ 100-20° subl.

Inorg. Synth., 1963, 7, 104 (*synth*)
Adams, R.D. *et al*, *J. Am. Chem. Soc.*, 1974, **96**, 749 (*cryst struct*)
Hackett, P. *et al*, *J. Chem. Soc., Dalton Trans.*, 1974, 1625 (*props*)
Birdwhistell, R. *et al*, *J. Organomet. Chem.*, 1978, **157**, 239 (*synth*)
Braunstein, P. *et al*, *Inorg. Chem.*, 1981, **20**, 3586 (*props*)
Lindsell, W.L., *J. Chem. Soc., Chem. Commun.*, 1981, 531 (*props*)

$C_{16}H_{11}CrNO_5$ Cr-00130

Pentacarbonyl[3-(dimethylamino)-3-phenyl-1,2-propadienylidene]chromium, 9CI
[60349-51-1]

$(OC)_5Cr{=}C{=}C{=}CPhNMe_2$

M 349.263
Red-violet cryst. (THF/pentane). Mp 125° dec.

Fischer, E.O. *et al*, *Angew. Chem., Int. Ed. Engl.*, 1976, **15**, 623 (*synth, struct*)

$C_{16}H_{12}CrO_3$ Cr-00131

Tricarbonyl[(1,2,3,4,5,6-η)-7-phenyl-1,3,5-cycloheptatriene]chromium, 8CI
[12147-95-4]

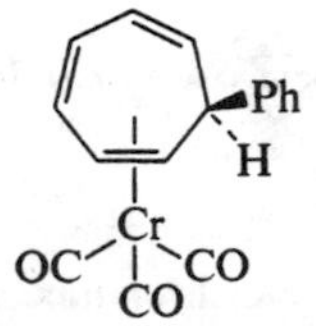

exo-form

M 304.265
exo-form
Red cryst. Mp 130-31°.
endo-form [12147-94-3]
Red needles. Mp 114-5°.

Pauson, P.L. *et al*, *J. Chem. Soc. (C)*, 1967, 1057, 1061; 1970, 2315 (*synth*)
Foreman, M.I. *et al*, *J. Chem. Soc., Perkin Trans. 2*, 1972, 1141.
Lloyd, M.K. *et al*, *J. Chem. Soc., Dalton Trans.*, 1973, 1768.

$C_{16}H_{16}Cr$ **Cr-00132**

($\eta^{1,2}$-[2,2]Paracyclophane)chromium, 10CI

[66901-71-1]

M 260.298

Lemon-yellow solid.

Elschenbroich, C. *et al*, *Angew. Chem.*, *Int. Ed. Engl.*, 1978, **17**, 531 (*synth*, *props*)

$C_{16}H_{18}Cl_2CrN_3$ **Cr-00133**

Dichloromethyltrispyridinechromium, 9CI

[36153-96-5]

$[MeCrCl_2(Py)_3]$

M 375.240

Green cryst.

Nishimura, U. *et al*, *J. Organomet. Chem.*, 1972, **37**, 317 (*synth*)

Yamamoto, A. *et al*, *J. Organomet. Chem.*, 1975, **102**, 57 (*synth*)

$C_{16}H_{18}CrO_4$ **Cr-00134**

Tetracarbonyl(1,2,3,4,5,6-hexamethylbicyclo[2.2.0]hexa-2,5-diene)chromium, 9CI

Tetracarbonyl(hexamethyldewarbenzene)chromium

[12242-79-4]

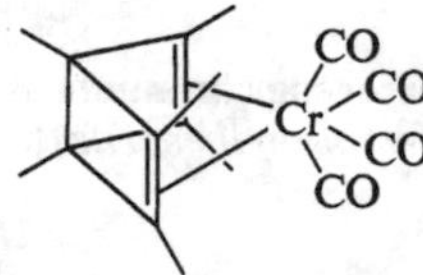

M 326.312

Yellow-orange cryst. Mp 115°.

Fischer, E.O. *et al*, *Chem. Ber.*, 1968, **101**, 824 (*synth*)

Huttner, E. *et al*, *J. Organomet. Chem.*, 1971, **29**, 275 (*cryst struct*)

$C_{16}H_{20}Cr$ **Cr-00135**

Bis[(1,2,3,4,5,6-η)ethylbenzene]chromium, 9CI

[12212-68-9]

As Bis(η^6-chlorobenzene)chromium, Cr-00097 with

$R = CH_2CH_3$

M 264.330

Brown-black liq.

Sorokin, A.Yu. *et al*, *J. Gen. Chem. USSR* (*Engl. Transl.*), 1965, **35**, 2123 (*synth*)

Suskina, I.A. *et al*, *Dokl. Akad. Nauk SSSR*, *Ser. Sci. Khim.*, 1971, 425 (*polarog*)

Graves, V. *et al*, *Inorg. Chem.*, 1976, **15**, 577 (*synth*, *nmr*)

Telnoi, V.I. *et al*, *Russ. Chem. Rev.* (*Engl. Transl.*), 1977, **46**, 689

$C_{16}H_{32}Cr_2^{\ominus\ominus\ominus\ominus}$ **Cr-00136**

Tetrakis(μ-tetramethylene)dichromate(4−), 8CI

M 328.421 (ion)

Tetra-Li salt: [31424-66-5].

$C_{16}H_{32}Cr_2Li_4$ M 356.185

Yellow air-sensitive cryst.

Krausse, J. *et al*, *J. Organomet. Chem.*, 1971, **27**, 59 (*synth*, *cryst struct*)

$C_{16}H_{40}Cr_2P_4$ **Cr-00137**

Tetrakis[(dimethylphosphonio)bis(methylene)]dichromium, 10CI

[53237-33-5]

M 460.379

Yellow cryst. Mp 163-5°.

Kurras, E. *et al*, *Z. Chem.*, 1974, **14**, 160 (*synth*)

Cotton, F.A. *et al*, *Angew. Chem.*, *Int. Ed. Engl.*, 1978, **17**, 953 (*synth*)

Cotton, F.A. *et al*, *Inorg. Chem.*, 1979, **18**, 2713 (*synth*, *cryst struct*)

$C_{16}H_{44}CrSi_4$ **Cr-00138**

Tetrakis[(trimethylsilyl)methyl]chromium, 9CI

[35394-18-4]

$[Cr(CH_2SiMe_3)_4]$ (T_d)

M 400.862

Purple air-sensitive needles. Sol. hydrocarbons. Mp 40°.

Yagupsky, G. *et al*, *J. Chem. Soc.*, *Chem. Commun.*, 1970, 1369 (*synth*)

Mowat, W. *et al*, *J. Chem. Soc.*, *Dalton Trans.*, 1972, 533 (*synth*)

$C_{16}H_{45}CrO_{16}P_5$ **Cr-00139**

Carbonylpentakis(trimethylphosphite-*P*)chromium, 8CI

[37478-32-3]

$Cr(CO)[P(OMe)_3]_5$

M 700.387

Pale-yellow cryst. Mp 175°.

Mathieu, R. *et al*, *Inorg. Chem.*, 1972, **11**, 1858 (*synth*)

$C_{17}H_{10}CrO_3$ Cr-00140

Anthracenetricarbonylchromium, 8CI

Tricarbonyl-η⁶-anthracenechromium

[12155-42-9]

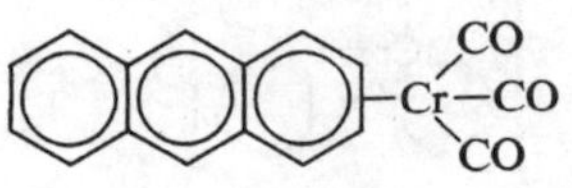

M 314.260

Dark-green cryst. Sol. C_6H_6, insol. hexane. Mp 189-92°.

Fischer, E.O. *et al*, *Chem. Ber.*, 1967, **100**, 3084 (*nmr*)
Mills, O.S. *et al*, *J. Organomet. Chem.*, 1968, **11**, 151 (*cryst struct*)
Muir, K.W. *et al*, *J. Chem. Soc.* (*B*), 1968, 467 (*cryst struct*)

$C_{17}H_{10}CrO_3$ Cr-00141

Tricarbonyl-η^6-phenanthrenechromium

[12094-55-2]

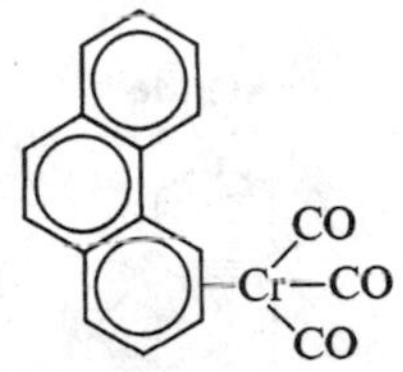

M 314.260

Bright-orange cryst. Mp 158-60°.

King, R.B. *et al*, *J. Am. Chem. Soc.*, 1960, **82**, 4557 (*synth*)
Fischer, E.O. *et al*, *Chem. Ber.*, 1967, **100**, 3084 (*synth*)
Muir, K.W. *et al*, *J. Chem. Soc., Dalton Trans.*, 1968, 467 (*cryst struct*)
Mills, O.S. *et al*, *J. Organomet. Chem.*, 1968, **11**, 151 (*cryst struct*)

$C_{17}H_{12}CrFeO_6$ Cr-00142

Pentacarbonyl(ferrocenylmethoxymethylene)chromium, 9CI

(*Methoxyferrocenylcarbene*)*pentacarbonylchromium*

[31725-05-0]

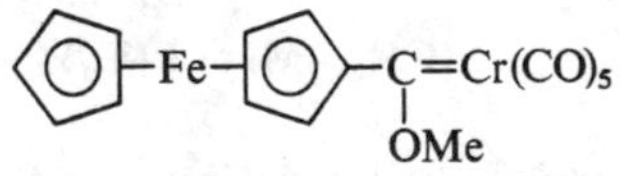

M 420.121

Purple cryst. (C_6H_6/heptane). Mp 170° dec. (softens at 135°).

Connor, J.A. *et al*, *J. Chem. Soc., Dalton Trans.*, 1972, 1470 (*synth, pmr, ir, uv, ms*)
Rausch, M.D. *et al*, *J. Organomet. Chem.*, 1973, **51**, 1 (*synth, pmr, ir*)

$C_{18}H_{10}CrO_5$ Cr-00143

Pentacarbonyl(diphenylmethylene)chromium, 9CI

[62589-11-1]

$(OC)_5Cr{=}CPh_2$

M 358.270

Air-sensitive black cryst.

Fischer, E.O. *et al*, *Chem. Ber.*, 1977, **110**, 656 (*synth*)
Doetz, K.H. *et al*, *Chem. Ber.*, 1977, **110**, 1555 (*props*)
Doetz, K.H. *et al*, *J. Organomet. Chem.*, 1978, **157**, C55 (*props*)
Cooper, N.J. *et al*, *J. Am. Chem. Soc.*, 1981, **103**, 238 (*synth, ir, nmr*)

$C_{18}H_{14}CrO_6Ru$ Cr-00144

Pentacarbonyl(ethoxyruthenocenylmethylene)chromium, 10CI

[75280-69-2]

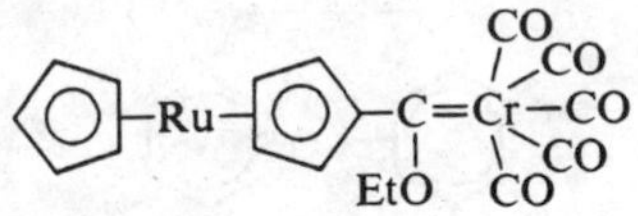

M 479.371

Orange-yellow cryst. (CH_2Cl_2/pentane). Mp 113°.

Fischer, E.O. *et al*, *J. Organomet. Chem.*, 1980, **191**, 261 (*synth, ir, ms, pmr, cmr*)

$C_{18}H_{23}CrNO_5Sn$ Cr-00145

[Di-*tert*-butyl(pyridine)tin]pentacarbonylchromium

[*Bis*(*1,1-dimethylethyl*)(*pyridine*)*tin*]*pentacarbonylchromium, 9CI.* [(*Pyridine*)*di*-tert-*butylstannanediyl*]-*pentacarbonylchromium*

[50982-07-5]

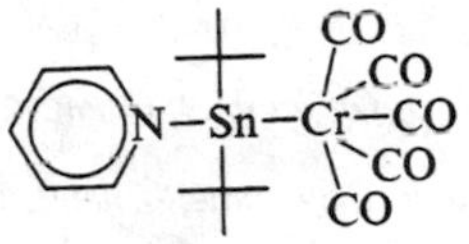

M 504.069

Pale yellow cryst. (hexane).

Marks, T.J., *J. Am. Chem. Soc.*, 1971, **93**, 7090 (*synth, ir, pmr*)
Brice, M.D. *et al*, *J. Am. Chem. Soc.*, 1973, **95**, 4529 (*struct*)
Grynkewich, G.W. *et al*, *Inorg. Chem.*, 1973, **12**, 2522 (*mössbauer, pe*)

$C_{19}H_{14}CrFeO_3$ Cr-00146

Tricarbonyl(phenylferrocene)chromium, 8CI

Tricarbonyl[*η⁶*-(*ferrocenylbenzene*)]*chromium*

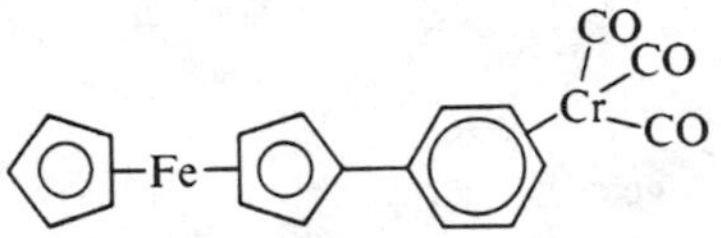

M 398.161

Orange cryst. (heptane). Mp 147-50° dec.

Gubin, S.P. *et al*, *J. Organomet. Chem.*, 1970, **22**, 449 (*synth, ir, pmr, polarog*)

$C_{19}H_{16}CrO_3$ Cr-00147

Tricarbonyl(3-8-η-[2.2]paracyclophane)chromium

Tricarbonyl[(*1,10,11,12,13,14-η*)-*tricyclo-*[*8.2.2.2*^{4,7}]*hexadeca-4,6,10,12,13,15-hexaene*]*chromium, 10CI*

[41354-64-7]

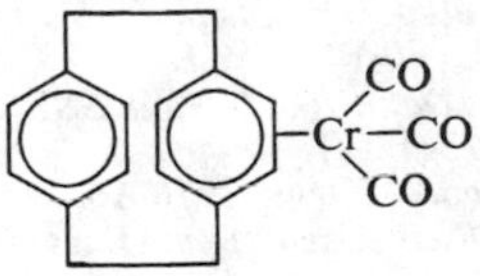

M 344.330

Air-stable cryst. (C_6H_6/pentane). Mp 253-5°.

Cram, D.J. *et al*, *J. Am. Chem. Soc.*, 1960, **82**, 5721 (*synth*)
Langer, E. *et al*, *Tetrahedron*, 1973, **29**, 375 (*props*)
Cristiani, F. *et al*, *Inorg. Chim. Acta*, 1975, **12**, 119 (*props*)
Kai, Y. *et al*, *Acta Crystallogr., Sect. B*, 1978, **34**, 2840 (*cryst struct*)
Langer, E. *et al*, *J. Organomet. Chem.*, 1979, **173**, 47 (*props*)

$C_{19}H_{38}CrO_5Si_4Sn$ Cr-00148

[Bis[bis(trimethylsilyl)methyl]tin]pentacarbonylchromium
Pentacarbonyl[[[(stannylenedimethylidyne)tetrakis[trimethylsilanato]](2−)-Sn]chromium, 9CI
[41772-92-3]

$[(Me_3Si)_2CH]_2SnCr(CO)_5$

M 629.534
Orange cryst. (hexane). Mp 120°.

Cotton, J.D. *et al*, *J. Chem. Soc., Dalton Trans.*, 1976, 2275 (*synth, struct*)

$C_{20}H_{16}Cr$ Cr-00149

Bis[(1,2,3,4,4*a*,8*a*-η)naphthalene]chromium, 10CI
[33085-81-3]

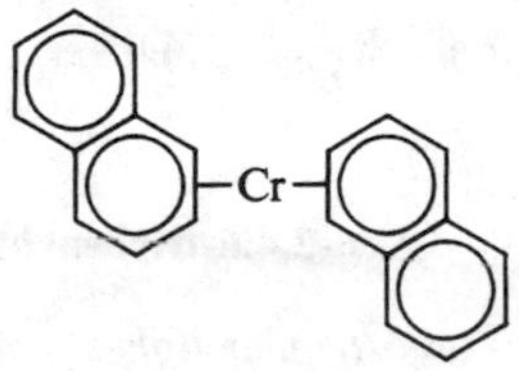

M 308.342
Black cryst.

Elschenbroich, C. *et al*, *Angew. Chem., Int. Ed. Engl.*, 1977, **16**, 870 (*props*)
Connor, J.A. *et al*, *J. Organomet. Chem.*, 1979, **179**, 331.
Kundig, E.P. *et al*, *J. Chem. Soc., Dalton Trans.*, 1980, 991 (*synth, props*)

$C_{20}H_{16}Cr_2GeO_6$ Cr-00150

(Dimethyldiphenylgermane)bis[tricarbonylchromium]
[75061-55-1]

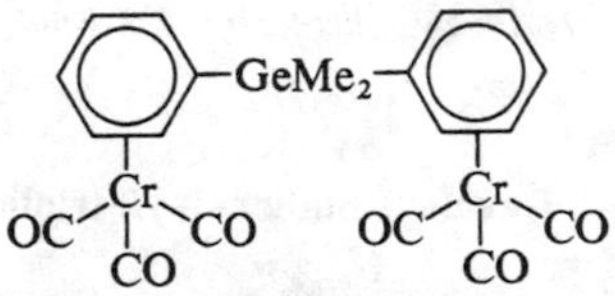

M 528.925
Yellow cryst. Mp 115-6°.

Nagelberg, S.B. *et al*, *Organometallics*, 1982, **1**, 851.

$C_{20}H_{44}Cr$ Cr-00151

Tetrakis(2,2-dimethylpropyl)chromium, 9CI
Tetrakis(neopentyl)chromium.
Tetraneopentylchromium
[37007-84-4]

$Cr[CH_2C(CH_3)_3]_4$

M 336.564
Maroon air-sensitive needles. Mp 110°.

Mowat, W. *et al*, *J. Chem. Soc., Dalton Trans.*, 1973, 770 (*synth*)

$C_{21}H_{14}CrO_3$ Cr-00152

Tricarbonyl(diphenylmethylenecyclopentadiene)chromium
Tricarbonyl[[[(2,3,4,5-η)-2,4-cyclopentadien-1-ylidene]phenylmethyl]benzene]chromium, 10CI. *Tricarbonyl(η⁶-6,6-diphenylfulvene)chromium*
[12248-79-2]

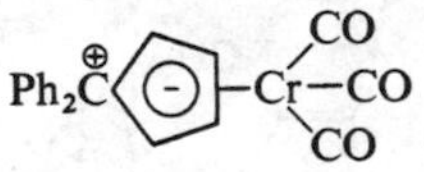

M 366.336
Brown needles. Mp 209°.

Andrianov, V.G. *et al*, *Zh. Strukt. Khim.*, 1977, **18**, 3118 (*struct*)
Edelmann, F. *et al*, *J. Organomet. Chem.*, 1977, **134**, 31.

$C_{21}H_{15}CrNO_6$ Cr-00153

Pentacarbonyl[[(diphenylmethylene)amino]ethoxymethylene]-chromium, 9CI
[54330-36-8]

$(OC)_5Cr{=}C(OEt)N{=}CPh_2$

M 429.349
Bright-red cryst. Mp 88-90°.

Doyle, M.J. *et al*, *J. Chem. Soc., Dalton Trans.*, 1974, 1494 (*synth, ir, nmr*)

$C_{21}H_{24}CrO_3$ Cr-00154

Tricarbonyl(η⁶-octamethylnaphthalene)chromium
[79391-56-3]

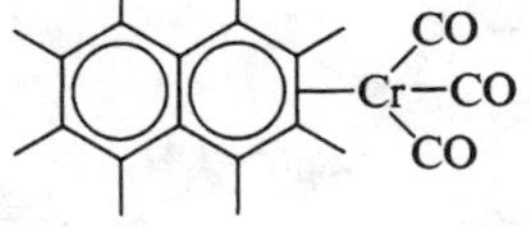

M 376.415
Dark-red cryst.

Hull, J.W. *et al*, *Organometallics*, 1982, **1**, 264 (*synth, struct*)

$C_{22}H_{62}Cr_2P_2Si_4$ Cr-00155

Bis(trimethylphosphine)bis[μ-[(trimethylsilyl)methyl]]bis[(trimethylsilyl)methyl]dichromium, 10CI
[64216-75-7]

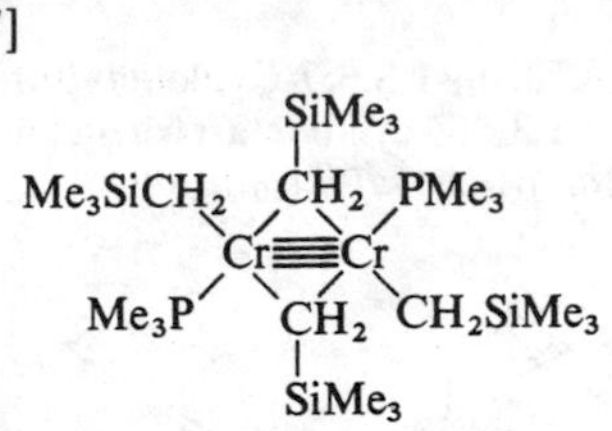

M 605.013
Deep-red air-sensitive cryst. (Et_2O). Mp 80-100° dec.

Anderson, R.A. *et al*, *J. Chem. Soc., Dalton Trans.*, 1978, 446 (*synth*)
Hursthouse, M.B. *et al*, *J. Chem. Soc., Dalton Trans.*, 1978, 1314 (*cryst struct*)

$C_{23}H_{22}CrGeO_5$ Cr-00156

[Bis(2,4,6-trimethylphenyl)germylene]pentacarbonylchromium, 10CI

(Dimesitylgermanediyl)pentacarbonylchromium

[63428-34-2]

M 503.010

Pale-yellow needles (pet. ether). Mp 114°.

Jutzi, P. *et al*, *Angew. Chem., Int. Ed. Engl.*, 1977, **16**, 639.

$C_{24}H_{20}CrO_4Tl_2$ Cr-00157

μ-(Chromato-*O,O'*)tetraphenyldithallium

Bis(diphenylthallium)chromate

$$(Ph_2Tl)_2CrO_4$$

M 833.182

Yellow needles. Sol. Py, insol. H_2O. Mp 290° (explodes).

▷Toxic, explodes on heating

Goddard, A.E., *J. Chem. Soc.*, 1922, 482 (*synth*)

$C_{24}H_{20}Cr_2O_6Sn$ Cr-00158

(Dibenzyldimethylstannane)bis[tricarbonylchromium]

[81496-90-4]

M 627.100

Yellow cryst. Mp 189-91°.

Nagelberg, S.B. *et al*, *Organometallics*, 1982, **1**, 851.

$C_{24}H_{24}Cr_2$ Cr-00159

[μ-[(1,2,3,4-η:5,6,7,8-η)-1,3,5,7-Cyclooctatetraene-][bis(1,2,3,4-η)-1,3,5,7-cyclooctatetraene]dichromium, 9CI

Tris(cyclooctatetraene)dichromium

[59982-80-8]

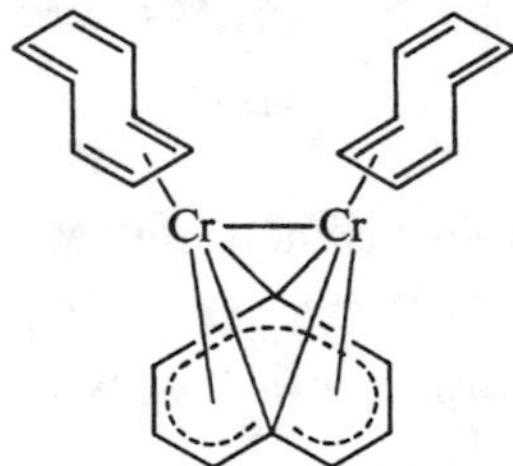

M 416.446

Air-sensitive black cryst.

Breil, H. *et al*, *Angew. Chem., Int. Ed. Engl.*, 1966, **5**, 898 (*synth*)

Brauer, D.J. *et al*, *Inorg. Chem.*, 1976, **15**, 2511 (*synth, cryst struct*)

$C_{24}H_{30}Cr_2O_4$ Cr-00160

Tetracarbonylbis[(1,2,3,4,5-η)-1,2,3,4,5-pentamethyl-2,4-cyclopentadien-1-yl]dichromium, 10CI

[37299-12-0]

M 486.491

Four "semi-bridging" carbonyls in cryst. Deep-green cryst. (hexane). Dec. >200°.

King, R.B. *et al*, *J. Organomet. Chem.*, 1979, **171**, 53 (*synth, ir, props*)

Robbins, J.L. *et al*, *Inorg. Chem.*, 1981, **20**, 1133 (*uv*)

$C_{26}H_{17}CrO_3P$ Cr-00161

Tricarbonyl[(1,2,3,4,5,6-η)-2,4,6-triphenylphosphorin]chromium, 9CI

Tricarbonyl(η⁶-2,4,6-triphenylphosphabenzene)chromium

[36153-02-3]

M 460.388

Ir, nmr and ms suggest π-bonded ring. Dark red-brown cryst. Mp 156-7°.

Deberitz, J. *et al*, *Chem. Ber.*, 1970, **103**, 2541 (*synth*)

Varenkamp, H. *et al*, *Chem. Ber.*, 1972, **105**, 1148 (*struct*)

Nöth, H. *et al*, *Chem. Ber.*, 1973, **106**, 2227 (*struct*)

Lückoff, M. *et al*, *Angew. Chem., Int. Ed. Engl.*, 1976, **15**, 503.

$C_{26}H_{20}CrO_3Pb$ Cr-00162

Tricarbonyl(η⁵-2,4-cyclopentadien-1-yl)(triphenylplumbyl)-chromium

(Triphenylplumbyl)(cyclopentadienyl)tricarbonylchromium

M 639.638

Yellow solid (CH_2Cl_2). Mp 195-7° dec.

Patil, H.R.H. *et al*, *Inorg. Chem.*, 1966, **5**, 1401.

$C_{26}H_{21}CrNO_5Si$ Cr-00163

Pentacarbonyl[(dimethylamino)(triphenylsilyl)methylene]chromium, 9CI

[60674-48-8]

$$(OC)_5Cr{=}C(NMe_2)SiPh_3$$

M 507.537

Yellow cryst.

Fischer, E.O. *et al*, *J. Organomet. Chem.*, 1976, **113**, C31 (*synth*)

Fischer, E.O. *et al*, *Chem. Ber.*, 1977, **110**, 3467 (*synth*)

Fischer, E.O. *et al*, *J. Organomet. Chem.*, 1979, **168**, 53 (*nmr*)

$C_{26}H_{21}CrOPS$ **Cr-00164**

(η^6-Benzene)(carbonothioyl)carbonyl(triphenylphosphine)chromium, 10CI

[63590-28-3]

M 464.481

Orange-brown cryst. Mp 120-4° dec.

Jaoven, G., *Tetrahedron Lett.*, 1973, 5159.
Jaoven, G. *et al*, *J. Organomet. Chem.*, 1974, **72**, 377.
Butler, I.S. *et al*, *J. Organomet. Chem.*, 1977, **133**, 59.

$C_{26}H_{30}Cr_2O_6$ **Cr-00165**

Hexacarbonylbis[(1,2,3,4,5-η)-1,2,3,4,5-pentamethyl-2,4-cyclopentadien-1-yl]dichromium, 10CI

[70605-18-4]

As Hexacarbonylbis(η^5-2,4-cyclopentadien-1-yl)dichromium, Cr-00129 with

R = CH_3

M 542.511

Deep-purple, v. air-sensitive cryst. (C_6H_6).

King, R.B., *Coord. Chem. Rev.*, 1976, **20**, 155 (*synth*)
King, R.B. *et al*, *J. Organomet. Chem.*, 1979, **171**, 53 (*synth, pmr*)

$C_{27}H_{36}CrN_3$ **Cr-00166**

Tris[[2-(dimethylamino)phenyl]methyl-*C,N*]chromium, 10CI

[64061-47-8]

M 454.598

Red cryst. (CH_2Cl_2). μ_{eff} = 3.6 BM.

▷Pyrophoric

Manzer, L.E., *J. Organomet. Chem.*, 1977, **135**, C6 (*synth, props*)
Lukehart, C.M. *et al*, *J. Organomet. Chem.*, 1979, **179**, C9 (*props*)

$C_{28}H_{38}CrN_4Si_2^{\oplus}$ **Cr-00167**

Bis(2,2′-bipyridine-*N,N′*)bis[(trimethylsilyl)methyl]chromium(1+), 9CI

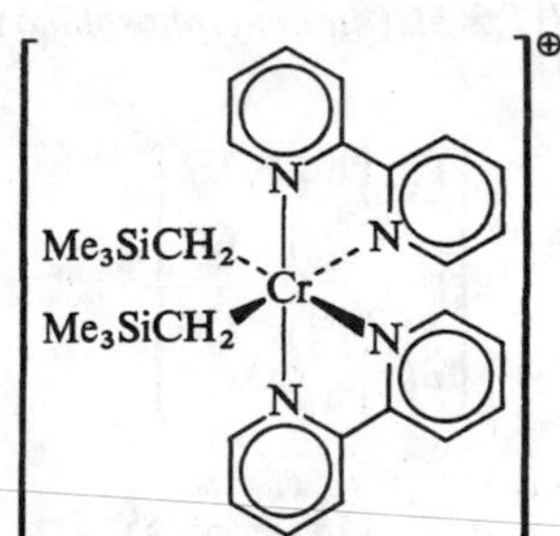

M 538.802 (ion)

μ_B = 3.64 BM.

Iodide: [41678-48-2].

$C_{28}H_{38}CrIN_4Si_2$ M 665.707

Red cryst.

Daly, J.J. *et al*, *Helv. Chim. Acta*, 1973, **56**, 503 (*synth, cryst struct*)
Daly, J.J. *et al*, *J. Chem. Soc., Dalton Trans.*, 1973, 1497.

$C_{29}H_{26}CrFeNO_4P$ **Cr-00168**

Tetracarbonyl[1-[(dimethylamino)methyl]-2-(diphenylphosphino)ferrocene-*N,P*]chromium, 9CI

[55960-25-3]

M 591.345

Cryst. (Me_2CO/pet. ether). Mp 160-4°.

Kotz, J.C. *et al*, *J. Organomet. Chem.*, 1975, **84**, 255 (*synth, ir, pmr*)

$C_{30}H_{24}CrO_4P_2$ **Cr-00169**

Tetracarbonyl[1,2-ethanediylbis[diphenylphosphine]-*P,P′*]-chromium, 10CI

[*1,2-Bis(diphenylphosphino)ethane*]*tetracarbonylchromium*

[29890-04-8]

M 562.461

Yellow cryst. Mp 211°.

Chatt, J. *et al*, *J. Chem. Soc.*, 1961, 4980 (*synth*)
Grim, S.O. *et al*, *Inorg. Chem.*, 1974, **13**, 1095 (*synth, nmr*)
Cohen, M.A. *et al*, *Inorg. Chem.*, 1976, **15**, 1417 (*synth, props*)
Brisdon, B.J. *et al*, *J. Mol. Struct.*, 1977, **41**, 99 (*ir*)
Gaebehein, H. *et al*, *J. Organomet. Chem.*, 1978, **156**, 389 (*ir*)
Treichel, P.M. *et al*, *Inorg. Chim. Acta*, 1979, **33**, 171 (*synth*)

$C_{30}H_{24}CrO_4P_2^{\oplus}$ **Cr-00170**

Tetracarbonyl[1,2-ethanediylbis[diphenylphosphine]-*P*,*P'*]-chromium(1+), 9CI

Tetracarbonyl[1,2-bis(diphenylphosphino)ethane]chromium(1+)

M 562.461 (ion)

Hexafluorophosphate: [37757-22-5].
$C_{30}H_{24}CrF_6O_4P_3$ M 707.425
Deep-purple cryst.

Johnson, B.F.G. *et al*, *J. Organomet. Chem.*, 1972, **40**, C36 (*synth*)

$C_{30}H_{39}CrO_3$ **Cr-00171**

Triphenyltris(tetrahydrofuran)chromium, 8CI

[13986-68-0]

$[CrPh_3(THF)_3]$

M 499.632

Red air-sensitive cryst.

Herwig, W. *et al*, *J. Am. Chem. Soc.*, 1957, **79**, 6561; 1959, **81**, 4798 (*synth*)
Whitesides, G.M. *et al*, *J. Am. Chem. Soc.*, 1970, **92**, 5625 (*props*)
Seidel, W. *et al*, *Z. Chem.*, 1971, **11**, 932 (*props*)

$C_{32}H_{26}CrN_4^{\oplus}$ **Cr-00172**

Bis(2,2'-bipyridine-*N*,*N'*)diphenylchromium(1+), 9CI

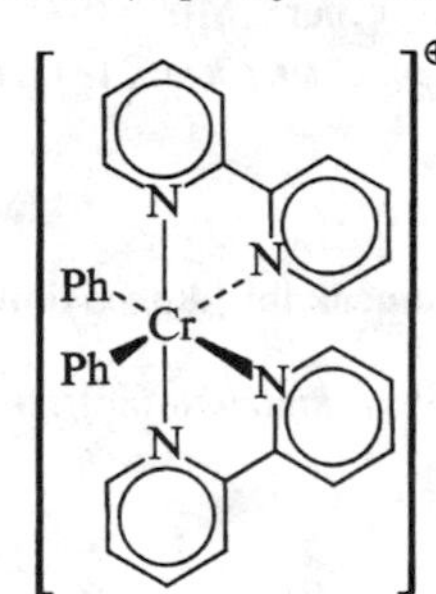

M 518.580 (ion)

Iodide: [25839-19-4].
$C_{32}H_{26}CrIN_4$ M 645.485
Red cryst. μ_B = 3.80 BM.

Muller, H., *Z. Chem.*, 1969, **9**, 311 (*synth*)
Sneeden, R.P.A. *et al*, *J. Organomet. Chem.*, 1973, **47**, 125 (*synth*)

Handle all chemicals with care

$C_{32}H_{28}CrFe_2$ **Cr-00173**

Bis[(η^6-phenyl)ferrocene]chromium, 10CI

(1,1'-Diferrocenyldi-π-benzene)chromium

[67574-52-1]

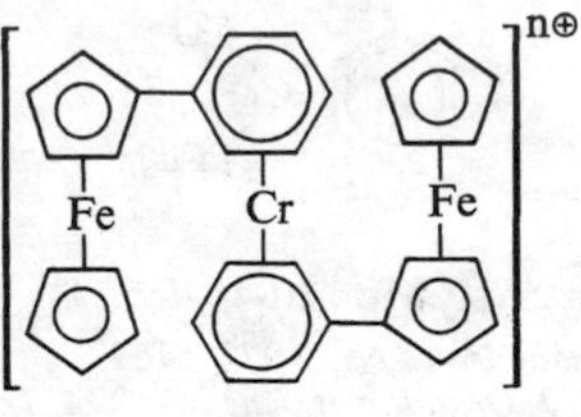

n = 0

M 576.263

Isol. as the oxidation product 89776-1.

Sheveler, Yu.A. *et al*, *Izv. Akad. Nauk SSSR, Ser. Khim.*, 1981, 1414 (*synth*)

$C_{32}H_{28}CrFe_2^{\oplus}$ **Cr-00174**

Bis[(η^6-phenyl)ferrocene]chromium(1+), 10CI

(1,1'-Diferrocenyldi-π-benzene)chromium(1+)

As Bis[(η^6-phenyl)ferrocene]chromium, Cr-00173 with

n = 1

M 576.263 (ion)

Iodide: [75826-53-8].
$C_{32}H_{28}CrFe_2I$ M 703.168
Red-brown lustrous cryst. (MeOH). Mp 132-4°.

Tetraphenylborate: [79102-83-3].
$C_{56}H_{48}BCrFe_2$ M 895.495
No phys. props. reported.

Nesmeyanov, A.N. *et al*, *Izv. Akad. Nauk SSSR, Ser. Khim.*, 1978, 1420 (*synth*, *esr*, *mössbauer*)
Shevelev, Yu.A. *et al*, *Izv. Akad. Nauk SSSR, Ser. Khim.*, 1980, 1414; 1981, 1414 (*synth*, *ir*, *uv*, *esr*)

$C_{32}H_{36}Cr_2O_8$ **Cr-00175**

Tetrakis[μ-(2,6-dimethoxyphenyl-*C*:*O*)]dichromium, 10CI

[64836-05-1]

M 652.624

Orange air-sensitive cryst. (THF/hexane).

Cotton, F.A. *et al*, *Inorg. Chim. Acta*, 1977, **25**, L105.
Cotton, F.A. *et al*, *J. Am. Chem. Soc.*, 1977, **99**, 7372.
Cotton, F.A. *et al*, *Inorg. Chem.*, 1978, **17**, 2087 (*synth*, *cryst struct*)

$C_{38}H_{30}CrINO_3P_2$ **Cr-00177**

Dicarbonyliodonitrosylbis(triphenylphosphine)chromium, 9CI

[60020-27-1]

$[Cr(CO)_2(PPh_3)_2(NO)I]$

M 789.508

Yellow solid. Mp >100° dec.

Connelly, N.G. *et al*, *J. Chem. Soc., Dalton Trans.*, 1976, 699 (*synth, props*)

$C_{40}H_{52}Cr$ **Cr-00178**

Tetrakis(2-methyl-2-phenylpropyl)chromium, 9CI

[37007-87-7]

$Cr[CH_2CPh(CH_3)_2]_4$ (T_d)

M 584.847

Purple air-sensitive prisms. Sol. hexane. Mp 120° dec.

Gramlich, V. *et al*, *J. Organomet. Chem.*, 1973, **61**, 247 (*cryst struct*)

Mowat, W. *et al*, *J. Chem. Soc., Dalton Trans.*, 1973, 770 (*synth*)

Dy Dysprosium

S. A. Cotton

Dysprosium (Fr., Ger.), Disprosio (Sp., Ital.), Диспрозий (Disprozii) (Russ.), ジスプロシウム (Japan.)

Atomic Number. 66

Atomic Weight. 162.50

Electronic Configuration. [Xe] $4f^{10}\,6s^2$

Oxidation State. +3

Coordination Number. 6 or greater.

Colour. Mainly yellow, but some are colourless.

Availability. Available starting materials are Dy_2O_3 and anhydrous $DyCl_3$. See also under Lanthanum.

Handling. See under Lanthanum.

Toxicity. See under Lanthanum.

Isotopic Abundance. ^{160}Dy, 2.29%; ^{161}Dy, 18.88%; ^{162}Dy, 25.53%; ^{163}Dy, 24.97%; ^{164}Dy, 28.18%.

Spectroscopy. ^1H nmr spectra of dysprosium compounds show very large paramagnetic shifts with very broad lines and as yet, no data on organodysprosium compounds are available. See also under Lanthanum.

Analysis. See under Lanthanum.

References. See under Lanthanum.

$C_5H_5Cl_2Dy$ Dy-00001

Dichloro(η^5-2,4-cyclopentadien-1-yl)dysprosium

Cyclopentadienyldysprosium dichloride

$(C_5H_5)DyCl_2$

M 298.501

Tris-THF adduct:
$C_{17}H_{19}Cl_2DyO_3$ M 504.741
Colourless cryst. Mp 85-90° dec.
Extremely air-sensitive. $\mu = 11.81\mu_B$. Struct. probably

Manastyrskyj, S. *et al*, *Inorg. Chem.*, 1963, **2**, 904 (*synth*)

$C_{10}H_{10}ClDy$ Dy-00002

Chlorobis(η^5-2,4-cyclopentadienyl)dysprosium

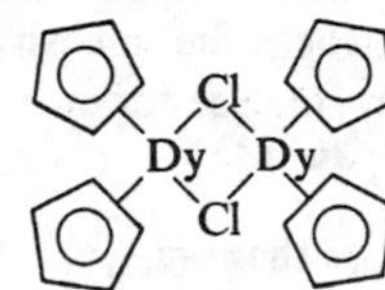

M 328.142
Dimeric.

Dimer: [60482-74-8]. *Di-μ-chlorotetrakis(η^5-2,4-cyclopentadien-1-yl)didysprosium*, 9CI.
$C_{20}H_{20}Cl_2Dy_2$ M 656.284
Yellow cryst. Sol. THF. Mp 343-6°. Dec. in air. Subl. *in vacuo*. $\mu_{eff} = 10.6\mu_B$.

Maginn, R.E. *et al*, *J. Am. Chem. Soc.*, 1963, **85**, 672 (*synth, ir*)

$C_{11}H_{13}Dy$ Dy-00003

Bis(η^5-2,4-cyclopentadien-1-yl)methyldysprosium

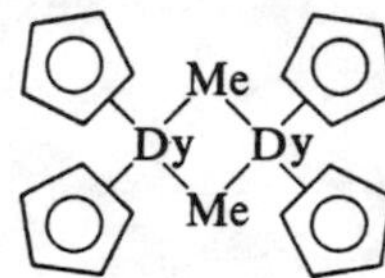

M 307.724
Dimeric.

Dimer: [60997-41-3]. *Tetrakis(η^5-2,4-cyclopentadienyl)-di-μ-methyldidysprosium*, 9CI.
$C_{22}H_{26}Dy_2$ M 615.447
Pale-yellow cryst. Sol. CH_2Cl_2, C_6H_6, toluene. Dec. >165°. Air-sensitive. $\mu_{eff} = 9.9\mu_B$.

Holton, J. *et al*, *J. Chem. Soc., Dalton Trans.*, 1979, 54 (*synth, ir*)

$C_{15}H_{15}Dy$ Dy-00004

Tris(η^5-2,4-cyclopentadien-1-yl)dysprosium, 9CI

Tricyclopentadienyldysprosium

[12088-04-9]

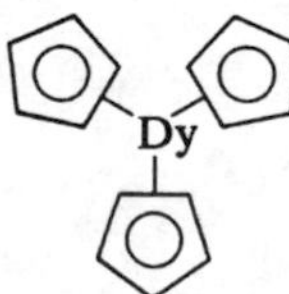

M 357.784
Yellow cryst. Sol. THF. Mp 302°. Hydrol. by H_2O. Subl. at 220° *in vacuo*. $\mu_{eff} = 10.0\mu_B$.

Isocyanocyclohexane adduct: [37298-98-9]. *Tris(2,4-cyclopentadien-1-yl)isocyanocyclohexanedysprosium.*
$C_{22}H_{26}DyN$ M 466.954
Sol. C_6H_6. Mp 163°.

Birmingham, J.M. *et al*, *J. Am. Chem. Soc.*, 1956, **78**, 42 (*synth*)
v. Ammon, R. *et al*, *Ber. Bunsenges. Phys. Chem.*, 1972, **76**, 995.
Aleksanyan, V.T. *et al*, *J. Raman. Spectrosc.*, 1974, **2**, 345 (*raman*)

$C_{20}H_{19}DyO_3W$ Dy-00005

[μ-(Carbonyl-*C,O*)]bis[(1,2,3,4,5-η)-1-methyl-2,4-cyclopentadien-1-yl][tricarbonyl(η^5-2,4-cyclopentadien-1-yl)tungsten]dysprosium, 9CI

$(H_3CC_5H_4)_2DyW(C_5H_5)(CO)_3$

M 653.718
Nitrosylated by *N*-methyl-*N*-nitroso-*p*-toluenesulfonamide to give $(C_5H_5)W(CO)_2NO$. Sol. THF, DMSO. Mp >220° dec. Air- and moisture-sensitive.

Crease, A.E. *et al*, *J. Chem. Soc., Dalton Trans.*, 1973, 1501 (*synth, ir*)

$C_{27}H_{21}Dy$ Dy-00006

Triindenyldysprosium

Tris(1,2,3,3a,7a-η)-1H-inden-1-yldysprosium

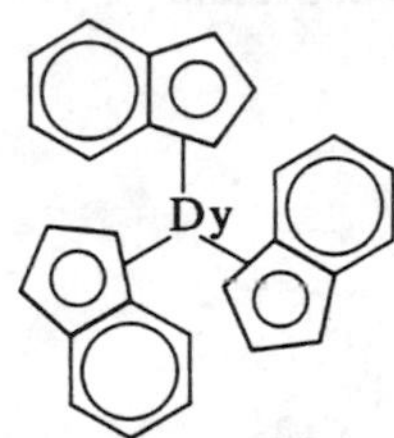

M 507.963
May be a mono- or trihapto complex.

THF adduct: [22696-75-9].
$C_{31}H_{29}DyO$ M 580.070
Pale-tan solid. Sol. THF. $\mu = 9.95\mu_B$.

Tsutsui, M. *et al*, *J. Am. Chem. Soc.*, 1969, **91**, 3175 (*synth*)

Er Erbium

S. A. Cotton

Erbium (Fr., Ger.), Erbio (Sp., Ital.), Эрбий Erbii) (Russ.), エルビウム (Japan.)

Atomic Number. 68

Atomic Weight. 167.26

Electronic Configuration. [Xe] $4f^{12}\,6s^2$

Oxidation State. +3

Coordination Number. 6 or greater, but less in compounds with bulky ligands such as $—CH_2SiMe_3$, $—C(CH_3)_3$.

Colour. Pink.

Availability. Available starting materials are Er_2O_3 and anhydrous $ErCl_3$. See also under Lanthanum.

Handling. See under Lanthanum.

Toxicity. See under Lanthanum.

Isotopic Abundance. ^{162}Er, 0.14%; ^{164}Er, 1.56%; ^{166}Er, 33.41%; ^{167}Er, 22.94%; ^{168}Er, 27.07%; ^{170}Er, 14.88%.

Spectroscopy. The resonances in ^1H nmr spectra of organoerbium compounds are very broad with large paramagnetic shifts.

Erbium compounds have electronic spectra with certain hypersensitive peaks, where the intensity in particular is affected by the environment. See for example Pappalardo, R., *J. Chem. Phys.*, 1968, **49,** 1545.

Analysis. See under Lanthanum.

References. See under Lanthanum for general references, and also, for ^1H nmr spectra, the following:

Schumann, H. *et al.*, *J. Organomet. Chem.*, 1978, **146,** C5.

$C_5H_5Cl_2Er$ **Er-00001**

Dichloro(η^5-2,4-cyclopentadien-1-yl)erbium

Cyclopentadienylerbium dichloride

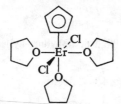

THF adduct

M 303.261

Tris-THF adduct: [81703-54-0].
$C_{17}H_{29}Cl_2ErO_3$ M 519.580
Pink cryst. Sol. THF, sl. sol. C_6H_6. Mp 91-4°. Dec. by air and moisture. $\mu = 9.68\mu_B$.

Manastyrskyj, S. *et al*, *Inorg. Chem.*, 1963, **2**, 904 (*synth, ir*)
Day, C.S. *et al*, *Organometallics*, 1982, **1**, 998 (*struct*)

$C_6H_{18}Er^{\ominus\ominus\ominus}$ **Er-00002**

Hexamethylerbate(3−)

$ErMe_6^{\ominus\ominus\ominus}$

M 257.468 (ion)

Tris[(tetramethylethylenediamine)lithium] salt: [66862-11-1].
$C_{24}H_{66}ErLi_3N_6$ M 626.909
Pink cryst. Sl. sol. Et_2O. Mp 138-9° dec. Extremely air- and moisture-sensitive.

Schumann, H. *et al*, *Angew. Chem., Int. Ed. Engl.*, 1978, **17**, 276; 1981, **20**, 120 (*synth, struct*)

C_8H_8ClEr **Er-00003**

(η^8-1,3,5,7-Cyclooctatetraene)chloroerbium

$(C_8H_8)ErCl$

M 306.864

THF adduct: (η^8-1,3,5,7-Cyclooctatetraene)-chloro(tetrahydrofuran)erbium.
$C_{12}H_{16}ClErO$ M 378.971
Pink powder. Sol. THF. Presumably dimeric by analogy with Chloro(cyclooctatetraene)cerium, Ce-00001.

Wayda, A.L., *Organometallics*, 1983, **2**, 565 (*synth, ir*)

$C_{10}H_{10}ClEr$ **Er-00004**

Chlorobis(η^5-2,4-cyclopentadien-1-yl)erbium

[53224-35-4]

M 332.902
Dimeric.

Dimer: Dichlorotetrakis(η^5-2,4-cyclopentadien-1-yl)-dierbium, 9CI.
$C_{20}H_{20}Cl_2Er_2$ M 665.804
Pink cryst. Sol. THF. Mp >200° dec. Dec. in air. Subl. *in vacuo*. $\mu_{eff} = 9.79\mu_B$.

Maginn, R.E. *et al*, *J. Am. Chem. Soc.*, 1963, **85**, 672 (*synth, ir*)
Marks, T.J. *et al*, *Inorg. Chem.*, 1976, **15**, 1302 (*synth*)

$C_{11}H_{13}Er$ **Er-00005**

Bis(η^5-2,4-cyclopentadien-1-yl)methylerbium

[61127-35-3]

M 312.484
Dimeric.

Dimer: Tetrakis(η^5-2,4-cyclopentadien-1-yl)di-μ-methyldierbium.
$C_{22}H_{26}Er_2$ M 624.967
Active catalyst for catalytic polymerisation of ethene. Convenient source for making alkynides. Pink cryst. Sol. THF, toluene, C_6H_6, CH_2Cl_2. Mp >150° dec. $\mu_{eff} = 9.41\mu_B$ (295°K), oxygen- and moisture-sensitive.

Ely, N.M. *et al*, *Inorg. Chem.*, 1975, **14**, 2680 (*synth, ir*)
Ballard, D.G.H. *et al*, *J. Chem. Soc., Chem. Commun.*, 1978, 994 (*use*)
Holton, J. *et al*, *J. Chem. Soc., Dalton Trans.*, 1979, 54 (*synth, uv, ir, ms*)
Evans, W.J. *et al*, *J. Organomet. Chem.*, 1980, **202**, C6 (*use*)

$C_{12}Co_3ErO_{12}$ **Er-00006**

Tris(tetracarbonylcobalto)erbium, 8CI

$Er[Co(CO)_4]_3$

M 680.184

Tetrakis-THF complex: [34216-63-2].
$C_{28}H_{32}Co_3ErO_{16}$ M 968.611
Struct. not yet established. Dark-red solid. Mp 138-40° dec. Air- and water-sensitive.

Marianelli, R.S. *et al*, *J. Organomet. Chem.*, 1971, **32**, C41 (*synth, ir*)

$C_{12}H_{14}ClEr$ **Er-00007**

Chlorobis(η^5-methylcyclopentadienyl)erbium

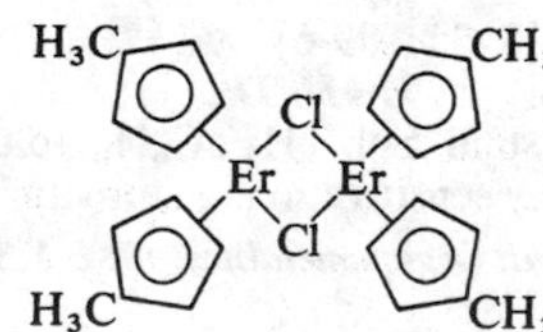

M 360.956
Dimeric.

Dimer: Di-μ-chlorotetrakis[(1,2,3,4,5-η)-1-methyl-2,4-cyclopentadien-1-yl]dierbium.
$C_{24}H_{28}Cl_2Er_2$ M 721.911
Pink cryst. Sol. C_6H_6, THF. Mp 119-22°. Dec. in air. Subl. *in vacuo*.

Maginn, R.E. *et al*, *J. Am. Chem. Soc.*, 1963, **85**, 672 (*synth*)

$C_{12}H_{33}ErSi_3$ **Er-00008**

Tris[(trimethylsilyl)methyl]erbium

$Er(CH_2SiMe_3)_3$

M 428.909

Bis-THF complex: [67483-83-4]. *Bis(tetrahydrofuran)-tris[(trimethylsilyl)methyl]erbium.*
$C_{20}H_{49}ErO_2Si_3$ M 573.122
Pink cryst. Sol. hexane. Mp 49-51°.

Tris-THF(?) complex: [66228-53-3]. *Tris(tetrahydrofuran)tris[(trimethylsilyl)methyl]erbium.*

$C_{24}H_{57}ErO_3Si_3$ M 645.229
Pink cryst. Sol. pentane, THF. Air-sensitive, unstable >30°. $\mu = 9.2\mu_B$. Possibly identical with the bis-complex above.

Atwood, J.L. *et al*, *J. Chem. Soc., Chem. Commun.*, 1978, 140 (*synth, ir*)
Schumann, H. *et al*, *J. Organomet. Chem.*, 1978, **146**, C5 (*synth, ir, pmr*)

$C_{14}H_{19}Er$ Er-00009

tert-Butylbis(η^5-cyclopentadienyl)erbium

(Bis-η^5-2,4-cyclopentadien-1-yl)(1,1-dimethylethyl)erbium, 10CI. *Bis(cyclopentadienyl)(tert-butyl)erbium*

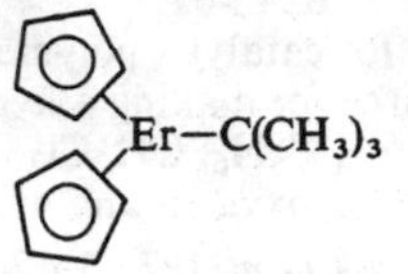

Er—C(CH₃)₃

M 354.564

THF complex: [78683-33-7].
$C_{18}H_{27}ErO$ M 426.671
Pink cryst. Sol. THF, toluene. Mp >40° dec. Extremely air- and moisture-sensitive.

Evans, W.J. *et al*, *J. Chem. Soc., Chem. Commun.*, 1981, 292 (*synth, ir*)
Schumann, H. *et al*, *Organometallics*, 1982, **1**, 1194 (*synth, ir*)

$C_{14}H_{21}ErSi$ Er-00010

Bis(η^5-2,4-cyclopentadien-1-yl)[(trimethylsilyl)methyl]erbium

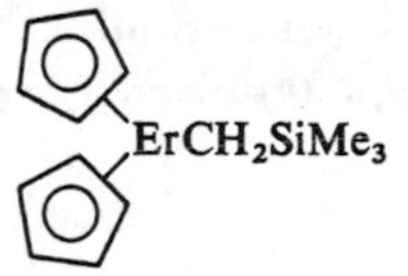

$ErCH_2SiMe_3$

M 384.665

THF complex: [82311-89-5].
$C_{18}H_{29}ErOSi$ M 456.772
Light-brown solid. Sol. THF, C_6H_6, toluene. Mp ~50° dec. Extremely sensitive to moisture and oxygen.

Schumann, H. *et al*, *Organometallics*, 1982, **1**, 1194 (*synth, ir*)

$C_{15}H_{15}Er$ Er-00011

Tris(η^5-2,4-cyclopentadien-1-yl)erbium, 9CI

Tricyclopentadienylerbium

[39330-74-0]

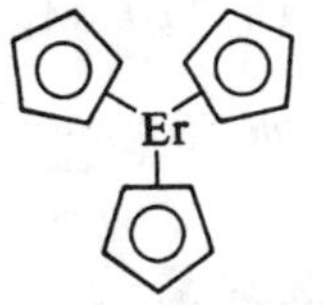

Er

M 362.544
Pink cryst. Sol. THF. Mp 285°. Hydrol. by H_2O. Subl. at 200° *in vacuo*. $\mu_{eff} = 9.44\mu_B$.

Isocyanocyclohexane complex: [37298-99-0]. *Tris(2,4-cyclopentadien-1-yl)isocyanocyclohexaneerbium.*
$C_{22}H_{26}ErN$ M 471.714
Sol. C_6H_6. Mp 168°.

Birmingham, J.M. *et al*, *J. Am. Chem. Soc.*, 1956, **78**, 42 (*synth*)
Pappalardo, R., *J. Chem. Phys.*, 1968, **49**, 1545 (*uv*)
v. Ammon, R. *et al*, *Ber. Bunsenges. Phys. Chem.*, 1972, **76**, 995.
Aleksanyan, V.T. *et al*, *J. Organomet. Chem.*, 1977, **131**, 251 (*raman*)

$C_{16}H_{19}Er$ Er-00012

Bis(2,4-cyclopentadien-1-yl)(3,3-dimethylbutynyl)erbium

[76747-86-9]

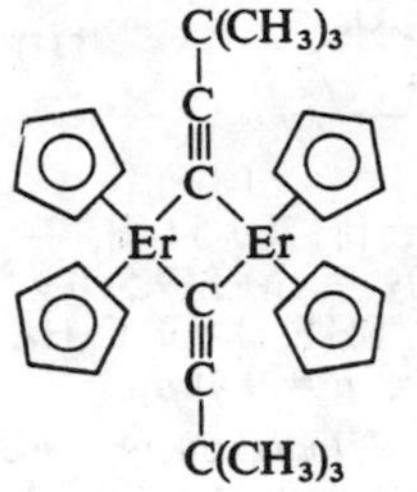

C(CH₃)₃ C C Er Er C C C(CH₃)₃

M 378.586
Dimeric.

Dimer: Tetrakis(η^5-2,4-cyclopentadien-1-yl)bis[μ-(3,3-dimethylbutynyl)]dierbium.
$C_{32}H_{38}Er_2$ M 757.172
Pink solid. Sol. THF. Extremely air- and moisture-sensitive.

Evans, W.J. *et al*, *J. Organomet. Chem.*, 1980, **202**, C6 (*synth, ir*)
Atwood, J.L. *et al*, *Inorg. Chem.*, 1981, **20**, 4115 (*synth, struct, ir, uv*)

$C_{16}H_{36}Er^{\ominus}$ Er-00013

Tetra-tert-butylerbate(1−)

Tetrakis(1,1-dimethylethyl)erbate(1−)

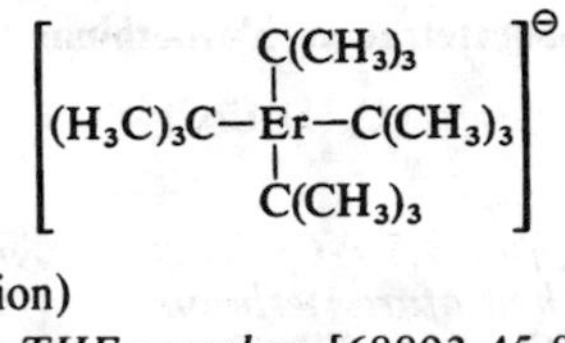

$[(H_3C)_3C-Er(C(CH_3)_3)_2-C(CH_3)_3]^{\ominus}$

M 395.720 (ion)

Li salt, tetra-THF complex: [68893-45-8].
$C_{32}H_{68}ErLiO_4$ M 691.088
Pink solid. Sol. Et_2O, THF. Mp 60° dec. (*in vacuo*). V. air- and moisture-sensitive. $\mu_{eff} = 9.7\mu_B$.

Wayda, A.L. *et al*, *J. Am. Chem. Soc.*, 1978, **100**, 7119 (*synth, ir, uv*)

$C_{18}H_{15}ErO_3W$ Er-00014

[μ-(Carbonyl-C,O)]bis(η^5-2,4-cyclopentadien-1-yl)[dicarbonyl(η^5-2,4-cyclopentadien-1-yl)tungsten]erbium

[39477-06-0]

$(C_5H_5)_2ErW(C_5H_5)(CO)_3$

M 630.425
Nitrosylated by N-methyl-N-nitroso-p-toluenesulfonamide to afford $(C_5H_5)W(CO)_2NO$. Sol. THF, DMSO. Air- and moisture-sensitive. Dec. >220°.

Crease, A.E. *et al*, *J. Chem. Soc., Dalton Trans.*, 1973, 1501 (*synth, ir*)

$C_{18}H_{30}Er_2$ Er-00015

Tris(3-hexyne)dierbium

$(H_3CCH_2CH_2CH_2C{\equiv}CH)_3Er_2$

M 580.955

Catalyst for homogeneous hydrogenation of alkenes and alkynes. Dark-brown solid. Sol. C_6H_6. Yields hexane and hexene on themolysis at 220°.

Evans, W.J. *et al*, *J. Chem. Soc., Chem. Commun.*, 1979, 1007 (*synth, ir, use*)

$C_{24}H_{36}Er^{\ominus}$ Er-00016

Tetrakis(3,3-dimethyl-1-butynyl)erbate(1–)

$$\left[\begin{array}{c} C(CH_3)_3 \\ | \\ C \\ ||| \\ C \\ | \\ (H_3C)_3C{-}C{\equiv}C{-}Er{-}C{\equiv}C{-}C(CH_3)_3 \\ | \\ C \\ ||| \\ C \\ | \\ C(CH_3)_3 \end{array}\right]^{\ominus}$$

M 491.808 (ion)

Li salt, THF complex: [76565-26-9].
$C_{28}H_{44}ErLiO$ M 570.856
Pink solid. Sol. THF, pentane. μ_{eff} 9.79μ_B.

Evans, W.J. *et al*, *J. Organomet. Chem.*, 1980, **202**, C6 (*synth, uv, ir*)

$C_{30}H_{33}ClEr_3^{\ominus}$ Er-00017

μ-Chlorohexakis(η⁵-2,4-cyclopentadien-1-yl)di-μ-hydro-μ₃-hydrotrierbate(1–)

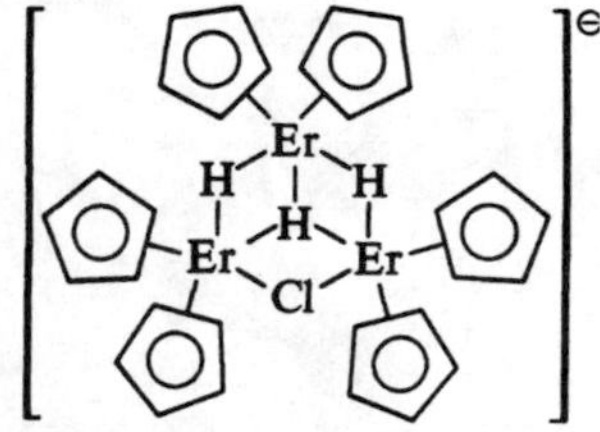

M 930.824 (ion)

Li salt, tetra-THF complex: [80795-37-5].
$C_{46}H_{65}ClEr_3LiO_4$ M 1226.191
Pink prisms. Extremely air- and moisture-sensitive.

Evans, W.J. *et al*, *J. Am. Chem. Soc.*, 1982, **104**, 2015 (*synth, struct, ir*)

$C_{32}H_{40}Er_2O_2$ Er-00018

Tris(1,3,5,7-cyclooctatetraene)bis(tetrahydrofuran)dierbium

[65101-97-5]

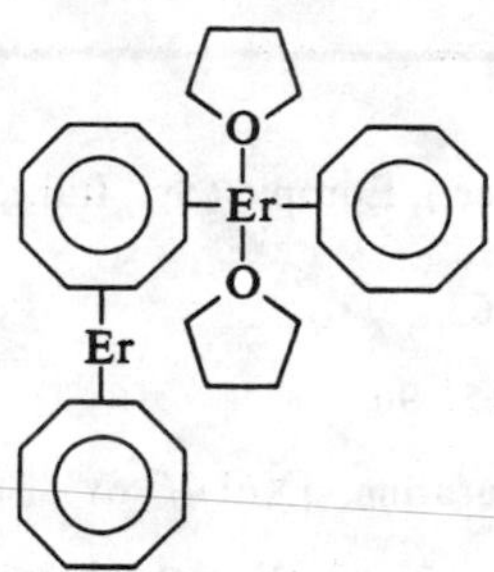

M 791.187

Struct. assigned by analogy with [μ-[(1,2-η:1,2,3,4,5,6,7,8-η)-1,3,5,7-Cyclooctatetraene]-[bis(η^8-1,3,5,7-cyclooctatetraene)bis(tetrahydrofuran)dineodymium, Nd-00012. Reddish-brown cryst. Sol. THF. Extremely air- and moisture-sensitive, inflames in air. $\mu_{eff} = 9.31\mu_B$.

Eu Europium

S. A. Cotton

Europium (Fr., Ger.), Europio (Sp., Ital.), Европий (Yevropii) (Russ.), ユーロピウム (Japan.)

Atomic Number. 63

Atomic Weight. 151.96

Electronic Configuration. [Xe] $4f^7\,6s^2$

Oxidation States. +2, +3, the latter being more common.

Coordination Number. 6 or greater.

Colour. Usually brown or yellow.

Availability. Available starting materials are Eu_2O_3 and anhydrous $EuCl_3$. See also under Lanthanum.

Handling. See under Lanthanum.

Toxicity. See under Lanthanum.

Isotopic Abundance. ^{151}Eu, 47.8%; ^{153}Eu, 52.18%.

Spectroscopy. The ^1H nmr spectra of Eu(III) ($4f^6$) compounds display large paramagnetic shifts. It is anticipated that lines in the ^1H nmr spectra of Eu(II) compounds would be too broad to give useful information, owing to the long relaxation times involved.

^{151}Eu, $I = \frac{5}{2}$, should be amenable to Mössbauer spectroscopy.

Analysis. See under Lanthanum.

References. See under Lanthanum.

$C_5H_5Cl_2Eu$ **Eu-00001**

Dichloro(η^5-2,4-cyclopentadien-1-yl)europium

Cyclopentadienyleuropium dichloride

$(C_5H_5)EuCl_2$

M 287.961

Tris-THF adduct:

$C_{17}H_{29}Cl_2EuO_3$ M 504.280

Purple cryst. Darkens >50°. Extremely air-sensitive. $\mu_{eff} = 4.24\mu_B$. Struct. probably resembles Dichloro(η^5-2,4-cyclopentadien-l-yl)erbium, Er-00001.

Manastyrskyj, S. *et al*, *Inorg. Chem.*, 1963, **2**, 904 (*synth*)

C_8H_8Eu **Eu-00002**

η^8-1,3,5,7-Cyclooctatetraeneeuropium

$(C_8H_8)Eu$

M 256.111

Orange solid. Sol. Py, DMF giving red solns. Extremely sensitive to air and water. Does not subl. Stable to 500°.

Hayes, R.G. *et al*, *J. Am. Chem. Soc.*, 1969, **91**, 6876 (*synth*, *epr*)

$C_{10}H_{10}Eu$ **Eu-00003**

Bis(η^5-2,4-cyclopentadien-1-yl)europium

Dicyclopentadienyleuropium. Europocene

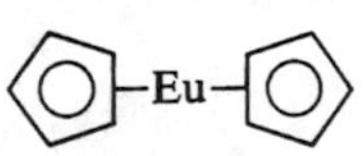

M 282.149

Yellow. Sol. THF. Subl. at 400° *in vacuo*. $\mu_{eff} = 7.63\mu_B$.

Fischer, E.O. *et al*, *J. Organomet. Chem.*, 1965, **3**, 181 (*synth*, *ir*)

Aleksanyan, V.T. *et al*, *J. Organomet. Chem.*, 1977, **131**, 113 (*ir*)

$C_{15}H_{15}Eu$ **Eu-00004**

Tris(η^5-2,4-cyclopentadien-1-yl)europium, 9CI

Tricyclopentadienyleuropium

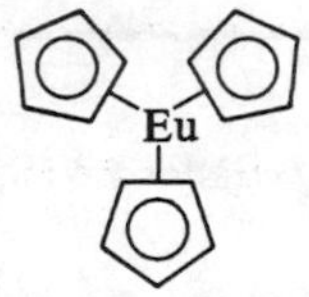

M 347.244

Brown cryst. Stable *in vacuo* up to 70°.

THF solvate: Mahogany-brown cryst. (THF).

Cyclohexylisocyanide adduct: [37299-00-6]. Sol. C_6H_6. Mp 150°.

Manastyrskyj, S. *et al*, *Inorg. Chem.*, 1964, **3**, 1647.

Tsutsui, M. *et al*, *Z. Naturforsch., B*, 1966, **21**, 1 (*synth*)

von Ammon, R. *et al*, *Ber. Bunsenges Phys. Chem.*, 1972, **76**, 995 (*nmr*, *ir*)

$C_{20}H_{30}Eu$ **Eu-00005**

Bis(pentamethylcyclopentadienyl)europium

Decamethyleuropocene

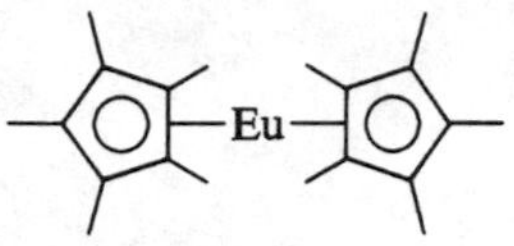

M 422.417

THF complex: [74282-45-4].

$C_{24}H_{38}EuO$ M 494.524

Red prisms. Sol. toluene. Mp 178-81°.

Complex with $2Et_2O + 1THF$: [74282-44-3].

$C_{32}H_{58}EuO_3$ M 642.768

Red cryst. Sol. Et_2O, hydrocarbons. Mp 181-2°. $\mu_{eff} = 7.99\mu_B$.

Tilley, T.D. *et al*, *Inorg. Chem.*, 1980, **19**, 2999 (*synth*, *ir*)

Gd Gadolinium

S. A. Cotton

Gadolinium (Fr., Ger.), Gadolinio (Sp., Ital.), Гадолиний (Gadolinii) (Russ.), ガドリニウム (Japan.)

Atomic Number. 64

Atomic Weight. 157.25

Electronic Configuration. [Xe] $4f^7\ 5d^1\ 6s^2$

Oxidation State. +3

Coordination Number. 6 or greater.

Colour. Colourless or yellow.

Availability. Available starting materials are Gd_2O_3 and anhydrous $GdCl_3$. See also under Lanthanum.

Handling. See under Lanthanum.

Toxicity. See under Lanthanum.

Isotopic Abundance. ^{154}Gd, 2.15%; ^{155}Gd, 14.73%; ^{156}Gd, 20.47%; ^{157}Gd, 15.68%; ^{158}Gd, 24.87%; ^{160}Gd, 21.90%.

Spectroscopy. See also under Lanthanum.

Owing to long relaxation times for Gd(III) ($4f^7$), ^1H nmr spectra are too broad to be observed. With the ^8S ground state, esr spectra should be capable of giving useful information about the environment of Gd(III) in organometallics.

Analysis. See under Lanthanum.

References. See under Lanthanum.

$C_5H_5Cl_2Gd$ Gd-00001

Dichloro(η^5-2,4-cyclopentadien-1-yl)gadolinium

Cyclopentadienylgadolinium dichloride

$(C_5H_5)GdCl_2$

M 293.251

Tris-THF adduct:
$C_{17}H_{29}Cl_2GdO_3$ M 509.570
Lavender cryst. Mp 82-6° dec. Extremely air-sensitive. Struct. probably resembles Dichloro(η^5-2,4-cyclopentadien-1-yl)erbium.

Manastyrskyj, S. *et al*, *Inorg. Chem.*, 1963, **2**, 904 (*synth*)

$C_{10}H_{10}ClGd$ Gd-00002

Chlorobis(η^5-2,4-cyclopentadien-1-yl)gadolinium

Chlorodicyclopentadienylgadolinium. Dicyclopentadienylgadolinium chloride

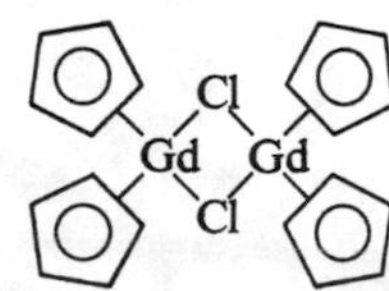

M 322.892
Dimeric.

Dimer: [11087-14-2]. *Di-μ-chlorotetrakis(η^5-2,4-cyclopentadien-1-yl)digadolinium*, 9CI.
$C_{20}H_{20}Cl_2Gd_2$ M 645.784
Colourless cryst. Sol. THF. Dec. >140°. Dec. by air. Subl. *in vacuo*. $\mu_{eff} = 8.86\mu_B$.

Maginn, R.E. *et al*, *J. Am. Chem. Soc.*, 1963, **85**, 672 (*synth, ir*)
Green, J.C. *et al*, *J. Organomet. Chem.*, 1981, **212**, 329 (*pe*)

$C_{12}H_{14}ClGd$ Gd-00003

Chlorobis(η^5-methylcyclopentadienyl)gadolinium

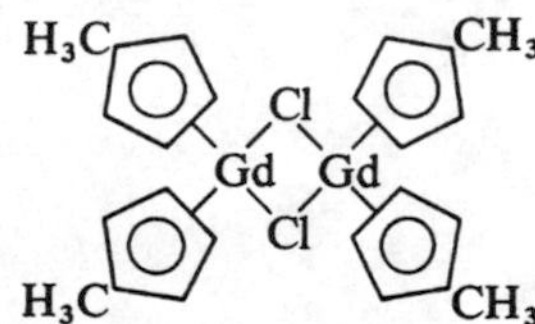

M 350.946
Dimeric.

Dimer: Di-μ-chlorotetrakis(η^5-1-methyl-2,4-cyclopentadien-1-yl)gadolinium, 9CI. Colourless cryst. Sol. C_6H_6, THF. Mp 188-97°. Subl. *in vacuo*, air-sensitive.

Maginn, R.E. *et al*, *J. Am. Chem. Soc.*, 1963, **85**, 672 (*synth*)

$C_{15}H_{15}Gd$ Gd-00004

Tris(η^5-2,4-cyclopentadien-1-yl)gadolinium, 9CI

Tri-π-cyclopentadienylgadolinium, 8CI

[1272-21-5]

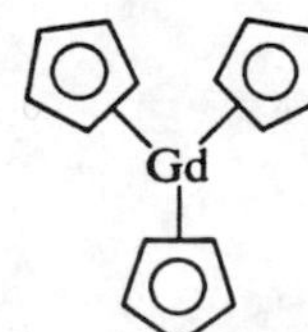

M 352.534

Pale-yellow cryst. Mp 350°. Subl. at 200-5° *in vacuo*, hydrol. by H_2O.

THF adduct: [75038-98-1]. White cryst. Sol. THF. Air-sensitive.

Cyclohexylisocyanide adduct: [37299-01-7]. Sol. C_6H_6. Mp 200°.

Wilkinson, G. *et al*, *J. Am. Chem. Soc.*, 1954, **76**, 6210 (*synth*)
Von Ammon, R. *et al*, *Ber. Bunsenges. Phys. Chem.*, 1972, **76**, 995 (*ir*)
Devyatykh, G.G. *et al*, *Dokl. Akad. Nauk. SSSR*, 1973, **208**, 1094 (*ms*)
Aleksanyan, V.T. *et al*, *J. Organomet. Chem.*, 1977, **131**, 251 (*raman*)
Rogers, R.D. *et al*, *J. Organomet. Chem.*, 1980, **192**, 65 (*struct*)

$C_{16}H_{15}Gd$ Gd-00005

Bis(η^5-2,4-cyclopentadien-1-yl)phenylgadolinium

[55672-18-9]

M 364.545

Lavender solid. Sol. THF, C_6H_6. Air- and moisture-sensitive. $\mu_{eff} = 7.69\mu_B$.

Ely, N.M. *et al*, *Inorg. Chem.*, 1975, **14**, 2680 (*synth, ir*)
Tsutsui, M. *et al*, *J. Am. Chem. Soc.*, 1975, **97**, 1280 (*synth, ir*)

$C_{18}H_{15}Gd$ Gd-00006

Bis(η^5-2,4-cyclopentadien-1-yl)(phenylethynyl)gadolinium

[53224-33-2]

Gd—C≡CPh

M 388.567

Pale-yellow solid. Sol. C_6H_6, THF. Mp 279-82° dec. Air- and moisture-sensitive. $\mu_{eff} = 7.98\mu_B$.

Ely, N.M. *et al*, *Inorg. Chem.*, 1975, **14**, 2680 (*synth, ir*)

Suggestions for new Entries are welcomed. Please write to the Editor, Dictionary of Organometallic Compounds, Chapman and Hall Ltd, 11 New Fetter Lane, London EC4P 4EE

$C_{27}H_{21}Gd$ **Gd-00008**

Triindenylgadolinium, 8CI

Tris(1,2,3,3a,7a-η)-1H-inden-1-ylgadolinium

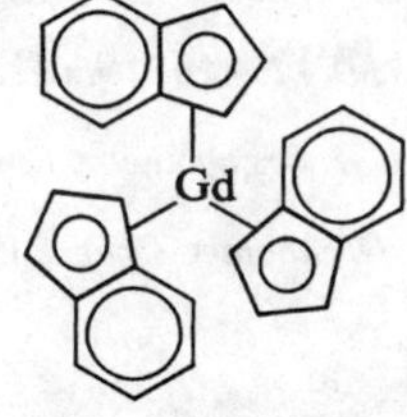

M 502.713

May be a mono- or a trihapto-complex.

THF complex: [22696-74-8].

$C_{31}H_{29}GdO$ M 574.820

Pale-green cryst. Sol. THF. $\mu = 7.89\mu_B$.

Tsutsui, M. *et al*, *J. Am. Chem. Soc.*, 1969, **91**, 3175 (*synth*)

Hf Hafnium

D. J. Cardin

Hafnium (Fr., Ger.), Hafnio (Sp.), Afnio (Ital.), Гафний (Gafnii) (Russ.), ハフニウム (Japan.)

Atomic Number. 72

Atomic Weight. 178.49 (Most commercial hafnium samples are contaminated with zirconium and so this value may not be appropriate).

Electronic Configuration. [Xe] $4f^{14}\ 5d^2\ 6s^2$

Oxidation States. The remarks for zirconium are applicable here with the general exception that there are markedly fewer examples of any type.

Coordination Number. Because of the extreme closeness of size of hafnium and zirconium the coordination numbers encountered are the same for analogous compounds with very few exceptions. Thus for some bulky alkyls the tetra-alkyl may be accessible for the lighter metal only. There is also a structural difference between the tetrakiscyclopentadienyls (q.v.). See under Zirconium.

Colour. See under Zirconium.

Availability. Common starting materials parallel those for zirconium, with fewer materials commercially available and at higher cost. Commercial Hf compounds are frequently contaminated with up to 9% of Zr.

Handling. Standard inert atmosphere techniques are required for air-sensitive compounds.

Toxicity. Hafnium compounds are not thought to be especially toxic but note that relatively little information is available and compounds should be treated appropriately.

Isotopic Abundance. ^{174}Hf, 0.18%; ^{176}Hf, 5.20%; ^{177}Hf, 18.50%; ^{178}Hf, 27.14%; ^{179}Hf, 13.75%; ^{180}Hf, 35.24%.

Spectroscopy. Neither of the two isotopes having nuclear spin (^{177}Hf, $I = \frac{7}{2}$ or ^{179}Hf, $I = \frac{9}{2}$), have been observed to date. The receptivities of these two isotopes relative to ^{13}C are 0.88 and 0.27 respectively.

Nmr of other nuclei is not generally useful for paramagnetic compounds of hafnium.

Esr is a valuable tool for d^1 compounds. Hyperfine couplings are not observed however.

Analysis. Because of the extreme similarity of the inorganic chemistry of these elements, methods used for zirconium are generally applicable. Difficulties may however arise for Zr/Hf ratios, for which chemical methods are not good (although available) especially when the hafnium content is low. In these cases emission spectroscopy, X-ray fluorescence or absorbance are preferred, and activation analysis is necessary for trace amounts.

References. General reviews are listed in the introduction to the *Sourcebook*.

Wailes, P. C. *et al.*, *Organometallic Chemistry of Titanium, Zirconium, and Hafnium*, Academic Press, London, 1974.

Larsen, E. M., *Adv. Inorg. Chem. Radiochem.*, 1970, **13,** 1.

Cardin, D. J. *et al.*, *The Organometallic Chemistry of Zirconium and Hafnium*, Ellis Horwood, in press (1984).

Hf-00001

Hf-00002

As Hf-00002 with
X = Cl

Hf-00003

As Hf-00002 with
X = F

Hf-00004

As Hf-00002 with
X = I

Hf-00005

Hf-00006

Hf-00007

Hf-00008

Hf-00009

Hf-00010

As Hf-00002 with
X = NCO

Hf-00011

As Hf-00002 with
X = CO

Hf-00012

As Hf-00002 with
X = Me

Hf-00013

Hf-00014

Hf-00015

Hf-00016

Hf-00017

As Hf-00002 with
X = NMe_2

Hf-00018

$Me_2Hf[N(SiMe_3)_2]_2$

Hf-00019

Hf-00020

$Hf(CH_2SiMe_3)_4$

Hf-00021

$Et_2Hf[N(SiMe_3)_2]_2$

Hf-00022

Hf-00023

As Hf-00002 with
R = NEt_2

Hf-00024

Hf-00025

Hf-00026

Hf-00027

Hf-00028

Hf-00029

$Hf[CH_2C(CH_3)_3]_4$

Hf-00030

Hf-00031

$ClHf[CH(SiMe_3)_2]_3$

Hf-00032

Hf-00033

As Hf-00002 with
X = Ph

Hf-00034

As Hf-00002 with
X = SePh

Hf-00035

Hf-00036

Hf-00037

Hf-00038

As Hf-00002 with
X = $-CH_2Ph$

Hf-00039

Hf-00040

Hf-00041

Hf-00042

Hf-00043

$Hf(CH_2Ph)_4$

Hf-00044

Hf-00045

Hf-00046

Hf-00047

Hf-00048

As Hf-00002 with
X = $—N{=}CPh_2$

Hf-00049

As Hf-00002 with
X = $—CHPh_2$

Hf-00050

Hf-00051

Hf-00052

$[(H_3C)_2C{=}CPh]_4Hf$

Hf-00053

$C_5H_5Cl_3Hf$ Hf-00001

Trichloro(η^5-2,4-cyclopentadien-1-yl)hafnium, 9CI

[61906-04-5]

M 349.944

Mp 125° dec. Forms a bis-THF adduct.

Renaut, P. *et al*, *J. Organomet. Chem.*, 1978, **148**, 35 (*nmr, ms*)

$C_{10}H_{10}Br_2Hf$ Hf-00002

Dibromobis(η^5-2,4-cyclopentadien-1-yl)hafnium, 9CI

Hafnocene dibromide

[37260-83-6]

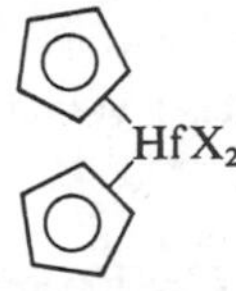

X = Br

M 468.487

Off-white solid. Mp 265-7°.

Druce, P.M. *et al*, *J. Chem. Soc.* (*A*), 1969, 2106, 2814 (*ir, ms, nmr*)
Samuel, E. *et al*, *Inorg. Chem.*, 1973, **12**, 881 (*ir*)
Cauletti, C. *et al*, *J. Electron Spectrosc. Relat. Phenom.*, 1980, **18**, 61 (*pe*)

$C_{10}H_{10}Cl_2Hf$ Hf-00003

Dichlorobis(η^5-2,4-cyclopentadien-1-yl)hafnium, 9CI

Dichlorodi-π-cyclopentadienylhafnium, 8CI. *Bis(cyclopentadienyl)hafnium dichloride. Hafnocene dichloride*

[12116-66-4]

As Dibromobis(η^5-2,4-cyclopentadien-1-yl)hafnium, Hf-00002 with

X = Cl

M 379.585

Colourless solid. Mp 234-5°, >300°. $Bp_{0.01}$ 155°.

▷MG4815000.

Druce, P.M. *et al*, *J. Chem. Soc.* (*A*), 1969, 2106, 2814 (*ir, ms, nmr, raman*)
Burmeister, J.L. *et al*, *Inorg. Chem.*, 1970, **9**, 58 (*ir, uv*)
Samuel, E. *et al*, *Inorg. Chem.*, 1973, **12**, 881 (*ir*)
Cauletti, C. *et al*, *J. Electron. Spectrosc. Relat. Phenom.*, 1980, **18**, 61 (*pe*)

$C_{10}H_{10}F_2Hf$ Hf-00004

Bis(η^5-2,4-cyclopentadien-1-yl)difluorohafnium, 9CI

Hafnocene difluoride

[37260-84-7]

As Dibromobis(η^5-2,4-cyclopentadien-1-yl)hafnium, Hf-00002 with

X = F

M 346.676

Colourless solid. Mp >200°.

Druce, P.M. *et al*, *J. Chem. Soc.* (*A*), 1969, 2106, 2814 (*ir, ms, nmr*)
Samuel, E. *et al*, *Inorg. Chem.*, 1973, **12**, 881 (*ir*)
Cauletti, C. *et al*, *J. Electron Spectrosc. Relat. Phenom.*, 1980, **18**, 61 (*pe*)

$C_{10}H_{10}HfI_2$ Hf-00005

Bis(η^5-2,4-cyclopentadien-1-yl)diiodohafnium, 9CI

Hafnocene diiodide

[37260-85-8]

As Dibromobis(η^5-2,4-cyclopentadien-1-yl)hafnium, Hf-00002 with

X = I

M 562.488

Yellow solid. Mp 303-5°.

Druce, P.M. *et al*, *J. Chem. Soc.* (*A*), 1969, 2106, 2814 (*ir, ms, nmr*)
Samuel, E. *et al*, *Inorg. Chem.*, 1973, **12**, 881 (*ir*)
Cauletti, C. *et al*, *J. Electron Spectrosc. Relat. Phenom.*, 1980, **18**, 61 (*pe*)

$C_{10}H_{10}HfS_5$ Hf-00006

Bis(η^5-2,4-cyclopentadien-1-yl)(pentathio)hafnium, 10CI

Bis(η^5-cyclopentadienyl)cyclo(pentasulfur-1,5-diyl)-hafnium

[75213-10-4]

M 468.979

Air-stable yellow cryst. (CH_2Cl_2). Mp 150-5° dec. E_{act} (ring inversion) 58 kJ. mol.$^{-1}$ (CD_2Cl_2).

McCall, J.M. *et al*, *J. Organomet. Chem.*, 1980, **193**, C37 (*synth, ir, nmr*)

$C_{10}H_{11}ClHf$ Hf-00007

Chlorobis(η^5-2,4-cyclopentadien-1-yl)hydrohafnium, 10CI

Chlorobis(η-cyclopentadienyl)hydridohafnium. Hafnocene hydridochloride

[72953-59-4]

M 345.140

'Hydrohafnation' reagent analogue to Chlorobis(η^5-2,4-cyclopentadien-1-yl)hydrozirconium.

Tolstikow, G.A. *et al*, *Izv. Akad. Nauk SSSR, Ser. Khim.*, 1979, 2576 (*synth, ir, use*)

$C_{10}H_{15}Cl_3Hf$ **Hf-00008**

Trichloro[(1,2,3,4,5-η)-1,2,3,4,5-pentamethyl-2,4-cyclopentadien-1-yl]hafnium, 10CI

[75181-08-7]

M 420.078

Green-yellow solid. $Bp_{0.001}$ 140° subl.

Blenkers, J. *et al*, *J. Organomet. Chem.*, 1981, **218**, 383 (*nmr*)

$C_{10}H_{18}B_2Hf$ **Hf-00009**

Bis(η^5-2,4-cyclopentadien-1-yl)bis[tetrahydroborato(1−)-*H,H'*]hafnium, 9CI

Hafnocene bis(borohydride)

[56420-26-9]

M 338.362

Colourless solid. Sol. Et_2O, aromatic hydrocarbons. Bp 110° subl. *in vacuo*. Dec. slowly in air. Deutero analogue also prepd.

Davies, N. *et al*, *J. Chem. Soc. (A)*, 1969, 2601 (*synth, ir, nmr*)
Marks, T.J. *et al*, *J. Am. Chem. Soc.*, 1975, **97**, 3397 (*nmr*)

$C_{11}H_{10}F_3HfOP$ **Hf-00010**

Carbonylbis(η^5-2,4-cyclopentadien-1-yl)(phosphorous trifluoride)hafnium

Carbonylbis(cyclopentadienyl)trifluorophosphinehafnium

[71763-65-0]

M 424.658

Degree of association uncertain. Yellow solid.

Sikora, D.J. *et al*, *J. Am. Chem. Soc.*, 1979, **101**, 5079 (*synth, nmr, ir*)

$C_{12}H_{10}HfN_2O_2$ **Hf-00011**

Bis(cyanato)di-π-cyclopentadienylhafnium, 8CI

Bis(η-cyclopentadienyl)dicyanatohafnium. Hafnocene dicyanate

As Dibromobis(η^5-2,4-cyclopentadien-1-yl)hafnium, Hf-00002 with

X = NCO

M 392.713

Mp 185°.

Burmeister, J.L. *et al*, *Inorg. Chem.*, 1970, **9**, 58 (*synth, ir, ms, uv*)

$C_{12}H_{10}HfO_2$ **Hf-00012**

Dicarbonylbis(η^5-2,4-cyclopentadien-1-yl)hafnium, 10CI

Hafnocene dicarbonyl

[59487-86-4]

As Dibromobis(η^5-2,4-cyclopentadien-1-yl)hafnium, Hf-00002 with

X = CO

M 364.700

Dark-red air-sensitive cryst. by subl. Sol. hydrocarbons.

Demerseman, B. *et al*, *J. Organomet. Chem.*, 1976, **107**, C19 (*synth, ir*)
Thomas, J.F. *et al*, *J. Organomet. Chem.*, 1976, **111**, 297 (*synth, ir, ms*)
Sikora, D.J. *et al*, *J. Am. Chem. Soc.*, 1979, **101**, 5079 (*cryst struct*)

$C_{12}H_{16}Hf$ **Hf-00013**

Bis(η^5-2,4-cyclopentadien-1-yl)dimethylhafnium, 10CI

Hafnocene dimethyl

[37260-88-1]

As Dibromobis(η^5-2,4-cyclopentadien-1-yl)hafnium, Hf-00002 with

X = Me

M 338.748

White cryst. by subl. Bp 90° subl. *in vacuo*.

Samuel, E. *et al*, *J. Am. Chem. Soc.*, 1973, **95**, 6263 (*synth*)
Alt, H. *et al*, *J. Am. Chem. Soc.*, 1974, **96**, 5936.
Alt, H.G. *et al*, *J. Organomet. Chem.*, 1976, **107**, 257.

$C_{12}H_{20}Hf$ **Hf-00014**

Tetrakis(η^3-2-propenyl)hafnium, 10CI

Tetraallylhafnium

[12241-95-1]

M 342.780

Red cryst. (pentane). $Bp_{0.0001}$ ~20° subl. Differs from Zr analogue Tetrakis(η^3-2-propenyl)zirconium.

Becconsall, J.K. *et al*, *J. Chem. Soc. (A)*, 1967, 423 (*synth, nmr, ms*)

$C_{13}H_{14}Cl_2Hf$ **Hf-00015**

Dichloro[1,3-propanediylbis[(1,2,3,4,5-η)-2,4-cyclopentadien-1-ylidene]]hafnium, 9CI

Dichloro(1,1'-trimethylenedicyclopentadienyl)hafnium

[38117-94-1]

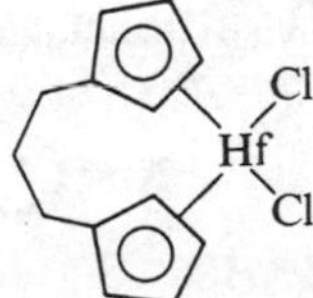

M 419.650

$Bp_{0.001-0.0001}$ 200° subl.

Hillman, M. *et al*, *J. Organomet. Chem.*, 1972, **42**, 123 (*synth*, *nmr*)
Saldarriaga-Molina, C.H. *et al*, *Inorg. Chem.*, 1974, **13**, 2880 (*cryst struct*)

$C_{14}H_{18}Hf$ Hf-00016

(η^8-1,3,5,7-Cyclooctatetraene)bis(η^3-2-propenyl)hafnium, 9CI
Bis(η-allyl)[η-cyclooctatetraene(2−)]hafnium
[41558-36-5]

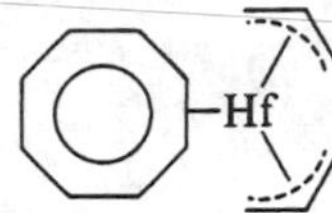

M 364.786
Yellow solid. In toluene shows dynamic allyl groups to low temp.

Kablitz, H.J. *et al*, *J. Organomet. Chem.*, 1972, **44**, C49 (*synth*, *nmr*)

$C_{14}H_{21}ClHfSi$ Hf-00017

Chlorobis(η^5-2,4-cyclopentadien-1-yl)[(trimethylsilyl)methyl]-hafnium, 10CI
[78205-96-6]

M 431.348
White solid.

Jeffery, J. *et al*, *J. Chem. Soc., Dalton Trans.*, 1981, 1593 (*synth*, *ir*, *nmr*)

$C_{14}H_{22}HfN_2$ Hf-00018

Bis(η^5-cyclopentadienyl)bis(dimethylamido)hafnium
Di-π-cyclopentadienylbis(dimethylaminato)hafnium, 8CI. *Hafnocene bis(dimethylamide)*
As Dibromobis(η^5-2,4-cyclopentadien-1-yl)hafnium, Hf-00002 with

$X = NMe_2$

M 396.831
Catalyses acrylonitrile polymerisation. Yellow solid by subl. $Bp_{0.02}$ 120-5° subl.

Chandra, G. *et al*, *J. Chem. Soc. (A)*, 1968, 1940 (*synth*, *nmr*, *ms*)
Jenkins, A.D. *et al*, *Polym. Lett.*, 1968, **6**, 865 (*use*)
Druce, P.M. *et al*, *J. Chem. Soc. (A)*, 1969, 2106 (*props*)

$C_{14}H_{42}HfN_2Si_4$ Hf-00019

Dimethylbis[bis(trimethylsilyl)amido]hafnium
Dimethylbis[1,1,1-trimethyl-N-(trimethylsilyl)silanaminato]hafnium, 10CI
[71000-84-5]

$Me_2Hf[N(SiMe_3)_2]_2$

M 529.331
Colourless liq. at r.t.; cryst. (pentane). Mp ca. 20°.

Andersen, R.A., *Inorg. Chem.*, 1979, **18**, 2928 (*synth*, *ir*, *pmr*, *cmr*)

$C_{15}H_{19}ClHfO_4$ Hf-00020

Chloro-π-cyclopentadienylbis(2,4-pentanedionato)hafnium, 8CI
Bis(1-acetyl-2-oxopropyl)chloro(η^5-2,4-cyclopentadien-1-yl)hafnium, 10CI. *Bis(acetylacetonato)chloro(η-cyclopentadienyl)hafnium*
[12147-78-3]

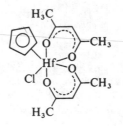

M 477.256
Mp 181-2°.

Minacheva, M.K. *et al*, *Izv. Akad. Nauk SSSR, Ser. Khim.*, 1969, 1104 (*nmr*)
Sudarikov, V.S. *et al*, *Zh. Strukt. Khim.*, 1969, **10**, 558 (*cryst struct*)
Kharlamova, E.N. *et al*, *Izv. Akad. Nauk SSSR, Ser. Khim.*, 1970, 2621 (*props*)
Pinnavaia, T.J. *et al*, *Inorg. Chem.*, 1971, **10**, 1388 (*synth*, *nmr*)

$C_{16}H_{44}HfSi_4$ Hf-00021

Tetrakis(trimethylsilylmethyl)hafnium, 9CI
[40334-04-1]

$Hf(CH_2SiMe_3)_4$

M 527.356
Polymerises olefins. Colourless liq. Mp 8-10°. $Bp_{0.001}$ ca. 50°. Shows markedly enhanced thermal stability relative to straight-chain alkyls.
▷Pyrophoric

Collier, M.R. *et al*, *J. Chem. Soc., Dalton Trans.*, 1973, 445 (*synth*, *nmr*, *ir*)
Lappert, M.F. *et al*, *J. Organomet. Chem.*, 1974, **66**, 271 (*pe*)
Ballard, D.G.H. *et al*, *J. Catal.*, 1976, **44**, 116 (*use*)

$C_{16}H_{46}HfN_2Si_4$ Hf-00022

Diethylbis[bis(trimethylsilyl)amido]hafnium
Diethylbis[1,1,1-trimethyl-N-(trimethylsilyl)silanaminatohafnium, 10CI
[70969-31-2]

$Et_2Hf[N(SiMe_3)_2]_2$

M 557.385
Colourless prisms (pentane). Mp 65-7°. Stable to β-H transfer, presumably for steric reasons.

Andersen, R.A., *Inorg. Chem.*, 1979, **18**, 2928 (*synth*, *ir*, *pmr*, *cmr*)

$C_{17}H_{25}HfP$ Hf-00023

Bis[(1,2,3,4,5,6-η)methylbenzene](trimethylphosphine)hafnium, 10CI

Ditoluene(trimethylphosphine)hafnium

[71361-39-2]

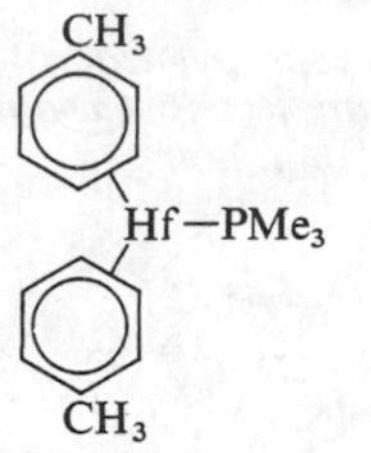

M 438.848

Dark-green cryst. Sol. aromatic solvs., less sol. aliphatic solvs. Solns. stable under Ar but not N_2.

Cloke, F.G.N. *et al*, *J. Chem. Soc.*, *Dalton Trans.*, 1981, 1938 (*synth*, *ms*, *nmr*)

$C_{18}H_{30}HfN_2$ Hf-00024

Bis(η^5-cyclopentadienyl)bis(diethylamido)hafnium

Di-π-cyclopentadienylbis(diethylaminato)hafnium, 8CI.
Hafnocene bis(diethylamide)

As Dibromobis(η^5-2,4-cyclopentadien-1-yl)hafnium, Hf-00002 with

R = NEt_2

M 452.938

Yellow cryst. by subl. $Bp_{0.03}$ 120-30° subl.

Chandra, G. *et al*, *J. Chem. Soc.* (*A*), 1968, 1940 (*synth*, *ms*)

$C_{18}H_{32}HfSi_2$ Hf-00025

Bis(η^5-2,4-cyclopentadien-1-yl)bis[(trimethylsilyl)methyl]-hafnium, 9CI

[11077-95-5]

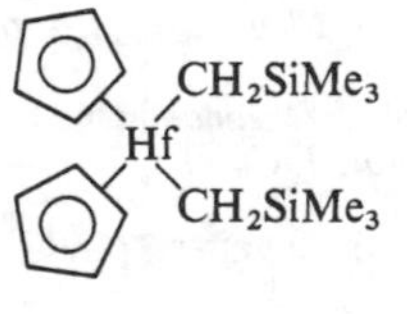

M 483.112

White needles. Mp 83°. More thermally stable than analogous alkyls. Air-sensitive, particularly in soln.

Collier, M.R. *et al*, *J. Organomet. Chem.*, 1970, **25**, C36 (*synth*)
Collier, M.R. *et al*, *J. Chem. Soc.*, *Dalton Trans.*, 1973, 445 (*synth*, *ir*, *nmr*)

$C_{20}H_{10}ClCo_3HfO_{10}$ Hf-00026

Chlorobis(η^5-2,4-cyclopentadien-1-yl)[μ_4-[methanolato(4−)-*C:C:C:O*]](nonacarbonyltricobalt)hafnium, 10CI

Chlorobis(η^5-cyclopentadienyl)[μ_3-(oxymethylidyne)-cyclotris(tricarbonylcobalt)hafnium

[67422-56-4]

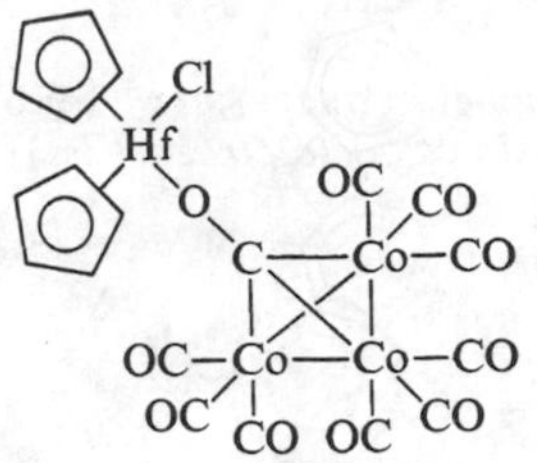

M 801.036

Dark-red cryst. (toluene). Spar. sol. aromatic, v. spar. sol. aliphatic solvs. Mp 139-42°.

Stutte, B. *et al*, *Chem. Ber.*, 1978, **111**, 1603 (*synth*, *ir*, *nmr*, *cryst struct*)
Ishii, M. *et al*, *Ber. Bunsenges Phys. Chem.*, 1979, **83**, 1026 (*synth*, *ir*, *nmr*)

$C_{20}H_{20}Cl_2Hf_2O$ Hf-00027

Dichlorotetrakis(η^5-2,4-cyclopentadien-1-yl)-μ-oxodihafnium

Dichlorotetra-π-cyclopentadienyloxodihafnium, 8CI.
μ-Oxobis[bis(η-cyclopentadienyl)chlorohafnium]

[12118-16-0]

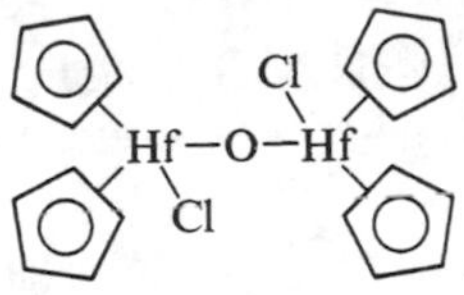

M 704.263

Cryst. Mp 265-9° dec.

Samuel, E., *Bull. Soc. Chim. Fr.*, 1966, 3548 (*synth*, *ir*, *uv*, *nmr*)
Minacheva, M.K., *Dokl. Akad. Nauk SSSR*, *Ser. Sci. Khim.*, 1967, **173**, 581 (*synth*)

$C_{20}H_{20}Hf$ Hf-00028

Bis[(1,2,3,3*a*,7*a*-η)-1*H*-inden-1-yl]dimethylhafnium, 9CI

[49596-06-7]

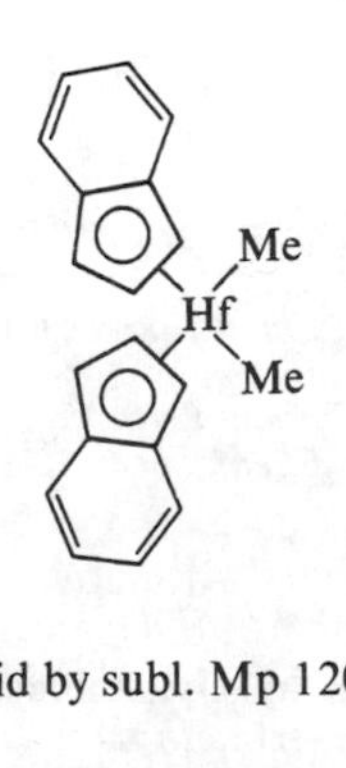

M 438.868

Straw-coloured solid by subl. Mp 120° dec. Bp ca. 120° subl. *in vacuo*.

Samuel, E. *et al*, *J. Am. Chem. Soc.*, 1973, **95**, 6263 (*synth*, *nmr*)
Atwood, J.L. *et al*, *Inorg. Chem.*, 1975, **14**, 1757 (*cryst struct*)

$C_{20}H_{24}HfSn_2$ **Hf-00029**

Bis[(1,2,3,4,5,6-η)methylbenzene]bis(trimethylstannyl)hafnium, 10CI

Ditoluenebis(trimethylstannyl)hafnium

[78379-40-5]

M 680.280

Black cryst. Solid only slowly dec. by air.

Cloke, F.G.N. *et al*, *J. Chem. Soc., Chem. Commun.*, 1981, 117 (*synth, nmr, ms, cryst struct*)

$C_{20}H_{44}Hf$ **Hf-00030**

Tetrakis(2,2-dimethylpropyl)hafnium, 9CI

Tetraneopentylhafnium

[50654-35-8]

$Hf[CH_2C(CH_3)_3]_4$

M 463.058

Colourless cryst. by subl. Mp 115-6°. $Bp_{0.001}$ 50° subl. Shows enhanced thermal stability over straight chain alkyls.

Davidson, P.J. *et al*, *J. Organomet. Chem.*, 1973, **57**, 269 (*synth, ir, raman, nmr, ms*)

Lappert, M.F. *et al*, *J. Organomet. Chem.*, 1974, **66**, 271 (*pe*)

Lappert, M.F. *et al*, *J. Chem. Soc., Chem. Commun.*, 1975, 830.

$C_{21}H_{24}HfO_5$ **Hf-00031**

η^5-2,4-Cyclopentadien-1-ylbis(2,4-pentanedionato-*O,O'*)phenoxyhafnium

Bis(acetylacetonato)(η-cyclopentadienyl)phenoxyhafnium

M 534.908

Mp 67.5-70°.

Brainina, E.M. *et al*, *Dokl. Akad. Nauk SSSR, Ser. Sci. Khim.*, 1971, **196**, 1085 (*synth, nmr*)

$C_{21}H_{57}ClHfSi_6$ **Hf-00032**

Tris[bis(trimethylsilyl)methyl]chlorohafnium, 9CI

Chlorotris[bis(trimethylsilyl)methyl]hafnium

[53668-85-2]

$ClHf[CH(SiMe_3)_2]_3$

M 692.137

White cryst.

Barker, G.K. *et al*, *J. Organomet. Chem.*, 1974, **76**, C45 (*synth, ir, nmr*)

$C_{22}H_{10}F_{10}Hf$ **Hf-00033**

Bis(η^5-2,4-cyclopentadien-1-yl)bis(pentafluorophenyl)hafnium, 9CI

[12636-90-7]

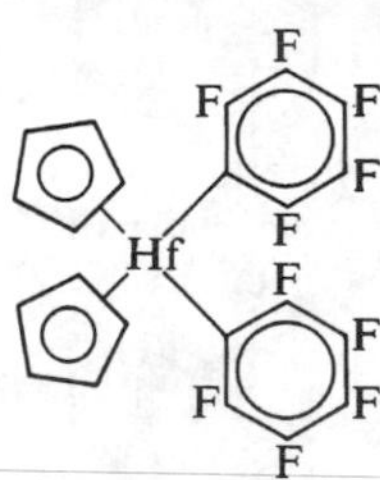

M 642.795

White microcryst. by subl. Mp 221-3°.

Rausch, M.D. *et al*, *J. Coord. Chem.*, 1971, **1**, 141 (*synth, ir, nmr, ms*)

$C_{22}H_{20}Hf$ **Hf-00034**

Bis(η^5-2,4-cyclopentadien-1-yl)diphenylhafnium, 10CI

Hafnocene diphenyl

[51231-81-3]

As Dibromobis(η^5-2,4-cyclopentadien-1-yl)hafnium, Hf-00002 with

X = Ph

M 462.890

White cryst. Mp 145° dec.

Samuel, E. *et al*, *J. Am. Chem. Soc.*, 1973, **95**, 6263 (*synth, ir, nmr*)

$C_{22}H_{20}HfSe_2$ **Hf-00035**

Bis(benzeneselenolato)bis(η^5-2,4-cyclopentadien-1-yl)hafnium, 10CI

Bis(η-cyclopentadienyl)bis(phenylselenato)hafnium

[78789-95-4]

As Dibromobis(η^5-2,4-cyclopentadien-1-yl)hafnium, Hf-00002 with

X = SePh

M 620.810

Golden-yellow air-stable cryst. (C_6H_6/pentane). Mp 160°. Deliquescent.

Gautheron, B. *et al*, *J. Organomet. Chem.*, 1981, **209**, C49 (*synth, ms, nmr*)

$C_{22}H_{30}HfO_2$ **Hf-00036**

Dicarbonylbis[(1,2,3,4,5-η)-1,2,3,4,5-pentamethyl-2,4-cyclopentadien-1-yl]hafnium

[76830-38-1]

M 504.968

Purple needles (heptane).

Sikora, D.J. *et al*, *J. Am. Chem. Soc.*, 1981, **103**, 1265 (*synth, cryst struct, ir, ms, nmr*)

$C_{23}H_{17}ClHfN_2O_2$ Hf-00037

Chloro-π-cyclopentadienylbis(8-quinolinolato)hafnium, 8CI
Chloro(cyclopentadienyl)hafnium bisoxinate

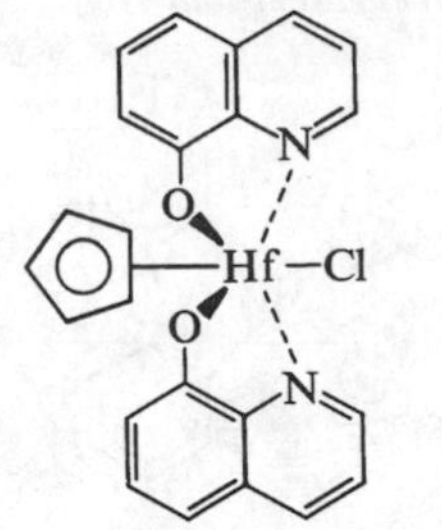

M 567.343
Yellow solid (Et_2O). Mp 248.5-251.5°.

Brainina, E.M. *et al*, *Izv. Akad. Nauk SSSR*, *Ser. Khim.*, 1968, 817 (*synth*, *ir*)
Charalambous, J. *et al*, *J. Chem. Soc.* (*A*), 1971, 2487 (*synth*, *ir*, *nmr*, *ms*)

$C_{23}H_{20}HfO$ Hf-00038

(η^2-Benzoyl)bis(η^5-2,4-cyclopentadien-1-yl)phenylhafnium, 10CI
[66152-54-3]

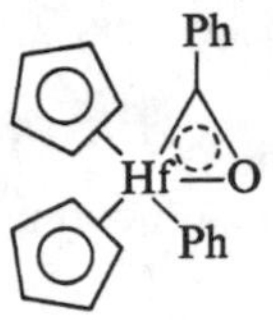

M 490.900
Characterized spectroscopically. Yellow-orange cryst. (toluene/pentane). Mp 22°. Stable to loss of CO both in solid state and in soln.

Fachinetti, G. *et al*, *J. Chem. Soc.*, *Dalton Trans.*, 1977, 1946 (*synth*, *ir*, *nmr*)

$C_{24}H_{24}Hf$ Hf-00039

Bis(benzyl)bis(η^5-cyclopentadienyl)hafnium
Bis(η^5-2,4-cyclopentadien-1-yl)bis(phenylmethyl)hafnium, 10CI. Hafnocene dibenzyl
[56770-59-3]

As Dibromobis(η^5-2,4-cyclopentadien-1-yl)hafnium, Hf-00002 with

$$X = -CH_2Ph$$

M 490.944
Yellow cryst. (toluene/heptane).

Fachinetti, G. *et al*, *J. Chem. Soc.*, *Dalton Trans.*, 1977, 1946 (*synth*, *nmr*)
Brindley, P.B. *et al*, *J. Chem. Soc.*, *Perkin Trans. 2*, 1981, 419 (*nmr*)

$C_{24}H_{34}HfSi_2$ Hf-00040

Bis(η^5-2,4-cyclopentadien-1-yl)[1,2-phenylenebis[(trimethylsilyl)methylene]]hafnium, 10CI
2,2-Bis(η^5-cyclopentadienyl)-1,3-bis(trimethylsilyl)-2-hafnaindane
[76933-96-5]

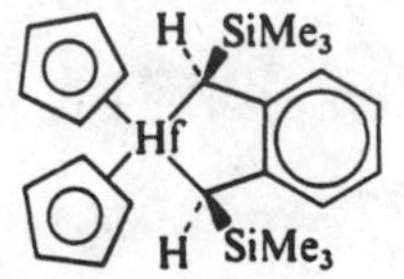

M 557.194

meso-form

Orange cryst. (pentane). Sol. org. solvs. Mp 144-6°. Limited air stability but thermally robust and sublimable.

Lappert, M.F. *et al*, *J. Chem. Soc.*, *Chem. Commun.*, 1980, 1284 (*synth*, *ir*, *ms*, *nmr*)

$C_{28}H_{25}ClGeHf$ Hf-00041

Chlorobis(η^5-2,4-cyclopentadien-1-yl)(triphenylgermyl)hafnium, 9CI
[12636-97-4]

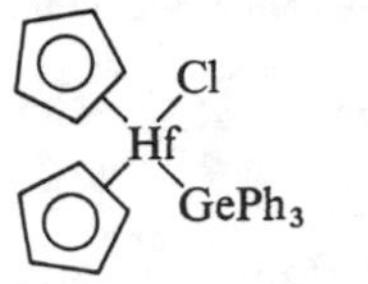

M 648.039
Yellow solid. Subl. at 200° *in vacuo*.

Kingston, B.M. *et al*, *J. Chem. Soc.*, *Dalton Trans.*, 1972, 69 (*synth*, *nmr*, *ms*)

$C_{28}H_{25}ClHfSi$ Hf-00042

Chlorobis(η^5-2,4-cyclopentadien-1-yl)(triphenylsilyl)hafnium, 9CI
[12636-99-6]

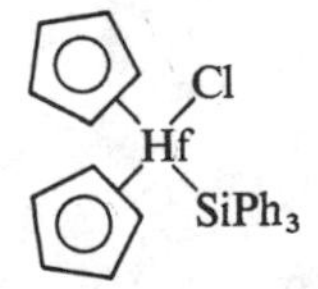

M 603.534
Yellow solid. Bp 180° subl. *in vacuo*.

Kingston, B.M. *et al*, *J. Chem. Soc.*, *Dalton Trans.*, 1972, 69 (*synth*, *nmr*, *ms*)

$C_{28}H_{25}ClHfSn$ Hf-00043

Chlorobis(η^5-2,4-cyclopentadien-1-yl)(triphenylstannyl)hafnium, 9CI
[12637-00-2]

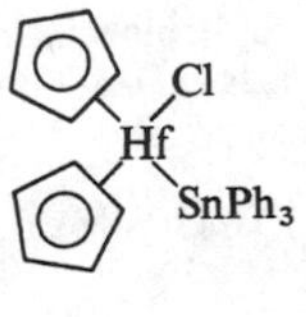

M 694.139
Yellow solid.

Kingston, B.M. *et al*, *J. Chem. Soc.*, *Dalton Trans.*, 1972, 69 (*synth*, *nmr*)

$C_{28}H_{28}Hf$ Hf-00044

Tetrabenzylhafnium

Tetrakis(phenylmethyl)hafnium, 9CI. *Hafnium tetrabenzyl*

[31406-67-4]

$Hf(CH_2Ph)_4$

M 543.019

Olefin polymerisation catalyst. Pale-yellow cryst. Mp 112-4°.

Davies, G.R. *et al*, *J. Chem. Soc., Chem. Commun.*, 1971, 1511 (*cryst struct*)

Felten, J.J. *et al*, *J. Organomet. Chem.*, 1972, **36**, 87; 1974, **82**, 375 (*synth, nmr*)

B.P., 1 428 108, (*1976*); *CA*, **85**, 63667y (*use*)

$C_{30}H_{10}Co_6HfO_{20}$ Hf-00045

Bis(η^5-2,4-cyclopentadien-1-yl)bis[μ_4-[methanolato(4−)-*C*:*C*:*C*:*O*]]bis(nonacarbonyltricobalt)dihafnium, 10CI

Bis(η^5-cyclopentadienyl)di[μ_3-(oxymethylidyne)cyclotris(tricarbonylcobalt)]hafnium

[67422-55-3]

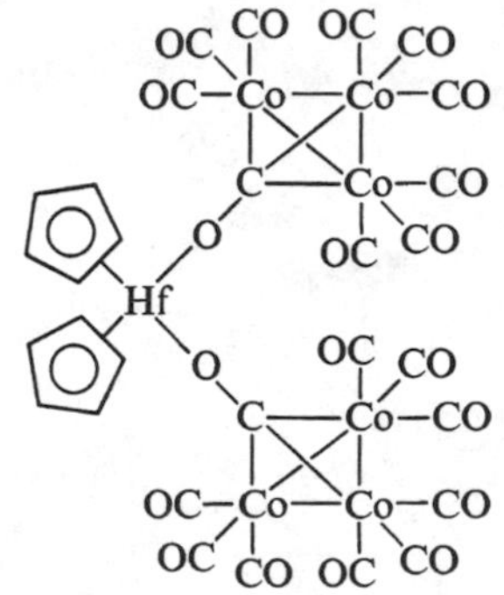

M 1222.486

Red cryst. (toluene).

Ishii, M. *et al*, *Ber. Bunsenges. Phys. Chem.*, 1978, **83**, 1026 (*cryst struct*)

Stutte, B. *et al*, *Chem. Ber.*, 1978, **111**, 1603 (*synth, ir, nmr, cryst struct*)

$C_{30}H_{36}HfS_2$ Hf-00046

Bis(η-*tert*-butylcyclopentadienyl)bis(phenylthiolato)hafnium

Bis(benzenethiolato)bis[(1,2,3,4,5-η)-1-(1,1-dimethylethyl)-2,4-cyclopentadien-1-yl]hafnium, 10CI

[75673-89-1]

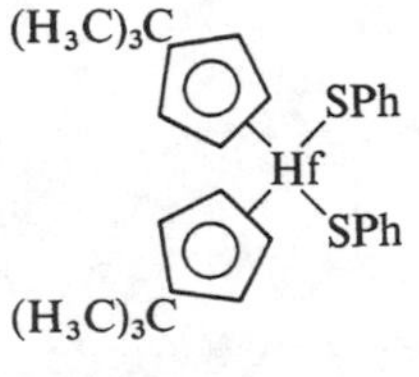

M 639.224

Yellow cryst. (hexane). Mp 460°.

Couturier, S. *et al*, *J. Organomet. Chem.*, 1980, **195**, 291 (*synth, ms, nmr*)

$C_{30}H_{38}Hf_2O_9$ Hf-00047

Bis(η^5-2,4-cyclopentadien-1-yl)-μ-oxotetrakis(2,4-pentanedionato-*O*,*O'*)dihafnium, 10CI

μ-Oxobis[bis(acetylacetonato)(η^5-cyclopentadienyl)hafnium]

[76124-85-1]

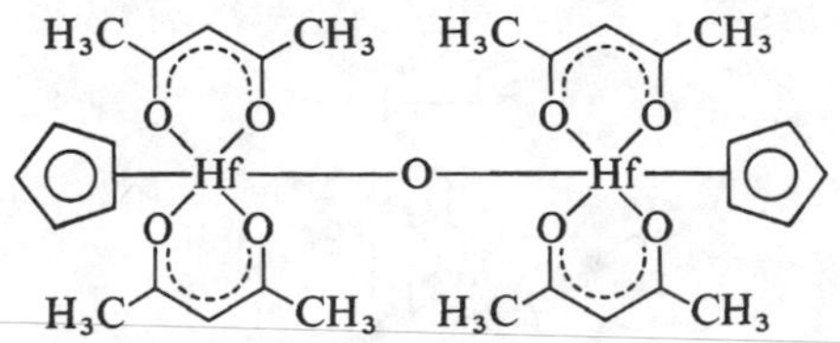

M 899.605

Mp 150-4°.

Brainina, E.M. *et al*, *Izv. Akad. Nauk SSSR, Ser. Khim.*, 1968, 817 (*synth, ir, props*)

$C_{32}H_{23}HfN_3O_3$ Hf-00048

π-Cyclopentadienyltris(8-quinolinolato)hafnium, 8CI

Tris(8-quinolinolato)cyclopentadienylhafnium

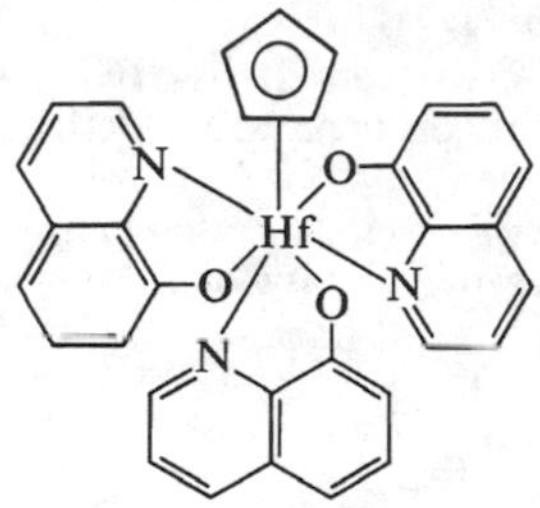

M 676.042

Solid. Mp 250° dec.

Minacheva, M.K. *et al*, *Dokl. Akad. Nauk SSSR, Ser. Sci. Khim.*, 1967, **173**, 581 (*synth*)

$C_{36}H_{30}HfN_2$ Hf-00049

Bis(η^5-cyclopentadienyl)bis(diphenylketimido)hafnium

Di-π-cyclopentadienylbis(1,1-diphenylmethyleneiminato)hafnium, 8CI. *Hafnocene bis(diphenylketimide)*

[11106-17-5]

As Dibromobis(η^5-2,4-cyclopentadien-1-yl)hafnium, Hf-00002 with

$X = —N{=}CPh_2$

M 669.136

Red-orange powder. $Bp_{0.001}$ ca. 150° subl.

Collier, M.R. *et al*, *Inorg. Nucl. Chem. Lett.*, 1971, **7**, 689 (*synth, ir, nmr, ms*)

$C_{36}H_{32}Hf$ Hf-00050

Bis(η^5-2,4-cyclopentadien-1-yl)bis(diphenylmethyl)hafnium, 10CI

[60175-23-7]

As Dibromobis(η^5-2,4-cyclopentadien-1-yl)hafnium, Hf-00002 with

$X = —CHPh_2$

M 643.139

Orange cryst. (C_6H_6 or toluene). Mp 192-4°. Sl. sensitive to air in solid state, dec. rapidly in soln.

Atwood, J.L. *et al*, *J. Am. Chem. Soc.*, 1977, **99**, 6645 (*synth, ir, nmr, cryst struct*)

$C_{38}H_{30}Hf$ Hf-00051

Bis(η^5-2,4-cyclopentadien-1-yl)(1,2,3,4-tetraphenyl-1,3-butadiene-1,4-diyl)hafnium, 10CI

1,1-Bis(η^5-cyclopentadienyl)-2,3,4,5-tetraphenylhafnole

[53433-59-3]

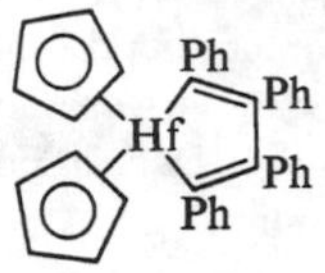

M 665.145

Yellow cryst. (pentane/CH_2Cl_2). Stable in solid state but air-sensitive in soln.

Atwood, J.L. *et al*, *J. Am. Chem. Soc.*, 1976, **98**, 2454 (*cryst struct*)

Sikora, D.J. *et al*, *J. Am. Chem. Soc.*, 1979, **101**, 5079 (*synth, ms, nmr*)

$C_{40}H_{20}Co_6Hf_2O_{21}$ Hf-00052

Tetrakis(η^5-2,4-cyclopentadien-1-yl)bis[μ_4-[methanolato(4−)-*C*:*C*:*C*:*O*]bis(nonacarbonyltricobalt)-μ-oxodihafnium, 10CI

μ-Oxobis[bis(η^5-cyclopentadienyl)[μ_3-(oxymethylidyne)cyclotris(tricarbonylcobalt)]hafnium]

[67422-50-8]

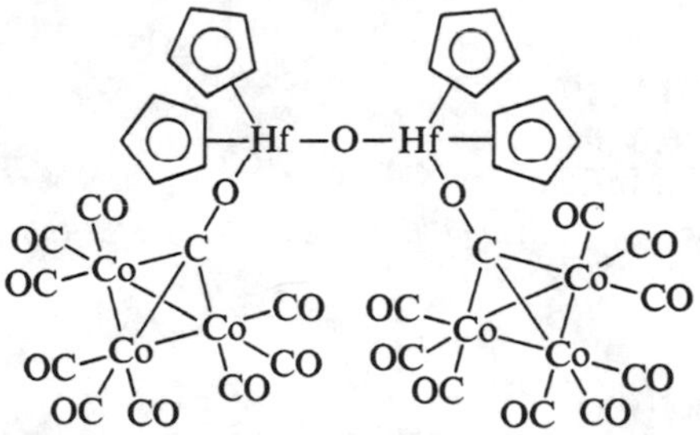

M 1547.165

Red cryst. (toluene). Mp 156-60°.

Stutte, B. *et al*, *Chem. Ber.*, 1978, **111**, 1603 (*synth, ir, nmr*)

$C_{40}H_{44}Hf$ Hf-00053

Tetrakis(2-methyl-1-phenyl-1-propenyl)hafnium, 10CI

Tetrakis(2,2-dimethyl-1-phenylethenyl)hafnium

[63422-31-1]

$[(H_3C)_2C{=}CPh]_4Hf$

M 703.278

Yellow cryst. Mp 70° dec.

Cardin, C.J. *et al*, *J. Organomet. Chem.*, 1977, **132**, C23 (*synth, nmr*)

Ho Holmium

S. A. Cotton

Holmium (Fr., Ger.), Holmio (Sp.), Olmio (Ital.), Гольмий (Gol'mii) (Russ.), ホルミウム (Japan.)

Atomic Number. 67

Atomic Weight. 164.9307

Electronic Configuration. [Xe] $4f^{11}\,6s^2$

Oxidation State. +3

Coordination Number. 6 or greater.

Colour. Usually yellow.

Availability. Available starting materials are the expensive Ho_2O_3 and anhydrous $HoCl_3$.

Handling. See under Lanthanum.

Toxicity. See under Lanthanum.

Isotopic Abundance. ^{165}Ho, 100%.

Spectroscopy. Organometallic compounds of the strongly paramagnetic Ho(III) show large paramagnetic shifts and very broad lines in their ^1H nmr spectra. For references see below. See also under Lanthanum.

Analysis. See under Lanthanum.

References. See under Lanthanum. For references to ^1H nmr spectroscopy, see also:

Von Ammon, R. *et al.*, *Inorg. Nucl. Chem. Lett.*, 1969, **5,** 315.

Schumann, H. *et al.*, *Z. Naturforsch.*, *B*, 1981, **36,** 1244.

$C_5H_5Cl_2Ho$ **Ho-00001**

Dichloro(η^5-2,4-cyclopentadien-1-yl)holmium

Cyclopentadienylholmium dichloride

$(C_5H_5)HoCl_2$

M 300.931

Tris-THF adduct: [52679-42-2].

$C_{17}H_{29}Cl_2HoO_3$ M 517.251

Yellow cryst. Mp 84-92°. Extremely air-sensitive. Struct. probably resembles Dichloro(η^5-2,4-cyclopentadien-1-yl)erbium.

Manastyrskyj, S. *et al*, *Inorg. Chem.*, 1963, **2**, 904 (*synth*)

$C_{10}H_{10}ClHo$ **Ho-00002**

Chlorobis(η^5-2,4-cyclopentadien-1-yl)holmium

Chlorodicyclopentadienylholmium. Dicyclopentadienylholmium chloride

[56200-26-1]

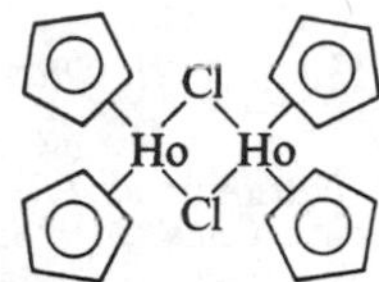

M 330.572

Dimeric.

Dimer: Di-μ-chlorotetrakis(η^5-2,4-cyclopentadien-1-yl-)diholmium, 9CI.

$C_{20}H_{20}Cl_2Ho_2$ M 661.145

Pale-yellow cryst. Sol. THF. Mp 340-3° dec. Subl. *in vacuo*. Air-sensitive. $\mu_{eff} = 10.3\mu_B$.

Maginn, R.E. *et al*, *J. Am. Chem. Soc.*, 1963, **85**, 672 (*synth*, *ir*)

$C_{11}H_{13}Ho$ **Ho-00003**

Bis(η^5-2,4-cyclopentadien-1-yl)methylholmium

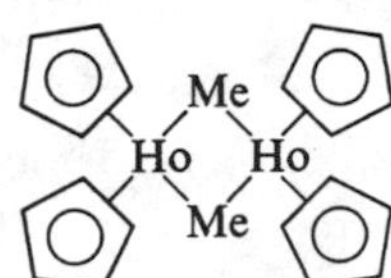

M 310.154

Dimeric.

Dimer: [60997-42-4]. *Tetrakis(η^5-2,4-cyclopentadien-1-yl)di-μ-methyldiholmium*, 9CI.

$C_{22}H_{26}Ho_2$ M 620.308

Straw cryst. Sol. CH_2Cl_2, C_6H_6, toluene. Dec. >160°. Air-sensitive. $\mu_{eff} = 10.0\mu_B$.

Holton, J. *et al*, *J. Chem. Soc., Dalton Trans.*, 1979, 54 (*synth*, *ir*)

$C_{13}H_{13}Ho$ **Ho-00004**

(η^8-1,3,5,7-Cyclooctatetraenyl)(η^5-2,4-cyclopentadien-1-yl)-holmium, 9CI

[52668-28-7]

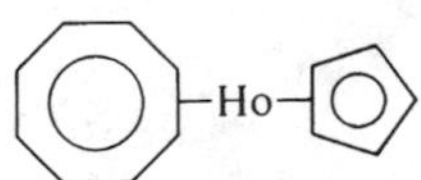

M 334.176

Cryst. Sol. THF. Extremely air-sensitive. Forms adducts with THF, NH_3, $C_6H_{11}NC$, Py.

Jamerson, J.D. *et al*, *J. Organomet. Chem.*, 1974, **65**, C33 (*synth*, *ms*, *ir*)

$C_{15}H_{15}Ho$ **Ho-00005**

Tris(η^5-2,4-cyclopentadien-1-yl)holmium, 9CI

Tricyclopentadienylholmium

[1272-22-6]

Ho

M 360.214

Yellow cryst. Sol. THF. Mp 295°. Hydrol. by H_2O. Subl. at 230° *in vacuo*. $\mu_{eff} = 10.2_{\mu B}$.

Isocyanocyclohexane adduct: [12098-32-7]. *Tris(η^5-2,4-cyclopentadien-1-yl)isocyanocyclohexaneholmium.*

$C_{22}H_{26}HoN$ M 469.385

Yellow cryst. Sol. C_6H_6. Mp 165°. Subl. at 150° *in vacuo*. $\mu_{eff} = 10.6\mu_B$.

Fischer, E.O. *et al*, *J. Organomet. Chem.*, 1965, **3**, 181 (*synth*, *ir*)

Fischer, E.O. *et al*, *J. Organomet. Chem.*, 1966, **6**, 141.

v. Ammon, R. *et al*, *Inorg. Nucl. Chem. Lett.*, 1969, **5**, 315 (*nmr*)

Aleksanyan, V.T. *et al*, *J. Organomet. Chem.*, 1977, **131**, 251 (*raman*)

$C_{20}H_{19}HoO_3W$ **Ho-00006**

[μ-(Carbonyl-*C,O*)]bis[(1,2,3,4,5-η)-1-methyl-2,4-cyclopentadien-1-yl][tricarbonyl(η^5-2,4-cyclopentadien-1-yl)tungsten]holmium, 9CI

[39477-08-2]

$(H_3CC_5H_4)_2HoW(C_5H_5)(CO)_3$

M 656.149

Nitrosylated by *N*-methyl-*N*-nitroso-*p*-toluenesulfonamide to give $(C_5H_5)_2W(CO)_2NO$. Sol. THF, DMSO. Mp >220° dec. Air- and moisture-sensitive.

Crease, A.E. *et al*, *J. Chem. Soc., Dalton Trans.*, 1973, 1501 (*synth*, *ir*)

La Lanthanum

S. A. Cotton

Lanthane (Fr.), Lanthan (Ger.), Lantano (Sp.), Lantanio (Ital.), Лантан (Lantan) (Russ.), ランタン (Japan.)

Atomic Number. 57

Atomic Weight. 138.905

Electronic Configuration. [Xe] $5d^1\,6s^2$

Oxidation State. +3

Coordination Number. Usually 6 or greater.

Colour. Colourless.

Availability. The most usual starting material for organolanthanum chemistry and for organolanthanides in general is the anhydrous chloride. Both this and the oxide are commercially available but the latter is invariably cheaper. $LaCl_3$ may be synthesized from La_2O_3 by treatment with NH_4Cl or by dissolving the oxide in hydrochloric acid, evaporating and treating the product with thionyl chloride. Commercial samples of anhydrous chlorides of the lanthanide may require refluxing with thionyl chloride. References to experimental procedures are given below.

Handling. Organolanthanides are extremely air- and moisture-sensitive and should be handled accordingly.

Toxicity. No special precautions are normally taken in handling lanthanum compounds. However, there have been few reports concerning their toxicity and they should be handled with caution.

Isotopic Abundance. ^{139}La, 99.91%.

Spectroscopy. ^{139}La, $I = \frac{7}{2}$, merits study as an nmr probe. Lanthanum compounds are diamagnetic and therefore give sharp, unshifted ^{1}H nmr spectra. There have to date been no studies reported of organolanthanum compounds. For references see below.

Mass spectra have been reported for relatively few organolanthanides—see *Gmelins Handbuch der Anorganischen Chemie.*

Lanthanum is amenable to Mössbauer spectroscopy and an account is given in *Organometallics of the f-Elements*—see below.

Analysis. Microanalytical results for organolanthanides are frequently lower than theoretical values, probably because of incomplete combustion and concomitant carbide formation.

Analysis for lanthanum may be carried out gravimetrically or complexometrically. Gravimetric determination involves precipitation of lanthanum as the hydroxide or oxalate, followed by ignition to the oxide. For complexometric determination, the compound is decomposed by ignition or by hydrolysis, the residue dissolved in acid and then titrated directly with Na_2EDTA at pH 5–6 using xylenol orange indicator. For details, see references below.

References

General

Marks, T. J., *Prog. Inorg. Chem.*, 1978, **24,** 51.
Cotton, S. A., *J. Organomet. Chem. Lib.*, 1977, **3,** 189.
Organometallics of the f-Elements, Marks, T. J. and Fischer, R. D. Eds, Reidel, Dortrecht, 1979.

Synthesis

Bravcr, G., *Handbuch der Präparativen Anorganischen Chemie*, Thieme, Stuttgart, 1962, **2,** 1002.
Kiess, N. H., *J. Res. Natl. Bur. Stand.*, 1963, **67A,** 1343.
Schumann, H. *et al.*, *Organometallics*, 1982, **1,** 1194.
Laubereau, P. G. *et al.*, *Inorg. Chem.*, 1970, **9,** 1091.
Saran, M. S. *et al.*, *J. Organomet. Chem.*, 1970, **21,** 147.
Tilley, T. D. *et al.*, *Inorg. Chem.*, 1981, **20,** 3267.

Nmr

Chemical Applications of Nmr in Paramagnetic Molecules, La Mar, G. N., Horrocks, W. D. and Holm, R. H. Eds, Academic Press, 1973.
Forsburg, J. H. *et al.*, *Gmelins Handbuch der Anorganischen Chemie*, 8th Edn, Part D6, Springer-Verlag, 1983.
Webb, G. A., *Ann. Rev. Nmr Spectrosc.*, 1976, **6A,** 1.
Marks, T. J., *Prog. Inorg. Chem.*, 1978, **24,** 51.
Keller, H. J. *et al.*, *Angew. Chem., Int. Ed. Engl.*, 1969, **5,** 315.
Von Ammon, R. *et al.*, *Ber. Bunsenges Phys. Chem.*, 1972, **76,** 995.
Von Ammon, R. *et al.*, *Inorg. Nucl. Chem. Lett.*, 1969, **5,** 315.

C_8H_8ClLa **La-00001**

(η^8-1,3,5,7-Cyclooctatetraene)chlorolanthanum

$(C_8H_8)LaCl$

M 278.510

Bis-THF adduct: *(η^8-1,3,5,7-Cyclooctatetraene)-bis(tetrahydrofuran)lanthanum.*
$C_{16}H_{24}ClLaO_2$ M 422.723
Pale-yellow powder. Sol. THF. Presumably dimeric by analogy with Chloro(cyclooctatetraene)cerium.

Wayda, A.L., *Organometallics*, 1983, **2**, 565 (*synth, ir, nmr*)

$C_{12}H_{30}LaP_3$ **La-00002**

Tris[(dimethylphosphinidenio)bis(methylene)]lanthanum

[61060-80-8]

M 406.196

Colourless solid. Sol. C_6H_6. Mp 195-205° dec.

Schumann, H. *et al*, *Chem. Ztg.*, 1976, **100**, 336 (*synth, nmr*)

$C_{13}H_{14}ClLa$ **La-00003**

Chloro[1,3-propanediylbis(η^5-2,4-cyclopentadien-1-ylidene)]-lanthanum

[75857-12-4]

M 344.612

Probably dimeric. Sol. THF. Mp >100°. Air- and moisture-sensitive.

John, J.N. *et al*, *J. Coord. Chem.*, 1980, **10**, 177 (*synth, nmr*)

$C_{15}H_{15}La$ **La-00004**

Tris(η^5-2,4-cyclopentadien-1-yl)lanthanum, 9CI

Tri-π-cyclopentadienyllanthanum, 8CI

[1272-23-7]

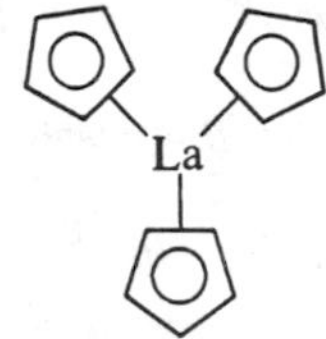

M 334.189

Cryst. Mp 395°. Subl. at 200-50° *in vacuo*, hydrol. by H_2O.

THF adduct: [79533-62-3]. White platelets. Sol. THF. Subl. *in vacuo* at 260°. Air-sensitive.

Cyclohexylisocyanide adduct: [37299-02-8]. Sol. C_6H_6. Mp 110°.

Wilkinson, G. *et al*, *J. Am. Chem. Soc.*, 1954, **76**, 6210 (*synth*)
Fritz, H.P., *Chem. Ber.*, 1959, **92**, 780 (*ir*)
Devyatykh, G.G. *et al*, *Dokl. Akad. Nauk. SSSR*, 1970, **193**, 1069 (*synth, ms*)
Von Ammon, R. *et al*, *Ber. Bunsenges. Phys. Chem.*, 1972, **76**, 995 (*nmr, ir*)
Aleksanyan, V.T. *et al*, *J. Raman Spec.*, 1974, **2**, 345 (*raman*)
Rogers, R.D. *et al*, *J. Organomet. Chem.*, 1981, **216**, 383 (*struct*)

$C_{16}H_{16}La^{\ominus}$ **La-00005**

Bis(η^8-1,3,5,7-cyclooctatetraene)lanthanate(1–), 9CI

Dicyclooctatetraenelanthanate(1–)

M 347.208 (ion)

K salt: [51187-42-9].
$C_{16}H_{16}KLa$ M 386.306
Green solid. Sol. dioxan, THF. Mp >345°. Extremely air- and moisture-sensitive.

Hodgson, K.O. *et al*, *J. Am. Chem. Soc.*, 1973, **95**, 8650 (*synth, ir*)

$C_{27}H_{21}La$ **La-00006**

Triindenyllanthanum

Tris(1,2,3,3a,7a-η)-1H-inden-1-yllanthanum

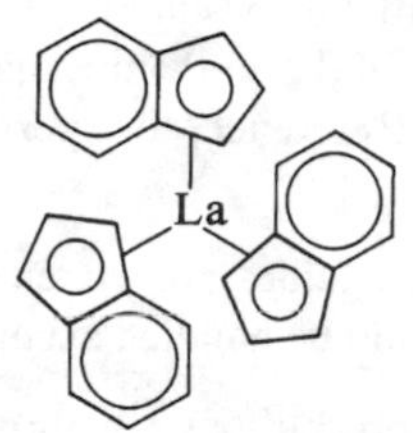

M 484.368

May be a monohapto- or a trihapto-complex.

THF adduct: [22696-73-7].
$C_{31}H_{29}LaO$ M 556.475
Pale-tan solid. Sol. THF. Diamagnetic.

Tsutsui, M. *et al*, *J. Am. Chem. Soc.*, 1969, **91**, 3175 (*synth, nmr*)

$C_{32}H_{40}La_2O_2$ **La-00007**

Tris(1,3,5,7-cyclooctatetraene)bis(tetrahydrofuran)dilanthanum

[65101-96-4]

M 734.478

Struct. assigned by analogy with [μ-[(1,2-η:1,2,3,4,5,6,7,8-η)-1,3,5,7-Cyclooctatetraene]-[bis(η^8-1,3,5,7-cyclooctatetraene)bis(tetrahydrofuran)dineodymium, Nd-00012. Gold cryst. Sol. THF. Extremely air- and moisture-sensitive, inflames in air.

DeKock, C.W. *et al*, *Inorg. Chem.*, 1978, **17**, 625 (*synth*)

Lu Lutetium

S. A. Cotton

Lutétium (Fr.), Lutetium (Ger.), Lutecio (Sp.), Lutezio (Ital.), Лютеций (Liutetsii) (Russ.), ルテチウム (Japan.)

Atomic Number. 71

Atomic Weight. 174.97

Electronic Configuration. [Xe] $4f^{14}\ 5d^{1}\ 6s^{2}$

Oxidation State. +3

Coordination Number. 6 or more, but less when bulky ligands such as $—CH_2SiMe_3$ and $—C(CH_3)_3$ are present.

Colour. Colourless.

Availability. Starting materials are Lu_2O_3 and anhydrous $LuCl_3$, which are expensive.

Handling. See under Lanthanum.

Toxicity. See under Lanthanum.

Isotopic Abundance. ^{175}Lu, 97.41%; ^{176}Lu, 2.59%.

Spectroscopy. Lu(III) compounds are diamagnetic ($4f^{14}$) and give sharp, unshifted ^{1}H nmr spectra of which a considerable number have now been reported. For these and for a discussion of ^{13}C nmr spectra see Forsburg, *loc cit*.

Analysis. See under Lanthanum

References. See under Lanthanum

$C_5H_5Cl_2Lu$ **Lu-00001**

Dichloro(η^5-2,4-cyclopentadien-1-yl)lutetium

Cyclopentadienyllutetium dichloride

$(C_5H_5)LuCl_2$

M 310.968

Tris-THF adduct:
$C_{17}H_{29}Cl_2LuO_3$ M 527.287
Colourless cryst. Mp 76-8°. Extremely air-sensitive. Struct. probably resembles Dichloro(η^5-2,4-cyclopentadien-1-yl)erbium.

Manastryrskyj, S. *et al*, *Inorg. Chem.*, 1963, **2**, 904 (*synth*)

$C_6H_{18}Lu^{\ominus\ominus\ominus}$ **Lu-00002**

Hexamethyllutetate(3–)

$LuMe_6^{\ominus\ominus\ominus}$

M 265.175 (ion)

Tris[(tetramethylethylenediamine)lithium] salt: [66862-13-3].
$C_{24}H_{66}Li_3LuN_6$ M 634.616
White cryst. Sl. sol. Et_2O. Mp 141-2° dec. V. sensitive to air and moisture.

Schumann, H. *et al*, *Angew. Chem., Int. Ed. Engl.*, 1978, **17**, 276 (*synth, nmr*)

C_8H_8ClLu **Lu-00003**

(η^8-1,3,5,7-Cyclooctatetraene)chlorolutetium

$(C_8H_8)LuCl$

M 314.571

THF adduct: (η^8-1,3,5,7-Cyclooctatetraene)-chloro(tetrahydrofuran)lutetium.
$C_{12}H_{16}ClLuO$ M 386.678
Off-white powder. Sol. THF. Probably dimeric by analogy with Chloro(cyclooctatetraene)cerium.

Wayda, A.L., *Organometallics*, 1983, **2**, 565 (*synth, ir, nmr*)

$C_{10}H_{10}ClLu$ **Lu-00004**

Chlorobis(η^5-2,4-cyclopentadien-1-yl)lutetium, 9CI

Chlorodicyclopentadienyllutetium. Dicyclopentadienyllutetium chloride

M 340.609

Dimeric.

Dimer: [76207-13-1]. *Di-μ-chlorotetrakis(η^5-2,4-cyclopentadien-1-yl)dilutetium.*
$C_{20}H_{20}Cl_2Lu_2$ M 681.218
Pale-green-white cryst. Sol. THF. Mp 318-20°. Diamagnetic. Air-sensitive. Subl. *in vacuo*.

Maginn, R.E. *et al*, *J. Am. Chem. Soc.*, 1963, **85**, 672 (*synth, ir*)

$C_{10}H_{11}Lu$ **Lu-00005**

Bis(η^5-2,4-cyclopentadien-1-yl)hydrolutetium

M 306.164

THF adduct: [79292-30-1]. *Bis(η^5-2,4-cyclopentadien-1-yl)hydro(tetrahydrofuran)lutetium,* 10CI.
$C_{14}H_{19}LuO$ M 378.271
Synthesized by reaction of $(C_5H_5)_2LuR$ (R = Me_3SiCH_2 or $C_6H_5CH_2$) with H_2. Colourless cryst. Does not react with alkenes.

Schumann, H. *et al*, *J. Organomet. Chem.*, 1981, **213**, C7 (*synth, nmr, ir*)

$C_{11}H_{13}Lu$ **Lu-00006**

Bis(η^5-2,4-cyclopentadien-1-yl)methyllutetium

M 320.191

THF complex: [76207-05-1].
$C_{15}H_{21}LuO$ M 392.297
White solid. Sol. THF, C_6H_6, toluene. Unstable at r.t. Extremely air- and moisture-sensitive.

Schumann, H. *et al*, *Angew. Chem., Int. Ed. Engl.*, 1981, **20**, 119 (*synth, nmr*)
Schumann, H. *et al*, *Organometallics*, 1982, **1**, 1194 (*synth, ir, nmr*)

$C_{12}H_{19}LuSi$ **Lu-00007**

(η^8-1,3,5,7-Cyclooctatetraene)(trimethylsilylmethyl)lutetium

$(C_8H_8)Lu(CH_2SiMe_3)$

M 366.335

Bis-THF adduct: [84582-80-9]. *(η^8-1,3,5,7-Cyclooctatetraene)(trimethylsilylmethyl)bis(tetrahydrofuran)-lutetium.*
$C_{20}H_{35}LuO_2Si$ M 510.548
Off-white powder. Sol. toluene, THF. Marginally stable at r.t.

Wayda, A.L. *et al*, *Organometallics*, 1983, **2**, 565 (*synth, ir, nmr*)

$C_{12}H_{30}LuP_3$ **Lu-00008**

Tris[(dimethylphosphinidenio)bis(methylene)]lutetium

[61060-87-5]

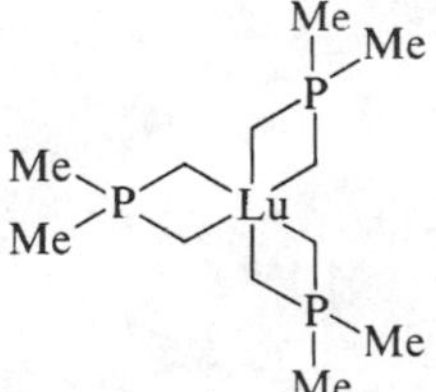

M 442.257

Colourless solid. Sol. C_6H_6. Mp 210° dec.

Schumann, H. *et al*, *Chem.-Ztg.*, 1976, **100**, 336 (*synth, nmr*)

$C_{14}H_{19}Lu$ **Lu-00009**

Butylbis(η^5-2,4-cyclopentadien-1-yl)lutetium

$Lu(CH_2)_3CH_3$

M 362.271

THF complex: [82293-70-7].

$C_{18}H_{27}LuO$ M 434.378

Colourless solid. Sol. THF, C_6H_6, toluene. Unstable at r.t., extremely sensitive to air and moisture.

Schumann, H. *et al*, *Angew. Chem., Int. Ed. Engl.*, 1981, **20**, 119 (*synth, nmr*)
Schumann, H. *et al*, *Organometallics*, 1981, **1**, 1194 (*synth, ir, nmr*)

$C_{14}H_{19}Lu$ **Lu-00010**

***tert*-Butylbis(η^5-cyclopentadienyl)lutetium**

Bis(η^5-2,4-cyclopentadien-1-yl)(1,1-dimethylethyl)lutetium

[76207-08-4]

$Lu{-}C(CH_3)_3$

M 362.271

THF complex:

$C_{18}H_{27}LuO$ M 434.378

$LuH(C_4H_8O)]_2$. Colourless cryst. + 1THF. Inserts CO at room temperature to form successively Bis(η^5-2,4-cyclopentadien-1-yl)(2,2-dimethyl-1-oxopropyl-*C,O*)lutetium, Lu-00013 and Tetrakis(η^5-2,4-cyclopentadien-1-yl)[μ-(2,2,7,7-tetramethyl-3,4,5,6-octanetetrone-O^3,O^5:O^4,O^6)]dilutetium, Lu-00024 Reacts with H_2 to afford $[(C_5H_5)_2$ 70-80 dec. Extremely air- and moisture-sensitive. $LuH(C_4H_8O)]_2$.

Evans, W.J. *et al*, *J. Chem. Soc., Chem. Commun.*, 1981, 292, 706 (*struct, ir, nmr, synth*)
Evans, W.J. *et al*, *J. Am. Chem. Soc.*, 1982, **104**, 2008 (*synth, nmr, cmr, ir*)
Schumann, H. *et al*, *Organometallics*, 1982, **1**, 1194 (*synth, ir, nmr, cmr*)

$C_{14}H_{21}LuSi$ **Lu-00011**

Bis(η^5-2,4-cyclopentadien-1-yl)[(trimethylsilyl)methyl]lutetium

(Trimethylsilylmethyl)dicyclopentadienyllutetium

$LuCH_2SiMe_3$

M 392.372

THF complex: [76207-10-8].

$C_{18}H_{29}LuOSi$ M 464.479

Yellow needles. Sol. THF, C_6H_6, toluene. Mp 100-10° dec.

Schumann, H. *et al*, *Angew. Chem., Int. Ed. Engl.*, 1981, **20**, 119 (*synth, struct, nmr*)
Schumann, H. *et al*, *Organometallics*, 1982, **1**, 1194 (*synth, struct, ir, nmr*)

$C_{15}H_{15}Lu$ **Lu-00012**

Tris(η^5-2,4-cyclopentadien-1-yl)lutetium

Tricyclopentadienyllutetium

[1272-24-8]

Lu

M 370.251

Colourless cryst. Mp 295°. Subl. at 260° *in vacuo*.

Isocyanocyclohexane adduct: [37299-03-9]. *Tris(η^5-2,4-cyclopentadien-1-yl)isocyanocyclohexanelutetium.*

$C_{22}H_{26}LuN$ M 479.421

Sol. C_6H_6. Mp 170°.

Fischer, E.O. *et al*, *J. Organomet. Chem.*, 1965, **3**, 181 (*synth, ir*)
v. Ammon, R. *et al*, *Ber. Bunsenges. Phys. Chem.*, 1972, **76**, 995.
Aleksanyan, V.T. *et al*, *J. Organomet. Chem.*, 1977, **131**, 251 (*raman*)

$C_{15}H_{19}LuO$ **Lu-00013**

Bis(η^5-2,4-cyclopentadien-1-yl)(2,2-dimethyl-1-oxopropyl-*C,O*)lutetium

[79503-47-2]

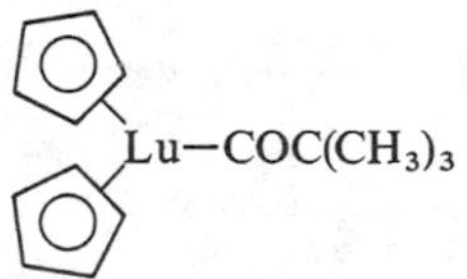

M 390.282

Formed by CO insertion reaction of *tert*-Butylbis(η^5-cyclopentadienyl)lutetium, Lu-00010 . Pale-yellow solid. Sol. THF, toluene. Air- and moisture-sensitive.

Evans, W.J. *et al*, *J. Chem. Soc., Chem. Commun.*, 1981, 706 (*synth, ir, pmr, cmr*)

$C_{15}H_{21}Lu$ **Lu-00014**

Bis(η^5-2,4-cyclopentadien-1-yl)(2,2-dimethylpropyl)lutetium

Dicyclopentadienylneopentyllutetium

[76207-09-5]

$LuCH_2C(CH_3)_3$

M 376.298

White cryst. + 1THF. Sol. C_6H_6, toluene. Mp 90-100° dec. Extremely air- and moisture-sensitive.

Schumann, H. *et al*, *Organometallics*, 1982, **1**, 1194 (*synth, ir, nmr, cmr*)

$C_{16}H_{44}LuSi^{\ominus}$ **Lu-00015**

Tetrakis[(trimethylsilyl)methyl]luteate(1−)

$[Lu(CH_2SiMe_3)_4]^{\ominus}$

M 439.576 (ion)

Li salt, tetrakis(tetrahydrofuranate): [70388-23-7].

$C_{32}H_{84}LiLuO_4Si$ M 743.007
Green-white solid. Sol. pentane. Mp 113-5° dec. Pyrophoric in air. Slowly dec. to Li[Lu(CH_2SiMe_3)$_2$($CHSiMe_3$)].

Schumann, H. *et al*, *J. Organomet. Chem.*, 1979, **169**, C1 (*synth, nmr*)

$C_{17}H_{17}Lu$ Lu-00016

Benzylbis(η^5-2,4-cyclopentadienyl)lutetium

Bis(η^5-2,4-cyclopentadien-1-yl)(phenylmethyl)lutetium, 9CI

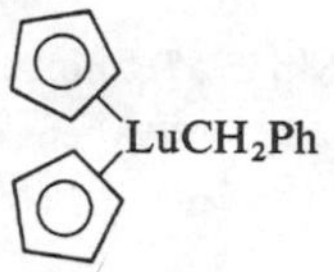

M 396.288

THF complex: [76207-11-9].
$C_{21}H_{25}LuO$ M 468.395
White cryst. Sol. THF, C_6H_6, toluene. Mp 100-10° dec. Extremely air- and moisture-sensitive.

Schumann, H. *et al*, *Angew. Chem., Int. Ed. Engl.*, 1981, **20**, 119.
Schumann, H. *et al*, *Organometallics*, 1982, **1**, 1194 (*synth, ir, nmr*)

$C_{17}H_{17}Lu$ Lu-00017

Bis(η^5-2,4-cyclopentadien-1-yl)(4-methylphenyl)lutetium

p-*Tolyldicyclopentadienyllutetium*

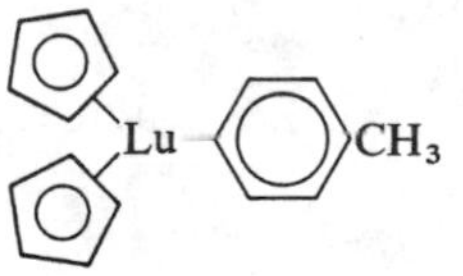

M 396.288

THF complex: [76207-12-0].
$C_{21}H_{25}LuO$ M 468.395
Colourless cryst. Sol. THF, C_6H_6, toluene. Mp 110-20° dec. V. sensitive to air and moisture.

Schumann, H. *et al*, *Angew. Chem., Int. Ed. Engl.*, 1981, **20**, 119 (*synth, nmr*)
Schumann, H. *et al*, *Organometallics*, 1982, **1**, 1194 (*synth, struct, nmr, ir*)

$C_{20}H_{31}Lu$ Lu-00018

Hydridobis[(1,2,3,4,5-η)-1,2,3,4,5-pentamethyl-2,4-cyclopentadien-1-yl]lutetium

[83463-59-6]

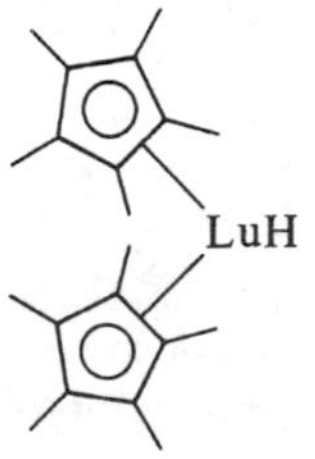

M 446.432

Reacts reversibly with C_6H_6 to form $(C_5Me_5)_2LuPh$ + H_2. Undergoes rapid H/D exhange. Ether adduct extrudes ethane to form $(C_5Me_5)_2LuOEt$. Undergoes metallation reactions with $SiMe_4$, pyridine, and ylides. White solid. Sol. pentane, hexane, C_6H_6, toluene. Highly reactive.

Watson, P.L. *et al*, *J. Am. Chem. Soc.*, 1982, **104**, 6471 (*synth, nmr*)
Waston, P.L., *J. Chem. Soc., Chem. Commun.*, 1983, 276 (*synth, nmr, ir*)

$C_{21}H_{33}Lu$ Lu-00019

Methylbis[(1,2,3,4,5-η)-1,2,3,4,5-pentamethyl-2,4-cyclopentadien-1-yl]lutetium

[85962-87-4]

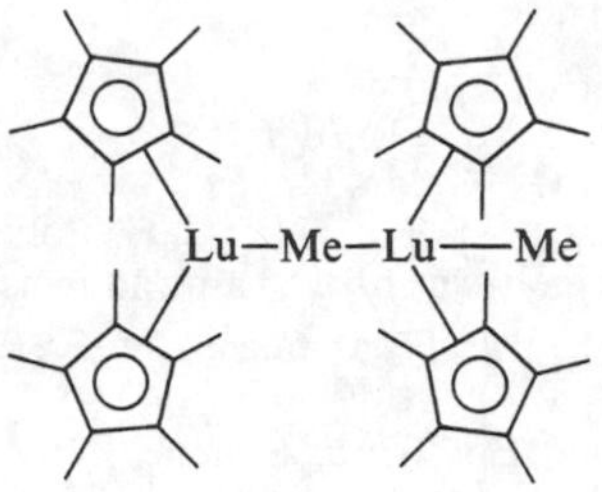

M 460.459

Asymmetric dimer in solid state, dissociates in soln. Activates methane in soln. Metallated by $SiMe_4$.

Dimer:
$C_{42}H_{66}Lu_2$ M 920.917
White solid. Sol. toluene, cyclohexane.

Et_2O complex: [80145-92-2].
$C_{25}H_{43}LuO$ M 534.581
Sol. toluene, hexane, Et_2O.

Watson, P.L., *J. Chem. Soc., Chem. Commun.*, 1983, 276 (*props*)
Watson, P.L., *J. Am. Chem. Soc.*, 1982, **104**, 337, 6471; 1983, **105**, 6491 (*struct, nmr, cmr, props, synth*)

$C_{24}H_{36}Lu^{\ominus}$ Lu-00020

Tetrakis(3,3-dimethyl-1-butynyl)lutetate(1−)

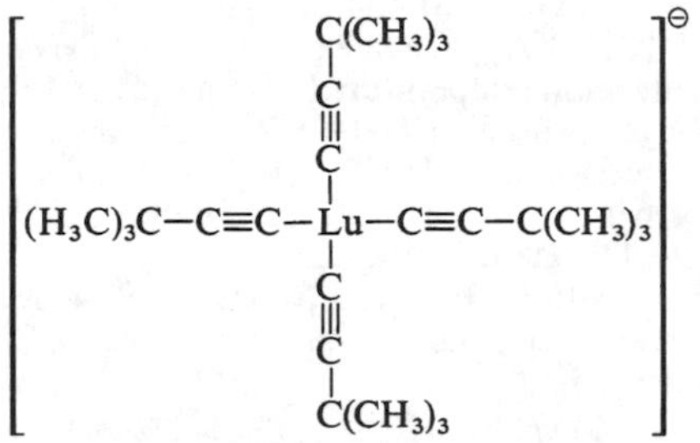

M 499.515 (ion)

Li salt, THF complex: [76565-30-5].
$C_{28}H_{44}LiLuO$ M 578.563
White solid. Sol. THF, pentane. Diamagnetic.

Evans, W.J. *et al*, *J. Organomet. Chem.*, 1980, **202**, C6 (*synth, ir*)

$C_{25}H_{35}LuN$ Lu-00021

[Bis(1,2,3,4,5-η)-1,2,3,4,5-pentamethyl-2,4-cyclopentadien-1-yl]-2-pyridinyllutetium

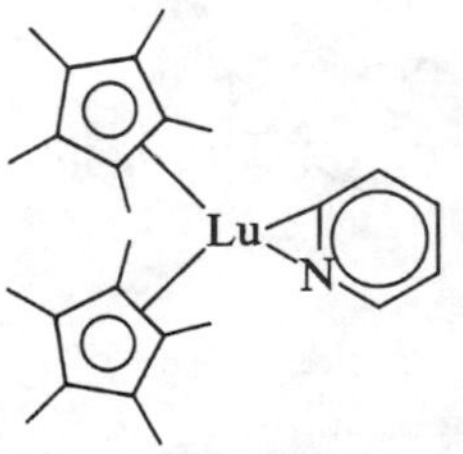

M 524.525

White cryst. Sol. cyclohexane. Formed by reaction of pyridine with Hydridobis[(1,2,3,4,5-η)-1,2,3,4,5-pentamethyl-2,4-cyclopentadien-1-yl]lutetium, Lu-00018.

Watson, P.L., *J. Chem. Soc., Chem. Commun.*, 1983, 276 (*synth, struct, nmr, cmr*)

$C_{29}H_{27}ClLuP$ Lu-00022

Chlorobis(η^5-2,4-cyclopentadien-1-yl)(triphenylphosphonium η-methylide)lutetium, 10CI

[78577-75-0]

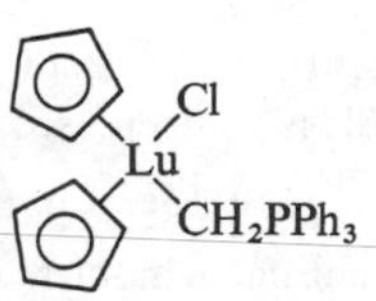

M 616.926

Colourless solid. Sol. C_6H_6, toluene, Et_2O, THF. Mp 172° dec. V. air- and moisture-sensitive.

Schumann, H. *et al*, *J. Organomet. Chem.*, 1981, **209**, C10 (*synth, nmr, ir*)

$C_{32}H_{36}Lu^{\ominus}$ Lu-00023

Tetrakis(2,6-dimethylphenyl)lutetate(1−)

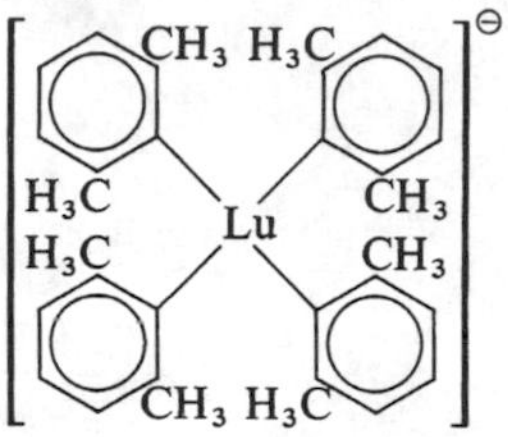

M 595.603 (ion)

Li salt, tetra-THF complex: [40844-50-6].

$C_{48}H_{68}LiLuO_4$ M 890.971

Colourless cryst. Sol. THF. Dec. by H_2O and air.

Cotton, S.A. *et al*, *J. Chem. Soc., Chem. Commun.*, 1972, 1225 (*synth, cryst struct*)

$C_{32}H_{38}Lu_2O_4$ Lu-00024

Tetrakis(η^5-2,4-cyclopentadien-1-yl)[μ-(2,2,7,7-tetramethyl-3,4,5,6-octanetetrone-O^3,O^5:O^4,O^6)]dilutetium

[79716-32-8]

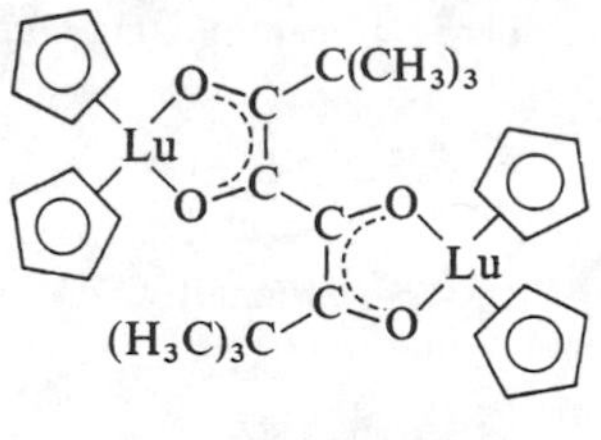

M 836.584

Formed by CO insertion reaction of *tert*-Butylbis(η^5-cyclopentadienyl)lutetium, Lu-00010 . Deep-purple cryst. Sol. pentane, toluene.

Evans, W.J. *et al*, *J. Chem. Soc., Chem. Commun.*, 1981, 706 (*synth, struct, cmr, pmr, ir*)

$C_{33}H_{36}LuP$ Lu-00025

Bis(η^5-2,4-cyclopentadien-1-yl)(*tert*-butyl)(triphenylphosphonium η-methylide)lutetium

*Dicyclopentadienyl-*tert*-butyllutetium(methylenetriphenylphosphorane)*

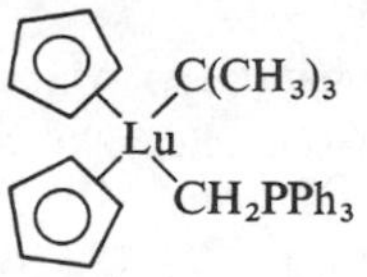

M 638.588

Colourless cryst. Sol. C_6H_6, toluene, Et_2O, THF. Mp 122° dec. V. sensitive to oxygen and moisture.

Schumann, H. *et al*, *J. Organomet. Chem.*, 1983, **255**, 305 (*synth, ir, nmr*)

$C_{33}H_{38}LuPSi$ Lu-00026

Bis(η^5-2,4-cyclopentadien-1-yl)[(trimethylsilyl)methyl](triphenylphosphonium η-methylide)lutetium

Dicyclopentadienyl[(trimethylsilyl)methyl]lutetium-(methylenetriphenylphosphorane)

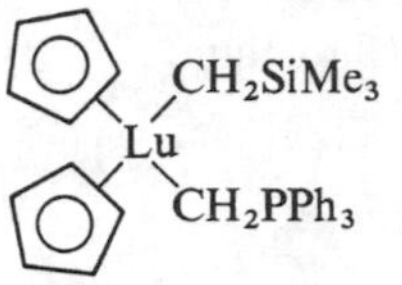

M 668.689

Colourless powder. Sol. C_6H_6, toluene, Et_2O, THF, CH_2Cl_2. Mp 103° dec. V. sensitive to oxygen and moisture.

Schumann, H. *et al*, *J. Organomet. Chem.*, 1983, **255**, 305 (*synth, ir, nmr*)

Mn Manganese

M. Green

Manganèse (Fr.), Mangan (Ger.), Manganeso (Sp.), Manganese (Ital.), Марганец (Marganets) (Russ.), マンガン (Japan.)

Atomic Number. 25

Atomic Weight. 54.9380

Electronic Configuration. [Ar] $3d^5\,4s^2$

Oxidation States. −3, −1, 0, +1, +2

Coordination Number. Most commonly 4, 5 or 6 with conventional ligands. Examples include $Mn(NO)_3CO$ (4), $[Mn(CO)_5]^-$ and $[Mn(CO)_4PPh_3]^-$ (5), $MnCl(CO)_5$ (6).

Colour. Variable.

Availability. Anhydrous $MnCl_2$ and $Mn_2(CO)_{10}$ are commonly available starting materials.

Handling. See under Chromium.

Toxicity. Mn compounds are not especially toxic except for the carbonyl complexes.

Isotopic Abundance. ^{55}Mn, 100%. Both ^{54}Mn ($t_{1/2}$ = 303d) and ^{56}Mn ($t_{1/2}$ = 2.57d) have been used in tracer studies.

Spectroscopy. ^{55}Mn has $I = \frac{5}{2}$ and is relatively easily observed by nmr although the quadrupole moment of 0.55×10^{-28} cm^2 gives rise to relatively broad lines. A range of chemical shifts has been reported for diamagnetic Mn(I) and Mn(−I) carbonyl-containing complexes. See for example Calderazzo, F. *et al.*, *J. Chem. Soc.*, (*A*), 1967, 154.

Analysis. For standard methods, see *Scott's Standard Methods of Chemical Analysis*, Furman, N. H. Ed., 6th Edn, Van Nostrand, 1962.

References. General reviews are listed in the introduction to the *Sourcebook*.

$Mn(NO)_3CO$

Mn-00001

Mn-00002

X = Br

Mn-00003

As Mn-00003 with
X = Cl

Mn-00004

Mn-00005

As Mn-00003 with
X = I

Mn-00006

Mn-00007

$[Mn(CO)_4]^{\ominus\ominus\ominus}$

Mn-00008

$MnBr(CO)_5$ (O_h)

Mn-00009

$MnCl(CO)_5$ (O_h)

Mn-00010

$HMn(CO)_5$(O_h)

Mn-00011

Mn-00012

$MnI(CO)_5$ (O_h)

Mn-00013

$Mn(CO)_5^{\ominus}$

Mn-00014

$[Mn(CO)_5(SO_2)]^{\oplus}$

Mn-00015

$MnMe(CO)_5$ (O_h)

Mn-00016

$[Mn(CO)_6]^{\oplus}$ (O_h)

Mn-00017

$Mn(OOCCF_3)(CO)_5$ (O_h)

Mn-00018

$Mn(CF_2CF_3)(CO)_5$ (O_h)

Mn-00019

$[Mn(CO)_5(NCMe)]^{\oplus}$

Mn-00020

$[Mn(CO)_5(CNMe)]^{\oplus}$ (O_h)

Mn-00021

Mn-00022

Mn-00023

L = N_2

Mn-00024

Mn-00025

As Mn-00024 with
L = $-NH_2NH_2$

Mn-00026

X = Br

Mn-00027

As Mn-00027 with
X = Cl

Mn-00028

$Mn(COCOCH_3)(CO)_5$

Mn-00029

Mn-00030

As Mn-00024 with
L = CS

Mn-00031

As Mn-00024 with
L = CSe

Mn-00032

Mn-00033

$H_2C{=}CHCH_2Mn(CO)_5$

Mn-00034

$[Mn(COOEt)(CO)_5]$ (O_h)

Mn-00035

$[(OC)_5Mn{=}CClNMe_2]^{\oplus}$ (O_h)

Mn-00036

Mn-00037

$Me_3PbMn(CO)_5$

Mn-00038

$Me_3SiMn(CO)_5$

Mn-00039

As Mn-00027 with
X = I

Mn-00040

$[(OC)_4Fe{-}Mn(CO)_5]^{\ominus}$

Mn-00041

Mn-00042

Mn-00043

Mn-00044

Mn-00045

Mn-00046

Mn-00047

R = CH_3

Mn-00048

$[Mn(CO)_3(CNMe)_3]^{\oplus}$

Mn-00049

$(MeCN)_3Mn(CO)_3^{\oplus}$

Mn-00050

$MnBr(CO)(CNMe)_4$

Mn-00051

Mn-00052

Mn-00053

Mn-00054

Mn-00055

Mn-00056

Mn-00057

Mn-00058

Mn-00059

R = CH_3

Mn-00060

Mn-00061

$(OC)_5Re{-}Mn(CO)_5$

Mn-00062

$(OC)_5Mn{-}Mn(CO)_5$

Mn-00063

$Mn(C_6F_5)(CO)_5$ (O_h)

Mn-00064

$(OC)_5Mn{-}Mn(CO)_4CNMe$ (*eq-*)

Mn-00065

$MnPh(CO)_5$ (O_h)

Mn-00066

Mn-00067

$[Mn(CO)(CNMe)_5]^{\oplus}$

Mn-00068

Mn-00069

Mn-00070

Mn-00071

Mn-00072

$MnOs_3H(CO)_{12}$

Mn-00073

Mn-00074

Mn-00075

Mn-00076

Mn-00077

$[(OC)_5Mn]_2PbMe_2$

Mn-00078

Mn-00079

Mn-00080

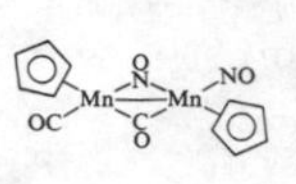

Mn-00081

Mn-00082

Mn-00083

Mn-00084

Mn-00085

$MnOs_3H_3(CO)_{13}$

Mn-00086

Mn-00087

Mn-00088

$MnBr(CO)[P(OMe)_3]_4$

Mn-00089

Mn-00090

$(OC)_5Mn—Fe(CO)_4—Re(CO)_5$

Mn-00091

$(OC)_5Mn—Fe(CO)_4—Mn(CO)_5$

Mn-00092

As Mn-00048 with
R = Ph

Mn-00093

Mn-00094

Mn-00095

$(OC)_5Mn–Os(CO)_4–Mn(CO)_5$

Mn-00096

$(OC)_5Mn—Ru(CO)_4—Mn(CO)_5$

Mn-00097

Mn-00098

Mn-00099

Mn-00100

Mn-00101

Mn-00102

As Mn-00060 with
R = Ph

Mn-00103

Mn-00104

$MnOs_3H(CO)_{16}$

Mn-00105

Mn-00106

Mn-00107

Mn-00108

Mn-00109

Mn-00110

$Ph_2TlMn(CO)_5$

Mn-00111

Mn-00112

Mn-00113

Mn-00114

Mn-00115

Mn-00116

Mn-00117

Mn-00118

Mn-00119

Mn-00120

Mn-00121

$U[Mn(CO)_5]_4$

Mn-00122

$Mn(O_3SCF_3)(CO)_2[P[OCH(CH_3)_2]_3]_2$

Mn-00123

$(OC)_4MnPPh_3^{\ominus}$

Mn-00124

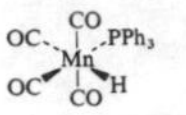

Mn-00125

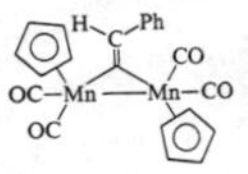

Mn-00126

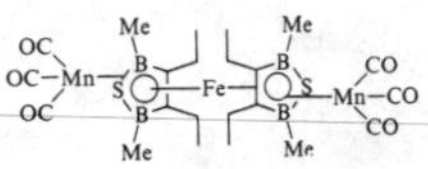

Mn-00127

$(OC)_5MnFe(NO)_2PPh_3$

Mn-00128

$[Mn(CO)_5(PPh_3)]^{\oplus}$ (O_h)

Mn-00129

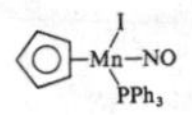

Mn-00130

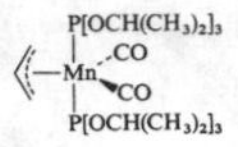

Mn-00131

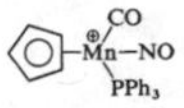

Mn-00132

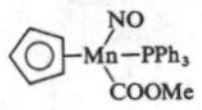

Mn-00133

$(OC)_5Mn—Mn(CO)_4(PPh_3)$(*ax*-)

Mn-00134

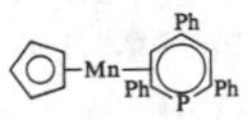

Mn-00135

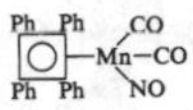

Mn-00136

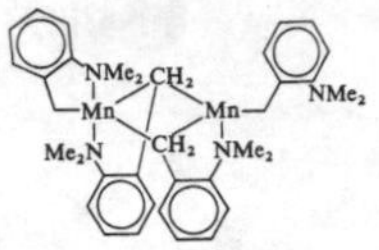

Mn-00137

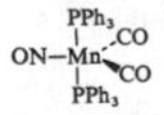

Mn-00138

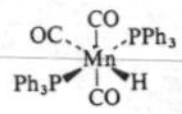

Mn-00139

$[Mn(CO)_4(PPh_3)_2]^{\oplus}$

Mn-00140

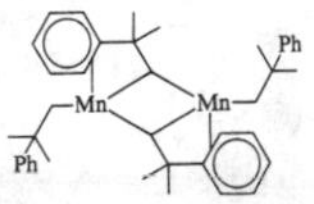

Mn-00141

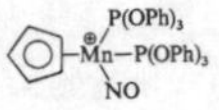

Mn-00142

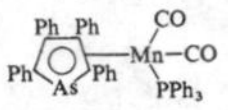

Mn-00143

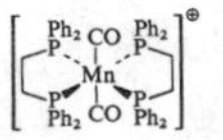

Mn-00144

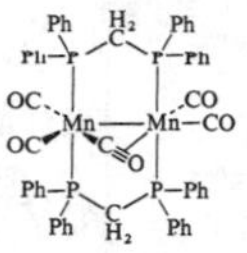

Mn-00145

$CMnN_3O_4$ **Mn-00001**

Carbonyltrinitrosylmanganese, 8CI

[14951-98-5]

$Mn(NO)_3CO$

M 172.967

Green solid. Mp 27°.

Barraclough, C.G. *et al*, *J. Chem. Soc.*, 1960, 4842 (*synth*)
Inorg. Synth., 1976, **16**, 4 (*synth*)

$C_3H_9MnN_3O_3^{\oplus}$ **Mn-00002**

Triamminetricarbonylmanganese(1+)

[14872-96-9]

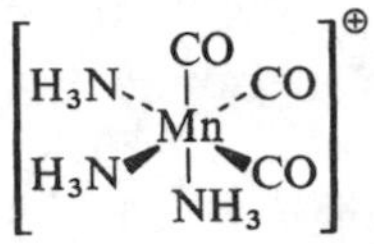

M 190.060 (ion)

fac-form

Pentacarbonylmanganate(1−):
$C_8H_9Mn_2N_3O_8$ M 385.050
Air-stable yellow cryst. Sol. Me_2CO, H_2O.

Tetraphenylborate: [14872-95-8].
$C_{27}H_{29}BMnN_3O_3$ M 509.292
Yellow cryst. (Me_2CO aq.). Sol. Me_2CO, insol. H_2O.

Behrens, H. *et al*, *Z. Anorg. Allg. Chem.*, 1967, **349**, 241 (*synth*)
Behrens, H. *et al*, *J. Organomet. Chem.*, 1972, **34**, 367 (*synth*)
Herberhold, M. *et al*, *J. Organomet. Chem.*, 1978, **152**, 329 (*pmr, cmr, ir, cryst struct, props*)

$C_4Br_2MnO_4^{\ominus}$ **Mn-00003**

Dibromotetracarbonylmanganate(1−)

[37176-07-1]

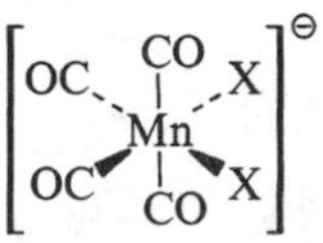

X = Br

M 326.788 (ion)

Tetraethylammonium salt: [43184-81-2].
$C_{12}H_{20}Br_2MnNO_4$ M 457.040
Cryst.

Abel, E.W. *et al*, *J. Chem. Soc.*, 1964, 434 (*synth, ir, nmr*)
Farona, M.F. *et al*, *J. Inorg. Nucl. Chem.*, 1967, **29**, 1814 (*props*)
Connor, J.A. *et al*, *J. Organomet. Chem.*, 1974, **73**, 351 (*props*)

$C_4Cl_2MnO_4^{\ominus}$ **Mn-00004**

Tetracarbonyldichloromanganate(1−)

As Dibromotetracarbonylmanganate(1−), Mn-00003 with

X = Cl

M 237.886 (ion)

Tetraethylammonium salt: [52901-72-1].
$C_{12}H_{20}Cl_2MnNO_4$ M 368.138
Cryst.

Abel, E.W. *et al*, *J. Chem. Soc.*, 1964, 434 (*synth, ir, nmr*)
Farona, M.F. *et al*, *J. Inorg. Nucl. Chem.*, 1967, **29**, 1814 (*props*)
Connor, J.A. *et al*, *J. Organomet. Chem.*, 1974, **73**, 351 (*props*)

$C_4H_3MnO_3S$ **Mn-00005**

Tricarbonyl(methanethiolato)manganese

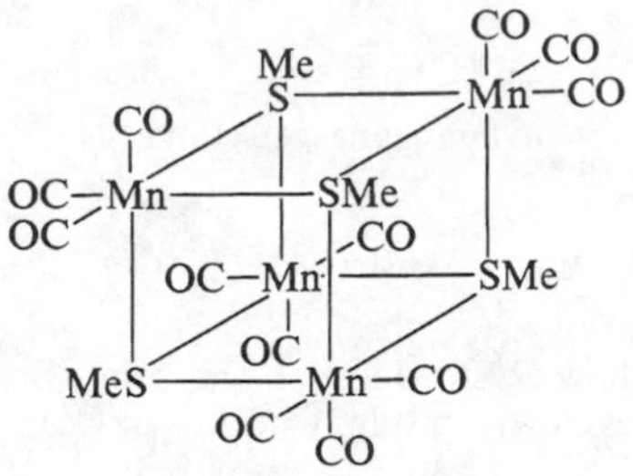

M 186.064

Tetrameric, originally thought to be trimeric.

Tetramer: [24819-01-0]. *Dodecacarbonyltetrakis(μ_3-methanethiolato)tetramanganese, 8CI.*
$C_{16}H_{12}Mn_4O_{12}S_4$ M 744.256
Yellow-orange cryst. Dec. at 210°.

Abel, E.W. *et al*, *J. Chem. Soc. (A)*, 1966, 1141 (*synth*)
Jenkins, C.R., *J. Organomet. Chem.*, 1968, **15**, 441.
Johnson, B.F.E. *et al*, *Inorg. Chem.*, 1968, **7**, 831 (*ms, synth*)

$C_4I_4MnO_4^{\ominus}$ **Mn-00006**

Tetracarbonyldiiodomanganate(1−)

[37176-05-9]

As Dibromotetracarbonylmanganate(1−), Mn-00003 with

X = I

M 674.598 (ion)

Tetramethylammonium salt: [53749-00-1].
$C_8H_{12}I_2MnNO_4$ M 494.934
Cryst.

Abel, E.W. *et al*, *J. Chem. Soc.*, 1964, 434 (*synth, ir, nmr*)
Ferona, M.F. *et al*, *J. Inorg. Nucl. Chem.*, 1967, **29**, 1814 (*props*)
Connor, J.A. *et al*, *J. Organomet. Chem.*, 1974, **73**, 351 (*props*)

C_4MnNO_5 **Mn-00007**

Tetracarbonylnitrosylmanganese, 8CI

[16104-17-9]

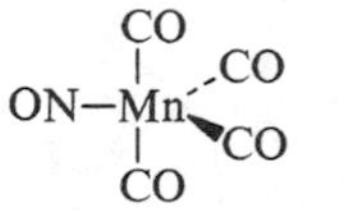

M 196.986

Red liq. Mp −1.5°.

Treichel, P.M. *et al*, *J. Am. Chem. Soc.*, 1961, **83**, 2593 (*synth*)
Organomet. Synth., 1965, **1**, 164 (*synth*)
Waersik, H. *et al*, *Inorg. Chem.*, 1967, **6**, 1066.
Keeton, D.P. *et al*, *Inorg. Chim. Acta*, 1972, **6**, 33 (*props*)
Herberhold, M. *et al*, *J. Organomet. Chem.*, 1974, **67**, 81.
Legzdins, P. *et al*, *Inorg. Chem.*, 1975, **14**, 1875.

$C_4MnO_4^{\ominus\ominus\ominus}$ **Mn-00008**

Tetracarbonylmanganate(3−), 10CI

[61769-24-2]

$[Mn(CO)_4]^{\ominus\ominus\ominus}$

M 166.980 (ion)

Tri-Na salt: [59123-04-5].
$C_4MnNa_3O_4$ M 235.949
Characterised spectroscopically.

Ellis, J.E. *et al*, *J. Am. Chem. Soc.*, 1977, **99**, 1801 (*synth, props*)

C_5BrMnO_5 Mn-00009

Bromopentacarbonylmanganese, 10CI, 9CI, 8CI
[14516-54-2]

$MnBr(CO)_5$ (O_h)

M 274.894
Orange-yellow cryst. Mod. sol. org. solvs. $Bp_{0.05}$ 50° subl. Dimerises slowly in soln. with loss of CO.

Abel, E.W. *et al*, *J. Chem. Soc.*, 1959, 1501 (*synth, props*)
Kaesz, H.D. *et al*, *J. Am. Chem. Soc.*, 1967, **89**, 2844 (*ir*)
Connor, J.A. *et al*, *J. Chem. Soc., Faraday Trans.*, 1972, **68**, 1754.
Aime, S. *et al*, *Transition Met. Chem.*, 1976, **1**, 177 (*cmr*)
Inorg. Synth., 1979, **19**, 160 (*synth*)

C_5ClMnO_5 Mn-00010

Pentacarbonylchloromanganese, 10CI
Chloropentacarbonylmanganese, 9CI, 8CI
[14100-30-2]

$MnCl(CO)_5$ (O_h)

M 230.443
Pale-yellow light-sensitive cryst. Mod. sol. org. solvs. $Bp_{0.1}$ 40° subl. Light sensitive. Dimerises slowly in soln. with loss of CO.

Abel, E.W. *et al*, *J. Chem. Soc.*, 1959, 1501 (*synth, props*)
Kaesz, H.D. *et al*, *J. Am. Chem. Soc.*, 1967, **89**, 2844 (*ir*)
Greene, P.T. *et al*, *J. Chem. Soc. (A)*, 1971, 1559 (*cryst struct*)
Connor, J.A. *et al*, *J. Chem. Soc., Faraday Trans.*, 1972, **68**, 1754.
Inorg. Synth., 1979, **19**, 159 (*synth*)

C_5HMnO_5 Mn-00011

Pentacarbonylhydromanganese, 9CI
Pentacarbonylmanganese hydride. Manganese hydrocarbonyl. Hydridopentacarbonylmanganese
[16972-33-1]

$HMn(CO)_5(O_h)$

M 195.998
Liq. Mp −20°. Sublimes *in vacuo*. Stable to 50°.
▷Poisonous vapour

Inorg. Synth., 1963, **7**, 196 (*synth*)
Organomet. Synth., 1965, **1**, 158 (*synth*)
Edgell, W.F. *et al*, *J. Am. Chem. Soc.*, 1966, **88**, 5451 (*synth*)
Calderazzo, F. *et al*, *J. Chem. Soc. (A)*, 1967, 154 (*nmr*)
Edgell, W.F. *et al*, *Inorg. Chem.*, 1969, **8**, 1103 (*ir*)
La Placa, S. *et al*, *Inorg. Chem.*, 1969, **8**, 1928 (*cryst struct*)
McNeill, E.A. *et al*, *J. Am. Chem. Soc.*, 1977, **99**, 6243 (*synth, struct*)
McKinney, R.J. *et al*, *Inorg. Chem.*, 1979, **18**, 3413.

$C_5H_3MnO_4S$ Mn-00012

Tetracarbonyl(methanethiolato)manganese

(Structure: dimer $[(OC)_4Mn(\mu\text{-SMe})_2Mn(CO)_4]$)

M 214.074
Dimeric.
Dimer: [21321-38-0]. *Octacarbonylbis(μ-methanethiolato)dimanganese*, 9CI, 8CI.
$C_{10}H_6Mn_2O_8S_2$ M 428.149
Golden-yellow cryst. (pentane).

Treichel, P.M. *et al*, *J. Chem. Soc.*, 1963, 720 (*synth, ir, pmr*)

C_5IMnO_5 Mn-00013

Pentacarbonyliodomanganese, 10CI
Iodopentacarbonylmanganese, 9CI, 8CI
[14879-42-6]

$MnI(CO)_5$ (O_h)

M 321.895
Red-orange cryst. (hexane). $Bp_{0.05}$ 50° subl. Iodine is liberated on prolonged standing at r.t., dimerises slowly in soln.

Abel, E.W. *et al*, *J. Chem. Soc.*, 1959, 1501 (*synth, props*)
Kaesz, H.D. *et al*, *J. Am. Chem. Soc.*, 1967, **89**, 2844 (*ir*)
Connor, J.A. *et al*, *J. Chem. Soc., Faraday Trans.*, 1972, **68**, 1754 (*props*)
Inorg. Synth., 1979, **19**, 160, 162 (*synth*)

$C_5MnO_5^{\ominus}$ Mn-00014

Pentacarbonylmanganate(1−), 10CI
[14971-26-7]

$Mn(CO)_5^{\ominus}$

M 194.990 (ion)
Li salt: [15689-01-7].
C_5LiMnO_5 M 201.931
Moisture-sensitive.
Na salt: [13859-41-1].
C_5MnNaO_5 M 217.980
Moisture-sensitive.
K salt:
C_5KMnO_5 M 234.088
Moisture-sensitive.
Bis(triphenylphosphine)iminium salt: [52542-59-3].
$C_{41}H_{30}MnNO_5P_2$ M 733.577
Orange cryst.

Inorg. Synth., 1974, **15**, 84 (*synth*)
Duffy, D.N. *et al*, *J. Organomet. Chem.*, 1979, **164**, 227 (*props*)
Gladysz, J.A. *et al*, *Inorg. Chem.*, 1979, **18**, 553 (*synth*)
Pearson, R.G. *et al*, *J. Am. Chem. Soc.*, 1980, **102**, 1541 (*props*)
Braunstein, P. *et al*, *J. Organomet. Chem.*, 1981, **213**, 79 (*props*)
Uson, R. *et al*, *J. Chem. Soc., Dalton Trans.*, 1981, 316 (*props*)

$C_5MnO_7S^{\oplus}$ Mn-00015

Pentacarbonyl(sulfur dioxide-*S*)manganese(1+)

$[Mn(CO)_5(SO_2)]^{\oplus}$

M 259.049 (ion)
Hexafluoroarsenate: [55853-06-0].

$C_5AsF_6MnO_7S$ M 447.961
Brownish-yellow solid.

Mews, R., *Angew. Chem., Int. Ed. Engl.*, 1975, **14**, 640; 1977, **16**, 56 (*synth*)

$C_6H_3MnO_5$ Mn-00016

Pentacarbonylmethylmanganese, 10CI
Methylpentacarbonylmanganese
[13601-24-6]

$MnMe(CO)_5$ (O_h)

M 210.025
Colourless cryst. (pet. ether).

Clossen, R.D. *et al*, *J. Org. Chem.*, 1957, **22**, 598 (*synth*, *ir*)
Hsieh, A.T.T. *et al*, *J. Organomet. Chem.*, 1970, **22**, 29 (*synth*)
Wojcicki, A., *Adv. Organomet. Chem.*, 1973, **11**, 87 (*synth*, *props*)
King, R.B. *et al*, *J. Am. Chem. Soc.*, 1978, **100**, 1687 (*props*)
Butts, S.B. *et al*, *J. Am. Chem. Soc.*, 1980, **102**, 5093 (*props*)
Butts, S.B. *et al*, *Inorg. Chem.*, 1981, **20**, 278.

$C_6MnO_6^{\oplus}$ Mn-00017

Hexacarbonylmanganese(1+), 9CI
Manganesehexacarbonyl(1+)
[21331-06-6]

$[Mn(CO)_6]^{\oplus}$ (O_h)

M 223.000 (ion)
Tetrafluoroborate: [15557-71-8].
$C_6BF_4MnO_6$ M 309.804
White solid.

Fischer, E.O. *et al*, *Chem. Ber.*, 1962, **95**, 249 (*synth*, *props*)
Kruck, T. *et al*, *Chem. Ber.*, 1964, **97**, 1693 (*synth*)
Beach, N.A. *et al*, *J. Am. Chem. Soc.*, 1968, **90**, 5713 (*synth*, *uv*)
DeKock, R.L. *et al*, *Inorg. Chem.*, 1971, **10**, 38 (*struct*)
Darensbourg, D.J. *et al*, *J. Am. Chem. Soc.*, 1977, **99**, 4726 (*props*)
Trogler, W.C., *J. Am. Chem. Soc.*, 1979, **101**, 6459 (*synth*)

$C_7F_3MnO_7$ Mn-00018

Pentacarbonyl(trifluoroacetato-*O*)manganese, 8CI
[15219-49-5]

$Mn(OOCCF_3)(CO)_5$ (O_h)

M 308.006
Bright-yellow volatile cryst.

Green, M.L.H. *et al*, *J. Organomet. Chem.*, 1967, **8**, 511 (*synth*)
King, R.B. *et al*, *J. Organomet. Chem.*, 1968, **15**, 457 (*synth*)
King, R.B. *et al*, *J. Inorg. Nucl. Chem.*, 1969, **31**, 2169 (*props*)
Garner, C.L. *et al*, *J. Chem. Soc., Dalton Trans.*, 1974, 735 (*props*)

$C_7F_5MnO_5$ Mn-00019

Pentacarbonyl(pentafluoroethyl)manganese, 8CI
[20791-80-4]

$Mn(CF_2CF_3)(CO)_5$ (O_h)

M 314.004
Liq. Mp 15-7°.

Kaesz, H.D. *et al*, *Z. Naturforsch., B*, 1960, **15**, 763 (*synth*, *props*)

$C_7H_3MnNO_5^{\oplus}$ Mn-00020

(Acetonitrile)pentacarbonylmanganese(1+), 9CI
Pentacarbonylacetonitrilemanganese. Pentacarbonyl(cyanomethane)manganese
[27674-37-9]

$[Mn(CO)_5(NCMe)]^{\oplus}$

M 236.042 (ion)
Hexafluorophosphate: [37504-44-2].
$C_7H_3F_6MnNO_5P$ M 381.007
Yellow cryst.

Drew, D. *et al*, *Inorg. Chem.*, 1975, **14**, 1579 (*synth*)
Darensbourg, D.J. *et al*, *J. Am. Chem. Soc.*, 1977, **99**, 4726, 5940 (*synth*, *props*)
Trogler, W.C., *J. Am. Chem. Soc.*, 1979, **101**, 6459 (*synth*, *props*)
Snow, M.R. *et al*, *Inorg. Chim. Acta Lett.*, 1980, **44**, 189 (*synth*)

$C_7H_3MnNO_5^{\oplus}$ Mn-00021

Pentacarbonyl(isocyanomethane)manganese(1+)
Pentacarbonyl(methylisocyanide)manganese(1+)
[45113-87-9]

$[Mn(CO)_5(CNMe)]^{\oplus}$ (O_h)

M 236.042 (ion)
Hexafluorophosphate: [38979-19-0].
$C_7H_3F_6MnNO_5P$ M 381.007
Cryst. (Me_2CO/Et_2O). Mp 192-4°.

Treichel, P.M. *et al*, *J. Organomet. Chem.*, 1972, **44**, 339 (*synth*, *props*)
Sarapu, A. *et al*, *Inorg. Chem.*, 1975, **14**, 247 (*synth*, *props*)

$C_7H_4MnNO_3$ Mn-00022

Tricarbonyl-π-pyrrolylmanganese, 8CI
Pyrrolylmanganese tricarbonyl
[32761-36-7]

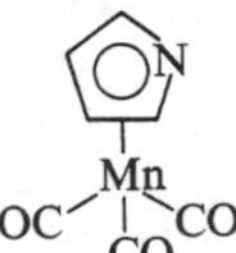

M 205.052
Cryst. (pentane). Mp 41°. Cryst. struct. of related 2-Me deriv. has been detd.

Joshi, K.K. *et al*, *J. Organomet. Chem.*, 1964, **1**, 471 (*synth*, *ir*)

$C_7H_5MnNO_3^{\oplus}$ Mn-00023

Dicarbonyl(η^5-2,4-cyclopentadien-1-yl)nitrosylmanganese(1+), 9CI
Dicarbonylnitrosyl(η^5-cyclopentadienyl)manganese(1+)
[31921-90-1]

M 206.059 (ion)
ir ν_{CO} 2125, 2075, ν_{NO} 1840 cm^{-1}. Dec. from 230°.
Hexafluorophosphate: [31921-90-1].
$C_7H_5F_6MnNO_3P$ M 351.024
Yellow cryst. (Me_2CO/Et_2O).

King, R.B. *et al*, *Inorg. Chem.*, 1964, **3**, 791 (*synth, ir, pmr*)
Brunner, H. *et al*, *J. Organomet. Chem.*, 1969, **19**, 135 (*props*)
King, R.B. *et al*, *Inorg. Chem.*, 1969, **8**, 2374 (*props*)
James, T.A. *et al*, *J. Chem. Soc. (A)*, 1970, 850 (*props*)
King, R.B. *et al*, *J. Organomet. Chem.*, 1973, **56**, 345 (*synth*)
Busetto, L. *et al*, *J. Organomet. Chem.*, 1974, **66**, 453 (*props*)

$C_7H_5MnN_2O_2$ Mn-00024

Dicarbonyl(η^5-2,4-cyclopentadien-1-yl)(dinitrogen)manganese, 9CI

[38600-58-7]

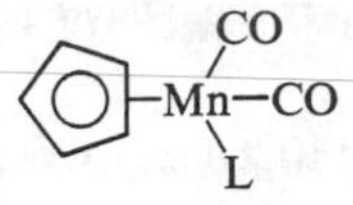

L = N_2

M 204.067

Air-stable red-brown cryst. by subl. Mp 60° dec. (gas evol.).

Sellmann, D. *et al*, *Angew. Chem., Int. Ed. Engl.*, 1971, **10**, 919; 1972, **11**, 534 (*synth, ir, ms, pmr*)
Sellmann, D., *J. Organomet. Chem.*, 1972, **44**, C46; 1978, **160**, 183 (*synth, props*)
Ziegler, M.L. *et al*, *Angew. Chem., Int. Ed. Engl.*, 1976, **15**, 695.

$C_7H_5MnO_4$ Mn-00025

Tetracarbonyl(η^3-2-propenyl)manganese, 9CI

(η^3-Allyl)tetracarbonylmanganese. π-Allylmanganese tetracarbonyl

[33307-28-7]

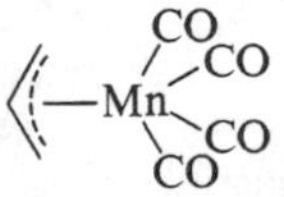

M 208.052

Pale-yellow solid or cryst. Mp 38-41°, 53°. Bp_9 60-2° subl.

Kaesz, H.D. *et al*, *Z. Naturforsch., B*, 1960, **18**, 763 (*synth, props*)
Davidson, G. *et al*, *J. Chem. Soc., Dalton Trans.*, 1972, 126 (*ir, raman*)
Abel, E.W. *et al*, *J. Chem. Soc., Dalton Trans.*, 1973, 1706 (*synth*)
Gibson, D.H. *et al*, *J. Organomet. Chem.*, 1979, **172**, C7 (*synth*)

$C_7H_9MnN_2O_2$ Mn-00026

Dicarbonyl(η^5-2,4-cyclopentadien-1-yl)(hydrazine-*N*)manganese, 10CI

[31921-91-2]

As Dicarbonyl(η^5-2,4-cyclopentadien-1-yl)(dinitrogen)-manganese, Mn-00024 with

L = $—NH_2NH_2$

M 208.098

Red-brown cryst.

Sellmann, D., *Z. Naturforsch., B*, 1970, **25**, 890 (*synth*)
Sellmann, D., *J. Organomet. Chem.*, 1972, **44**, C46.
Sellmann, D. *et al*, *J. Organomet. Chem.*, 1978, **160**, 183; 1979, **178**, 433.

$C_8Br_2Mn_2O_8$ Mn-00027

Di(μ-bromo)octacarbonyldimanganese, 9CI, 8CI

Octacarbonyldi(μ-bromo)dimanganese

[18535-44-9]

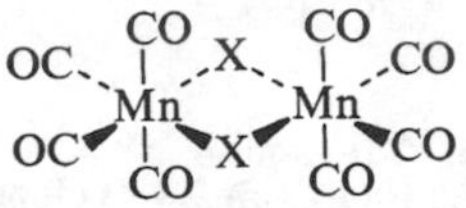

X = Br

M 493.767

Brown cryst. (C_6H_6 or CCl_4). Subl. at 150°. Converted by CO into $BrMn(CO)_5$, Mn-00009.

Monomer: Bromotetracarbonylmanganese.
C_4BrMnO_4 M 246.884

Abel, E.W. *et al*, *J. Chem. Soc.*, 1959, 1501 (*synth, props*)
Dahl, L.F. *et al*, *Acta Crystallogr., Sect. B*, 1963, **16**, 611 (*cryst struct*)
Abel, E.W. *et al*, *Trans. Faraday Soc.*, 1967, **63**, 45 (*ir*)
Farona, M.F. *et al*, *J. Inorg. Nucl. Chem.*, 1967, **29**, 1814 (*synth*)
Bamford, C.H. *et al*, *J. Chem. Soc., Dalton Trans.*, 1972, 1846 (*synth, uv*)

$C_8Cl_2Mn_2O_8$ Mn-00028

Di(μ-chloro)octacarbonyldimanganese, 9CI, 8CI

Octacarbonyldi(μ-chloro)dimanganese

[18535-43-8]

As Di(μ-bromo)octacarbonyldimanganese, Mn-00027 with

X = Cl

M 404.865

Monomer: Tetracarbonylchloromanganese. Chlorotetracarbonylmanganese.
C_4ClMnO_4 M 202.433

Abel, E.W. *et al*, *J. Chem. Soc.*, 1959, 1501 (*synth, props*)
El Sayed, M.A. *et al*, *Inorg. Chem.*, 1963, **2**, 158 (*ir*)
Farona, M.F. *et al*, *J. Inorg. Nucl. Chem.*, 1967, **29**, 1814 (*synth*)
Zingalas, F. *et al*, *Inorg. Chem.*, 1967, **6**, 1243 (*props*)
Bamford, C.H. *et al*, *J. Chem. Soc., Dalton Trans.*, 1972, 1846 (*synth*)

$C_8H_3MnO_7$ Mn-00029

Pentacarbonyl(1,2-dioxopropyl)manganese, 9CI

[33678-85-2]

$Mn(COCOCH_3)(CO)_5$

M 266.046

Red cryst.

Casey, C.P. *et al*, *J. Am. Chem. Soc.*, 1976, **98**, 1166 (*synth, props, cryst struct*)

$C_8H_4MnO_7^{\oplus}$ Mn-00030

Pentacarbonyl-1,3-dioxalan-2-ylidenemanganese(1+)

[40901-26-6]

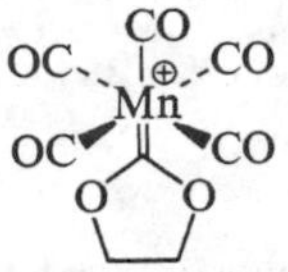

M 267.053 (ion)

Hexafluorophosphate:
$C_5H_4F_6MnO_7P$ M 375.985
Yellow cryst.

Bowen, D.H. *et al*, *J. Chem. Soc., Dalton Trans.*, 1974, 1189 (*synth, ir, nmr, props*)

$C_8H_5MnO_2S$ Mn-00031

(Carbonothioyl)dicarbonyl(η^5-2,4-cyclopentadien-1-yl)manganese, 9CI

[31741-76-1]

As Dicarbonyl(η^5-2,4-cyclopentadien-1-yl)(dinitrogen)-manganese, Mn-00024 with

L = CS

M 220.124

Bright-yellow cryst. (pentane). Mp 56°. $Bp_{0.001}$ 28° subl.

Fenster, A.E. *et al*, *Inorg. Chem.*, 1974, **13**, 915 (*synth, ir, nmr*)
Inorg. Synth., 1976, **16**, 53 (*synth*)
Butler, I.S. *et al*, *Inorg. Chem.*, 1977, **16**, 1779 (*synth*)
Butler, I.S. *et al*, *J. Organomet. Chem.*, 1977, **133**, 59 (*synth, ir, nmr*)

$C_8H_5MnO_2Se$ Mn-00032

(Carbonoselenoyl)dicarbonyl(η^5-2,4-cyclopentadien-1-yl)-manganese, 10CI

[55987-17-2]

As Dicarbonyl(η^5-2,4-cyclopentadien-1-yl)(dinitrogen)-manganese, Mn-00024 with

L = CSe

M 267.024

Golden cryst. Mp 64-5°. $Bp_{0.005}$ 25° (subl.).

Butler, I.S. *et al*, *Inorg. Chem.*, 1977, **16**, 1779 (*synth, nmr, ir*)

$C_8H_5MnO_3$ Mn-00033

Cyclopentadienyltricarbonylmanganese

Tricarbonylcyclopentadienylmanganese. Cymantrene

[12079-65-1]

(cyclopentadienyl)–$Mn(CO)_3$

M 204.064

Pale-yellow cryst. with camphoraceous odour. Mp 76.8-77.1°. Solns. in org. solvs. slowly oxidised by air. Undergoes electrophilic substitution reactions. Sublimes at 60°.

▷Highly toxic, TLV (skin) 0.1. OO9720000.

Piper, T.S. *et al*, *J. Inorg. Nucl. Chem.*, 1955, **1**, 165 (*synth, ir*)
Kozikowski, J. *et al*, *J. Am. Chem. Soc.*, 1959, **81**, 2995.
Organomet. Synth., 1965, **1**, 111.
Müller, J. *et al*, *J. Organomet. Chem.*, 1967, **10**, 313 (*ms*)
Abel, E.W. *et al*, *Angew. Chem., Int. Ed. Engl.*, 1971, **10**, 339 (*synth*)
Lokshin, B.V. *et al*, *J. Organomet. Chem.*, 1974, **77**, 69 (*ir, raman*)
Sax, N.I., *Dangerous Properties of Industrial Materials*, 5th Ed., Van Nostrand-Reinhold, 1979, 787.
Hazards in the Chemical Laboratory, (Bretherick, L., Ed.), 3rd Ed., Royal Society of Chemistry, London, 1981, 264.

$C_8H_5MnO_5$ Mn-00034

Pentacarbonyl-2-propenylmanganese, 9CI

Allylpentacarbonylmanganese

[14057-83-1]

$H_2C{=}CHCH_2Mn(CO)_5$

M 236.063

Lemon-yellow liq. Bp_{15} 45°. Air sensitive. Dec. at 80° → $CO + \eta^3\text{-}C_3H_5Mn(CO)_4$.

Kaesz, H.D. *et al*, *Z. Naturforsch., B*, 1960, **15**, 682 (*synth*)
McClellan, W.R. *et al*, *J. Am. Chem. Soc.*, 1961, **83**, 1601 (*synth*)
Clarke, H.L. *et al*, *J. Organomet. Chem.*, 1972, **40**, 379 (*ir, raman*)

$C_8H_5MnO_7$ Mn-00035

Pentacarbonyl(ethoxycarbonyl)manganese, 9CI

[28300-66-5]

$[Mn(COOEt)(CO)_5]$ (O_h)

M 268.061

Cryst. Mp 59.5°.

Kruck, T. *et al*, *Chem. Ber.*, 1964, **97**, 1693 (*synth*)
Beach, N.A. *et al*, *J. Am. Chem. Soc.*, 1968, **90**, 5713 (*props*)
Drew, D. *et al*, *Inorg. Chem.*, 1975, **14**, 1579 (*synth*)

$C_8H_6ClMnNO_5^{\oplus}$ Mn-00036

Pentacarbonyl[chloro(dimethylamino)methylene]manganese(1+)

[58938-56-0]

$[(OC)_5Mn{=}CClNMe_2]^{\oplus}$ (O_h)

M 286.530 (ion)

Mp 121-2° dec.

Perchlorate: [58938-57-1].
$C_8H_6Cl_2MnNO_9$ M 385.981
Pale-yellow solid. Mp 121-2° dec.

Hartshorn, A.J. *et al*, *J. Chem. Soc., Chem. Commun.*, 1975, 929 (*synth, props*)

$C_8H_6MnO_6^{\ominus}$ Mn-00037

Diacetyltetracarbonylmanganate(1−), 10CI

$$\left[\begin{array}{c} \text{CO} \\ \text{OC}\diagdown\ |\ \diagup\text{COCH}_3 \\ \text{Mn} \\ \text{OC}\diagup\ |\ \diagdown\text{COCH}_3 \\ \text{CO} \end{array}\right]^{\ominus}$$

M 253.070 (ion)

cis-Diacetyl moiety is electronically equiv. to the acetoacetonate anion.

Li salt: [57891-59-5].
$C_8H_6LiMnO_6$ M 260.011
Not. isol., deep-yellow soln. in Et_2O.

Bis(triphenylphosphine)iminium salt: [66964-10-1]. No phys. props. reported.

Al salt: [58034-11-0]. *Hexakis[μ-(acetyl-C:O)](aluminum)dodecacarbonyltrimanganese, 9CI.*
$C_{24}H_{18}AlMn_3O_{18}$ M 786.191
Pale-yellow hexagonal needles. Sol. CH_2Cl_2, $CHCl_3$, mod. sol. toluene. Dec. from 140°.

Lukehart, C.M. *et al*, *J. Am. Chem. Soc.*, 1975, **97**, 6903 (*synth, ir, pmr, cryst struct*)

$C_8H_9MnO_5Pb$ Mn-00038

Pentacarbonyl(trimethylplumbyl)manganese, 10CI

(Trimethylplumbyl)pentacarbonylmanganese

[36527-71-6]

$Me_3PbMn(CO)_5$

M 447.294

Yellow. Disproportionates in THF soln. to Me_4Pb and Bis(pentacarbonylmanganese)dimethyllead, Mn-00078

Haupt, H.J. *et al*, *J. Organomet. Chem.*, 1973, **54**, 231 (*synth*)
Kodel, W. *et al*, *Inorg. Chim. Acta*, 1981, **49**, 209.

$C_8H_9MnO_5Si$ **Mn-00039**

Pentacarbonyl(trimethylsilyl)manganese, 9CI, 8CI
Trimethylsilylmanganese pentacarbonyl
[26500-16-3]

$Me_3SiMn(CO)_5$

M 268.180
Mp 26.5°.

Malisch, W. *et al*, *Chem. Ber.*, 1974, **107**, 979 (*synth*, *ir*)
McLean, R.A.N., *J. Chem. Soc., Dalton Trans.*, 1974, 1568 (*uv*)
Spalding, T.R., *J. Organomet. Chem.*, 1978, **149**, 371 (*ms*)
Gladysz, J.A. *et al*, *Inorg. Chem.*, 1979, **18**, 553 (*synth*)
Li, S. *et al*, *J. Organomet. Chem.*, 1979, **166**, 317 (*nmr*)

$C_8I_2Mn_2O_8$ **Mn-00040**

Di(μ-iodo)octacarbonyldimanganese, 9CI, 8CI
Octacarbonyldi(μ-iodo)dimanganese
As Di(μ-bromo)octacarbonyldimanganese, Mn-00027 with

X = I

M 587.768
Monomer: Tetracarbonyliodomanganese. Iodotetracarbonylmanganese.
C_4IMnO_4 M 293.884

Abel, E.W. *et al*, *J. Chem. Soc.*, 1959, 1501 (*synth*, *props*)
El Sayad, M.A. *et al*, *Inorg. Chem.*, 1963, **2**, 158 (*ir*)
Farona, M.F. *et al*, *J. Inorg. Nucl. Chem.*, 1967, **29**, 1814 (*synth*)
Zingeles, F. *et al*, *Inorg. Chem.*, 1967, **6**, 1243 (*props*)
Bamford, C.H. *et al*, *J. Chem. Soc., Dalton Trans.*, 1972, 1846 (*synth*)

$C_9FeMnO_9^{\ominus}$ **Mn-00041**

Nonacarbonylferratemanganate(1−), 8CI

$[(OC)_4Fe—Mn(CO)_5]^{\ominus}$

M 362.879 (ion)
Bis(triphenylphosphine)iminium salt:
$C_{45}H_{30}FeMnNO_9P_2$ M 901.466
Dark-red cryst. (pentane/Et_2O). Mp 134-5°.

Ruff, J.K., *Inorg. Chem.*, 1968, **7**, 1818 (*synth*, *ir*)

$C_9H_5ClMnO_3^{\oplus}$ **Mn-00042**

(Chlorobenzene)tricarbonylmanganese(1+)
Tricarbonyl(η⁶-chlorobenzene)manganese(1+), 9CI
[57812-91-6]

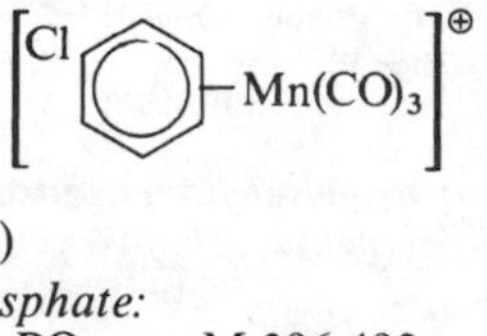

$[Cl-C_6H_4-Mn(CO)_3]^{\oplus}$

M 251.528 (ion)
Hexafluorophosphate:
$C_9H_5ClF_6MnPO_3$ M 396.492
Pale-yellow cryst. (Me_2CO/EtOH).

Pauson, P.L. *et al*, *J. Chem. Soc., Dalton Trans.*, 1975, 1677.

$C_9H_5MnN_3OS_2^{\ominus}$ **Mn-00043**

π-Cyclopentadienyl[dimercaptomaleonitrilato(2−)]nitrosylmanganate(1−)
[32629-44-0]

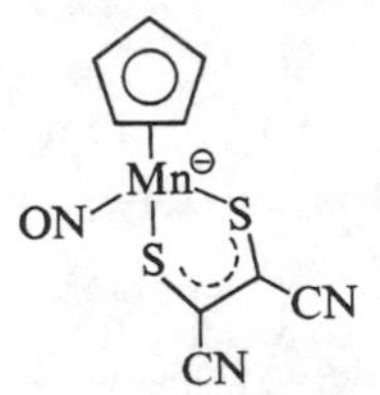

M 290.216 (ion)

McCleverty, J.A. *et al*, *Inorg. Chem.*, 1969, **8**, 1340 (*synth*, *props*)
McCleverty, J.A. *et al*, *J. Chem. Soc. (A)*, 1970, 3315 (*synth*, *props*)

$C_9H_5MnO_5$ **Mn-00044**

Tricarbonyl(carboxy-π-cyclopentadienyl)manganese, 8CI
(Carboxycyclopentadienyl)tricarbonylmanganese
[12082-07-4]

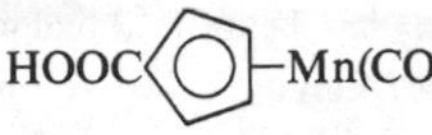

$HOOC-C_5H_4-Mn(CO)_3$

M 248.074
Yellow cryst. ($ClCH_2CH_2Cl$). Mp 190-9° (195-7°).
Me ester:
$C_{10}H_7MnO_5$ M 262.100
Yellow oil. $Bp_{2.5}$ 127-8°.
Amide:
$C_9H_6MnNO_4$ M 247.089
Mp 152-4°.

Cais, M. *et al*, *J. Am. Chem. Soc.*, 1960, **82**, 5667 (*synth*)
Riemschneider, R. *et al*, *Z. Naturforsch., B*, 1960, **15**, 627 (*synth*)
Tirosh, N. *et al*, *J. Organomet. Chem.*, 1966, **5**, 357.
Shew, L.M.C. *et al*, *J. Organomet. Chem.*, 1966, **5**, 362 (*deriv*)
Nesmeyanov, A.N. *et al*, *Izv. Akad. Nauk SSSR, Ser. Khim.*, 1969, 1945 (*deriv*)
Brill, T.B. *et al*, *Inorg. Chem.*, 1971, **10**, 74 (*nqr*)

$C_9H_6MnO_3^{\oplus}$ **Mn-00045**

(η⁶-Benzene)tricarbonylmanganese(1+), 9CI
Tricarbonyl(η⁶-benzene)manganese(1+)

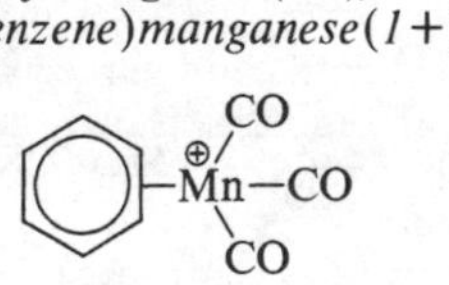

M 217.083 (ion)
Hexafluorophosphate: [38834-51-4].
$C_9H_6F_6MnO_3P$ M 362.047
Pale-yellow cryst.

Jones, D. *et al*, *J. Chem. Soc.*, 1962, 4458 (*props*)
Walker, P.J.C. *et al*, *J. Chem. Soc., Dalton Trans.*, 1973, 662 (*props*)
Pauson, P.L. *et al*, *J. Chem. Soc., Dalton Trans.*, 1975, 1677 (*synth*, *ir*, *nmr*)
Bachmann, P. *et al*, *Z. Naturforsch., B*, 1977, **32**, 471 (*synth*)
Kane-Maguire, L.A.P. *et al*, *Inorg. Chem.*, 1979, **18**, 700.

$C_9H_7MnO_3$ **Mn-00046**

(Methylcyclopentadienyl)tricarbonylmanganese

Tricarbonyl(methylcyclopentadienyl)manganese. Methylcymantrene

[12108-13-3]

H_3C–C_5H_4–$Mn(CO)_3$

M 218.091

Antiknock additive for petrol. Yellow liq. Almost insol. H_2O, misc. nonpolar solvs. d_4^{20} 1.388. Mp 1.5°. Bp 233°, Bp_{10} 102°.

▷Highly toxic, TLV (skin) 0.2. Highly inflammable, flash p. 111°. OP1450000.

Piper, T.S. *et al*, *J. Inorg. Nucl. Chem.*, 1955, **1**, 165 (*synth*)
Gibson, H.J. *et al*, *Angew. Chem.*, 1959, **71**, 629 (*synth*)
Schlögl, K. *et al*, *Monatsh. Chem.*, 1967, **98**, 82 (*pmr*)
Gansow, O.A. *et al*, *J. Chem. Soc., Chem. Commun.*, 1972, 456 (*cmr*)
Parker, D.J., *Spectrochim. Acta, Part A*, 1975, **31**, 1789 (*ir, raman*)
Sax, N.I., *Dangerous Properties of Industrial Materials*, 5th Ed., Van Nostrand-Reinhold, 1979, 816.
Hazards in the Chemical Laboratory, (Bretherick, L., Ed.), 3rd Ed., Royal Society of Chemistry, London, 1981, 395.

$C_9H_7MnO_3$ **Mn-00047**

Tricarbonyl[(1,2,3,4,5-η)-2,4-cyclohexadien-1-yl]manganese, 9CI

Cyclohexadienyltricarbonylmanganese

[12108-14-4]

M 218.091

Yellow plates (pet. ether at −78°). Mp 79°.

Winkhaus, G. *et al*, *J. Chem. Soc.*, 1961, 3807 (*synth, ir, pmr*)
King, R.B. *et al*, *Org. Mass. Spectrom.*, 1974, **9**, 195 (*ms*)
Munro, G.A.M. *et al*, *Isr. J. Chem.*, 1976/77, **15**, 258 (*synth*)

$C_9H_8MnO_2^{\oplus}$ **Mn-00048**

Dicarbonyl(η⁵-2,4-cyclopentadien-1-yl)ethylidynemanganese(1+), 10CI

[59831-15-1]

C_5H_5–$Mn^{\oplus}(CO)_2{\equiv}CR$

R = CH_3

M 203.099 (ion)

Tetrachloroborate: [59831-16-2].
$C_9H_8BCl_4MnO_2$ M 355.721
Characterised spectroscopically.

Fischer, E.O. *et al*, *Chem. Ber.*, 1977, **110**, 1140 (*synth*)
Kreissl, F.R. *et al*, *Chem. Ber.*, 1977, **110**, 3040 (*props*)
Fischer, E.O. *et al*, *J. Organomet. Chem.*, 1979, **169**, C27 (*props*)
Fischer, E.O. *et al*, *Z. Naturforsch., B*, 1979, **34**, 1168.

$C_9H_9MnN_3O_3^{\oplus}$ **Mn-00049**

Tricarbonyltris(isocyanomethane)manganese(1+)

Tricarbonyltris(methylisocyanide)manganese(1+)

[45167-07-5]

$[Mn(CO)_3(CNMe)_3]^{\oplus}$

M 262.126 (ion)

Hexafluorophosphate: [38833-22-6].
$C_9H_9F_6MnN_3O_3P$ M 407.091
White cryst. (Me_2CO/Et_2O). Mp 175-7°.

Treichel, P.M. *et al*, *J. Organomet. Chem.*, 1972, **44**, 339 (*synth, ir, nmr*)
Sawai, T. *et al*, *J. Organomet. Chem.*, 1974, **80**, 91 (*synth*)
Sarapu, A. *et al*, *Inorg. Chem.*, 1975, **14**, 247 (*synth*)
Treichel, P.M. *et al*, *Isr. J. Chem.*, 1977, **15**, 253.

$C_9H_9MnN_3O_3^{\oplus}$ **Mn-00050**

Tris(acetonitrile)tricarbonylmanganese(1+)

Tricarbonyltris(cyanomethane)manganese(1+)

[55029-83-9]

$(MeCN)_3Mn(CO)_3^{\oplus}$

M 262.126 (ion)

Hexafluorophosphate: [54039-60-0].
$C_9H_9F_6MnN_3O_3P$ M 407.091
Pale-yellow plate-like cryst. ($CH_2Cl_2/EtOH$). Mp 131-3° dec.

Reimann, R.H. *et al*, *J. Chem. Soc., Dalton Trans.*, 1974, 808 (*synth, ir, nmr*)
Edwards, D.A. *et al*, *J. Organomet. Chem.*, 1977, **131**, 73 (*synth, props*)
Kane-Maguire, L.A.P. *et al*, *Inorg. Chem.*, 1979, **18**, 700 (*synth*)

$C_9H_{12}BrMnN_4O$ **Mn-00051**

Bromocarbonyltetrakis(isocyanomethane)manganese, 8CI

Bromocarbonyltetrakis(methyl isocyanide)manganese

[38833-19-1]

$MnBr(CO)(CNMe)_4$

M 327.062

Bright-yellow cryst. (CH_2Cl_2/hexane). Mp 176-8°.

Treichel, P.M. *et al*, *J. Organomet. Chem.*, 1972, **44**, 339 (*synth, ir, nmr*)
Adams, R.D. *et al*, *J. Organomet. Chem.*, 1975, **87**, C48 (*synth*)
Treichel, P.M. *et al*, *J. Organomet. Chem.*, 1976, **122**, 229 (*synth*)

$C_9H_{12}MnO$ **Mn-00052**

Bis(η⁴-1,3-butadiene)carbonylmanganese, 9CI

[54688-87-8]

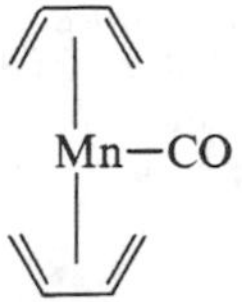

M 191.131

Long blue-green needles (pentane). Paramagnetic with 1 unpaired electron. Dec. at 150°, ir ν_{CO} 1968 cm^{-1} (pentane).

Herberhold, M. *et al*, *Angew. Chem., Int. Ed. Engl.*, 1975, **14**, 351 (*synth, ir, ms, esr*)
Huttner, G. *et al*, *Angew. Chem., Int. Ed. Engl.*, 1975, **87**, 352 (*cryst struct*)
Harlow, R.L. *et al*, *Organometallics*, 1982, **1**, 1506 (*synth*)

$C_9H_{18}BrMnO_9P_2$ **Mn-00053**

Bromotricarbonylbis(trimethylphosphite-*P*)manganese, 8CI

[52841-89-1]

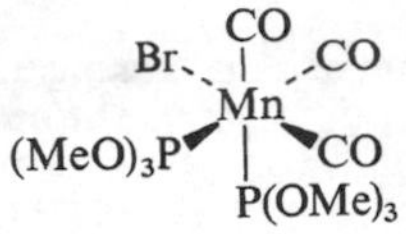

M 467.025

Orange cryst. Mp 110°.

Treichel, P.M. *et al*, *J. Organomet. Chem.*, 1969, **17**, P37 (*synth*)
Reimann, R.H., *J. Chem. Soc., Dalton Trans.*, 1973, 841 (*synth*)

$C_{10}H_7MnO_3$ **Mn-00054**

Tricarbonyl[(1,2,3,4,5-η)-2,4,6-cycloheptatrien-1-yl]manganese, 9CI

[53011-14-6]

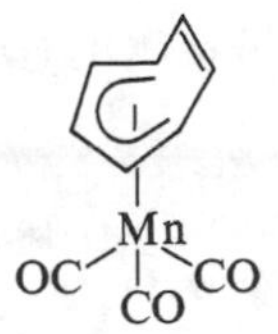

M 230.102

Bright-orange solid. Mp 64°. $Bp_{0.04}$ 35° subl.

Whitesides, T.H. *et al*, *Inorg. Chem.*, 1976, **15**, 874 (*synth, ir, nmr*)

$C_{10}H_7MnO_4$ **Mn-00055**

(Acetylcyclopentadienyl)tricarbonylmanganese

[(1,2,3,4,5-η)-1-Acetyl-2,4-cyclopentadien-1-yl]tricarbonylmanganese, 9CI. *Tricarbonyl(acetyl-π-cyclopentadienyl)manganese*, 8CI. *Acetylcymantrene*

[12116-28-8]

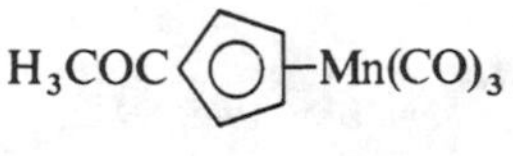

H_3COC–C_5H_4–$Mn(CO)_3$

M 246.101

Yellow cryst. (pet. ether). Mp 41.5-42.5°.

2,4-Dinitrophenylhydrazone: Mp 223-4°.

Cotton, F.A. *et al*, *Chem. Ind. (London)*, 1958, 1368 (*synth, ir*)
Fischer, E.O. *et al*, *Chem. Ber.*, 1958, **91**, 2719 (*synth, ir*)
Kozikowski, J. *et al*, *J. Am. Chem. Soc.*, 1959, **81**, 2995 (*synth, ir*)
Khotsyanova, J.L. *et al*, *J. Organomet. Chem.*, 1975, **88**, 351 (*struct*)

$C_{10}H_9MnO_3$ **Mn-00056**

Tricarbonyl[(1,2,3,4,5-η)-2,4-cycloheptadien-1-yl]manganese, 9CI

[32798-86-0]

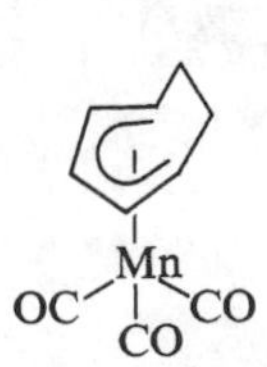

M 232.117

Yellow cryst. (hexane). Mp 66-8°.

King. R.B. *et al*, *Inorg. Chem.*, 1974, **13**, 637 (*synth, ir, nmr*)

$C_{10}H_9MnO_4$ **Mn-00057**

(η[6]-Benzene)dicarbonyl(methoxycarbonyl)manganese, 8CI

[53127-56-3]

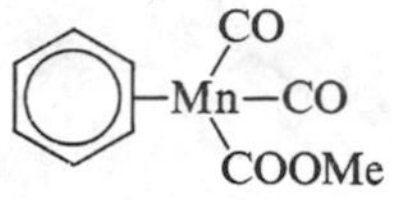

M 248.117

Unstable yellow cryst.

Walker, P.J.C. *et al*, *Inorg. Chim. Acta*, 1973, **7**, 621 (*synth, ir, nmr*)

$C_{10}H_{10}Mn$ **Mn-00058**

Manganocene, 9CI

Di-π-Cyclopentadienylmanganese, 8CI. *Biscyclopentadienylmanganese*

[1271-27-8]

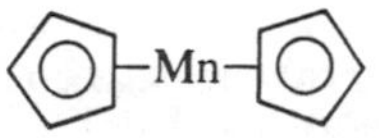

M 185.127

Polymeric in solid state. Amber-brown cryst. becoming pinkish white at 159-60°. Sol. Py, THF, mod. sol. C_6H_6. Mp 173°. Sublimes *in vacuo* at 100-30°. Stable to 350°.

▷Inflames expolsively in air when finely divided

Wilkinson, G. *et al*, *J. Inorg. Nucl. Chem.*, 1956, **2**, 95 (*synth*)
Reid, A.F. *et al*, *Aust. J. Chem.*, 1966, **19**, 309 (*synth*)
Parker, D.J. *et al*, *J. Chem. Soc. (A)*, 1970, 480 (*ir*)
Ammeter, J.H. *et al*, *J. Am. Chem. Soc.*, 1974, **96**, 7833.
Switzer, M.E. *et al*, *J. Am. Chem. Soc.*, 1974, **96**, 7669.
U.S.P., 3 880 743, (*1975*); *CA*, **83**, 97557 (*synth*)
Binder, W. *et al*, *Z. Naturforsch., B*, 1978, **33**, 1235 (*cryst struct*)
Haaland, A., *Acc. Chem. Res.*, 1979, **12**, 415 (*struct*)
Haaland, A., *Inorg. Nucl. Chem. Lett.*, 1979, **15**, 267 (*ed*)
Chipperfield, J.R. *et al*, *J. Organomet. Chem.*, 1979, **178**, 177.

$C_{10}H_{11}MnO_2$ **Mn-00059**

Dicarbonyl(η[5]-2,4-cyclopentadien-1-yl)(1-methylethylidene)manganese, 9CI

[59831-13-9]

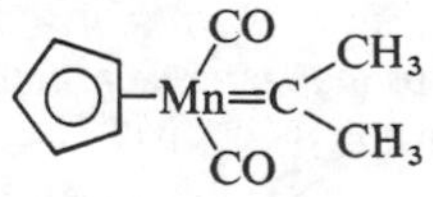

M 218.134

Herrmann, W.A., *Chem. Ber.*, 1975, **108**, 486 (*synth*)
Friedrich, P. *et al*, *J. Organomet. Chem.*, 1977, **139**, C68 (*cryst struct*)

$C_{10}H_{11}MnO_3$ **Mn-00060**

Dicarbonyl(η[5]-2,4-cyclopentadien-1-yl)(1-methoxyethylidene)manganese, 9CI

[12244-94-9]

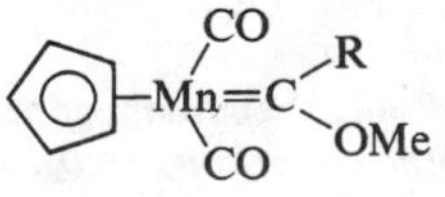

R = CH_3

M 234.133

Red-brown liq.

Fischer, E.O. *et al*, *Chem. Ber.*, 1967, **100**, 2445 (*synth, props*)

$C_{10}H_{15}MnN_2OS_2$ Mn-00061

π-Cyclopentadienyl(diethyldithiocarbamato)nitrosylmanganese, 8CI

[32680-07-2]

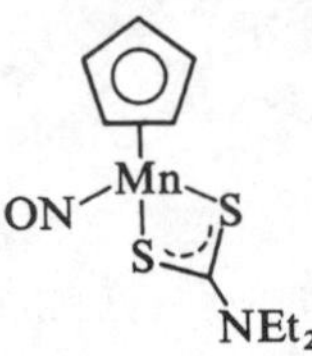

M 298.299

McCleverty, J.A. *et al*, *Inorg. Chem.*, 1969, **8**, 1340 (*synth, props*)
McCleverty, J.A. *et al*, *J. Chem. Soc.*, 1970, 3315 (*synth, props*)

$C_{10}MnO_{10}Re$ Mn-00062

Pentacarbonyl(pentacarbonylmanganese)rhenium, 8CI

Manganese rhenium decacarbonyl. Decacarbonylmanganeserhenium

[14693-30-2]

$(OC)_5Re{-}Mn(CO)_5$

M 521.249

White, air-stable cryst.

Nesmeyanov, A.N. *et al*, *Izv. Akad. Nauk SSSR, Ser. Khim.*, 1963, 194 (*synth*)
Struchkov, Y.T. *et al*, *Dokl. Akad. Nauk SSSR, Ser. Sci. Khim.*, 1967, **172**, 107 (*cryst struct*)
Junk, G.A. *et al*, *J. Chem. Soc.* (*A*), 1970, 2102.
Spiro, T.G., *Prog. Inorg. Chem.*, 1970, **11**, 1 (*spectra*)
Abel, E.W. *et al*, *Inorg. Nucl. Chem. Lett.*, 1971, **7**, 587 (*synth*)
Knox, S.A.R. *et al*, *Inorg. Chem.*, 1971, **10**, 2636 (*synth, struct*)

$C_{10}Mn_2O_{10}$ Mn-00063

Decacarbonyldimanganese, 8CI

Manganese carbonyl. Dimanganese decacarbonyl

[10170-69-1]

$(OC)_5Mn{-}Mn(CO)_5$

M 389.980

Starting material for manganese carbonyl chemistry. Yellow cryst. Sol. org. solvs. Mp 152-4°. $Bp_{0.1}$ 80° subl.

Podall, H.E. *et al*, *J. Am. Chem. Soc.*, 1960, **82**, 1325 (*synth*)
Calderazzo, F. *et al*, *Inorg. Chem.*, 1965, **4**, 293 (*synth*)
Lewis, J. *et al*, *J. Chem. Soc.* (*A*), 1966, 1663 (*ms*)
Haas, H. *et al*, *J. Chem. Phys.*, 1967, **47**, 2996 (*ir*)
King, R.B., *J. Organomet. Chem.*, 1968, **11**, 641 (*synth*)
Almenningen, A. *et al*, *Acta Chem. Scand.*, 1969, **23**, 865 (*cryst struct*)
Brill, T.B. *et al*, *Inorg. Chem.*, 1974, **13**, 470 (*synth*)
Ireland, P.S. *et al*, *J. Magn. Reson.*, 1976, **23**, 485 (*nmr*)

$C_{11}F_5MnO_5$ Mn-00064

Pentacarbonyl(pentafluorophenyl)manganese, 8CI

Pentafluorophenylpentacarbonylmanganese

[14122-94-2]

$Mn(C_6F_5)(CO)_5$ (O_h)

M 362.048

White cryst. Mp 123-4°.

Jolly, P.W. *et al*, *J. Chem. Soc.*, 1965, 5830 (*synth, nmr*)
Booth, B.L. *et al*, *J. Chem. Soc.* (*A*), 1970, 1974 (*synth*)
Oliver, A.J. *et al*, *Inorg. Chem.*, 1970, **99**, 2578 (*props*)

$C_{11}H_3Mn_2NO_9$ Mn-00065

Nonacarbonyl(isocyanomethane)dimanganese, 9CI

Nonacarbonyl(methyl isocyanidè)dimanganese

[57956-59-9]

$(OC)_5Mn{-}Mn(CO)_4CNMe$ (*eq-*)

M 403.022

Yellow cryst. (Et_2O/pentane). Mp 95°.

Ziegler, M.L. *et al*, *Chem. Ber.*, 1965, **98**, 2454 (*synth*)
Fischer, E.O. *et al*, *Chem. Ber.*, 1969, **102**, 2449 (*synth, props*)
Koelle, U., *J. Organomet. Chem.*, 1978, **155**, 53 (*synth*)

$C_{11}H_5MnO_5$ Mn-00066

Pentacarbonylphenylmanganese, 9CI

Phenylpentacarbonylmanganese

[13985-77-8]

$MnPh(CO)_5$ (O_h)

M 272.096

White cryst. Mp 52°.

Booth, B.L. *et al*, *J. Chem. Soc.* (*A*), 1969, 920 (*props*)
Wojcicki, A., *Adv. Organomet. Chem.*, 1973, **11**, 87 (*synth, props*)
Blumer, D.J. *et al*, *J. Organomet. Chem.*, 1979, **173**, 71 (*props*)
Stewart, R.P., *Inorg. Chem.*, 1979, **18**, 2083 (*props*)

$C_{11}H_{11}Mn$ Mn-00067

(η^6-Benzene)(η^5-2,4-cyclopentadien-1-yl)manganese, 9CI

[1271-43-8]

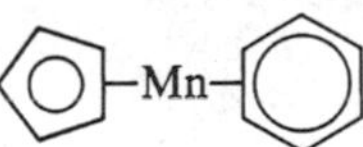

M 198.146

Deep-red cryst.

Denning, R.G. *et al*, *J. Am. Chem. Soc.*, 1966, **88**, 4619 (*synth*)
Fischer, E.O. *et al*, *Chem. Ber.*, 1966, **99**, 2213 (*synth, props*)

$C_{11}H_{15}MnN_5O^{\oplus}$ Mn-00068

Carbonylpentakis(isocyanomethane)manganese(1+)

Carbonylpentakis(methyl isocyanide)manganese(1+)

[45211-82-3]

$[Mn(CO)(CNMe)_5]^{\oplus}$

M 288.210 (ion)

Hexafluorophosphate: [39044-46-7].
$C_{11}H_{15}F_6MnN_5OP$ M 433.175
Cryst. (CH_2Cl_2/Et_2O). Mp 215°.

Treichel, P.M. *et al*, *J. Organomet. Chem.*, 1972, **44**, 339 (*synth, ir, nmr*)
Sarapu, A. *et al*, *Inorg. Chem.*, 1975, **14**, 247.
Treichel, P.M. *et al*, *Isr. J. Chem.*, 1977, **15**, 253.

$C_{11}H_{21}MnO_3P^{\ominus}$ Mn-00069

Bis(η^4-1,3-butadiene)(trimethyl phosphite)manganate(1−)

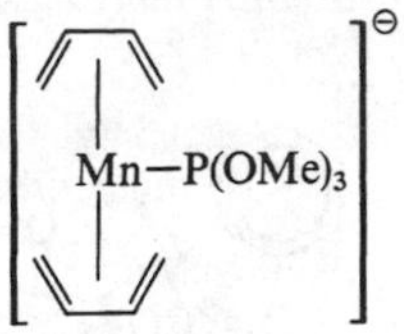

M 287.197 (ion)

Tetrabutylammonium salt: [82963-81-3].
$C_{27}H_{57}MnNO_3P$ M 529.664
Red needles (THF/Et_2O).

Harlow, R.L. *et al*, *Organometallics*, 1982, **1**, 1506 (*synth, props*)

$C_{11}H_{21}MnO_3P$ Mn-00070

Bis(η^4-1,3-butadiene)(trimethyl phosphite)manganese
[82963-77-7]

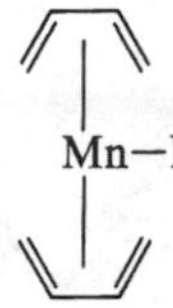

M 287.197
Green paramagnetic cryst.

Harlow, R.L. *et al*, *Organometallics*, 1982, **1**, 1506 (*synth, cryst struct*)

$C_{11}H_{27}BrMnO_{11}P_3$ Mn-00071

Bromodicarbonyltris(trimethylphosphite-*P*)manganese, 8CI
[39604-51-8]

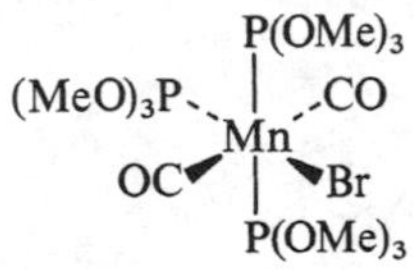

M 563.091
Orange needles. Mp 125° dec., 161°.

Treichel, P.M. *et al*, *J. Organomet. Chem.*, 1969, **17**, P37 (*synth*)
Reimann, R.H. *et al*, *J. Chem. Soc., Dalton Trans.*, 1973, 841, 2658 (*synth, ir, nmr*)

$C_{12}Fe_2MnO_{12}^{\ominus}$ Mn-00072

Tetracarbonyl(di-μ-carbonylhexacarbonyldiferrate)manganate(1−), 10CI
Dodecacarbonyldiferratemanganate(1−), 9CI

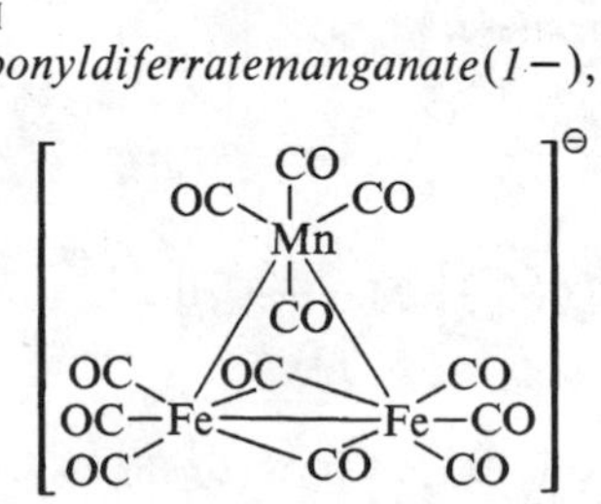

M 502.757 (ion)

Tetraethylammonium salt: [53433-49-1].
$C_{20}H_{20}Fe_2MnNO_{12}$ M 633.010
Intensely blue-black cryst. (EtOH).

Bis(triphenylphosphine)iminium salt: [68520-37-6].
$C_{48}H_{30}Fe_2MnNO_{12}P_2$ M 1041.344
Dark-blue cryst. (CH_2Cl_2/Et_2O). Mp 171-3°.

Anders, U. *et al*, *J. Chem. Soc., Chem. Commun.*, 1966, 291 (*synth, ir, uv*)
Ruff, J.K., *Inorg. Chem.*, 1968, **7**, 1818 (*synth, ir*)
Knight, J.A. *et al*, *J. Chem. Soc., Dalton Trans.*, 1970, 1006 (*ir*)
Lindauer, M.W. *et al*, *Inorg. Chem.*, 1970, **9**, 1694 (*mössbauer*)
Tyler, D.R. *et al*, *J. Am. Chem. Soc.*, 1978, **100**, 7888 (*uv*)

$C_{12}HMnO_{12}Os_3$ Mn-00073

Dodecacarbonylhydromanganesediosmium, 8CI
[12560-43-9]

$MnOs_3H(CO)_{12}$

M 962.671
No phys. props. reported.

Knight, J. *et al*, *J. Chem. Soc., Chem. Commun.*, 1971, 62 (*synth, ir*)

$C_{12}H_3Mn_3O_{12}$ Mn-00074

Dodecacarbonyltrihydrotrimanganese
Tris(hydridomanganese tetracarbonyl)
[18444-56-9]

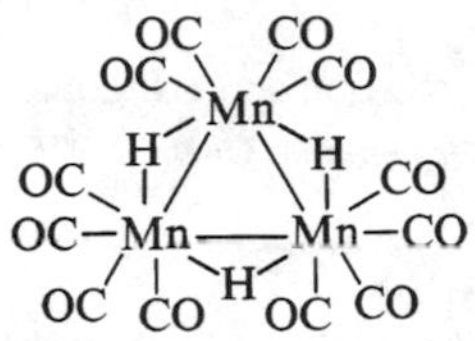

M 503.963
Orange-red cryst., stable to air. Bp_1 60° subl. Dec. at 123-5°.

Johnson, B.F.G. *et al*, *J. Organomet. Chem.*, 1967, **10**, 105 (*synth, ir, nmr*)
Inorg. Synth., 1968, **12**, 43 (*synth*)
Kirtley, S.W. *et al*, *J. Am. Chem. Soc.*, 1973, **95**, 4532 (*cryst struct*)

$C_{12}H_5FeMnO_7$ Mn-00075

Pentacarbonyl[dicarbonyl(η^5-2,4-cyclopentadien-1-yl)iron]-manganese, 9CI
Dicarbonyl(η^5-2,4-cyclopentadien-1-yl)pentacarbonylmanganioiron. Pentacarbonyl(dicarbonyl-π-cyclopentadienylferrio)manganese, 8CI. Cyclopentadienylirondicarbonylpentacarbonylmanganese. (η^5-Cyclopentadienyl)dicarbonylironpentacarbonylmanganese. Heptacarbonyl(η^5-cyclopentadienyl)ironmanganese
[12088-73-2]

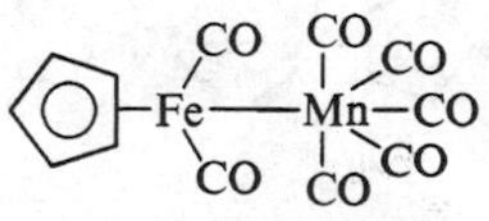

M 371.952
Dark-red plates. Mp 76°. Bp 40° *in vacuo* subl.

King, R.B. *et al*, *Chem. Ind. (London)*, 1961, 747 (*synth*)
Herber, R.H. *et al*, *Inorg. Chem.*, 1964, **3**, 101 (*mössbauer*)
Dessy, R.E. *et al*, *J. Am. Chem. Soc.*, 1966, **88**, 5117, 5124.
Hansen, P.J. *et al*, *J. Organomet. Chem.*, 1966, **6**, 389 (*cryst struct*)
Madach, T. *et al*, *Chem. Ber.*, 1980, **113**, 2675 (*synth, pmr, ir, uv*)
Busetto, L. *et al*, *J. Chem. Soc., Dalton Trans.*, 1982, 1631 (*ir*)

$C_{12}H_6MnO_{10}Re$ Mn-00076

Pentacarbonyl[tetracarbonyl(1-methoxyethylidene)]manganeserhenium, 9CI

Nonacarbonyl(1-methoxyethylidene)manganeserhenium

[49547-64-0]

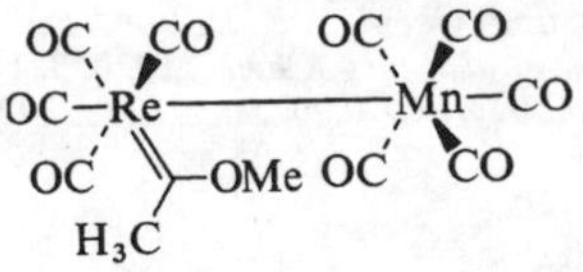

M 551.318

Red-orange cryst. Mp 107-8°.

Casey, C.P. *et al*, *J. Am. Chem. Soc.*, 1975, **97**, 3053 (*synth, struct*)

$C_{12}H_6Mn_2N_2O_8$ Mn-00077

Octacarbonylbis(isocyanomethane)dimanganese, 9CI

Octacarbonylbis(methyl isocyanide)dimanganese

[58034-14-3]

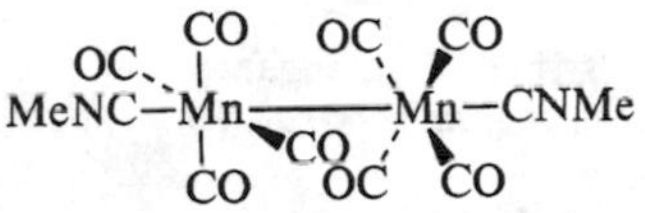

M 416.064

Fluxional. Characterised spectroscopically.

Grant, S. *et al*, *J. Organomet. Chem.*, 1975, **96**, C11 (*synth*)

$C_{12}H_6Mn_2O_{10}Pb$ Mn-00078

Bis(pentacarbonylmanganese)dimethyllead

Decacarbonyl[μ-(dimethylplumbylene)]dimanganese, 10CI

[42167-74-8]

$[(OC)_5Mn]_2PbMe_2$

M 627.249

Yellow solid.

Haupt, H.J. *et al*, *J. Organomet. Chem.*, 1973, **54**, 231.
Kodel, W. *et al*, *Inorg. Chim. Acta*, 1981, **49**, 209.

$C_{12}H_7FeMnO_6$ Mn-00079

μ-Carbonylpentacarbonyl(η^5-2,4-cyclopentadienyl)iron-μ-methylenemanganese

(μ-Carbonyl)(μ-methylene)[(cyclopentadienylcarbonyliron)tetracarbonylmanganese]. Hexacarbonyl(η^5-cyclopentadienyl)iron-μ-methylenemanganese

[83350-15-6]

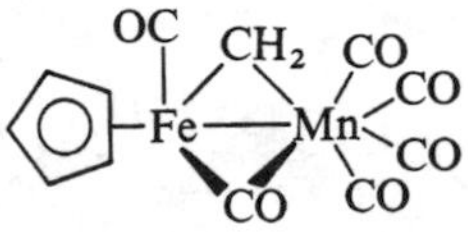

M 357.969

Cryst. (pentane or by subl. *in vacuo*) (dimorph.). $Bp_{0.05}$ 55° subl.

Gadol, S.M. *et al*, *Organometallics*, 1982, **1**, 1607 (*struct*)

$C_{12}H_7MnO_5$ Mn-00080

(2-Acetylphenyl-*C,O*)tetracarbonylmanganese, 9CI

Acetophenonetetracarbonylmanganese

[50831-23-7]

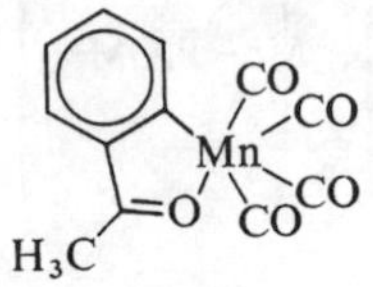

M 286.122

Yellow cryst.

Knobler, C.B. *et al*, *Inorg. Chem.*, 1975, **14**, 2062 (*struct*)
McKinney, R.J. *et al*, *Inorg. Chem.*, 1975, **14**, 2057 (*synth, ir, ms, nmr*)

$C_{12}H_{10}Mn_2N_2O_4$ Mn-00081

μ-Carbonylcarbonylbis(η^5-2,4-cyclopentadien-1-yl)[μ-(nitrosyl-*N:N*)]nitrosyldimanganese, 9CI

Dicarbonyldicyclopentadienyldinitrosyldimanganese

[38999-59-6]

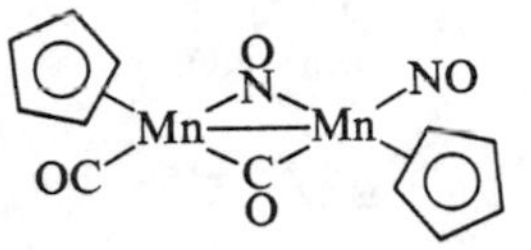

M 356.098

Purple-brown cryst. (CH_2Cl_2/hexane). Dec. at ca. 200°.

King, R.B. *et al*, *Inorg. Chem.*, 1964, **3**, 791 (*synth, ir*)
Adam, R.D. *et al*, *J. Am. Chem. Soc.*, 1973, **95**, 6589 (*nmr*)
Kirchner, R.M. *et al*, *J. Am. Chem. Soc.*, 1973, **95**, 6602 (*cryst struct*)

$C_{12}H_{12}Mn^{\oplus}$ Mn-00082

(η^7-Cycloheptatrienylium)(η^5-2,4-cyclopentadien-1-yl)manganese(1+), 9CI

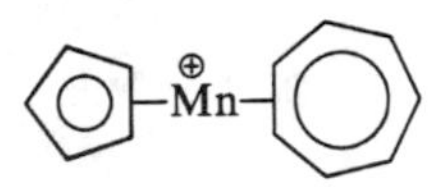

M 211.165 (ion)

Hexafluorophosphate: [52308-73-3].
$C_{12}H_{12}F_6MnP$ M 356.129
Cryst. (Me_2CO/EtOH).

Pauson, P.L. *et al*, *J. Chem. Soc., Dalton Trans.*, 1975, 2387 (*synth, props*)

$C_{12}H_{13}Mn$ Mn-00083

[(1,2,3,4,5,6-η)-1,3,5-Cycloheptatriene](η^5-2,4-cyclopentadien-1-yl)manganese, 9CI

[52308-70-0]

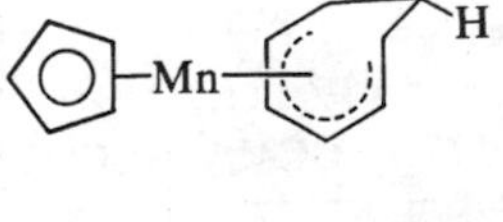

M 212.173

Red-black needle-shaped cryst. $Bp_{0.01}$ 60° subl.

Pauson, P.L. *et al*, *J. Chem. Soc., Dalton Trans.*, 1975, 2387 (*synth, props*)

$C_{12}H_{23}MnO_6P_2S$ **Mn-00084**

Carbonothioyl(η^5-2,4-cyclopentadien-1-yl)bis(trimethylphosphite-*P*)manganese, 9CI

[51831-93-7]

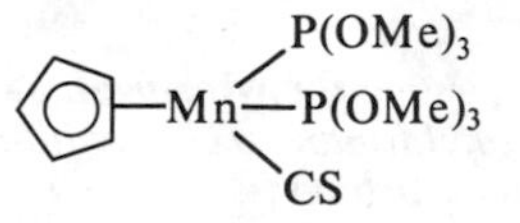

M 412.256

Yellow cryst. (CH_2Cl_2/hexane). Dec. at ca. 180°.

Coville, N.J. *et al*, *J. Organomet. Chem.*, 1974, **64**, 101 (*synth*)
Butler, I.S. *et al*, *J. Organomet. Chem.*, 1974, **80**, 235.
Fenster, A.E. *et al*, *Inorg. Chem.*, 1974, **13**, 915 (*synth, ir, nmr*)

$C_{12}MnO_{12}Os_2^{\ominus}$ **Mn-00085**

Dodecacarbonylmanganatediosmate(1–), 8CI

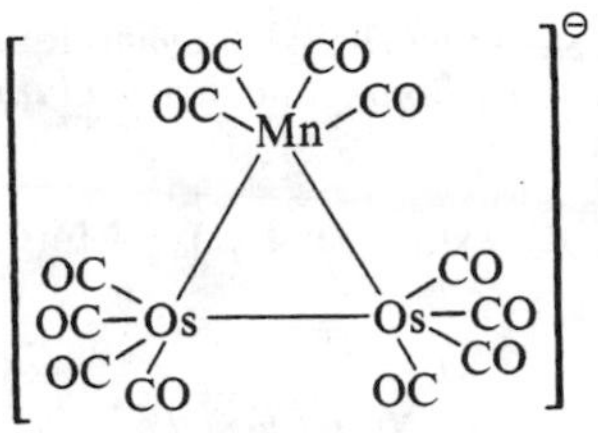

M 771.463 (ion)

Tetramethylammonium salt: [12563-65-4].
$C_{16}H_{12}MnNO_{12}Os_2$ M 845.608
Characterised spectroscopically.

Knight, J. *et al*, *J. Chem. Soc., Chem. Commun.*, 1971, 62 (*synth, ir*)

$C_{13}H_3MnO_{13}Os_3$ **Mn-00086**

Tridecacarbonyltrihydromanganesetriosmium, 8CI

[12563-75-6]

$MnOs_3H_3(CO)_{13}$

M 992.697

No phys. props. reported.

Knight, J. *et al*, *J. Chem. Soc., Chem. Commun.*, 1971, 62 (*synth, ir*)

$C_{13}H_7FeMnO_6$ **Mn-00087**

Tricarbonyl[μ-[(1,2,3-η:4,5,6,7-η)-2,4,6-cycloheptatrien-1-yl]](tricarbonyliron)manganese, 9CI

μ-[(1,2,3-η:4,5,6,7-η)-Cycloheptatrienyl]tricarbonylirontricarbonylmanganese

[60767-46-6]

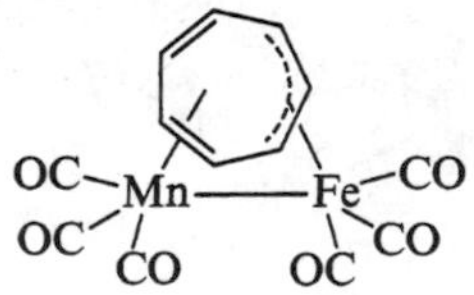

M 369.980

Fluxional molecule. Orange cryst. (pentane). Mp 141-2°.

Bennett, M.J. *et al*, *J. Am. Chem. Soc.*, 1976, **98**, 4810 (*synth, pmr, cmr, ms, ir*)

$C_{13}H_{12}MnNO_4$ **Mn-00088**

Tetracarbonyl[2-[(dimethylamino)methyl]phenyl-*C,N*]manganese, 9CI

[38162-89-9]

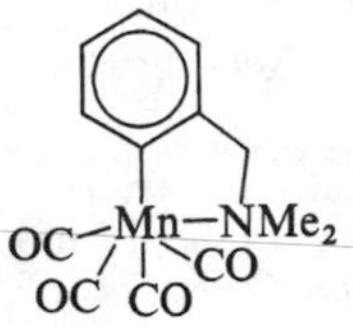

M 301.180

Bright-yellow needles (pet. ether). Mp 73°. $Bp_{0.01}$ 100° subl.

Little, R.G. *et al*, *Inorg. Chem.*, 1973, **12**, 844 (*cryst struct*)
Bennett, R.L. *et al*, *Aust. J. Chem.*, 1975, **28**, 1265 (*synth, ir, nmr*)

$C_{13}H_{36}BrMnO_{13}P_4$ **Mn-00089**

Bromocarbonyltetrakis(trimethylphosphite-*P*)manganese, 8CI

[52646-26-1]

$MnBr(CO)[P(OMe)_3]_4$

M 659.157

Orange cryst. Mp 131°.

Treichel, P.M. *et al*, *J. Organomet. Chem.*, 1969, **17**, P37 (*synth*)
Reimann, R.H. *et al*, *J. Chem. Soc., Dalton Trans.*, 1973, 841.

$C_{13}H_{37}MnO_{13}P_4$ **Mn-00090**

Carbonylhydrotetrakis(trimethylphosphite-*P*)manganese, 9CI

Hydridocarbonyltetrakis(trimethylphosphite)manganese

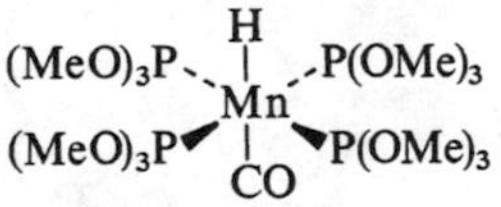

M 580.261

trans-form [66562-91-2]

Pale-brown-pink cryst. (hexane). Sol. C_6H_6, polar solvs.

Stuhl, L.S. *et al*, *Inorg. Chem.*, 1978, **17**, 2148 (*synth, ir, nmr*)

$C_{14}FeMnO_{14}Re$ **Mn-00091**

Pentacarbonyl(pentacarbonylmanganese)(tetracarbonyliron)-rhenium, 9CI

Tetradecacarbonylmanganeserheniumiron, 8CI

[33958-72-4]

$(OC)_5Mn—Fe(CO)_4—Re(CO)_5$

M 689.138

Orange needles (Me_2CO).

Evans, G.O. *et al*, *J. Inorg. Nucl. Chem.*, 1968, **30**, 2862 (*synth, ir, ms*)
Evans, G.O. *et al*, *Inorg. Chem.*, 1970, **9**, 979 (*ir*)
Tyler, D.R. *et al*, *J. Am. Chem. Soc.*, 1978, **100**, 7888 (*struct*)

$C_{14}FeMn_2O_{14}$ Mn-00092

Tetracarbonylbis(pentacarbonylmanganio)iron, 8CI
Decacarbonyl(tetracarbonyliron)dimanganese, 10CI
[15668-57-2]

$(OC)_5Mn—Fe(CO)_4—Mn(CO)_5$

M 557.869
Air-stable red needles (Me_2CO). Sol. org. solvs. Mp 112°. Bp 75° subl. *in vacuo*.

Schubert, E.H. *et al*, *Z. Naturforsch., B*, 1965, **20**, 1306 (*synth, ir*)
Agron, P.A. *et al*, *Acta Crystallogr.*, 1967, **23**, 1079 (*cryst struct*)
Evans, G.O. *et al*, *Inorg. Chem.*, 1970, **9**, 979 (*ir*)
Tyler, D.R. *et al*, *J. Am. Chem. Soc.*, 1978, **100**, 7888 (*uv*)

$C_{14}H_{10}MnO_2^{\oplus}$ Mn-00093

Dicarbonyl(η^5-2,4-cyclopentadien-1-yl)(phenylmethylidyne)manganese(1+), 9CI
[59831-17-3]
As Dicarbonyl(η^5-2,4-cyclopentadien-1-yl)ethylidynemanganese(1+), Mn-00048 with

R = Ph

M 265.170 (ion)
Tetrachloroborate: [59831-18-4].
$C_{14}H_{10}BCl_4MnO_2$ M 417.792
Yellow cryst.

Fischer, E.O. *et al*, *Chem. Ber.*, 1977, **110**, 1140 (*synth*)
Fischer, E.O. *et al*, *J. Organomet. Chem.*, 1977, **129**, 197 (*props*)
Kreissl, F.R. *et al*, *Chem. Ber.*, 1977, **110**, 3040 (*props*)
Fischer, E.O. *et al*, *J. Organomet. Chem.*, 1979, **169**, C27 (*props*)
Fischer, E.O. *et al*, *Z. Naturforsch., B*, 1979, **34**, 1168 (*synth*)

$C_{14}H_{10}MnO_4Re$ Mn-00094

(η^5-2,4-Cyclopentadien-1-yl)[μ-(η:η^5-2,4-cyclopentadien-1-ylidene)]hydrotetracarbonylmanganeserhenium, 10CI
[71486-19-6]

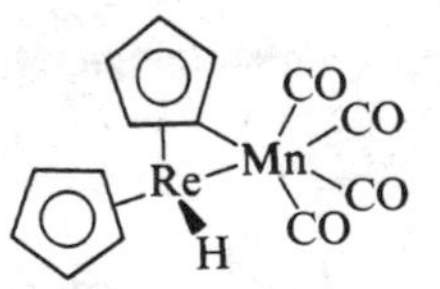

Re Mn CO CO CO CO H

M 483.376
Air-stable red cryst. Air-sensitive in soln.

Hoxmeier, R.J. *et al*, *Inorg. Chem.*, 1979, **18**, 3453 (*synth, nmr, props*)

$C_{14}H_{38}MnP_4$ Mn-00095

Bis[1,2-ethanediylbis[dimethylphosphine]]-*P,P'*]dimethylmanganese
Bis[(1,2-dimethylphosphino)ethane]dimethylmanganese
[87450-50-8]

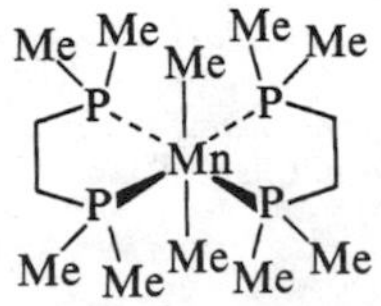

Me Me Me Me Me P P Mn P P Me Me Me Me Me Me

M 385.287
Red paramagnetic cryst.

Wilkinson, G. *et al*, *J. Am. Chem. Soc.*, 1983, **105**, 6752 (*synth, struct*)

$C_{14}Mn_2O_{14}Os$ Mn-00096

Tetracarbonylbis(pentacarbonylmanganio)osmium, 8CI
Tetradecacarbonyldimanganeseosmium. Dimanganese osmium tetradecacarbonyl
[33292-90-9]

$(OC)_5Mn-Os(CO)_4-Mn(CO)_5$

M 692.222
No phys. props. reported. ir ν_{CO} 2068, 2020, 1987 cm^{-1}.

Abel, E.W. *et al*, *Inorg. Nucl. Chem. Lett.*, 1971, **7**, 587 (*synth, ir*)

$C_{14}Mn_2O_{14}Ru$ Mn-00097

Tetracarbonylbis(pentacarbonylmanganio)ruthenium, 8CI
Tetradecacarbonyldimanganeseruthenium
[33152-40-8]

$(OC)_5Mn—Ru(CO)_4—Mn(CO)_5$

M 603.092
Yellow.

Abel, E.W. *et al*, *Inorg. Nucl. Chem. Lett.*, 1971, **7**, 587 (*synth, ir*)

$C_{14}Mn_3O_{14}^{\ominus}$ Mn-00098

Tetradecacarbonyltrimanganate(1−)
Trimanganesetetradecacarbonyl(1−)

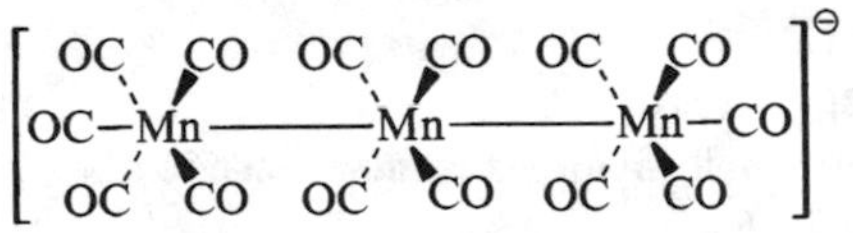

[OC CO OC CO OC CO / OC–Mn——Mn——Mn–CO / OC CO OC CO OC CO]$^{\ominus}$

M 556.960 (ion)
Tetraphenylarsonium salt: [52340-90-6].
$C_{38}H_{20}AsMn_3O_{14}$ M 940.303
Red cryst.

Curtis, M.D., *Inorg. Chem.*, 1972, **11**, 802 (*synth, ir*)
Bau, R. *et al*, *J. Am. Chem. Soc.*, 1974, **96**, 988 (*synth, cryst struct*)

$C_{15}H_9MnMoO_5$ Mn-00099

Molybdenumcarbonyl(η^5-2,4-cyclopentadien-1-yl)[μ-(η:η^5-2,4-cyclopentadien-1-ylidene)]tetracarbonylmanganese, 10CI
[71411-08-0]

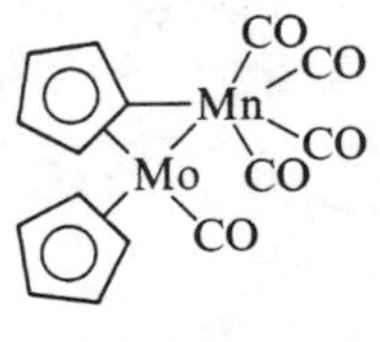

CO CO Mn CO Mo CO CO

M 420.111
Red cryst.

Hoxmeier, R.J., *J. Am. Chem. Soc.*, 1971, **93**, 536 (*synth*)
Hoxmeier, R.J., *Inorg. Chem.*, 1979, **18**, 3453, 3462 (*synth, struct*)

$C_{15}H_{11}MnO_2$ **Mn-00100**

Dicarbonyl(η^5-2,4-cyclopentadien-1-yl)(2-phenylethenylidene)manganese, 9CI

[60039-69-2]

M 278.189

Dark-red cryst. Mp 64°.

Nesmeyanov, A.N. *et al*, *J. Organomet. Chem.*, 1976, **110**, C36 (*synth*)

Antonova, A.B. *et al*, *J. Organomet. Chem.*, 1977, **137**, 55.

$C_{15}H_{12}BMnO_3$ **Mn-00101**

Tricarbonyl[(1,2,3,4,5,6-η)-4-methyl-1-phenylboratabenzene]manganese, 10CI

[62980-54-5]

M 306.006

Yellow cryst. Mp 73-5°.

Herberich, G.E. *et al*, *Chem. Ber.*, 1977, **110**, 1167 (*synth, nmr*)

$C_{15}H_{12}Mn_2O_4$ **Mn-00102**

Tetracarbonylbis(η^5-2,4-cyclopentadien-1-yl)-μ-methylenedimanganese, 10CI

[57603-41-5]

M 366.133

Cryst. Mp 135-6°.

Herrmann, W.A. *et al*, *J. Organomet. Chem.*, 1975, **97**, 245 (*synth*)

Creswick, M. *et al*, *J. Organomet. Chem.*, 1979, **172**, C39 (*cryst struct*)

$C_{15}H_{13}MnO_3$ **Mn-00103**

Dicarbonyl(η^5-2,4-cyclopentadien-1-yl)(2-methoxy-1-phenylethylidene)manganese, 9CI

[12245-61-3]

As Dicarbonyl(η^5-2,4-cyclopentadien-1-yl)(1-methoxyethylidene)manganese, Mn-00060 with

R = Ph

M 296.204

Red-brown liq.

Fischer, E.O. *et al*, *Chem. Ber.*, 1967, **100**, 2445 (*synth, props*)

Hadicke, E. *et al*, *Acta Crystallogr., Sect. B*, 1971, **27**, 270.

$C_{15}H_{15}Mn_3N_4O_4$ **Mn-00104**

Tris(η^5-2,4-cyclopentadien-1-yl)-μ_3-nitrosyltri-μ-nitrosyltrimanganese, 9CI

[12312-71-9]

M 480.122

Needle-shaped cryst.

King, R.B. *et al*, *Inorg. Chem.*, 1964, **3**, 791 (*synth*)

Elder, R.C., *Inorg. Chem.*, 1974, **13**, 1037 (*cryst struct*)

$C_{16}HMnO_{16}Os_3$ **Mn-00105**

Hexadecacarbonylhydromanganesetriosmium, 8CI

[12566-57-3]

$MnOs_3H(CO)_{16}$

M 1074.712

No phys. props. reported.

Knight, J. *et al*, *J. Chem. Soc., Chem. Commun.*, 1971, 62 (*synth, ir*)

$C_{16}H_9MnN_2O_4$ **Mn-00106**

Tetracarbonyl[2-(phenylazo)phenyl]manganese, 9CI

(*Azobenzene*)*tetracarbonylmanganese*

[19528-32-6]

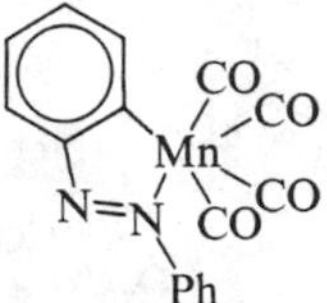

M 348.196

Red needle-shaped cryst. $Bp_{0.01}$ 90-100° subl.

Bruce, M.I. *et al*, *J. Chem. Soc.* (*A*), 1970, 3204 (*synth*)

Bennett, R.L. *et al*, *Aust. J. Chem.*, 1974, **27**, 2131 (*synth*)

$C_{16}H_{11}FeMnO_5$ **Mn-00107**

(2-Acetylferrocenyl-*C,O*)tetracarbonylmanganese, 9CI

[56708-99-7]

M 394.045

Red cryst. (hexane), dec. after some days in air. Soln. dec. overnight in air. Mp 126-127.5°.

Crawford, S.S. *et al*, *Inorg. Chem.*, 1977, **16**, 3193 (*synth, ms*)

$C_{16}H_{11}FeMnO_5$ Mn-00108

Pentacarbonyl(ferrocenylmethyl)manganese, 9CI

[60349-59-9]

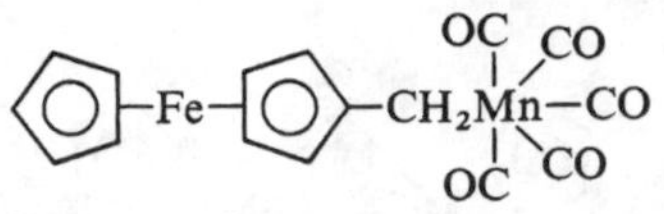

M 394.045

Mp 74-6°.

Pannell, K.H. *et al*, *Inorg. Chem.*, 1976, **15**, 2671 (*synth, pmr, ir*)

$C_{16}H_{21}B_2FeMnO_3S$ Mn-00109

Tricarbonyl[(η^5-2,4-cyclopentadien-1-yl)iron][μ-[(1,2,3,4,5-η:1,2,3,4,5-η)-3,4-diethyl-2,5-dihydro-2,5-dimethyl-1,2,5-thiadiborole]]manganese, 10CI

[66921-71-9]

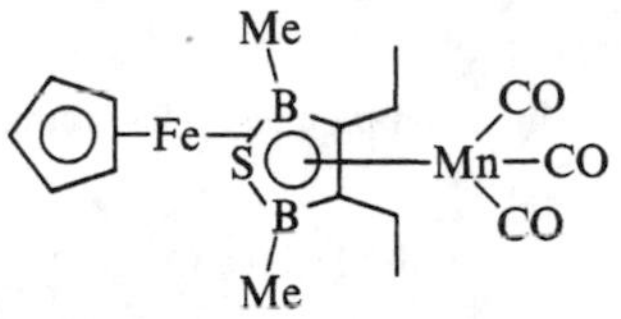

M 425.805

Green air-stable cryst. Mp 194°.

Siebert, W. *et al*, *Angew. Chem., Int. Ed. Engl.*, 1978, **17**, 527 (*synth, pmr, nmr, ir, ms*)

$C_{17}H_{10}MnNO_4$ Mn-00110

Tetracarbonyl[2-[(phenylimino)methyl]phenyl-*C,N*]manganese, 9CI

Tetracarbonyl-2-(N*-phenylformimidoyl)phenylmanganese*

[36594-09-9]

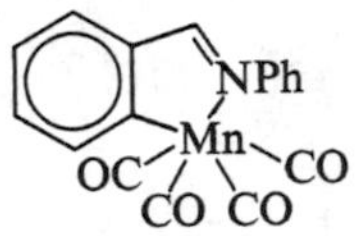

M 347.208

Yellow cryst.

Bruce, M.I. *et al*, *J. Chem. Soc., Chem. Commun.*, 1971, 1595 (*synth, pmr, struct*)

Little, R.G. *et al*, *Inorg. Chem.*, 1973, **12**, 840 (*cryst struct*)

$C_{17}H_{10}MnO_5Tl$ Mn-00111

Pentacarbonyl(diphenylthallium)manganese, 9CI

[41654-98-2]

$Ph_2TlMn(CO)_5$

M 553.584

▷Toxic

Haupt, H.J. *et al*, *J. Organomet. Chem.*, 1973, **50**, 63 (*synth, ir, nmr*)

$C_{18}H_{13}FeMnO_3$ Mn-00112

Tricarbonyl[(1,2,3,4,5-η)-ferrocenyl-2,4-cyclopentadien-1-yl]manganese, 9CI

Cymantrenylferrocene

[37048-11-6]

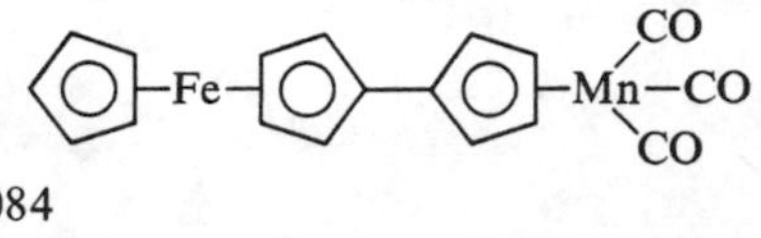

M 388.084

Orange cryst. (EtOH aq.). Mp 108-109.5°.

Nesmeyanov, A.N. *et al*, *Izv. Akad. Nauk SSSR, Ser. Khim.*, 1972, 735; 1980, 1171 (*synth*)

Nesmeyanov, A.N. *et al*, *Zh. Strukt. Khim.*, 1973, **14**, 49 (*cmr*)

$C_{18}H_{18}BMnO$ Mn-00113

Carbonyl(η^5-2,4-cyclopentadien-1-yl)[(1,2,3,6,7-η)-4,5-dihydro-1-phenyl-1*H*-borepin]manganese

[67158-14-9]

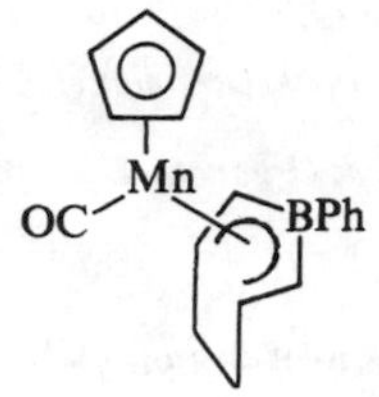

M 316.088

Violet oil, characterised by ir and nmr.

Koelle, U. *et al*, *J. Organomet. Chem.*, 1978, **152**, 7 (*synth, ir, nmr*)

$C_{19}H_{15}FeMnO_3$ Mn-00114

Tricarbonyl[(1,2,3,4,5-η)-1-(ferrocenylmethyl)-2,4-cyclopentadien-1-yl]manganese

(Ferrocenylmethyl)cymantrene. (Ferrocenylmethyl)cyclopentadienylmanganese tricarbonyl

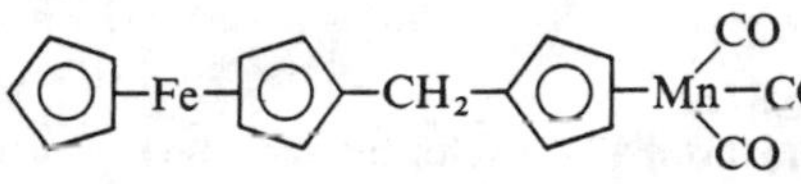

M 402.111

Orange cryst. Mp 121.5-122°.

Cais, M. *et al*, *Tetrahedron Lett.*, 1961, 440 (*synth*)

Maoz, N. *et al*, *Tetrahedron Lett.*, 1965, 2087, 2095 (*ms*)

$C_{19}H_{19}Mn$ Mn-00115

[(1,2,3,4,5-η)-1-methyl-2,4-cyclopentadien-1-yl][(1,2,3,4,5,6-η)-7-phenyl-1,3,5-cycloheptatriene]manganese

[58355-39-8]

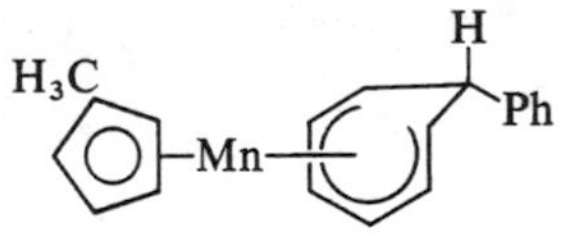

M 302.297

Cryst. $Bp_{0.0001}$ 80° subl.

Pauson, P.L. *et al*, *J. Chem. Soc., Dalton Trans.*, 1975, 2387 (*synth, ir, nmr, props*)

Jeffreys, J.A.D. *et al*, *J. Chem. Soc., Dalton Trans.*, 1978, 144 (*cryst struct*)

$C_{20}H_{10}FeMnO_8P$ Mn-00116

Tetracarbonyl[μ-(diphenylphosphino)](tetracarbonylmanganio)iron, 8CI

[32649-18-6]

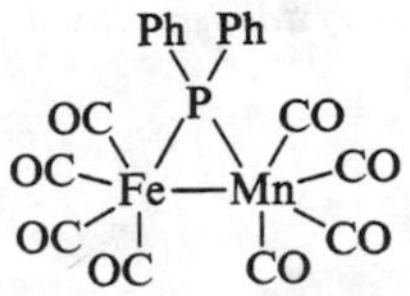

M 520.053

Orange cryst. Mp 164-5°.

Benson, B.C. *et al*, *J. Chem. Soc., Chem. Commun.*, 1968, 1506 (*synth*)
Yasufuku, K. *et al*, *J. Organomet. Chem.*, 1971, **28**, 415 (*synth, ir*)

$C_{20}H_{11}Mn_2O_8P$ Mn-00117

Octacarbonyl-μ-(diphenylphosphino)-μ-hydrodimanganese, 8CI

[19663-11-7]

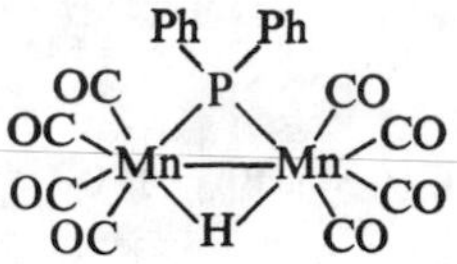

M 520.152

Mod. air-sensitive yellow cryst. (pet. ether). Mp 154-5° dec. Diamagnetic.

Green, M.L.H. *et al*, *Z. Naturforsch, B*, 1962, **17**, 783 (*synth, ir, pmr*)
Hayter, R.E., *Z. Naturforsch, B*, 1963, **18**, 581 (*synth*)
Doedens, R.J. *et al*, *J. Am. Chem. Soc.*, 1967, **89**, 4323 (*cryst struct*)

$C_{20}H_{19}FeMnO_3$ Mn-00118

Dicarbonyl[(1,2,3,4,5-η)-1-methyl-2,4-cyclopentadien-1-yl-](ferrocenylmethoxymethylene)manganese, 9CI

[Ferrocenyl(methoxy)carbene]methylcyclopentadienyldicarbonylmanganese

[35226-88-1]

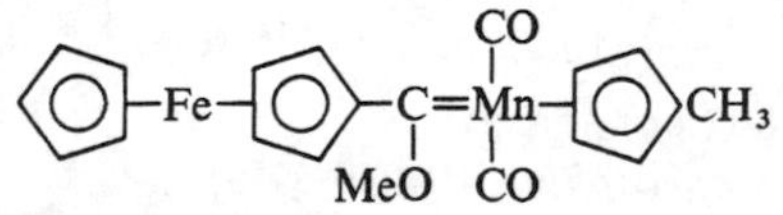

M 418.153

Red-brown cryst. (pentane). Mp 53-4°.

Connor, J.A. *et al*, *J. Chem. Soc., Dalton Trans.*, 1972, 1470 (*synth, pmr, ir, uv, ms*)

$C_{20}H_{30}Mn$ Mn-00119

Decamethylmanganocene, 10CI

Bis(η⁵-pentamethylcyclopentadienyl)manganese

[67506-86-9]

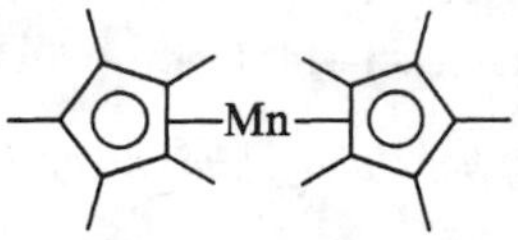

M 325.395

Red-orange cryst. (hexane); air-sensitive. Mp 292°.

Robbins, J.L. *et al*, *J. Am. Chem. Soc.*, 1979, **101**, 3853 (*synth, esr*)
Freyberg, D.P. *et al*, *J. Am. Chem. Soc.*, 1979, **101**, 892 (*cryst struct*)

Handle all chemicals with care

$C_{20}H_{30}Mn^{\oplus}$ Mn-00120

Decamethylmanganocenium

Bis(pentamethylcyclopentadienyl)manganese(1+)

[71163-80-9]

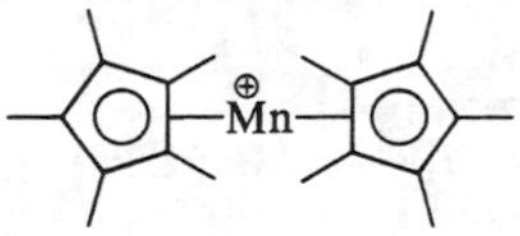

M 325.395 (ion)

Hexafluorophosphate: [71163-81-0]. $C_{20}H_{30}F_6MnP$ M 470.359
Red cryst.

Robbins, J.L. *et al*, *J. Am. Chem. Soc.*, 1979, **101**, 3853 (*synth, props*)

$C_{20}H_{30}Mn_2O_8P_2$ Mn-00121

Octacarbonylbis(triethylphosphine)dimanganese, 8CI

[15529-60-9]

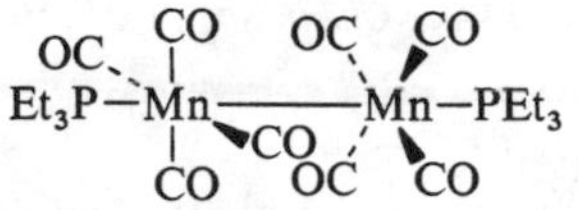

M 570.276

Yellow cryst. Mp 153-4°.

Lewis, T. *et al*, *Chem. Ind. (London)*, 1963, 1398 (*synth*)
Osborne, A.G. *et al*, *J. Chem. Soc.*, 1964, 634 (*synth*)
Lewis, J. *et al*, *J. Chem. Soc. (A)*, 1966, 845.
Bennett, M.J. *et al*, *J. Chem. Soc. (A)*, 1968, 75 (*cryst struct*)

$C_{20}Mn_4O_{20}U$ Mn-00122

Tetrakis(pentacarbonylmanganio)uranium, 8CI

[31135-22-5]

$U[Mn(CO)_5]_4$

M 1017.989

Struct. not yet established. Bright-orange solid. Sol. Me_2CO, THF. Mp 180° dec. Extremely air-sensitive; air oxidn. affords $Mn_2(CO)_{10}$, reaction with bromine yields $UBr_4 + Mn(CO)_5Br$. Reacts quickly with cold chloroform forming $Mn(CO)_5Cl$.

Bennett, R.L. *et al*, *J. Organomet. Chem.*, 1971, **26**, 355 (*synth, ir, ms*)

$C_{21}H_{42}F_3MnO_{11}P_2S$ Mn-00123

Dicarbonyl(trifluoromethanesulfonate-*O,O'*)bis(triisopropylphosphite-*P*)manganese

*Dicarbonyl(trifluoromethanesulfonate-*O,O'*)bis[tris(1-methylethyl)phosphite-*P*]manganese, 10CI*

[66562-90-1]

$Mn(O_3SCF_3)(CO)_2[P[OCH(CH_3)_2]_3]_2$

M 676.497

Orange-red cryst. (hexane at low temp.). Sol. hexane. Mp ~91° dec. Struct. not fully elucidated.

Stuhl, L.J. *et al*, *Inorg. Chem.*, 1978, **17**, 2148 (*synth, ir, nmr, props*)

$C_{22}H_{15}MnO_4P^{\ominus}$ Mn-00124

Tetracarbonyl(triphenylphosphine)manganate(1−), 10CI

[53418-18-1]

$(OC)_4MnPPh_3^{\ominus}$

M 429.270 (ion)

Na salt: [19457-74-0].
$C_{22}H_{15}MnNaO_4P$ M 452.260
Moisture-sensitive.

K salt: [67204-49-3].
$C_{22}H_{15}KMnO_4P$ M 468.368
No phys. props. reported.

Tetramethylammonium salt: [74365-56-3].
$C_{26}H_{27}MnNO_4P$ M 503.415
Moisture-sensitive orange cryst.

Bis(triphenylphosphine)iminium salt: [67047-38-5].
$C_{58}H_{45}MnNO_4P_3$ M 967.857
Orange cryst.

Müller, R. *et al*, *J. Organomet. Chem.*, 1978, **150**, C11 (*props*)
Wrighton, M.S. *et al*, *J. Am. Chem. Soc.*, 1978, **100**, 2701 (*synth*)
Darensbourg, M.Y. *et al*, *Inorg. Chem.*, 1979, **18**, 3286 (*props*)
Casey, C.P. *et al*, *J. Am. Chem. Soc.*, 1980, **102**, 2728 (*synth*)

$C_{22}H_{16}MnO_4P$ Mn-00125

Tetracarbonylhydro(triphenylphosphine)manganese, 9CI

Hydridotetracarbonyltriphenylphosphinemanganese

[16925-29-4]

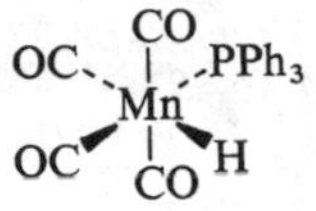

M 430.278

Cryst. (Et_2O/pet. ether). Mp 137-8°.

Hieber, W. *et al*, *Z. Anorg. Allg. Chem.*, 1962, **314**, 125 (*synth, ir, nmr*)
Hieber, W. *et al*, *Chem. Ber.*, 1965, **98**, 2933 (*synth*)
Booth, B.L. *et al*, *J. Chem. Soc. (A)*, 1966, 157 (*synth, props*)
Froelich, J.A. *et al*, *Inorg. Chem.*, 1977, **16**, 960 (*props*)

$C_{22}H_{16}Mn_2O_4$ Mn-00126

Tetracarbonylbis(η^5-2,4-cyclopentadien-1-yl)[μ-(phenylethenylidene)dimanganese, 10CI

[60429-52-9]

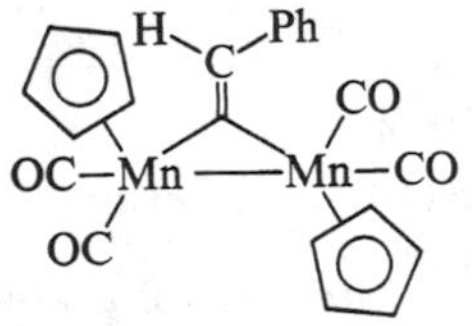

M 454.242

Violet cryst. Mp 144°.

Nesmeyanov, A.N. *et al*, *J. Organomet. Chem.*, 1976, **110**, C36 (*synth*)
Antonova, A.B. *et al*, *J. Organomet. Chem.*, 1977, **137**, 55 (*cryst struct*)

$C_{22}H_{32}B_4FeMn_2O_6S_2$ Mn-00127

Hexacarbonylbis[μ-[(1,2,3,4,5-η:1,2,3,4,5-η)-3,4-diethyl-2,5-dihydro-2,5-dimethyl-1,2,5-thiadiborole]](iron)-dimanganese, 10CI

[66921-70-8]

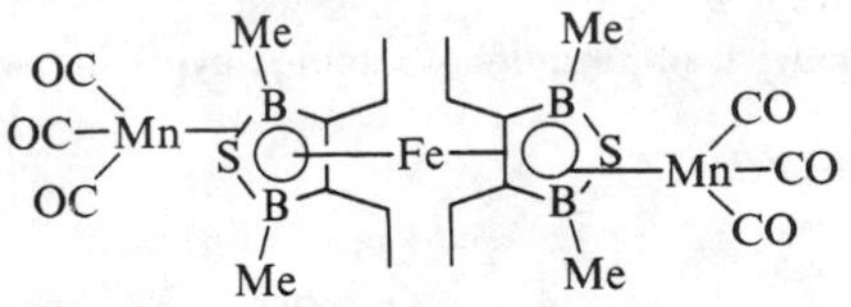

M 665.574

Orange-red cryst. (CS_2). $Bp_{0.1}$ 140-60° subl. Darkens >155°, dec. at 210°.

Siebert, W. *et al*, *Angew. Chem., Int. Ed. Engl.*, 1978, **17**, 527 (*synth, ir, ms, struct*)

$C_{23}H_{15}FeMnN_2O_7P$ Mn-00128

Dinitrosyl(pentacarbonylmanganio)(triphenylphosphine)iron, 8CI

[28939-17-5]

$(OC)_5MnFe(NO)_2PPh_3$

M 573.139

Characterised spectroscopically.

Hieber, W. *et al*, *Z. Naturforsch., B*, 1970, **25**, 663 (*synth, ir*)

$C_{23}H_{15}MnO_5P^{\oplus}$ Mn-00129

Pentacarbonyl(triphenylphosphine)manganese(1+)

$[Mn(CO)_5(PPh_3)]^{\oplus}$ (O_h)

M 457.280 (ion)

Hexafluorophosphate: [54039-57-5].
$C_{23}H_{15}F_6MnO_5P_2$ M 602.244
Yellow cryst.

Kruck, T. *et al*, *Chem. Ber.*, 1963, **96**, 3035 (*synth, props*)
Darensbourg, D.J. *et al*, *J. Organomet. Chem.*, 1975, **85**, 73 (*synth*)
Darensbourg, D.T. *et al*, *J. Am. Chem. Soc.*, 1977, **99**, 5940 (*synth*)
Froelich, J.A. *et al*, *Inorg. Chem.*, 1977, **16**, 960.
Behrens, H. *et al*, *J. Organomet. Chem.*, 1978, **159**, 201.
Usón, R. *et al*, *J. Chem. Soc., Dalton Trans.*, 1979, 996.

$C_{23}H_{20}IMnNOP$ Mn-00130

(η^5-2,4-Cyclopentadien-1-yl)iodonitrosyl(triphenylphosphine)manganese, 10CI

Iodonitrosyltriphenylphosphine(η^5-cyclopentadienyl)-manganese

[69120-57-6]

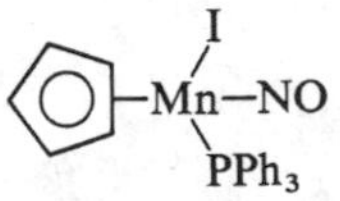

M 539.233

James, T.A. *et al*, *J. Chem. Soc. (A)*, 1970, 850 (*synth*)
Haines, B.W. *et al*, *Inorg. Chem.*, 1981, **20**, 650 (*props*)

$C_{23}H_{47}MnO_8P_2$ **Mn-00131**

Dicarbonyl(η^3-2-propenyl)bis(triisopropylphosphite-*P*)manganese

Dicarbonyl(η^3-2-propenyl)bis[tris(1-methylethyl)phosphite-P]manganese, 10CI. *Dicarbonyl(η^3-allyl)bis(triisopropylphosphite)manganese*

[66562-85-4]

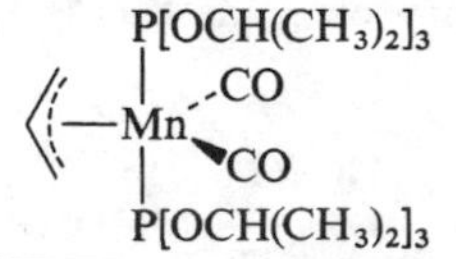

M 568.505

Catalyst for hydrogenation of alkenes. Air-sensitive yellow cryst. (hexane at −40°). Mp 104.6-106°.

Stuhl, L.S. *et al*, *Inorg. Chem.*, 1978, **17**, 2148 (*synth, ir, nmr, props*)

$C_{24}H_{20}MnNO_2P^{\oplus}$ **Mn-00132**

Carbonyl(η^5-2,4-cyclopentadien-1-yl)nitrosyl(triphenylphosphine)manganese(1+), 9CI

Carbonylnitrosyl(triphenylphosphine)(η^5-cyclopentadienyl)manganese(1+)

[34691-61-7]

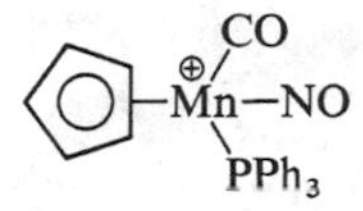

M 440.339 (ion)

Hexafluorophosphate: [32880-58-3]. $C_{24}H_{20}F_6MnNO_2P_2$ M 585.303 Characterised spectroscopically.

Brunner, H., *Z. Anorg. Allg. Chem.*, 1969, **368**, 120 (*props*)
Brunner, H. *et al*, *J. Organomet. Chem.*, 1970, **24**, C7 (*props*)
James, T.A. *et al*, *J. Chem. Soc. (A)*, 1970, 850 (*synth, props*)
Brunner, H. *et al*, *J. Organomet. Chem.*, 1973, **54**, 221; 1975, **87**, 223; 1976, **104**, 347 (*props*)

$C_{25}H_{23}MnNO_3P$ **Mn-00133**

η^5-2,4-Cyclopentadien-1-yl(methoxycarbonyl)nitrosyl(triphenylphosphine)manganese, 9CI

[35886-41-0]

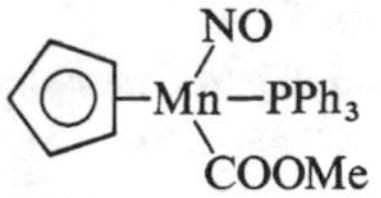

M 471.373

Known in opt. active forms.

(+)-*form*

$[\alpha]^{25}_{579}$+630°.

(−)-*form*

$[\alpha]^{25}_{579}$−645°.

(±)-*form*

Red cryst., air-sensitive in soln. Mod. sol. C_6H_6, CH_2Cl_2, spar. sol. hexane, MeOH. Dec. at 125°.

Brunner, H., *Z. Anorg. Allg. Chem.*, 1969, **368**, 120 (*synth*)
Brunner, H. *et al*, *Z. Naturforsch., B*, 1971, **26**, 1220.
Brunner, H., *Adv. Organomet. Chem.*, 1980, **18**, 151 (*rev*)

$C_{27}H_{15}Mn_2O_9P$ **Mn-00134**

Nonacarbonyl(triphenylphosphine)dimanganese, 8CI

[14592-26-8]

$(OC)_5Mn—Mn(CO)_4(PPh_3)$(*ax*-)

M 624.260

Orange cryst. (hexane). Mp 145° dec.

Ziegler, M.L. *et al*, *Chem. Ber.*, 1965, **98**, 2454 (*synth*)
Wawersik, H. *et al*, *J. Chem. Soc., Chem. Commun.*, 1966, 366 (*synth*)
Fischer, E.O. *et al*, *Chem. Ber.*, 1969, **102**, 2449 (*synth, props*)

$C_{28}H_{22}MnP$ **Mn-00135**

(η^5-2,4-Cyclopentadien-1-yl)[(1,2,3,4,5,6-η)-2,4,6-triphenylphosphorin]manganese, 10CI

[73752-27-9]

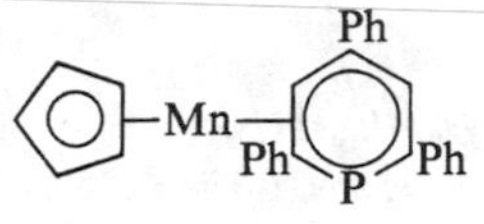

M 444.394

Cryst. (pentane). Mp 145-50° dec.

Nief, F. *et al*, *J. Organomet. Chem.*, 1980, **187**, 277 (*synth, props*)

$C_{30}H_{20}MnNO_3$ **Mn-00136**

Dicarbonylnitrosyl(η^4-tetraphenylcyclobutadiene)manganese

[83378-77-2]

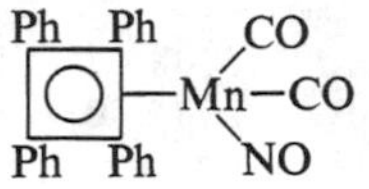

M 497.431

Orange cryst. (CH_2Cl_2/hexane). Mp 192° dec.

Rausch, M.D. *et al*, *Organometallics*, 1982, **1**, 1567 (*synth, struct*)

$C_{36}H_{48}Mn_2N_4$ **Mn-00137**

Bis[μ-[[2-(dimethylamino)phenyl]methyl-*C*:*C*,*N*]][[2-(dimethylamino)phenyl]methyl][[2-(dimethylamino)phenyl]methyl-*C*,*N*]dimanganese, 10CI

[65186-91-6]

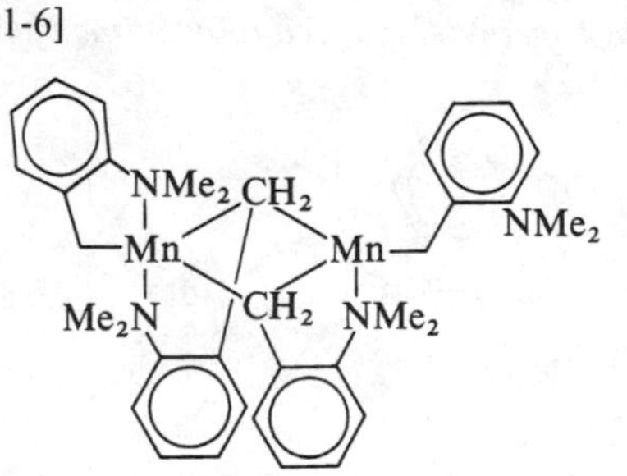

M 646.678

Manzer, L.E. *et al*, *J. Organomet. Chem.*, 1977, **139**, C34 (*synth, cryst struct*)

$C_{38}H_{30}MnNO_3P_2$ **Mn-00138**

Dicarbonylnitrosylbis(triphenylphosphine)manganese, 8CI

[14653-43-1]

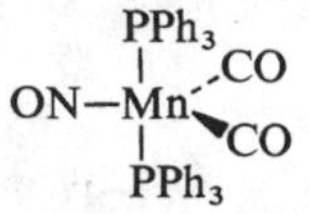

M 665.545

Orange cryst.

Hieber, W. *et al*, *Z. Anorg. Allg. Chem.*, 1962, **318**, 136 (*synth*)

Enemark, J.H. *et al*, *Inorg. Chem.*, 1967, **6**, 1575 (*struct*)
Keeton, D.P. *et al*, *Inorg. Chim. Acta*, 1972, **6**, 33 (*synth*)

$C_{39}H_{31}MnO_3P_2$ Mn-00139

Tricarbonylhydrobis(triphenylphosphine)manganese, 10CI
Hydridotricarbonylbis(triphenylphosphine)manganese
[17030-38-5]

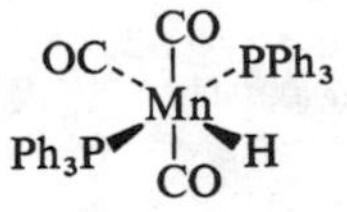

M 664.558
Cream-white cryst. (C_6H_6/pet. ether). Mp 237-9° dec.

Ugo, R. *et al*, *J. Organomet. Chem.*, 1967, **8**, 189 (*synth, ir, nmr*)
Hayakawa, H. *et al*, *Bull. Chem. Soc. Jpn.*, 1978, **51**, 2041 (*cryst struct*)

$C_{40}H_{30}MnO_4P_2^{\oplus}$ Mn-00140

Tetracarbonylbis(triphenylphosphine)manganese(1+)

$[Mn(CO)_4(PPh_3)_2]^{\oplus}$

M 691.560 (ion)
Hexafluorophosphate: [70083-76-0].
$C_{40}H_{30}F_6MnO_4P_3$ M 836.524
No phys. props. reported.

Kruck, T. *et al*, *Chem. Ber.*, 1963, **96**, 3035 (*synth, props*)
Green, M.L.H. *et al*, *J. Organomet. Chem.*, 1967, **8**, 511 (*synth*)
Behrens, H. *et al*, *J. Organomet. Chem.*, 1972, **34**, 367 (*props*)
Brink, R.W. *et al*, *Inorg. Chem.*, 1973, **12**, 1067 (*synth, ir*)
Behrens, H. *et al*, *J. Organomet. Chem.*, 1978, **159**, 201 (*props*)

$C_{40}H_{52}Mn_2$ Mn-00141

Bis[μ-(2-methyl-2-phenylpropyl)]bis(2-methyl-2-phenylpropyl)dimanganese, 9CI
Tetrakis(2-methyl-2-phenylpropyl)dimanganese
[62450-90-2]

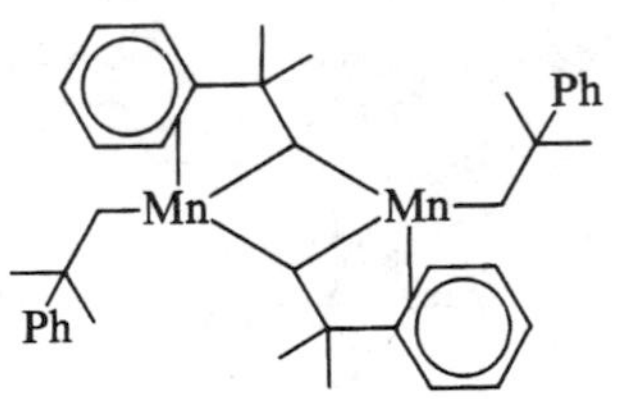

M 642.727

Anderson, R.A. *et al*, *J. Chem. Soc., Dalton Trans.*, 1976, 2204.

$C_{41}H_{35}MnNO_7P_2^{\oplus}$ Mn-00142

(η^5-2,4-Cyclopentadien-1-yl)nitrosylbis(triphenyl phosphite-*P*)manganese(1+)
Nitrosylbis(triphenyl phosphite)(η^5-cyclopentadienyl)-manganese(1+)

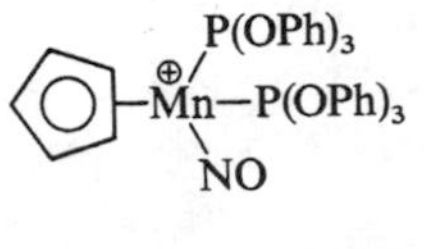

M 770.616 (ion)
Hexafluorophosphate: [36223-58-2].
$C_{41}H_{35}F_6MnNO_7P_3$ M 915.580
Pale-yellow cryst. Mp 134-7°.

James, T.A. *et al*, *J. Chem. Soc. (A)*, 1970, 850 (*synth, props*)

$C_{48}H_{35}AsMnO_2P$ Mn-00143

Dicarbonyl[(1,2,3,4,5-η)-2,3,4,5-tetraphenyl-1*H*-arsol-1-yl-](triphenylphosphine)manganese, 10CI
[72415-82-8]

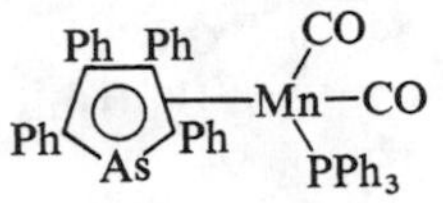

M 804.637
Yellow cryst. (hexane/$CHCl_3$). Mp 244-50°.

Abel, E.W. *et al*, *J. Chem. Soc., Chem. Commun.*, 1973, 258 (*synth, cryst struct*)
Abel, E.W. *et al*, *J. Chem. Soc., Dalton Trans.*, 1979, 814.
Abel, E.W. *et al*, *J. Chem. Soc., Dalton Trans.*, 1979, 1552 (*synth, props*)

$C_{54}H_{48}MnO_2P_4^{\oplus}$ Mn-00144

Dicarbonylbis[1,2-ethanediylbis[diphenylphosphine]-*P,P'*]-manganese(1+)

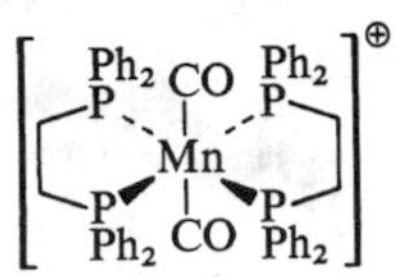

M 907.805 (ion)
Air-stable.

trans-form

Perchlorate: [14836-77-2].
$C_{54}H_{48}ClMnO_6P_4$ M 1007.256
Yellow cryst. (MeOH). Sol. org. solvs. except Et_2O, pet. ether.
Chloride:
$C_{54}H_{48}ClMnO_2P_4$ M 943.258
Yellow cryst. + 2MeOH (MeOH) or + $2CHCl_3$ ($CHCl_3$). Sol. org. solvs., except pet. ether, Et_2O.

Saeco, A., *Gazz. Chim. Ital.*, 1963, **93**, 698 (*synth*)
Osborne, A.G. *et al*, *J. Chem. Soc.*, 1965, 700 (*synth*)
Wimmer, F.L. *et al*, *Inorg. Chem.*, 1974, **13**, 1617 (*props*)

$C_{55}H_{44}Mn_2O_5P_4$ Mn-00145

μ-Carbonyltetracarbonylbis[μ-[methylenebis[diphenylphosphine]-*P,P'*]]dimanganese, 10CI
[56665-73-7]

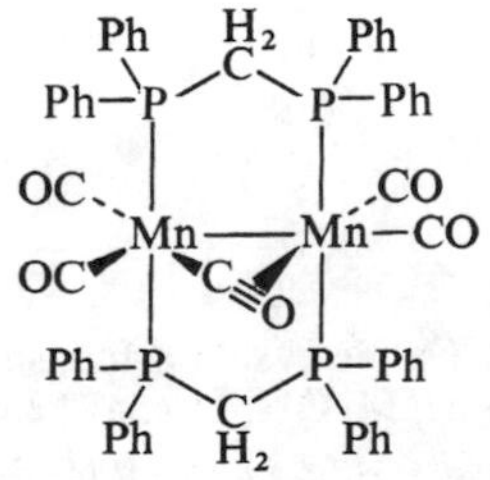

M 1018.721
Red diamagnetic cryst.

Colton, R. *et al*, *Aust. J. Chem.*, 1975, **28**, 1673 (*synth, ir*)
Commons, C.J. *et al*, *Aust. J. Chem.*, 1975, **28**, 1663 (*struct*)
Caulton, K.G. *et al*, *J. Organomet. Chem.*, 1976, **114**, C11.
Caulton, K.G. *et al*, *Organometallics*, 1982, **1**, 274 (*nmr*)

Mo Molybdenum

M. Green

Molybdène (Fr.), Molybdän (Ger.), Molibdeno (Sp., Ital.), Молибден (Molibden) (Russ.), モリブデン (Japan.)

Atomic Number. 42

Atomic Weight. 95.94

Electronic Configuration. [Kr] $4d^5\ 5s^1$

Oxidation States. −4, −2, 0, +2, +4

Coordination Number. This ranges from 4 in $[Mo(CO)_4]^{4-}$ upwards.

Colour. Variable including colourless, yellow, red, green and blue.

Availability. Readily available starting materials include $Mo(CO)_6$ and anhydrous $MoCl_5$.

Handling. See under Chromium.

Toxicity. Molybdenum compounds, except for the carbonyl complexes, are generally not regarded as highly toxic.

Isotopic Abundance. ^{92}Mo, 15.86%; ^{94}Mo, 9.12%; ^{95}Mo, 15.78%; ^{96}Mo, 16.50%; ^{97}Mo, 9.45%; ^{98}Mo, 23.75%; ^{100}Mo, 9.62%. Thirteen radioisotopes are also known. ^{99}Mo ($t_{1/2} \sim$ 66h) is commonly used in tracer work.

Spectroscopy. Both ^{95}Mo and ^{97}Mo have $I = \frac{5}{2}$ and although fundamental studies have been carried out, few chemical shift data are available.

Analysis. For standard methods see *Scott's Standard Methods of Chemical Analysis*, Furman, N. H. Ed., 6th Edn, Van Nostrand, 1962.

References. General reviews are listed in the introduction to the *Sourcebook*.

$[Mo(CO)_4]^{\ominus\ominus\ominus\ominus}$

Mo-00001

Mo-00002

Mo-00003

$(OC)_5MoI^{\ominus}$

Mo-00004

Mo-00005

$[MoCN(CO)_5]^{\ominus}$

Mo-00006

$Mo(CO)_6$ (O_h)

Mo-00007

Mo-00008

Mo-00009

Mo-00010

L = Br

Mo-00011

As Mo-00011 with
L = Cl

Mo-00012

As Mo-00011 with
L = I

Mo-00013

Mo-00014

As Mo-00011 with
L = Tl

Mo-00015

As Mo-00011 with
L = H

Mo-00016

$(OC)_5Mo{=}C(OMe)CH_3$

Mo-00017

Mo-00018

Mo-00019

Mo-00020

Mo-00021

Mo-00022

Mo-00023

Mo-00024

Mo-00025

$(H_3CCN)_3Mo(CO)_3$

Mo-00026

Mo-00027

Mo-00028

Mo-00029

Mo-00030

Mo-00031

Mo-00032

Mo-00033

Mo-00034

Mo-00035

Mo-00036

Mo-00037

Mo-00038

Mo-00039

$(C_5H_5)_2Mo(H)Li$

Mo-00040

Mo-00041

Mo-00042

Mo-00043

Mo-00044

$[(OC)_5Mo{-}Mo(CO)_5]^{\ominus\ominus}$

Mo-00045

Mo-00046

Mo-00047

Mo-00048

Mo-00049

Mo-00050

Mo-00051

Mo-00052

Mo-00053

Mo-00054

Mo-00055

As Mo-00011 with
L = $GeMe_3$

Mo-00056

Mo-00057

As Mo-00011 with
L = $SiMe_3$

Mo-00058

As Mo-00011 with
L = $SnMe_3$

Mo-00059

As Mo-00011 with
L = $Co(CO)_4$

Mo-00060

Mo-00061

Mo-00062

Mo-00063

Mo-00064

Mo-00065

Mo-00066

Mo-00067

Mo-00068

Mo-00069

Mo-00070

$Mo(S_2CNEt_2)_2(CO)_2$

Mo-00071

Mo-00072

X = Cl

Mo-00073

As Mo-00073 with
X = I

Mo-00074

Mo-00075

Mo-00076

Mo-00077

Mo-00078

R = H

Mo-00079

Mo-00080

X = Cl

Mo-00081

Mo-00082

R = R' = CH_3

Mo-00083

cis-form

Mo-00084

Mo-00085

Mo-00086

Mo-00087

Mo-00088

Mo-00089

Mo-00090

Mo-00091

Mo-00092

Mo-00093

Mo-00094

Mo-00095

Mo-00096

Mo-00097

Mo-00098

Mo-00099

As Mo-00081 with
X = OMe

Mo-00100

$Ph_2C{=}Mo(CO)_5$

Mo-00101

Mo-00102

Mo-00103

Mo-00104

Mo-00105

Mo-00106

Mo-00107

As Mo-00109 with
R = OMe

Mo-00108

R = Me

Mo-00109

$[(Me_3Si)_2CH]_2Pb{-}Mo(CO)_5$

Mo-00110

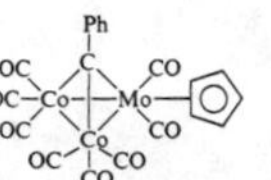

Mo-00111

Mo-00112

Mo-00113

Mo-00114

Mo-00115

Mo-00116

Mo-00117

Mo-00118

$(OC)_5MoPPh_3$ (O_h)

Mo-00119

Mo-00120

As Mo-00083 with
R = Ph, R′ = CH_3

Mo-00121

Mo-00122

Mo-00123

As Mo-00079 with
R = CH_3

Mo-00124

Mo-00125

Mo-00126

trans-form

Mo-00127

Mo-00128

Mo-00129

Mo-00130

$(OC)_5Mo{=}C(OEt)SiPh_3$

Mo-00131

Mo-00132

Mo-00133

$[Mo(CO)_3(NO)(Ph_2PCH_2CH_2PPh_2)]^{\oplus}$

Mo-00134

Mo-00135

Mo-00136

Mo-00137

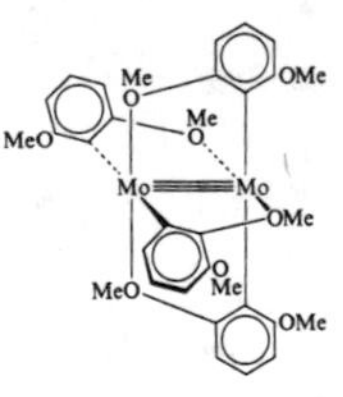

Mo-00138

Mo-00139

As Mo-00083 with
R = R′ = Ph

Mo-00140

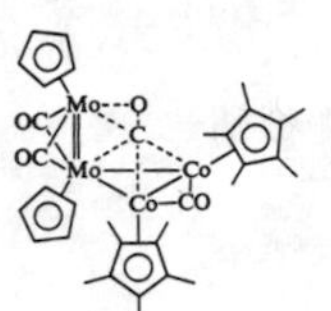

Mo-00141

Mo-00142

Mo-00143

Mo-00144

Mo-00145

Mo-00146

$C_4MoO_4^{\ominus\ominus\ominus\ominus}$ **Mo-00001**

Tetracarbonylmolybdate(4–)

$[Mo(CO)_4]^{\ominus\ominus\ominus\ominus}$

M 207.982 (ion)

Tetra-Na salt: [67202-63-5].
$C_4MoNa_4O_4$ M 299.941
Pale-orange solid. Insol. THF, HMPA, liq. NH_3.
▷Extremely pyrophoric, reacting explosively with fluorolube, water or air. Deflagrates on percussion

Lin, J.T. *et al*, *J. Am. Chem. Soc.*, 1983, **105**, 2296 (*synth*)

$C_5H_5ClMoN_2O_2$ **Mo-00002**

Chloro(η^5-2,4-cyclopentadien-1-yl)dinitrosylmolybdenum, 10CI

[12305-00-9]

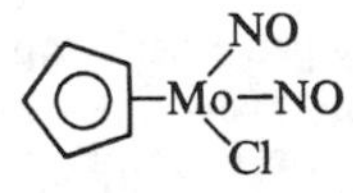

M 256.500

Green cryst. Mp 116°. $Bp_{0.005}$ 40-50° subl.

Hoyano, J.K. *et al*, *J. Chem. Soc., Dalton Trans.*, 1975, 1022.
Inorg. Synth., 1978, **18**, 126 (*synth*)
Botto, R.E. *et al*, *Inorg. Chem.*, 1979, **18**, 2049 (*nmr*)
Legzdins, P. *et al*, *Inorg. Chem.*, 1979, **18**, 1250.

$C_5H_5Cl_2MoNO$ **Mo-00003**

Dichloro(η^5-2,4-cyclopentadien-1-yl)nitrosylmolybdenum

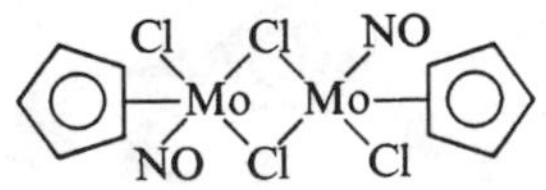

M 261.947

Dimeric.

Dimer: [41395-41-9]. *Di-μ-chlorodichlorobis(η^5-2,4-cyclopentadien-1-yl)dinitrosyldimolybdenum, 10CI. Tetrachlorobis(cyclopentadienyl)dinitrosyldimolybdenum.*
$C_{10}H_{10}Cl_4Mo_2N_2O_2$ M 523.893

Inorg. Synth., 1976, **16**, 24 (*synth*)
Inorg. Synth., 1978, **18**, 126 (*synth*)

$C_5IMoO_5^{\ominus}$ **Mo-00004**

Pentacarbonyliodomolybdate(1–), 9CI

$(OC)_5MoI^{\ominus}$

M 362.897 (ion)

Tetraethylammonium salt: [14781-00-1].
$C_{13}H_{20}IMoNO_5$ M 493.149
Yellow cryst.

Bis(triphenylphosphine)iminium salt: [39004-14-3].
$C_{41}H_{30}IMoNO_5P_2$ M 901.484
Yellow cryst.

Abel, E.W. *et al*, *J. Chem. Soc.*, 1963, 2068 (*synth*)
Ruff, J.K. *et al*, *Synth. React. Inorg. Metal-Org. Chem.*, 1972, **1**, 215 (*synth*)
Bond, A.M. *et al*, *Inorg. Chem.*, 1974, **13**, 602 (*props*)

$C_6H_{18}MoO_6P$ **Mo-00005**

Diperoxooxohexamethylphosphoramidomolybdenum

(Hexamethylphosphoric triamide-O)oxodiperoxymolybdenum

[25377-12-2]

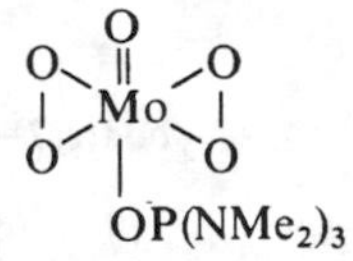

$OP(NMe_2)_3$

M 313.118

Reagent for efficient stereospecific epoxidn. of olefins, also for enolate hydroxylation and for oxidn. of alkyl bromides and nitriles. Yellow-orange needles. Mp 101-2° dec.

Himoun, H. *et al*, *Tetrahedron*, 1970, **26**, 37 (*synth*)
Sharpless, K.B. *et al*, *J. Am. Chem. Soc.*, 1972, **94**, 295 (*use*)
Schmitt, G. *et al*, *J. Organomet. Chem.*, 1978, **152**, 271 (*use*)
Vedjs, E. *et al*, *J. Org. Chem.*, 1978, **43**, 188 (*use*)
Matlin, S.A. *et al*, *J. Chem. Soc., Perkin Trans. 1*, 1979, 2481 (*synth, use*)
Fieser, M. *et al*, *Reagents for Organic Synthesis*, Wiley, 1967-83, **8**, 206 (*use*)

$C_6MoNO_5^{\ominus}$ **Mo-00006**

Pentacarbonyl(cyano-*C*)molybdate(1–), 8CI

$[MoCN(CO)_5]^{\ominus}$

M 262.010 (ion)

Bis(triphenylphosphine)iminium salt: [39048-39-0].
$C_{42}H_{30}MoN_2P_2O_5$ M 800.597
White cryst.

King, R.B., *Inorg. Chem.*, 1967, **6**, 25 (*synth, props*)
Ruff, J.K. *et al*, *Synth. React. Inorg. Metal-Org. Chem.*, 1971, **1**, 215.

C_6MoO_6 **Mo-00007**

Molybdenum carbonyl, 10CI

Molybdenum hexacarbonyl. Hexacarbonylmolybdenum

[13939-06-5]

$Mo(CO)_6$ (O_h)

M 264.002

Important starting material. Colourless, odourless, air-stable, diamagnetic solid. Sol. org. solvs.

Junk, G.A. *et al*, *Z. Naturforsch., B*, 1968, **23**, 1 (*ms*)
Jones, L.H. *et al*, *Inorg. Chem.*, 1969, **11**, 2349 (*ir*)
Lloyd, D.R. *et al*, *J. Chem. Soc., Faraday Trans. 2*, 1973, **69**, 1659 (*pe*)
Mann, B.E., *J. Chem. Soc., Dalton Trans.*, 1973, 2012 (*nmr*)
Hagan, A.P. *et al*, *Inorg. Chem.*, 1978, **17**, 1369 (*synth*)
Michels, G.D. *et al*, *Inorg. Chem.*, 1980, **19**, 479.

$C_8Cl_4Mo_2O_8$ **Mo-00008**

Octacarbonyltetrachlorodimolybdenum, 9CI

Octacarbonyldi-μ-chlorodichlorodimolybdenum

[12655-17-3]

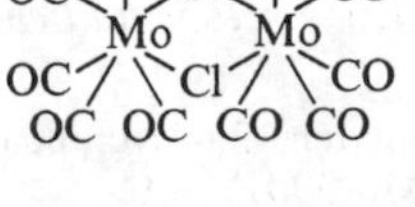

M 557.775

Yellow powder. Sol. CH_2Cl_2, $CHCl_3$ (part. dec.). Thermally unstable. ir ν_{CO} 2100w, 2050m, 1980s, 1956w cm^{-1} (nujol).

Monomer: Tetracarbonyldichloromolybdenum.
$C_4Cl_2MoO_4$ M 278.888
Unknown.

Colton, R. *et al*, *Aust. J. Chem.*, 1966, **19**, 1143 (*synth*, *ir*)
Davis, R. *et al*, *J. Chem. Soc., Dalton Trans.*, 1972, 508 (*props*)

$C_8H_2Mo_2O_8^{\ominus\ominus}$ Mo-00009

Octacarbonyl-μ-dihydrodimolybdate(2−)

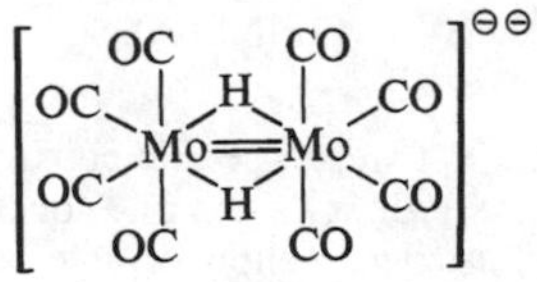

M 417.979 (ion)
Tetraethylammonium salt: [84850-68-0].
$C_{24}H_{42}Mo_2N_2O_8$ M 678.484
Orange air-sensitive cryst.

Ellis, J.E. *et al*, *J. Am. Chem. Soc.*, 1983, **105**, 2296 (*synth*)

$C_8H_5AsMoO_3$ Mo-00010

(η^6-Arsenin)tricarbonylmolybdenum, 10CI
Tricarbonyl-η^6-arsabenzenemolybdenum. Arsabenzenetricarbonylmolybdenum
[65198-26-7]

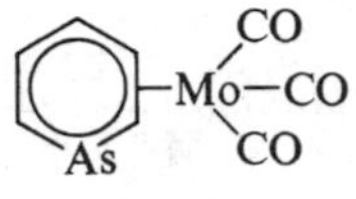

M 319.987
Air-stable red cryst.

Ashe, A.J. *et al*, *J. Am. Chem. Soc.*, 1977, **99**, 8099 (*synth*, *props*)

$C_8H_5BrMoO_3$ Mo-00011

Bromotricarbonyl(η^5-2,4-cyclopentadien-1-yl)molybdenum, 9CI
Cyclopentadienyltricarbonylmolybdenum bromide
[12079-79-7]

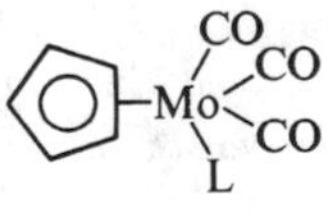

L = Br

M 324.970
Red cryst. (CH_2Cl_2). Mp 150-1°.

Piper, T.S. *et al*, *J. Inorg. Nucl. Chem.*, 1956, **3**, 104 (*synth*)
Burkett, A.R. *et al*, *J. Organomet. Chem.*, 1974, **67**, 67 (*synth*, *uv*, *pmr*, *ir*)
King, R.B., *Coord. Chem. Rev.*, 1976, **20**, 155.

$C_8H_5ClMoO_3$ Mo-00012

Tricarbonylchloro(η^5-2,4-cyclopentadien-1-yl)molybdenum, 9CI
Chlorotricarbonyl(η^5-cyclopentadienyl)molybdenum
[12128-23-3]
As Bromotricarbonyl(η^5-2,4-cyclopentadien-1-yl)-molybdenum, Mo-00011 with

L = Cl

M 280.519
Orange-red cryst. Dec. at 145° without melting.

Piper, T.S. *et al*, *J. Inorg. Nucl. Chem.*, 1956, **3**, 104 (*synth*)
Parker, D.J. *et al*, *J. Chem. Soc.* (*A*), 1970, 480, 1382 (*ir*)
Wrighton, M.S. *et al*, *J. Am. Chem. Soc.*, 1975, **97**, 4246 (*synth*)
King, R.B., *Coord. Chem. Rev.*, 1976, **20**, 155 (*rev*)
Todd, L.J. *et al*, *J. Organomet. Chem.*, 1978, **154**, 151 (*cmr*)

$C_8H_5IMoO_3$ Mo-00013

Tricarbonyl(η^5-2,4-cyclopentadien-1-yl)iodomolybdenum, 9CI
Iodotricarbonyl(η^5-cyclopentadienyl)molybdenum
[12287-61-5]
As Bromotricarbonyl(η^5-2,4-cyclopentadien-1-yl)-molybdenum, Mo-00011 with

L = I

M 371.970
Red cryst. (CH_2Cl_2/pet. ether). Mp 134-134.5°.

Piper, T.S. *et al*, *J. Inorg. Nucl. Chem.*, 1956, **3**, 104 (*synth*)
White, C. *et al*, *J. Chem. Soc.* (*A*), 1971, 940.
King, R.B., *Coord. Chem. Rev.*, 1976, **20**, 155.
St. Laurent, J.C.T.R.B. *et al*, *J. Organomet. Chem.*, 1978, **153**, C19.
Todd, L.J. *et al*, *J. Organomet. Chem.*, 1978, **154**, 151 (*struct*)
Colville, N.J., *J. Organomet. Chem.*, 1980, **190**, C84.

$C_8H_5MoO_3^{\ominus}$ Mo-00014

Tricarbonyl(η^5-2,4-cyclopentadien-1-yl)molybdate(1−), 9CI
[12126-18-0]

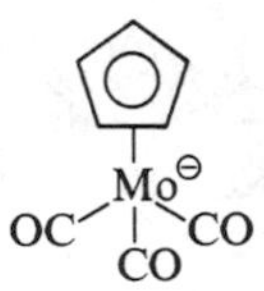

M 245.066 (ion)
Li salt: [68550-41-4].
$C_8H_5LiMoO_3$ M 252.007
Characterised spectroscopically.
Tetrabutylammonium salt:
$C_{24}H_{41}MoNO_3$ M 487.533
Cryst. Characterised crystallographically.

King, R.B. *et al*, *Inorg. Chem.*, 1965, **4**, 475.
Ellis, J.E. *et al*, *J. Organomet. Chem.*, 1975, **99**, 263.
King, R.B., *Coord. Chem. Rev.*, 1976, **20**, 155.
Crotty, D.E. *et al*, *Inorg. Chem.*, 1977, **16**, 920 (*cryst struct*)
Gladysz, J.A. *et al*, *Inorg. Chem.*, 1979, **18**, 553 (*synth*)

$C_8H_5MoO_3Tl$ Mo-00015

Tricarbonyl(η^5-2,4-cyclopentadien-1-yl)thalliummolybdenum, 10CI
Thalliumtricarbonyl(η^5-cyclopentadienyl)molybdenum
[52720-93-1]
As Bromotricarbonyl(η^5-2,4-cyclopentadien-1-yl)-molybdenum, Mo-00011 with

L = Tl

M 449.449
Light-yellow cryst. Mp 230° dec.

Burlitch, J.M. *et al*, *J. Chem. Soc., Dalton Trans.*, 1974, 828 (*synth*, *ir*)

$C_8H_6MoO_3$ Mo-00016

Tricarbonyl(η^5-2,4-cyclopentadien-1-yl)hydromolybdenum, 10CI

Hydridotricarbonyl(η^5-cyclopentadienyl)molybdenum

[12176-06-6]

As Bromotricarbonyl(η^5-2,4-cyclopentadien-1-yl)-molybdenum, Mo-00011 with

L = H

M 246.074

Yellow cryst.; air-sensitive, darkens immediately on exposure to air. $Bp_{0.1}$ 50° subl.

Inorg. Synth., 1963, **7**, 99 (*synth*)
Dub, M., *Organometallic Compounds*, Springer-Verlag, Berlin, 2nd Ed., 1966, Vol. 1.
Treichel, P.M. *et al*, *Inorg. Chem.*, 1967, **6**, 1328 (*synth*)
Keppie, S.A. *et al*, *J. Organomet. Chem.*, 1969, **19**, P5 (*synth*)
Faller, J.W. *et al*, *J. Am. Chem. Soc.*, 1970, **92**, 5852.
Kalck, P. *et al*, *J. Organomet. Chem.*, 1970, **24**, 445.

$C_8H_6MoO_6$ Mo-00017

Pentacarbonyl(1-methoxyethylidene)molybdenum, 10CI

Pentacarbonyl(methoxymethylcarbene)molybdenum.
Pentacarbonyl[methoxy(methyl)methylene]molybdenum

[12365-46-7]

$(OC)_5Mo{=}C(OMe)CH_3$

M 294.072

Unstable.

Fischer, E.O. *et al*, *Chem. Ber.*, 1967, **100**, 2445 (*synth*, *props*)

$C_8H_9MoN_2O_3^{\oplus}$ Mo-00018

Tricarbonyl(η^5-2,4-cyclopentadien-1-yl)(hydrazine-*N*)molybdenum(1+), 10CI

[Cp–Mo(CO)(CO)(CO)(NH₂NH₂)]⊕

M 277.111 (ion)

Chloride: [68643-99-2].
$C_8H_9ClMoN_2O_3$ M 312.564
Unstable.

Fischer, E.O. *et al*, *J. Organomet. Chem.*, 1964, **2**, 230 (*synth*)
Nolte, M.J. *et al*, *J. Chem. Soc., Dalton Trans.*, 1978, 932 (*synth*, *props*)

$C_8H_{10}IMoNO$ Mo-00019

(η^5-2,4-Cyclopentadien-1-yl)iodonitrosyl(η^3-2-propenyl)-molybdenum, 10CI

η^3-Allyliodonitrosyl(η^5-cyclopentadienyl)molybdenum

[66979-94-0]

NO, Mo–I

M 359.018

Faller, J.W. *et al*, *J. Organomet. Chem.*, 1980, **187**, 227 (*struct*, *synth*)

$C_8H_{24}Mo_2^{\ominus\ominus\ominus\ominus}$ Mo-00020

Octamethyldimolybdenum(4−), 9CI

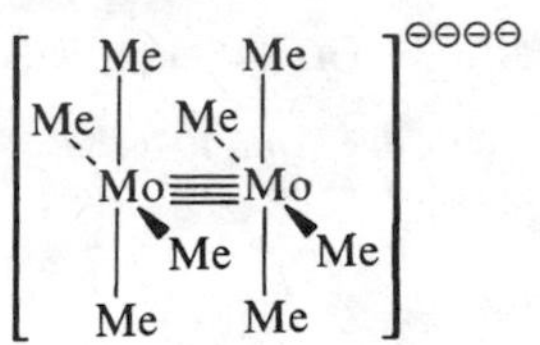

M 312.158 (ion)

Tetra-Li salt, tetrakis-THF complex: [53307-60-1].
$C_{24}H_{56}Li_4Mo_2O_4$ M 628.348
Mo—Mo 2.147(3) Å. Diamagnetic.

Heyn, B. *et al*, *Z. Chem.*, 1972, **12**, 338 (*synth*)
Cotton, F.A. *et al*, *J. Am. Chem. Soc.*, 1974, **96**, 3824 (*synth*, *cryst struct*)

$C_8I_2Mo_2O_8$ Mo-00021

Octacarbonyldi-μ-iododimolybdenum, 9CI

[55463-48-4]

(OC)₄Mo(μ-I)₂Mo(CO)₄

M 669.772

Black cryst. (hexane). Mo—Mo bond 3.16Å.

Schmid, C. *et al*, *Chem. Ber.*, 1975, **108**, 260 (*synth*)
Boese, R. *et al*, *Acta Crystallogr., Sect. B*, 1976, **32**, 582 (*cryst struct*)

$C_9H_5MoO_4^{\oplus}$ Mo-00022

Tetracarbonyl(η^5-2,4-cyclopentadien-1-yl)molybdenum(1+), 10CI

[Cp–Mo(CO)₄]⊕

M 273.076 (ion)

Air-stable.

Hexafluorophosphate: [41618-11-5].
$C_9H_5F_6MoO_4P$ M 418.040
Yellow cryst.

Tetrafluoroborate: [67251-59-6].
$C_9H_5BF_4MoO_4$ M 359.880
Yellow cryst.

Treichel, P.M. *et al*, *Inorg. Chem.*, 1967, **6**, 1328; 1971, **10**, 1183 (*props*)
Abel, E.W. *et al*, *Adv. Organomet. Chem.*, 1970, **8**, 117 (*synth*, *props*)
Behrens, H. *et al*, *Z. Anorg. Allg. Chem.*, 1977, **438**, 61 (*props*)
Pfister, A. *et al*, *Z. Anorg. Allg. Chem.*, 1977, **438**, 53 (*props*)

$C_9H_6MoO_3$ Mo-00023

η^6-Benzenetricarbonylmolybdenum, 10CI

Tricarbonyl(η^6-benzene)molybdenum

[12287-81-9]

C₆H₆–Mo(CO)₃

M 258.085

Yellow cryst., air-stable in solid state. Mp 120-5°.

Fischer, E.O. *et al*, *Chem. Ber.*, 1958, **91**, 2763 (*synth*)

Mann, B.E., *J. Chem. Soc., Dalton Trans.*, 1973, 2012 (*nmr*)
Kane-Maguire, L.A.P. *et al*, *J. Chem. Soc., Dalton Trans.*, 1974, 428.
Nesmeyanov, A.N. *et al*, *J. Organomet. Chem.*, 1975, **102**, 185 (*synth*)
Barbeau, C. *et al*, *Can. J. Chem.*, 1976, **54**, 1612 (*ir*)
Muetterties, E.L. *et al*, *J. Organomet. Chem.*, 1979, **178**, 197.

$C_9H_7ClMoO_2$ Mo-00024

Dicarbonylchloro(η^7-cycloheptatrienylium)molybdenum

Dicarbonylchlorotropyliummolybdenum

[55411-20-6]

M 278.546

Dark-green solid. Mp ca. 185°.

King, R.B. *et al*, *Inorg. Chem.*, 1964, **3**, 785 (*synth*)
Beall, T.W. *et al*, *Inorg. Chem.*, 1972, **11**, 915 (*props*)
Isaacs, E.E. *et al*, *Can. J. Chem.*, 1975, **53**, 975 (*props*)
Ziegler, M.L. *et al*, *Z. Naturforsch., B*, 1975, **30**, 26 (*cryst struct*)
Bochmann, M. *et al*, *J. Chem. Soc., Dalton Trans.*, 1977, 714 (*synth*)
Bowerbank, R. *et al*, *J. Chem. Soc., Chem. Commun.*, 1977, 245 (*props*)

$C_9H_8MoO_3$ Mo-00025

Tricarbonyl(η^5-2,4-cyclopentadien-1-yl)methylmolybdenum, 9CI

Cyclopentadienyl(methyl)tricarbonylmolybdenum

[12082-25-6]

M 260.100

Yellow cryst. by subl. *in vacuo*.

Piper, T.S. *et al*, *J. Inorg. Nucl. Chem.*, 1956, **3**, 104.
King, R.B. *et al*, *Can. J. Chem.*, 1969, **47**, 2959 (*ir*)
Samuel, E. *et al*, *J. Organomet. Chem.*, 1979, **172**, 309 (*uv*)

$C_9H_9MoN_3O_3$ Mo-00026

Tris(acetonitrile)tricarbonylmolybdenum, 9CI, 8CI

Tricarbonyltris(acetonitrile)molybdenum

[15038-48-9]

$(H_3CCN)_3Mo(CO)_3$

M 303.128

Bright-yellow cryst. (MeCN/hexane). Mp 140° dec.

Tate, D.P. *et al*, *Inorg. Chem.*, 1962, **1**, 433 (*synth*)
Werner, H. *et al*, *Helv. Chim. Acta*, 1970, **53**, 2002 (*synth, props, ir*)
Graham, J.M., *J. Organomet. Chem.*, 1974, **77**, 247.

$C_9H_{10}MoNO_2^{\oplus}$ Mo-00027

Carbonyl(η^5-2,4-cyclopentadien-1-yl)nitrosyl(η^3-2-propenyl)-molybdenum(1+), 10CI

η^3-Allylcarbonylcyclopentadienylnitrosylmolybdenum-(1+)

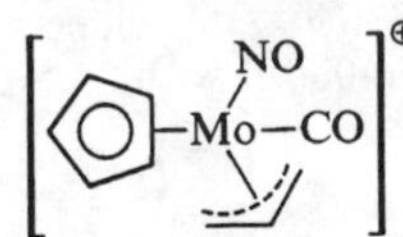

M 260.124 (ion)

Hexafluorophosphate: [54438-52-7].
$C_9H_{10}F_6MoNO_2P$ M 405.088
Yellow cryst.

Faller, J.W. *et al*, *Ann. N.Y. Acad. Sci.*, 1977, **295**, 186 (*synth, props*)
Adams, R.D. *et al*, *J. Am. Chem. Soc.*, 1979, **101**, 2570 (*props*)
McCleverty, J.A. *et al*, *Transition Met. Chem.*, 1979, **4**, 273 (*synth, nmr*)

$C_9H_{11}ClMo$ Mo-00028

(η^6-Benzene)chloro(η^3-2-propenyl)molybdenum

Allylbenzenechloromolybdenum

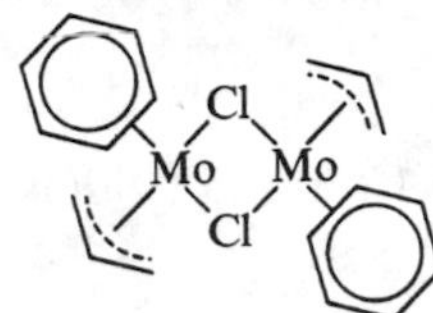

M 250.579

Dimeric.

Dimer: [12701-35-8]. *Bis(η^6-benzene)di-μ-chloro-bis(η^3-2-propenyl)dimolybdenum,* 9CI. *Bis(allyl)-bis(η^6-benzene)di-μ-chlorodimolybdenum. Diallyl-bis-η^6-benzenedichloromolybdenum.*
$C_{18}H_{22}Cl_2Mo_2$ M 501.158
Purple cryst., sensitive to oxygen.

Green, M.L.H. *et al*, *J. Chem. Soc., Dalton Trans.*, 1974, 311; 1976, 213 (*synth*)
Inorg. Synth., 1977, **17**, 57 (*synth*)

$C_9H_{11}ClMoN_2O_2$ Mo-00029

Bis(acetonitrile)dicarbonylchloro(η^3-2-propenyl)molybdenum, 9CI

Bis(acetonitrile)η^3-allyldicarbonylchloromolybdenum

[33221-75-9]

M 310.591

Yellow cryst.

Hayter, R.G., *J. Organomet. Chem.*, 1968, **13**, P1 (*synth*)
Tom Dieck, H. *et al*, *J. Organomet. Chem.*, 1968, **14**, 375 (*synth, ir, nmr*)
Brisdon, B.J., *J. Organomet. Chem.*, 1977, **125**, 225.

$C_9H_{11}IMoN_2O_2$ Mo-00030

Bis(acetonitrile)dicarbonyliodo(η^3-2-propenyl)molybdenum, 10CI

Bis(acetonitrile)(η^3-allyl)dicarbonyliodomolybdenum

[33221-77-1]

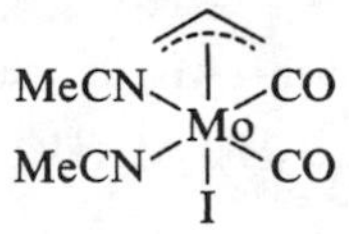

M 402.043

Orange cryst.

Hayter, R.E., *J. Organomet. Chem.*, 1968, **13**, P1.
Dieck, H.T. *et al*, *J. Organomet. Chem.*, 1968, **14**, 375 (*synth*)
Brisdon, B.J. *et al*, *Inorg. Chim. Acta*, 1979, **35**, L381; **36**, 127.
Clark, D.A. *et al*, *J. Chem. Soc., Dalton Trans.*, 1980, 565 (*react*)

$C_9H_{11}MoNO_2$ Mo-00031

Carbonyl(η^5-2,4-cyclopentadien-1-yl)nitrosyl(1,2-η-1-propene)molybdenum, 10CI

[54438-48-1]

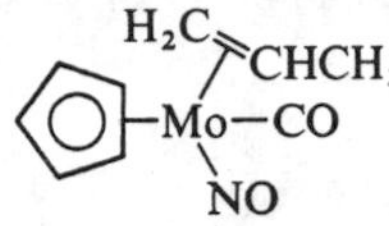

M 261.131

Yellow cryst.

Bailey, N.A. *et al*, *J. Chem. Soc., Chem. Commun.*, 1974, 592.
Faller, J.W. *et al*, *Ann. N.Y. Acad. Sci.*, 1977, **295**, 186.
Adams, R.D. *et al*, *J. Am. Chem. Soc.*, 1979, **101**, 2570 (*props*)
Faller, J.W. *et al*, *J. Organomet. Chem.*, 1980, **187**, 227 (*props*)

$C_{10}H_7MoO_3^{\oplus}$ Mo-00032

Tricarbonyl(η^7-cycloheptatrienylium)molybdenum(1+), 9CI

Tricarbonyl(η^7-tropylium)molybdenum(1+)

[46238-45-3]

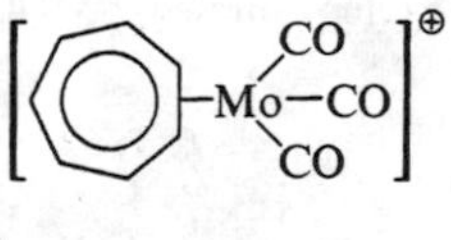

M 271.104 (ion)

Tetrafluoroborate: [12170-21-7].
$C_{10}H_7BF_4MoO_3$ M 357.907
Orange air-stable cryst. Mp >270°.

Dauben, H.J. *et al*, *J. Am. Chem. Soc.*, 1958, **80**, 5570 (*synth*)
Clark, G.R. *et al*, *J. Organomet. Chem.*, 1973, **50**, 185 (*struct*)
Isaacs, E.E. *et al*, *Can. J. Chem.*, 1975, **53**, 975 (*props*)
Isaacs, E.E. *et al*, *J. Organomet. Chem.*, 1975, **90**, 319 (*props*)
Salzer, A. *et al*, *J. Organomet. Chem.*, 1975, **87**, 101 (*synth*)
Tremmel, P.O. *et al*, *Z. Naturforsch., B*, 1975, **30**, 699 (*props*)
Olah, G. *et al*, *J. Org. Chem.*, 1976, **41**, 1694 (*cmr*)
Green, M. *et al*, *J. Chem. Soc., Dalton Trans.*, 1977, 1755 (*props*)

$C_{10}H_8MoO_3$ Mo-00033

Tricarbonyl[(1,2,3,4,5,6η)-1,3,5-cycloheptatriene]molybdenum, 9CI

(η^6-1,3,5-Cycloheptatriene)tricarbonylmolybdenum

[12125-77-8]

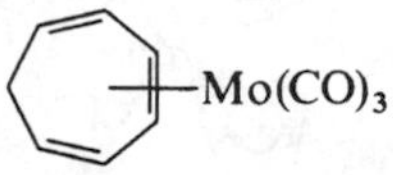

M 272.111

Red hexagonal prisms (hexane), air-stable in solid state. Mp 100.5-101.5°, 95° dec. Sensitive to light, dec. by CCl_4. Sublimes *in vacuo* at 85°.

▷QA4700000.

Abel, E.W. *et al*, *J. Chem. Soc.*, 1958, 4559 (*synth, ir*)
Dunitz, J.D. *et al*, *Helv. Chim. Acta*, 1960, **43**, 2188 (*cryst struct*)
Inorg. Synth., 1967, **9**, 121.
Guenther, H. *et al*, *Z. Naturforsch., B*, 1967, **22**, 389 (*pmr*)
Olah, G.A. *et al*, *J. Org. Chem.*, 1976, **41**, 1694 (*cmr*)
Gower, M. *et al*, *Inorg. Chim. Acta*, 1979, **37**, 79.

$C_{10}H_8MoO_3$ Mo-00034

Tricarbonyl[(1,2,3,4,5,6-η)methylbenzene]molybdenum, 9CI

Tricarbonyl(η^6-toluene)molybdenum

[12083-34-0]

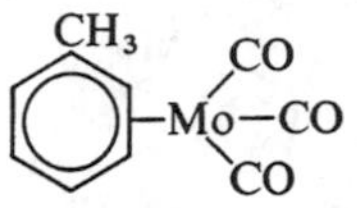

M 272.111

Mp 118-20° dec.

Strohmeier, W., *Chem. Ber.*, 1961, **94**, 3337 (*synth, props*)
Zingales, F. *et al*, *J. Am. Chem. Soc.*, 1966, **88**, 2707.
Nesmeyanov, A.N. *et al*, *J. Organomet. Chem.*, 1975, **102**, 185 (*synth, props*)
Holtzapple, G.M. *et al*, *Inorg. Nucl. Chem. Lett.*, 1976, **12**, 623.
Muetterties, E.L. *et al*, *J. Organomet. Chem.*, 1979, **178**, 197.

$C_{10}H_9MoO_3^{\oplus}$ Mo-00035

Tricarbonyl(η^5-2,4-cyclopentadien-1-yl)(η^2-ethene)molybdenum(1+)

[46238-41-9]

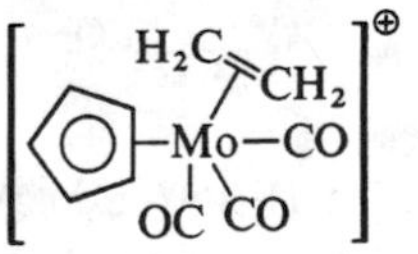

M 273.119 (ion)

Tetrafluoroborate: [62866-16-4].
$C_{10}H_9BF_4MoO_3$ M 359.923
Yellow cryst.

Dub, M., *Organometallic Compounds*, Springer-Verlag, Berlin, 2nd. Ed., 1966, **1**.
Beck, W. *et al*, *Z. Naturforsch., B*, 1978, **33**, 1214.
Beck, W. *et al*, *Chem. Ber.*, 1981, **114**, 867 (*synth*)

$C_{10}H_{10}Cl_2Mo$ Mo-00036

Dichlorobis(η^5-2,4-cyclopentadien-1-yl)molybdenum, 9CI

[12184-22-4]

M 297.035

Green cryst. Dec. at 270°.

Dub, M., *Organometallic Compounds*, Springer-Verlag, Berlin, 2nd Ed., 1966, Vol. 1 (*synth*)

Cooper, R.L. *et al*, *J. Chem. Soc.* (*A*), 1967, 1155 (*synth*, *ir*, *nmr*, *uv*, *esr*)

Prout, K. *et al*, *Acta Crystallogr.*, *Sect. B*, 1974, **30**, 2290 (*struct*)

Lindsell, W.E., *J. Chem. Soc.*, *Dalton Trans.*, 1975, 2508.

Thomas, J.L., *Inorg. Chem.*, 1978, **17**, 1507.

Aviles, T. *et al*, *J. Chem. Soc.*, *Dalton Trans.*, 1979, 1367.

$C_{10}H_{10}IMoNO$ Mo-00037

Bis(η^5-2,4-cyclopentadien-1-yl)iodonitrosylmolybdenum, 9CI

[12300-89-9]

M 383.040

Green-black cryst. Mp 157-9°.

King, R.B., *Inorg. Chem.*, 1968, **7**, 90 (*synth*)

Cotton, F.A. *et al*, *J. Am. Chem. Soc.*, 1972, **94**, 402 (*props*)

$C_{10}H_{10}MoN_2O_5$ Mo-00038

Pentacarbonyl(1,3-dimethyl-2-imidazolidinylidene)molybdenum, 10CI

[57560-35-7]

M 334.139

Air-stable white cryst. (Et_2O/hexane). Mp 108-9°.

Lappert, M.F. *et al*, *J. Less-Common Met.*, 1977, **54**, 191 (*synth*, *ir*, *cmr*)

$C_{10}H_{10}MoO_2$ Mo-00039

Dicarbonyl(η^5-2,4-cyclopentadien-1-yl)(η^3-2-propenyl)molybdenum, 10CI

η^3-Allyldicarbonyl(η^5-cyclopentadienyl)molybdenum

[12128-87-9]

M 258.128

Yellow cryst. Mp 175-6°.

Davison, A. *et al*, *Inorg. Chem.*, 1967, **6**, 2124 (*ir*)

Faller, J.W. *et al*, *Inorg. Chem.*, 1968, **7**, 840 (*nmr*)

Faller, J.W. *et al*, *J. Organomet. Chem.*, 1971, **31**, C75 (*nmr*)

Abel, E.W. *et al*, *J. Chem. Soc.*, *Dalton Trans.*, 1973, 1706 (*synth*)

Gibson, D.H. *et al*, *J. Organomet. Chem.*, 1979, **172**, C7 (*synth*)

$C_{10}H_{11}LiMo$ Mo-00040

Bis(η^5-2,4-cyclopentadien-1-yl)hydromolybdenumlithium

$(C_5H_5)_2Mo(H)Li$

M 234.078

Tetrameric.

Tetramer: [51837-59-3]. *Tetrakis[bis(η^5-2,4-cyclopentadien-1-yl)hydromolybdenum]tetralithium*, 9CI.

$C_{40}H_{44}Li_4Mo_4$ M 936.312

Yellow-orange cryst.

Francis, B.R. *et al*, *J. Chem. Soc.*, *Dalton Trans.*, 1976, 1339 (*synth*, *props*)

Meunier, B. *et al*, *Acta Crystallogr.*, *Sect. B*, 1979, **35**, 2558.

Berry, M. *et al*, *J. Chem. Soc.*, *Dalton Trans.*, 1980, 29 (*props*)

$C_{10}H_{11}MoNO_3$ Mo-00041

Dicarbonyl(η^5-2,4-cyclopentadien-1-yl)(2-propanoneoximato-*N*,*O*)molybdenum, 10CI

[61491-36-9]

M 289.142

Purple cryst. (CH_2Cl_2/hexane). Mp 100-1° dec.

Doedens, R.J. *et al*, *Inorg. Chem.*, 1977, **16**, 907 (*cryst struct*)

King, R.B. *et al*, *Inorg. Chem.*, 1977, **16**, 1164 (*synth*, *nmr*)

$C_{10}H_{11}MoO_3Tl$ Mo-00042

Tricarbonyl(η^5-2,4-cyclopentadien-1-yl)(dimethylthallium)-molybdenum, 10CI

M 479.518

Mp 68-9°.

▷Toxic

Walther, B. *et al*, *J. Organomet. Chem.*, 1978, **145**, 285 (*pmr*, *ir*, *synth*)

$C_{10}H_{12}Mo$ Mo-00043

Bis(η^5-2,4-cyclopentadien-1-yl)dihydromolybdenum, 10CI

Dihydridobis(η^5-cyclopentadienyl)molybdenum. Bis(cyclopentadienyl)molybdenum dihydride

[1291-40-3]

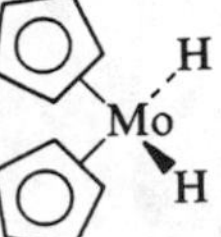

M 228.145

Yellow cryst. Mp 183-5° (163-5°, 174°). Subl. *in vacuo* at 50°.

Green, M.L.H. *et al*, *J. Chem. Soc.*, 1961, 4854 (*synth, ir, nmr*)
Dub, H., *Organometallic Compounds*, Springer-Verlag, Berlin, 2nd Ed., 1966, Vol. 1 (*synth*)
Macosek, F. *et al*, *Collect. Czech. Chem. Commun.*, 1970, **35**, 993 (*ms*)
Van Dam, E.M. *et al*, *J. Am. Chem. Soc.*, 1975, **97**, 465 (*synth*)
Schultz, A.J. *et al*, *Inorg. Chem.*, 1977, **16**, 3303 (*struct*)
Calado, J.C.E. *et al*, *J. Organomet. Chem.*, 1979, **174**, 77.
Hoxmeier, R.J. *et al*, *Inorg. Chem.*, 1979, **18**, 3462.

$C_{10}H_{30}Mo_2N_4$ Mo-00044

Dimethyltetrakis(*N*-dimethylamino)dimolybdenum

Dimethyltetrakis(N-methylmethaneaminato)dimolybdenum, 9CI

[67030-82-4]

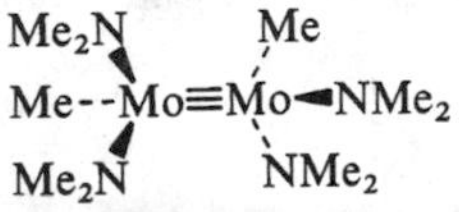

Me_2N Me
Me--Mo≡Mo◄NMe_2
Me_2N NMe_2

M 398.254

Thermally stable yellow cryst. $Bp_{0.01}$ 100° subl.

Chisholm, M.H. *et al*, *Inorg. Chem.*, 1976, **15**, 2244; 1977, **16**, 320.
Chisholm, M.H. *et al*, *Inorg. Chem.*, 1978, **17**, 2338 (*cryst struct, synth, ir, nmr*)
Chisholm, M.H. *et al*, *J. Am. Chem. Soc.*, 1979, **101**, 6784.

$C_{10}Mo_2O_{10}^{\ominus\ominus}$ Mo-00045

Decarbonyldimolybdate(2−), 9CI

[57348-62-6]

$[(OC)_5Mo—Mo(CO)_5]^{\ominus\ominus}$

M 471.984 (ion)

Staggered conformation. Mo—Mo distance 3.123(7) Å.

Di-K salt: [57348-62-6].
$C_{10}K_2Mo_2O_{10}$ M 550.181
Yellow cryst.

Bis[bis(triphenylphosphine)iminium]salt: [49736-19-8].
$C_{82}H_{60}Mo_2N_2O_{10}P_4$ M 1549.158
Yellow cryst.

Hayter, R.G., *J. Am. Chem. Soc.*, 1966, **88**, 4376 (*synth*)
Dahl, L.F. *et al*, *J. Am. Chem. Soc.*, 1970, **92**, 7312 (*cryst struct*)
Ungurenasu, C. *et al*, *J. Chem. Soc., Chem. Commun.*, 1975, 388 (*synth*)

$C_{11}H_5F_9MoOS$ Mo-00046

Carbonyl(η^5-2,4-cyclopentadien-1-yl)[(2,3-η)-1,1,1,4,4,4-hexafluoro-2-butyne](trifluoromethanethiolato)molybdenum, 9CI

[Bis(trifluoromethyl)acetylene]carbonyl(η^5-cyclopentadienyl)(trifluoromethanethiolato)molybdenum

[59464-63-0]

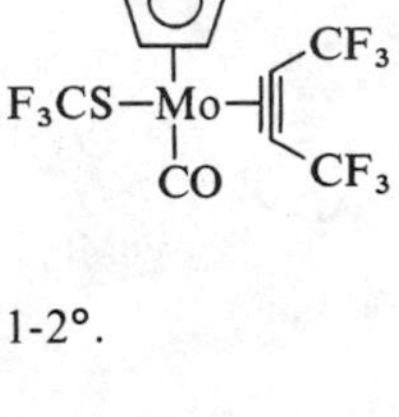

F_3CS—Mo—(CF₃C≡CCF₃), CO

M 452.146

Blue cryst. Mp 111-2°.

Braterman, P.S. *et al*, *J. Chem. Soc., Dalton Trans.*, 1976, 241 (*synth*)

$C_{11}H_8MoO_3$ Mo-00047

Tricarbonyl[(1,2,3,4,5,6-η)-1,3,5,7-cyclooctatetraene]molybdenum, 9CI

[12108-93-9]

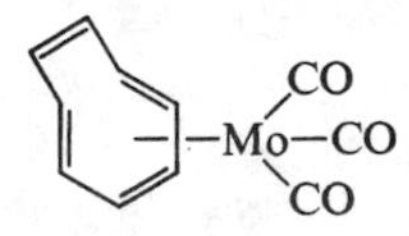

M 284.122

Red or yellow cryst.

Winstein, S. *et al*, *J. Am. Chem. Soc.*, 1965, **87**, 3267 (*synth*)
Kaesz, H.D. *et al*, *J. Am. Chem. Soc.*, 1966, **88**, 1319 (*props*)
Paul, I.C. *et al*, *J. Am. Chem. Soc.*, 1966, **88**, 5927 (*cryst struct*)
King, R.B., *Inorg. Chem.*, 1967, **8**, 139 (*synth*)
Cotton, F.A. *et al*, *J. Am. Chem. Soc.*, 1974, **96**, 4723, 7926 (*synth, nmr*)
Mann, B.E. *et al*, *J. Chem. Soc., Dalton Trans.*, 1979, 1021 (*nmr*)

$C_{11}H_8MoO_3$ Mo-00048

Tricarbonyl[(1,2,3,4,5,6-η)ethenylbenzene]molybdenum, 9CI

Tricarbonyl(η^6-styrene)molybdenum

[53041-79-5]

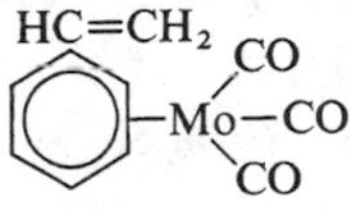

M 284.122

Mp 80-2°.

Nesmeyanov, A.N. *et al*, *J. Organomet. Chem.*, 1975, **102**, 185 (*synth*)

$C_{11}H_8MoO_4$ Mo-00049

[(2,3,5,6-η)-Bicyclo[2.2.1]hepta-2,5-diene]tetracarbonylmolybdenum, 9CI

Tetracarbonylnorbornadienemolybdenum

[12146-37-1]

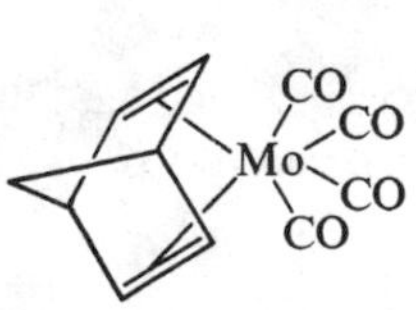

M 300.122

Pale-yellow cryst. Mp 77-8°. $Bp_{0.1}$ 100° subl.

Bennett, M.A. *et al*, *J. Chem. Soc.*, 1961, 2037 (*synth, pmr*)
Lauterbur, P.C., *J. Am. Chem. Soc.*, 1965, **87**, 3266 (*cmr*)
Werner, H. *et al*, *J. Organomet. Chem.*, 1966, **5**, 79 (*props*)
Mann, B.E., *J. Chem. Soc., Dalton Trans.*, 1973, 2012 (*cmr, ir*)
Skinner, H.A. *et al*, *J. Organomet. Chem.*, 1977, **142**, 321 (*props*)

$C_{11}H_{10}MoO$ **Mo-00050**

Carbonylbis(η^5-2,4-cyclopentadien-1-yl)molybdenum, 10CI

[12701-85-8]

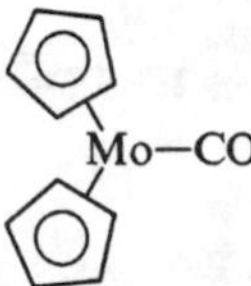

M 254.139

Bright-green cryst. $Bp_{0.001}$ 35° subl.

Thomas, J.L., *J. Am. Chem. Soc.*, 1973, **95**, 1838 (*synth*)
Geoffrey, G.L. *et al*, *Inorg. Chem.*, 1978, **17**, 2410.
Grebenik, P. *et al*, *J. Chem. Soc., Chem. Commun.*, 1979, 742.

$C_{11}H_{10}MoO_3$ **Mo-00051**

Tricarbonyl[(1,2,3,4,5,6-η)-1,3,5-cyclooctatriene]molybdenum, 9CI

[12093-14-0]

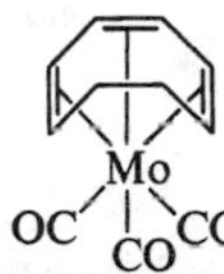

M 286.138

Orange-red cryst.

Fischer, E.O. *et al*, *Chem. Ber.*, 1959, **92**, 2645 (*synth*)
Aumann, R. *et al*, *Tetrahedron Lett.*, 1970, 903 (*synth, props*)
Kreiter, C.E. *et al*, *Chem. Ber.*, 1975, **108**, 1502 (*nmr*)
Kaesz, H.D. *et al*, *J. Am. Chem. Soc.*, 1966, **88**, 1319 (*props*)

$C_{11}H_{10}MoO_3$ **Mo-00052**

Tricarbonyl[η^6-5-(1-methylethylidene)-1,3-cyclopentadiene]molybdenum, 10CI

Tricarbonyl-η^6-6,6-dimethylfulvenemolybdenum

[63835-71-2]

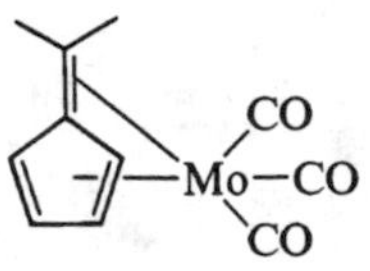

M 286.138

Dark-red cryst. (hexane). Dec. ca. 130°.

Edelmann, F. *et al*, *J. Organomet. Chem.*, 1977, **134**, 31 (*synth, props*)

$C_{11}H_{11}Mo$ **Mo-00053**

(η^6-Benzene)(η^5-2,4-cyclopentadien-1-yl)molybdenum, 9CI

[12153-25-2]

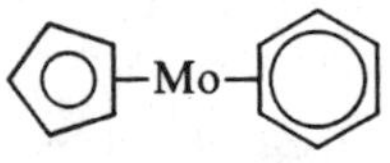

M 239.148

Red cryst., paramagnetic, air-sensitive. Sol. polar solvs., spar. sol. C_6H_6. $Bp_{0.01}$ 85°.

Wehner, W.H. *et al*, *Chem. Ber.*, 1970, **103**, 2258.
Green, M.L.H. *et al*, *J. Chem. Soc., Dalton Trans.*, 1974, 311; 1976, 213.
Inorg. Synth., 1980, **20**, 196 (*synth*)

$C_{11}H_{11}MoN_2O_2^{\oplus}$ **Mo-00054**

Bis(acetonitrile)dicarbonyl(η^5-2,4-cyclopentadien-1-yl)molybdenum(1+), 9CI

[12109-14-7]

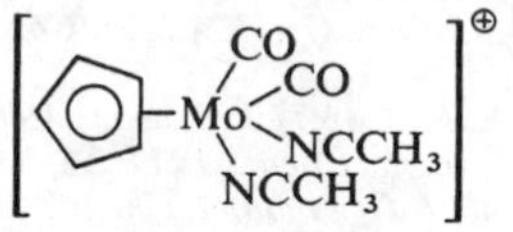

M 299.160 (ion)

Hexafluorophosphate:
$C_{11}H_{11}F_6MoN_2O_2P$ M 444.124
Orange-red cryst. Sol. CH_2Cl_2.
Tetraphenylborate: [68868-08-6].
$C_{35}H_{31}BMoN_2O_2$ M 618.392
Orange-red cryst.

Treichel, P.M. *et al*, *J. Organomet. Chem.*, 1967, **7**, 449 (*synth*)
Nolte, M.J. *et al*, *J. Chem. Soc., Dalton Trans.*, 1978, 932 (*synth*)

$C_{11}H_{11}MoO_2^{\oplus}$ **Mo-00055**

(η^4-1,3-Butadiene)dicarbonyl(η^5-2,4-cyclopentadien-1-yl)molybdenum(1+), 9CI

Dicarbonyl(η^4-1,3-butadiene)(η^5-cyclopentadienyl)molybdenum(1+)

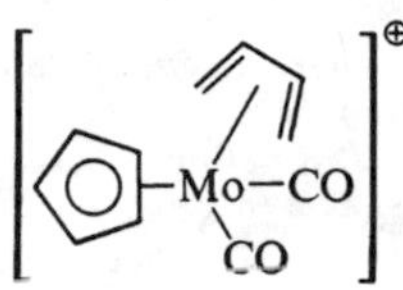

M 271.147 (ion)

Tetrafluoroborate: [63947-89-7].
$C_{11}H_{11}BF_4MoO_2$ M 357.950
Yellow cryst. Reacts with nucleophiles.

Faller, J.W. *et al*, *J. Am. Chem. Soc.*, 1977, **99**, 4858 (*synth*)
Faller, J.W. *et al*, *Ann. N.Y. Acad. Sci.*, 1977, **295**, 186.

$C_{11}H_{14}GeMoO_3$ **Mo-00056**

Tricarbonyl(η^5-2,4-cyclopentadien-1-yl)(trimethylgermyl)molybdenum, 9CI

Trimethylgermyltricarbonyl(η^5-cyclopentadienyl)molybdenum

[33306-91-1]

As Bromotricarbonyl(η^5-2,4-cyclopentadien-1-yl)molybdenum, Mo-00011 with

L = $GeMe_3$

M 362.760

Pale-yellow cryst. Mp 84°. $Bp_{0.001}$ 70° subl.

Carrick, A. *et al*, *J. Chem. Soc. (A)*, 1968, 913 (*synth*)
Cardin, D.J. *et al*, *J. Chem. Soc. (A)*, 1970, 2594 (*synth, ir, nmr, uv*)
Dean, W.K. *et al*, *Inorg. Chem.*, 1977, **16**, 1061 (*synth, props*)

$C_{11}H_{14}MoO_3Pb$ **Mo-00057**

Tricarbonyl(η^5-2,4-cyclopentadien-1-yl)(trimethylplumbyl)molybdenum

(Trimethylplumbyl)(cyclopentadienyl)tricarbonylmolybdenum

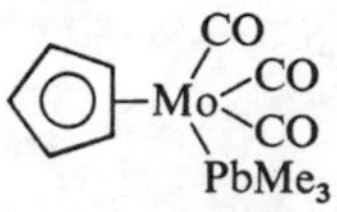

M 497.370

Pale-orange solid (CH_2Cl_2/pentane). Mp 93-5°.

Patil, H.R.H. *et al*, *Inorg. Chem.*, 1966, **5**, 1401.

$C_{11}H_{14}MoO_3Si$ Mo-00058

Tricarbonyl(η^5-2,4-cyclopentadien-1-yl)(trimethylsilyl)molybdenum, 9CI

Trimethylsilyltricarbonyl(η^5-cyclopentadienyl)molybdenum

[12282-32-5]

As Bromotricarbonyl(η^5-2,4-cyclopentadien-1-yl)molybdenum, Mo-00011 with

L = $SiMe_3$

M 318.255

Orange-yellow solid, air- and light-sensitive. $Bp_{0.001}$ 100° subl.

Cardin, D.J. *et al*, *J. Chem. Soc. (A)*, 1970, 2594 (*synth*)
Hagen, A.P. *et al*, *Inorg. Chem.*, 1971, **10**, 1657 (*synth*)
Malisch, W. *et al*, *Chem. Ber.*, 1974, **107**, 979 (*synth*)

$C_{11}H_{14}MoO_3Sn$ Mo-00059

Tricarbonyl(η^5-2,4-cyclopentadien-1-yl)(trimethylstannyl)molybdenum, 9CI

Trimethylstannyltricarbonyl(η^5-cyclopentadienyl)molybdenum

[12214-92-5]

As Bromotricarbonyl(η^5-2,4-cyclopentadien-1-yl)molybdenum, Mo-00011 with

L = $SnMe_3$

M 408.860

Pale-yellow cryst. (hexane). Mp 99-100°. $Bp_{0.001}$ 80° subl.

Abel, E.W. *et al*, *J. Organomet. Chem.*, 1970, **22**, C31.
Inorg. Synth., 1970, **12**, 60 (*synth*, *ir*, *nmr*)
Cardin, D.J. *et al*, *J. Chem. Soc. (A)*, 1970, 2594.
Abel, E.W. *et al*, *J. Organomet. Chem.*, 1973, **49**, 435 (*synth*)

$C_{12}H_5CoMoO_7$ Mo-00060

Tricarbonyl(η^5-2,4-cyclopentadien-1-yl)(tetracarbonylcobalt)molybdenum, 10CI

[62015-65-0]

As Bromotricarbonyl(η^5-2,4-cyclopentadien-1-yl)molybdenum, Mo-00011 with

L = $Co(CO)_4$

M 416.041

Cryst. (pentane). Mp 63-5° (sealed under Ar).

Schmid, E. *et al*, *Z. Naturforsch., B*, 1977, **32**, 1277 (*synth*, *props*)
Abrahamson, H.B. *et al*, *Inorg. Chem.*, 1978, **17**, 1003 (*synth*, *props*)

$C_{12}H_{10}Cl_2Mo$ Mo-00061

Bis(η^6-chlorobenzene)molybdenum, 9CI

[52346-34-6]

Cl Cl
—Mo—

M 321.057

Light green, air-sensitive powder. Dec. at 92° (under N_2 sealed tube).

▷Pyrophoric

Inorg. Synth., 1979, **19**, 80 (*synth*)
Green, M.L.H., *J. Organomet. Chem.*, 1980, **200**, 119 (*props*)

$C_{12}H_{10}MoO_3$ Mo-00062

[(2,3,4,5,6,7-η)-Bicyclo[6.1.0]nona-2,4,6-triene]tricarbonylmolybdenum, 9CI

Tricarbonyl-η^6-bicyclo[6.1.0]nona-2,4,6-trienemolybdenum

[41354-56-7]

—Mo(CO)(CO)(CO)

M 298.149

Orange cryst.

Grimme, W., *Chem. Ber.*, 1967, **100**, 113 (*synth*, *props*)
Brookhart, M. *et al*, *Inorg. Chem.*, 1974, **13**, 1540 (*props*)
Darensbourg, D.J. *et al*, *J. Am. Chem. Soc.*, 1978, **100**, 4119 (*props*)

$C_{12}H_{12}Mo$ Mo-00063

Bis(η^6-benzene)molybdenum, 9CI

[12129-68-9]

—Mo—

M 252.167

Green cryst. Mp 115° dec. Air-sensitive. Sublimes (high vac.) at 110-5°.

Fischer, E.O. *et al*, *Chem. Ber.*, 1956, **89**, 1805; 1960, **93**, 2065 (*synth*)
Fritz, H.P. *et al*, *Spectrochim. Acta*, 1961, **17**, 1068 (*ir*)
Schneider, R. *et al*, *Naturwissenchaften*, 1962, **48**, 452 (*cryst struct*)
Benfield, F.W.S. *et al*, *J. Chem. Soc., Chem. Commun.*, 1973, 866 (*synth*)
Silvon, M.P. *et al*, *J. Am. Chem. Soc.*, 1974, **96**, 1945 (*synth*)
Green, M.L.H., *J. Organomet. Chem.*, 1980, **200**, 119 (*synth*, *props*)

$C_{12}H_{12}MoO_3$ Mo-00064

(η^6-1,3,5-Trimethylbenzene)tricarbonylmolybdenum

Mesitylenetricarbonylmolybdenum. Tricarbonyl-1,3,5-trimethylbenzenemolybdenum

[12089-15-5]

CH_3
—$Mo(CO)_3$
H_3C CH_3

M 300.165

Yellow cryst. Mp 140° dec.

Price, J.T. *et al*, *Can. J. Chem.*, 1968, **46**, 515 (*pmr*)
U.S.P., 3 382 263, (*1968*); *CA*, **69**, 59376 (*synth*)
Lokshin, B.V. *et al*, *Zh. Strukt. Khim.*, 1975, **16**, 592; *CA*, **84**, 51628 (*ir*, *raman*)
Nesmeyanov, A.N. *et al*, *J. Organomet. Chem.*, 1975, **102**, 185 (*synth*)
Koskland, D.E. *et al*, *Acta Crystallogr., Sect. B*, 1977, **53**, 2013 (*struct*)

$C_{12}H_{12}MoO_4$ Mo-00065

Tetracarbonyl[(1,2,5,6-η)-1,5-cyclooctadiene]molybdenum, 9CI

[12109-74-9]

M 316.164

Yellow-brown solid. Mp 125-9° dec.

Leigh, G.J. *et al*, *J. Organomet. Chem.*, 1965, **4**, 461 (*synth*)
Zingales, F. *et al*, *J. Organomet. Chem.*, 1967, **7**, 461 (*props*)
King, R.B., *J. Organomet. Chem.*, 1967, **8**, 139 (*synth*)
Zingales, F. *et al*, *J. Am. Chem. Soc.*, 1967, **89**, 256 (*props*)
Tayim, H.A. *et al*, *Inorg. Nucl. Chem. Lett.*, 1972, **8**, 231.

$C_{12}H_{14}Mo$ Mo-00066

Bis(η^5-cyclopentadienyl)(η^2-ethylene)molybdenum

Bis(η^5-2,4-cyclopentadien-1-yl)(η^2-ethene)molybdenum, 9CI

[37343-05-8]

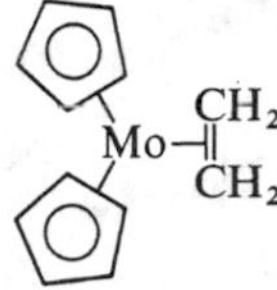

M 254.183

Bright-red needles. $Bp_{0.0001}$ 50° subl.

Thomas, J.L., *J. Am. Chem. Soc.*, 1973, **95**, 1838 (*synth, nmr, ms*)

$C_{12}H_{14}MoO_3Sn$ Mo-00067

Tricarbonyl(η^6-trimethylstannylbenzene)molybdenum

Tricarbonyl(trimethylphenylstannane)molybdenum, 8CI

[31870-76-5]

M 420.871

Shiny-yellow needles (pet. ether). Mp 103-5°.

Poeth, T.P. *et al*, *Inorg. Chem.*, 1971, **10**, 522 (*synth, ir, nmr, ms, uv*)

$C_{12}H_{15}MoO_3Tl$ Mo-00068

Tricarbonyl(η^5-2,4-cyclopentadien-1-yl)(diethylthallium)-molybdenum

[66125-44-8]

M 507.572

Mp 86-7°.

▷Toxic

Walther, B. *et al*, *J. Organomet. Chem.*, 1978, **145**, 285 (*synth, ir, nmr*)

$C_{12}H_{16}Mo$ Mo-00069

(η^6-Benzene)bis(η^3-2-propenyl)molybdenum, 9CI

Bis(η^3-allyl)(η^6-benzene)molybdenum. Diallyl-η^6-benzenemolybdenum

[35625-70-8]

M 256.198

Green, M.L.H. *et al*, *J. Chem. Soc., Dalton Trans.*, 1974, 311; 1976, 213.

$C_{12}H_{18}Mo$ Mo-00070

Tris(η^4-1,3-butadiene)molybdenum, 9CI

[51733-17-6]

M 258.214

Yellow cryst. (heptane). $Bp_{0.001}$ 50-60° subl.

Skell, P.S. *et al*, *J. Am. Chem. Soc.*, 1974, **96**, 626 (*synth, ir, nmr, ms*)
Skell, P.S. *et al*, *Angew. Chem., Int. Ed. Engl.*, 1975, **14**, 195 (*synth, struct*)

$C_{12}H_{20}MoN_2O_2S_4$ Mo-00071

Dicarbonylbis(diethylcarbamodithioato-*S,S'*)molybdenum, 10CI

[18947-43-8]

$Mo(S_2CNEt_2)_2(CO)_2$

M 448.482

16-Electron complex.

Colton, R. *et al*, *Aust. J. Chem.*, 1968, **21**, 15 (*synth*)
McDonald, J.W. *et al*, *J. Organomet. Chem.*, 1975, **92**, C25.
Inorg. Synth., 1976, **16**, 235 (*synth*)

$C_{12}H_{20}Mo_2$ Mo-00072

Bis[μ-[(1-η:2,3-η)-2-propenyl)]bis(η^3-2-propenyl)dimolybdenum, 9CI

Tetraallyldimolybdenum

[79739-15-4]

M 356.170

Dark-green cryst., air- and moisture-sensitive.

Oberkirch, W., *Diss. Techn. Hochschule, Aachen*, 1963 (*synth*)
Wilke, E. *et al*, *Angew. Chem., Int. Ed. Engl.*, 1966, **5**, 151 (*synth*)
Ramey, K.C. *et al*, *J. Am. Chem. Soc.*, 1968, **90**, 4275 (*nmr*)
Cotton, F.A. *et al*, *J. Am. Chem. Soc.*, 1971, **93**, 5441 (*synth, struct*)
Benn, R. *et al*, *J. Organomet. Chem.*, 1979, **174**, C38 (*props*)

$C_{13}H_5ClF_{12}Mo$ Mo-00073

Chloro(η^5-2,4-cyclopentadien-1-yl)bis[(2,3-η)-1,1,1,4,4,4-hexafluoro-2-butyne]molybdenum, 10CI

[54721-00-5]

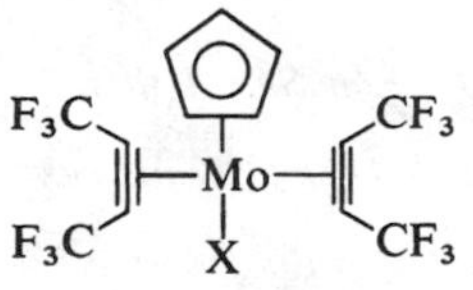

X = Cl

M 520.556

Yellow cryst.

Davidson, J.L. *et al*, *J. Chem. Soc., Dalton Trans.*, 1975, 2531; 1977, 287 (*props, synth*)
Davidson, J.L., *J. Organomet. Chem.*, 1980, **186**, C19 (*props*)

$C_{13}H_5F_{12}IMo$ Mo-00074

(η^5-2,4-Cyclopentadien-1-yl)bis[(2,3-η)-1,1,1,4,4,4-hexafluoro-2-butyne]iodomolybdenum, 10CI

[59172-20-2]

As Chloro(η^5-2,4-cyclopentadien-1-yl)bis[(2,3-η)-1,1,1,4,4,4-hexafluoro-2-butyne]molybdenum, Mo-00073 with

X = I

M 612.008

Red-brown cryst.

Davidson, J.L. *et al*, *J. Chem. Soc., Dalton Trans.*, 1975, 2531.
Davidson, J.L. *et al*, *J. Chem. Soc., Chem. Commun.*, 1980, 113 (*synth, nmr*)

$C_{13}H_8MoN_3O_4^{\oplus}$ Mo-00075

(2,2′-Bipyridine-*N,N′*)tricarbonylnitrosylmolybdenum(1+)

[63890-69-7]

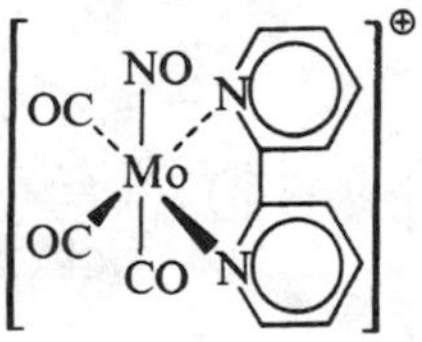

M 366.164 (ion)

fac-form

Tetrafluoroborate: [63890-70-0].
$C_{13}H_8BF_4MoN_3O_4$ M 452.968
Red-brown solid. Sol. CH_2Cl_2.

Condon, D. *et al*, *J. Chem. Soc., Dalton Trans.*, 1977, 925 (*synth, props*)

$C_{13}H_{13}Mo^{\oplus}$ Mo-00076

(η^8-1,3,5,7-Cyclooctatetraene)(η^5-2,4-cyclopentadien-1-yl)molybdenum(1+), 10CI

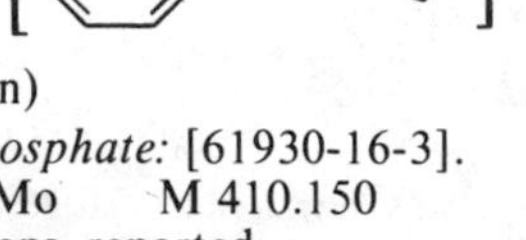

M 265.186 (ion)

Hexafluorophosphate: [61930-16-3].
$C_{13}H_{13}F_6PMo$ M 410.150
No phys. props. reported.

Segel, J.A. *et al*, *J. Chem. Soc., Chem. Commun.*, 1976, 766 (*synth, props*)

$C_{13}H_{15}Mo^{\oplus}$ Mo-00077

Bis(η^5-2,4-cyclopentadien-1-yl)(η^3-2-propenyl)molybdenum(1+), 10CI

[53449-94-8]

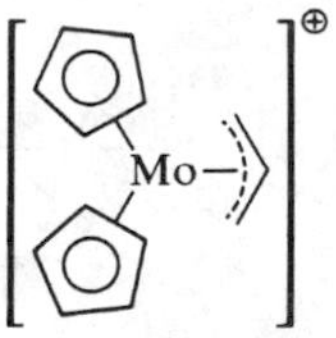

M 267.202 (ion)

Hexafluorophosphate: [53449-95-9].
$C_{13}H_{15}F_6MoP$ M 412.166
Red cryst.

Green, M.L.H. *et al*, *J. Chem. Soc., Dalton Trans.*, 1976, 1993 (*synth*)
Ephritikhine, M. *et al*, *J. Chem. Soc., Dalton Trans.*, 1977, 1131 (*props*)
Davies, S. *et al*, *J. Chem. Soc., Dalton Trans.*, 1978, 1510 (*synth*)

$C_{14}H_5F_{12}IMoO$ Mo-00078

Carbonyl(η^5-2,4-cyclopentadien-1-yl)iodo[(1,2,3,4-η)-1,2,3,4-tetrakis(trifluoromethyl)cyclobutadiene]molybdenum, 10CI

[74890-12-3]

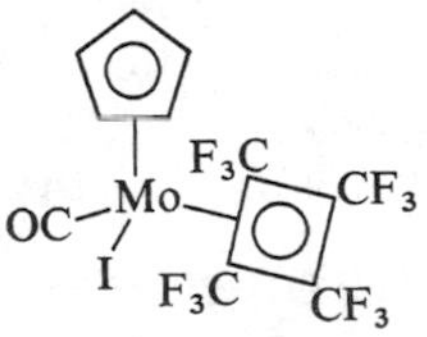

M 640.018

Dynamic nmr reported.

Davidson, J.L. *et al*, *J. Chem. Soc., Chem. Commun.*, 1980, 113 (*synth, nmr*)

$C_{14}H_{10}Mo_2O_4$ Mo-00079

Tetracarbonylbis(η^5-2,4-cyclopentadien-1-yl)dimolybdenum, 10CI

Dicyclopentadienyltetracarbonyldimolybdenum

[56200-27-2]

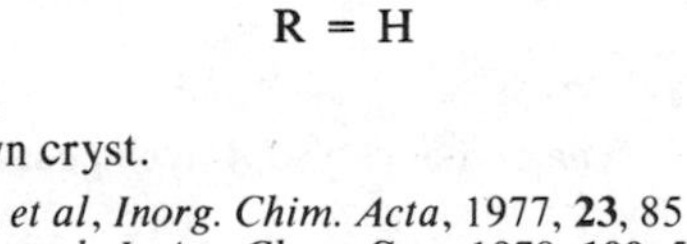

R = H

M 434.111

Dark-brown cryst.

Ginley, D.S. *et al*, *Inorg. Chim. Acta*, 1977, **23**, 85 (*synth*)
Bailey, W.I. *et al*, *J. Am. Chem. Soc.*, 1978, **100**, 5764.
Chisholm, M.H. *et al*, *J. Am. Chem. Soc.*, 1978, **100**, 807.
Klinger, R.J. *et al*, *J. Am. Chem. Soc.*, 1978, **100**, 5034 (*synth*)
King, R.B. *et al*, *J. Organomet. Chem.*, 1979, **171**, 53.
Jemmis, E.D. *et al*, *J. Am. Chem. Soc.*, 1980, **102**, 2576.

$C_{14}H_{12}MoO_2$ Mo-00080

Benzyldicarbonyl-π-cyclopentadienylmolybdenum, 8CI

Dicarbonyl(η⁵-2,4-cyclopentadien-1-yl)(phenylmethyl)-molybdenum

[12126-95-3]

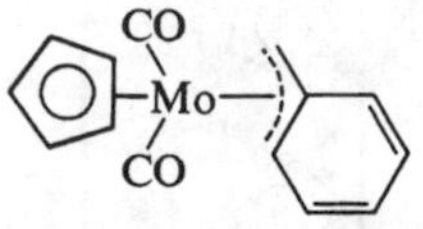

M 308.188

Dark-red cryst., blackens on storage at r.t. in closed vials. Mp 83-5°. $Bp_{0.1}$ 80° subl.

King, R.B. *et al*, *J. Am. Chem. Soc.*, 1966, **88**, 709 (*synth*, *ir*, *nmr*)
Cotton, F.A. *et al*, *J. Am. Chem. Soc.*, 1968, **90**, 5418 (*struct*)
Cotton, F.A. *et al*, *J. Am. Chem. Soc.*, 1969, **91**, 1339.

$C_{14}H_{14}Cl_3Mo_2$ Mo-00081

Trichlorobis(η^7-2,4,6-cycloheptatrien-1-yl)dimolybdenum, 9CI

Bis(η^7-2,4,6-cycloheptatrien-1-yl)tri-μ-chlorodimolybdenum

[51809-70-2]

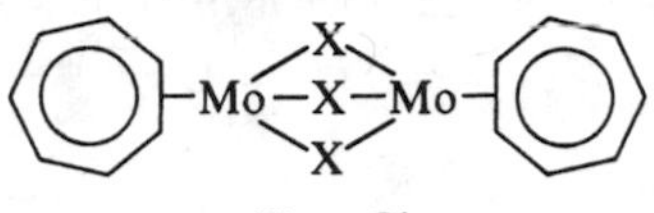

X = Cl

M 480.504

Allen, G.C. *et al*, *J. Chem. Soc., Chem. Commun.*, 1976, 794 (*pe*)
Bochmann, M. *et al*, *J. Chem. Soc., Dalton Trans.*, 1977, 714 (*synth*)
Ashworth, E.F. *et al*, *J. Chem. Soc., Dalton Trans.*, 1977, 1693 (*synth*, *props*)

$C_{14}H_{16}Mo$ Mo-00082

Bis[(1,2,3,4,5,6-η)methylbenzene]molybdenum, 9CI

Bis(η^6-toluene)molybdenum

[12131-22-5]

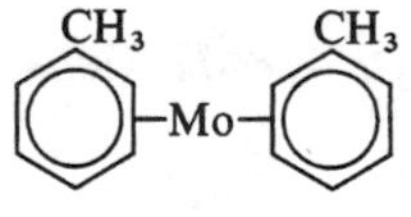

M 280.220

Green air-sensitive cryst.

Green, M.L.H. *et al*, *J. Chem. Soc. (A)*, 1971, 2929.
Green, M.L.H. *et al*, *J. Chem. Soc., Dalton Trans.*, 1973, 301.
Green, M.L.H., *J. Organomet. Chem.*, 1980, **200**, 119 (*synth*, *props*)

$C_{14}H_{17}MoO^{\oplus}$ Mo-00083

Bis[(2,3-η)-2-butyne]carbonyl(η^5-2,4-cyclopentadien-1-yl-)molybdenum(1+), 10CI

[66615-14-3]

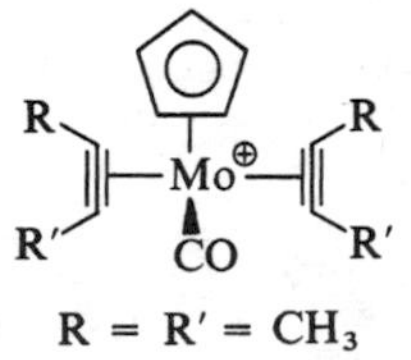

R = R′ = CH_3

M 297.228 (ion)

Tetrafluoroborate: [66615-15-4].
$C_{14}H_{17}BF_4MoO$ M 384.031
Yellow cryst. (CH_2Cl_2/Et_2O).
Hexafluorophosphate: [74380-92-0].
$C_{14}H_{17}F_6MoOP$ M 442.192
Yellow cryst.

Bottrill, M. *et al*, *J. Chem. Soc., Dalton Trans.*, 1977, 2365 (*synth*)
Bergmann, R.E. *et al*, *J. Am. Chem. Soc.*, 1980, **102**, 2698 (*synth*)
Allen, J.R. *et al*, *J. Chem. Soc., Dalton Trans.*, 1981, 873 (*props*)

$C_{14}H_{20}MoN_4O_4$ Mo-00084

Tetracarbonylbis(1,3-dimethyl-2-imidazolidinylidene)molybdenum, 10CI

[64161-94-0]

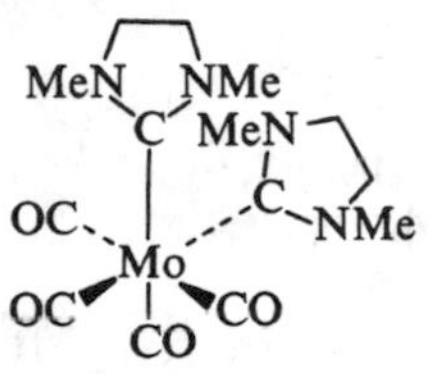

cis-form

M 404.276

Dicarbene complex.

cis-form

Pale-yellow microcryst. Sol. Me_2CO, $CHCl_3$. Mp 274°.

trans-form

By irradiation of *cis-form* in Me_2CO. Deep-orange microcryst. Spar. sol. Me_2CO. Readily reverts to *cis-form* on heating to 140-60° or on dissolution in $CHCl_3$ (30°).

Lappert, M.F. *et al*, *J. Chem. Soc., Dalton Trans.*, 1977, 1272 (*ir*, *pmr*, *cmr*, *cryst struct*, *synth*, *props*)

$C_{15}H_9BrFeMoO_4$ Mo-00085

Bromotetracarbonyl(ferrocenylmethylidyne)molybdenum, 10CI

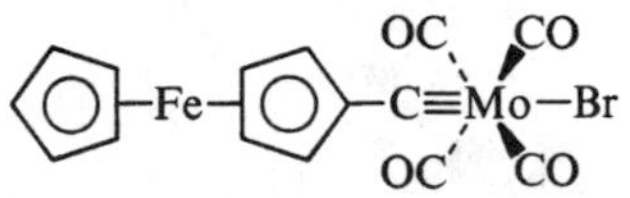

M 484.925

trans-form [68830-12-6]

Red microcryst. Mp >40° dec.

Fischer, E.O. *et al*, *Chem. Ber.*, 1978, **111**, 3530 (*synth*, *pmr*, *cmr*, *ir*, *polarog*)

$C_{15}H_9MnMoO_5$ Mo-00086

Molybdenumcarbonyl(η^5-2,4-cyclopentadien-1-yl)[μ-(η:η^5-2,4-cyclopentadien-1-ylidene)]tetracarbonylmanganese, 10CI

[71411-08-0]

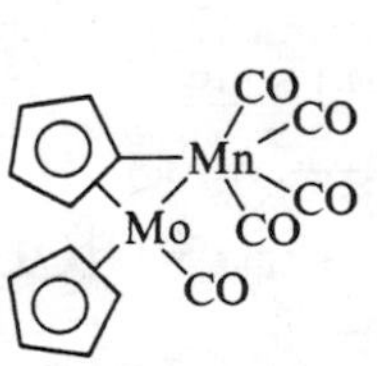

M 420.111

Red cryst.

Hoxmeier, R.J., *J. Am. Chem. Soc.*, 1971, **93**, 536 (*synth*)
Hoxmeier, R.J., *Inorg. Chem.*, 1979, **18**, 3453, 3462 (*synth*, *struct*)

$C_{15}H_{10}FeMoO_5$ Mo-00087

Tricarbonyl(η^5-2,4-cyclopentadien-1-yl)[dicarbonyl(η^5-2,4-cyclopentadien-1-yl)iron]molybdenum, 10CI

Tricarbonyl-π-cyclopentadienyl(dicarbonyl-π-cyclopentadienylferrio)molybdenum, 8CI. Pentacarbonylbis(η^5-2,4-cyclopentadien-1-yl)ironmolybdenum

[12130-13-1]

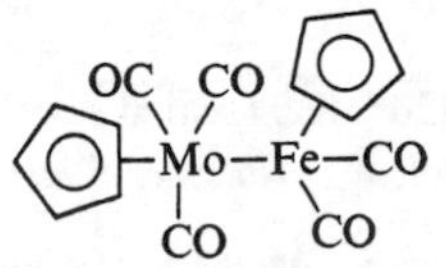

M 422.028

Red-violet cryst. (pentane or C_6H_6). Mp 209° (205°).

King, R.B. *et al*, *Chem. Ind.* (*London*), 1961, 747 (*synth, ir*)
Abrahamson, H.B. *et al*, *Inorg. Chem.*, 1978, **17**, 1003 (*uv, ir, synth, pmr*)
Madach, T. *et al*, *Chem. Ber.*, 1980, **113**, 2675 (*pmr, ir*)

$C_{15}H_{13}MoN_2O_2^{\oplus}$ Mo-00088

Bis(acetonitrile)dicarbonyl[(1,2,3,3*a*,7*a*-η)-1*H*-inden-1-yl-]molybdenum(1+), 9CI

[66615-32-5]

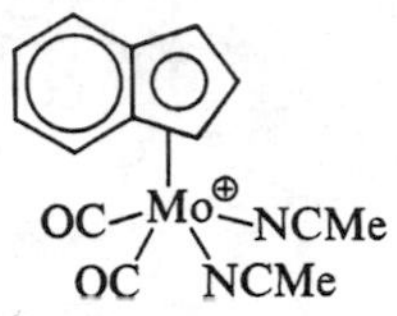

M 349.220 (ion)

Tetrafluoroborate: [66615-33-6].
$C_{15}H_{13}BF_4MoN_2O_2$ M 436.024
Orange cryst. Characterised spectroscopically.

Bottrill, M. *et al*, *J. Chem. Soc., Dalton Trans.*, 1977, 2365 (*synth, ir, nmr*)

$C_{15}H_{13}MoO_2^{\oplus}$ Mo-00089

(η^4-1,3-Butadiene)dicarbonyl[(1,2,3,3*a*,7*a*-η)-1*H*-inden-1-yl]molybdenum(1+)

[66615-20-1]

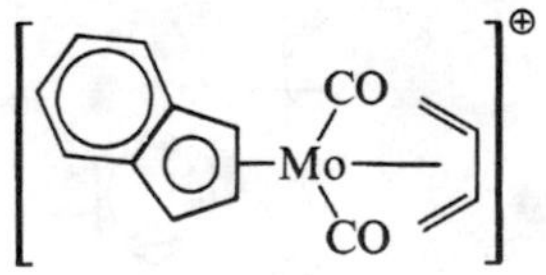

M 321.207 (ion)

Tetrafluoroborate: [66615-21-2].
$C_{15}H_{13}BF_4MoO_2$ M 408.010
Yellow cryst. (CH_2Cl_2/Et_2O).

Bottrill, M. *et al*, *J. Chem. Soc., Dalton Trans.*, 1977, 2365 (*synth, props*)

$C_{15}H_{16}MoO_3$ Mo-00090

Tricarbonyl(η^6-1,3,5,7-tetramethyl-1,3,5,7-cyclooctatetraene)molybdenum, 8CI

[12111-76-1]

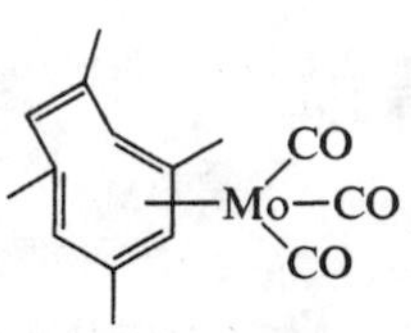

M 340.230

Cotton, F.A. *et al*, *J. Am. Chem. Soc.*, 1968, **90**, 1438 (*synth, nmr*)

$C_{15}H_{17}BMoN_6O_2$ Mo-00091

Dicarbonyl[hydrotris(1*H*-pyrazolato-N^1)borato(1−)-N^2,N^2,N^2][(1,2,3-η)-2-methyl-2-propenyl]molybdenum, 9CI

[38834-21-8]

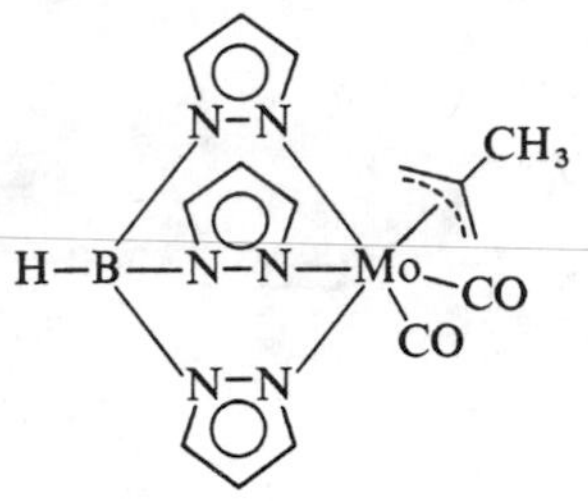

M 420.088

Yellow cryst.

Trofimenko, S. *et al*, *J. Am. Chem. Soc.*, 1969, **91**, 588, 3183 (*synth, nmr*)
Meekin, P. *et al*, *J. Am. Chem. Soc.*, 1972, **94**, 5677 (*nmr*)
Holt, E.M. *et al*, *J. Chem. Soc., Dalton Trans.*, 1973, 2444 (*cryst struct*)

$C_{16}H_8Mo_2O_6$ Mo-00092

[μ-[(1,2,3,3*a*,8*a*-η:4,5,6,7,8-η)azulene]]hexacarbonyldimolybdenum, 10CI

Hexacarbonylazulenedimolybdenum. Azulenedimolybdenum hexacarbonyl

[60295-03-6]

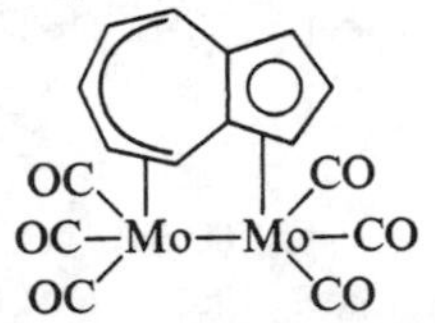

M 488.116

Blackish-red cryst.; dec. >150°. Diamagnetic. Fluxional with CO scrambling.

Burton, R. *et al*, *J. Chem. Soc.*, 1960, 4290 (*synth, nmr*)
King, R.B. *et al*, *Inorg. Chem.*, 1965, **4**, 475 (*synth, ir, nmr*)
Churchill, M.R. *et al*, *Inorg. Chem.*, 1968, **7**, 1545 (*cryst struct*)
Cotton, F.A. *et al*, *Inorg. Chem.*, 1976, **15**, 1866 (*cryst struct*)
Cotton, F.A. *et al*, *Inorg. Chem.*, 1976, **15**, 2806 (*cmr*)

$C_{16}H_{10}Mo_2O_6$ Mo-00093

Hexacarbonylbis(η^5-2,4-cyclopentadien-1-yl)dimolybdenum, 10CI

Bis(tricarbonyl-π-cyclopentadienylmolybdenum)

[12091-64-4]

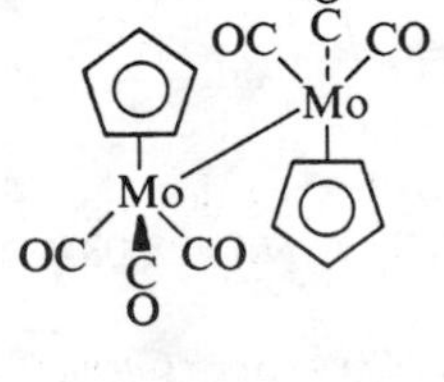

M 490.131

Red-violet cryst. (pentane/CH_2Cl_2). $Bp_{0.1}$ 150° subl.

Inorg. Synth., 1963, **7**, 99 (*synth*)
Dub, M., *Organometallic Compounds*, Springer-Verlag, Berlin, 2nd Ed., 1966, Vol. 1 (*props*)

Curtis, M.D., *Inorg. Nucl. Chem. Lett.*, 1970, 859.
Haines, R.J. *et al*, *J. Organomet. Chem.*, 1971, **28**, 97.
Birdwhistell, R. *et al*, *J. Organomet. Chem.*, 1978, **157**, 239 (*synth*)

$C_{16}H_{14}ClF_{12}MoP$ Mo-00094

Chloro(η^5-2,4-cyclopentadien-1-yl)bis[(2,3-η)-1,1,1,4,4,4-hexafluoro-2-butyne](trimethylphosphine)molybdenum, 10CI

[66118-38-5]

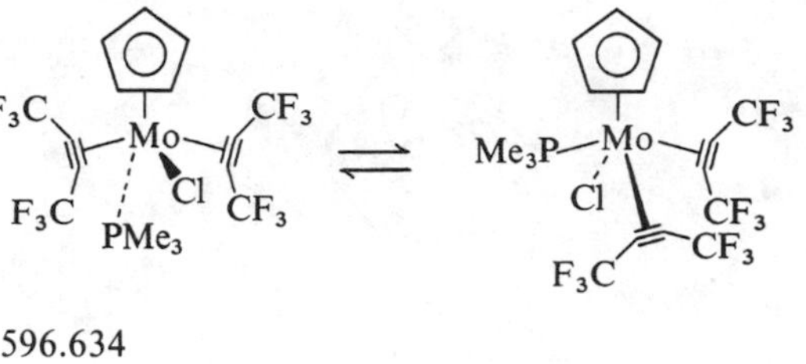

M 596.634

Mixt. of isomers in soln., with suggested structs. as shown. Pale-yellow cryst. (CH_2Cl_2/hexane). Mp 175° dec.

Davidson, J.L. *et al*, *J. Chem. Soc., Dalton Trans.*, 1977, 2246 (*synth, ir, pmr, nmr*)

$C_{16}H_{17}Mo_2O_4P$ Mo-00095

Tetracarbonylbis(η^5-2,4-cyclopentadien-1-yl)[μ-(dimethylphosphino)]-μ-hydrodimolybdenum, 10CI

[12092-01-2]

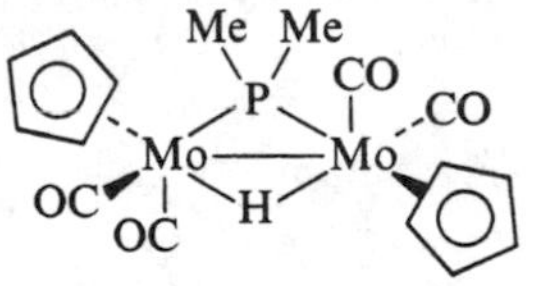

M 496.162

Orange-red rod-like cryst. (C_6H_6/hexane). Mp 215°.

Hayter, R.G., *Inorg. Chem.*, 1962, **2**, 1031 (*synth, ir, nmr*)
Petersen, J.L. *et al*, *Inorg. Chem.*, 1978, **17**, 1308 (*struct*)
Petersen, J.L. *Inorg. Chem.*, 1980, **19**, 186 (*struct*)

$C_{16}H_{21}MoO_2^{\oplus}$ Mo-00096

Dicarbonyl(η^4-2-methylene-1,3-propanediyl)[(1,2,3,4,5-η)-1,2,3,4,5-pentamethyl-2,4-cyclopentadien-1-yl]molybdenum(1+), 10CI

Dicarbonyl(η^4-trimethylenemethane)(η^5-cyclopentadienyl)molybdenum(1+)

[75593-05-4]

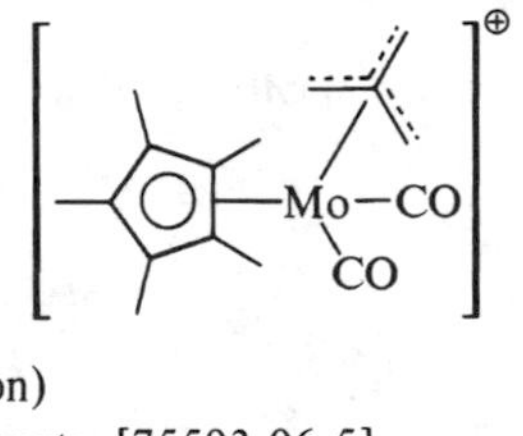

M 341.281 (ion)

Tetrafluoroborate: [75593-06-5].
$C_{16}H_{21}BF_4MoO_2$ M 428.084
Yellow cryst.

Barnes, S.G. *et al*, *J. Chem. Soc., Chem. Commun.*, 1980, 267 (*synth, props*)

$C_{16}H_{48}Mo_2P_4$ Mo-00097

Tetramethyltetrakis(trimethylphosphine)dimolybdenum, 9CI

[64376-69-8]

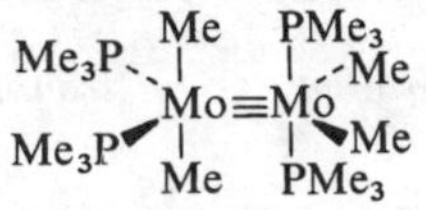

M 556.330

Blue cryst. Mp 220° dec. (sealed tube).

Anderson, R.A. *et al*, *J. Chem. Soc., Chem. Commun.*, 1977, 283 (*synth*)
Anderson, R.A. *et al*, *J. Chem. Soc., Dalton Trans.*, 1978, 446 (*synth, nmr, struct*)
Anderson, R.A. *et al*, *J. Am. Chem. Soc.*, 1981, **103**, 3953 (*cryst struct*)

$C_{17}H_{14}MoO_4$ Mo-00098

Tetracarbonylbis(η^5-2,4-cyclopentadien-1-yl)[μ-[(1,2-η:2,3-η)-1,2-propadiene]]dimolybdenum, 10CI

Allenetetracarbonylbiscyclopentadienyldimolybdenum

[62573-31-3]

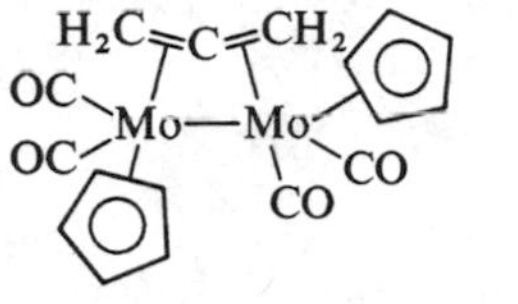

M 378.235

Red cryst. (toluene/pentane), mod. air stable in solid state. Sol. aromatic solvs. Sol. CH_2Cl_2, $CHCl_3$ but dec. after several hrs. V-shaped allene moiety.

Bailey, W.I., *J. Am. Chem. Soc.*, 1978, **100**, 802 (*ir, nmr, ms, synth, cryst struct*)

$C_{17}H_{19}Mo^{\oplus}$ Mo-00099

(η^6-Benzene)[(1,2,3,4-η)-1,3,5,7-cyclooctatetraene](η^3-2-propenyl)molybdenum(1+)

η^3-Allyl-η^6-benzene-η^4-cyclooctatetraenemolybdenum(1+)

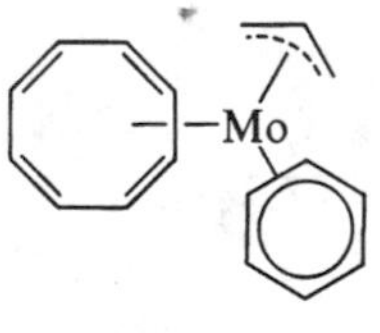

M 319.277 (ion)

Hexafluorophosphate: [59464-56-1].
$C_{17}H_{19}F_6PMo$ M 464.241
Dark-red cryst. (CH_2Cl_2, THF).

Green, M.L.H. *et al*, *J. Chem. Soc., Dalton Trans.*, 1974, 311; 1976, 213 (*synth, nmr*)

$C_{17}H_{23}Mo_2O_3$ Mo-00100

Bis(η^7-2,4,6-cycloheptatrien-1-yl)tri-μ-methoxydimolybdenum, 10CI

[53321-77-0]

As Trichlorobis(η^7-2,4,6-cycloheptatrien-1-yl)dimolybdenum, Mo-00081 with

X = OMe

M 467.247

Ashworth, E.F. *et al*, *J. Chem. Soc., Dalton Trans.*, 1977, 1693 (*synth*)

$C_{18}H_{10}MoO_5$ Mo-00101

Pentacarbonyl(diphenylmethylene)molybdenum, 10CI

[76346-28-6]

$Ph_2C{=}Mo(CO)_5$

M 402.214

Dark-red cryst.

Beatty, R.P. *et al*, *J. Am. Chem. Soc.*, 1981, **103**, 238 (*synth, props*)

$C_{18}H_{14}FeMoO_3$ Mo-00102

Tricarbonyl(η^5-2,4-cyclopentadien-1-yl)ferrocenylmolybdenum, 9CI

[60349-56-6]

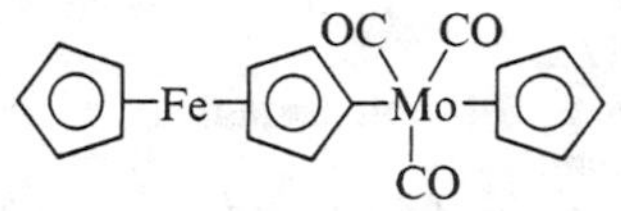

M 430.094

Mp 140° dec.

Pannell, K.H. *et al*, *Inorg. Chem.*, 1976, **15**, 2671 (*synth, ir, pmr*)

$C_{18}H_{14}FeMoO_6$ Mo-00103

Pentacarbonyl(ethoxyferrocenylmethylene)molybdenum, 10CI

[68830-11-5]

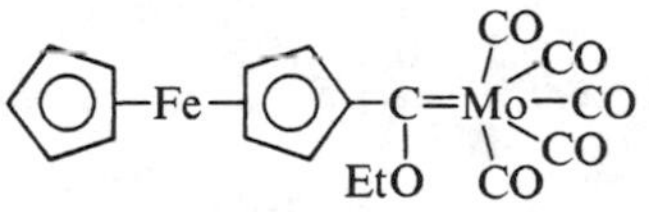

M 478.092

Black-red cryst. (pentane). Mp >85° dec.

Fischer, E.O. *et al*, *Chem. Ber.*, 1978, **111**, 3530 (*synth, ir, pmr, cmr*)

Fischer, E.O. *et al*, *J. Organomet. Chem.*, 1980, **191**, 261 (*ir, pmr, cmr*)

$C_{18}H_{14}MoO_6Ru$ Mo-00104

Pentacarbonyl(ethoxyruthenocenylmethylene)molybdenum, 10CI

[75280-68-1]

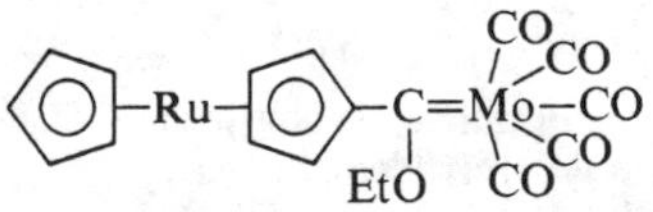

M 523.315

Orange-yellow cryst. (CH_2Cl_2/pentane). Mp 108°.

Fischer, E.O. *et al*, *J. Organomet. Chem.*, 1980, **191**, 261 (*synth, ir, ms, pmr, cmr*)

$C_{18}H_{16}Mo_2O_4$ Mo-00105

[η-[(2,3-η:2,3-η)-2-Butyne]]tetracarbonylbis(η^5-2,4-cyclopentadien-1-yl)dimolybdenum, 10CI

μ-Dimethylacetylenebis(cyclopentadienyldicarbonylmolybdenum)

[68079-61-8]

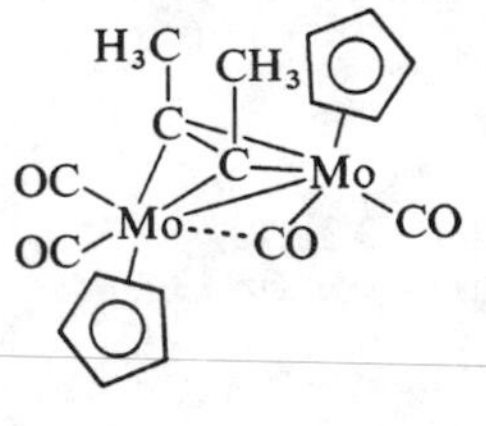

M 488.202

Dark-red cryst. Sol. C_6H_6, CH_2Cl_2.

Ginley, D.S. *et al*, *Inorg. Chim. Acta*, 1977, **23**, 85 (*synth*)

Curtis, M.D. *et al*, *J. Organomet. Chem.*, 1978, **161**, 23 (*synth*)

Bailey, W.I. *et al*, *J. Am. Chem. Soc.*, 1978, **100**, 5764 (*synth, cryst struct*)

Knox, S.A.R. *et al*, *J. Chem. Soc., Chem. Commun.*, 1978, 221; 1979, 934 (*props*)

$C_{18}H_{24}Mo$ Mo-00106

Bis[(1,2,3,4,5,6-η)-1,3,5-trimethylbenzene]molybdenum, 9CI

Bis(η^6-mesitylene)molybdenum

[12131-50-9]

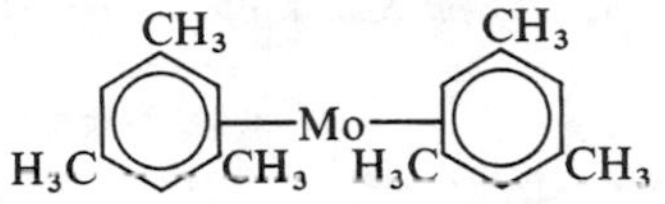

M 336.328

Green air-sensitive cryst.

Benfield, F.W.S. *et al*, *J. Chem. Soc., Chem. Commun.*, 1973, 866 (*synth*)

Green, M.L.H. *et al*, *J. Chem. Soc., Dalton Trans.*, 1973, 301 (*synth*)

$C_{19}H_{18}MoO_3Zr$ Mo-00107

[μ-(Acetyl-*C*:*O*)][μ-(η^2-carbonyl)][carbonyl(η^5-2,4-cyclopentadien-1-yl)molydenum]bis(η^5-2,4-cyclopentadien-1-yl)zirconium

*μ-[Acetyl-*C(Mo):O(Zr)*]μ-[carbonyl-*C(Mo):O(Zr)*]bis-η-cyclopentadienylzirconium(carbonyl)η-cyclopentadienylmolybdenum*

[76671-79-9]

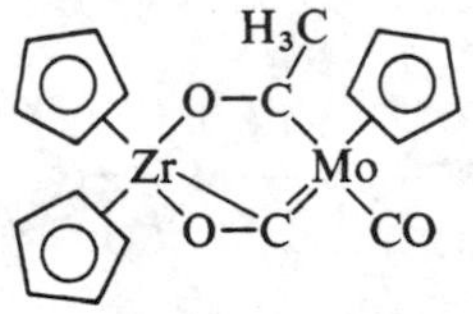

M 481.509

Air-stable solid. Yellow cryst. Unreactive to most donor ligands.

Longato, B. *et al*, *J. Am. Chem. Soc.*, 1981, **103**, 209 (*synth, ir, nmr, cryst struct*)

$C_{19}H_{31}MoO_6P_2^{\oplus}$ Mo-00108

[(2,3-η)-2-Butyne][(1,2,3,3*a*,7*a*-η)-1*H*-inden-1-yl]bis(trimethylphosphite)molybdenum(1+), 10CI

[64347-67-7]

As [(2,3-η)-2-Butyne][(1,2,3,3*a*,7*a*-η)-1*H*-inden-1-yl]bis(trimethylphosphine)molybdenum(1+), Mo-00109 with

R = OMe

M 513.338 (ion)

Tetrafluoroborate: [64347-68-8].
$C_{19}H_{31}BF_4MoO_6P_2$ M 600.141
Yellow cryst.

Bottrill, M. *et al*, *J. Chem. Soc., Dalton Trans.*, 1977, 2365 (*synth*)
Allen, S.R. *et al*, *J. Chem. Soc., Dalton Trans.*, 1983, 927 (*props*)

$C_{19}H_{31}MoP_2^{\oplus}$ Mo-00109

[(2,3-η)-2-Butyne][(1,2,3,3a,7a-η)-1H-inden-1-yl]bis(trimethylphosphine)molybdenum(1+), 10CI

[78091-05-1]

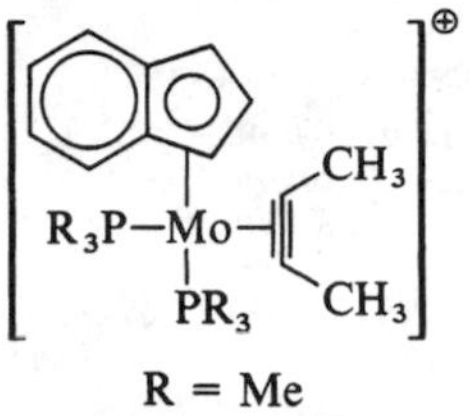

R = Me

M 417.341 (ion)

Tetrafluoroborate: [78091-06-2].
$C_{19}H_{31}BF_4MoP_2$ M 504.145
Blue air-sensitive cryst.

Allen, S.R. *et al*, *J. Chem. Soc., Dalton Trans.*, 1981, 873 (*synth*)

$C_{19}H_{38}MoO_5PbSi_4$ Mo-00110

[Bis[bis(trimethylsilyl)methyl]plumbylene]pentacarbonylmolybdenum, 10CI

[62789-86-0]

$[(Me_3Si)_2CH]_2Pb—Mo(CO)_5$

M 761.988
Yellow complex.

Cotton, J.D. *et al*, *J. Chem. Soc., Dalton Trans.*, 1976, 2275.

$C_{20}H_{10}Co_2MoO_8$ Mo-00111

Dicarbonyl(η⁵-2,4-cyclopentadien-1-yl)(hexacarbonyldicobalt)[μ₃-(phenylmethylidyne)]molybdenum, 10CI

[68185-45-5]

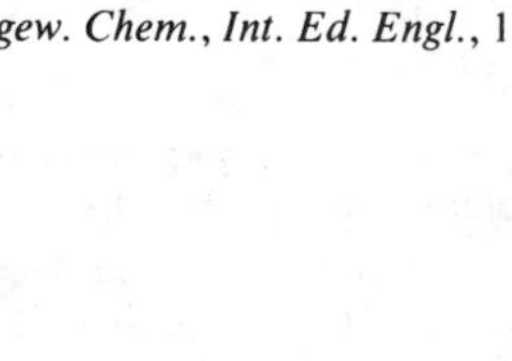

M 592.101
Black cryst. Mp 165° dec. pmr δ 5.22 (s, C_5H_5) 7.21 (m, C_6H_5—)ppm (CCl_4).

Beurich, H., *Angew. Chem., Int. Ed. Engl.*, 1978, **17**, 863 (*synth, ir, pmr*)

$C_{20}H_{10}MoO_5$ Mo-00112

Pentacarbonyl(2,3-diphenyl-2-cyclopropen-1-ylidene)molybdenum, 9CI

[37766-71-5]

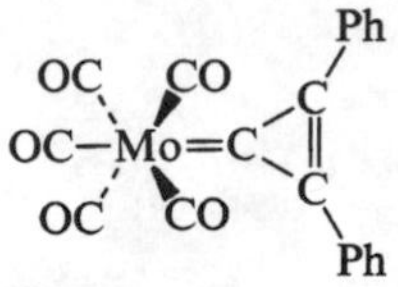

M 426.236
Yellow cryst. Mp 180° dec.

Rees, C.W. *et al*, *J. Chem. Soc., Chem. Commun.*, 1972, 428 (*synth, props*)

$C_{20}H_{15}ClMoO$ Mo-00113

Carbonylchloro(η⁵-2,4-cyclopentadien-1-yl)(diphenylacetylene)molybdenum

Carbonylchloro(η⁵-2,4-cyclopentadien-1-yl)[1,1′-(η²-1,2-ethynediyl)bis[benzene]]molybdenum, 10CI

[54720-88-6]

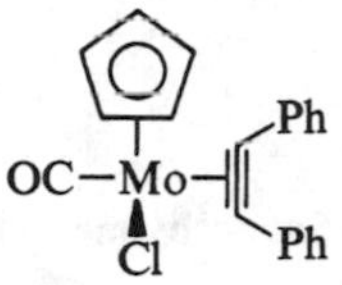

M 402.731
Green cryst.

Davidson, J.L. *et al*, *J. Chem. Soc., Dalton Trans.*, 1977, 287, 2246 (*props*)

$C_{20}H_{16}Mo$ Mo-00114

Bis[(1,2,3,4,4a,8a-η)naphthalene]molybdenum, 10CI

[67422-84-8]

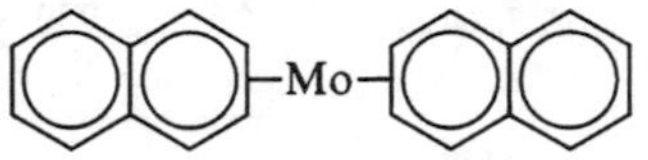

M 352.286
V. air-sensitive deep-red solid.

Kündig, E.P. *et al*, *J. Chem. Soc., Chem. Commun.*, 1977, 912 (*synth, props*)

$C_{20}H_{18}Mo_2$ Mo-00115

Bis(η⁵-2,4-cyclopentadien-1-yl)bis[μ-(η:η⁵-2,4-cyclopentadien-1-ylidene)]dimolybdenum, 10CI

Dehydromolybdenocene

[66705-66-6]

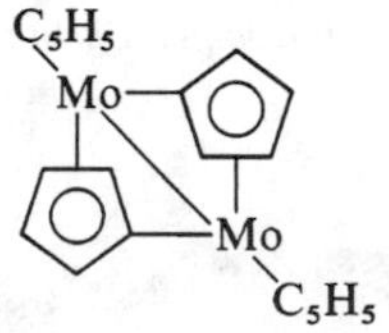

M 450.242
Starting material for $(C_5H_5)_2Mo$ chemistry. Dark-red cryst.

Smart, J.C. *et al*, *Inorg. Chem.*, 1978, **17**, 3290.
Meunier, B. *et al*, *Acta Crystallogr., Sect. B*, 1979, **35**, 2558 (*cryst struct*)
Berry, M. *et al*, *J. Chem. Soc., Dalton Trans.*, 1980, 29 (*synth, nmr*)

$C_{20}H_{28}MoON_2O_4$ Mo-00116

Dicarbonylbis(2-methyl-2-propanolato)bis(pyridine)molybdenum, 10CI

Bis(tert-*butoxy*)*dicarbonylbis*(*pyridine*)*molybdenum*

[72202-89-2]

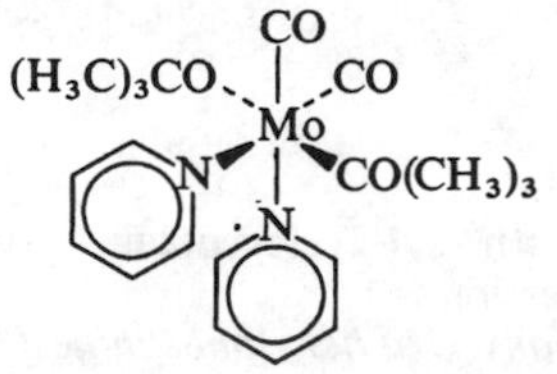

M 472.392

Green, air-sensitive cryst. Sol. toluene with dec., insol. hexane. ir ν_{CO} 1908, 1768 cm^{-1} (nujol).

Chisholm, M.H. *et al*, *J. Am. Chem. Soc.*, 1979, **101**, 7615 (*synth*, *ir*, *cryst struct*, *ms*)

$C_{20}H_{48}Mo_2N_4$ Mo-00117

Bis[μ-[2-methyl-2-propanaminato(2−)]]bis[2-methyl-2-propanaminato(2−)]tetramethyldimolybdenum, 10CI

Tetrakis(tert-*butylaminato*)*tetramethyldimolybdenum*

[73448-06-3]

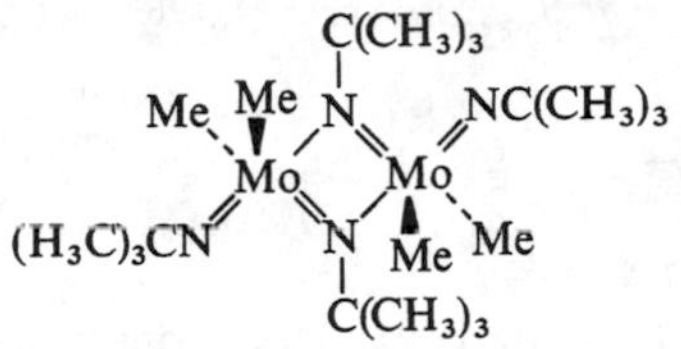

M 536.506

Air-sensitive, violet cryst. Sol. hexane.

Nugent, W.A. *et al*, *J. Am. Chem. Soc.*, 1980, **102**, 1759 (*synth*)
Thorn, D.L. *et al*, *J. Am. Chem. Soc.*, 1981, **103**, 357 (*struct*)

$C_{21}H_{24}MoO_3$ Mo-00118

Tricarbonyl(η^6-octamethylnaphthalene)molybdenum

[79391-57-4]

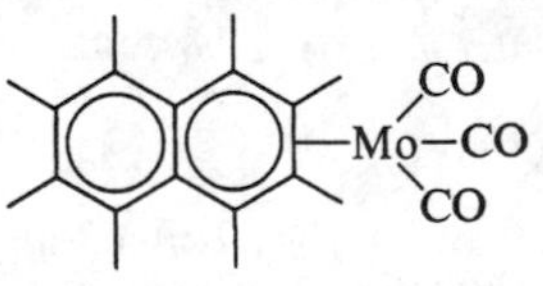

M 420.359

Bright-orange cryst.

Hull, J.W. *et al*, *Organometallics*, 1982, **1**, 264 (*synth*, *struct*)

$C_{23}H_{15}MoO_5P$ Mo-00119

Pentacarbonyl(triphenylphosphine)molybdenum, 10CI

[14971-42-7]

$(OC)_5MoPPh_3$ (O_h)

M 498.282

Pale-yellow cryst. Mp 127-8°.

Magee, T.A. *et al*, *J. Am. Chem. Soc.*, 1961, **83**, 3200 (*synth*)
Cotton, F.A. *et al*, *Inorg. Chem.*, 1981, **20**, 580 (*synth*, *cryst struct*)

$C_{23}H_{16}MoO_3$ Mo-00120

Tricarbonyl[(1,2,3,4,5,6-η)-7-(diphenylmethylene)-1,3,5-cycloheptatriene]molybdenum, 9CI

Tricarbonyl η^6-8,8-diphenylheptafulvenemolybdenum

[52445-38-2]

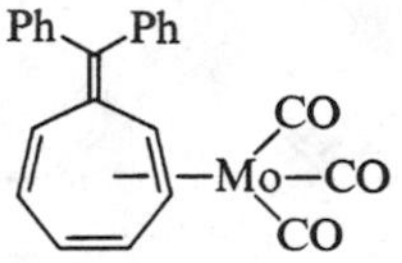

M 436.318

Red cryst.

Howell, J.A.S., *J. Chem. Soc., Dalton Trans.*, 1974, 293 (*synth*, *props*)

$C_{23}H_{21}ClMo$ Mo-00121

Chlorobis(1-phenyl-1-propyne)(η^5-cyclopentadienyl)molybdenum

Chloro(η^5-*2,4-cyclopentadien-1-yl*)*bis*[[(*1,2-η*)-*1-propynyl*]*benzene*]*molybdenum*, *10CI*

[60109-61-7]

As Bis[(2,3-η)-2-butyne]carbonyl(η^5-2,4-cyclopentadien-1-yl)molybdenum(1+), Mo-00083 with

R = Ph, R′ = CH_3

M 428.812

Yellow cryst. (CH_2Cl_2/hexane). Mp 160-2°.

Davidson, J.L. *et al*, *J. Chem. Soc., Dalton Trans.*, 1976, 738 (*synth*)
Faller, J.W. *et al*, *J. Organomet. Chem.*, 1979, **172**, 171 (*pmr*, *isom*)

$C_{24}H_{17}MoO_4P$ Mo-00122

Tetracarbonyl[[2-(η^2-ethenyl)phenyl]diphenylphosphine-*P*]-molybdenum, 9CI

Tetracarbonyl(o-*styryldiphenylphosphine*)*molybdenum*

[12290-82-3]

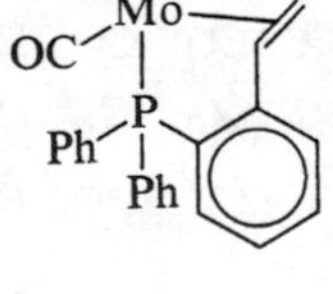

M 496.310

Air-stable yellow cryst. Mp 155-65° dec. Rotational isomers thought to be present in soln.

Bennett, M.A. *et al*, *J. Organomet. Chem.*, 1967, **10**, 301 (*synth*, *props*, *nmr*, *ir*)
Bennett, M.A. *et al*, *J. Organomet. Chem.*, 1973, **51**, 289 (*ir*, *synth*, *props*)

$C_{24}H_{24}Mo_2$ Mo-00123

[μ-[(1,2,3,4-η:5,6,7,8-η)-1,3,5,7-cyclooctatetraene-]]bis[(1,2,3,4-η)-1,3,5,7-cyclooctatetraene]dimolybdenum, 10CI

Tris(cyclooctatetraene)dimolybdenum

[66719-22-0]

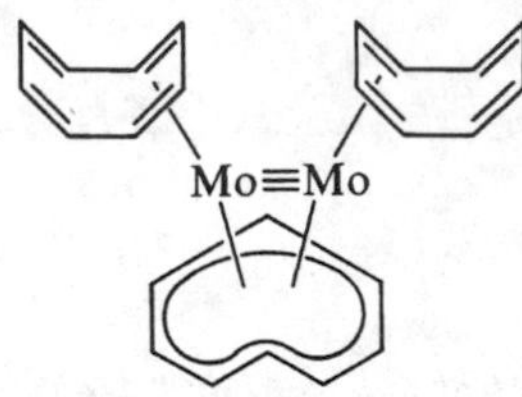

M 504.334

Black cryst. (toluene).

Cotton, F.A. *et al*, *Inorg. Chem.*, 1978, **17**, 2093 (*synth*, *struct*)

$C_{24}H_{30}Mo_2O_4$ Mo-00124

Tetracarbonylbis[(1,2,3,4,5-η)-1,2,3,4,5-pentamethyl-2,4-cyclopentadien-1-yl]dimolybdenum, 10CI

[12132-04-6]

As Tetracarbonylbis(η^5-2,4-cyclopentadien-1-yl)dimolybdenum, Mo-00079 with

R = CH_3

M 574.379

Deep-red cryst. Mp 219° dec. ir ν_{CO} 1867, 1842 cm^{-1}.

Monomer: Dicarbonyl(pentamethylcyclopentadienyl)-molybdenum.

$C_{12}H_{15}MoO_2$ M 287.189

Unknown.

King, R.B., *Coord. Chem. Rev.*, 1976, **20**, 155 (*rev*)
Curtis, M.D. *et al*, *J. Organomet. Chem.*, 1978, **161**, 23 (*synth*)
King, R.B. *et al*, *J. Organomet. Chem.*, 1979, **171**, 53 (*synth*)

$C_{24}H_{54}Mo_2O_6$ Mo-00125

Hexakis(2-methyl-2-propanolato)dimolybdenum, 10CI

*Hexa-*tert*-butoxydimolybdenum*

[60764-63-8]

$(H_3C)_3CO$ $CO(CH_3)_3$
$(H_3C)_3CO-Mo \equiv Mo-CO(CH_3)_3$
$(H_3C)_3CO$ $CO(CH_3)_3$

M 630.567

Has no bridging ligands. Red-orange volatile cryst. (pentane). Sol. hydrocarbons, THF. $Bp_{0.001}$ 100-20° subl. O_2- and moisture-sensitive.

Chisholm, M.H. *et al*, *Inorg. Chem.*, 1977, **16**, 1801 (*synth*, *ms*, *ir*, *pmr*, *cmr*)
Chisholm, M.H. *et al*, *J. Am. Chem. Soc.*, 1981, **103**, 1305 (*props*)

$C_{24}H_{66}Mo_2Si_6$ Mo-00126

Hexakis[(trimethylsilyl)methyl]dimolybdenum, 9CI

[34439-17-3]

Me_3SiCH_2 CH_2SiMe_3
$Me_3SiCH_2-Mo \equiv Mo-CH_2SiMe_3$
Me_3SiCH_2 CH_2SiMe_3

M 715.178

Air-stable yellow plate-like cryst. Mp 99°. Bp 120° subl. Mo-Mo distance 2.167Å.

Huq, F. *et al*, *J. Chem. Soc.*, *Chem. Commun.*, 1971, 1079 (*cryst struct*)
Mowat, W. *et al*, *J. Chem. Soc.*, *Dalton Trans.*, 1972, 533 (*synth*)
Wilkinson, G. *et al*, *Pure Appl. Chem.*, 1972, **30**, 627 (*synth*, *props*)

$C_{25}H_{20}ClMoO_2P$ Mo-00127

Dicarbonylchloro(η^5-2,4-cyclopentadien-1-yl)(triphenylphosphine)molybdenum, 10CI

Chlorodicarbonyl(triphenylphosphine)(η^5-cyclopentadienyl)molybdenum

[12115-01-4]

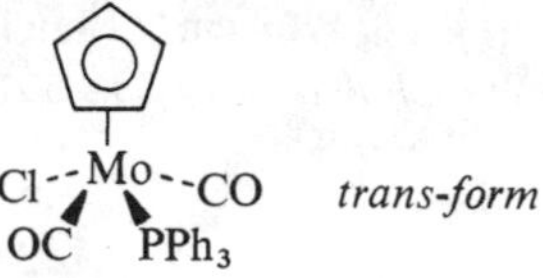

M 514.799

Exists as mixt. of *cis-* and *trans-*forms, not separated. Orange cryst. (C_6H_6/2,2,4-trimethylpentane). Sol. $CHCl_3$. Mp 170° dec. Equilibrium ratio of *cis/trans* in $CHCl_3$ ~50:1.

Haines, R.J. *et al*, *J. Chem. Soc. (A)*, 1967, 94 (*synth*, *ir*, *pmr*, *uv*, *props*)
Faller, J.W. *et al*, *J. Am. Chem. Soc.*, 1970, **92**, 5852 (*pmr*)
Blumer, D.J. *et al*, *J. Organomet. Chem.*, 1979, **173**, 71 (*synth*)

$C_{25}H_{21}MoO_5P$ Mo-00128

Tetracarbonyl(1-methoxyethylidene)(triphenylphosphine)-molybdenum

H_3C OMe
C
OC ‖ CO
Mo
OC | PPh_3
CO

M 528.352

***cis*-form** [23606-69-1]

Stable yellow-orange cryst. Mp 104-6°.

Fischer, E.O. *et al*, *Chem. Ber.*, 1969, **102**, 1495 (*synth*, *pmr*, *ir*)

$C_{25}H_{34}MoO_6P_2$ Mo-00129

(η^5-2,4-Cyclopentadien-1-yl)[(1,2-η)-1,2-diphenylethenyl]bis(trimethylphosphite-*P*)molybdenum, 10CI

[76791-31-6]

MeO Ph
MeO—P—Mo —H
MeO
MeO—P Ph
MeO OMe

M 588.428

Green cryst.

Green, M. *et al*, *J. Am. Chem. Soc.*, 1981, **103**, 1267 (*synth*, *cryst struct*)

$C_{25}H_{54}Mo_2O_7$ **Mo-00130**

Hexa-*tert*-butoxy-μ-carbonyldimolybdenum

[67063-52-9]

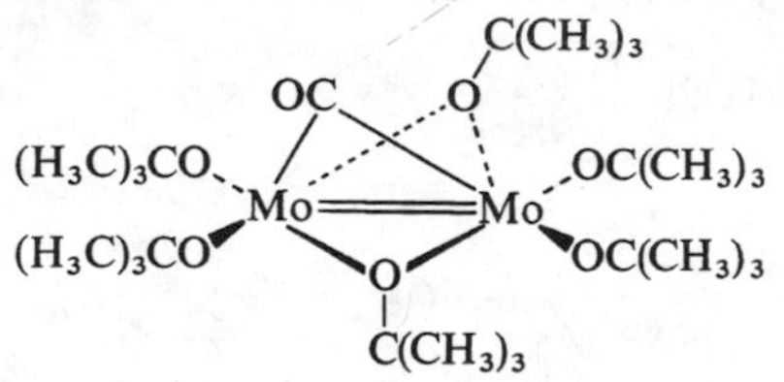

M 658.577

Example of a complex with an Mo-Mo double bond. Purple cryst. (hexane at −15°). Sol. alkanes. Dec. on warming with loss of CO.

Chisholm, M.H. *et al*, *J. Am. Chem. Soc.*, 1979, **101**, 7645 (*synth, cmr, pmr, ir, cryst struct*)
Chisholm, M.H. *et al*, *J. Am. Chem. Soc.*, 1980, **102**, 3451.

$C_{26}H_{20}MoO_6Si$ **Mo-00131**

Pentacarbonyl[ethoxy(triphenylsilyl)methylene]molybdenum, 10CI

Pentacarbonyl[ethoxy(triphenylsilyl)carbene]molybdenum

[65573-54-8]

$(OC)_5Mo{=}C(OEt)SiPh_3$

M 552.466

Orange cryst. Mp 77-9°.

Fischer, E.O. *et al*, *Chem. Ber.*, 1977, **110**, 3467 (*synth, props*)

$C_{27}H_{26}MoO_2PTl$ **Mo-00132**

Dicarbonyl(η^5-2,4-cyclopentadien-1-yl)(dimethylthallium)(triphenylphosphine)molybdenum

[66125-47-1]

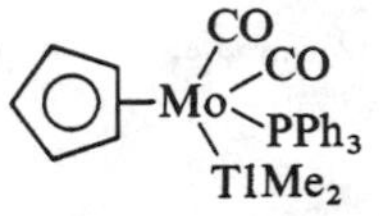

M 713.798

Mp 146-8°.

Walther, B. *et al*, *J. Organomet. Chem.*, 1978, **145**, 285.

$C_{28}H_{20}Mo_2O_4$ **Mo-00133**

μ-Carbonyltricarbonylbis(η^5-2,4-cyclopentadien-1-yl)[μ-[1,1′-(η^2:η^2-1,2-ethynediylbisbenzene]]dimolybdenum, 10CI

[67507-24-8]

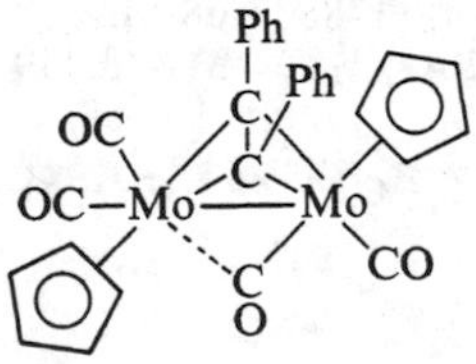

M 612.344

Ginley, D.S. *et al*, *Inorg. Chim. Acta*, 1977, **23**, 85 (*synth*)
Bailey, W.I. *et al*, *J. Am. Chem. Soc.*, 1978, **100**, 5764 (*synth, cryst struct*)
Curtis, M.D. *et al*, *J. Organomet. Chem.*, 1978, **161**, 23 (*synth, ir, nmr*)
Knox, S.A.R. *et al*, *J. Chem. Soc., Chem. Commun.*, 1978, 221; 1979, 934 (*props*)

$C_{29}H_{24}MoNO_4P_2^{\oplus}$ **Mo-00134**

Tricarbonyl[1,2-ethanediylbis[diphenylphosphine]-*P,P′*]nitrosylmolybdenum(1+), 9CI

$[Mo(CO)_3(NO)(Ph_2PCH_2CH_2PPh_2)]^{\oplus}$

M 608.400 (ion)

Stereochemistry not reported.

Hexafluorophosphate: [37757-20-3].
$C_{29}H_{24}F_6MoNO_4P_3$ M 753.365
Yellow cryst. (CH_2Cl_2/hexane). Sol. polar org. solvs. Mp 160° dec. Solns. slowly dec. in air.

Johnson, B.F.G. *et al*, *J. Organomet. Chem.*, 1972, **40**, C36 (*synth*)
Connelly, N.G., *J. Chem. Soc., Dalton Trans.*, 1973, 2183 (*synth*)

$C_{29}H_{26}FeMoNO_4P$ **Mo-00135**

Tetracarbonyl[1-[(dimethylamino)methyl]-2-(diphenylphosphino)ferrocene-*N,P*]molybdenum, 9CI

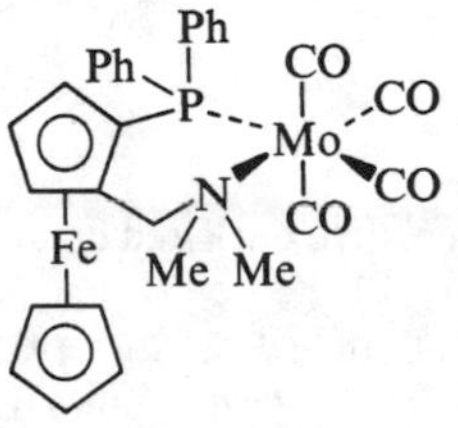

M 635.289

Cryst. (Me_2CO/pet. ether). Mp 175-8° dec.

Kotz, J.C. *et al*, *J. Organomet. Chem.*, 1975, **84**, 255 (*synth, ir, pmr*)

$C_{30}H_{66}Mo_2$ **Mo-00136**

Hexakis(2,2-dimethylpropyl)dimolybdenum, 9CI

Hexakisneopentyldimolybdenum

[42077-05-4]

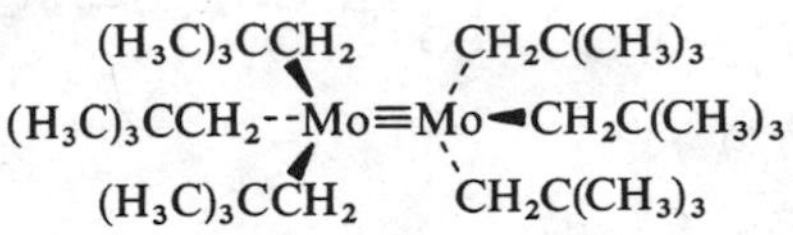

M 618.731

Yellow plate-like cryst. Mp 135-8°. Subl. *in vacuo* at 130° with dec.

Mowat, W. *et al*, *J. Chem. Soc., Dalton Trans.*, 1973, 1120 (*synth, props*)

$C_{31}H_{29}ClMoO_2P_2$ **Mo-00137**

Dicarbonylchloro[1,2-ethanediylbis[diphenylphosphine]-*P,P′*](η^3-2-propenyl)molybdenum, 10CI

Bis(η^3-allyl)dicarbonylchloro[1,2-bis(diphenylphosphino)ethane]molybdenum

[33135-95-4]

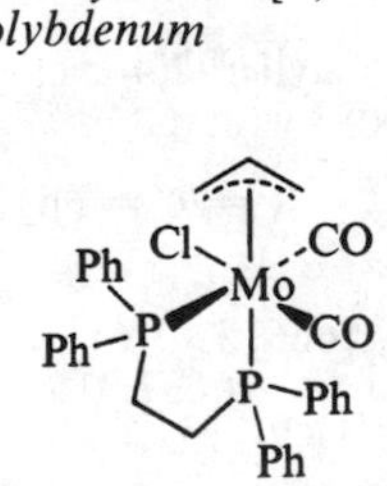

M 626.909

Orange cryst. ($CHCl_3$/pet. ether).

Hayter, R.G., *J. Organomet. Chem.*, 1968, **13**, P1.
Tom Dieck, H. *et al*, *J. Organomet. Chem.*, 1968, **14**, 375 (*synth*)
Graham, A.J. *et al*, *Cryst. Struct. Commun.*, 1976, **5**, 891 (*cryst struct*)

Brisdon, B.J., *J. Organomet. Chem.*, 1977, **125**, 225 (*synth, ir, nmr*)
Faller, J.W. *et al*, *J. Am. Chem. Soc.*, 1979, **101**, 865 (*nmr*)

$C_{32}H_{36}Mo_2O_8$ Mo-00138

Tetrakis[μ-(2,6-dimethoxyphenyl)-*C*:*O*)]dimolybdenum, 10CI

[64836-04-0]

M 740.512

Extremely air-sensitive dark-red cryst. (THF/hexane). Mo-Mo 2.065(1)Å.

Cotton, F.A. *et al*, *J. Am. Chem. Soc.*, 1977, **99**, 7372 (*synth*)
Cotton, F.A. *et al*, *Inorg. Chem.*, 1978, **17**, 2087 (*cryst struct*)

$C_{33}H_{62}MoO_2P_2$ Mo-00139

[(2,3,5,6-η)-Bicyclo[2.2.1]hepta-2,5-diene]dicarbonylbis(tributylphosphine)molybdenum, 10CI

[64146-27-6]

M 648.739

Yellow cryst.

Tom Dieck, H. *et al*, *J. Organomet. Chem.*, 1977, **132**, 255 (*synth, ir, nmr*)

$C_{34}H_{25}MoO^{\oplus}$ Mo-00140

Carbonyl(η^5-cyclopentadienyl)bis(diphenylacetylene)molybdenum(1+)

Carbonyl(η^5-2,4-cyclopentadien-1-yl)bis[1,1′-(η^2-1,2-ethynediyl)bis[benzene]]molybdenum(1+), 10CI

[66615-16-5]

As Bis[(2,3-η)-2-butyne]carbonyl(η^5-2,4-cyclopentadien-1-yl)molybdenum(1+), Mo-00083 with

R = R′ = Ph

M 545.511 (ion)

Tetrafluoroborate: [66615-17-6].
$C_{34}H_{25}BF_4MoO$ M 632.315
Yellow cryst. (CH_2Cl_2/Et_2O).

Bottrill, M. *et al*, *J. Chem. Soc., Dalton Trans.*, 1977, 2365 (*synth*)
Allen, J.R. *et al*, *J. Chem. Soc., Dalton Trans.*, 1981, 873 (*props*)

$C_{34}H_{40}Co_2Mo_2O_4$ Mo-00141

Di-μ-carbonyl[μ_4-(η^2-carbonyl-*C*:*C*:*C*)][μ-carbonylbis[(1,2,3,4,5-η)-1,2,3,4,5-pentamethyl-2,4-cyclopentadien-1-yl]dicobalt]bis(η^5-2,4-cyclopentadien-1-yl)dimolybdenum

[84474-41-9]

M 822.434

Model for CO activation on a stepped metal surface. Isol. from toluene. ir ν_{CO} 1750 s, 1711 vs, 1696 vs, 1633 vw cm^{-1} (THF). Bond length of μ_4—CO group 1.283(3) Å.

Brun, P. *et al*, *J. Chem. Soc., Chem. Commun.*, 1982, 926 (*synth, cryst struct, nmr, ir*)

$C_{37}H_{30}MoO_2PTl$ Mo-00142

Dicarbonyl(η^5-2,4-cyclopentadien-1-yl)(diphenylthallium)-(triphenylphosphine)molybdenum

[66125-48-2]

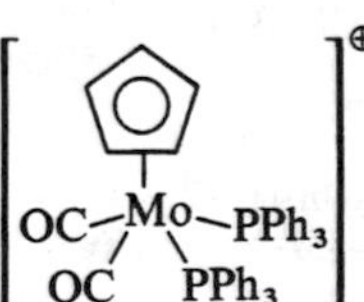

M 837.940

Mp 173-5°.

Walther, B. *et al*, *J. Organomet. Chem.*, 1978, **145**, 285 (*synth, ir, nmr*)

$C_{43}H_{35}MoO_2P_2^{\oplus}$ Mo-00143

Dicarbonyl(η^5-2,4-cyclopentadien-1-yl)bis(triphenylphosphine)molybdenum(1+)

M 741.636 (ion)

Tetrafluoroborate: [78833-68-8].
$C_{43}H_{35}BF_4MoO_2P_2$ M 828.439
Cryst.

Suenkel, K. *et al*, *Z. Naturforsch., B*, 1981, **36**, 474 (*synth, ir, nmr*)

$C_{45}H_{45}MoP_3$ Mo-00144

(η^6-Benzene)tris(methyldiphenylphosphine)molybdenum, 9CI

[35004-32-1]

M 774.712

Orange cryst. (C_6H_6).

Green, M.L.H. *et al*, *J. Chem. Soc.* (*A*), 1971, 2929 (*synth*)
Green, M.L.H. *et al*, *J. Chem. Soc.*, *Dalton Trans.*, 1974, 1361.
Inorg. Synth., 1977, **17**, 59 (*synth*)

$C_{52}H_{48}MoN_4P_4$ **Mo-00145**

Bis(dinitrogen)bis[1,2-bis(diphenylphosphino)ethane]molybdenum

Bis(dinitrogen)bis[1,2-ethanediylbis(diphenylphosphine)-P,P′]molybdenum, 9CI. *Bis(dinitrogen)bis[ethylenebis(diphenylphosphine)]molybdenum*, 8CI

[41700-58-7]

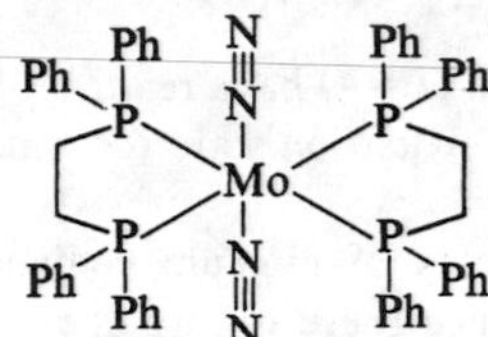

M 948.813

Orange cryst. (C_6H_6/MeOH). Mp 165° dec.

Atkinson, L.K. *et al*, *J. Chem. Soc.*, *Chem. Commun.*, 1971, 157 (*synth*)
Chatt, J. *et al*, *J. Organomet. Chem.*, 1971, **27**, C15 (*synth*)
George, T.A. *et al*, *J. Organomet. Chem.*, 1971, **30**, C13 (*synth*, *ir*)
Uchida, T. *et al*, *Bull. Chem. Soc. Jpn.*, 1971, **44**, 2883 (*struct*)
Inorg. Synth., 1980, **20**, 119 (*synth*)

$C_{56}H_{56}MoP_4$ **Mo-00146**

Bis[1,2-ethanediylbis[diphenylphosphine]-*P*,*P*′]bis(η^2-ethene)molybdenum, 10CI

[Bis(1,2-diphenylphosphino)ethane]bis(ethylene)molybdenum

[56307-56-3]

M 948.893

Orange cryst.

Bryne, J.W. *et al*, *J. Am. Chem. Soc.*, 1975, **97**, 3871 (*synth*)
Bryne, J.W. *et al*, *J. Chem. Soc.*, *Chem. Commun.*, 1977, 662 (*ir*, *nmr*)

Nb Niobium

J. A. Labinger

Niobium (Fr.), Niob or Niobium (Ger.), Niobio (Sp., Ital.), Ниобий (Niobii) (Russ.), ニオブ (Japan.)

Atomic Number. 41

Atomic Weight. 92.9064 (1)

Electronic Configuration. [Kr] $4d^4\,5s^1$

Oxidation States. These range from −3 (for $Nb(CO)_5^{3-}$) to +5, the more common values being −1, +1, +3, +4 and +5.

Coordination Number. 3 − 8 for conventional ligands.

Colour. Organoniobium compounds of oxidation state $+4(d^1)$ are highly coloured, often black. Those of even-electron oxidation states range from colourless (rare) through yellow to red and brown.

Availability. The commonest starting material is $NbCl_5$.

Handling. All organoniobium compounds require standard techniques for handling air-sensitive materials.

Toxicity. Niobium compounds are not considered to be especially toxic except for the carbonyls.

Isotopic Abundance. ^{93}Nb, 100%.

Spectroscopy. ^{93}Nb ($I = \frac{9}{2}$) has a relatively high sensitivity but Nb(V) is the only oxidation state for which data are available.

Mass spectrometry of niobium compounds is relatively straightforward since there is only one naturally occurring isotope.

Esr is a useful technique for compounds of oxidation state $+4(d^1)$.

Analysis. Niobium is usually determined gravimetrically. Tantalum interferes with the determination but this is not a problem when assaying organometallic compounds.

References. General reviews are listed in the introduction to the *Sourcebook*.

$MeNbCl_4$

Nb-00001

Me_2NbCl_3

Nb-00002

$Me_2NbCl_4^{\ominus}$

Nb-00003

Me_3NbCl_2

Nb-00004

$Me_3NbCl_3^{\ominus}$

Nb-00005

Nb-00006

$Nb(CO)_6^{\ominus}$

Nb-00007

Nb-00008

Nb-00009

X = Br

Nb-00010

X = Cl

Nb-00011

Nb-00012

X = Y = I

Nb-00013

As Nb-00010 with
X = H

Nb-00014

Nb-00015

X = Cl

Nb-00016

As Nb-00013 with
X = Cl, Y = CO

Nb-00017

As Nb-00013 with
X = CO, Y = H

Nb-00018

As Nb-00013 with
X = Cl, Y = Me

Nb-00019

As Nb-00011 with
X = Me

Nb-00020

Nb-00021

As Nb-00013 with
X = Y = CN

Nb-00022

Nb-00023

Nb-00024

Nb-00025

As Nb-00013 with
X = CO, Y = Me

Nb-00026

R = H

Nb-00027

As Nb-00013 with
X = Y = Me

Nb-00028

Nb-00029

As Nb-00016 with
X = CO

Nb-00030

As Nb-00027 with
R = Et

Nb-00031

Nb-00032

R = Me

Nb-00033

As Nb-00013 with
X = Me, Y = PEt_3

Nb-00034

Nb-00035

As Nb-00013 with
X = Br, Y = PMe_2Ph

Nb-00036

As Nb-00013 with
X = —PMe_2Ph, Y = H

Nb-00037

Nb-00038

$MeNbCl_4(PPh_3)$

Nb-00039

Nb-00040

Nb-00041

Nb-00042

As Nb-00013 with
X = Y = SPh

Nb-00043

As Nb-00033 with
R = Ph

Nb-00044

$Ph_3SnNb(CO)_6$

Nb-00045

Nb-00046

Nb-00047

Nb-00048

Nb-00049

Nb-00050

Nb-00051

CH_3Cl_4Nb **Nb-00001**

Tetrachloromethylniobium, 9CI
Methyltetrachloroniobium
[41453-01-4]

$MeNbCl_4$

M 249.753
Orange cryst. (CH_2Cl_2/pentane). Sol. CH_2Cl_2, toluene. Mp 65° dec.

Fowles, G.W.A. *et al*, *J. Chem. Soc., Dalton Trans.*, 1972, 2313 (*pmr*)
Santini-Scampucci, C. *et al*, *J. Chem. Soc., Dalton Trans.*, 1973, 2436 (*synth, pmr, ir*)

$C_2H_6Cl_3Nb$ **Nb-00002**

Trichlorodimethylniobium, 9CI
Dimethyltrichloroniobium
[41453-02-5]

Me_2NbCl_3

M 229.335
Yellow-orange solid (pentane).

Fowles, G.W.A. *et al*, *J. Chem. Soc., Dalton Trans.*, 1972, 2313 (*pmr*)

$C_2H_6Cl_4Nb^{\ominus}$ **Nb-00003**

Tetrachlorodimethylniobate(1−), 9CI
Dimethyltetrachloroniobate(1−)

$Me_2NbCl_4^{\ominus}$

M 264.788 (ion)
Tetraethylammonium salt: [38856-62-1].
$C_{10}H_{26}Cl_4NNb$ M 395.041
Violet solid (MeCN/C_6H_6).

Fowles, G.W.A. *et al*, *J. Chem. Soc., Dalton Trans.*, 1972, 2313 (*pmr, ir*)

$C_3H_9Cl_2Nb$ **Nb-00004**

Dichlorotrimethylniobium, 9CI
Trimethyldichloroniobium
[2948-60-9]

Me_3NbCl_2

M 208.917
Yellow cryst. (pentane), thermally unstable. Sol. aliphatic hydrocarbons. Subl. at 20° *in vacuo*.

Juvinall, G.L., *J. Am. Chem. Soc.*, 1964, **86**, 4202 (*synth*)
Fowles, G.W.A. *et al*, *J. Chem. Soc., Dalton Trans.*, 1973, 961 (*pmr*)

$C_3H_9Cl_3Nb^{\ominus}$ **Nb-00005**

Trichlorotrimethylniobate(1−), 9CI
Trimethyltrichloroniobate(1−)

$Me_3NbCl_3^{\ominus}$

M 244.370 (ion)
Tetraethylammonium salt: [53470-46-5].
$C_{11}H_{29}Cl_3NNb$ M 374.622
Cryst. (CH_2Cl_2/pentane).

Fowles, G.W.A. *et al*, *J. Chem. Soc., Dalton Trans.*, 1974, 1080 (*synth, ir, pmr*)

$C_5H_5Cl_4Nb$ **Nb-00006**

Tetrachloro(η^5-2,4-cyclopentadien-1-yl)niobium, 9CI
Cyclopentadienyltetrachloroniobium
[33114-15-7]

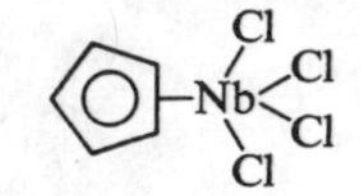

M 299.813
Red cryst. (CH_2Cl_2). Sl. sol. org. solvs. Mp 180° dec. Forms adducts with nitriles.

Cardoso, A.M. *et al*, *J. Chem. Soc., Dalton Trans.*, 1980, 1156 (*synth*)
Bunker, M.J. *et al*, *J. Chem. Soc., Dalton Trans.*, 1980, 2155 (*synth*)

$C_6NbO_6^{\ominus}$ **Nb-00007**

Hexacarbonylniobate(1−), 9CI

$Nb(CO)_6^{\ominus}$

M 260.969 (ion)
Tetraphenylarsonium salt: [60119-18-8].
$C_{30}H_{20}AsNbO_6$ M 644.312
Yellow cryst. (diglyme aq.). Sol. polar org. solvs. Dec. >160°.

Inorg. Synth., 1976, **16**, 68 (*synth, ir*)

$C_8H_5Cl_2NbO_3$ **Nb-00008**

Tricarbonyldichloro(η^5-2,4-cyclopentadien-1-yl)niobium, 9CI
Cyclopentadienyltricarbonyldichloroniobium
[73394-64-6]

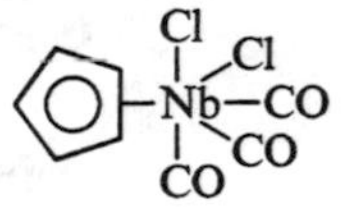

M 312.938
Orange-red cryst. (THF). Mp 150° dec.

Cardoso, A.M. *et al*, *J. Organomet. Chem.*, 1980, **186**, 237 (*ir, pmr*)

$C_9H_5NbO_4$ **Nb-00009**

Tetracarbonyl(η^5-2,4-cyclopentadien-1-yl)niobium, 9CI
Cyclopentadienyltetracarbonylniobium
[12108-03-1]

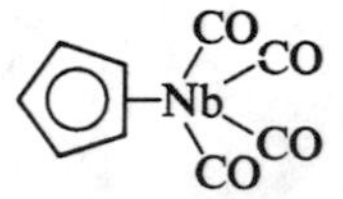

M 270.043
Orange-red cryst. by subl. *in vacuo*. Mp 144-6°. Dec. >120°. Sublimes at 60-80°/0.1 mm.

Dem'yanchuk, V.V. *et al*, *Zh. Obshch. Khim.*, 1970, **40**, 961; *CA*, **73**, 70470 (*synth*)
Nesmeyanov, A.N. *et al*, *Zh. Strukt. Khim.*, 1972, **13**, 1033; *CA*, **78**, 90759 (*cmr*)
Lokshin, B.V. *et al*, *J. Organomet. Chem.*, 1973, **55**, 315 (*ir*)
Fredericks, S. *et al*, *J. Am. Chem. Soc.*, 1978, **100**, 350 (*synth*)
Herrmann, W.A. *et al*, *Chem. Ber.*, 1979, **112**, 3942; *J. Organomet. Chem.*, 1980, **191**, 397 (*synth*)
Herrmann, W.A. *et al*, *Chem. Ber.*, 1981, **114**, 3558 (*struct*)

$C_{10}H_{10}Br_3Nb$ **Nb-00010**

Tribromobis(η^5-2,4-cyclopentadien-1-yl)niobium, 9CI

Bis(cyclopentadienyl)tribromoniobium

[53522-49-9]

X = Br

M 462.807

Sensitizer for electrophotography. Red-brown cryst. ($CHCl_3$), stable to dry air. Sol. polar org. solvs., H_2O with partial hydrolysis. Mp 260° dec.

Wilkinson, G. *et al*, *J. Am. Chem. Soc.*, 1954, **76**, 4281 (*synth, ir*)

$C_{10}H_{10}ClNbO$ **Nb-00011**

Chlorobis(η^5-2,4-cyclopentadien-1-yl)oxoniobium, 9CI

Bis(cyclopentadienyl)oxochloroniobium

[59412-84-9]

X = Cl

M 274.548

Yellow cryst. (THF).

Limonovskii, D.A. *et al*, *Dokl. Akad. Nauk SSSR, Ser. Sci. Khim.*, 1976, **226**, 65 (*synth, nmr, ir*)

$C_{10}H_{10}Cl_2Nb$ **Nb-00012**

Dichlorobis(η^5-2,4-cyclopentadien-1-yl)niobium, 9CI

Bis(cyclopentadienyl)dichloroniobium

[12793-14-5]

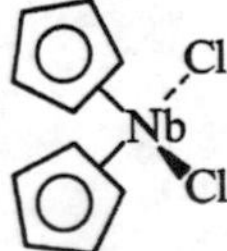

M 294.001

Black cryst. Sl. sol. toluene, CH_2Cl_2. Bp$_{0.25}$ 270° subl. Paramagnetic, μ = 1.63 BM.

▷QU0400000.

Green, M.L.H. *et al*, *J. Organomet. Chem.*, 1974, **74**, 245 (*synth*)

Prout, K. *et al*, *Acta Crystallogr., Sect. B*, 1974, **30**, 2290 (*struct*)

Inorg. Synth., 1976, **16**, 107 (*synth, ir*)

$C_{10}H_{10}I_2Nb$ **Nb-00013**

Bis(η^5-2,4-cyclopentadien-1-yl)diiodoniobium

[56028-53-6]

X = Y = I

M 476.904

Black cryst. (C_6H_6). Mp 155° dec. Paramagnetic, μ = 1.8 BM.

Treichel, P.M. *et al*, *J. Organomet. Chem.*, 1968, **12**, 479.

$C_{10}H_{13}Nb$ **Nb-00014**

Bis(η^5-2,4-cyclopentadien-1-yl)trihydroniobium, 9CI

Dicyclopentadienyltrihydroniobium

[11105-67-2]

As Tribromobis(η^5-2,4-cyclopentadien-1-yl)niobium, Nb-00010 with

X = H

M 226.119

Yellow cryst. (toluene). Sol. aromatic hydrocarbons. Dec. at 80° in soln.

Tebbe, F.N. *et al*, *J. Am. Chem. Soc.*, 1971, **93**, 3793 (*synth, pmr*)

Wilson, R.D. *et al*, *J. Am. Chem. Soc.*, 1977, **99**, 1775 (*struct*)

Labinger, J.A. *et al*, *J. Organomet. Chem.*, 1979, **170**, 373 (*synth*)

$C_{10}H_{14}BNb$ **Nb-00015**

Bis(η^5-2,4-cyclopentadien-1-yl)[tetrahydroborato(1–)-*H,H'*-]niobium, 9CI

Bis(cyclopentadienyl)niobium borohydride

[37298-41-2]

M 237.937

Dark-green solid (toluene), highly air-sensitive. Sol. Et_2O, aromatic hydrocarbons. Bp 80° subl. (part. dec.).

▷Pyrophoric in air

Kirillova, N.I. *et al*, *Zh. Strukt. Khim.*, 1974, **15**, 718 (*struct*)

Inorg. Synth., 1976, **16**, 107 (*synth, ir*)

$C_{10}H_{15}Cl_4Nb$ **Nb-00016**

Tetrachloro[(1,2,3,4,5-η)-1,2,3,4,5-pentamethyl-2,4-cyclopentadien-1-yl]niobium

Pentamethylcyclopentadienyltetrachloroniobium

[80432-35-5]

X = Cl

M 369.947

Deep-brown waxy solid (C_6H_6). Sol. aromatic hydrocarbons; insol. aliphatic hydrocarbons.

Herrmann, W.A. *et al*, *Chem. Ber.*, 1981, **114**, 3558 (*synth*)

$C_{11}H_{10}ClNbO$ Nb-00017

Carbonylchlorobis(η^5-2,4-cyclopentadien-1-yl)niobium

Bis(cyclopentadienyl)chlorocarbonylniobium

[63782-46-7]

As Bis(η^5-2,4-cyclopentadien-1-yl)diiodoniobium, Nb-00013 with

X = Cl, Y = CO

M 286.559

Grey cryst. (THF). Sol. C_6H_6, THF, Me_2CO; insol. Et_2O, alcohols, pentane. Dec. >180°.

Limonovski, D.A. *et al*, *J. Organomet. Chem.*, 1977, **132**, C14 (*ir, pmr, synth*)
Fredericks, S. *et al*, *J. Am. Chem. Soc.*, 1978, **100**, 350 (*synth, pmr, ir*)
Otto, E.E.H. *et al*, *J. Organomet. Chem.*, 1978, **148**, 29 (*synth, pmr, ir*)

$C_{11}H_{11}NbO$ Nb-00018

Carbonylbis(η^5-2,4-cyclopentadien-1-yl)hydroniobium, 9CI

Bis(cyclopentadienyl)hydro(carbonyl)niobium

[11105-68-3]

As Bis(η^5-2,4-cyclopentadien-1-yl)diiodoniobium, Nb-00013 with

X = CO, Y = H

M 252.114

Pale-violet cryst. by subl. Sol. aromatic hydrocarbons.

Tebbe, F.N. *et al*, *J. Am. Chem. Soc.*, 1971, **93**, 3793 (*synth, pmr, ir*)
Kirillova, N.I. *et al*, *Zh. Strukt. Khim.*, 1972, **13**, 473 (*struct*)
Otto, E.E.H. *et al*, *J. Organomet. Chem.*, 1978, **148**, 29 (*synth*)

$C_{11}H_{13}ClNb$ Nb-00019

Chlorobis(η^5-2,4-cyclopentadien-1-yl)methylniobium, 9CI

Bis(cyclopentadienyl)methylchloroniobium

[70616-56-7]

As Bis(η^5-2,4-cyclopentadien-1-yl)diiodoniobium, Nb-00013 with

X = Cl, Y = Me

M 273.583

Brown cryst. (toluene). Sol. toluene.

Otto, E.E.H. *et al*, *J. Organomet. Chem.*, 1979, **170**, 209 (*synth*)

$C_{11}H_{13}NbO$ Nb-00020

Bis(cyclopentadienyl)methyloxoniobium

[67552-61-8]

As Chlorobis(η^5-2,4-cyclopentadien-1-yl)oxoniobium, Nb-00011 with

X = Me

M 254.130

White cryst. (toluene).

Otto, E.E.H. *et al*, *J. Organomet. Chem.*, 1979, **170**, 209 (*ms, pmr*)

Middleton, A.R. *et al*, *J. Chem. Soc., Dalton Trans.*, 1980, 1888 (*synth, ir, pmr*)

$C_{11}H_{31}NbP_2$ Nb-00021

[1,2-Ethanediylbis[dimethylphosphine]-*P,P'*]pentamethylniobium, 9CI

Pentamethyl[bis(dimethylphosphino)ethane]niobium

[53437-03-9]

M 318.220

Yellow cryst. (Et_2O).

Schrock, R.R. *et al*, *J. Am. Chem. Soc.*, 1974, **96**, 5288 (*nmr*)

$C_{12}H_{10}N_2Nb$ Nb-00022

Bis(cyano-*C*)bis(η^5-2,4-cyclopentadien-1-yl)niobium, 9CI

Bis(cyclopentadienyl)dicyanoniobium

[39333-45-4]

As Bis(η^5-2,4-cyclopentadien-1-yl)diiodoniobium, Nb-00013 with

X = Y = CN

M 275.131

No phys. props. reported.

Stewart, C.P. *et al*, *J. Chem. Soc., Dalton Trans.*, 1973, 722 (*synth, esr*)

$C_{12}H_{11}NbO_2$ Nb-00023

Dicarbonyl(cyclopentadiene)-π-cyclopentadienylniobium, 8CI

[31869-16-6]

M 280.124

Yellow needles (pentane).

Nesmeyanov, A.N. *et al*, *Izv. Akad. Nauk SSSR, Ser. Khim.*, 1970, 727 (*synth, ir, pmr*)

$C_{12}H_{12}Nb$ Nb-00024

Bis(η^6-benzene)niobium

[68088-94-8]

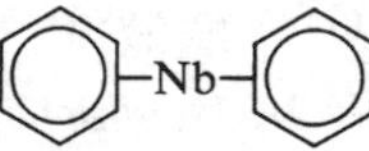

M 249.133

Obt. by metal-vapour synth. Red-purple cryst. (pet. ether). Bp 80° subl. (part. dec.).

Cloke, F.G.N. *et al*, *J. Chem. Soc., Chem. Commun.*, 1978, 431 (*synth, pe*)

$C_{12}H_{12}Nb$ **Nb-00025**

(η^7-Cycloheptatrienylium)(η^5-2,4-cyclopentadien-1-yl)niobium, 9CI

(Cycloheptatrienyl)(cyclopentadienyl)niobium

[54360-38-2]

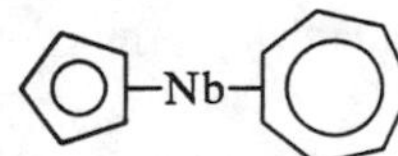

M 249.133

Solid by subl. $Bp_{0.1}$ 120° subl.

van Oven, H.O. *et al*, *J. Organomet. Chem.*, 1974, **81**, 379 (*synth, ir, ms, esr*)

Groenenboom, C.J. *et al*, *J. Organomet. Chem.*, 1975, **97**, 73 (*pe*)

$C_{12}H_{13}NbO$ **Nb-00026**

Carbonylbis(η^5-2,4-cyclopentadien-1-yl)methylniobium, 9CI

Bis(cyclopentadienyl)methyl(carbonyl)niobium

[65652-98-4]

As Bis(η^5-2,4-cyclopentadien-1-yl)diiodoniobium, Nb-00013 with

X = CO, Y = Me

M 266.141

Green solid. Sol. toluene. Bp 50-60° subl.

Otto, E.E.H. *et al*, *J. Organomet. Chem.*, 1979, **170**, 209 (*synth, ir*)

Foust, D.F. *et al*, *J. Organomet. Chem.*, 1980, **193**, 209 (*synth, pmr*)

Threlkel, R.S. *et al*, *J. Am. Chem. Soc.*, 1981, **103**, 2560 (*synth*)

$C_{12}H_{15}Nb$ **Nb-00027**

Bis(η^5-2,4-cyclopentadien-1-yl)(η^2-ethene)hydroniobium, 9CI

Bis(cyclopentadienyl)hydro(ethylene)niobium

[11105-70-7]

R = H

M 252.157

Sol. aromatic hydrocarbons.

Tebbe, F.N. *et al*, *J. Am. Chem. Soc.*, 1971, **93**, 3793 (*synth*)

Guggenberger, L.J. *et al*, *J. Am. Chem. Soc.*, 1974, **96**, 5420 (*nmr, struct*)

$C_{12}H_{16}Nb$ **Nb-00028**

Bis(η^5-2,4-cyclopentadien-1-yl)dimethylniobium, 9CI

[54373-66-9]

As Bis(η^5-2,4-cyclopentadien-1-yl)diiodoniobium, Nb-00013 with

X = Y = Me

M 253.165

Red-brown cryst. (pentane). Sol. Et_2O, pentane. $Bp_{0.0001}$ 80° subl. Paramagnetic, μ = 1.5 BM.

▷Explodes at 128°

Elson, I.H. *et al*, *J. Am. Chem. Soc.*, 1974, **96**, 7374 (*esr*)

Manzer, L.E., *Inorg. Chem.*, 1977, **16**, 525 (*synth*)

$C_{13}H_{15}Nb$ **Nb-00029**

Bis(η^5-2,4-cyclopentadien-1-yl)(η^3-2-propenyl)niobium, 9CI

Allylbis(cyclopentadienyl)niobium

[39413-65-5]

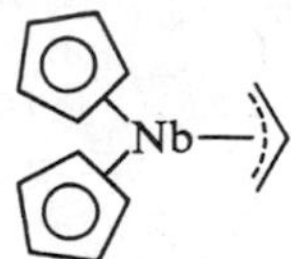

M 264.168

Green-black cryst. (pentane). Sol. pentane. Dec. at 118°.

van Baalen, A. *et al*, *J. Organomet. Chem.*, 1974, **74**, 245 (*synth, ir, pmr, ms*)

Green, J.C. *et al*, *J. Chem. Soc., Dalton Trans.*, 1975, 403 (*synth, pe*)

$C_{14}H_{15}NbO_4$ **Nb-00030**

Tetracarbonyl[(1,2,3,4,5-η)-1,2,3,4,5-pentamethyl-2,4-cyclopentadien-1-yl]niobium

Pentamethylcyclopentadienyltetracarbonylniobium

[80432-28-6]

As Tetrachloro[(1,2,3,4,5-η)-1,2,3,4,5-pentamethyl-2,4-cyclopentadien-1-yl]niobium, Nb-00016 with

X = CO

M 340.177

Orange-red cryst. (pentane). Sol. pentane, C_6H_6. Mp 128-30°.

Herrmann, W.A. *et al*, *Chem. Ber.*, 1981, **114**, 3558 (*synth, ir, pmr*)

$C_{14}H_{19}Nb$ **Nb-00031**

Bis(η^5-2,4-cyclopentadien-1-yl)(η^2-ethene)ethylniobium, 9CI

Bis(cyclopentadiene)ethyl(ethylene)niobium

[54244-99-4]

As Bis(η^5-2,4-cyclopentadien-1-yl)(η^2-ethene)-hydroniobium, Nb-00027 with

R = Et

M 280.211

Tebbe, F.N. *et al*, *J. Am. Chem. Soc.*, 1971, **93**, 3793 (*synth*)

Guggenberger, L.J. *et al*, *J. Am. Chem. Soc.*, 1974, **96**, 5420 (*nmr, struct*)

$C_{16}H_{41}NbP_4$ **Nb-00032**

Bis[1,2-ethanediylbis[dimethylphosphine]-*P*,*P'*]bis(η^2-ethene)hydroniobium, 9CI

Hydrobis(ethylene)bis[bis(dimethylphosphino)ethane]-niobium

[62651-63-2]

M 450.301

Yellow cryst. (pentane). Sol. aliphatic hydrocarbons.

Schrock, R.R., *J. Organomet. Chem.*, 1976, **121**, 373 (*synth, nmr*)

$C_{17}H_{19}Nb$ **Nb-00033**

Bis(1,3,5,7-cyclooctatetraene)methylniobium, 9CI

[59296-78-5]

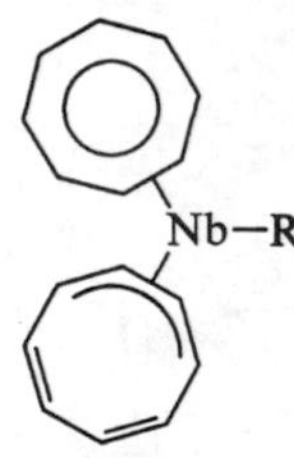

R = Me

M 316.244

Proposed struct. shown. Red-brown cryst. (THF). Sol. THF, CH_2Cl_2. V. air- and moisture-sensitive.

Schrock, R.R. *et al*, *J. Am. Chem. Soc.*, 1976, **98**, 903 (*synth*, *pmr*)

$C_{17}H_{28}NbP$ **Nb-00034**

Bis(cyclopentadienyl)methyl(triethylphosphine)niobium

[70616-59-0]

As Bis(η^5-2,4-cyclopentadien-1-yl)diiodoniobium, Nb-00013 with

X = Me, Y = PEt_3

M 356.288

Sol. aliphatic hydrocarbons.

Otto, E.E.H. *et al*, *J. Organomet. Chem.*, 1979, **170**, 209 (*synth*, *pmr*, *ms*)

$C_{17}H_{37}NbP_4$ **Nb-00035**

(η^5-2,4-Cyclopentadien-1-yl)bis[1,2-ethanediylbis[dimethylphosphine]-*P*,*P'*]niobium, 9CI

Cyclopentadienylbis[bis(dimethylphosphino)ethane]niobium

[77299-51-5]

M 458.281

Yellow needles (THF).

Bunker, M.J. *et al*, *J. Chem. Soc., Dalton Trans.*, 1981, 85 (*synth*, *pmr*)

$C_{18}H_{21}BrNbP$ **Nb-00036**

Bromobis(η^5-2,4-cyclopentadien-1-yl)(dimethylphenylphosphine)niobium

Bis(cyclopentadienyl)bromo(dimethylphenylphosphine)niobium

[37298-77-4]

As Bis(η^5-2,4-cyclopentadien-1-yl)diiodoniobium, Nb-00013 with

X = Br, Y = PMe_2Ph

M 441.148

Green cryst. (THF). Sol. aromatic solvs. THF, Me_2CO.

Inorg. Synth., 1976, **16**, 107 (*synth*, *ir*, *pmr*)

$C_{18}H_{22}NbP$ **Nb-00037**

Bis(η^5-2,4-cyclopentadien-1-yl)hydro(dimethylphenylphosphine)niobium, 9CI

As Bis(η^5-2,4-cyclopentadien-1-yl)diiodoniobium, Nb-00013 with

X = —PMe_2Ph, Y = H

M 362.252

Red cryst. (pet. ether). Sol. org. solvs. Reacts with halogenated solvs. Solns. dec. >40°.

Inorg. Synth., 1976, **16**, 107 (*synth*, *ir*, *pmr*)

$C_{18}H_{25}Nb$ **Nb-00038**

Bis(cyclopentadienyl)hydro(η^2-4-octyne)niobium

[55534-71-9]

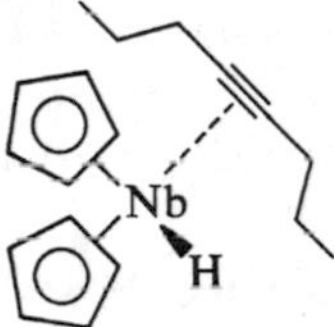

M 334.302

Sol. org. solvs.

Labinger, J.A. *et al*, *J. Am. Chem. Soc.*, 1975, **97**, 1596 (*synth*, *ir*, *pmr*)

$C_{19}H_{18}Cl_4NbP$ **Nb-00039**

Tetrachloromethyl(triphenylphosphine)niobium, 9CI

[51877-88-4]

$MeNbCl_4(PPh_3)$

M 512.043

Purple powder (CH_2Cl_2/pet. ether). Mp 85° dec.

Santini-Scampucci, C. *et al*, *J. Chem. Soc., Dalton Trans.*, 1973, 2436 (*synth*, *nmr*, *ir*)

$C_{20}H_{20}Nb_2$ **Nb-00040**

Bis(η^5-2,4-cyclopentadien-1-yl)bis[μ-[(1,η:1,2,3,4,5-η)-2,4-cyclopentadien-1-ylidene]]dihydrodiniobium, 9CI

Niobocene dimer

[11105-93-4]

M 446.191

Yellow cryst. (toluene). Sol. aromatic hydrocarbons.

Tebbe, F.N. *et al*, *J. Am. Chem. Soc.*, 1971, **93**, 3793 (*synth*, *pmr*)

Guggenberger, L.J., *Inorg. Chem.*, 1973, **12**, 294 (*struct*)

$C_{21}H_{15}NbO_2$ Nb-00041

Dicarbonyl(η^5-2,4-cyclopentadien-1-yl)(diphenylacetylene)-niobium

M 392.255

Dimeric in solid state, monomeric in soln.

Dimer: Tetracarbonylbis(cyclopentadienyl)bis(diphenylacetylene)diniobium.

$C_{42}H_{30}Nb_2O_4$ M 784.509

Air-stable violet cryst. (Et_2O). Sol. org. solvs. Loses CO to form $[(C_5H_5)Nb(CO)(PhC{\equiv}CPh)]_2$ at 145°.

Nesmeyanov, A.N. *et al*, *Izv. Akad. Nauk SSSR, Ser. Khim.*, 1969, 100; *Bull. Acad. Sci. USSR, Div. Chem. Sci.*, 87 (*synth, ir*)

$C_{22}H_{15}Nb_3O_7$ Nb-00042

Heptacarbonyltris(cyclopentadienyl)triniobium

[72228-91-2]

M 670.076

Has a doubly semibridging CO group. Black needles (hexane). Sol. CH_2Cl_2, THF, Et_2O. Dissolves to give brown, air-sensitive solns. Dec. >100°.

Herrmann, W.A. *et al*, *Angew. Chem., Int. Ed. Engl.*, 1979, **18**, 960 (*synth, ir, nmr, struct*)

Lewis, L.N. *et al*, *Inorg. Chem.*, 1980, **19**, 3201 (*nmr*)

$C_{22}H_{20}NbS_2$ Nb-00043

Bis(benzenethiolato)bis(η^5-2,4-cyclopentadien-1-yl)niobium, 9CI

Bis(cyclopentadienyl)bis(benzenethiolato)niobium

[37328-40-8]

As Bis(η^5-2,4-cyclopentadien-1-yl)diiodoniobium, Nb-00013 with

X = Y = SPh

M 441.426

Green cryst. (CH_2Cl_2/pet. ether). Mp 135-40°.

Douglas, W.E. *et al*, *J. Chem. Soc., Dalton Trans.*, 1972, 1796 (*synth, esr*)

$C_{22}H_{21}Nb$ Nb-00044

Bis(1,3,5,7-cyclooctatetraene)phenylniobium, 9CI

[59296-77-4]

As Bis(1,3,5,7-cyclooctatetraene)methylniobium, Nb-00033 with

R = Ph

M 378.314

Ethylene dimerisation catalyst. Red-brown plates (THF). Sol. THF, CH_2Cl_2. V. air- and moisture-sensitive. Dec. at *ca* 100°.

Schrock, R.P. *et al*, *J. Am. Chem. Soc.*, 1976, **98**, 903 (*synth, pmr*)

U.S.P., 3 932 477; *CA*, **85**, 33190 (*use*)

$C_{24}H_{15}NbO_6Sn$ Nb-00045

Hexacarbonyl(triphenylstannyl)niobium

[36669-56-4]

$Ph_3SnNb(CO)_6$

M 610.975

Red plates (CH_2Cl_2). Sol. CH_2Cl_2, Et_2O, aromatic solvs. Dec. in polar solvs., thermally labile.

Davison, A. *et al*, *J. Organomet. Chem.*, 1972, **36**, 113 (*synth, ir*)

Ellis, J.E. *et al*, *Inorg. Chem.*, 1976, **15**, 3168.

$C_{24}H_{62}Nb_2Si_6$ Nb-00046

Tetrakis[(trimethylsilyl)methyl]bis[μ-(trimethylsilyl)methylidyne]diniobium, 9CI

[34844-61-6]

M 705.080

Red-brown prisms (pet. ether), highly air- and moisture-sensitive. Sol. Et_2O, hydrocarbons. Mp 152° dec.

Huq, F. *et al*, *J. Chem. Soc., Chem. Commun.*, 1971, 1477 (*struct*)

Mowat, W. *et al*, *J. Chem. Soc., Dalton Trans.*, 1973, 1120 (*synth, pmr*)

$C_{26}H_{20}NbO_3P$ Nb-00047

Tricarbonyl(η^5-2,4-cyclopentadien-1-yl)(triphenylphosphine)-niobium, 9CI

Cyclopentadienyltricarbonyl(triphenylphosphine)niobium

M 504.322

Red needles. Sol. C_6H_6, CH_2Cl_2. Mp 186° dec.

Herrmann, W.A. *et al*, *J. Organomet. Chem.*, 1980, **191**, 397 (*synth, pmr, ir*)

$C_{34}H_{25}NbO$ Nb-00048

Carbonyl-η^5-2,4-cyclopentadien-1-ylbis(diphenylacetylene)niobium

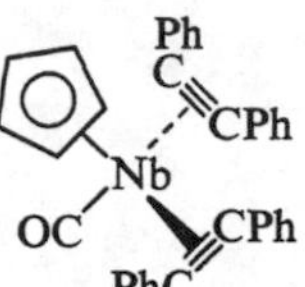

M 542.477

Air-stable yellow prisms (Et_2O). Sol. org. solvs. Mp 135-6°. Bp 200° subl. *in vacuo* (part. dec.).

Nesmeyanov, A.N. *et al*, *Izv. Akad. Nauk SSSR, Ser. Khim.*, 1969, 100 (*synth, pmr*)
Nesmeyanov, A.N. *et al*, *J. Chem. Soc., Chem. Commun.*, 1969, 277 (*struct*)

$C_{36}H_{54}Cl_6Nb_3^{\oplus}$ Nb-00049

Hexa-μ-chlorotris(hexamethylbenzene)triniobium(1+), 9CI

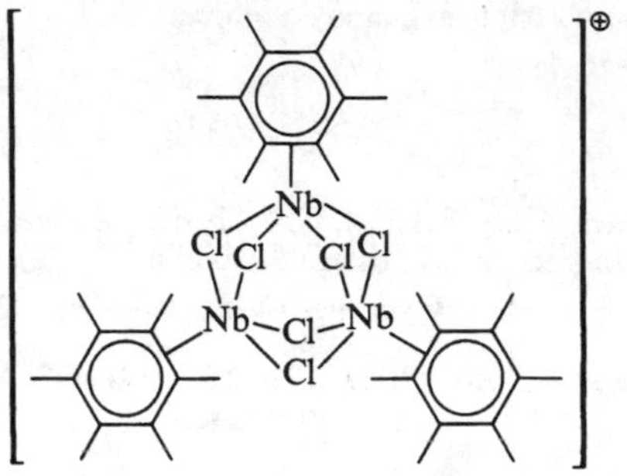

M 978.260 (ion)

Chloride: [12103-45-6].
$C_{36}H_{54}Cl_7Nb_3$ M 1013.713
Dark-green needles ($CHCl_3$/pentane). Mp 80° dec.

Fischer, E.O. *et al*, *J. Organomet. Chem.*, 1966, **6**, 53 (*synth, ir, pmr*)
Churchill, M.R. *et al*, *J. Chem. Soc., Chem. Commun.*, 1974, 248 (*struct*)

$C_{40}H_{30}Nb_2O_2$ Nb-00050

Dicarbonylbis(η[5]-cyclopentadien-1-yl)bis[1,1′-(1,2-ethynediyl)bis[benzene]]diniobium, 9CI

Dicarbonylbis(cyclopentadienyl)bis(diphenylacetylene)-diniobium

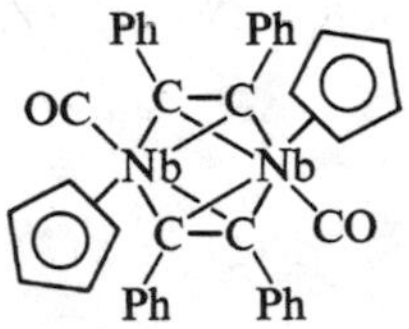

M 728.489

Air-stable, dark-lilac cryst. (toluene). Sol. aromatic solvs., Et_2O, $CHCl_3$. Mp 248-50° dec.

Nesmeyanov, A.N. *et al*, *Dokl. Akad. Nauk SSSR, Ser. Sci. Khim.*, 1968, **181**, 736 (*synth, ir, pmr*)
Nesmeyanov, A.N. *et al*, *J. Chem. Soc., Chem. Commun.*, 1968, 1365 (*struct*)

$C_{48}H_{35}NbO$ Nb-00051

Cyclopentadienylcarbonyl(diphenylacetylene)(tetraphenylcyclobutadiene)niobium

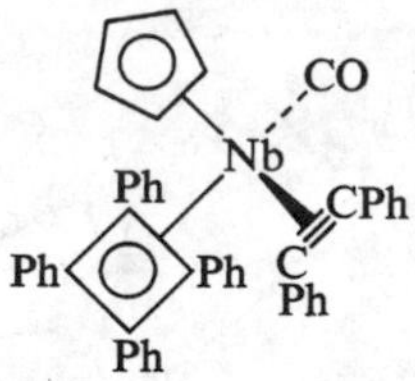

M 720.710

Red cryst. (C_6H_6). Sol. org. solvs. Mp 120° dec. Dec. thermally to C_6Ph_6.

Nesmeyanov, A.N. *et al*, *J. Chem. Soc., Chem. Commun.*, 1969, 739 (*synth, ir, pmr, struct*)

Nd Neodymium

S. A. Cotton

Néodyme (Fr.), Neodym (Ger.), Neodimio (Sp., Ital.), Неодим (Nieodim) (Russ.), ネオジム (Japan.)

Atomic Number. 60

Atomic Weight. 144.24

Electronic Configuration. [Xe] $4f^4\,6s^2$

Oxidation State. +3

Coordination Number. Usually 6 or greater.

Colour. Generally blue.

Availability. Common starting materials are Nd_2O_3 and anhydrous $NdCl_3$. For general remarks, see under Lanthanum.

Handling. See under Lanthanum.

Toxicity. See under Lanthanum.

Isotopic Abundance. ^{142}Nd, 27.11%; ^{143}Nd, 12.17%; ^{144}Nd, 23.85%; ^{145}Nd, 8.30%; ^{146}Nd, 17.62%; ^{148}Nd, 5.73%; ^{150}Nd, 5.62%.

Spectroscopy. See the Lanthanum section for general references. The limited ^{1}H nmr data published for Nd(III) ($4f^3$) compounds show relatively small shifts and broadened lines.

Electronic spectra of $Nd(C_5H_5)_3$ show hypersensitive transitions (at 588 and 594 nm) with greater extinction coefficients than usual.

Analysis. See under Lanthanum.

References. See under Lanthanum.

$C_6H_{18}Nd^{\ominus\ominus\ominus}$ **Nd-00001**

Hexamethylneodymate(3−)

$NdMe_6^{\ominus\ominus\ominus}$

M 234.448 (ion)

Tris[(tetramethylethylenediamine)lithium] salt: [76206-88-7].
$C_{24}H_{66}Li_3N_6Nd$ M 603.889
Blue solid. Sol. Et_2O. Mp 78-83° dec.

Schumann, H. *et al*, *Angew. Chem., Int. Ed. Engl.*, 1981, **20**, 120 (*synth*)

C_8H_8ClNd **Nd-00002**

Chloro(cyclooctatetraene)neodymium

Di-μ-chlorobis(η[8]-1,3,5,7-cyclooctatetraene)dineodymium, 9CI

$(C_8H_8)NdCl$

M 283.844

THF adduct is dimeric, struct. may resemble Chloro(cyclooctatetraene)cerium.

Bis-THF adduct: [51177-49-2].
$C_{32}H_{48}Cl_2Nd_2O_4$ M 856.115
Bright-green cryst. Sol. dioxan, THF. Mp >345°. Extremely air- and moisture-sensitive. $\mu = 3.37\mu_B$.

Hodgson, K.O. *et al*, *J. Am. Chem. Soc.*, 1973, **95**, 8650 (*synth, ir uv*)

$C_{13}H_{13}Nd$ **Nd-00003**

(η[8]-1,3,5,7-Cyclooctatetraene)(η[5]-2,4-cyclopentadien-1-yl)-neodymium, 9CI

[52668-26-5]

M 313.486

Cryst. Sol. THF. Extremely air-sensitive. Forms THF adduct.

Jamerson, J.D. *et al*, *J. Organomet. Chem.*, 1974, **65**, C33 (*synth*)

$C_{15}H_{15}Nd$ **Nd-00004**

Tris(η[5]-2,4-cyclopentadien-1-yl)neodymium, 9CI

Tri-π-cyclopentadienylneodymium, 8CI

[1273-98-9]

M 339.524

Pale-blue cryst. Sol. THF. Mp 380°. Subl. at 200-50° *in vacuo*, hydrol. by H_2O.

Cyclohexylisocyanide adduct: [12098-33-8]. Violet cryst. Sol. C_6H_6. Mp 147°. $\mu = 3.4\mu_B$; Subl. *in vacuo* at 150°.

Wilkinson, G. *et al*, *J. Am. Chem. Soc.*, 1954, **76**, 6210 (*synth*)
Reid, A.F. *et al*, *Inorg. Chem.*, 1966, **5**, 1213 (*synth*)
Devyatykh, G.G. *et al*, *Dokl. Akad. Nauk. SSSR*, 1970, **193**, 1069 (*synth, ms*)
Thomas, J.L. *et al*, *J. Organomet. Chem.*, 1970, **23**, 487 (*ms*)
Aleksanyan, V.T. *et al*, *J. Raman Spectrosc.*, 1974, **2**, 345 (*ir, raman*)
Von Ammon, R. *et al*, *Inorg. Nucl. Chem. Lett.*, 1969, **5**, 315 (*nmr*)
Fischer, E.O. *et al*, *J. Organomet. Chem.*, 1966, **6**, 141 (*synth*)
Von Ammon, R. *et al*, *Ber. Bunsenges. Phys. Chem.*, 1972, **76**, 995 (*ir, nmr*)

$C_{16}H_{16}Nd^{\ominus}$ **Nd-00005**

Bis(η[8]-1,3,5,7-cyclooctatetraene)neodymate(1−), 9CI

Dicyclooctatetraeneneodymate(1−)

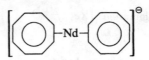

M 352.542 (ion)

K salt: [51177-54-9].
$C_{16}H_{16}KNd$ M 391.641
Pale-green solid. Sol. dioxan, THF. Mp >345°. Extremely air- and moisture-sensitive. $\mu = 2.98\mu_B$.

Mares, F. *et al*, *J. Organomet. Chem.*, 1970, **24**, C68 (*synth*)
Hodgson, K.O. *et al*, *J. Am. Chem. Soc.*, 1973, **95**, 8650 (*synth, ir, nmr, uv*)

$C_{18}H_{21}Nd$ **Nd-00006**

Tris[(1,2,3,4,5-η)-1-methyl-2,4-cyclopentadien-1-yl]neodymium, 9CI

[39470-13-8]

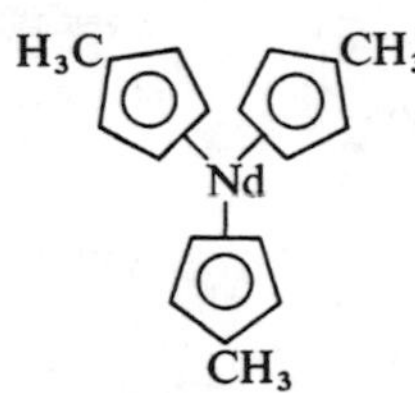

M 381.604

Blue-violet cryst. Sol. THF. Mp 165°. Air- and moisture-sensitive. Subl. *in vacuo* at 200°.

Reynolds, L.T. *et al*, *J. Inorg. Nucl. Chem.*, 1959, **9**, 86 (*synth*)
Burns, J.H. *et al*, *Inorg. Chem.*, 1974, **13**, 1916 (*struct*)

$C_{20}H_{30}Cl_2Nd^{\ominus}$ **Nd-00008**

Dichlorobis[(1,2,3,4,5-η)-pentamethyl-2,4-cyclopentadien-1-yl]neodymate(1−)

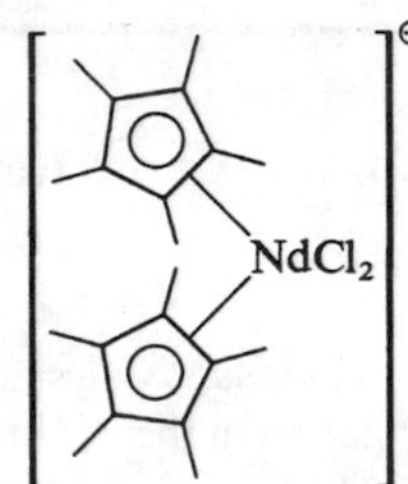

M 485.603 (ion)

Li salt, THF complex: [73597-12-3].
$C_{28}H_{46}Cl_2LiNdO_2$ M 636.757
Blue-violet cryst. Sol. toluene, pentane. Mp 270° dec. Extremely air- and moisture-sensitive. $\mu_{eff} = 3.50\mu_B$.

Wayda, A.L. *et al*, *Inorg. Chem.*, 1980, **19**, 2190 (*synth*, *ir*, *nmr*, *uv*)

$C_{21}H_{33}Nd$ **Nd-00009**

Tris[(1,2,3,4,5-η)-2,4-dimethyl-2,4-pentadienyl]neodymium, 10CI

[80502-52-9]

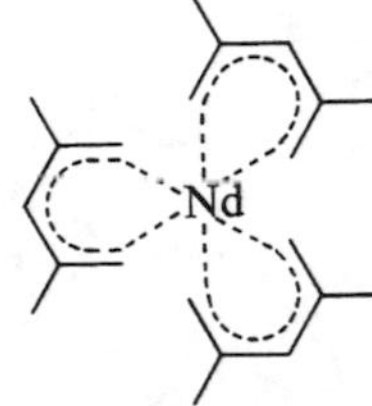

M 429.732

Bright-green cryst. (red by transmittance). Sol. THF, Et_2O. Extremely air- and water-sensitive. $\mu_{eff} = 3.58\mu_B$.

Ernst, R.D. *et al*, *Organometallics*, 1982, **1**, 708 (*synth*, *struct*, *uv*, *ir*, *nmr*)

$C_{22}H_{42}Cl_2NdSi_4^{\ominus}$ **Nd-00010**

Dichlorobis[(1,2,3,4,5-η)-1,3-bis(trimethylsilyl)-2,4-cyclopentadien-1-yl]neodymate(1−)

M 634.062 (ion)

Li salt, Bis-THF complex: [81507-33-7].
$C_{30}H_{58}Cl_2LiNdO_2Si_4$ M 785.216
Blue-green or blue-pink cryst. Sol. pentane.

Lappert, M.F. *et al*, *J. Chem. Soc., Chem. Commun.*, 1981, 1191 (*synth*, *struct*)

$C_{26}H_{50}Cl_2Nd^{\ominus}$ **Nd-00011**

Bis[(1,2,3,4,5-η)-1,3-bis(trimethylsilylmethyl)-2,4-cyclopentadien-1-yl]dichloroneodymium(1−)

M 577.827 (ion)

Tetraphenylarsonium salt:
$C_{50}H_{70}AsCl_2NdSi_4$ M 1073.513
Relatively air-stable violet cryst. Sol. Et_2O, CH_2Cl_2. Mp ~165-75°.

Benzyltriphenylphosphonium salt:
$C_{51}H_{72}Cl_2NdP$ M 931.250
Relatively air-stable violet cryst. Sol. Et_2O, CH_2Cl_2.

Lappert, M.F. *et al*, *J. Chem. Soc., Chem. Commun.*, 1983, 69 (*synth*, *struct*)

$C_{32}H_{40}Nd_2O_2$ **Nd-00012**

[μ-[(1,2-η:1,2,3,4,5,6,7,8-η)-1,3,5,7-Cyclooctatetraene-]][bis(η⁸-1,3,5,7-cyclooctatetraene)bis(tetrahydrofuran)dineodymium

Cyclooctatetraenylbis(tetrahydrofuran)neodymiumbis-(cyclooctatetraenyl)neodymate

[59458-55-8]

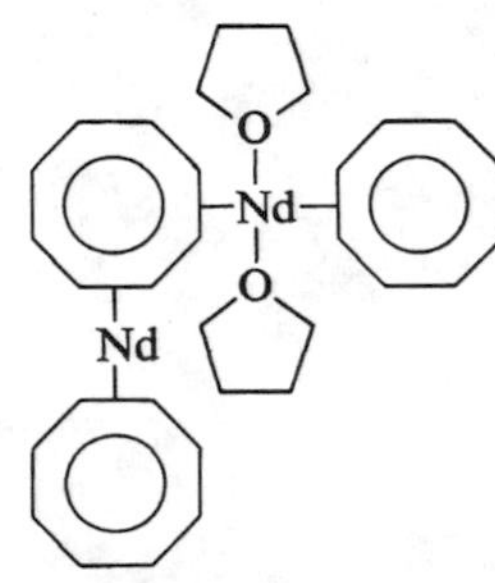

M 745.147

Bright-green cryst. Sol. THF. Air-sensitive. $\mu_{eff} = 3.20\mu_B$.

Ely, S.R. *et al*, *J. Am. Chem. Soc.*, 1976, **98**, 1624 (*synth*, *struct*)

deKock, C.W. *et al*, *Inorg. Chem.*, 1978, **17**, 625 (*synth*, *struct*, *ir*, *raman*)

Np Neptunium

S. A. Cotton

Neptunium (Fr., Ger.), Neptunio (Sp.), Nettunio (Ital.), Нептуний (Nieptunii) (Russ.), ネプツニウム (Japan.)

Atomic Number. 93

Atomic Weight. 237.0482

Electronic Configuration. [Rn] $5f^5\ 7s^2$

Oxidation States. +3, +4

Coordination Number. See under Uranium.

Colour. Variable, including yellow, brown, red and green.

Availability. ^{237}Np is the most stable isotope ($t_{1/2} = 2.20 \times 10^6$y). Neptunium compounds are not generally available. $NpCl_4$ is the most useful starting material, prepared by the reaction of NpO_2 with CCl_4. See under Uranium for references.

Handling. Neptunium is a highly radioactive α-emitter and is highly toxic. Neptunium compounds require scrupulously careful handling in glove boxes at slightly below atmospheric pressure.

Toxicity. See above.

Spectroscopy. Np(IV) compounds, despite being paramagnetic, give useful information in their ^1H nmr spectra. ^{239}Np Mössbauer spectra have been analysed giving valuable information on covalency. See references under Uranium and also Karraker, D. G. *et al.*, *Inorg. Chem.*, 1979, **18,** 2205.

Analysis. Following destruction of the original compound and redissolution in acid, neptunium content may be analysed by α-counting the solution. Isomorphism with corresponding uranium compounds may also be used as proof of identity.

References. See under Uranium and also:

Karraker, D. G. *et al.*, *Inorg. Chem.*, 1979, **18,** 2205 (*Mössbauer*, *analysis*)

Alvey, P. J. *et al.*, *J. Chem. Soc., Dalton Trans.*, 1973, 2308 (*analysis*)

$C_{15}H_{15}ClNp$ Np-00001

Chlorotris(η^5-2,4-cyclopentadien-1-yl)neptunium

Tricyclopentadienylneptunium chloride

[1317-00-6]

M 467.785

Brown solid. Sol. C_6H_6. Mp >330° dec. Subl. at 100° *in vacuo*. $\mu_{eff} = 2.20\mu_B$.

Baumgartner, F. *et al*, *Naturwissenschaften*, 1965, **52**, 560 (*synth*)
Fischer, E.O. *et al*, *J. Organomet. Chem.*, 1966, **5**, 583 (*synth*)
Fischer, R.D. *et al*, *Z. Naturforsch., A*, 1969, **24**, 616.
Karraker, D.G. *et al*, *Inorg. Chem.*, 1972, **11**, 1742 (*synth*, *mössbauer*)

$C_{15}H_{15}FNp$ Np-00002

Fluorotris(η^5-2,4-cyclopentadien-1-yl)neptunium

Tricyclopentadienylneptunium fluoride

[70445-59-9]

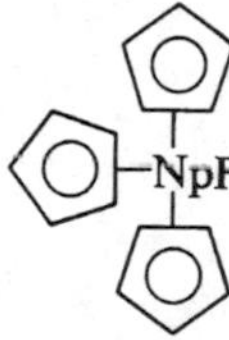

M 451.330

Green solid. Sol. THF. Subl. at 170° *in vacuo*.

Fischer, E.O. *et al*, *J. Organomet. Chem.*, 1966, **5**, 583 (*synth*)
Stollenwerk, A.H., *J. Phys. Colloq.*, 1979, 179 (*synth*)

$C_{16}H_{16}Np^{\ominus}$ Np-00003

Bis(η^8-1,3,5,7-cyclooctatetraene)neptunate(1−), 9CI

$[Np(C_8H_8)_2]^{\ominus}$

M 445.351 (ion)

K salt, bis-THF complex: [54326-81-7].
$C_{24}H_{32}KO_2Np$ M 628.662
Burgundy cryst. Extremely air- and water-sensitive. Easily oxid. to Bis(η^8-1,3,5,7-cyclooctatetraene)neptunium, Np-00004 .

Karraker, D.G. *et al*, *J. Am. Chem. Soc.*, 1974, **96**, 6685 (*synth*, *mössbauer*)

$C_{16}H_{16}Np$ Np-00004

Bis(η^8-1,3,5,7-cyclooctatetraene)neptunium, 9CI

Neptunocene

[37281-22-4]

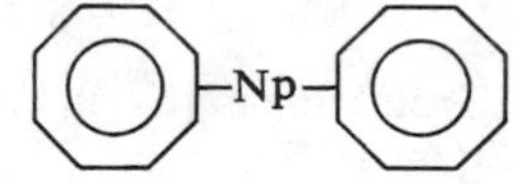

M 445.351

Orange solid. Sol. THF, toluene, C_6H_6, CCl_4. Air-sensitive. $\mu_{eff} = 1.81\mu_B$.

Karraker, D.G. *et al*, *J. Am. Chem. Soc.*, 1970, **92**, 4841 (*synth*, *mossbauer*, *ir*, *uv*)

$C_{20}H_{20}Np$ Np-00005

Tetrakis(η^5-2,4-cyclopentadien-1-yl)neptunium, 9CI

[37216-56-1]

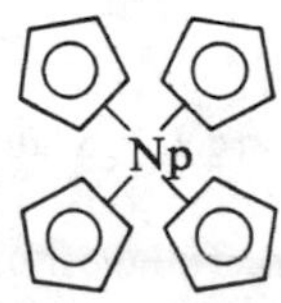

M 497.426

Brown-red solid. Sol. C_6H_6, THF. Subl. with dec. at 200-20° *in vacuo*. $\mu_{eff} = 2.43\mu_B$.

Baumgartner, F. *et al*, *Angew. Chem., Int. Ed. Engl.*, 1968, **7**, 634 (*synth*, *ir*)
Fischer, R.D. *et al*, *Z. Naturforsch., A*, 1969, **24**, 616 (*synth*)
Karraker, D.G. *et al*, *Inorg. Chem.*, 1972, **11**, 1742 (*mössbauer*)
Kanellakopulos, B. *et al*, *J. Chem. Phys.*, 1980, **72**, 6311 (*nmr*)

Pa Protactinium

S. A. Cotton

Protactinium (Fr.), Protaktinium (Ger.), Protactinio (Sp.), Protattinio (Ital.), Протактиний (Protactinii) (Russ.), プロトアクチニウム (Japan.)

Atomic Number. 91

Atomic Weight. 231.0359

Electronic Configuration. [Rn] $5f^2\ 6d^1\ 7s^2$ or [Rn] $5f^1\ 6d^2\ 7s^2$

Oxidation State. +4

Coordination Number. See under Uranium.

Colour. Orange-yellow.

Availability. Not commercially available. The commonest isotope, ^{231}Pa is an α-emitter ($t_{1/2} = 3.25 \times 10^4$y) obtained from uranium extraction. The most useful starting material is $PaCl_4$ obtained by the reduction of $PaCl_5$ with Al powder. $PaCl_5$ is in turn obtained from Pa_2O_5 by reaction with thionyl chloride. See Brown, D. *et al.*, *J. Chem. Soc.* (*A*), 1966, 874; 1967, 719.

Handling. See under Uranium.

Toxicity. Protactinum is extremely toxic, requiring handling procedures similar to those used for plutonium.

Isotopic Abundance. ^{231}Pa, 100%.

Spectroscopy. There are as yet no ^{1}H nmr data on the paramagnetic Pa(IV) ($5f^1$) organometallics.

Analysis. The identity of protactinium organometallics has largely been established by showing them to be isomorphous with corresponding compounds of other actinides.

References. See under Uranium.

$C_{16}H_{16}Pa$ **Pa-00001**

Bis(η^8-1,3,5,7-cyclooctatetraene)protactinium

[51056-18-9]

M 439.338

Golden-yellow cryst. Sol. C_6H_6, THF. Subl. at 155° *in vacuo*. V. air-sensitive.

Goffart, J. *et al*, *Inorg. Nucl. Chem. Lett.*, 1974, **10**, 413 (*synth*, *ir*)

Starks, D.F. *et al*, *Inorg. Chem.*, 1974, **13**, 1307 (*synth*)

$C_{20}H_{20}Pa$ **Pa-00002**

Tetrakis(η^5-2,4-cyclopentadien-1-yl)protactinium, 9CI

Tetra-π-cyclopentadienylprotactinium, 8CI

[79063-52-8]

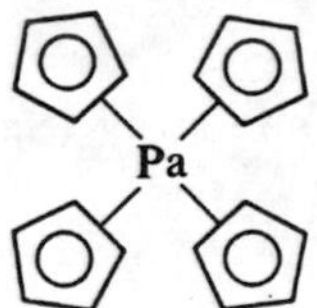

M 491.414

Orange-yellow cryst. Subl. with dec. at 220° *in vacuo*.

Baumgärtner, F. *et al*, *Angew. Chem., Int. Ed. Engl.*, 1969, **8**, 202 (*synth*, *ir*, *ms*)

Pm Promethium

S. A. Cotton

Promethium (Fr., Ger.), Prometio (Sp.), Promezio (Ital.), Прометий (Prometii) (Russ.), プロメチウム (Japan.)

Atomic Number. 61

Atomic Weight. All isotopes are radioactive, not naturally occurring.

Electronic Configuration. [Xe] $4f^5\,6s^2$

Oxidation State. +3

Coordination Number. Expected to be 6 or greater.

Colour. The one organopromethium compound so far reported is orange.

Availability. Extremely limited owing to the radioactivity and short half-life of the element. The stablest isotope, ^{147}Pm has $t_{1/2}$ = 17.7 y. The most useful isotope, ^{145}Pm, which has $t_{1/2}$ = 2.5 y, is a soft β-emitter.

Handling. See under Lanthanum.

Toxicity. Because of their radioactivity, promethium compounds should be handled with great care.

Spectroscopy. See the section under Lanthanum. The ^{1}H nmr spectra would be expected to display shifted and broadened lines compared with diamagnetic analogues. The optical spectra exhibit a range of sharp transitions. See Laubereau, P. G. *et al.*, *Inorg. Chem.*, 1970, **9**, 1091.

Analysis. The identity of the only known complex was established by comparison of powder photographs with those of analogous compounds.

References. See under Lanthanum.

$C_{15}H_{15}Pm$ **Pm-00001**

Tris(η^5-2,4-cyclopentadien-1-yl)promethium

Tricyclopentadienylpromethium

[12277-27-9]

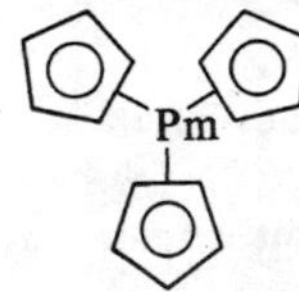

M 340.284

Yellow-orange cryst. Sol. THF. Subl. at 140-260° *in vacuo*. Hydrol. by H_2O.

Baumgärtner, F. *et al*, *Radiochim. Acta.*, 1967, **7**, 188 (*synth*)
Kopune, C.R. *et al*, *Radiochim. Radioanal. Letters*, 1969, **1**, 117 (*synth*)
Laubereau, P.G. *et al*, *Inorg. Chem.*, 1970, **9**, 1091 (*synth*)

Pr Praseodymium

S. A. Cotton

Praséodyme (Fr.), Praseodym (Ger.), Praseodimio (Sp., Ital.), Празеодимий (Prazeodimii) (Russ.), プラセオジム (Japan.)

Atomic Number. 59

Atomic Weight. 140.9077

Electronic Configuration. [Xe] $4f^3\ 6s^2$

Oxidation State. +3

Coordination Number. Usually 6 or greater.

Colour. With few exceptions (violet, yellow), organopraseodymium compounds are pale green.

Availability. Usual starting materials are Pr_6O_{11} and anhydrous $PrCl_3$. See also the section under Lanthanum.

Handling. Sec under Lanthanum.

Toxicity. See under Lanthanum.

Isotopic Abundance. ^{141}Pr, 100%.

Spectroscopy. ^{141}Pr has $I = \frac{5}{2}$ and has been little studied as an nmr probe.

The ^1H nmr spectra of Pr(III) compounds are reasonably sharp in spite of the fact that Pr(III) is paramagnetic ($4f^2$). See Von Ammon, R. *et al.*, *Chem. Ber.*, 1971, **104,** 1072, and also under Lanthanum.

Analysis. See under Lanthanum.

References. See under Lanthanum.

$C_6H_{18}Pr^{\ominus\ominus\ominus}$ Pr-00001

Hexamethylpraesodymate(3−)

$PrMe_6^{\ominus\ominus\ominus}$

M 231.116 (ion)

Tris[(tetramethylethylenediamine)lithium] salt: [76206-86-5].

$C_{24}H_{66}Li_3N_6Pr$ M 600.556

Green solid. Sol. Et_2O. Mp 59-62° dec. Extremely air- and moisture-sensitive.

Schumann, H. *et al*, *Angew. Chem., Int. Ed. Engl.*, 1981, **20**, 120 (*synth*)

C_8H_8ClPr Pr-00002

Chloro(cyclooctatetraene)praesodymium

Di-μ-chlorobis(η⁸-1,3,5,7-cycloctatetraene)dipraesodymium, 9CI

$(C_8H_8)PrCl$

M 280.512

Bis-THF complex probably dimeric; see Chloro(cyclooctatetraene)cerium.

Bis-THF complex: [51177-48-1].

$C_{32}H_{48}Cl_2O_4Pr_2$ M 849.450

Pale-green cryst. Sol. dioxan, THF. Mp >345°. Extremely air- and moisture-sensitive. $\mu = 3.39\mu_B$.

Hodgson, K.O. *et al*, *J. Am. Chem. Soc.*, 1973, **95**, 8650 (*synth*, *ir*, *uv*)

$C_{15}H_{15}Pr$ Pr-00003

Tris(η-⁵-2,4-cyclopentadien-1-yl)praseodymium, 9CI

Tri-π-cyclopentadienylpraseodymium, 8CI

[11077-59-1]

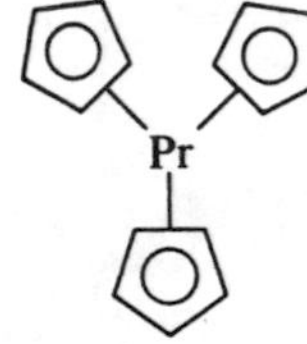

M 336.191

Pale-green cryst. Sol. THF. Mp 420°. Subl. at 200-50° *in vacuo*, hydrol. by H_2O.

Cyclohexylisocyanide adduct: [11057-26-4]. Green needles. Sol. C_6H_6, sl sol. pentane. Air- and moisture-sensitive.

Wilkinson, G. *et al*, *J. Am. Chem. Soc.*, 1954, **76**, 6210 (*synth*)
Mueller, J., *Chem. Ber.*, 1969, **102**, 152 (*ms*)
Von Ammon, R. *et al*, *Inorg. Nucl. Chem. Lett.*, 1969, **5**, 315 (*nmr*)
Devyatykh, G.G. *et al*, *Dokl. Akad. Nauk. SSSR*, 1970, **193**, 1069 (*synth*, *ms*)
Krasnova, S.G. *et al*, *Zh. Neorg. Khim.*, 1971, **16**, 1733 (*synth*)
Von Ammon, R. *et al*, *Ber. Bunsenges. Phys. Chem.*, 1972, **76**, 995 (*nmr*, *ir*)
Burns, J.H. *et al*, *J. Organomet. Chem.*, 1976, **120**, 361 (*struct*)
Aleksanyan, V.T. *et al*, *J. Organomet. Chem.*, 1977, **131**, 251 (*raman*)

$C_{16}H_{16}Pr^{\ominus}$ Pr-00004

Bis(η⁸-1,3,5,7-cyclooctatetraene)praesodymate(1−), 9CI

Dicyclooctatetraenepraesodymate(1−)

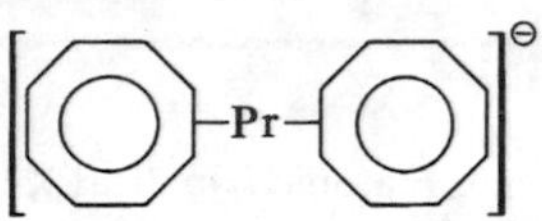

M 349.210 (ion)

K salt: [51177-53-8].

$C_{16}H_{16}KPr$ M 388.308

Yellow-gold solid. Sol. dioxan, THF. Mp >345°. Extremely air- and moisture-sensitive. $\mu = 2.84\mu_B$.

Mares, F. *et al*, *J. Organomet. Chem.*, 1970, **24**, C68 (*synth*)
Hodgson, K.O. *et al*, *J. Am. Chem. Soc.*, 1973, **95**, 8650 (*synth*, *ir*, *uv*)

$C_{22}H_{42}ClPrSi_4$ Pr-00006

Bis[1,3-bis(trimethylsilyl)cyclopentadienyl)chloropraesodymium

[81507-56-4]

M 595.277

Dimeric.

Dimer: Tetrakis[(1,2,3,4,5-η)-1,3-bis(trimethylsilyl-)-2,4-cyclopentadien-1-yl]di-μ-chlorodipraesodymium.

$C_{44}H_{84}Cl_2Pr_2Si_8$ M 1190.553

Yellow cryst. Sol. toluene, THF.

Lappert, M.F. *et al*, *J. Chem. Soc., Chem. Commun.*, 1981, 1190 (*synth*, *struct*)

Pu Plutonium

S. A. Cotton

Plutonium (Fr., Ger.), Plutonio (Sp., Ital.), Плутоний (Plutonii) (Russ.), プルトニウム (Japan.)

Atomic Number. 94

Atomic Weight. 244 (stablest isotope, $t_{1/2} = 8 \times 10^7$ y)

Electronic Configuration. [Rn] $5f^6\,7s^2$

Oxidation States. +3, +4

Coordination Number. See under Uranium.

Colour. Pu(IV) compounds are usually red, and Pu(III) compounds are green.

Availability. Plutonium compounds are not generally available. The most useful starting material is Cs_2PuCl_6 (see for example Bibler, J. P. *et al.*, *Inorg. Chem.*, 1968, **7**, 982. For the use of thc trihalides see Karraker, D. G. *et al.*, *J. Am. Chem. Soc.*, 1974, **96**, 6885).

Handling. Extreme care is required in the handling of plutonium, due to its acute toxicity. Remote control techniques are preferably used.

Toxicity. Plutonium is both highly radioactive and a potent carcinogen. Microgram quantities are potentially lethal and accumulate in the body.

Spectroscopy. Nmr data on organoplutonium complexes are sparse. Spectra reported for $Pu(Cp)_3$ show lines which are an order of magnitude broader than for U(IV) compounds.

Analysis. See under Neptunium.

References. See under Uranium.

$C_{15}H_{15}ClPu$ **Pu-00001**

Chlorotris(η^5-2,4-cyclopentadien-1-yl)plutonium

Tricyclopentadienylplutonium chloride

[81033-04-7]

M 474.737

Dark-brown powder. Sol. MeCN, THF, v. sl. sol. pentane. Mp >230°. V. sensitive to hydrolysis, particularly in soln. Does not subl. *in vacuo* up to 230°.

Bagnall, K.W. *et al*, *J. Organomet. Chem.*, 1982, **224**, 263 (*synth, ir, uv*)

$C_{15}H_{15}Pu$ **Pu-00002**

Tris(η^5-2,4-cyclopentadien-1-yl)plutonium

Tricyclopentadienylplutonium

[12216-08-9]

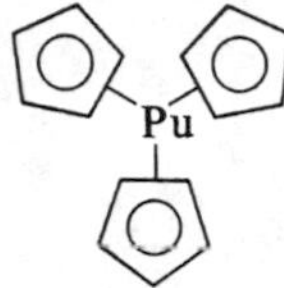

M 439.284

Moss-green solid. Sol. THF. Mp >195° dec. Subl. at 140-65° *in vacuo*. Extremely sensitive to air.

THF complex: Tris(η^5-2,4-cyclopentadienyl)-tetrahydrofuranplutonium.
$C_{19}H_{23}OPu$ M 511.390
Green cryst.

Nicotine complex: Tris(η^5-2,4-cyclopentadien-1-yl)-(nicotine)plutonium.
$C_{25}H_{29}N_2Pu$ M 601.518
Green cryst.

Baumgartner, F. *et al*, *Angew. Chem., Int. Ed. Engl.*, 1965, **4**, 878 (*synth, ir*)
Crisler, L.R. *et al*, *J. Inorg. Nucl. Chem.*, 1974, **36**, 1424 (*synth, uv, ms*)

$C_{16}H_{16}Pu^{\ominus}$ **Pu-00003**

Bis(η^8-1,3,5,7-cyclooctatetraene)plutonate(1−), 9CI

$[Pu(C_8H_8)_2]^{\ominus}$

M 452.302 (ion)

K salt, bis-THF complex: [54326-82-8].
$C_{24}H_{32}KO_2Pu$ M 635.614
Turquoise-green cryst. Extremely air- and water-sensitive. Easily oxid. to Bis(η^8-1,3,5,7-cyclooctatetraene)-plutonium, Pu-00004 μ_{eff} = 1.25μ_B.

Karraker, D.G. *et al*, *J. Am. Chem. Soc.*, 1974, **96**, 6885 (*synth*)

$C_{16}H_{16}Pu$ **Pu-00004**

Bis(η^8-1,3,5,7-cyclooctatetraene)plutonium, 9CI

Plutonocene

[37281-23-5]

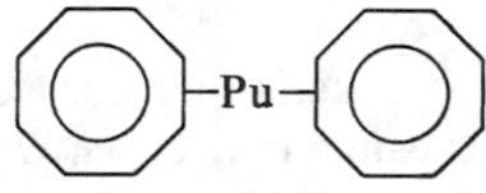

M 452.302

Cherry-red solid. Sol. THF, toluene, C_6H_6, CCl_4. Air-sensitive. Diamagnetic.

Karraker, D.G. *et al*, *J. Am. Chem. Soc.*, 1970, **92**, 4841 (*synth, ir, uv*)

Re Rhenium

M. Green

Rhénium (Fr.), Rhenium (Ger.), Renio (Sp., Ital.), Рений (Renii) (Russ.), レニウム (Japan.)

Atomic Number. 75

Atomic Weight. 186.207

Electronic Configuration. [Xe] $4f^{14}\,5d^5\,6s^2$

Oxidation States. −1, +1, +2, +3

Coordination Number. Commonly 6.

Colour. Variable.

Availability. Starting materials include $Re_2(CO)_{10}$, $KReO_4$, $ReCl_3$ and ReO_3. Rhenium is an expensive metal.

Handling. See under Chromium.

Toxicity. Rhenium compounds, with the exception of the carbonyl complexes, are not regarded as highly toxic.

Isotopic Abundance. ^{185}Re, 37.07%; ^{187}Re, 62.93%.

Spectroscopy. Both ^{185}Re and ^{187}Re have $I = \frac{5}{2}$ and only fundamental nmr studies have been carried out on these nuclei.

Analysis. For standard methods see *Scott's Standard methods of Chemical Analysis*, Furman, N. H. Ed., 6th Edn, Van Nostrand, 1962.

References. General reviews are listed in the introduction to the *Sourcebook*.

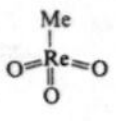

Re-00001

$ReCl_3(CO)_3^{\ominus\ominus}$

Re-00002

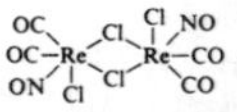

Re-00003

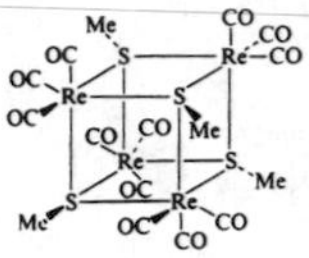

Re-00004

$Me_4Re{=}O$

Re-00005

$Re(CO)_5Br$

Re-00006

$Re(CO)_5Cl$ (O_h)

Re-00007

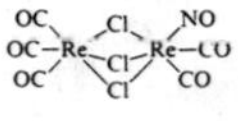

Re-00008

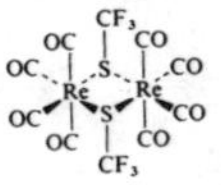

Re-00009

$HRe(CO)_5$ (O_h)

Re-00010

$(OC)_5ReI$ (O_h)

Re-00011

$Re(CO)_5^{\ominus}$

Re-00012

$[(OC)_5ReSO_2]^{\oplus}$ (O_h)

Re-00013

$MeRe(CO)_5$ (O_h)

Re-00014

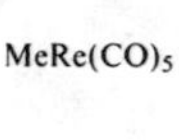

X = H

Re-00015

$ReMe_6$ (O_h)

Re-00016

$Re(CO)_6^{\oplus}$ (O_h)

Re-00017

$H_3CCNRe(CO)_5^{\oplus}$ (O_h)

Re-00018

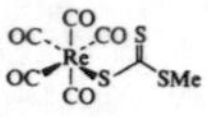

Re-00019

$(OC)_5ReCOCH_3$ (O_h)

Re-00020

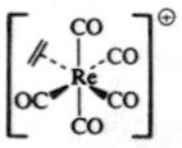

Re-00021

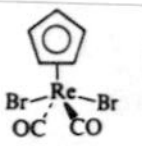

trans-form

Re-00022

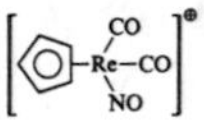

Re-00023

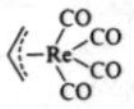

Re-00024

As Re-00015 with X = CHO

Re-00025

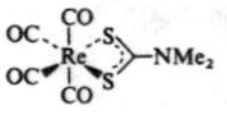

Re-00026

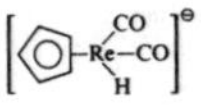

Re-00027

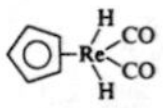

Re-00028

As Re-00015 with X = Me

Re-00029

As Re-00015 with X = CH_2OH

Re-00030

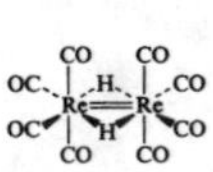

Re-00031

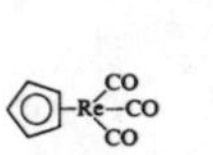

Re-00032

Re-00033

$(OC)_5ReCOOEt$ (O_h)

Re-00034

$Me_3PbRe(CO)_5$

Re-00035

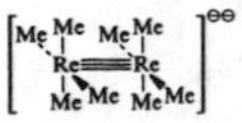

Re-00036

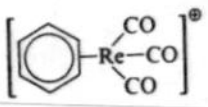

Re-00037

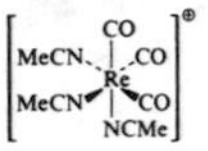

Re-00038

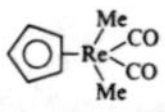

Re-00039

$Re_2(CO)_9^{\ominus\ominus}$

Re-00040

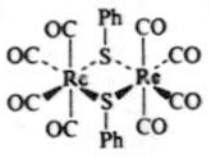

Re-00041

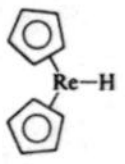

Re-00042

$(OC)_5Re{-}Mn(CO)_5$

Re-00043

$(OC)_5Re{-}Re(CO)_5$

Re-00044

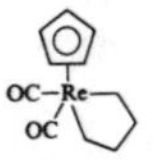

Re-00045

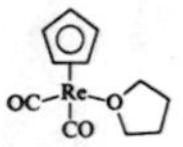

Re-00046

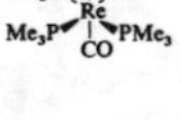

Re-00047

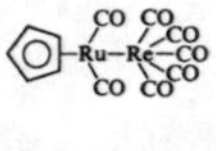

Re-00048

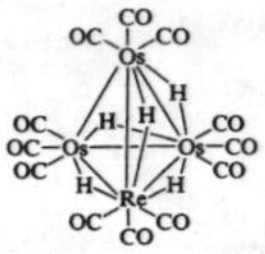

Re-00049

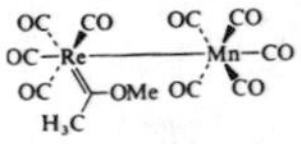

Re-00050

$Me_2Pb[Re(CO)_5]_2$

Re-00051

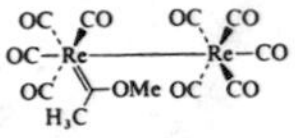

Re-00052

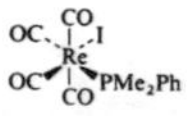

cis-form

Re-00053

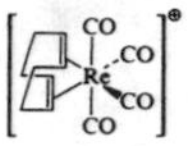

Re-00054

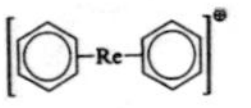

Re-00055

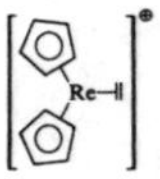

Re-00056

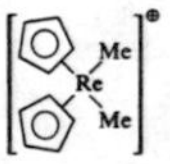

Re-00057

Re-00058

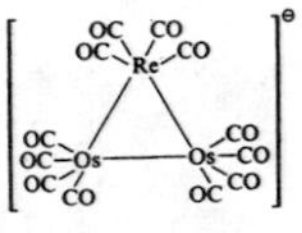

Re-00059

$ReOs_3H_3(CO)_{13}$

Re-00060

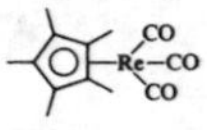

Re-00061

$(OC)_5Mn{-}Fe(CO)_4{-}Re(CO)_5$

Re-00062

$(OC)_5Re—Fe(CO)_4—Re(CO)_5$

Re-00063

$Re_3H(CO)_{14}$

Re-00064

Re-00065

Re-00066

Re-00067

$(OC)_5Re—Os(CO)_4—Re(CO)_5$

Re-00068

Re-00069

$ReOs_3H(CO)_{16}$

Re-00070

Re-00071

Re-00072

Re-00073

Re-00074

Re-00075

Re-00076

Re-00077

Re-00078

Re-00079

Re-00080

Re-00081

Re-00082

Re-00083

Re-00084

Re-00085

Re-00086

Re-00087

Re-00088

Re-00089

Re-00090

$Ph_3P—Re(CO)_4—Re(CO)_4—PPh_3$ (*diax-*)

Re-00091

Re-00092

Re-00093

CH_3O_3Re **Re-00001**

Methyltrioxorhenium, 9CI

[70197-13-6]

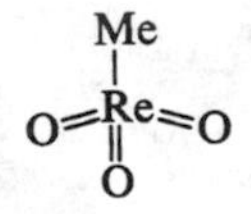

M 249.240

Colourless cryst. Sol. $CHCl_3$, C_6H_6, Et_2O.

Mertis, K. *et al*, *J. Chem. Soc.*, *Dalton Trans.*, 1975, 1488.
Beattie, I.R. *et al*, *Inorg. Chem.*, 1979, **18**, 2318 (*synth*, *spectra*)

$C_3Cl_3O_3Re^{\ominus\ominus}$ **Re-00002**

Tricarbonyltrichlororhenate(2−), 9CI

$ReCl_3(CO)_3^{\ominus\ominus}$

M 376.597 (ion)

Di-Cs salt: [35084-05-0].
$C_3CsCl_3O_3Re$ M 509.503
White cryst.

Bistetraethylammonium salt: [25908-19-4].
$C_{19}H_{40}Cl_3N_2O_3Re$ M 637.103
No phys. props. reported.

Ginsberg, A.P. *et al*, *Inorg. Chem.*, 1969, **8**, 2189 (*synth*)
Colton, R. *et al*, *Aust. J. Chem.*, 1972, **25**, 9 (*synth*)

$C_4Cl_4N_2O_6Re_2$ **Re-00003**

Tetracarbonyldi-μ-chlorodichlorodinitrosyldirhenium, 9CI

[25360-92-3]

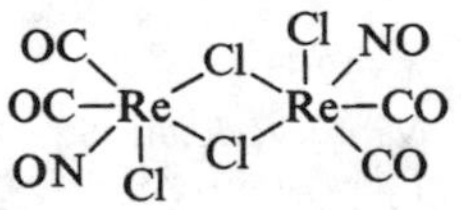

M 686.280

Starting material for rhenium nitrosyl chemistry. Stable yellow cryst. (CH_2Cl_2/hexane). Sol. Me_2CO, C_6H_6, $PhNO_2$, chlorinated hydrocarbons, insol. aliphatic hydrocarbons.

Monomer: Dicarbonyldichloronitrosylrhenium.
$C_2Cl_2NO_3Re$ M 343.140
Unknown.

Zingales, F. *et al*, *Inorg. Chem.*, 1971, **10**, 507 (*synth*)
Zingales, F. *et al*, *Inorg. Chem.*, 1971, **10**, 510 (*props*)
Trovati, A. *et al*, *Inorg. Chem.*, 1971, **10**, 851 (*props*)
Norton, J.R. *et al*, *Inorg. Chem.*, 1973, **12**, 485 (*synth*, *props*)
Inorg. Synth., 1976, **16**, 35 (*synth*, *ir*, *ms*)

$C_4H_3O_3ReS$ **Re-00004**

Tricarbonyl(methanethiolato)rhenium

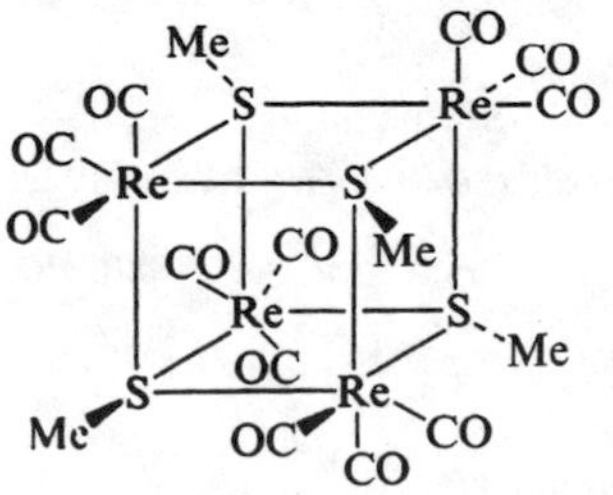

M 317.333

Tetrameric.

Tetramer: [23591-73-3]. *Dodecacarbonyltetrakis(μ_3-methanethiolato)tetrarhenium*, 8CI.
$C_{16}H_{12}O_{12}Re_4S_4$ M 1269.332
Yellow cryst.

Abel, E.W. *et al*, *J. Chem. Soc. (A)*, 1966, 1141 (*synth*, *ir*, *nmr*)
Osborne, A.G. *et al*, *J. Chem. Soc. (A)*, 1966, 1143.
Harrison, W. *et al*, *J. Chem. Soc.*, *Dalton Trans.*, 1972, 1009 (*cryst struct*)

$C_4H_{12}ORe$ **Re-00005**

Tetramethyloxorhenium

[53022-70-1]

$Me_4Re{=}O$

M 262.345

Red-purple cryst. by subl. Mp ca. 45°.

Mertis, K. *et al*, *J. Chem. Soc.*, *Dalton Trans.*, 1975, 607 (*synth*)
Gibson, J.F. *et al*, *J. Chem. Soc.*, *Dalton Trans.*, 1975, 1073 (*synth*)
Mertis, K. *et al*, *J. Chem. Soc.*, *Dalton Trans.*, 1976, 1488 (*synth*)

C_5BrO_5Re **Re-00006**

Bromopentacarbonylrhenium, 9CI, 8CI

Rhenium pentacarbonyl bromide

[14220-21-4]

$Re(CO)_5Br$

M 406.163

Yellow cryst. by subl.

Kirkham, W.J. *et al*, *J. Chem. Soc.*, 1965, 550 (*synth*)
Junk, G.A. *et al*, *J. Am. Chem. Soc.*, 1968, **90**, 5758 (*ms*)
Keeling, G. *et al*, *J. Chem. Soc. (A)*, 1971, 3143 (*ir*)
Colton, R. *et al*, *Aust. J. Chem.*, 1972, **25**, 9 (*synth*)
Hall, M.B., *J. Am. Chem. Soc.*, 1975, **97**, 2057 (*uv*)
Webb, M.J. *et al*, *J. Organomet. Chem.*, 1975, **93**, 119 (*cmr*)

C_5ClO_5Re **Re-00007**

Pentacarbonylchlororhenium, 9CI, 8CI

Chloropentacarbonylrhenium. Rhenium pentacarbonyl chloride

[14099-01-5]

$Re(CO)_5Cl$ (O_h)

M 361.712

Cryst. (C_6H_6). Sublimes at 140°.

Kirkham, W.J. *et al*, *J. Chem. Soc.*, 1965, 550 (*synth*)
Junk, G.A. *et al*, *J. Am. Chem. Soc.*, 1968, **90**, 5758 (*ms*)
Colton, R. *et al*, *Aust. J. Chem.*, 1972, **25**, 9 (*synth*)
Higginson, B.R. *et al*, *J. Chem. Soc.*, *Faraday Trans. 2*, 1975, **71**, 1913.
Wrighton, M.S. *et al*, *J. Am. Chem. Soc.*, 1976, **98**, 1111 (*uv*)

$C_5Cl_3NO_6Re_2$ **Re-00008**

Pentacarbonyltri-μ-chloronitrosyldirhenium, 9CI

[37402-69-0]

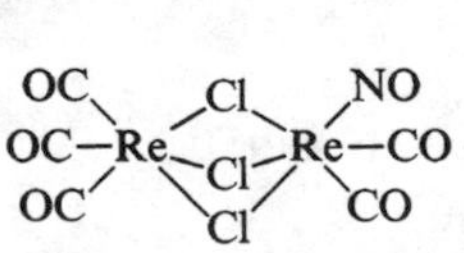

M 648.831

Mod. air-stable light-orange cryst. (CH_2Cl_2/hexane). Sol. CH_2Cl_2, CCl_4, C_6H_6, reacts with EtOH.

Zingales, F. *et al*, *Inorg. Chem.*, 1971, **10**, 507 (*props*)
Norton, J.R. *et al*, *Inorg. Chem.*, 1973, **12**, 485 (*synth*, *ir*)
Inorg. Synth., 1976, **15**, 35 (*synth*, *ir*)

$C_5F_3O_4ReS$ Re-00009

Tetracarbonyl(trifluoromethanethiolato)rhenium

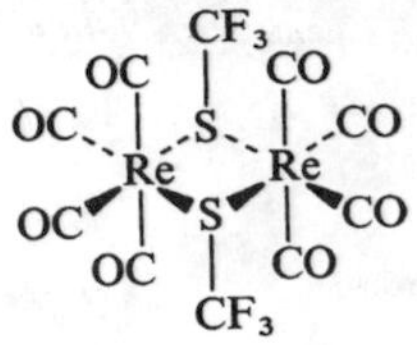

M 399.315

Dimeric.

Dimer: [25031-36-1]. *Octacarbonylbis[μ-(trifluoromethanethiolato)]dirhenium*, 8CI.

$C_{10}F_6O_8Re_2S_2$ M 798.630

White cryst. by fractional subl. Mp 129°. $Bp_{0.1}$ 70° subl.

King, R.B. *et al*, *Inorg. Chem.*, 1969, **8**, 2540 (*synth, ir, nmr*)
Grobe, J. *et al*, *J. Fluorine Chem.*, 1976, **8**, 145 (*synth, ir, nmr*)

C_5HO_5Re Re-00010

Pentacarbonylhydrorhenium, 9CI, 8CI

Hydridopentacarbonylrhenium. Pentacarbonylrhenium(I) hydride. Hydrogen pentacarbonyl rhenate(−1)

[16457-30-0]

$HRe(CO)_5$ (O_h)

M 327.267

Liq. Bp ca. 100° (extrapolated). With radical initiators gives ·$Re(CO)_5$ useful in synth. of substituted complexes. Dec. at ca. 100° to $H_2 + Re_2(CO)_{10}$. Air and moisture sensitive.

Hieber, W. *et al*, *Z. Naturforsch., B*, 1959, **14**, 132 (*synth, props*)
Hileman, J.C. *et al*, *Inorg. Chem.*, 1962, **1**, 933 (*synth, ir, nmr*)
Braterman, P.S. *et al*, *J. Am. Chem. Soc.*, 1967, **89**, 2851 (*ir*)
Byers, B.H. *et al*, *J. Am. Chem. Soc.*, 1977, **99**, 2527.
Miles, W.J. *et al*, *J. Organomet. Chem.*, 1977, **131**, 93 (*synth*)

C_5IO_5Re Re-00011

Pentacarbonyliodorhenium, 9CI

Iodopentacarbonylrhenium

[13821-00-6]

$(OC)_5ReI$ (O_h)

M 453.164

Yellow cryst.

Abel, E.W. *et al*, *J. Chem. Soc.*, 1958, 3149 (*props*)
Higginson, B.R. *et al*, *J. Chem. Soc., Faraday Trans. 2*, 1975, 1912 (*pe*)
Dunn, J.G. *et al*, *J. Organomet. Chem.*, 1974, **102**, 199 (*props*)
Burgess, J. *et al*, *J. Chem. Soc., Dalton Trans.*, 1976, 1158 (*props*)
Lindner, E. *et al*, *J. Organomet. Chem.*, 1976, **114**, C23 (*props*)
Wrighton, M.S. *et al*, *J. Am. Chem. Soc.*, 1976, **98**, 1111.

$C_5O_5Re^{\ominus}$ Re-00012

Pentacarbonylrhenate(1−), 10CI

[14971-38-1]

$Re(CO)_5^{\ominus}$

M 326.259 (ion)

Salts generally prepd. and used *in situ* due to extreme sensitivity to air and moisture.

Na salt: [33634-75-2].

C_5NaO_5Re M 349.249

Yellow-green soln. in THF.

K salt: [63139-42-4].

C_5KO_5Re M 365.357

Moisture-sensitive.

Inorg. Synth., 1964, **7**, 196 (*synth*)
Dessy, R.E. *et al*, *J. Am. Chem. Soc.*, 1966, **88**, 5121 (*props*)
Boerner, M. *et al*, *J. Chem. Res. (S)*, 1977, 74 (*props*)
Benson, I.B. *et al*, *J. Chem. Soc., Dalton Trans.*, 1978, 1240 (*props*)
Curtis, M.D. *et al*, *J. Organomet. Chem.*, 1979, **172**, 177 (*props*)
Beck, W. *et al*, *J. Organomet. Chem.*, 1981, **214**, 81 (*props*)

$C_5O_7ReS^{\oplus}$ Re-00013

Pentacarbonyl(sulfur dioxide-*S*)rhenium(1+)

$[(OC)_5ReSO_2]^{\oplus}$ (O_h)

M 390.318 (ion)

Hexafluoroarsenate: [55853-08-2].

$C_5AsF_6O_7ReS$ M 579.230

Pale-yellow solid. Dec. at 40-50° *in vacuo* with loss of SO_2 reversibly forming $Re(CO_5)F. AsF_5$.

Mews, R., *Angew. Chem., Int. Ed. Engl.*, 1975, **14**, 640 (*synth, ir*)

$C_6H_3O_5Re$ Re-00014

Pentacarbonylmethylrhenium, 9CI

Methylpentacarbonylrhenium

[14524-92-6]

$MeRe(CO)_5$ (O_h)

M 341.294

Colourless cryst. Mp 120°.

Hieber, W. *et al*, *Z. Naturforsch., B*, 1959, **14**, 132 (*synth*)
Beck, W. *et al*, *Chem. Ber.*, 1961, **94**, 862 (*synth*)
Tattershall, B.W. *et al*, *J. Chem. Soc. (A)*, 1968, 899 (*props*)
Bruce, M.I. *et al*, *J. Chem. Soc. (C)*, 1970, 3204 (*props*)
Bennett, R.J. *et al*, *J. Chem. Soc., Dalton Trans.*, 1972, 1787 (*props*)
King, R.B. *et al*, *J. Am. Chem. Soc.*, 1978, **100**, 1687 (*props*)

$C_6H_6NO_2Re$ Re-00015

Carbonyl(η[5]-2,4-cyclopentadien-1-yl)hydronitrosylrhenium, 9CI

Carbonyl(η[5]-cyclopentadienyl)hydridonitrosylrhenium

[38814-46-9]

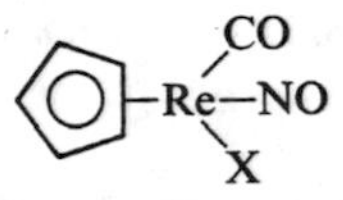

X = H

M 310.326

Orange thermally stable liq.

Stewart, R.P. *et al*, *J. Organomet. Chem.*, 1972, **42**, C32 (*synth, nmr*)
Casey, C.P. *et al*, *J. Am. Chem. Soc.*, 1980, **102**, 1927 (*synth*)

$C_6H_{18}Re$ Re-00016

Hexamethylrhenium, 9CI

Rhenium hexamethyl

[56090-02-9]

$ReMe_6$ (O_h)

M 276.415

Green cryst. Sol. hexane. Mp 10-2° dec.

▷Explosion hazard

Mertis, K., *J. Chem. Soc., Dalton Trans.*, 1975, 607 (*synth, props*)
Gibson, J.F. *et al, J. Chem. Soc., Dalton Trans.*, 1975, 1073 (*props*)
Gibson, J.F. *et al, J. Chem. Soc., Dalton Trans.*, 1976, 1492 (*esr*)
Green, J.C. *et al, J. Chem. Soc., Dalton Trans.*, 1978, 1403 (*pe*)

$C_6O_6Re^{\oplus}$ Re-00017

Hexacarbonylrhenium(1+)

Rhenium hexacarbonyl(1+)

[41944-00-7]

$Re(CO)_6^{\oplus}$ (O_h)

M 354.269 (ion)

Hexafluorophosphate: [38656-75-6].

$C_6F_6O_6PRe$ M 499.234

White air-stable cryst.

Hieber, W. *et al, Angew. Chem.*, 1961, **73**, 586 (*synth*)
Hieber, W. *et al, Z. Naturforsch., B*, 1961, **16**, 709 (*synth*)
Kruck, T. *et al, Chem. Ber.*, 1963, **96**, 3028 (*synth*)
Kruck, T. *et al, Chem. Ber.*, 1966, **99**, 1153 (*synth*)
Abel, E.W. *et al, J. Mol. Spectrosc.*, 1969, **30**, 29 (*ir*)
Bruce, D.M. *et al, J. Chem. Soc., Dalton Trans.*, 1978, 1627 (*cryst struct*)

$C_7H_3NO_5Re^{\oplus}$ Re-00018

(Acetonitrile)pentacarbonylrhenium(1+)

[29421-17-8]

$H_3CCNRe(CO)_5^{\oplus}$ (O_h)

M 367.311 (ion)

Hexafluorophosphate: [55057-83-5].

$C_7H_3F_6NO_5PRe$ M 512.276

Off-white powder. Sol. polar solvs. Mp 163-70°. Air-stable as solid, but air-sensitive in soln.

Connolly, N.G. *et al, J. Chem. Soc., Chem. Commun.*, 1970, 880 (*synth, ir*)
Drew, D. *et al, Inorg. Chem.*, 1975, **14**, 1579 (*synth, ir, nmr*)

$C_7H_3O_5ReS_3$ Re-00019

Pentacarbonyl(monomethylcarbonotrithioato-*S*)rhenium, 10CI

[55190-76-6]

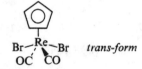

M 449.485

Yellow cryst. Mp 83°. Readily decarbonylates on warming.

Benson, I.B. *et al, J. Chem. Soc., Dalton Trans.*, 1978, 1240 (*synth, ir, nmr*)

$C_7H_3O_6Re$ Re-00020

Acetylpentacarbonylrhenium, 10CI

Pentacarbonylacetylrhenium

[23319-44-0]

$(OC)_5ReCOCH_3$ (O_h)

M 369.304

Sl. volatile, pale-yellow air-stable solid. Mp 79.5-80.5°.

Nesmeyanov, A.N. *et al, Dokl. Akad. Nauk SSSR, Ser. Sci. Khim.*, 1967, **175**, 1293 (*props*)
Inorg. Synth., 1980, **20**, 201 (*synth*)

$C_7H_4O_5Re$ Re-00021

Pentacarbonyl(ethene)rhenium(1+)

Pentacarbonyl(ethylene)rhenium(1+), 8CI

M 354.313

Hexafluorophosphate: [31922-27-7].

$C_7H_4F_6O_5ReP$ M 499.277

White cryst. (Me_2CO/Et_2O).

Fischer, E.O. *et al, Angew. Chem., Int. Ed. Engl.*, 1962, **1**, 52 (*synth, nmr*)
Brodie, A.M. *et al, J. Organomet. Chem.*, 1970, **24**, 201 (*synth*)

$C_7H_5Br_2O_2Re$ Re-00022

Dibromodicarbonyl(η^5-2,4-cyclopentadien-1-yl)rhenium, 9CI

[32966-00-0]

trans-form

M 467.130

Exists as *cis/trans* isomers in solid state, in soln. converts to the *trans* isomer shown.

cis-form

lateral-form

Dark-brown microcryst. Mp 217-20°. ir ν_{CO} 2055, 1987 cm^{-1} ($CHCl_3$).

trans-form

diagonal-form

Maroon plates. Mp 221-3°. ir ν_{CO} 2070, 2008 cm^{-1} ($CHCl_3$).

Nesmeyanov, A.N. *et al, Bull. Akad. Sci. USSR, Div. Chem. Sci.*, 1969, 1687 (*synth*)
King, R.B. *et al, J. Organomet. Chem.*, 1975, **93**, C23 (*synth*)
Sizoi, V.F. *et al, J. Organomet. Chem.*, 1975, **94**, 425 (*props*)

$C_7H_5NO_3Re^{\oplus}$ Re-00023

Dicarbonyl(η^5-2,4-cyclopentadien-1-yl)nitrosylrhenium(1+), 10CI

[45978-17-4]

M 337.328 (ion)

Tetrafluoroborate: [31960-40-4].

$C_7H_5BF_4NO_3Re$ M 424.132

Characterised spectroscopically.

Stewart, R.P. *et al, J. Organomet. Chem.*, 1972, **42**, C32 (*synth, props*)
Casey, C.P. *et al, J. Am. Chem. Soc.*, 1979, **101**, 741 (*props*)
Sweet, J.R. *et al, J. Organomet. Chem.*, 1979, **173**, C9.
Tam, W. *et al, J. Am. Chem. Soc.*, 1979, **101**, 1589 (*props*)
Wong, W.K. *et al, J. Am. Chem. Soc.*, 1979, **101**, 5440 (*props*)

Casey, C.P. *et al*, *J. Am. Chem. Soc.*, 1980, **102**, 1927 (*props*)

$C_7H_5O_4Re$ Re-00024

Tetracarbonyl(η^3-2-propenyl)rhenium, 9CI

(η^3-*Allyl*)*tetracarbonylrhenium*

[33307-29-8]

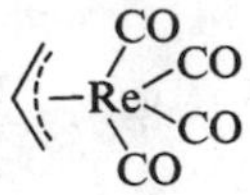

M 339.321

Yellow cryst. (hydrocarbons). Mp 32-5°.

Abel, E.W. *et al*, *J. Chem. Soc., Dalton Trans.*, 1973, 1706 (*synth, nmr*)

$C_7H_6NO_3Re$ Re-00025

Carbonyl(η^5-2,4-cyclopentadien-1-yl)formyl(nitrosyl)rhenium, 10CI

[69621-06-3]

As Carbonyl(η^5-2,4-cyclopentadien-1-yl)hydronitrosylrhenium, Re-00015 with

X = CHO

M 338.336

Orange oil.

Casey, C.P. *et al*, *J. Am. Chem. Soc.*, 1979, **101**, 741, 3371 (*synth, nmr*)
Tam, W. *et al*, *J. Am. Chem. Soc.*, 1979, **101**, 1589; 1982, **104**, 141 (*synth, props*)
Sweet, J.R. *et al*, *J. Organomet. Chem.*, 1979, **173**, C9 (*synth, nmr*)
Casey, C.P. *et al*, *J. Am. Chem. Soc.*, 1980, **102**, 1927 (*synth, props*)
Casey, C.P. *et al*, *Pure Appl. Chem.*, 1980, **52**, 625 (*props*)

$C_7H_6NO_4ReS_2$ Re-00026

Tetracarbonyl(dimethylcarbamodithioato-*S,S'*)rhenium, 9CI

[68795-84-6]

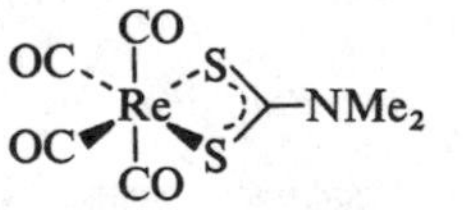

M 418.456

White cryst. Sol. CH_2Cl_2, toluene. Mp 120° dec.

Rowbottom, J.F. *et al*, *J. Chem. Soc., Dalton Trans.*, 1974, 684 (*synth*)
Calderazzo, F. *et al*, *J. Organomet. Chem.*, 1978, **160**, 207 (*synth, ir, nmr, cryst struct*)
Nakamoto, M. *et al*, *J. Chem. Soc., Dalton Trans.*, 1979, 87 (*synth*)

$C_7H_6O_2Re^{\ominus}$ Re-00027

Dicarbonyl(η^5-2,4-cyclopentadien-1-yl)hydrorhenate(1–)

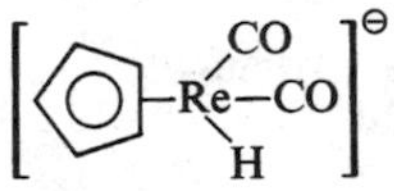

M 308.330 (ion)

K salt: [87145-51-5].

$C_7H_6KO_2Re$ M 347.429

Slightly yellow cryst.

Tetraethylammonium salt: [87145-52-6].

$C_{15}H_{26}NO_2Re$ M 438.583

Yellow cryst.

Bergmann, R.G. *et al*, *J. Am. Chem. Soc.*, 1983, **105**, 6500 (*synth, props*)

$C_7H_7O_2Re$ Re-00028

Dicarbonyl(η^5-2,4-cyclopentadien-1-yl)dihydrorhenium

[80952-45-0]

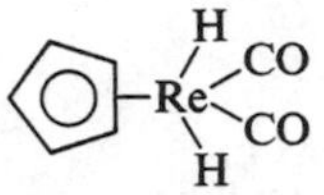

M 309.338

Useful starting material.

Hoyano, J.K. *et al*, *Organometallics*, 1982, **1**, 783 (*synth*)
Bergmann, R.G. *et al*, *J. Am. Chem. Soc.*, 1983, **105**, 6500 (*synth, props*)

$C_7H_8NO_2Re$ Re-00029

Carbonyl(η^5-2,4-cyclopentadien-1-yl)methylnitrosylrhenium, 9CI

[38814-45-8]

As Carbonyl(η^5-2,4-cyclopentadien-1-yl)hydronitrosylrhenium, Re-00015 with

X = Me

M 324.353

Red air-stable cryst. Mp 75°.

Stewart, R.P. *et al*, *J. Organomet. Chem.*, 1972, **42**, C32 (*synth, ir, nmr*)

$C_7H_8NO_3Re$ Re-00030

Carbonyl(η^5-2,4-cyclopentadien-1-yl)(hydroxymethyl)nitrosylrhenium, 10CI

[31960-41-5]

As Carbonyl(η^5-2,4-cyclopentadien-1-yl)hydronitrosylrhenium, Re-00015 with

X = CH_2OH

M 340.352

Air-stable orange powder. Mp 105°.

Sweet, J.R. *et al*, *J. Organomet. Chem.*, 1979, **173**, C9 (*synth, props, nmr*)
Tam, W. *et al*, *J. Am. Chem. Soc.*, 1979, **101**, 1589 (*synth, props, nmr*)
Casey, C.P. *et al*, *J. Am. Chem. Soc.*, 1980, **102**, 1927.

$C_8Cl_2O_8Re_2$ Re-00031

Octacarbonyldi-μ-chlorodirhenium, 10CI

[15189-52-3]

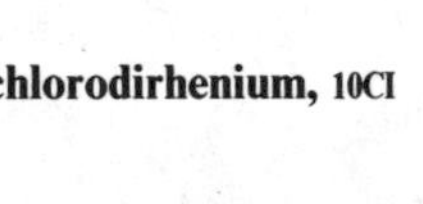

M 667.403

Bridged dimer. Off-white cryst.

Monomer: Tetracarbonylchlororhenium. Tetracarbonylrhenium chloride.

C_4ClO_4Re M 333.702

Unknown.

Rowbottom, J.F. *et al*, *J. Chem. Soc., Dalton Trans.*, 1974, 684 (*props*)
Swanson, B.I. *et al*, *Inorg. Chem.*, 1975, **14**, 1737 (*raman*)

Inorg. Synth., 1976, **16**, 35 (*synth*)
McCleverty, J.A. *et al*, *J. Organomet. Chem.*, 1979, **169**, 289 (*props*)
Lausarot, P.M. *et al*, *J. Organomet. Chem.*, 1980, **201**, 459 (*props*)

$C_8H_2O_8Re_2$ Re-00032

Octacarbonyldi-μ-hydrodirhenium, 10CI

[38887-05-7]

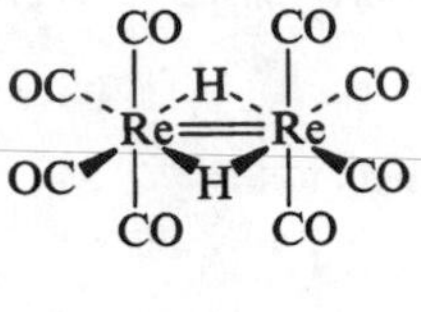

M 598.513

Yellow cryst., mod. air-stable.

Bennett, M.J. *et al*, *J. Am. Chem. Soc.*, 1972, **94**, 6232 (*synth, ir, nmr, cryst struct*)
Andrews, M.A. *et al*, *Inorg. Chem.*, 1977, **16**, 1556 (*synth, ir, nmr*)
Epstein, R.A. *et al*, *J. Am. Chem. Soc.*, 1979, **101**, 3847 (*synth*)

$C_8H_5O_3Re$ Re-00033

Tricarbonyl(η⁵-2,4-cyclopentadien-1-yl)rhenium, 9CI

Cyclopentadienyltricarbonylrhenium

[12079-73-1]

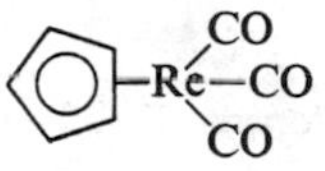

M 335.333

Cryst. (hexane or by subl. *in vacuo*). Mp 112°.

Fischer, E.O. *et al*, *J. Organomet. Chem.*, 1963, **1**, 191 (*synth*)
Nesmeyanov, A.N. *et al*, *Izv. Akad. Nauk SSSR*, 1963, 193; *CA*, **58**, 12588 (*synth*)
Lokshin, B.V. *et al*, *Spectrochim. Acta, Part A*, 1972, **28**, 2209; *Izv. Akad. Nauk SSSR*, 1974, 710; *CA*, **81**, 18816 (*ir*)
Ignatov, B.G. *et al*, *Izv. Akad. Nauk SSSR, Ser. Fiz.*, 1975, **39**, 2630; *CA*, **84**, 128662 (*nqr*)
Cozak, D. *et al*, *Spetrosc. Lett.*, 1976, **9**, 673 (*cmr*)
Lichtenberger, D.L. *et al*, *J. Am. Chem. Soc.*, 1976, **98**, 50.

$C_8H_5O_7Re$ Re-00034

Pentacarbonyl(ethoxycarbonyl)rhenium

Pentacarbonyl(hydroformatoethyl ester)rhenium, 8CI

[15692-05-4]

$(OC)_5ReCOOEt$ (O_h)

M 399.330

White cryst. Mp 69-70°.

Kruck, T. *et al*, *Chem. Ber.*, 1966, **99**, 1153 (*synth*)
Brodie, A.M. *et al*, *J. Organomet. Chem.*, 1970, **24**, 201 (*synth, props*)
Angelic, R.J., *Acc. Chem. Res.*, 1972, **5**, 335 (*props*)

$C_8H_9O_5PbRe$ Re-00035

Pentacarbony(trimethylplumbyl)rhenium, 10CI

(Trimethylplumbyl)pentacarbonylrhenium

[55318-19-9]

$Me_3PbRe(CO)_5$

M 578.563

Orange solid. Mp 48-50°. Disproportionates in Me_2CO to $Me_2Pb[Re(CO)_5]_2$ and Me_4Pb.

Schubert, W. *et al*, *Z. Naturforsch., B*, 1974, **29**, 694 (*synth*)
Kodel, W. *et al*, *Inorg. Chim. Acta*, 1981, **49**, 209.

$C_8H_{24}Re_2^{\ominus\ominus}$ Re-00036

Octamethyldirhenate(2−), 9CI

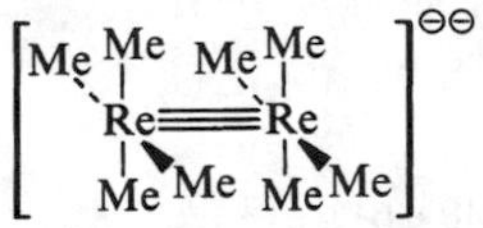

M 492.692 (ion)

Bis[tetrakis(tetrahydrofuran)lithium(1+)]salt: [62171-36-2].
$C_{40}H_{88}Li_2O_8Re_2$ M 1083.426
Red cryst., air- and water-sensitive, but thermally stable.

Cotton, F.A. *et al*, *J. Am. Chem. Soc.*, 1976, **98**, 6922 (*synth, cryst struct*)

$C_9H_6O_3Re^{\oplus}$ Re-00037

(η⁶-Benzene)tricarbonylrhenium(1+)

[68927-89-9]

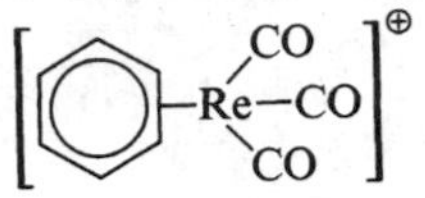

M 348.352 (ion)

Hexafluorophosphate: [68927-90-2].
$C_9H_6F_6O_3PRe$ M 493.316
Yellow cryst.

Kane-Maguire, L.A.P. *et al*, *Inorg. Chem.*, 1979, **18**, 700 (*synth, nmr, props*)

$C_9H_9N_3O_3Re^{\oplus}$ Re-00038

Tris(acetonitrile)tricarbonylrhenium(1+), 9CI

[66610-17-1]

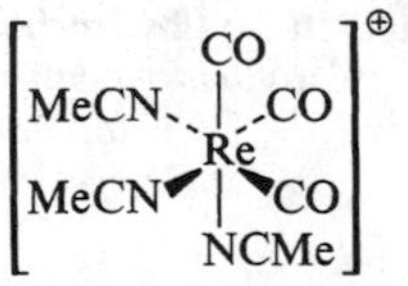

M 393.395 (ion)

Hexafluorophosphate: [66610-18-2].
$C_9H_9F_6N_3O_3PRe$ M 538.360
Yellow air-stable cryst.

Reimann, R.H. *et al*, *J. Organomet. Chem.*, 1973, **59**, C24 (*synth*)
Chan, L.Y.Y. *et al*, *Can. J. Chem.*, 1977, **55**, 111 (*synth, cryst struct*)
Zolanovitch, V.I. *et al*, *J. Organomet. Chem.*, 1978, **146**, 63 (*synth*)
Kane-Maguire, L.A.P. *et al*, *Inorg. Chem.*, 1979, **18**, 700 (*synth, ir, nmr*)

$C_9H_{11}O_2Re$ Re-00039

Dicarbonyl(η⁵-2,4-cyclopentadien-1-yl)dimethylrhenium

[80952-46-1]

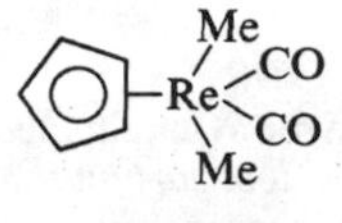

M 337.392

Pale-yellow cryst. Sol. C_6H_6. Mp 150-5°.

Hoyano, J.K. *et al*, *Organometallics*, 1982, **1**, 783 (*synth, props*)

$C_9O_9Re_2^{\ominus\ominus}$ Re-00040

μ-Carbonyloctacarbonyldirhenium(2−), 10CI

Nonacarbonyldirhenium(2−)

$Re_2(CO)_9^{\ominus\ominus}$

M 624.508 (ion)

Di-K salt: [66940-26-9].

$C_9H_9K_2Re_2$ M 567.781

Orange powder. Dec. at 120-40°.

Beck, W. *et al*, *Z. Anorg. Allg. Chem.*, 1961, **308**, 23 (*synth*)
Ellis, J.E. *et al*, *J. Organomet. Chem.*, 1975, **99**, 263.
Gladysz, J.A. *et al*, *J. Am. Chem. Soc.*, 1978, **100**, 2545 (*synth, ir*)

$C_{10}H_5O_4ReS$ Re-00041

(Benzenethiolato)tetracarbonylrhenium

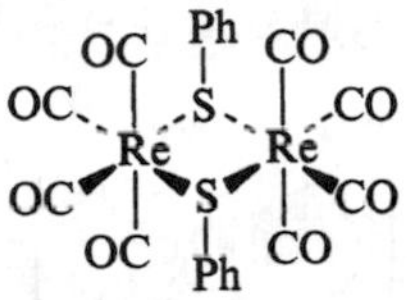

M 407.414

Dimeric.

Dimer: [15680-94-1]. *Bis(μ-benzenethiolato)octacarbonyldirhenium*, 8CI.

$C_{20}H_{10}O_8Re_2S_2$ M 814.828

Pale-yellow cryst. (C_6H_6/pentane). Mp 270° dec. ir ν_{CO} 2105m, 2030s, 2000m, 1967s cm^{-1} (cyclohexane).

Osborne, A.G. *et al*, *J. Chem. Soc. (A)*, 1966, 1143.

$C_{10}H_{11}Re$ Re-00042

Bis(η[5]-2,4-cyclopentadien-1-yl)hydrorhenium, 9CI

Bis(cyclopentadienyl)hydridorhenium. Bis(cyclopentadienyl)rhenium hydride. Dicyclopentadienylhydridorhenium

[1271-32-5]

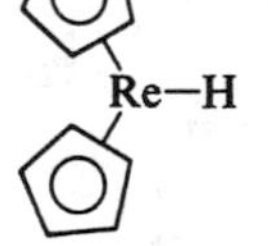

M 317.404

Yellow air-sensitive cryst. by subl. Sol. org. solvs., insol. H_2O. Mp 161-2°. $Bp_{0.01}$ ca. 80° subl.

Wilkinson, G. *et al*, *J. Am. Chem. Soc.*, 1955, **77**, 3421 (*nmr*)
Organomet. Synth., 1965, **1**, 80 (*synth*)
Cooper, R.L. *et al*, *J. Chem. Soc. (A)*, 1967, 1155 (*props*)
Shriver, D.F., *Acc. Chem. Res.*, 1970, **3**, 231 (*props*)
Hoxmeier, R.J., *Inorg. Chem.*, 1979, **18**, 3453 (*props*)
Casey, C.P. *et al*, *J. Am. Chem. Soc.*, 1980, **102**, 1927 (*props*)

$C_{10}MnO_{10}Re$ Re-00043

Pentacarbonyl(pentacarbonylmanganese)rhenium, 8CI

Manganese rhenium decacarbonyl. Decacarbonylmanganeserhenium

[14693-30-2]

$(OC)_5Re{-}Mn(CO)_5$

M 521.249

White, air-stable cryst.

Nesmeyanov, A.N. *et al*, *Izv. Akad. Nauk SSSR, Ser. Khim.*, 1963, 194 (*synth*)
Struchkov, Y.T. *et al*, *Dokl. Akad. Nauk SSSR, Ser. Sci. Khim.*, 1967, **172**, 107 (*cryst struct*)
Junk, G.A. *et al*, *J. Chem. Soc. (A)*, 1970, 2102.
Spiro, T.G., *Prog. Inorg. Chem.*, 1970, **11**, 1 (*spectra*)
Abel, E.W. *et al*, *Inorg. Nucl. Chem. Lett.*, 1971, **7**, 587 (*synth*)
Knox, S.A.R. *et al*, *Inorg. Chem.*, 1971, **10**, 2636 (*synth, struct*)

$C_{10}O_{10}Re_2$ Re-00044

Decacarbonyldirhenium, 10CI

Dirhenium decacarbonyl

[14285-68-8]

$(OC)_5Re{-}Re(CO)_5$

M 652.518

White air-stable cryst. by subl. Spar. sol. org. solvs., insol. H_2O. Mp 177°.

Organomet. Synth., 1965, **1**, 92 (*synth*)
Quicksall, C.O. *et al*, *Inorg. Chem.*, 1969, **8**, 2363 (*ir, raman*)
Swenson, B.I. *et al*, *Prog. Inorg. Chem.*, 1970, **11**, 1 (*ir*)
Bor, G. *et al*, *J. Chem. Soc., Dalton Trans.*, 1974, 440 (*electron diff*)
Jackson, R.A. *et al*, *Inorg. Chem.*, 1978, **17**, 997.
Churchill, M.R. *et al*, *Inorg. Chem.*, 1981, **20**, 1609 (*cryst struct*)

$C_{11}H_{13}O_2Re$ Re-00045

Dicarbonyl(η[5]-2,4-cyclopentadien-1-yl)tetramethylenerhenium

[87145-50-4]

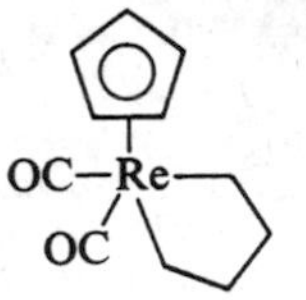

M 363.430

Yellow air-stable cryst.

Bergmann, R.G. *et al*, *J. Am. Chem. Soc.*, 1983, **105**, 6500 (*synth, props*)

$C_{11}H_{13}O_3Re$ Re-00046

Dicarbonyl(η[5]-2,4-cyclopentadien-1-yl)(tetrahydrofuran)rhenium, 9CI

[59423-86-8]

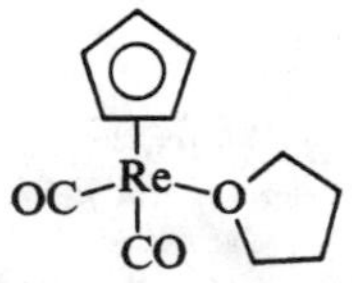

M 379.429

Stable yellow cryst. (THF). Mp 75-8° dec.

Sellmann, D. *et al*, *Z. Naturforsch., B*, 1977, **32**, 795 (*synth, nmr*)
Caulton, K.G., *Coord. Chem. Rev.*, 1981, **38**, 1 (*rev*)

$C_{11}H_{27}BrO_2P_3Re$ **Re-00047**

Bromodicarbonyltris(trimethylphosphine)rhenium, 9CI

[49742-48-5]

M 550.365

mer,cis-form

White cryst. Mp 175-8°.

Reimann, R.H. *et al*, *J. Organomet. Chem.*, 1973, **59**, 309 (*synth, ir, nmr, props*)

$C_{12}H_5O_7ReRu$ **Re-00048**

Dicarbonyl(η^5-2,4-cyclopentadien-1-yl)(pentacarbonylrhenio)ruthenium

Cyclopentadienylrutheniumdicarbonylpentacarbonylrhenium. (η^5-Cyclopentadienyl)dicarbonylrutheniumpentacarbonylrhenium. Heptacarbonyl(η^5-cyclopentadienyl)rheniumruthenium

M 548.444

Yellow solid. $Bp_{0.1}$ 60° subl.

Blackmore, T. *et al*, *J. Chem. Soc. (A)*, 1968, 2931 (*synth, ir*)
Bruce, M.I., *Int. J. Mass Spectrom. Ion Phys.*, 1968, **1**, 141 (*ms*)

$C_{12}H_5O_{12}Os_3Re$ **Re-00049**

Tricarbonyltri-μ-hydro(nonacarbonyldi-μ-hydrotriosmium-)rhenium, 10CI

Dodecacarbonylpenta-μ-hydrotriosmiumrhenium

[77424-05-6]

M 1097.971

Yellow cryst.

Churchill, M.R. *et al*, *J. Am. Chem. Soc.*, 1981, **103**, 2430 (*synth, struct, ir, pmr, ms*)

$C_{12}H_6MnO_{10}Re$ **Re-00050**

Pentacarbonyl[tetracarbonyl(1-methoxyethylidene)]manganeserhenium, 9CI

Nonacarbonyl(1-methoxyethylidene)manganeserhenium

[49547-64-0]

M 551.318

Red-orange cryst. Mp 107-8°.

Casey, C.P. *et al*, *J. Am. Chem. Soc.*, 1975, **97**, 3053 (*synth, struct*)

$C_{12}H_6O_{10}PbRe_2$ **Re-00051**

Bis(pentacarbonylrhenium)dimethyllead

Decacarbonyl[μ-(dimethylplumbylene)]dirhenium, 10CI

[55318-17-7]

$Me_2Pb[Re(CO)_5]_2$

M 889.787

Yellow solid. Mp 124-6°.

Schubert, W. *et al*, *Z. Naturforsch., B*, 1974, **29**, 694.

$C_{12}H_6O_{10}Re_2$ **Re-00052**

Nonacarbonyl(1-methoxyethylidene)dirhenium, 8CI

[49547-63-9]

M 682.587

Colourless cryst. Mp 110°. Subl. *in vacuo*.

Fischer, E.O. *et al*, *Chem. Ber.*, 1972, **105**, 3027 (*synth, nmr*)

$C_{12}H_{11}IO_4PRe$ **Re-00053**

Tetracarbonyl(dimethylphenylphosphine)iodorhenium, 8CI

[26901-29-1]

cis-form

M 563.302

cis-form

Pale-yellow cryst. Mp 90°.

trans-form [26901-38-2]

Pale-yellow cryst. Mp 122°.

Singleton, E. *et al*, *J. Organomet. Chem.*, 1970, **21**, 449 (*synth, ir, nmr*)

$C_{12}H_{12}O_4Re^{\oplus}$ **Re-00054**

Tetracarbonyl[(1,2,5,6-η)-1,5-cyclooctadiene)]rhenium(1+), 10CI

[66531-51-9]

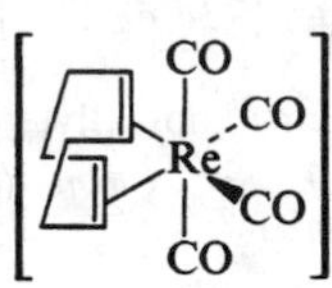

M 406.431 (ion)

Tetrafluoroborate: [66531-52-0].
$C_{12}H_{12}BF_4O_4Re$ M 493.235
Magnolia cryst. (Me_2CO/Et_2O).

Harris, P.J. *et al*, *J. Organomet. Chem.*, 1978, **148**, 327 (*synth, ir, nmr*)

$C_{12}H_{12}Re^{\oplus}$ **Re-00055**

Bis(η^6-benzene)rhenium(1+), 8CI

[11077-50-2]

M 342.434 (ion)

Dec. at 300°.

Hexafluorophosphate:

$C_{12}H_{12}F_6PRe$ M 487.398

Yellow cryst. (Me_2CO/Et_2O).

Fischer, E.O. *et al*, *Chem. Ber.*, 1966, **99**, 2206 (*synth*, *nmr*)

$C_{12}H_{14}Re^{\oplus}$ **Re-00056**

Bis(η^5-2,4-cyclopentadien-1-yl)(η^2-ethene)rhenium(1+), 10CI

Bis(η^5-cyclopentadienyl)(ethylene)rhenium(1+)

[72316-87-1]

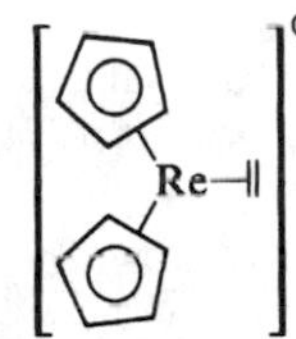

M 344.450 (ion)

Tetrafluoroborate: [72316-87-1].

$C_{12}H_{14}BF_4Re$ M 431.253

Deep-maroon cryst. (Me_2CO/Et_2O).

Mink, R.I. *et al*, *J. Am. Chem. Soc.*, 1979, **101**, 6928 (*synth*, *nmr*)

$C_{12}H_{16}Re^{\oplus}$ **Re-00057**

Bis(η^5-2,4-cyclopentadien-1-yl)dimethylrhenium(1+), 10CI

Dimethylbis(η^5-cyclopentadienyl)rhenium(1+)

[73449-72-6]

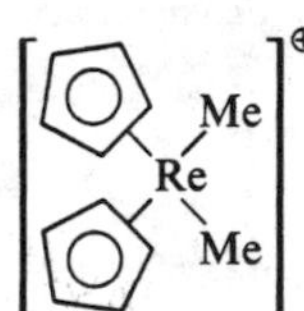

M 346.465 (ion)

Hexafluorophosphate: [73449-73-7].

$C_{12}H_{16}F_6PRe$ M 491.430

Characterised spectroscopically.

Baudry, D. *et al*, *J. Chem. Soc.*, *Chem. Commun.*, 1979, 895; 1980, 249 (*synth*, *props*)

$C_{12}H_{20}Re_2$ **Re-00058**

Tetrakis(η^3-2-propenyl)dirhenium, 10CI

Tetrakis(η^3-allyl)dirhenium

[67178-24-9]

M 536.704

Orange cryst. (pentane). Mp 120° dec.

Masters, A.F. *et al*, *Nouv. J. Chem.*, 1977, **1**, 389 (*synth*, *ir*, *nmr*)

Cotton, F.A. *et al*, *J. Am. Chem. Soc.*, 1978, **100**, 3788 (*cryst struct*)

$C_{12}O_{12}Os_2Re^{\ominus}$ **Re-00059**

Dodecacarbonylrhenatediosmate(1−), 8CI

M 902.732 (ion)

Tetramethylammonium salt: [12563-67-6].

$C_{16}H_{12}NO_{12}Os_2Re$ M 976.877

Characterised spectroscopically.

Knight, J. *et al*, *J. Chem. Soc.*, *Chem. Commun.*, 1971, 62 (*synth*, *ir*)

$C_{13}H_3O_{13}Os_3Re$ **Re-00060**

Tridecacarbonyltrihydrorheniumtriosmium, 8CI

[12563-76-7]

$ReOs_3H_3(CO)_{13}$

M 1123.966

No phys. props. reported.

Knight, J. *et al*, *J. Chem. Soc.*, *Chem. Commun.*, 1971, 62 (*synth*, *ir*)

$C_{13}H_{15}O_3Re$ **Re-00061**

Tricarbonyl[(1,2,3,4,5-η)-1,2,3,4,5-pentamethyl-2,4-cyclopentadien-1-yl)]rhenium, 9CI

[12130-88-0]

M 405.467

White cryst. (hexane). Mp 151°.

King, R.B. *et al*, *J. Organomet. Chem.*, 1967, **8**, 287 (*synth*, *nmr*)

King, R.B. *et al*, *J. Organomet. Chem.*, 1979, **171**, 53 (*synth*, *props*)

$C_{14}FeMnO_{14}Re$ **Re-00062**

Pentacarbonyl(pentacarbonylmanganese)(tetracarbonyliron)-rhenium, 9CI

Tetradecacarbonylmanganeserheniumiron, 8CI

[33958-72-4]

$(OC)_5Mn—Fe(CO)_4—Re(CO)_5$

M 689.138

Orange needles (Me_2CO).

Evans, G.O. *et al*, *J. Inorg. Nucl. Chem.*, 1968, **30**, 2862 (*synth*, *ir*, *ms*)

Evans, G.O. *et al*, *Inorg. Chem.*, 1970, **9**, 979 (*ir*)

Tyler, D.R. *et al*, *J. Am. Chem. Soc.*, 1978, **100**, 7888 (*struct*)

$C_{14}FeO_{14}Re_2$ **Re-00063**

Tetracarbonylbis(pentacarbonylrhenio)iron, 8CI

Decacarbonyl(tetracarbonyliron)dirhenium, 10CI

[16040-31-6]

$(OC)_5Re—Fe(CO)_4—Re(CO)_5$

M 820.407

Stable yellow solid. Mp 163° dec.

Evans, G.O. *et al*, *J. Chem. Soc., Chem. Commun.*, 1967, 186 (*synth, ir*)
Evans, G.O. *et al*, *Inorg. Chem.*, 1971, **10**, 1598 (*ir, raman*)
Tyler, D.R. *et al*, *J. Am. Chem. Soc.*, 1978, **100**, 7888 (*uv*)

$C_{14}HO_{14}Re_3$ Re-00064

Tetradecacarbonylhydrotrirhenium, 9CI
[12086-80-5]

$Re_3H(CO)_{14}$

M 951.775
Yellow cryst.

Fellmann, W. *et al*, *Inorg. Nucl. Chem. Lett.*, 1966, **2**, 63 (*synth, ir, nmr*)
Curtis, M.D., *Inorg. Nucl. Chem. Lett.*, 1970, **6**, 859 (*synth*)
Curtis, M.D., *Inorg. Chem.*, 1972, **11**, 802 (*synth, ir*)

$C_{14}H_{10}MnO_4Re$ Re-00065

(η^5-2,4-Cyclopentadien-1-yl)[μ-(η:η^5-2,4-cyclopentadien-1-ylidene)]hydrotetracarbonylmanganeserhenium, 10CI
[71486-19-6]

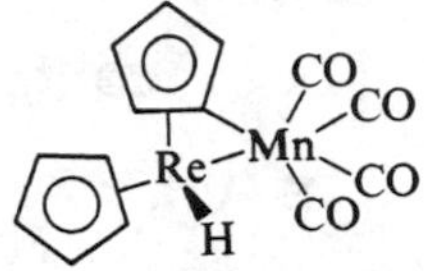

M 483.376
Air-stable red cryst. Air-sensitive in soln.

Hoxmeier, R.J. *et al*, *Inorg. Chem.*, 1979, **18**, 3453 (*synth, nmr, props*)

$C_{14}H_{10}O_2Re^{\oplus}$ Re-00066

Dicarbonyl(η^5-2,4-cyclopentadien-1-yl)(phenylmethylidyne)-rhenium(1+)
[61993-63-3]

M 396.439 (ion)
Tetrachloroborate: [61993-64-4].
$C_{14}H_{10}BCl_4O_2Re$ M 549.061
Yellow cryst. Mp 78° dec.

Fischer, E.O. *et al*, *J. Organomet. Chem.*, 1976, **120**, C6 (*synth, nmr*)

$C_{14}H_{30}Cl_2O_2P_2Re$ Re-00067

Dicarbonyldichlorobis(triethylphosphine)rhenium, 10CI
[61818-15-3]

M 549.450
trans-form
Green cryst.

Hertzer, C.A. *et al*, *Inorg. Chem.*, 1978, **17**, 2383 (*synth, ir, nmr, props*)

$C_{14}O_{14}OsRe_2$ Re-00068

Tetracarbonylbis(pentacarbonylrhenio)osmium, 8CI
Tetradecacarbonylosmiumdirhenium. Osmium dirhenium tetradecacarbonyl
[33153-70-7]

$(OC)_5Re{-}Os(CO)_4{-}Re(CO)_5$

M 954.760
No phys. props. reported. ir ν_{CO} 2094, 2021, 1982 cm^{-1}.

Abel, E.W. *et al*, *Inorg. Nucl. Chem. Lett.*, 1971, **7**, 587 (*synth, ir*)

$C_{15}H_5BrO_8Re_2$ Re-00069

μ-Bromooctacarbonyl[μ-(phenylmethylidyne)dirhenium
[58384-14-8]

M 765.518
Black cryst. (CH_2Cl_2/Et_2O). Mp 108° dec.

Fischer, E.O. *et al*, *Angew. Chem., Int. Ed. Engl.*, 1976, **15**, 231 (*synth, nmr, cryst struct*)

$C_{16}HO_{16}Os_3Re$ Re-00070

Hexadecacarbonylhydrorheniumtriosmium, 8CI
[12566-58-4]

$ReOs_3H(CO)_{16}$

M 1205.981
Yellow solid, not obt. pure.

Knight, J. *et al*, *J. Chem. Soc., Chem. Commun.*, 1971, 62 (*synth*)
Shapley, J.R. *et al*, *J. Am. Chem. Soc.*, 1977, **99**, 8064 (*synth, ms, pmr*)

$C_{16}H_9N_2O_4Re$ Re-00071

Tetracarbonyl[2-(phenylazo)phenyl]rhenium, 8CI
[19529-33-0]

M 479.465
Red needles by subl. Mp 121-2°. $Bp_{0.01}$ 100° subl.

Bruce, M.I. *et al*, *J. Chem. Soc. (A)*, 1970, 3204 (*synth, nmr*)

$C_{16}H_{30}N_3OReS_6$ Re-00072

Carbonyltris(diethylcarbamodithioato-*S,S'*)rhenium, 9CI

[52193-18-7]

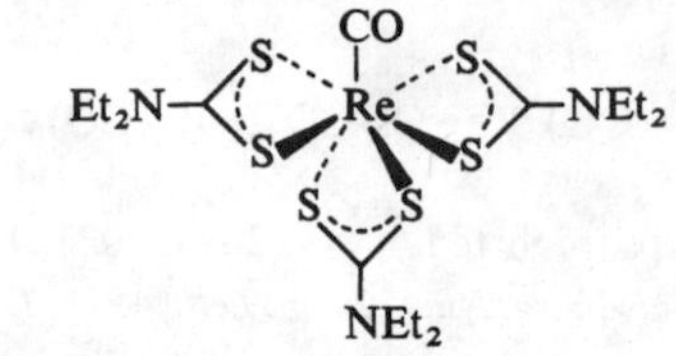

M 659.000

Distorted pentagonal bipyramidal geometry. Wine-red plates (Me_2CO).

Fletcher, S.R. *et al, J. Chem. Soc., Dalton Trans.*, 1974, 486 (*cryst struct*)

Rowbottom, J.F. *et al, J. Chem. Soc., Dalton Trans.*, 1974, 684 (*synth, ir, nmr*)

$C_{16}O_{16}Re_4^{\ominus\ominus}$ Re-00073

Hexadecacarbonyltetrarhenate(2−)

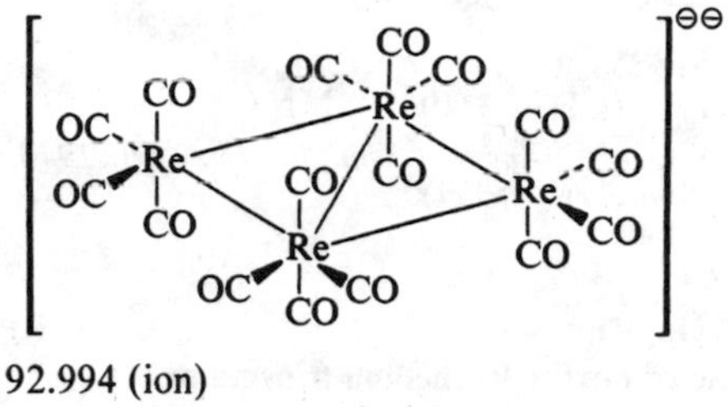

M 1192.994 (ion)

Di-Na salt: [19497-41-7].
$C_{16}Na_2O_{16}Re_4$ M 1238.974
No phys. props. reported.

Bis(tetraethylammonium) salt:
$C_{32}H_{40}N_2O_{16}Re_4$ M 1453.500
Dark-red cryst. (Me_2CO/EtOH).

Bau, R. *et al, J. Am. Chem. Soc.*, 1967, **89**, 6374 (*synth*)

Churchill, M.R. *et al, Inorg. Chem.*, 1968, 7, 2606 (*cryst struct*)

Ciani, G. *et al, J. Organomet. Chem.*, 1978, **157**, 199 (*synth, cryst struct*)

$C_{17}H_4NO_{15}Os_3Re$ Re-00074

[(Acetonitrile)decacarbonyl-μ-hydrotriosmium]pentacarbonylrhenium, 10CI

(Acetonitrile)pentadecacarbonyl-μ-hydrotriosmiumrhenium

[79255-11-1]

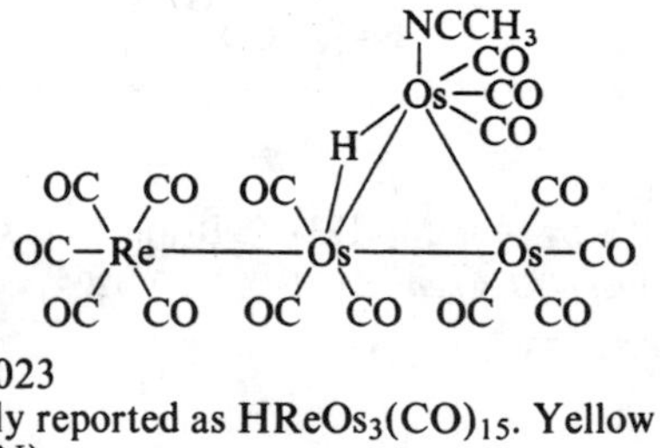

M 1219.023

Previously reported as $HReOs_3(CO)_{15}$. Yellow cryst. (MeCN).

Shapley, J.R. *et al, J. Am. Chem. Soc.*, 1977, **99**, 8064 (*synth, ir, ms*)

Churchill, M.R. *et al, J. Am. Chem. Soc.*, 1981, **103**, 2430 (*struct*)

$C_{17}H_{10}NO_4Re$ Re-00075

Tetracarbonyl[2-[(phenylimino)methyl]phenyl-*C,N*]rhenium, 9CI

[38883-46-4]

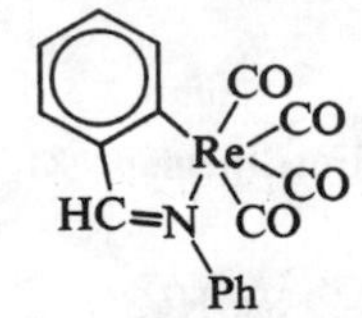

M 478.477

Yellow needles (pet. ether). Sol. cyclohexane, Me_2CO. Mp 115.5-116°.

Bennett, R.L. *et al, J. Chem. Soc., Dalton Trans.*, 1972, 1787 (*synth, ir, nmr*)

$C_{20}H_2O_{20}Os_3Re_2$ Re-00076

Decacarbonyl(decacarbonyldi-μ-hydrotriosmium)diruthenium, 10CI

Icosacarbonyldi-μ-hydrotriosmiumdiruthenium. Eicosacarbonyldi-μ-hydrotriosmiumdiruthenium

[64885-82-1]

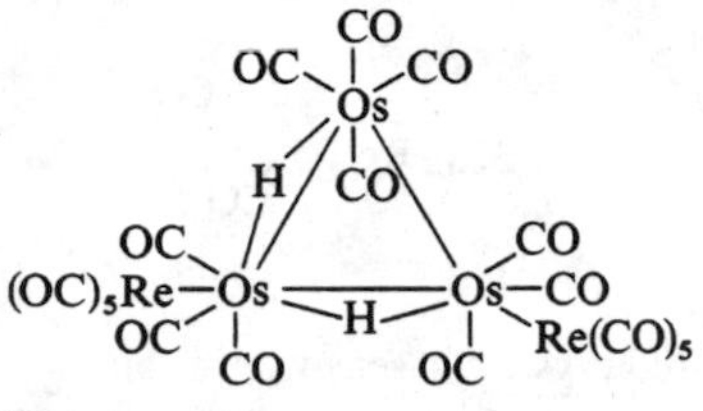

M 1505.238

Clear-yellow plates.

Churchill, M.R. *et al, Inorg. Chem.*, 1978, **17**, 3546 (*struct*)

Churchill, M.R. *et al, J. Am. Chem. Soc.*, 1981, **103**, 2430 (*synth, ir, pmr, ms*)

$C_{20}H_{11}O_5Re$ Re-00077

Pentacarbonyl(1,2-diphenyl-2-cyclopropen-1-yl)rhenium, 10CI

[79643-33-7]

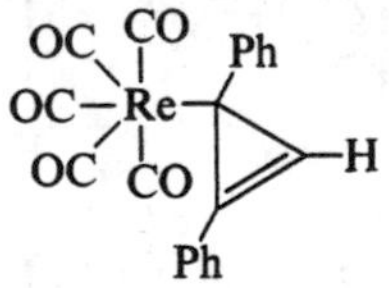

M 517.511

Pale-yellow cryst. Mp 100-3°.

Desrosiers, P.J. *et al, J. Am. Chem. Soc.*, 1981, **103**, 5593 (*synth, nmr*)

$C_{21}H_{11}O_6Re$ Re-00078

Pentacarbonyl[(2,3-diphenyl-2-cyclopropen-1-yl)]carbonylrhenium, 10CI

[79643-31-5]

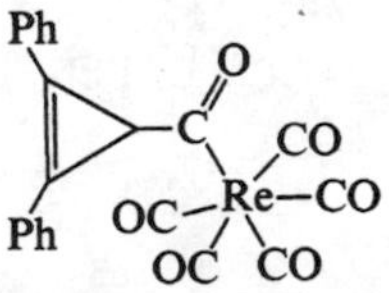

M 545.521

Pale-yellow cryst. Mp 92-4°.

Desrosiers, P.J. *et al, J. Am. Chem. Soc.*, 1981, **103**, 5593 (*synth, nmr*)

$C_{22}H_{14}O_4PRe$ Re-00079

Tetracarbonyl[2-(diphenylphosphinophenyl-*C,P*)]rhenium, 9CI

[56483-30-8]

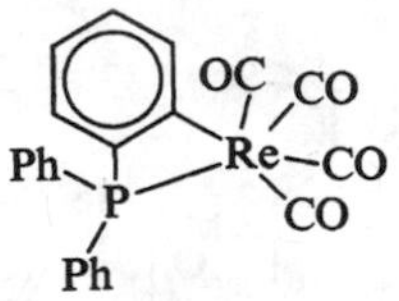

M 559.531

Red cryst.

McKinney, R.J. *et al*, *J. Am. Chem. Soc.*, 1975, **97**, 3066 (*synth, nmr*)
Huie, B.T. *et al*, *J. Am. Chem. Soc.*, 1977, **99**, 7852 (*props*)

$C_{24}H_{22}NOPRe^{\oplus}$ Re-00080

(η^5-2,4-Cyclopentadien-1-yl)methylenenitrosyl(triphenylphosphine)rhenium(1+), 10CI

Methylenenitrosyl(triphenylphosphine)(η^5-cyclopentadienyl)rhenium(1+)

[71763-23-0]

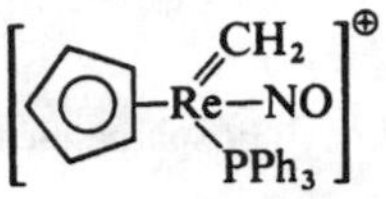

M 557.625 (ion)

Tetrafluoroborate: [71763-22-9].
$C_{24}H_{22}BF_4NOPRe$ M 644.428
Unstable, characterised by nmr.

Wong, W.K. *et al*, *J. Am. Chem. Soc.*, 1979, **101**, 5441 (*synth*)
Kiel, W.A. *et al*, *J. Am. Chem. Soc.*, 1980, **102**, 3299 (*props*)

$C_{26}H_{20}O_2ReSi^{\oplus}$ Re-00081

Dicarbonyl(η^5-2,4-cyclopentadien-1-yl)[(triphenylsilyl)methylidyne]rhenium(1+)

[76096-93-0]

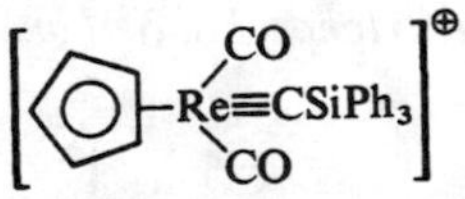

M 578.735 (ion)

Tetrafluoroborate: [76096-94-1].
$C_{26}H_{20}BF_4O_2ReSi$ M 665.539
Bright-yellow cryst. Mp 140° dec.

Fischer, E.O. *et al*, *Z. Naturforsch., B*, 1980, **35**, 1083 (*synth, nmr*)

$C_{26}H_{33}ClO_2P_3Re$ Re-00082

Dicarbonylchlorotris(dimethylphenylphosphine)rhenium, 9CI

[26566-78-9]

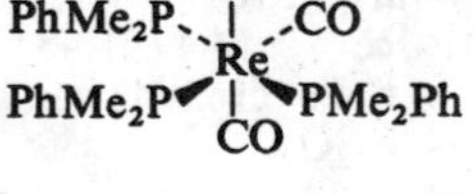

M 692.127

Yellow prisms.

Douglas, P.G. *et al*, *J. Chem. Soc. (A)*, 1969, 1491.
Chatt, J. *et al*, *J. Organomet. Chem.*, 1974, **64**, 245 (*synth, ir, nmr*)

$C_{31}H_{33}O_7P_3Re_2$ Re-00083

Heptacarbonyltris(dimethylphenylphosphine)dirhenium, 10CI

[26901-33-7]

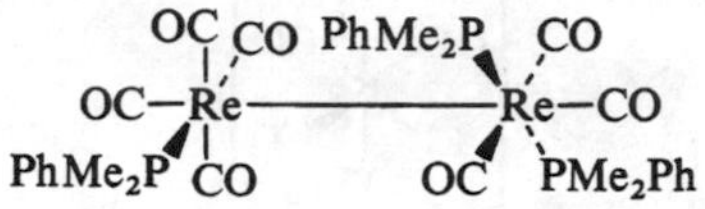

M 982.933

White prisms (EtOAc/hexane). Mp 148°.

Singleton, E. *et al*, *J. Organomet. Chem.*, 1971, **21**, 449 (*synth*)
Moelwyn-Hughes, J.T. *et al*, *J. Organomet. Chem.*, 1971, **26**, 373 (*synth*)
Cox, D.J. *et al*, *J. Organomet. Chem.*, 1980, **186**, 347 (*synth, ir*)

$C_{33}H_{44}ClOP_4Re$ Re-00084

Carbonylchlorotetrakis(dimethylphenylphosphine)rhenium, 9CI

[25259-88-5]

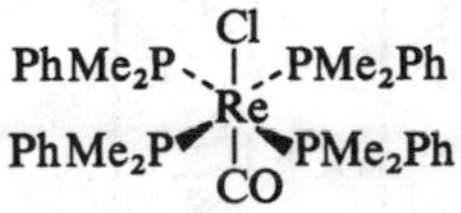

M 802.265

White cryst. Mp 166-79° dec.

Douglas, P.G. *et al*, *J. Chem. Soc. (A)*, 1969, 1491 (*synth, ir, nmr*)
Chatt, J. *et al*, *J. Organomet. Chem.*, 1974, **64**, 245 (*synth, ir, nmr*)

$C_{38}H_{30}NO_3P_2Re$ Re-00085

Dicarbonylnitrosylbis(triphenylphosphine)rhenium, 9CI

[53183-18-9]

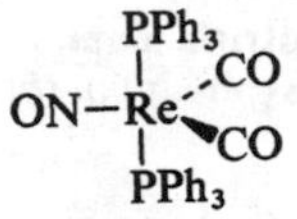

M 796.814

Brick-red cryst. Mp 180° dec.

La Monica, G. *et al*, *J. Organomet. Chem.*, 1974, **71**, 57 (*synth*)
Giusto, D. *et al*, *J. Organomet. Chem.*, 1976, **105**, 91 (*synth, ir, nmr, props*)

$C_{38}H_{45}P_2Re$ Re-00086

Bis(diethylphenylphosphine)triphenylrhenium, 9CI

[15390-66-6]

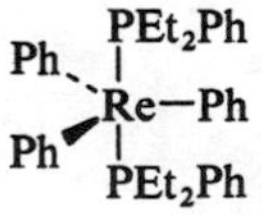

M 749.928

Deep-blue cryst. (hexane).

Chatt, J. *et al*, *J. Chem. Soc. (A)*, 1966, 1834 (*synth*)
Gibson, J.F. *et al*, *J. Chem. Soc., Dalton Trans.*, 1975, 1073 (*esr*)
Carroll, W.E. *et al*, *J. Chem. Soc., Chem. Commun.*, 1978, 825 (*struct*)
Green, J.C. *et al*, *J. Chem. Soc., Dalton Trans.*, 1978, 1403 (*pe*)

$C_{39}H_{30}ClO_3P_2Re^{\oplus}$ Re-00087

Tricarbonylchlorobis(triphenylphosphine)rhenium(1+), 9CI

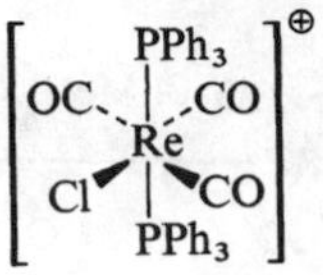

M 830.272 (ion)

mer,trans-form

Tetrafluoroborate: [42584-07-6].
$C_{39}H_{30}BClF_4O_3P_2Re$ M 917.075
White solid. Mp 173-8°.

Eaborn, C. *et al*, *J. Chem. Soc., Dalton Trans.*, 1976, 58 (*synth, ir, nmr*)

$C_{40}H_{30}O_4P_2Re^{\oplus}$ Re-00088

Tetracarbonylbis(triphenylphosphine)rhenium(1+)

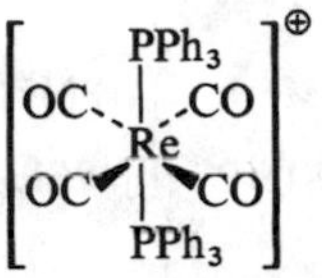

M 822.829 (ion)

trans-form

Hexafluorophosphate: [38496-54-7].
$C_{40}H_{30}F_6O_4P_3Re$ M 967.793
Characterised spectroscopically.

Angelici, R.J. *et al*, *Inorg. Chem.*, 1973, **12**, 1067 (*synth, ir*)
Behrens, H. *et al*, *J. Organomet. Chem.*, 1978, **159**, 201 (*synth, ir, nmr*)

$C_{42}H_{42}Cl_3Re_3$ Re-00089

Hexabenzyltri-μ-chlorotrirhenium

Tri-μ-chlorohexakis(phenylmethyl)trirhenium, 10CI
[73682-46-9]

M 1211.774
Brown-black cryst. Sol. hydrocarbons. Mp 110° dec.

Edwards, P. *et al*, *J. Chem. Soc., Dalton Trans.*, 1980, 334 (*synth, props*)

$C_{42}H_{42}O_6Re_2$ Re-00090

Tetrakis[μ-(2-methoxyphenyl-*C*:*O*)]bis(2-methoxyphenyl-*C*,*O*)dirhenium, 10CI

Hexakis(2-methoxyphenyl)dirhenium
[71110-90-2]

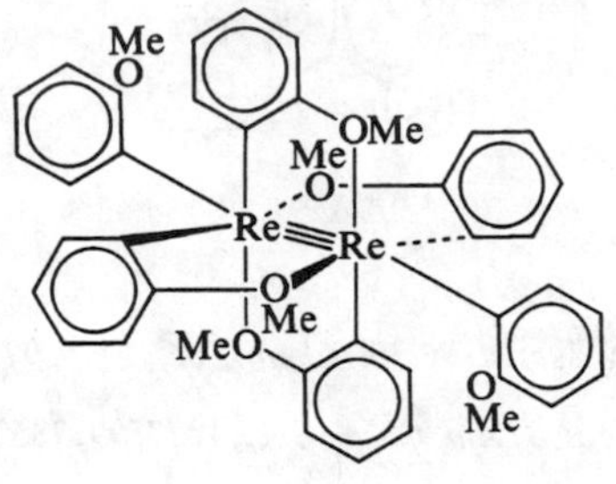

M 1015.204
Dark-green cryst. (CH_2Cl_2). Mod. sol. Me_2CO, chlorinated solvs., spar. sol. hydrocarbons. Mp 180-200° dec. Fluxional in soln. with exchange between end and bridging ligand positions.

Jones, R.A. *et al*, *J. Chem. Soc., Dalton Trans.*, 1979, 472 (*synth, pmr, cmr*)

$C_{44}H_{30}O_8P_2Re_2$ Re-00091

Octacarbonylbis(triphenylphosphine)dirhenium, 10CI

[14172-94-2]

$Ph_3P—Re(CO)_4—Re(CO)_4—PPh_3$ (*diax-*)

M 1121.078
White plates (C_6H_6/hexane). Mp 239-41°.

Dewit, D.G. *et al*, *J. Chem. Soc., Dalton Trans.*, 1976, 528 (*synth*)
Cox, D.J. *et al*, *J. Organomet. Chem.*, 1980, **186**, 339, 347 (*synth, ir, nmr*)

$C_{53}H_{48}ClOP_4Re^{\oplus}$ Re-00092

Carbonylchlorobis[1,2-ethanediylbis(diphenylphosphine-*P*,*P*′)]rhenium(1+)

Bis[1,2-bis(diphenylphosphino)ethane]carbonylchlororhenium

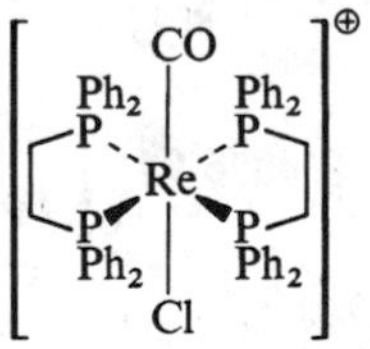

M 1046.517 (ion)

trans-form

Tribromide:
$C_{53}H_{48}Br_3ClOP_4Re$ M 1286.229
By oxidn. of $Re(CO)Cl(PPh_2CH_2CH_2PPh_2)_2$ with Br_2. Green needles ($CHCl_3/Et_2O$).

Hexafluorophosphate: [51886-84-1].
$C_{53}H_{48}ClF_6OP_5Re$ M 1191.481
By metathesis of the tribromide complex. Green needles ($CHCl_3/Et_2O$).

Chatt, J. *et al*, *J. Organomet. Chem.*, 1974, **64**, 245 (*synth, ir, nmr*)

$C_{54}H_{47}NOP_3Re$ **Re-00093**

Dihydronitrosyltris(triphenylphosphine)rhenium, 9CI

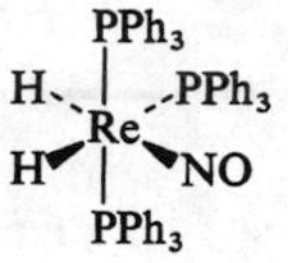

PPh₃ / H, PPh₃ / Re / H, NO / PPh₃

M 1005.100

mer,cis-form [58694-74-9]

Yellow cryst. (C_6H_6/EtOH).

Adams, R.W. *et al*, *J. Chem. Soc., Dalton Trans.*, 1974, 1075 (*synth*, *ir*, *nmr*)
Ciani, E. *et al*, *J. Chem. Soc., Dalton Trans.*, 1976, 1943 (*cryst struct*)
Giusto, D. *et al*, *J. Organomet. Chem.*, 1976, **105**, 91 (*synth*)

Sc Scandium

S. A. Cotton

Scandium (Fr., Ger.), Escandio (Sp.), Scandio (Ital.), Скандий (Scandii) (Russ.), スカンジウム (Japan.)

Atomic Number. 21

Atomic Weight. 44.9559

Electronic Configuration. [Ar] $3d^1\ 4s^2$

Oxidation State. +3

Coordination Number. Usually 6 but may be lower in compounds containing bulky ligands such as $—CH_2SiMe_3$.

Colour. Generally colourless but may be yellow or brown.

Availability. Usual starting materials are Sc_2O_3 and anhydrous $ScCl_3$. For the synthesis of the latter, see Saran, M. S. *et al.*, *J. Organomet. Chem.*, 1970, **21,** 147. See also under Lanthanum.

Handling. See under Lanthanum.

Toxicity. Little is known about the toxicity of scandium and its compounds.

Isotopic Abundance. ^{45}Sc, 100%.

Spectroscopy. ^{45}Sc has $I = \frac{7}{2}$ and has been little studied. Sc(III) compounds are diamagnetic and their ^1H nmr spectra are sharp. Coupling has not been resolved between ^1H and ^{45}Sc.

Analysis. See the corresponding Lanthanum section. Gravimetric analysis is known to the Editor to have given good results.

References. See under Lanthanum.

$C_{10}H_{10}ClSc$ **Sc-00001**

Chlorobis(η^5-2,4-cyclopentadien-1-yl)scandium

Chlorodicyclopentadienylscandium. Dicyclopentadienylscandium chloride

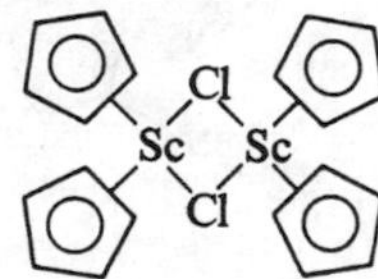

M 210.598

Dimeric.

Dimer: [37205-32-6]. *Di-μ-chlorotetrakis(η^5-2,4-cyclopentadien-1-yl)discandium*, 9CI.

$C_{20}H_{20}Cl_2Sc_2$ M 421.196

Yellow-green cryst. Sol. C_6H_6, THF. Mp 313-5°. V. sensitive to moisture.

Coutts, R.S.P. *et al*, *J. Organomet. Chem.*, 1970, **25**, 117 (*synth*, *nmr*)

Atwood, J.L. *et al*, *J. Chem. Soc., Dalton Trans.*, 1973, 2487 (*struct*)

$C_{11}H_{13}Sc$ **Sc-00002**

Bis(η^5-2,4-cyclopentadien-1-yl)methylscandium, 9CI

M 190.180

Py complex: [60936-88-1].

$C_{16}H_{18}NSc$ M 269.281

Pale-yellow solid. Sol. toluene. Py removable *in vacuo*; Py-free product not fully characterised.

Holton, J. *et al*, *J. Chem. Soc., Chem. Commun.*, 1976, 480 (*synth*, *ir*)

Holton, J. *et al*, *J. Chem. Soc., Dalton Trans.*, 1979, 54 (*synth*, *ir*, *nmr*)

$C_{13}H_{13}Sc$ **Sc-00003**

(η^8-1,3,5,7-Cyclooctatetraene)(η^5-2,4-cyclopentadien-1-yl)scandium, 9CI

[60830-03-7]

M 214.202

Yellowish-white cryst. Sol. Et_2O, THF. $Bp_{0.1}$ 120° subl. Rather air-sensitive.

Westerhof, A. *et al*, *J. Organomet. Chem.*, 1976, **116**, 319 (*synth*, *ir*, *ms*)

$C_{14}H_{22}AlSc$ **Sc-00004**

Bis(2,4-cyclopentadien-1-yl)(dimethylaluminum)di-μ-methylscandium

[60475-11-8]

M 262.265

Pale-yellow cryst. Sol. C_6H_6, toluene, CH_2Cl_2. Mp 110-8°. Air-sensitive.

Holton, J. *et al*, *J. Chem. Soc., Dalton Trans.*, 1979, 45 (*synth*, *ir*, *nmr*, *cmr*)

$C_{15}H_{15}Sc$ **Sc-00005**

Tris(η^5-2,4-cyclopentadien-1-yl)scandium

Tricyclopentadienylscandium

[1298-54-0]

M 240.239

Straw cryst. Sol. Py, THF, dioxan. Mp 240°. Dec. by water. Subl. at 200-50° *in vacuo*.

Birmingham, J.M. *et al*, *J. Am. Chem. Soc.*, 1956, **78**, 42 (*synth*)

Reid, A.F. *et al*, *Inorg. Chem.*, 1966, **5**, 1213 (*synth*)

Atwood, J.L. *et al*, *J. Am. Chem. Soc.*, 1973, **95**, 1488 (*struct*)

$C_{18}H_{15}Sc$ **Sc-00006**

Triphenylscandium, 8CI

Scandium triphenyl

[21500-33-4]

Ph_3Sc

M 276.272

Yellow-brown powder. Sol. THF. Mp >140°. Pyrophoric in air.

Hart, F.A. *et al*, *J. Organomet. Chem.*, 1970, **21**, 147 (*synth*, *ir*)

$C_{19}H_{22}NSc$ **Sc-00007**

Bis(η^5-2,4-cyclopentadien-1-yl)[2-[(dimethylamino)methyl]-phenyl-*C*,*N*]scandium

[68878-72-8]

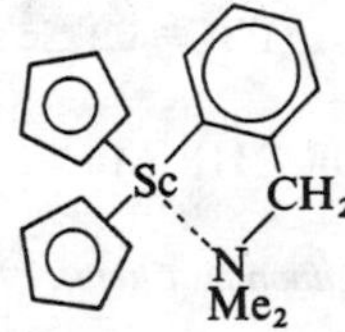

M 309.345

White cryst. Sol. Et_2O. Mp 150°. Very air-sensitive.

Manzer, L.E., *J. Am. Chem. Soc.*, 1978, **100**, 8068 (*synth*, *nmr*)

$C_{21}H_{57}ScSi_6$ **Sc-00008**

Tris[bis(trimethylsilyl)methyl]scandium

$Sc[CH(SiMe_3)_2]_3$

M 523.150

Bis-THF complex: [53659-74-8].

$C_{29}H_{73}O_2ScSi_6$ M 667.363

White cryst. Air-sensitive.

Barker, G.K. *et al*, *J. Organomet. Chem.*, 1974, **76**, C45 (*synth*, *pmr*)

$C_{22}H_{26}BScSi_4$ **Sc-00009**

Bis[(1,2,3,4,5-η)-1,3-bis(trimethylsilyl)-2,4-cyclopentadien-1-yl]tetrahydroboratoscandium

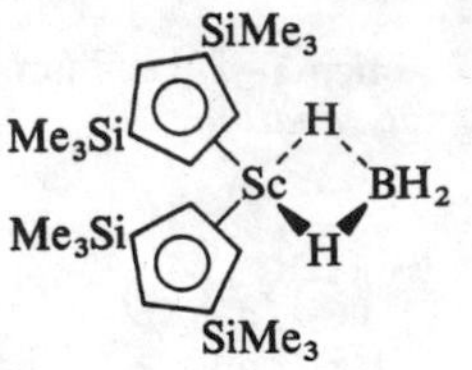

M 458.555

White cryst. Sol. pentane, THF. Mp 82-4°.

Lappert, M.F. *et al*, *J. Chem. Soc., Chem. Commun.*, 1983, 206 (*synth, cryst struct, cmr, nmr, ir*)

$C_{22}H_{42}ClScSi_4$ **Sc-00010**

Bis[1,3-bis(trimethylsilyl)cyclopentadienyl]chloroscandium

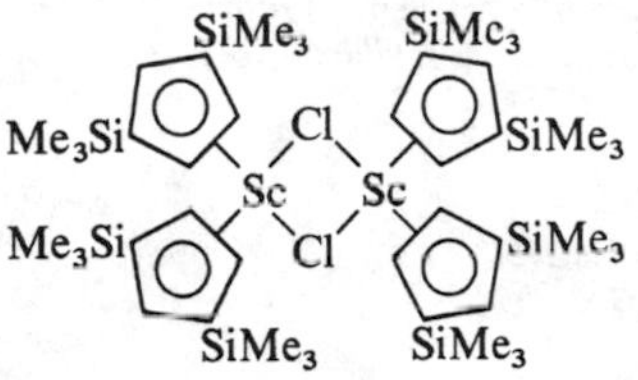

M 499.325

Bridged dimer.

Dimer: [81507-53-1].

$C_{44}H_{84}Cl_2Sc_2Si_8$ M 998.649

Colourless cryst. Sol. toluene, THF.

Lappert, M.F. *et al*, *J. Chem. Soc., Chem. Commun.*, 1981, 1190 (*synth, struct*)

$C_{24}H_{15}Sc$ **Sc-00011**

Tris(phenylethynyl)scandium, 8CI

[21500-34-5]

$(PhC{\equiv}C)_3Sc$

M 348.338

Dark-brown solid. Sol. THF. Mp >250°. Pyrophoric in air.

Hart, F.A. *et al*, *J. Organomet. Chem.*, 1970, **21**, 147 (*synth, ir*)

$C_{24}H_{24}PSc$ **Sc-00012**

Bis(η^5-2,4-cyclopentadien-1-yl)[(diphenylphosphinidenio)bis-(methylene)]scandium, 9CI

[59982-77-3]

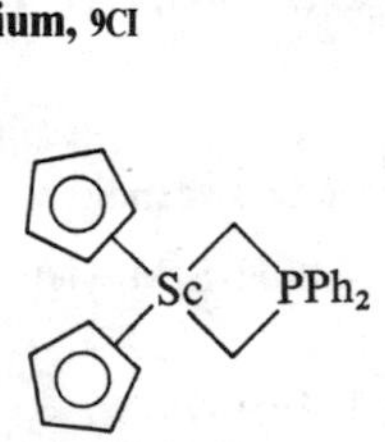

M 388.383

Reacts with cyclohexanone to give methylenecyclohexane quantitatively (more reactive than Ti,V analogues). Pale-yellow cryst. Sol. Et_2O. Very air-sensitive.

Manzer, L.E., *Inorg. Chem.*, 1976, **15**, 2567 (*synth, nmr*)

$C_{27}H_{36}N_3Sc$ **Sc-00013**

Tris[2-[(dimethylamino)methyl]phenyl-*C,N*]scandium

[65521-47-3]

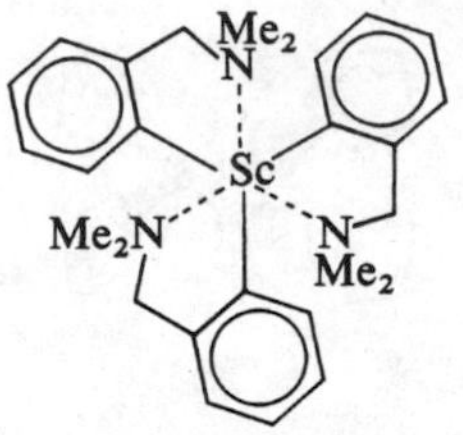

M 447.557

White solid. Sol. hot THF. Mp 180-2°. Pyrophoric in air. Dec. by MeOH and CH_2Cl_2.

Manzer, L.E., *J. Am. Chem. Soc.*, 1978, **100**, 8068 (*synth, nmr*)

$C_{27}H_{36}N_3Sc$ **Sc-00014**

[Tris(2-(dimethylamino)phenyl]methyl-*C,N*)scandium

Tris(2-dimethylaminobenzyl)scandium

[64061-48-9]

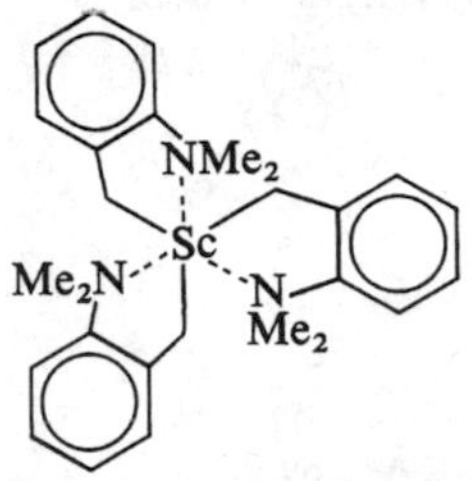

M 447.557

Ethylene polymerisation catalyst. Pale-yellow cryst. Sol. THF, toluene. Mp 115-8°. Extremely air-sensitive.

U.S.P., 597 981, (*1977*); *CA*, **88**, 62468 (*use*)

Manzer, L.E., *J. Am. Chem. Soc.*, 1978, **100**, 8068 (*synth, nmr*)

Sm Samarium

S. A. Cotton

Samarium (Fr., Ger.), Samario (Sp., Ital.), Самарий (Samarii) (Russ.), サマリウム (Japan.)

Atomic Number. 62

Atomic Weight. 150.4

Electronic Configuration. [Xe] $4f^6\,6s^2$

Oxidation States. +2 (unusual), +3 (common).

Coordination Number. 6 or greater.

Colour. Sm(III) organometallics are yellow or orange whilst those of Sm(II) are purple.

Availability. Commonly available materials are Sm_2O_3 and anhydrous $SmCl_3$. See also under Lanthanum.

Handling. See under Lanthanum.

Toxicity. As for Lanthanum.

Isotopic Abundance. ^{144}Sm, 3.1%; ^{150}Sm, 7.4%; ^{152}Sm, 26.7%; ^{154}Sm, 22.7%.

Spectroscopy. ^1H nmr spectra of Sm(III) compounds ($4f^5$) show lines with paramagnetic shifts. Mössbauer spectroscopy is potentially a tool for studying compounds of ^{149}Sm ($t_{1/2} \sim 4 \times 10^{14}$ y).

Analysis. See under Lanthanum.

References. See under Lanthanum.

$C_5H_5Cl_2Sm$ **Sm-00001**

Dichloro(η^5-2,4-cyclopentadien-1-yl)samarium, 9CI

Cyclopentadienylsamarium dichloride

$(C_5H_5)SmCl_2$

M 286.361

Tris-THF adduct: [52679-41-1].

$C_{17}H_{29}Cl_2O_3Sm$ M 502.680

Beige cryst. Darkens >50°. Extremely air-sensitive. Struct. probably resembles Dichloro(η^5-2,4-cyclopentadien-1-yl)erbium.

Manastyrskyj, S. *et al*, *Inorg. Chem.*, 1963, **2**, 904 (*synth*)

$C_6H_{18}Sm^{\ominus\ominus\ominus}$ **Sm-00002**

Hexamethylsamarate(3−)

$SmMe_6^{\ominus\ominus\ominus}$

M 240.568 (ion)

Tris[(tetramethylethylenediamine)lithium] salt: [76206-90-1].

$C_{24}H_{66}Li_3N_6Sm$ M 610.009

Yellow solid. Sol. Et_2O. Mp 85-8° dec. Extremely air- and moisture-sensitive.

Schumann, H. *et al*, *Angew. Chem., Int. Ed. Engl.*, 1981, **20**, 120 (*synth*)

C_8H_8ClSm **Sm-00003**

Chloro(cyclooctatetraene)samarium

Di-μ-chlorobis(η^8-1,3,5,7-cyclooctatetraene)disamarium

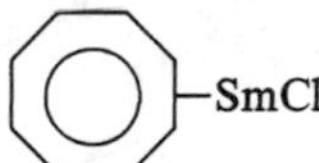

M 289.964

THF complex: [52679-44-4].

$C_{32}H_{48}Cl_2O_4Sm_2$ M 868.355

Purple cryst. Sol. dioxan, THF. Mp >345°. Dimeric; struct. may resemble Chloro(cyclooctatetraene)-cerium.

Hodgson, K.O. *et al*, *J. Am. Chem. Soc.*, 1973, **95**, 8650 (*synth, ir, uv, props*)

$C_{10}H_{10}ClSm$ **Sm-00004**

Chlorobis(η^5-2,4-cyclopentadien-1-yl)samarium, 9CI

Chlorodicyclopentadienylsamarium. Dicyclopentadienylsamarium chloride

M 316.002

Dimeric.

Dimer: [56200-25-0]. *Di-μ-chlorotetrakis(η^5-2,4-cyclopentadien-1-yl)disamarium.*

$C_{20}H_{20}Cl_2Sm_2$ M 632.004

Yellow solid. Sol. THF. Mp >200° dec. Subl. *in vacuo*. Air-sensitive. $\mu_{eff} = 1.62\mu_B$.

Maginn, R.E. *et al*, *J. Am. Chem. Soc.*, 1963, **85**, 672 (*synth, ir*)
Marks, T.J. *et al*, *Inorg. Chem.*, 1976, **15**, 1302 (*synth*)

$C_{10}H_{10}Sm$ **Sm-00005**

Bis(η^5-2,4-cyclopentadien-1-yl)samarium

Dicyclopentadienylsamarium

M 280.549

THF complex:

$C_{14}H_{18}OSm$ M 352.656

Purple solid. Pyrophoric in air. $\mu_{eff} = 3.6\mu_B$.

Watt, G.W. *et al*, *J. Am. Chem. Soc.*, 1969, **91**, 775 (*synth, ir*)

$C_{12}H_{14}ClSm$ **Sm-00006**

Chlorobis[(1,2,3,4,5-η)-1-methyl-2,4-cyclopentadien-1-yl]samarium

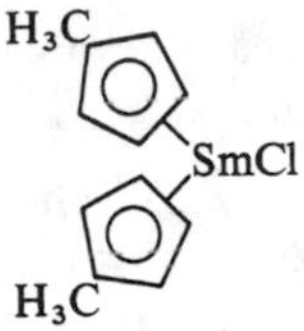

M 344.056

THF adduct: [84623-27-8].

$C_{16}H_{22}ClOSm$ M 416.162

Orange cryst. Sol. toluene, THF. Extremely air- and moisture-sensitive.

Evans, W.J. *et al*, *Organometallics*, 1983, **2**, 709 (*synth, nmr, cmr, ir*)

$C_{13}H_{13}Sm$ **Sm-00007**

(η^8-1,3,5,7-Cyclooctatetraene)(η^5-2,4-cyclopentadien-1-yl)-samarium, 9CI

[52668-27-6]

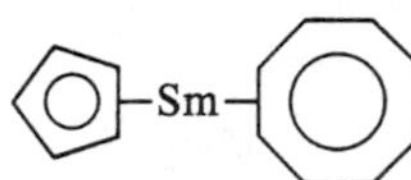

M 319.606

Cryst. Sol. THF. Extremely air-sensitive. Forms THF adducts.

Jamerson, J.D. *et al*, *J. Organomet. Chem.*, 1974, **65**, C33 (*synth*)

$C_{14}H_{21}SiSm$ **Sm-00008**

Bis(η^5-2,4-cyclopentadien-1-yl)[(trimethylsilyl)methyl]samarium

[82293-73-0]

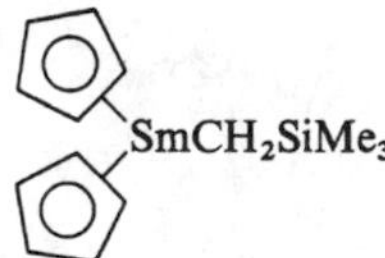

M 367.765

Colourless cryst. Sol. THF, toluene, C_6H_6. Mp <50° dec. Extremely sensitive to moisture and oxygen.

Schumann, H. *et al*, *Organometallics*, 1982, **1**, 1194 (*synth, ir*)

$C_{15}H_{15}Sm$ **Sm-00009**

Tris(η^5-2,4-cyclopentadien-1-yl)samarium, 9CI

Tri-π-cyclopentadienylsamarium, 8CI

[1298-55-1]

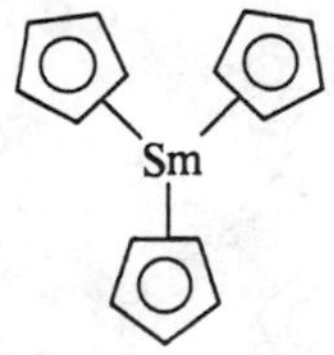

M 345.644

Orange cryst. Sol. THF. Mp 365°. Subl. at 200-50° *in vacuo*, hydrol. by H_2O.

Cyclohexylisocyanide adduct: [37299-04-0]. Sol. C_6H_6. Mp 148°.

Wilkinson, G. *et al*, *J. Am. Chem. Soc.*, 1954, **76**, 6210 (*synth*)
Reid, A.F. *et al*, *Inorg. Chem.*, 1966, **5**, 1213 (*synth*)
Wong, C.-H. *et al*, *Acta. Crystallogr., Sect. B*, 1969, **25**, 2580 (*struct*)
Thomas, J.L. *et al*, *J. Organomet. Chem.*, 1970, **23**, 487 (*ms*)
Von Ammon, R. *et al*, *Ber. Bunsenges. Phys. Chem.*, 1972, **76**, 995 (*ir*)

$C_{16}H_{16}Sm^{\ominus}$ **Sm-00010**

Bis(η^8-1,3,5,7-cyclooctatetraene)samarate(1−), 9CI

Dicyclooctatetraenesamarate(1−)

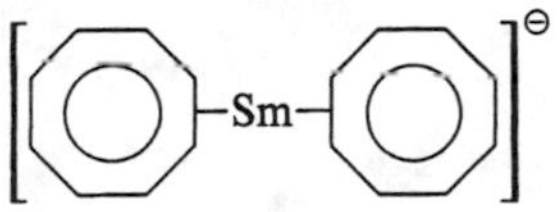

M 358.662 (ion)

K salt: [51177-55-0].
$C_{16}H_{16}KSm$ M 397.761
Brown solid. Sol. dioxan, THF. Mp >345°. Extremely air- and moisture-sensitive. $\mu = 1.42\mu_B$.

Mares, F. *et al*, *J. Organomet. Chem.*, 1970, **24**, C68 (*synth*)
Hodgson, K.O. *et al*, *J. Am. Chem. Soc.*, 1973, **95**, 8650 (*synth, ir, nmr, uv*)

$C_{16}H_{36}Sm^{\ominus}$ **Sm-00011**

Tetra-*tert*-butylsamarate(1−)

Tetrakis(1,1-dimethylethyl)samarate(1−)

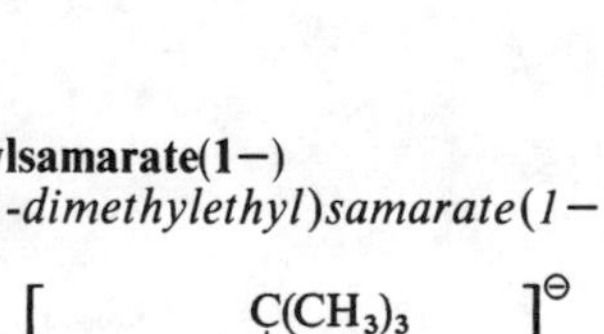

$[(H_3C)_3C-Sm(C(CH_3)_3)_3]^{\ominus}$

M 378.820 (ion)

Li salt, Tetra-THF complex: [68868-92-8].
$C_{32}H_{68}LiO_4Sm$ M 674.188
Dark-gold solid. Sol. THF, Et_2O. V. air- and moisture-sensitive. $\mu_{eff} = 2.1\mu_B$.

Wayda, A.L. *et al*, *J. Am. Chem. Soc.*, 1978, **100**, 7119 (*synth, ir, pmr*)

$C_{20}H_{30}Sm$ **Sm-00012**

Bis[(1,2,3,4,5-η)-1,2,3,4,5-pentamethyl-2,4-cyclopentadien-1-yl]samarium

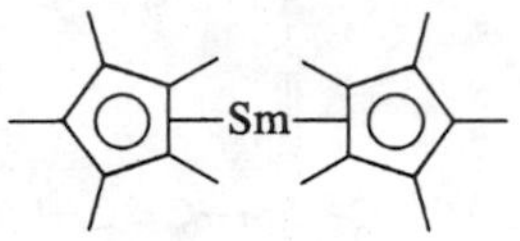

M 420.817

Bis-THF complex: [79372-14-8].
$C_{28}H_{46}O_2Sm$ M 565.030
Polymerises ethylene. Catalyses hydrogenation of 3-hexyne → (*Z*)-3-hexene (>99%). Reacts with CO, NO. Purple cryst. $\mu_{eff} = 3.6\mu_B$.

Evans, W.J. *et al*, *J. Am. Chem. Soc.*, 1981, **103**, 6507 (*synth, ir, uv, nmr, cmr, struct*)

$C_{20}H_{31}Sm$ **Sm-00013**

Hydrobis(pentamethylcyclopentadienyl)samarium

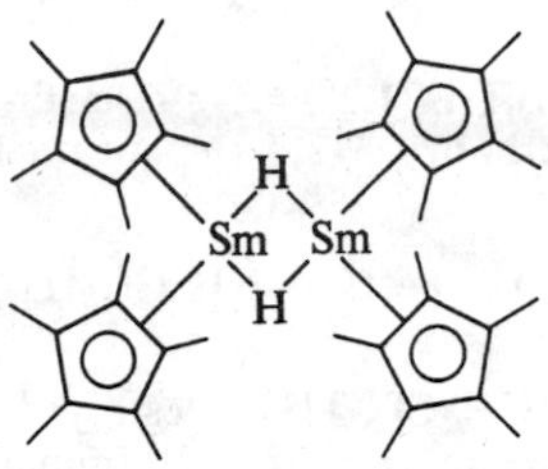

M 421.825

Dimeric.

Dimer: Di-μ-hydrotetrakis[(1,2,3,4,5-η)-1,2,3,4,5-pentamethyl-2,4-cyclopentadien-1-yl]disamarium.
$C_{40}H_{62}Sm_2$ M 843.650
Red-orange prisms. Sol. toluene. $\mu_{eff} = 1.4\mu_B$. Dec. after a few days in solid state.

Evans, W.J. *et al*, *J. Am. Chem. Soc.*, 1983, **105**, 1401 (*synth, struct, ir, nmr*)

$C_{22}H_{21}CoO_4Sm$ **Sm-00014**

[μ-(Carbonyl-*C,O*)]tris[(1,2,3,4,5-η)-1-methyl-2,4-cyclopentadien-1-yl](tricarbonylcobalt)samarium

$Co(CO)_4 \cdot [H_3CC_5H_4]_3Sm$

M 558.699

Dimeric. Bright-yellow solid. Sol. Et_2O, THF, insol. nonpolar solvs. Extremely sensitive to air and moisture.

Crease, A.E. *et al*, *J. Chem. Soc., Dalton Trans.*, 1973, 1501 (*synth, ir*)

$C_{24}H_{36}Sm^{\ominus}$ **Sm-00015**

Tetrakis(3,3-dimethyl-1-butynyl)samarate(1−)

$$\left[\begin{array}{c} C(CH_3)_3 \\ | \\ C \\ ||| \\ C \\ | \\ (H_3C)_3C-C\equiv C-Sm-C\equiv C-C(CH_3)_3 \\ | \\ C \\ ||| \\ C \\ | \\ C(CH_3)_3 \end{array}\right]^{\ominus}$$

M 474.908 (ion)

Li salt, THF complex: [76565-28-1].

$C_{28}H_{44}LiOSm$ M 553.956

Off-white solid. Sol. THF, pentane. $\mu_{eff} = 1.66\mu_B$.

Evans, W.J. *et al*, *J. Organomet. Chem.*, 1980, **202**, C6 (*synth, ir, cmr*)

$C_{25}H_{26}FeO_2Sm$ **Sm-00016**

[μ-(Carbonyl-*C,O*)]Tris[(1,2,3,4,5-η)-1-methyl-2,4-cyclopentadien-1-yl][carbonyl(η⁵-2,4-cyclopentadien-1-yl)iron]-samarium

$$[(C_5H_5)Fe(CO)_2](H_3CC_5H_4)_3Sm$$

M 564.686

Red solid. Sol. CH_2Cl_2, THF. Destroyed by air and moisture. Unstable above 120° at 10^{-1}mm. Cannot be subl.

Crease, A.E. *et al*, *J. Chem. Soc., Dalton Trans.*, 1973, 1501 (*synth, ir*)

$C_{27}H_{21}Sm$ **Sm-00017**

Triindenylsamarium

Tris(1,2,3,3a,7a-η)-1H-inden-1-ylsamarium, 9CI

[39358-33-3]

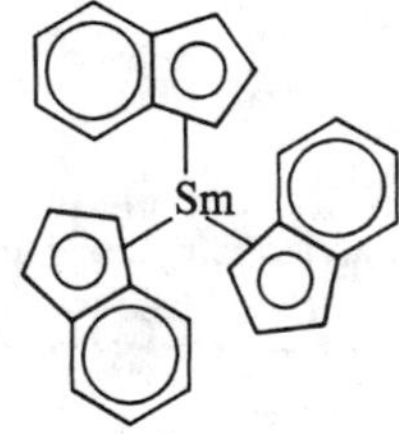

M 495.823

Deep-red cryst. Sol. C_6H_6. Air-sensitive.

THF adduct: [22533-62-6].

$C_{31}H_{29}OSm$ M 567.930

Deep-red cryst. Sol. THF. Mp 185-200°. $\mu = 1.55\mu_B$.

Tsutsui, M. *et al*, *J. Am. Chem. Soc.*, 1969, **91**, 3175 (*nmr*)

Atwood, J.L. *et al*, *J. Am. Chem. Soc.*, 1973, **95**, 1830 (*synth, struct*)

$C_{36}H_{46}Sm_2$ **Sm-00018**

Bis[μ-(3,3-dimethyl-1-butyn-1-yl)]tetrakis(1,2,3,4,5-η-1-methyl-2,4-cyclopentadien-1-yl)disamarium

[84642-15-9]

C(CH₃)₃ CH₃ C C Sm Sm CH₃ H₃C C C H₃C C(CH₃)₃

M 779.479

Reacts with $HPPh_2$ to form $(MeC_5H_4)_2SmPPh_2$. Yellow rectangular prisms (toluene). Sol. toluene, THF. V. air- and moisture-sensitive.

Evans, W.J. *et al*, *Organometallics*, 1983, **2**, 709 (*synth, ir, nmr, cmr, uv, cryst struct*)

Ta Tantalum

J. A. Labinger

Tantale (Fr.), Tantal (Ger.), Tántalo (Sp.), Tantalio (Ital.), Тантал (Tantal) (Russ.), タンタル (Japan.)

Atomic Number. 73

Atomic Weight. 180.7479(3)

Electronic Configuration. [Xe] $5d^3\,6s^2$

Oxidation States. These range from −3 (for $Ta(CO)_5{}^{3-}$) to +5, the more common values being −1, +1, +3, +4 and +5.

Coordination Number. 3–8 for conventional ligands.

Colour. Organotantalum compounds of oxidation state +4 (d^1) are highly coloured, often black. Those of even-electron oxidation states range from colourless (rare) through yellow to red and brown.

Availability. The commonest starting material is $TaCl_5$.

Handling. All organotantalum compounds require standard techniques for handling air-sensitive materials.

Toxicity. Tantalum compounds in general are not especially toxic although the carbonyls must be considered highly toxic as are those of other metals.

Isotopic Abundance. ^{181}Ta, 99.988%.

Spectroscopy. ^{181}Ta ($I = \frac{7}{2}$) has relative receptivity 0.036. The nucleus has been studied with little success since the quadrupole moment is so high at 3×10^{-24} cm^2.

Mass spectrometry of Ta compounds is straightforward as there is essentially only one naturally occurring isotope.

Esr is a useful technique for the +4 (d^1) oxidation state.

Analysis. Tantalum may be determined gravimetrically. Niobium interferes but this is not a problem when assaying organometallic compounds.

References. General reviews are listed in the introduction to the *Sourcebook*.

$MeTaCl_4$

Ta-00001

Me_2TaCl_3

Ta-00002

Me_3TaCl_2

Ta-00003

X = Cl

Ta-00004

$TaMe_5$

Ta-00005

$Ta(CO)_6^{\ominus}$

Ta-00006

As Ta-00004 with
X = Me

Ta-00007

Ta-00008

Ta-00009

X = Y = Cl

Ta-00010

R = H

Ta-00011

X = Cl

Ta-00012

As Ta-00010 with
X = CO, Y = Cl

Ta-00013

As Ta-00010 with
X = CO, Y = H

Ta-00014

Ta-00015

$(OC)_6TaHgPh$

Ta-00016

Ta-00017

Ta-00018

Ta-00019

Ta-00020

Ta-00021

As Ta-00011 with
R = Me

Ta-00022

$PhCH{=}TaCl_3(PMe_3)_2$

Ta-00023

As Ta-00012 with
X = CO

Ta-00024

As Ta-00012 with
X = Me

Ta-00025

R = H

Ta-00026

Ta-00027

Ta-00028

As Ta-00026 with
R = Me

Ta-00029

Ta-00030

Ta-00031

$[(H_3C)_3CO]_3Ta[CHC(CH_3)_3]$

Ta-00032

Ta-00033

Ta-00034

$(H_3C)_3CCH{=}Ta[CH_2C(CH_3)_3]_3$

Ta-00035

Ta-00036

Ta-00037

Ta-00038

$(PhCH_2)_4TaCl$

Ta-00039

$Ta(CH_2Ph)_5$

Ta-00040

CH_3Cl_4Ta **Ta-00001**

Tetrachloromethyltantalum

Methyltetrachlorotantalum

[41453-04-7]

$MeTaCl_4$

M 337.795

Yellow cryst. by subl. Sol. CH_2Cl_2. Mp 50° dec. $Bp_{0.1}$ 20° subl. Disproportionates in soln. to Me_2TaCl_3 and $TaCl_5$.

Fowles, G.W.A. *et al*, *J. Chem. Soc., Dalton Trans.*, 1973, 961 (*synth, pmr*)

Santini-Scampucci, C. *et al*, *J. Chem. Soc., Dalton Trans.*, 1973, 2436 (*synth, ir, pmr*)

$C_2H_6Cl_3Ta$ **Ta-00002**

Trichlorodimethyltantalum

Dimethyltrichlorotantalum

[41453-05-8]

Me_2TaCl_3

M 317.376

Yellow soln. in pentane. Sol. aliphatic hydrocarbons. Disproportionates in soln. to $MeTaCl_4$ and Me_3TaCl_2.

Fowles, G.W.A. *et al*, *J. Chem. Soc., Dalton Trans.*, 1973, 961 (*synth, pmr*)

$C_3H_9Cl_2Ta$ **Ta-00003**

Dichlorotrimethyltantalum, 9CI

Trimethyldichlorotantalum

[3020-02-8]

Me_3TaCl_2

M 296.958

Air- and moisture-sensitive volatile yellow solid. V. sol. org. solvs.

Schrock, R.R. *et al*, *J. Am. Chem. Soc.*, 1978, **100**, 2389 (*synth, pmr*)

$C_5H_5Cl_4Ta$ **Ta-00004**

Tetrachloro(η^5-2,4-cyclopentadien-1-yl)tantalum, 9CI

Cyclopentadienyltetrachlorotantalum

[62927-98-4]

X = Cl

M 387.854

Yellow solid (CH_2Cl_2). Sl. sol. org. solvs., HMPA with some dec. Mp 190° dec. $Bp_{0.05}$ 230° subl.

Burt, R.J. *et al*, *J. Organomet. Chem.*, 1977, **129**, C33 (*synth, ir, ms*)

Cardoso, A.M. *et al*, *J. Chem. Soc., Dalton Trans.*, 1980, 1156 (*synth*)

Bunker, M.J. *et al*, *J. Chem. Soc., Dalton Trans.*, 1980, 2155 (*synth*)

$C_5H_{15}Ta$ **Ta-00005**

Pentamethyltantalum

[53378-72-6]

$TaMe_5$

M 256.121

Yellow cryst. (Et_2O), dec. at r.t. Mp 0°.

▷May dec. violently

Galyer, L. *et al*, *J. Chem. Soc., Chem. Commun.*, 1975, 497 (*pe*)

Schrock, R.R., *J. Organomet. Chem.*, 1976, **122**, 209 (*synth, pmr, ms, haz*)

$C_6O_6Ta^{\ominus}$ **Ta-00006**

Hexacarbonyltantalate(1−)

$Ta(CO)_6^{\ominus}$

M 349.010 (ion)

Tetraphenylarsonium salt: [57288-89-8].

$C_{30}H_{20}AsO_6Ta$ M 732.354

Yellow cryst. (diglyme aq.). Sol. polar org. solvs. Mp 193-4° dec.

Wrighton, M.S. *et al*, *Inorg. Chem.*, 1976, **15**, 434 (*uv*)

Inorg. Synth., 1976, **16**, 68 (*synth, ir*)

$C_8H_{14}ClTa$ **Ta-00007**

Chloro(η^5-2,4-cyclopentadien-1-yl)trimethyltantalum, 9CI

Cyclopentadienyltrimethylchlorotantalum

[49598-13-2]

As Tetrachloro(η^5-2,4-cyclopentadien-1-yl)tantalum, Ta-00004 with

X = Me

M 326.600

Orange needles (toluene). Sol. aromatic hydrocarbons. Sl. dec. at 25°.

Schrock, R.R. *et al*, *J. Am. Chem. Soc.*, 1978, **100**, 2389 (*synth, pmr*)

$C_8H_{22}Cl_3P_2Ta$ **Ta-00008**

Trichloro(η^2-ethene)bis(trimethylphosphine)tantalum, 9CI

(*Ethylene*)*trichlorobis*(*trimethylphosphine*)*tantalum*

[71860-94-1]

M 467.516

Royal-blue solid (Et_2O). Sol. aromatic hydrocarbons, Et_2O.

Sattelberger, A.P. *et al*, *J. Am. Chem. Soc.*, 1980, **102**, 7111 (*synth, nmr*)

Rocklage, S.M. *et al*, *J. Am. Chem. Soc.*, 1981, **103**, 1440 (*synth, nmr, uv*)

$C_9H_5O_4Ta$ **Ta-00009**

Tetracarbonyl(η^5-2,4-cyclopentadien-1-yl)tantalum, 9CI

Cyclopentadienyltetracarbonyltantalum

[32628-95-8]

M 358.084

Ruby-red cryst. by subl. Mp 171-3°. Sublimes at 30-40°/0.0001 mm.

Werner, R.P.M. *et al*, *Inorg. Chem.*, 1964, **3**, 298 (*synth*, *ir*)
Nesmeyanov, A.N. *et al*, *Zh. Strukt. Khim.*, 1972, **13**, 1033; *CA*, **78**, 90759 (*cmr*)

$C_{10}H_{10}Cl_2Ta$ Ta-00010

Dichlorobis(η^5-2,4-cyclopentadien-1-yl)tantalum, 9CI
Bis(cyclopentadienyl)dichlorotantalum
[54039-37-1]

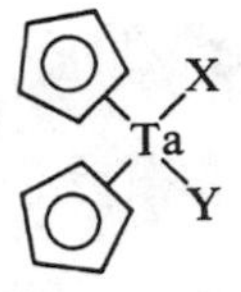

X = Y = Cl

M 382.043
Brown cryst., mod. air-stable. Sol. toluene, CH_2Cl_2, Me_2CO. $Bp_{0.001}$ 222° subl.

Green, M.L.H. *et al*, *J. Organomet. Chem.*, 1978, **161**, C25 (*synth*)
Bunker, M.J. *et al*, *J. Chem. Soc., Dalton Trans.*, 1980, 2155 (*esr*)

$C_{10}H_{13}Ta$ Ta-00011

Bis(cyclopentadienyl)trihydrotantalum
Dicyclopentadienyltrihydrotantalum
[12117-02-1]

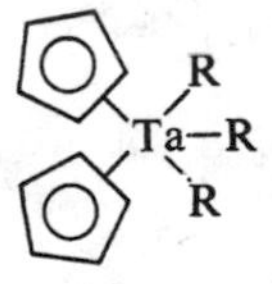

R = H

M 314.161
White cryst. Sol. C_6H_6. Mp 187-9° dec. $Bp_{0.01}$ 120° subl.

Green, M.L.H. *et al*, *J. Chem. Soc. (A)*, 1961, 4854 (*synth*, *pmr*, *ir*)
Wilson, R.D. *et al*, *J. Am. Chem. Soc.*, 1977, **99**, 1775 (*struct*)
Green, M.L.H. *et al*, *J. Organomet. Chem.*, 1978, **161**, C25 (*synth*)

$C_{10}H_{15}Cl_4Ta$ Ta-00012

Tetrachloro[(1,2,3,4,5-η)-1,2,3,4,5-pentamethyl-2,4-cyclopentadien-1-yl]tantalum
Pentamethylcyclopentadienyltetrachlorotantalum
[71414-47-6]

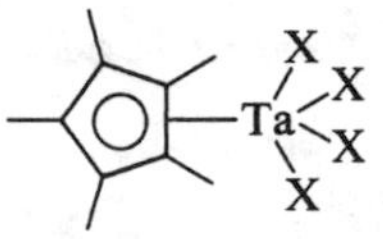

X = Cl

M 457.988
Yellow needles (C_6H_6). Sol. aromatic hydrocarbons, insol. aliphatic hydrocarbons. Dec. >250°.

McClain, S.J. *et al*, *J. Am. Chem. Soc.*, 1979, **101**, 4558 (*synth*, *pmr*, *ir*)
Herrmann, W.A. *et al*, *Chem. Ber.*, 1981, **114**, 3558 (*synth*)

$C_{11}H_{10}ClOTa$ Ta-00013

Carbonylchlorobis(η^5-2,4-cyclopentadien-1-yl)tantalum
Bis(cyclopentadienyl)chlorocarbonyltantalum
[69593-26-6]
As Dichlorobis(η^5-2,4-cyclopentadien-1-yl)tantalum, Ta-00010 with

X = CO, Y = Cl

M 374.600
Grey-green solid (toluene). Sol. toluene, THF. Mp 221°.

Klazinga, A.H. *et al*, *J. Organomet. Chem.*, 1979, **165**, 31 (*synth*)

$C_{11}H_{11}OTa$ Ta-00014

Carbonylbis(η^5-2,4-cyclopentadien-1-yl)hydrotantalum, 9CI
Bis(cyclopentadienyl)hydro(carbonyl)tantalum
[11105-69-4]
As Dichlorobis(η^5-2,4-cyclopentadien-1-yl)tantalum, Ta-00010 with

X = CO, Y = H

M 340.155
Green cryst. (toluene). Sol. toluene, THF. Mp 157°. Dec. at 219°.

Klazinga, A.H. *et al*, *J. Organomet. Chem.*, 1979, **165**, 31 (*synth*, *ir*, *pmr*)
Klazinga, A.H. *et al*, *J. Organomet. Chem.*, 1980, **194**, 309.

$C_{11}H_{28}Cl_3P_2Ta$ Ta-00015

Trichloro(2,2-dimethylpropylidene)bis(trimethylphosphine)-tantalum, 10CI
Bis(trimethylphosphine)neopentylidenetrichlorotantalum
[70083-62-4]

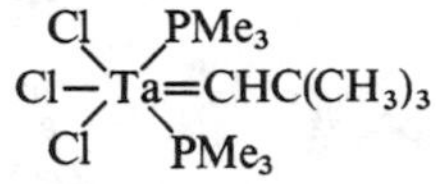

$Cl_3Ta(PMe_3)_2{=}CHC(CH_3)_3$

M 509.597
Mixt. of isomers. Purple-pink fibres (Et_2O). Sl. sol. pentane, Et_2O.

Rupprecht, G.A. *et al*, *J. Am. Chem. Soc.*, 1980, **102**, 6236 (*synth*, *nmr*, *ir*)

$C_{12}H_5HgO_6Ta$ Ta-00016

Hexacarbonyl(phenylmercury)tantalum

$(OC)_6TaHgPh$

M 626.706
Orange-red plates. Sol. CH_2Cl_2. Mp 65-8° dec.

Keblys, K.A. *et al*, *Inorg. Chem.*, 1964, **3**, 1646 (*synth*)
Davison, A. *et al*, *J. Organomet. Chem.*, 1972, **36**, 113 (*ir*)

$C_{12}H_{15}Ta$ **Ta-00017**

Bis(η^5-2,4-cyclopentadien-1-yl)methyl(methylidene)tantalum, 9CI

[57913-16-3]

M 340.198

First isol. methylidene complex. Greenish-white cryst. (toluene), dec. slowly at r.t. Sol. aromatic hydrocarbons.

Guggenberger, L.J. *et al, J. Am. Chem. Soc.*, 1975, **97**, 6578 (*struct*)

Schrock, R.R. *et al, J. Am. Chem. Soc.*, 1978, **100**, 2389 (*synth, ir, pmr*)

$C_{12}H_{19}Cl_2Ta$ **Ta-00018**

Dichloro(η^2-ethene)(η^5-1,2,3,4,5-pentamethyl-2,4-cyclopentadien-1-yl)tantalum, 9CI

(Pentamethylcyclopentadienyl)dichloro(ethylene)tantalum

[71414-49-8]

M 415.136

Olefin dimerisation catalyst. Purple cryst. (toluene). Sol. aromatic hydrocarbons.

McLain, S.J. *et al, J. Am. Chem. Soc.*, 1979, **101**, 4558 (*synth, nmr, struct*)

McLain, S.J. *et al, J. Am. Chem. Soc.*, 1980, **102**, 5610 (*use*)

$C_{13}H_{15}Ta$ **Ta-00019**

Bis(η^5-2,4-cyclopentadien-1-yl)(η^3-2-propenyl)tantalum, 9CI

Allylbis(cyclopentadienyl)tantalum

[54039-40-6]

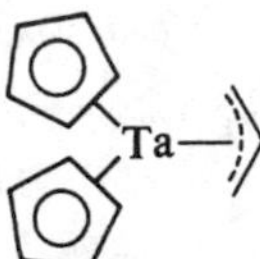

M 352.209

Green cryst. by subl. Sol. pentane. $Bp_{0.1}$ 130° subl.

van Baalen, A. *et al, J. Organomet. Chem.*, 1974, **74**, 245 (*synth, ir, pmr, ms*)

$C_{13}H_{17}Cl_2N_2Ta$ **Ta-00020**

(2,2′-Bipyridine-*N,N′*)dichlorotrimethyltantalum, 9CI

Trimethyldichloro(2,2′-bipyridyl)tantalum

[50589-81-6]

M 453.145

Yellow cryst. (CH_2Cl_2).

Drew, M.G.B. *et al, J. Chem. Soc., Dalton Trans.*, 1973, 1830 (*struct*)

Fowles, G.W.A. *et al, J. Chem. Soc., Dalton Trans.*, 1974, 1080 (*synth, ir, pmr*)

$C_{13}H_{17}Ta$ **Ta-00021**

Bis(cyclopentadienyl)hydropropenetantalum

[68586-68-5]

M 354.225

Yellow cryst. (pentane). Mp 109°. Dec. >125°.

Klazinga, A.H. *et al, J. Organomet. Chem.*, 1978, **157**, 413 (*synth, nmr*)

Klazinga, A.H. *et al, J. Organomet. Chem.*, 1979, **165**, 31.

$C_{13}H_{19}Ta$ **Ta-00022**

Bis(η^5-2,4-cyclopentadien-1-yl)trimethyltantalum, 9CI

[61649-18-1]

As Bis(cyclopentadienyl)trihydrotantalum, Ta-00011 with

R = Me

M 356.241

White cryst. (toluene), thermally unstable. Sol. aromatic hydrocarbons.

Schrock, R.R. *et al, J. Am. Chem. Soc.*, 1978, **100**, 2389 (*synth, pmr*)

$C_{13}H_{24}Cl_3P_2Ta$ **Ta-00023**

Benzylidenetrichlorobis(trimethylphosphine)tantalum

Trichloro(phenylmethylidene)bis(trimethylphosphine)-tantalum

[77026-98-3]

$PhCH{=}TaCl_3(PMe_3)_2$

M 529.587

Exists as mixt. of isomers. Green cryst. (C_6H_6). Sol. polar org. solvs.

Rupprecht, G.A. *et al, J. Am. Chem. Soc.*, 1980, **102**, 6236 (*synth, nmr, ir*)

$C_{14}H_{15}O_4Ta$ **Ta-00024**

Tetracarbonyl[(1,2,3,4,5-η)-1,2,3,4,5-pentamethyl-2,4-cyclopentadien-1-yl]tantalum

Pentamethylcyclopentadienyltetracarbonyltantalum

As Tetrachloro[(1,2,3,4,5-η)-1,2,3,4,5-pentamethyl-2,4-cyclopentadien-1-yl]tantalum, Ta-00012 with

X = CO

M 428.218

Orange-yellow needles (C_6H_6). Sol. aromatic hydrocarbons. Mp 136-9°.

Herrmann, W.A. *et al*, *Chem. Ber.*, 1981, **114**, 3558 (*synth*, *ir*, *pmr*)

$C_{14}H_{27}Ta$ Ta-00025

[(1,2,3,4,5-η)-1,2,3,4,5-Pentamethyl-2,4-cyclopentadien-1-yl]tetramethyltantalum, 9CI

[71763-35-4]

As Tetrachloro[(1,2,3,4,5-η)-1,2,3,4,5-pentamethyl-2,4-cyclopentadien-1-yl]tantalum, Ta-00012 with

X = Me

M 376.315

Yellow cryst. (pentane). Sol. pentane, Et_2O.

Wood, C.D. *et al*, *J. Am. Chem. Soc.*, 1979, **101**, 5421 (*synth*)

$C_{14}H_{33}TaO_2P_4$ Ta-00026

Dicarbonylbis[1,2-ethanediylbis[dimethylphosphine]-*P*,*P'*]-hydrotantalum, 9CI

Dicarbonylhydrobis[bis(dimethylphosphino)ethane]tantalum

[62050-78-6]

R = H

M 538.256

Orange cryst. (hexane). Mp 140-41°.

Tebbe, F.N., *J. Am. Chem. Soc.*, 1973, **95**, 5823 (*synth*, *ir*)
Meakin, P. *et al*, *Inorg. Chem.*, 1974, **13**, 1025 (*nmr*, *struct*)
Bond, A.M. *et al*, *Inorg. Chem.*, 1980, **19**, 1760.

$C_{15}H_{20}NTa$ Ta-00027

Bis(η⁵-2,4-cyclopentadien-1-yl)(isocyanomethane)propyltantalum

Bis(cyclopentadienyl)propyl(methyl isocyanide)-tantalum

[74909-91-4]

M 395.278

Red-brown cryst. (pentane). Mp 104°. Dec. at 111°.

Klazinga, A.H. *et al*, *J. Organomet. Chem.*, 1980, **192**, 75 (*synth*, *ir*, *nmr*)

$C_{15}H_{25}Cl_2Ta$ Ta-00028

Dichloro(2,2-dimethylpropylidene)[(1,2,3,4,5-η)-1,2,3,4,5-pentamethyl-2,4-cyclopentadien-1-yl]tantalum, 9CI

(Pentamethylcyclopentadienyl)dichloro(neopentylidene)-tantalum

[71414-58-9]

M 457.216

Red oil (pentane).

Wood, C.D. *et al*, *J. Am. Chem. Soc.*, 1979, **101**, 3210 (*synth*, *nmr*)

$C_{15}H_{35}TaO_2P_4$ Ta-00029

Dicarbonylbis[1,2-ethanediylbis[dimethylphosphine]-*P*,*P'*]-methyltantalum, 9CI

Methyldicarbonylbis[bis(dimethylphosphino)ethane]-tantalum

[61916-37-8]

As Dicarbonylbis[1,2-ethanediylbis[dimethylphosphine]-*P*,*P'*]hydrotantalum, Ta-00026 with

R = Me

M 552.283

Yellow cryst. (toluene).

Datta, S. *et al*, *Inorg. Chem.*, 1977, **16**, 1134 (*synth*, *nmr*, *ir*)
Bond, A.M. *et al*, *Inorg. Chem.*, 1980, **19**, 1760.

$C_{16}H_{26}PTa$ Ta-00030

Bis(η⁵-2,4-cyclopentadien-1-yl)hydro(triethylphosphine)tantalum, 10CI

[11105-81-0]

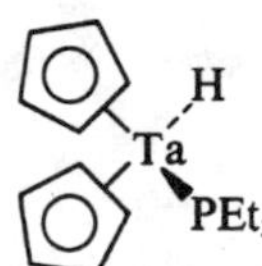

M 430.303

Red cryst. (pentane). Sol. aromatic hydrocarbons.

Barefield, E.K. *et al*, *J. Am. Chem. Soc.*, 1970, **92**, 5234 (*synth*, *pmr*)
Klazinga, A.H. *et al*, *J. Organomet. Chem.*, 1978, **157**, 413 (*synth*)

$C_{16}H_{26}P_2Ta^{\oplus}$ Ta-00031

Bis(η⁵-2,4-cyclopentadien-1-yl)[1,2-ethanediylbis[dimethylphosphine]-*P*,*P'*]tantalum(1+), 9CI

Bis(cyclopentadienyl)[bis(dimethylphosphino)ethane]-tantalum(1+)

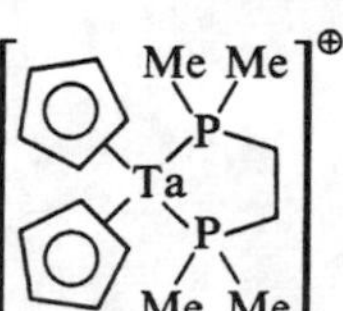

M 461.277 (ion)

Chloride: [66632-54-0].

$C_{16}H_{26}ClP_2Ta$ M 496.730

Red cryst. (CH_2Cl_2/heptane).

Foxman, B.M. *et al*, *Inorg. Chem.*, 1978, **17**, 2311 (*nmr, struct*)

$C_{17}H_{37}O_3Ta$ Ta-00032

(2,2-Dimethylpropylidene)tris(2-methyl-2-propanolato)tantalum, 9CI

Tris(tert-*butoxy*)(*2,2-dimethylpropylidene*)*tantalum*

[76992-41-1]

$[(H_3C)_3CO]_3Ta[CHC(CH_3)_3]$

M 470.425

Olefin metathesis catalyst. Orange oil (Et_2O).

Rocklage, S.M. *et al*, *J. Am. Chem. Soc.*, 1981, **103**, 1440 (*synth, cmr, use*)

$C_{18}H_{25}Ta$ Ta-00033

(η^2-Benzyne)dimethyl[(1,2,3,4,5-η)-1,2,3,4,5-pentamethyl-2,4-cyclopentadien-1-yl]tantalum, 9CI

(*Pentamethylcyclopentadienyl*)*dimethyl*(η^2-*benzyne*)-*tantalum*

[69302-77-8]

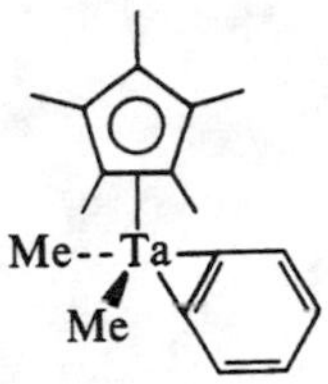

M 422.343

Yellow cryst. (pentane), mod. air-stable.

McLain, S.J. *et al*, *J. Am. Chem. Soc.*, 1979, **101**, 263 (*synth, nmr*)

Churchill, M.R. *et al*, *Inorg. Chem.*, 1979, **18**, 1697 (*struct*)

$C_{20}H_{38}PTa$ Ta-00034

(2,2-Dimethylpropylidene)-η^2-ethene(η^5-pentamethyl-2,4-cyclopentadien-1-yl)(trimethylphosphine)tantalum

(*Pentamethylcyclopentadienyl*)(*neopentylidene*)(*ethylene*)(*trimethylphosphine*)*tantalum*

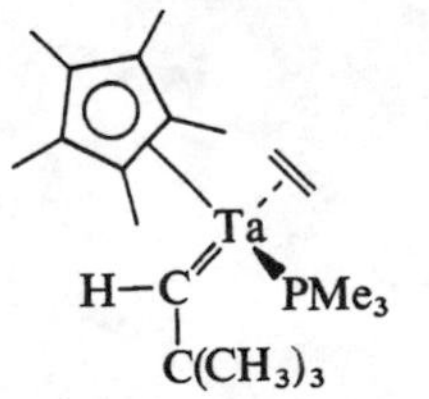

M 490.442

No phys. details recorded.

Schultz, A.J. *et al*, *J. Am. Chem. Soc.*, 1981, **103**, 169 (*nd, struct*)

$C_{20}H_{43}Ta$ Ta-00035

Tris(2,2-dimethylpropyl)(2,2-dimethylpropylidene)tantalum, 9CI

Tris(*neopentyl*)*neopentylidenetantalum*

[54294-45-0]

$(H_3C)_3CCH{=}Ta[CH_2C(CH_3)_3]_3$

M 464.508

First non-heteroatom stabilised alkylidene complex isol. Orange cryst. (Et_2O); highly air-sensitive. Sol. nonpolar org. solvs. Mp 71°.

Schrock, R.R. *et al*, *J. Am. Chem. Soc.*, 1978, **100**, 3359 (*synth, nmr*)

$C_{22}H_{40}ClP_4Ta$ Ta-00036

Chlorobis[1,2-ethanediylbis[dimethylphosphine]-*P,P'*](η^4-naphthalene)tantalum, 9CI

(η^4-*Naphthalene*)*chlorobis*[*bis*(*dimethylphosphino*)*ethane*]*tantalum*

[64367-79-9]

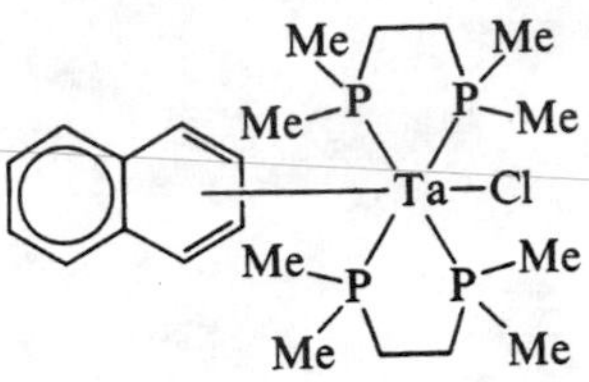

M 644.854

Red cryst. (toluene). Sol. toluene, sl. sol. pentane.

Albright, J.O. *et al*, *J. Am. Chem. Soc.*, 1979, **101**, 611 (*synth, nmr, struct*)

Bond, A.M. *et al*, *Inorg. Chem.*, 1980, **19**, 1760 (*props*)

$C_{24}H_{25}Cl_2Ta$ Ta-00037

Dichloro(diphenylacetylene)[(1,2,3,4,5-η)-1,2,3,4,5-pentamethyl-2,4-cyclopentadien-1-yl]tantalum

(*Pentamethylcyclopentadienyl*)*dichloro*(*diphenylacetylene*)*tantalum*

[75522-28-0]

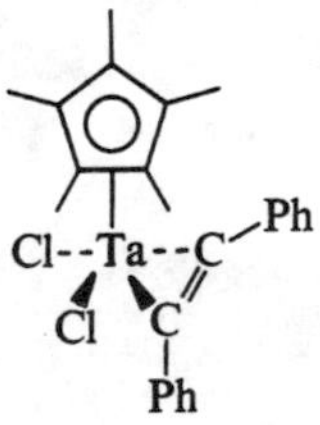

M 565.315

Orange cryst. (toluene). Sol. toluene, CH_2Cl_2, insol. pentane.

Smith, G. *et al*, *Inorg. Chem.*, 1981, **20**, 387 (*synth, nmr, struct*)

$C_{25}H_{49}P_2Ta$ Ta-00038

Bis(2,2-dimethylpropylidene)(2,4,6-trimethylphenyl)bis(trimethylphosphine)tantalum, 9CI

Mesitylbis(*neopentylidene*)*bis*(*trimethylphosphine*)*tantalum*

[69552-43-8]

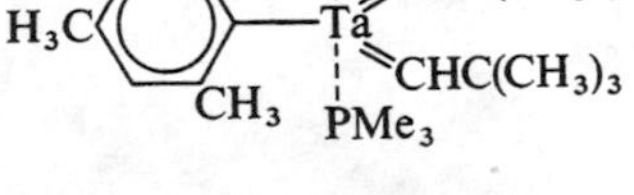

M 592.558

Yellow cryst. (Et_2O). Sol. pentane, Et_2O.

Churchill, M.R. *et al*, *Inorg. Chem.*, 1979, **18**, 1930 (*struct*)

Fellmann, J.D. *et al*, *J. Am. Chem. Soc.*, 1981, **103**, 5752 (*synth, nmr*)

$C_{28}H_{28}ClTa$ **Ta-00039**

Tetrabenzylchlorotantalum

Chlorotetrakis(phenylmethyl)tantalum

[74920-40-4]

$(PhCH_2)_4TaCl$

M 580.930

Red cryst. (C_6H_6), thermally unstable. Sol. aromatic hydrocarbons.

Messerle, L.W. *et al*, *J. Am. Chem. Soc.*, 1980, **102**, 6744 (*synth*, *nmr*)

$C_{35}H_{35}Ta$ **Ta-00040**

Pentabenzyltantalum

Pentakis(phenylmethyl)tantalum

[62044-68-2]

$Ta(CH_2Ph)_5$

M 636.609

Red cryst. (toluene). Sol. aromatic hydrocarbons. Dec. at 80° in soln., subl. with dec.

Schrock, R.R., *J. Organomet. Chem.*, 1976, **122**, 209 (*synth*, *pmr*, *ms*)

Tb Terbium

S. A. Cotton

Terbium (Fr., Ger.), Terbio (Sp., Ital.), Тербий (Terbii) (Russ.), テルビウム (Japan.)

Atomic Number. 65

Atomic Weight. 158.9254

Electronic Configuration. [Xe] $4f^9\ 6s^2$

Oxidation State. +3

Coordination Number. 6 or greater.

Colour. Colourless or yellow.

Availability. The available starting materials are Tb_4O_7 and anhydrous $TbCl_3$ and are expensive. See also the section under Lanthanum.

Handling. See under Lanthanum.

Toxicity. See under Lanthanum.

Isotopic Abundance. ^{159}Tb, 100%.

Spectroscopy. Very few data are available for ^1H nmr spectra, but compounds of the paramagnetic Tb(III) species give large paramagnetic shifts with considerable line-broadening. See also under Lanthanum.

Analysis. See under Lanthanum.

References. See under Lanthanum. Also, for ^1H nmr spectra:

Schumann, H. *et al.*, *Z. Anorg. Allg. Chem.*, 1976, **426,** 127.

Schumann, H. *et al.*, *Z. Naturforsch.*, *B*, 1981, **366,** 1244.

$C_{12}H_{33}Tb$ Tb-00001

Tris[(trimethylsilyl)methyl]terbium

$Tb(CH_2SiMe_3)_3$

M 336.318

Bis-THF complex: [67483-82-3]. *Bis(tetrahydrofuran)-tris[(trimethylsilyl)methyl]terbium.*

$C_{20}H_{49}O_2Si_3Tb$ M 564.788

Colourless cryst. Sol. hexane. Mp 50-4°. Air-sensitive in soln. but stable as large cryst. μ_{eff} 9.12μ_B.

Atwood, J.L. *et al*, *J. Chem. Soc., Chem. Commun.*, 1978, 141 (*synth, ir*)

$C_{15}H_{15}Tb$ Tb-00002

Tris(η^5-2,4-cyclopentadien-1-yl)terbium, 9CI

Tricyclopentadienylterbium

[1272-25-9]

M 354.209

Colourless cryst. Sol. THF. Mp 316°. Subl. at 230° *in vacuo*. Hydrol. by H_2O. μ_{eff} = 8.9μ_B.

Isocyanocyclohexane adduct: [12098-34-9]. *Tris(η^5-2,4-cyclopentadien-1-yl)isocyanocyclohexaneterbium.*

$C_{22}H_{26}NTb$ M 463.380

Colourless cryst. Sol. C_6H_6. Mp 162°. μ_{eff} = 10.1μ_B. Subl. *in vacuo* at 150°.

Fischer, E.O. *et al*, *Angew. Chem.*, 1964, **76**, 52 (*synth*)
Fischer, E.O. *et al*, *J. Organomet. Chem.*, 1965, **3**, 181; 1966, **6**, 141 (*synth*)
Laubereau, P.G. *et al*, *Inorg. Chem.*, 1970, **9**, 1091 (*synth*)
Aleksanyan, V.T. *et al*, *J. Organomet. Chem.*, 1977, **131**, 251 (*raman*)

$C_{16}H_{16}Tb^{\ominus}$ Tb-00003

Bis(η^8-1,3,5,7-cyclooctatetraene)terbate(1−), 9CI

Dicyclooctatetraeneterbate(1−)

$[Tb(C_8H_8)_2]^{\ominus}$

M 367.228 (ion)

K salt: [51177-57-2].

$C_{16}H_{16}KTb$ M 406.326

Yellow-brown solid. Sol. THF. Mp >345°. Extremely air- and moisture-sensitive. μ = 9.86μ_B.

Mares, F. *et al*, *J. Organomet. Chem.*, 1970, **24**, C68 (*synth*)
Hodgson, K.O. *et al*, *J. Am. Chem. Soc.*, 1973, **95**, 8650 (*synth, ir, nmr, uv*)

$C_{27}H_{21}Tb$ Tb-00005

Triindenylterbium, 8CI

Tris(1,2,3,3a,7a-η)-1H-inden-1-ylterbium

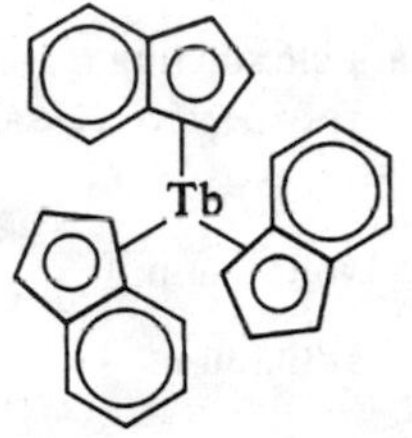

M 504.388

May be a mono- or a trihapto complex.

THF adduct: [22877-87-8].

$C_{31}H_{29}OTb$ M 576.495

Pale yellow. μ = 9.43μ_B.

Tsutsui, M. *et al*, *J. Am. Chem. Soc.*, 1969, **91**, 3175 (*synth*)

Tc Technetium

M. Green

Technétium (Fr.), Technetium (Ger.), Technecio (Sp.), Tecnezio (Ital.), Технеций (Technietsii) (Russ.), テクネチウム (Japan.)

Atomic Number. 43

Atomic Weight. 99 (most common isotope)

Electronic Configuration. [Kr] $4d^6\ 5s^1$

Oxidation States. −1, +1

Coordination Number. Most often 6.

Colour. Physical appearance and reactivity of technetium complexes are similar to those of manganese and rhenium.

Availability. Technetium is not naturally occurring and all isotopes are radioactive. Of the long-lived isotopes only ^{99}Tc ($t_{1/2} = 2.12 \times 10^5$ y) is readily available. The short-lived isotopes of greatest practical importance are ^{95}Tc ($t_{1/2} = 60$ d), ^{97}Tc ($t_{1/2} = 90.5$ d) and ^{99m}Tc ($t_{1/2} = 6.02$ h).

Handling. The handling of ^{99}Tc on a small scale (<0.05 g) does not present a serious health hazard provided elementary precautions are taken. Intending workers should first consult the current *Code of Practice for the Protection of Persons Exposed to Ionizing Radiation* (HMSO).

Toxicity. See above.

Analysis. ^{99}Tc compounds may be determined by conventional chemical and spectroscopic methods. The percentage of ^{99}Tc in a compound is determined by direct counting of the β-decay.

References. General reviews are listed in the introduction to the *Sourcebook*.

C_5BrO_5Tc Tc-00001

Bromopentacarbonyltechnetium, 10CI

[15841-97-1]

$(OC)_5TcBr$ (O_h)

M 317.956

Yellow cryst. by subl. Sublimes *in vacuo*.

Bor, G. *et al*, *Inorg. Nucl. Chem. Lett.*, 1973, **9**, 1073 (*props*)
Behrens, R.G. *et al*, *J. Less-Common Met.*, 1978, **61**, 321 (*props*)
Kruglov, A.A. *et al*, *Zh. Neorg. Khim.*, 1981, **26**, 960 (*cryst struct*)
Michels, E.D. *et al*, *Inorg. Chem.*, 1981, **20**, 3445 (*synth*)

C_5ClO_5Tc Tc-00002

Pentacarbonylchlorotechnetium, 10CI

Chloropentacarbonyltechnetium

[15841-98-2]

$(OC)_5TcCl$ (O_h)

M 273.505

Colourless cryst.

Hileman, J.C. *et al*, *J. Am. Chem. Soc.*, 1961, **83**, 2453 (*synth*)
Behrens, R.G., *J. Less-Common Met.*, 1978, **61**, 321 (*props*)
Kruglov, A.A. *et al*, *Zh. Neorg. Khim.*, 1981, **26**, 960 (*cryst struct*)

C_5IO_5Tc Tc-00003

Pentacarbonyliodotechnetium, 10CI

[15841-96-0]

$(OC)_5TcI$ (O_h)

M 364.957

No phys. props. recorded.

Hileman, J.C. *et al*, *J. Am. Chem. Soc.*, 1961, **83**, 2953 (*synth*)
DeJong, I.G. *et al*, *Inorg. Chem.*, 1973, **12**, 2519 (*synth*)
Behrens, R.G., *J. Less-Common Met.*, 1978, 321.

$C_8H_5O_3Tc$ Tc-00004

Cyclopentadienyltricarbonyltechnetium

Tricarbonyl(η^5-2,4-cyclopentadien-1-yl)technetium, 9CI

[60184-30-7]

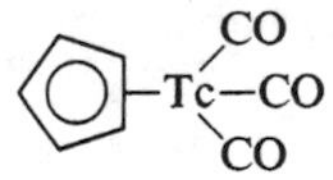

M 247.126

Cryst. by subl. in high vac. Mp 87.5°.

Palm, C. *et al*, *Naturwissenschaften*, 1962, **49**, 279 (*synth, uv, ir, pmr*)
Fischer, E.O. *et al*, *J. Organomet. Chem.*, 1963, **1**, 191.
de Jong, I.G. *et al*, *Inorg. Chem.*, 1976, **15**, 2588 (*synth*)

$C_{10}H_{10}Tc_2$ Tc-00005

Decacarbonylditechnetium, 10CI

Ditechnetium decacarbonyl

[14837-15-1]

$(OC)_5Tc{-}Tc(CO)_5$

M 326.189

White cryst.

Hileman, J.C., *Prep. Inorg. React.*, Jolly, W.J., Ed., Interscience, New York, 1964, **1**, 102 (*synth*)
Bailey, M.F. *et al*, *Inorg. Chem.*, 1965, **4**, 1140 (*cryst struct*)
Junk, G.A. *et al*, *J. Chem. Soc.* (*A*), 1970, 2102 (*ms*)
Spiro, T.G., *Prog. Inorg. Chem.*, 1970, **11**, 1 (*ir, raman*)
Jackson, R.A. *et al*, *Inorg. Chem.*, 1978, **17**, 997.
Kidd, D.R. *et al*, *J. Am. Chem. Soc.*, 1978, **100**, 4095 (*props*)

$C_{10}H_{11}Tc$ Tc-00006

Bis(η^5-2,4-cyclopentadien-1-yl)hydrotechnetium, 9CI

Bis(cyclopentadienyl)hydridotechnetium. Bis(cyclopentadienyl)technetium hydride. Dicyclopentadienylhydridotechnetium

[12116-92-6]

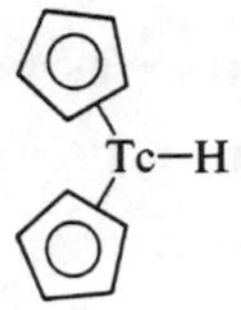

M 229.197

Yellow cryst. (toluene).

Fischer, E.O. *et al*, *Chem. Ber.*, 1969, **102**, 1954 (*synth, nmr*)

$C_{12}H_6O_{10}Tc_2$ Tc-00007

Nonacarbonyl(1-methoxyethylidene)ditechnetium, 8CI

[38855-80-0]

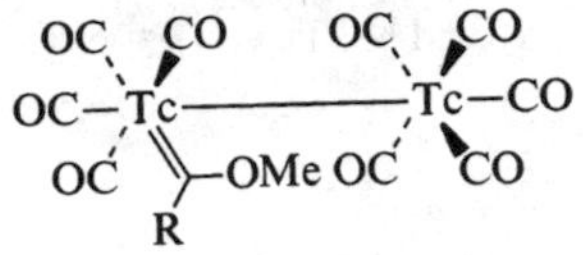

R = CH_3

M 506.173

Yellow cryst. Mp 103°.

Fischer, E.O. *et al*, *Chem. Ber.*, 1972, **105**, 3027 (*synth, nmr*)

$C_{12}H_{12}Tc^{\oplus}$ Tc-00008

Bis(η^6-benzene)technetium(1+)

[11077-51-3]

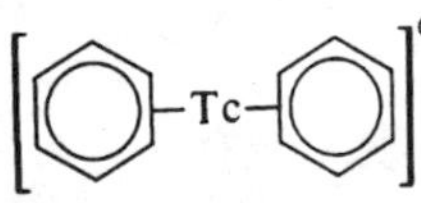

M 254.227 (ion)

Hexafluorophosphate:

$C_{12}H_{12}F_6PTe$ M 428.791

Yellow-green cryst.

Palm, C. *et al*, *Tetrahedron Lett.*, 1962, 253 (*synth, nmr*)

$C_{17}H_8O_{10}Tc_2$ Tc-00009

Nonacarbonyl(methoxyphenylmethylene)ditechnetium, 8CI

[38855-79-7]

As Nonacarbonyl(1-methoxyethylidene)ditechnetium, Tc-00007 with

R = Ph

M 568.244

Yellow-orange cryst. by subl. Mp 88° dec.

Fischer, E.O. *et al*, *Chem. Ber.*, 1972, **105**, 3027 (*synth, nmr*)

$C_{25}H_{33}Cl_3OP_3Tc$ Tc-00010

Carbonyltrichlorotris(dimethylphenylphosphine)technetium, 10CI

[64347-45-1]

$[TcCl_3(CO)(PMe_2Ph)_3]$

M 646.815

Pale-yellow cryst.

Bandoli, G. *et al*, *J. Chem. Soc., Dalton Trans.*, 1978, 373 (*synth, ir, nmr*)

$C_{29}H_{39}ClN_2OP_3Tc^{\oplus\oplus}$ Tc-00011

Bis(acetonitrile)carbonylchlorotris(dimethylphenylphosphine)technetium(2+)

[74847-93-1]

$[TcCl(CO)(NCMe)_2(PMe_2Ph)_3]^{\oplus\oplus}$

M 658.014 (ion)

Diperchlorate: [74847-94-2].

$C_{29}H_{39}Cl_3N_2O_9P_3Tc$ M 856.915

Yellow cryst.

Mazzi, U. *et al*, *Inorg. Chim. Acta*, 1980, **41**, 95 (*synth, ir, nmr, props*)

$C_{39}H_{30}ClO_3P_2Tc$ Tc-00012

Tricarbonylchlorobis(triphenylphosphine)technetium, 10CI

[64396-16-3]

PPh3
OC, CO
Tc
Cl, CO
PPh3

M 742.065

mer,trans-form

White cryst.

Mazzi, U. *et al*, *J. Organomet. Chem.*, 1977, **135**, 177 (*synth, ir, nmr*)

$C_{44}H_{30}O_8P_2Tc_2$ Tc-00013

Octacarbonylbis(triphenylphosphine)ditechnetium, 10CI

[61663-90-9]

$Ph_3P—Tc(CO)_4—Tc(CO)_4—PPh_3$ (*diax-*)

M 944.664

Characterised spectroscopically in soln.

Fawcett, J.P. *et al*, *J. Chem. Soc., Dalton Trans.*, 1976, 2039 (*synth, ir*)

Jackson, R.A. *et al*, *Inorg. Chem.*, 1979, **18**, 3331 (*synth, ir*)

Th Thorium

S. A. Cotton

Thorium (Fr.), Thor (Ger.), Torio (Sp., Ital.), Торий (Torii) (Russ.), トリウム (Japan.)

Atomic Number. 90

Atomic Weight. 232.0382 (one natural isotope)

Electronic Configuration. [Rn] $6d^2\,7s^2$

Oxidation State. +3 (very rare), +4

Coordination Number. See under Uranium.

Colour. Colourless or yellow.

Availability. Materials available (at medium cost) include ThO_2, anhydrous $ThCl_4$ and thorium metal. The most convenient starting material is the anhydrous chloride, prepared either by direct reaction of the elements or by the reaction of ThO_2 with CCl_4 at 450°. References to experimental procedures are given under Uranium.

Handling. See under Uranium.

Toxicity. ^{232}Th is an α-emitter with a long half-life (1.41×10^{10} y) and its decay generates gaseous ^{220}Rn.

Isotopic Abundance. ^{232}Th, 100%.

Spectroscopy. Th(IV) compounds are diamagnetic and give sharp unshifted ^1H nmr spectra.

Analysis. The compound is first destroyed with acid and then titrated against EDTA using xylenol orange indicator at pH 2.5–3.5. Alternatively, thorium may be precipitated either as the hydroxide or oxalate followed by ignition to the oxide which is determined gravimetrically.

Lower than predicted values for C and H microanalysis may be obtained and this is probably due to carbide formation and incomplete combustion.

References. See under Uranium.

Th-00001

Th-00002

Th-00003

Th-00004

Th-00005

Th-00006

Th-00007

Th-00008

Th-00009

Th-00010

Th-00011

Th-00012

$MeTh[N(SiMe_3)_2]_3$

Th-00013

Th-00014

Th-00015

Th-00016

Th-00017

Th-00018

Th-00019

Th-00020

Th-00021

Th-00022

Th-00023

Th-00024

Th-00025

Th-00026

Th-00027

Th-00028

Th-00029

$Th(CH_2Ph)_4$

Th-00030

Th-00031

Th-00032

Th-00033

Th-00034

Th-00035

Th-00036

Th-00037

Th-00038

Th-00039

$C_8H_8Cl_2Th$ Th-00001

Dichloro(η^8-1,3,5,7-cyclooctatetraene)thorium

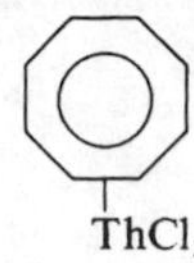

M 407.095

Di-THF complex: [73652-04-7].
$C_{16}H_{24}Cl_2O_2Th$ M 551.309
White cryst. Very air-sensitive.

LeVanda, C. *et al*, *J. Am. Chem. Soc.*, 1980, **102**, 2128 (*synth, cmr, pmr, ir*)
Zalkin, A. *et al*, *Inorg. Chem.*, 1980, **19**, 2560 (*struct*)

$C_{10}H_{15}Cl_3Th$ Th-00002

[(1,2,3,4,5-η)-1,2,3,4,5-Pentamethyl-2,4-cyclopentadien-1-yl]trichlorothorium

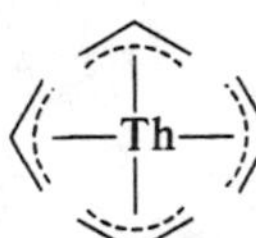

M 473.626

Bis-THF complex: [82511-71-5].
$C_{18}H_{31}Cl_3O_2Th$ M 617.839
Colourless cryst. Sol. THF, dioxan.

Mintz, E.A. *et al*, *J. Am. Chem. Soc.*, 1982, **104**, 4692 (*synth, nmr, ir*)

$C_{12}H_{20}Th$ Th-00003

Tetrakis(η^3-2-propenyl)thorium
Tetraallylthorium

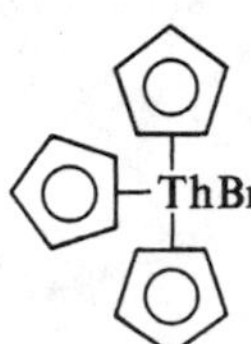

M 396.328
Dark-yellow solid. Sol. C_6H_6. Mp 0° dec.

Wilke, G. *et al*, *Angew. Chem., Int. Ed. Engl.*, 1966, **5**, 151 (*synth, ir, nmr*)

$C_{15}H_{15}BrTh$ Th-00004

Bromotris(η^5-2,4-cyclopentadien-1-yl)thorium

M 507.226
Pale-yellow solid. Subl. at 180° *in vacuo*.

Kanellakopulos, B., *MTP International Review of Science, Inorg. Chem.*, Series One, 1972, **7**, 302.

$C_{15}H_{15}ClTh$ Th-00005

Chlorotris(η^5-2,4-cyclopentadien-1-yl)thorium, 9CI
[1284-82-8]

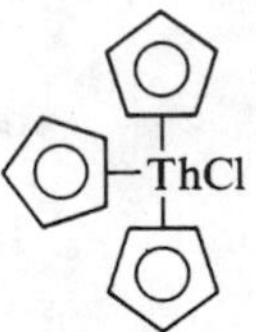

M 462.775
Colourless solid. Sol. THF, $CHCl_3$, toluene, CH_2Cl_2. Subl. at 200° *in vacuo*. Air-sensitive.

Ter Haan, G.L. *et al*, *Inorg. Chem.*, 1964, **3**, 1648 (*synth, nmr, ir*)
Marks, T.J. *et al*, *Inorg. Synth.*, 1976, **16**, 147 (*synth, nmr, ir*)

$C_{15}H_{15}FTh$ Th-00006

Fluorotris(η^5-2,4-cyclopentadien-1-yl)thorium
Tricyclopentadienylthorium fluoride

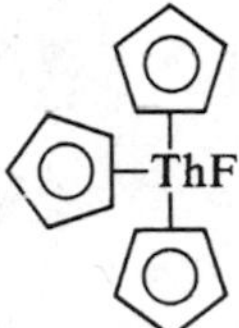

M 446.320
Pale-yellow solid. Subl. at 200° *in vacuo*.

Mueller, J., *Chem. Ber.*, 1969, **102**, 152 (*ms*)
Kanellakopulos, B., *MTP International Review of Science, Inorg. Chem.*, Series one, 1972, **7**, 302.

$C_{15}H_{15}ITh$ Th-00007

Iodotris(η^5-2,4-cyclopentadien-1-yl)thorium
Tricyclopentadienylthorium iodide
[80410-12-4]

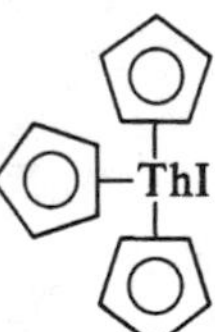

M 554.226
Pale-yellow solid. Sol. toluene. Subl. at 190° *in vacuo*.

Mueller, J., *Chem. Ber.*, 1969, **102**, 152 (*ms*)
Bruno, J.W. *et al*, *J. Am. Chem. Soc.*, 1982, **104**, 1860 (*synth, nmr, ms*)

$C_{15}H_{15}Th$ Th-00008

Tris(η^5-2,4-cyclopentadien-1-yl)thorium
Tricyclopentadienylthorium
[52550-20-6]

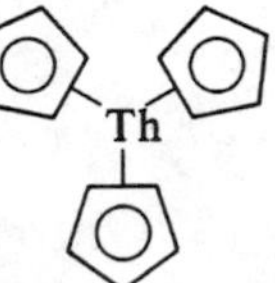

M 427.322

Purple-form

Purple solid. Sol. THF, insol. C_6H_6, pentane. Dec. >170°. Forms adduct with $C_6H_{11}NC$. μ_{eff} 0.40μ_B. Possibly dimeric.

Green-form

Prod. by thermolysis and photolysis of $(C_5H_5)_3ThC_3H_7$. Dark-green cryst. Generally insol. (Dec. by MeOH). Highly air-sensitive. $\mu_{eff} = 2.10\mu_B$.

Kanellakopulos, B. *et al*, *Inorg. Nucl. Chem. Lett.*, 1974, **10**, 155 (*synth, ir, ms*)

Kalina, D.G. *et al*, *J. Am. Chem. Soc.*, 1977, **99**, 3877 (*synth, ir, ms, raman, uv*)

$C_{16}H_{16}Th$ Th-00009

Bis(η^8-1,3,5,7-cyclooctatetraene)thorium

Thoracene

[12702-09-9]

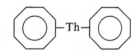

M 440.341

Bright-yellow solid. Sol. DMSO. $Bp_{0.03}$ 160° subl. Dec. >190°. Air-sensitive.

▷Explodes when red-hot.

Streitweiser, A. *et al*, *J. Am. Chem. Soc.*, 1969, **91**, 7528 (*synth, nmr, ms*)

Ardeef, A. *et al*, *Inorg. Chem.*, 1972, **11**, 1083 (*struct*)

$C_{16}H_{18}Th$ Th-00010

Tris(η^5-2,4-cyclopentadien-1-yl)methylthorium

[80410-05-5]

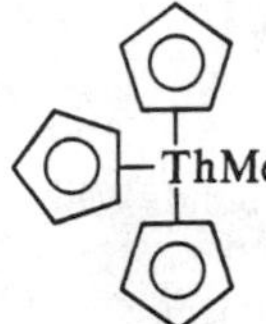

M 442.356

White microcrystalline solid. Sol. toluene, C_6H_6.

Bruno, J.W. *et al*, *J. Am. Chem. Soc.*, 1982, **104**, 1860 (*synth, nmr, ir*)

$C_{19}H_{24}OTh$ Th-00011

Butoxytris(η^5-2,4-cyclopentadien-1-yl)thorium

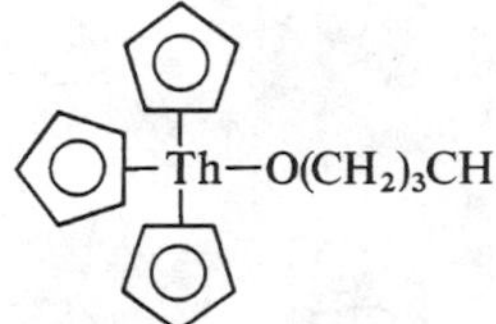

M 500.436

White cryst. Sol. dimethoxyethane, THF, CH_2Cl_2, $CHCl_3$. Mp 148-50°. Diamagnetic. Subl. at 135° *in vacuo*. Oxidized and hydrolysed in air.

Ter Haar, G.L. *et al*, *Inorg. Chem.*, 1964, **3**, 1648 (*synth, ir, nmr*)

$C_{19}H_{24}Th$ Th-00012

Butyltris(η^5-2,4-cyclopentadien-1-yl)thorium, 9CI

Tricyclopentadienylbutylthorium

[54067-92-4]

M 484.437

White cryst. Sol. toluene. Mp 210° dec. Dec. by air.

Marks, T.J. *et al*, *J. Am. Chem. Soc.*, 1976, **98**, 703 (*synth, ir, nmr, raman*)

$C_{19}H_{57}N_3Si_6Th$ Th-00013

Methyltris[(hexamethyldisilyl)amino]thorium

*Methyltris[1,1,1-trimethyl-*N*-(trimethylsilyl)silylanaminato]thorium*

[69517-43-7]

$MeTh[N(SiMe_3)_2]_3$

M 728.231

White needles (pentane). Sol. Et_2O, pentane. Mp 133-5°. Air- and moisture-sensitive.

Turner, H.W. *et al*, *Inorg. Chem.*, 1979, **18**, 1221 (*synth, ir, nmr*)

$C_{20}H_{20}Th$ Th-00014

Tetrakis(η^5-2,4-cyclopentadien-1-yl)thorium

[1298-75-5]

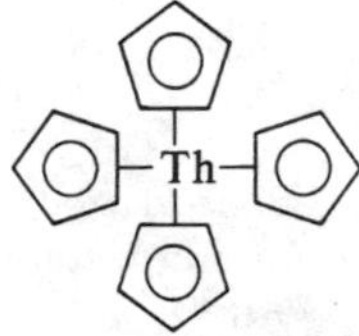

M 492.416

Colourless solid. Sol. Et_2O. Mp >170° dec. Subl. at 250-90° *in vacuo*.

Fischer, E.O. *et al*, *Z. Naturforsch., B*, 1962, **17**, 276 (*synth, ir, nmr*)

Green, J.C. *et al*, *J. Organomet. Chem.*, 1981, **212**, 329 (*pe*)

$C_{20}H_{26}Th$ Th-00015

Tris(η^5-2,4-cyclopentadien-1-yl)(2,2-dimethylpropyl)thorium, 9CI

Tricyclopentadienylneopentylthorium

[58920-14-2]

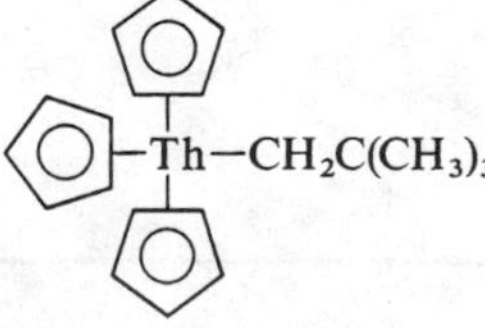

M 498.464

White microcryst. solid. Sol. toluene. Mp 200° dec. Smokes on exposure to air.

Marks, T.J. *et al*, *J. Am. Chem. Soc.*, 1976, **98**, 703 (*synth, ir, nmr*)

$C_{20}H_{30}Cl_2Th$ **Th-00016**

Dichlorobis[(1,2,3,4,5-η)-1,2,3,4,5-pentamethyl-2,4-cyclopentadien-1-yl]thorium

Bis(pentamethylcyclopentadienyl)thorium dichloride

[67506-88-1]

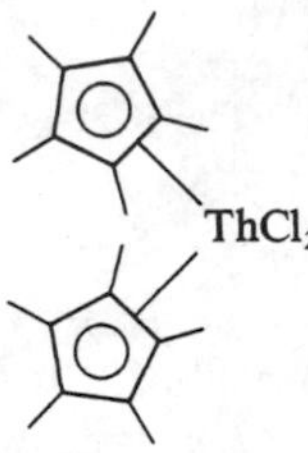

M 573.401

White cryst. Sol. toluene, THF. Air-sensitive.

Manriquez, J.M. *et al*, *J. Am. Chem. Soc.*, 1978, **100**, 3939 (*synth, ir, pmr*)
Fagan, P.J. *et al*, *J. Am. Chem. Soc.*, 1981, **103**, 6650 (*synth, ir, pmr*)

$C_{20}H_{32}Th$ **Th-00017**

Dihydrobis(pentamethylcyclopentadienyl)thorium

Bis(pentamethylcyclopentadienyl)thorium dihydride

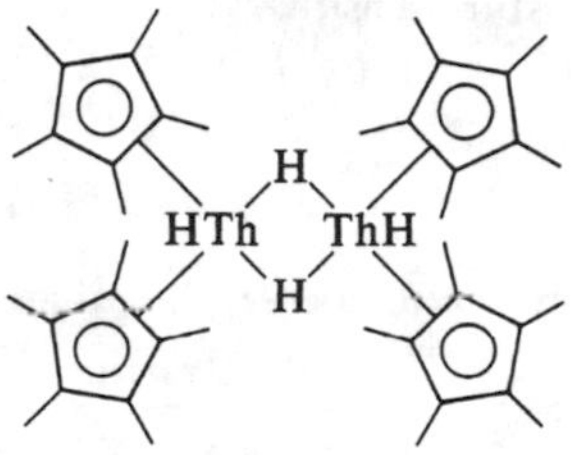

M 504.511

Dimeric.

Dimer: [67506-92-7]. *Di-μ-hydrodihydrotetrakis[(1,2,3,4,5-η)-1,2,3,4,5-pentamethyl-2,4-cyclopentadien-1-yl]dithorium*, 9CI.

$C_{40}H_{64}Th_2$ M 1009.022

Inserts ethylene to form diethyl complex; reacts rapidly with MeCl, acetone; catalyses hydrogenation of hex-1-ene; inserts CO; catalyses hydrogenation of coordinated CO in Chloro(3,3-dimethyl-1-oxobutyl-*C,O*)bis[(1,2,3,4,5-η)-1,2,3,4,5-pentamethyl-2,4-cyclopentadien-1-yl]thorium, Th-00027 and (Benzoyl-*C,O*)chlorobis[(1,2,3,4,5-η)-1,2,3,4,5-pentamethyl-2,4-cyclopentadien-1-yl]uranium. Pale-yellow cryst. Sol. toluene, almost insol. pentane. Mp >80°. Air-sensitive.

Broach, R.W. *et al*, *Science*, 1979, **203**, 172 (*cryst struct*)
Maata, E.A. *et al*, *J. Am. Chem. Soc.*, 1981, **103**, 3578 (*use*)
Fagan, P.J. *et al*, *J. Am. Chem. Soc.*, 1981, **103**, 6650, 6959 (*synth, ir, nmr, use*)

The first digit of the Entry number defines the Supplement in which the Entry is found. 0 indicates the Main Work

$C_{21}H_{33}ClTh$ **Th-00018**

Chloro(methyl)bis[(1,2,3,4,5-η)-1,2,3,4,5-pentamethyl-2,4-cyclopentadien-1-yl]thorium

[79301-20-5]

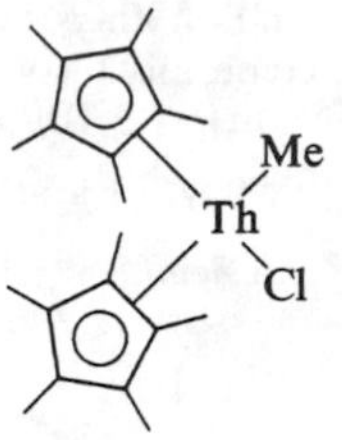

M 552.983

Reacts additively with alcohols and carbonyl compounds. White cryst. Sol. toluene, C_6H_6. Air- and moisture-sensitive.

Fagan, P.J. *et al*, *J. Am. Chem. Soc.*, 1981, **103**, 6650 (*synth, nmr, props*)

$C_{22}H_{36}ClNTh$ **Th-00019**

Chloro(dimethylamino)bis(pentamethylcyclopentadienyl)thorium

Chloro(N-methylmethanaminato)bis[(1,2,3,4,5η)-1,2,3,4,5-pentamethyl-2,4-cyclopentadien-1-yl]thorium

[77260-84-5]

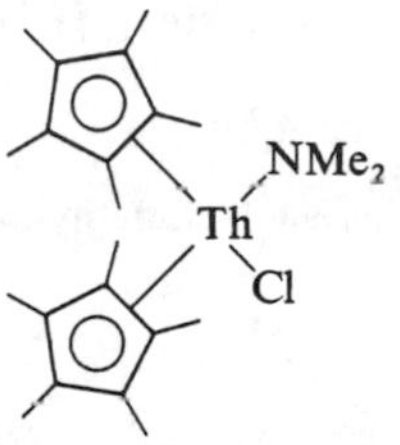

M 582.024

Inserts CO into Th-N bond. White cryst. Sol. toluene. Air- and moisture-sensitive.

Fagan, P.J. *et al*, *J. Am. Chem. Soc.*, 1981, **103**, 2206 (*synth, ir, nmr*)

$C_{22}H_{36}Th$ **Th-00020**

Dimethylbis[(1,2,3,4,5-η)-1,2,3,4,5-pentamethyl-2,4-cyclopentadien-1-yl]thorium

[67506-90-5]

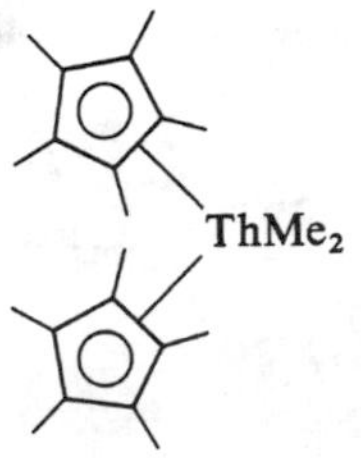

M 532.565

When supported on alumina, very active catalyst for hydrogenation of propene and polymerisation of ethylene. Inserts CO. Reacts with H_2 to give Dihydrobis(pentamethylcyclopentadienyl)thorium, Th-00017 . White cryst. Sol. toluene, pentane. V. air-sensitive.

Fagan, P.J. *et al*, *J. Am. Chem. Soc.*, 1978, **100**, 3939 (*synth*)
Manriquez, J.M. *et al*, *J. Am. Chem. Soc.*, 1978, **100**, 7112.
Bowman, R.G. *et al*, *J. Chem. Soc., Chem. Commun.*, 1981, 257.
Fagan, P.J. *et al*, *J. Am. Chem. Soc.*, 1981, **103**, 6050 (*synth, ir, pmr*)

$C_{23}H_{39}NTh$ **Th-00021**

Methyl(*N*-methylmethanaminato)bis[(1,2,3,4,5-η)-1,2,3,4,5-pentamethyl-2,4-cyclopentadien-1-yl]thorium, 10CI

[77260-87-8]

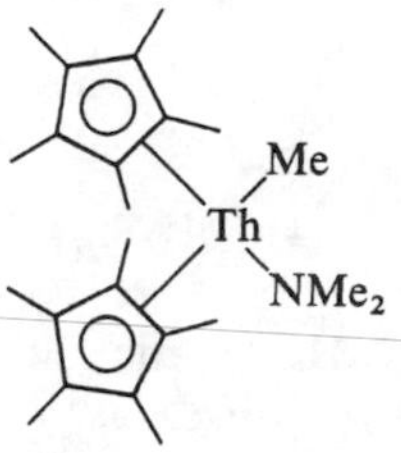

M 561.606

Inserts CO preferentially at Th—C bond to give $(C_5Me_5)_2Th(\eta^2-COMe)(NMe_2)$. White cryst. Sol. pentane, Et_2O. Air- and moisture-sensitive.

Fagan, P.J. *et al*, *J. Am. Chem. Soc.*, 1981, **103**, 2206 (*synth, ir, nmr*)

$C_{24}H_{40}ClNTh$ **Th-00022**

Chloro(diethylamino)bis(pentamethylcyclopentadienyl)thorium

Chloro(N-*ethylethanaminato*)*bis*[(*1,2,3,4,5-η*)*-1,2,3,4,5-pentamethyl-2,4-cyclopentadien-1-yl*]*thorium*

[77260-81-2]

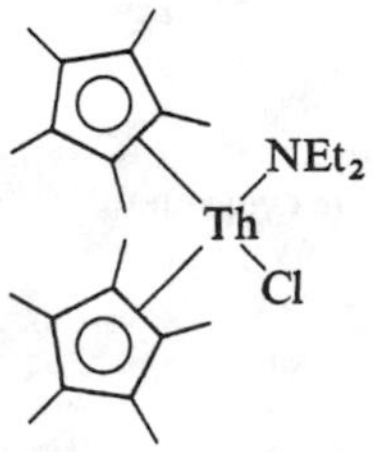

M 610.078

Inserts CO into Th-N bond. Colourless cryst. Sol. toluene. Air- and moisture-sensitive.

Fagan, P.J. *et al*, *J. Am. Chem. Soc.*, 1981, **103**, 2206 (*synth, ir, nmr*)

$C_{24}H_{42}N_2Th$ **Th-00023**

Bis(dimethylamino)bis(pentamethylcyclopentadienyl)thorium

Bis(N-*methylmethanaminato*)*bis*[(*1,2,3,4,5-η*)*-1,2,3,4,5-pentamethyl-2,4-cyclopentadien-1-yl*]*thorium*

[77431-02-8]

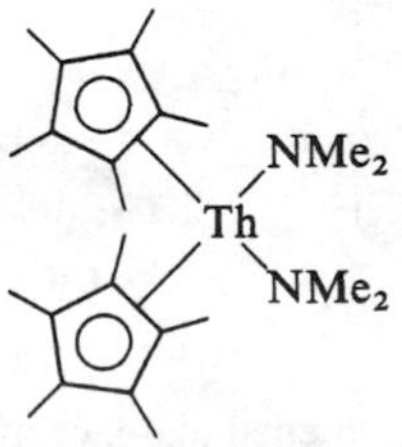

M 590.647

Inserts CO into one Th-N bond (into both at 60°). Colourless cryst. Sol. Et_2O, sl. sol. pentane. Air- and moisture-sensitive.

Fagan, P.J. *et al*, *J. Am. Chem. Soc.*, 1981, **103**, 2206 (*synth, ir, nmr*)

$C_{25}H_{40}ClNOTh$ **Th-00024**

Chloro[(diethylamino)carbonyl-*C,O*]bis[(1,2,3,4,5-η)-1,2,3,4,5-pentamethyl-2,4-cyclopentadien-1-yl]thorium

[77260-90-3]

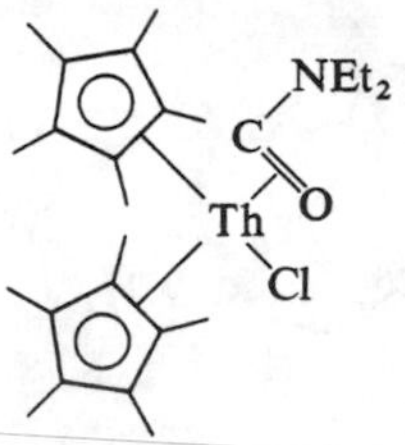

M 638.088

Formed by CO insertion reaction of Chloro(diethylamino)bis(pentamethylcyclopentadienyl)thorium, Th-00022 . Colourless cryst. Sol. toluene, pentane. Air- and moisture-sensitive.

Fagan, P.J. *et al*, *J. Am. Chem. Soc.*, 1981, **103**, 2206 (*synth, ir, nmr, cryst struct*)

$C_{25}H_{41}ClOSiTh$ **Th-00025**

Chloro[(1,2,3,4,5-η)-1,2,3,4,5-pentamethyl-2,4-cyclopentadien-1-yl][1-(trimethylsilyl)ethenolato]thorium

[68963-89-3]

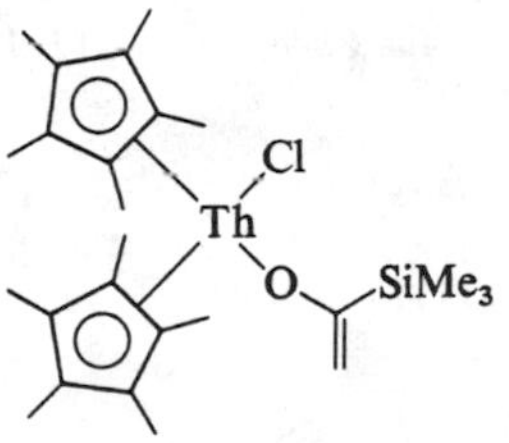

M 653.175

Formed by migratory CO insertion reactions of $(C_5Me_5)_2ThCl(CH_2SiMe_3)$. Colourless needles. Sol. C_6H_6, toluene.

Manriquez, J.M. *et al*, *J. Am. Chem. Soc.*, 1978, **100**, 7112 (*synth, ir, nmr*)

$C_{25}H_{41}ClTh$ **Th-00026**

Chloro(2,2-dimethylpropyl)bis[(1,2,3,4,5-η)-1,2,3,4,5-pentamethyl-2,4-cyclopentadien-1-yl]thorium

[74587-39-6]

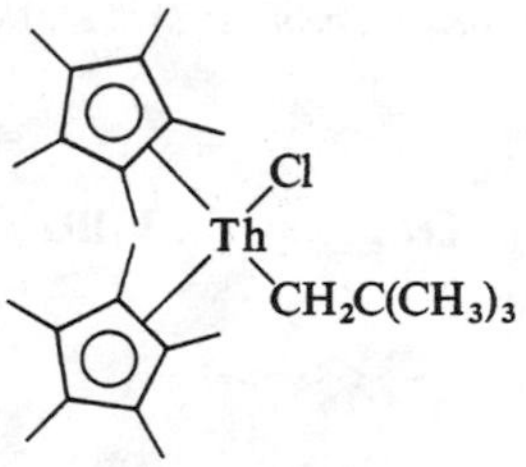

M 609.090

Th-C bond undergoes CO insertion reactions. White solid. Sol. toluene.

Fagan, P.J. *et al*, *J. Am. Chem. Soc.*, 1980, **102**, 5393 (*synth, props*)

$C_{26}H_{41}ClOTh$ **Th-00027**

Chloro(3,3-dimethyl-1-oxobutyl-*C,O*)bis[(1,2,3,4,5-η)-1,2,3,4,5-pentamethyl-2,4-cyclopentadien-1-yl]thorium

[74587-36-3]

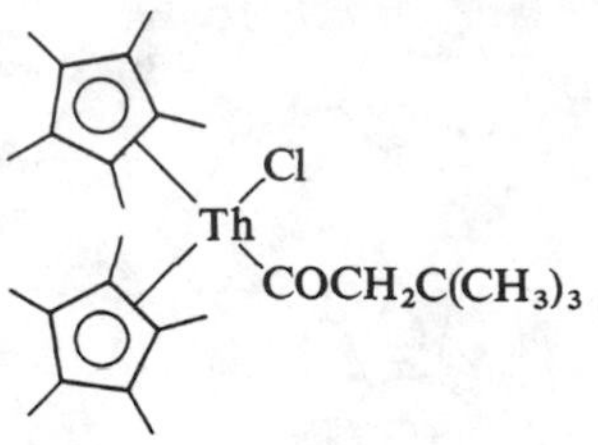

M 637.100

Formed by CO insertion of Chloro(2,2-dimethylpropyl)-bis[(1,2,3,4,5-η)-1,2,3,4,5-pentamethyl-2,4-cyclopentadien-1-yl]thorium, Th-00026 . Coordinated CO– reduced to $-CH_2O-$ by H_2 with catalyst of Dihydrobis(pentamethylcyclopentadienyl)thorium, Th-00017 Rearranges by H migration at 100°, reacts with excess CO to give Dichlorotetrakis[(1,2,3,4,5-η)-1,2,3,4,5-pentamethyl-2,4-cyclopentadien-1-yl][μ-(2,2,9,9-tetramethyl-4,5,6,7-decanetetrone-O^4,O^5,O^6,O^7)]dithorium, Th-00039 . Pale-yellow plates. Sol. toluene.

Fagan, P.J. *et al*, *J. Am. Chem. Soc.*, 1980, **102**, 5393 (*synth, ir, struct, cmr*)

Matta, E.A. *et al*, *J. Am. Chem. Soc.*, 1981, **103**, 3576.

$C_{27}H_{21}ClTh$ **Th-00028**

Chlorotris[(1,2,3,3*a*,7*a*-η)-1*H*-inden-1-yl)]thorium, 9CI

Triindenylthorium chloride

[11133-05-4]

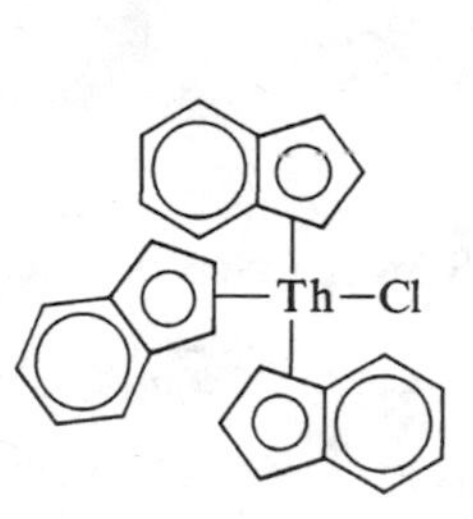

M 612.954

Sol. THF, sl. sol. pentane. Mp 181°. Air- and moisture-sensitive. Subl. *in vacuo* at 160°.

Laubereau, P.G. *et al*, *Inorg. Chem.*, 1971, **10**, 2274 (*synth, ir, ms, nmr*)

Fragala, I.E. *et al*, *Inorg. Chem.*, 1983, **22**, 216 (*pe*)

$C_{28}H_{24}Th$ **Th-00029**

Tris[(1,2,3,3*a*,7*a*-η)-1*H*-indenyl]methylthorium

[63643-54-9]

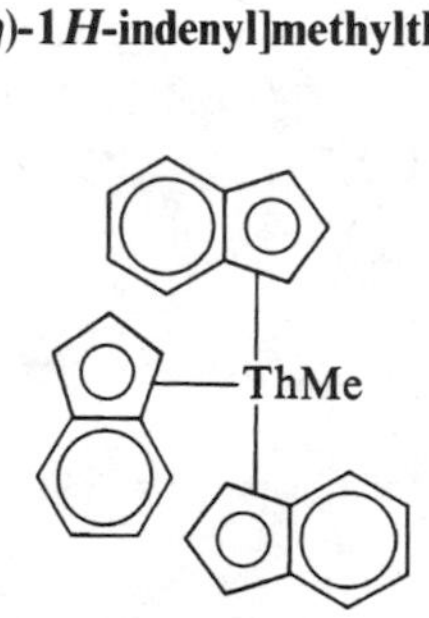

M 592.536

Yellow solid. Sol. C_6H_6, THF, sl. sol. pentane. Mp 213-5°. Subl. at 190-5° *in vacuo*.

Goffart, J. *et al*, *Inorg. Nucl. Chem. Lett.*, 1977, **13**, 189 (*synth, ir, raman, ms*)

$C_{28}H_{28}Th$ **Th-00030**

Tetrabenzylthorium

Tetrakis(phenylmethyl)thorium, 9CI. *Thorium tetrabenzyl*

[54008-63-8]

$Th(CH_2Ph)_4$

M 596.567

Yellow cryst. Sol. C_6H_6, toluene. Air- and light-sensitive.

Köhler, E. *et al*, *J. Organomet. Chem.*, 1974, **76**, 235 (*synth, ir, pmr*)

$C_{28}H_{52}Si_2Th$ **Th-00031**

Bis[(1,2,3,4,5-η)-1,2,3,4,5-pentamethyl-2,4-cyclopentadien-1-yl]bis[(trimethylsilyl)methyl]thorium

[69040-88-6]

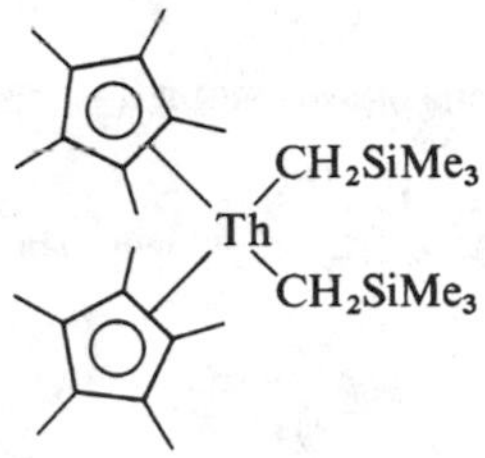

M 676.928

Inserts two CO groups into Th-C bonds at 25°. Undergoes intermolecular cyclometallation on heating to 80°, eliminating $SiMe_4$. White cryst. Sol. heptane, pentane. Air- and moisture-sensitive.

Manriquez, J.M. *et al*, *J. Am. Chem. Soc.*, 1978, **100**, 7112 (*props*)

Fagan, P.J. *et al*, *J. Am. Chem. Soc.*, 1981, **103**, 6650 (*synth, ir, nmr*)

Bruno, J.W. *et al*, *J. Organomet. Chem.*, 1983, **250**, 237 (*synth, struct*)

$C_{29}H_{50}OTh$ **Th-00032**

Hydrobis[(1,2,3,4,5-η)-1,2,3,4,5-pentamethyl-2,4-cyclopentadien-1-yl](2,2,4,4-tetramethyl-3-pentanolato)thorium, 10CI

[79932-08-4]

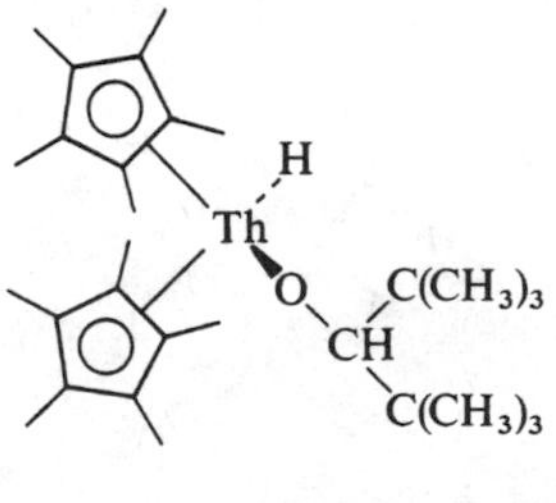

M 646.752

Carbonylated to yield enediolate, or in presence of H_2, methoxide $Th(C_5Me_5)_2(OMe)[OCH(CMe_3)_2]$. Colourless cryst. Sol. toluene.

Fagan, P.J. *et al*, *J. Am. Chem. Soc.*, 1981, **103**, 6959 (*synth*)

Katahira, D.A. *et al*, *Organometallics*, 1982, **1**, 1723.

$C_{30}H_{28}Th_2$ Th-00033

Tetrakis(η^5-2,4-cyclopentadien-1-yl)bis[μ-(1-η: 1,2,3,4,5-η)-2,4-cyclopentadien-1-ylidene]dithorium, 9CI

[54067-93-5]

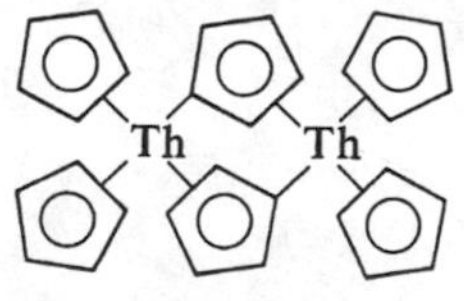

M 852.627

Formed by thermal decomposition of Butyltris(η^5-2,4-cyclopentadien-1-yl)thorium, Th-00012 . Colourless cryst. Sol. toluene. Air- and moisture-sensitive.

Baker, E.C. *et al*, *J. Am. Chem. Soc.*, 1974, **96**, 7586 (*synth, struct*)

Marks, T.J. *et al*, *J. Am. Chem. Soc.*, 1976, **98**, 703 (*synth, ms*)

$C_{30}H_{50}ClOPTh$ Th-00034

Chloro[1-hydroxy-4,4-dimethyl-1-(trimethylphosphoranylidene)-2-pentanonato-*O,O'*]bis[(1,2,3,4,5-η)-1,2,3,4,5-pentamethyl-2,4-cyclopentadien-1-yl]thorium

[86543-76-2]

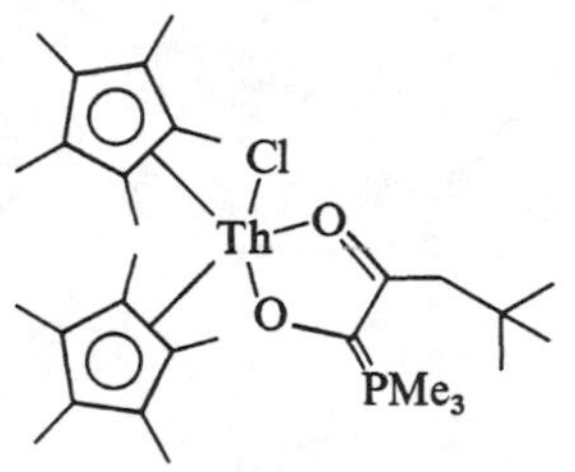

M 725.189

Prepd. by reaction of Chloro(3,3-dimethyl-1-oxobutyl-*C,O*)bis[(1,2,3,4,5-η)-1,2,3,4,5-pentamethyl-2,4-cyclopentadien-1-yl]thorium, Th-00027 and PMe_3 with CO. Cryst. (toluene). Sol. toluene.

Moloy, K.G. *et al*, *J. Am. Chem. Soc.*, 1983, **105**, 5696 (*synth, struct, nmr*)

$C_{31}H_{30}Th$ Th-00035

Butyltris[(1,2,3,3*a*,7*a*-η)-1*H*-inden-1-yl]thorium

Tris(η^5-indenyl)butylthorium

[63643-43-6]

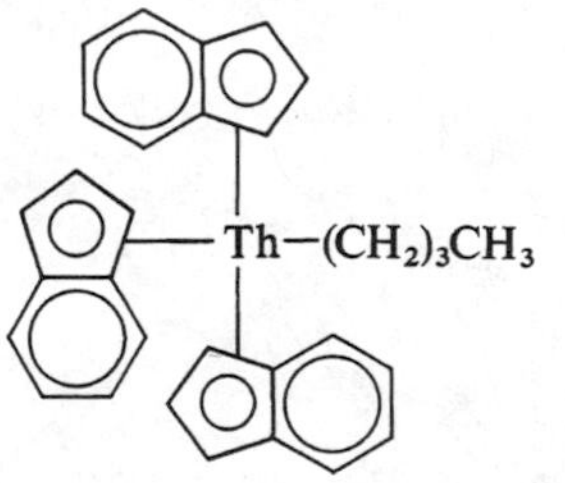

M 634.616

Yellow solid. Sol. THF, pentane. Mp 125-6°. Subl. at 185-90° *in vacuo*.

Goffart, J. *et al*, *Inorg. Nucl. Chem. Lett.*, 1977, **13**, 189 (*synth, ir, raman, ms*)

$C_{31}H_{36}Th$ Th-00036

Tribenzyl(pentamethylcyclopentadienyl)thorium

[(1,2,3,4,5-η)-1,2,3,4,5-Pentamethyl-2,4-cyclopentadien-1-yl]tris(phenylmethyl)thorium. (Pentamethylcyclopentadienyl)tribenzylthorium

[82511-73-7]

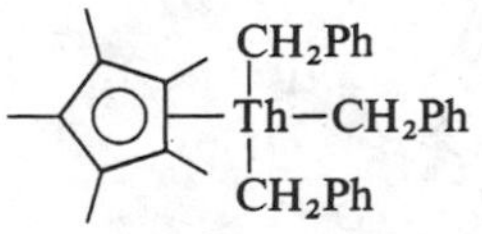

M 640.664

Reacts with H_2 to form toluene, inserts formaldehyde to yield alkoxide. Yellow cryst. Sol. toluene.

Mintz, E.A. *et al*, *J. Am. Chem. Soc.*, 1982, **104**, 4692 (*synth, struct, nmr, cmr, ir*)

$C_{36}H_{39}ClTh$ Th-00037

Chlorotris[(1,2,3,3*a*,7*a*-η)-1,4,7-trimethyl-1*H*-inden-1-yl]thorium

Tris(1,4,7-trimethylindenyl)thorium chloride

[78599-84-5]

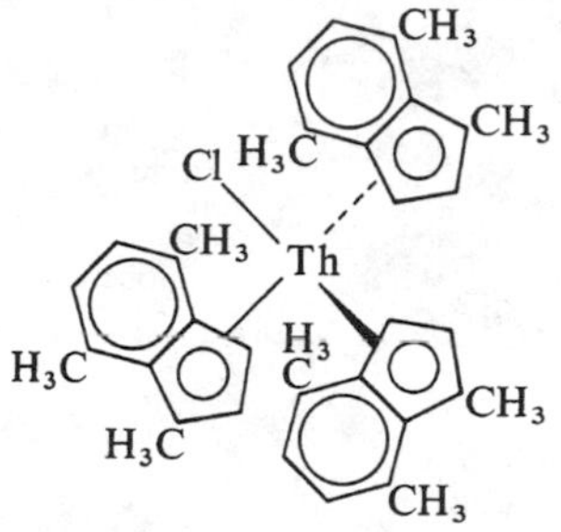

M 739.195

Yellow cryst. Sol. THF, sl. sol. pentane. Air- and moisture-sensitive. Struct. by analogy with the corresponding U complex.

Goffart, J. *et al*, *J. Organomet. Chem.*, 1981, **209**, 281 (*synth, ir, ms, raman, nmr*)

$C_{48}H_{72}Th_2O_4$ Th-00038

Bis[μ-(2-butene-2,3-diolato(2-)-*O,O'*]tetrakis[(1,2,3,4,5-η)-1,2,3,4,5-pentamethyl-2,4-cyclopentadien-1-yl]dithorium

[69021-32-5]

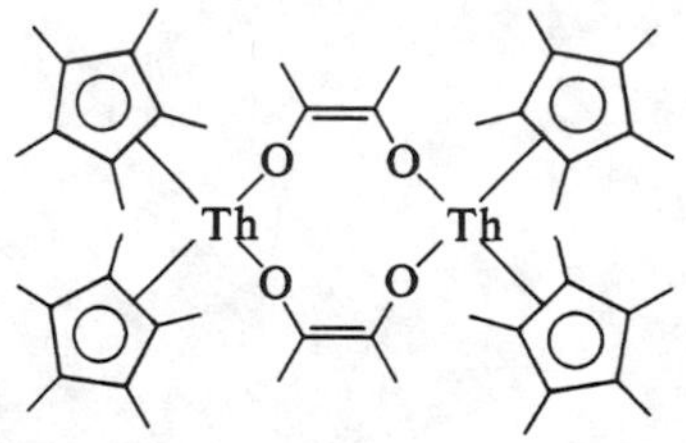

M 1177.171

Formed by CO insertion of $(C_5Me_5)_2ThMe_2$. Colourless cryst. Sol. toluene.

Manriquez, J.M. *et al*, *J. Am. Chem. Soc.*, 1978, **100**, 7112 (*synth, ir, nmr*)

$C_{54}H_{82}Cl_2O_4Th_2$ **Th-00039**

Dichlorotetrakis[(1,2,3,4,5-η)-1,2,3,4,5-pentamethyl-2,4-cyclopentadien-1-yl][μ-(2,2,9,9-tetramethyl-4,5,6,7-decanetetrone-O^4,O^5,O^6,O^7)]dithorium

[74587-38-5]

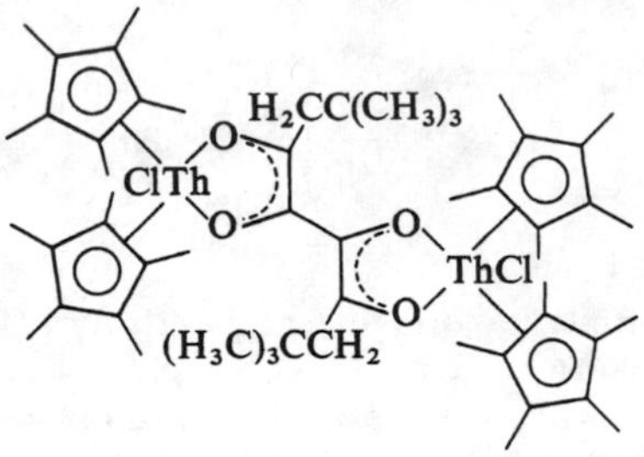

M 1330.222

Dark-purple cryst. Sol. toluene.

Fagan, P.J. *et al*, *J. Am. Chem. Soc.*, 1980, **102**, 5393 (*synth, struct*)

Ti Titanium

D. J. Cardin

Titane (Fr.), Titan (Ger.), Titanio (Sp., Ital.), Титан (Teetan) (Russ.), チタン (Japan.)

Atomic Number. 22

Atomic Weight. 47.90

Electronic Configuration. [Ar] $3d^2\ 4s^2$

Oxidation States. By far the largest number of organotitanium compounds prepared so far are derivatives of Ti(IV). There is also a fairly substantial organometallic chemistry of Ti(III). Ti(II) is a much smaller class, represented for example by bis(cyclopentadienyl) derivatives with neutral donors, and some arene complexes. Arene and butadiene compounds of Ti(0) are known.

Coordination Number. (Note, all ligands in this section are treated as monohapto, including cyclic hydrocarbons, which may or, more frequently, may not be). For Ti(IV), and especially the bis(cyclopentadienyl) derivatives, tetrahedral geometry predominates. However, with chelating ligands higher coordination numbers are known, eg. 7 in $Ti(S_2CNMe_2)_3(\eta^5\text{-}C_5H_5)$. In lower oxidation states, coordination numbers vary from 2 in bis(arene)metal(0) compounds to 5 in the Ti(II) complex $[Ti_3(C_6Me_6)_3Cl_6]^+Cl^-$, and in the $Ti(\eta^5\text{-}C_5H_5)(S_2CNR_2)_2$ species of Ti(III). For metals in oxidation states (II) and (III) four coordination is still common.

Colour. Despite the d^0 configuration of titanium (IV) the vast majority of organic derivatives are coloured, presumably owing to charge transfer transitions. Derivatives with electron withdrawing substituents and a single cyclopentadienyl ligand are frequently yellow or pale in colour. Bis(cyclopentadienyl) compounds are generally deeply coloured, often red or orange. Compounds of the lower oxidation states are generally intensely coloured. Red (and similar coloured) compounds are common, but colours are highly ligand-dependent.

Availability. The prime (low cost) starting material is $TiCl_4$; the element is unreactive with regard to organylation except in the vapour phase. $TiCl_4$ is readily converted to $TiCl_2(\eta^5\text{-}C_5H_5)_2$, an important departure point for the huge range of cyclopentadienyl derivatives. Other halogens are usually incorporated metathetically from the chlorides. In some cases, (eg. alkyls) alkoxides have been recommended over the tetrachloride. $TiCl_3$ has been employed for metal(III) alkyls, but for cyclopentadienyl Ti(III) compounds reduction from $TiX_3(\eta^5\text{-}C_5H_5)$ (X = halide) has been widely employed. Apart from these, the compounds $TiCl_3(\eta^5\text{-}C_5H_5)$, $TiCl_2(C_9H_7)_2$, (C_9H_7 = indenyl), $TiCl_2(\eta^5\text{-}C_5H_5)_2$, $Ti(OMe)_4$, $Ti(OEt)_4$, $Ti(NMe_2)_4$, $Ti(acac)_2Cl_2$, and TiX_4 (X = F, Br, I) are readily obtainable commercially.

Handling. Standard inert atmosphere techniques are required for those compound of titanium which are air-sensitive.

Toxicity. Organotitanium compounds are not associated with any known specific toxicity.

Isotopic Abundance. ^{46}Ti, 7.93%; ^{47}Ti, 7.28%; ^{48}Ti, 73.94%; ^{49}Ti, 5.51%; ^{50}Ti, 5.34%.

Spectroscopy. ^{47}Ti and ^{49}Ti have nuclear spins $I = \frac{5}{2}$ and $\frac{7}{2}$ and receptivities relative to ^{13}C of 0.864 and 1.18 respectively. In their nmr spectra the two resonances always occur 270 ppm apart as the frequencies lie extremely close, eg. at 5.6390 and 5.6375 MHz referred to ^{1}H TMS at 100 MHz.

Nmr is not generally useful for paramagnetic compounds. Esr data are reported for many titanium(III) compounds but hyperfine splitting is not always observed.

Analysis. For precise analysis the organic compounds are generally decomposed by oxidation, alkaline fusion, or hydrolysis, after which the aqueous solution can be estimated titrimetrically, colorimetrically, or gravimetrically. Experimental details are given in standard analytical works. X-ray fluorescence and polarographic methods have also been employed.

For detection of the element, after conversion to an aqueous medium, the orange colour produced by hydrogen peroxide in the presence of sulfuric acid has been the most widely used. This colour is bleached by fluoride ions (distinction from vanadium). Other reagents for aqueous media include pyrocatechol, morphine, and chromotropic acid.

References. In addition to references given in the introduction to the *Sourcebook*, the following provide further reading:

Wailes, P. C. *et al.*, *Organometallic Chemistry of Titanium, Zirconium and Hafnium*, Academic Press, London, 1974.

Clark, R. J. H. *et al.*, *J. Organomet. Chem. Lib.*, 1977, **3**, 223.

$MeTiBr_3$

Ti-00001

$MeTiCl_3$

Ti-00002

$EtTiCl_3$

Ti-00003

$Me_3SiCH_2TiCl_3$

Ti-00004

$Ti(OMe)_4$

Ti-00005

$TiMe_4$

Ti-00006

X = Br

Ti-00007

Ti-00008

As Ti-00007 with
X = Cl

Ti-00009

As Ti-00007 with
X = F

Ti-00010

As Ti-00007 with
X = I

Ti-00011

$(C_6F_5)TiCl_3$

Ti-00012

$Me_2Ti(NMe_2)_2$

Ti-00013

$MeTi(NMe_2)_3$

Ti-00014

As Ti-00007 with
X = NCS

Ti-00015

As Ti-00007 with
R = OMe

Ti-00016

As Ti-00007 with
X = Me

Ti-00017

$Ti(NMe_2)_4$

Ti-00018

Ti-00019

Ti-00020

Ti-00021

Ti-00022

X = Br

Ti-00023

X = Br

Ti-00024

As Ti-00023 with
X = Cl

Ti-00025

As Ti-00024 with
X = Cl

Ti-00026

Ti-00027

As Ti-00023 with
X = F

Ti-00028

As Ti-00024 with
X = F

Ti-00029

As Ti-00023 with
X = I

Ti-00030

As Ti-00024 with
X = I

Ti-00031

As Ti-00024 with
$X = N_3$

Ti-00032

X = S

Ti-00033

As Ti-00033 with
X = Se

Ti-00034

Ti-00035

$(H_3C)_2C{=}CPhTiCl_3$

Ti-00036

As Ti-00023 with
X = H

Ti-00037

Ti-00038

Ti-00039

Ti-00040

Ti-00041

$MeTi[OCH(CH_3)_2]_3$

Ti-00042

$Me_3SiCH_2Ti(NMe_2)_3$

Ti-00043

Ti-00044

X = Me

Ti-00045

Ti-00046

As Ti-00007 with
X = OAc

Ti-00047

As Ti-00007 with
X = OEt

Ti-00048

As Ti-00007 with
$X = NMe_2$

Ti-00049

As Ti-00024 with
X = NCO

Ti-00050

Ti-00051

$TiPh_2$

Ti-00052

Ti-00053

Ti-00054

Ti-00055

Ti-00056

As Ti-00045 with
$X = -CH{=}CH_2$

Ti-00057

Ti-00058

As Ti-00024 with
X = SMe

Ti-00059

Ti-00060

Ti-00061

$(Me_3SiCH_2)_2Ti(NMe_2)_2$

Ti-00062

As Ti-00045 with
X = —N=C$(CF_3)_2$

Ti-00063

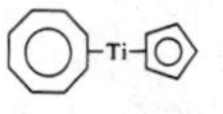

Ti-00064

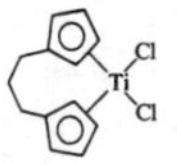

Ti-00065

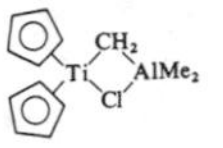

Ti-00066

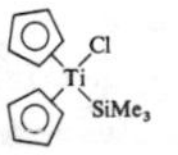

Ti-00067

$MeTi(NEt_2)_3$

Ti-00068

As Ti-00024 with
X = —OOCCF$_3$

Ti-00069

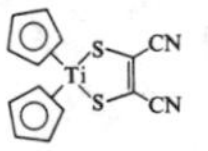

Ti-00070

As Ti-00024 with
X = OAc

Ti-00071

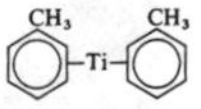

Ti-00072

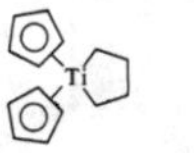

Ti-00073

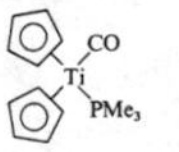

Ti-00074

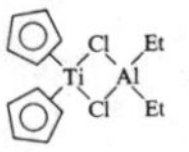

Ti-00075

As Ti-00024 with
X = NMe_2

Ti-00076

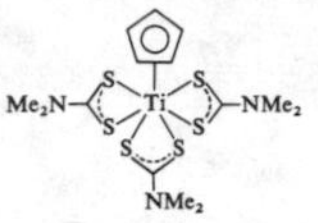

Ti-00077

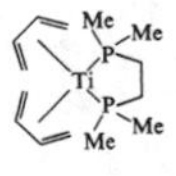

Ti-00078

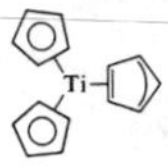

Ti-00079

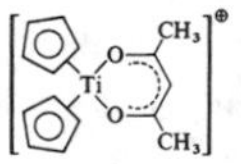

Ti-00080

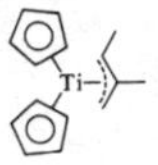

Ti-00081

As Ti-00045 with
X = —$CH_2C(CH_3)_3$

Ti-00082

$PhTi[OCH(CH_3)_2]_3$

Ti-00083

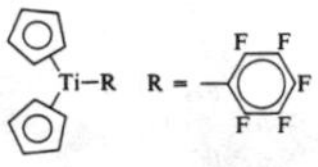

Ti-00084

As Ti-00024 with
X = —C≡CCF_3

Ti-00085

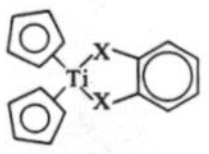

X = O

Ti-00086

As Ti-00086 with
X = S

Ti-00087

As Ti-00045 with
X = Ph

Ti-00088

As Ti-00084 with
R = Ph

Ti-00089

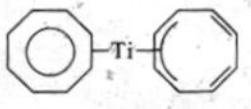

Ti-00090

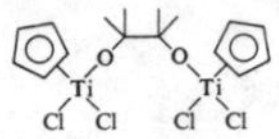

Ti-00091

As Ti-00024 with
X = $P(OMe)_3$

Ti-00092

$Ti(CH_2SiMe_3)_4$

Ti-00093

As Ti-00084 with
R = CH_2Ph

Ti-00094

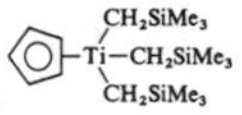

Ti-00095

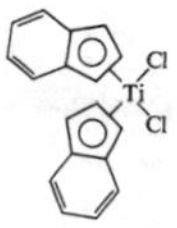

Ti-00096

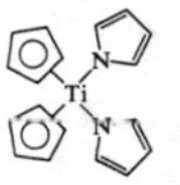

Ti-00097

As Ti-00084 with
R = 2,6-Dimethylphenyl

Ti-00098

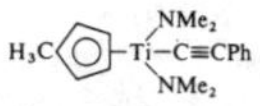

Ti-00099

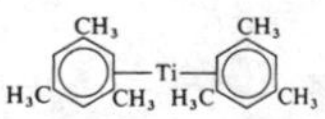

Ti-00100

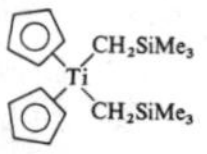

Ti-00101

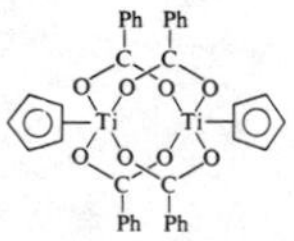

Ti-00102

As Ti-00045 with
X = —N=C$[C(CH_3)_3]_2$

Ti-00103

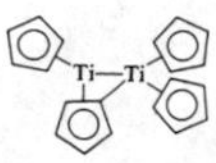

Ti-00104

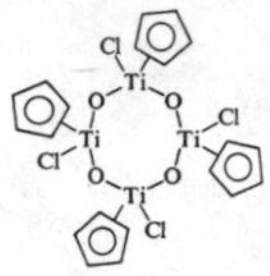

Ti-00105

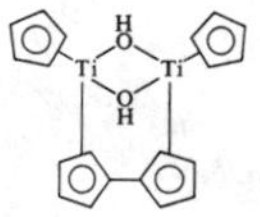

Ti-00106

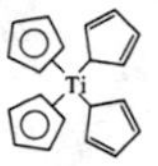

Ti-00107

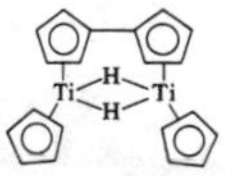

Ti-00108

As Ti-00045 with
X = —CPh=C$(CH_3)_2$

Ti-00109

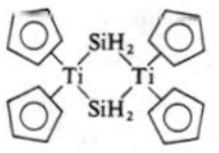

Ti-00110

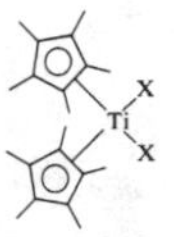

X = Cl

Ti-00111

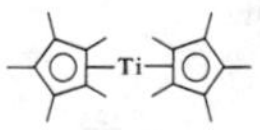

Ti-00112

$Ti[CH_2C(CH_3)_3]_4$

Ti-00113

$Ti[CH(SiMe_3)_2]_3$

Ti-00114

As Ti-00024 with
X = C_6F_5

Ti-00115

As Ti-00024 with
X = OPh

Ti-00116

As Ti-00024 with
X = —SPh

Ti-00117

As Ti-00024 with
X = Ph

Ti-00118

As Ti-00111 with
X = Me

Ti-00119

Ti-00120

As Ti-00045 with
X = —N=CPh_2

Ti-00121

As Ti-00007 with
R = OPh

Ti-00122

Ti-00123

$TiPh_4$

Ti-00124

As Ti-00024 with
X = CH_2Ph

Ti-00125

Ti-00126

Ti-00127

Ti-00128

As Ti-00024 with
X = —C≡CPh

Ti-00129

Ti-00130

Ti-00131

Ti-00132

THF complex

Ti-00133

Ti-00134

THF complex

Ti-00135

$Ti(CH_2Ph)_4$

Ti-00136

Ti-00137

Ti-00138

Ti-00139

Ti-00140

Ti-00141

Ti-00142

Ti-00143

Ti-00144

R = H

Ti-00145

Ti-00146

As Ti-00145 with
R = CH_3

Ti-00147

$[(H_3C)_2C{=}CPh]_4Ti$

Ti-00148

Ti-00149

Ti-00150

Ti-00151

Ti-00152

Ti-00153

CH_3Br_3Ti Ti-00001

Tribromomethyltitanium, 9CI

[30043-33-5]

$MeTiBr_3$

M 302.627

Dark-violet cryst. melting to yellow liq., dec. slowly at r.t., extremely air-sensitive. Sol. hydrocarbons, Et_2O, THF. Mp 2-3°. Forms a complex with bipyridine.

Thiele, K.H. *et al*, *Z. Anorg. Allg. Chem.*, 1968, **356**, 195; 1970, **378**, 62 (*synth*)
Clark, R.J.H. *et al*, *J. Chem. Soc., Dalton Trans.*, 1972, 2454 (*nmr*)

CH_3Cl_3Ti Ti-00002

Trichloromethyltitanium, 9CI

[2747-38-8]

$MeTiCl_3$

M 169.274

Monomeric in soln. Violet cryst. Mp 28.5°.

Beerman, C. *et al*, *Angew. Chem.*, 1959, **71**, 618 (*synth*)
Karavinka, G.L. *et al*, *J. Polym. Sci.*, 1961, **50**, 143 (*synth, ir*)
Dijkgraaf, C. *et al*, *Spectrochim. Acta, Part A*, 1969, **25**, 1455 (*uv*)
Hawlan, J.F. *et al*, *Can. J. Chem.*, 1972, **50**, 747 (*nmr*)
Inorg. Synth., 1976, **16**, 120 (*synth*)
Basso-Bert, M. *et al*, *J. Organomet. Chem.*, 1977, **136**, 201 (*uv*)

$C_2H_5Cl_3Ti$ Ti-00003

Trichloroethyltitanium, 9CI

[4984-20-7]

$EtTiCl_3$

M 183.301

Violet solid.

Thiele, K.H. *et al*, *Z. Anorg. Allg. Chem.*, 1968, **356**, 195 (*synth*)
Hanlaw, J.F. *et al*, *Can. J. Chem.*, 1972, **50**, 747 (*synth, nmr*)

$C_4H_{11}Cl_3SiTi$ Ti-00004

Trichloro(trimethylsilylmethyl)titanium, 9CI

[54943-97-4]

$Me_3SiCH_2TiCl_3$

M 241.455

Red-orange liq.

Beilin, S.I. *et al*, *Dokl. Akad. Nauk SSSR, Ser. Sci. Khim. (Engl. Transl.)*, 1974, **218**, 718 (*synth, ir, nmr*)
Nametkin, N.S. *et al*, *Bull. Acad. Sci. USSR, Div. Chem. Sci.*, 1974, 2767 (*ms*)
Sonnek, G. *et al*, *Z. Anorg. Allg. Chem.*, 1976, **426**, 232 (*synth*)

$C_4H_{12}O_4Ti$ Ti-00005

Tetramethyl titanate

Titanium(IV) methoxide. Methyl titanate(IV). Methyl orthotitanate

$Ti(OMe)_4$

M 172.016

Cocatalyst with organo-Al compds. for polymerisations. Cryst. or liq. Mp 209-10° (tetramer).

B.P., 512 452, (*1941*); *CA*, **35**, 463 (*synth*)
Kriegsmann, H. *et al*, *Z. Elektrochem.*, 1958, **62**, 1163 (*ir, raman*)
Deluzarche, A., *Ann. Chim. (Paris)*, 1961, **6**, 661; *CA*, **56**, 12531 (*synth*)
Adams, R.W. *et al*, *Aust. J. Chem.*, 1966, **19**, 207 (*struct*)
Wright, D.A. *et al*, *Acta Crystallogr., Sect. B*, 1968, **24**, 1108 (*cryst struct*)

$C_4H_{12}Ti$ Ti-00006

Tetramethyltitanium, 10CI

Titanium tetramethyl

[2371-70-2]

$TiMe_4$

M 108.019

Yellow cryst. (<−78°) (hexane). Sol. pet. ether. Dec. autocatalytically above −78°, more stable in Et_2O. Adducts substantially more stable.

▷Dioxan adduct explodes at 0°

Clauss, K. *et al*, *Angew. Chem.*, 1959, **71**, 627 (*synth, props*)
Berthold, H.J. *et al*, *Z. Anorg. Allg. Chem.*, 1963, **319**, 230 (*synth*)
Eysel, H.H. *et al*, *Spectrochim. Acta, Part A*, 1970, **26**, 1595 (*ir*)
Tabacchi, R. *et al*, *Helv. Chim. Acta*, 1970, **53**, 1977 (*nmr*)

$C_5H_5Br_3Ti$ Ti-00007

Tribromo(η^5-2,4-cyclopentadien-1-yl)titanium, 10CI

Cyclopentadienyltitanium tribromide

[12240-42-5]

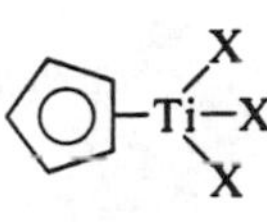

X = Br

M 352.687

Orange solid. Mp 197-9°. Readily hydrol.

Sloan, C.L. *et al*, *J. Am. Chem. Soc.*, 1959, **81**, 1364 (*synth, ir*)
Ronova, I.A. *et al*, *Dokl. Chem. (Engl. Transl.)*, 1969, **185**, 334 (*ed*)
Semin, G.K. *et al*, *Dokl. Chem. (Engl. Transl.)*, 1970, **194**, 635 (*nqr*)
Nesmayanov, A.N. *et al*, *Dokl. Chem. (Engl. Transl.)*, 1972, **205**, 632 (*cmr*)
Samuel, E. *et al*, *Inorg. Chem.*, 1973, **12**, 881 (*ir*)
Cardoso, A.M. *et al*, *J. Chem. Soc., Dalton Trans.*, 1980, 1156 (*synth*)

$C_5H_5Cl_2Ti$ Ti-00008

Dichloro(η^5-2,4-cyclopentadien-1-yl)titanium, 10CI

Cyclopentadienyltitanium dichloride

[31781-62-1]

$TiCl_2$

M 183.881

Degree of association of unsolvated material uncertain. THF complex shown to be monomeric in soln. Cocatalyst for ethylene polymerisation. Purple solid. Sol. polar org. solvs., insol. C_6H_6, Ph_2O, heptane.

THF complex:

$C_9H_{13}Cl_2OTi$ M 255.987

Blue.

Bartlett, P.D. *et al*, *J. Am. Chem. Soc.*, 1961, **83**, 581 (*synth, uv, esr, nmr, use*)
Coutts, R.S.P. *et al*, *Aust. J. Chem.*, 1971, **24**, 1079, 2533 (*synth, nmr, ir, uv*)
Floriani, C. *et al*, *J. Chem. Soc., Dalton Trans.*, 1973, 1954 (*synth*)

$C_5H_5Cl_3Ti$ Ti-00009

Trichloro(η^5-2,4-cyclopentadien-1-yl)titanium, 10CI

Cyclopentadienyltitanium trichloride

[1270-98-0]

As Tribromo(η^5-2,4-cyclopentadien-1-yl)titanium, Ti-00007 with

X = Cl

M 219.334

Orange-yellow cryst. Mp 208-11°. Sensitive to hydrol.

Gorsich, R.D., *J. Am. Chem. Soc.*, 1958, **80**, 4744 (*synth, ir*)
Ganis, P. *et al*, *Atti. Accad. Nazl. Lincei Rend. Classe Sci. Fis. Mat. Nat.*, 1962, **33**, 303 (*cryst struct*)
Reid, A.F. *et al*, *Spectrochim. Acta*, 1964, **20**, 1257 (*ir*)
Nesmayanov, A.N. *et al*, *Dokl. Akad. Nauk SSSR, Ser. Sci. Khim.* (*Engl. Transl.*), 1972, **205**, 632 (*cmr*)
Glivicky, A. *et al*, *Can. J. Chem.*, 1973, **51**, 2609 (*pmr*)
Inorg. Synth., 1976, **16**, 237 (*synth, ir*)
Cardoso, A.M. *et al*, *J. Chem. Soc., Dalton Trans.*, 1980, 1156 (*synth, ms, nmr*)
Chandra, K. *et al*, *Chem. Ind.* (*London*), 1980, 288 (*synth, nmr*)

$C_5H_5F_3Ti$ Ti-00010

(η^5-2,4-Cyclopentadien-1-yl)trifluorotitanium, 10CI

Cyclopentadienyltitanium trifluoride

[63166-76-7]

As Tribromo(η^5-2,4-cyclopentadien-1-yl)titanium, Ti-00007 with

X = F

M 169.970

Yellow cryst. Spar. sol. org. solvs. (less sol. than other $C_5H_5TiX_3$ halides). Mp 140° dec. Readily hydrol.

Nesmayanov, A.N. *et al*, *Dokl. Chem.* (*Engl. Transl.*), 1968, **182**, 874 (*ir*)
Nesmayanov, A.N. *et al*, *Izv. Akad. Nauk SSSR, Ser. Khim.*, 1968, 514 (*synth, ir*)

$C_5H_5I_3Ti$ Ti-00011

(η^5-Cyclopentadien-1-yl)triiodotitanium, 10CI

Cyclopentadienyltitanium triiodide

[12240-43-6]

As Tribromo(η^5-2,4-cyclopentadien-1-yl)titanium, Ti-00007 with

X = I

M 493.688

Dark-red cryst. Mp 148-50°. Hydrolytically sensitive.

Sloan, C.L. *et al*, *J. Am. Chem. Soc.*, 1959, **81**, 1364 (*synth, ir*)
Nesmayanov, A.N. *et al*, *Izv. Akad. Nauk SSSR, Ser. Khim.*, 1968, 514 (*synth, ir*)
Bryukhova, E.V. *et al*, *J. Organomet. Chem.*, 1974, **81**, 195 (*nqr*)
Cardoso, A.M. *et al*, *J. Chem. Soc., Dalton Trans.*, 1980, 1156 (*synth*)

$C_6Cl_3F_5Ti$ Ti-00012

Trichloro(pentafluorophenyl)titanium, 9CI

[37759-22-1]

$(C_6F_5)TiCl_3$

M 321.297

Yellow solid. Sol. polar solvs. including H_2O, dioxan, THF. Mp 114-8°. $Bp_{0.0001}$ 100° subl. Forms a bispyridine adduct. Ti—C bond stable to hydrolysis by acid or alkali.

Lahuerta Pena, P. *et al*, *Rev. Acad. Cieac. Exactas, Fis. Quim. Natur. Zaragoza*, 1972, **27**, 75; *CA*, **77**, 164813 (*synth*)
Razuvaev, G.A. *et al*, *Dokl. Chem.*, 1972, **203**, 220 (*synth*)

$C_6H_{18}N_2Ti$ Ti-00013

Bis(dimethylamido)dimethyltitanium

Dimethylbis(N-*methylmethanaminato*)*titanium*, 9CI

[59512-77-5]

$Me_2Ti(NMe_2)_2$

M 166.102

Yellow solid. Dec. at −35°.

Bürger, H. *et al*, *J. Organomet. Chem.*, 1976, **108**, 69 (*synth, ir, nmr*)

$C_7H_{21}N_3Ti$ Ti-00014

Tris(dimethylaminato)methyltitanium, 8CI

Tris(*dimethylamido*)*methyltitanium*

[25483-53-8]

$MeTi(NMe_2)_3$

M 195.143

Yellow liq. Mp 8-9°. $Bp_{0.001}$ 20° subl. Dec. at 80°. $TiCD_3$ analogue also prepd. (Mp reported 9-10°).

Bürger, H. *et al*, *J. Organomet. Chem.*, 1969, **20**, 129 (*synth, ir, nmr, raman*)
Bürger, H. *et al*, *J. Organomet. Chem.*, 1970, **21**, 381 (*uv*)

$C_8H_5N_3S_3Ti$ Ti-00015

(η^5-Cyclopentadienyl)trithiocyanatotitanium, 8CI

As Tribromo(η^5-2,4-cyclopentadien-1-yl)titanium, Ti-00007 with

X = NCS

M 287.208

Red solid. Mp 162-5°. $Bp_{0.001}$ 150° subl.

Köpf, H. *et al*, *Z. Naturforsch., B*, 1968, **23**, 1534 (*synth, ir*)

$C_8H_{14}O_3Ti$ Ti-00016

(η^5-Cyclopentadienyl)trimethoxytitanium, 8CI

[12145-64-1]

As Tribromo(η^5-2,4-cyclopentadien-1-yl)titanium, Ti-00007 with

R = OMe

M 206.077

Monomeric in C_6H_6. White solid. Mp 50-2°. $Bp_{1.5}$ 88°.

Nesmayanov, A.N. *et al*, *Izv. Akad. Nauk SSSR, Ser. Khim.*, 1967, 808 (*synth, props*)

$C_8H_{14}Ti$ Ti-00017

(η^5-2,4-Cyclopentadien-1-yl)trimethyltitanium, 9CI

[38386-55-9]

As Tribromo(η^5-2,4-cyclopentadien-1-yl)titanium, Ti-00007 with

X = Me

M 158.079
Yellow cryst. Mp ca. 20° dec.

Giannini, U. *et al*, *Tetrahedron Lett.*, 1960, 19 (*synth*)
Clauss, K. *et al*, *Justus Liebigs Ann. Chem.*, 1962, **654**, 8 (*synth*)
Samuel, E. *et al*, *Inorg. Chem.*, 1973, **12**, 881 (*raman, nmr*)
Green, M.L.H. *et al*, *J. Organomet. Chem.*, 1974, **73**, 259 (*ir, nmr*)

$C_8H_{24}N_4Ti$ Ti-00018

Tetrakis(dimethylamino)titanium

$Ti(NMe_2)_4$

M 224.184
Reagent for conversion of hindered ketones to enamines. Yellow liq. $Bp_{0.05}$ 50°.

Bradley, D.C. *et al*, *J. Chem. Soc.*, 1960, 3857 (*synth*)
Buerger, H. *et al*, *Inorg. Nucl. Chem. Lett.*, 1966, **3**, 209 (*ir*)
Fieser, M. *et al*, *Reagents for Organic Synthesis*, Wiley, 1967-83, **2**, 399.

$C_8H_{32}B_{20}Ti^{\ominus\ominus}$ Ti-00019

Bis[η^6-decahydro-*C,C'*-dimethyldicarbadodecaborato(2−)]-titanate(2−), 9CI
*4,4'-Bis(decahydro-1,6-dimethyl-1,6-dicarba-4-titana-*closo*-tridecaborate)(2−)*

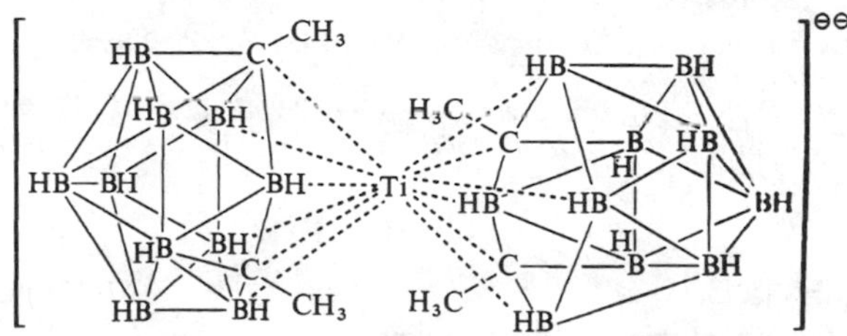

M 392.421 (ion)
Bis(tetramethylammonium) salt: [55266-92-7].
$C_{16}H_{56}B_{20}N_2Ti$ M 540.712
Red cryst. + $2Me_2CO$ (Me_2CO/EtOH), air-stable in solid state.

Salentine, C.G. *et al*, *J. Am. Chem. Soc.*, 1975, **97**, 426 (*synth, pmr, nmr, props*)
Yo, F.Y. *et al*, *J. Am. Chem. Soc.*, 1975, **97**, 428 (*cryst struct*)

$C_9H_{11}Cl_3Ti$ Ti-00020

Trichloro(η^5-tetrahydroindenyl)titanium
(η^5-Tetrahydroindenyl)titanium trichloride

M 273.425
Characterised by hydrol. prods.

Samuel, E., *J. Organomet. Chem.*, 1969, **19**, 87 (*synth, ir, nmr*)

$C_9H_{15}O_2Ti$ Ti-00021

Cyclopentadienyldiethoxytitanium

M 203.096
Dimeric.
Dimer: Bis(η^5-2,4-cyclopentadien-1-yl)diethoxy-di-μ-ethoxydititanium. Dicyclopentadienyldi-μ-ethoxy-diethoxydititanium. Bis[(η-cyclopentadienyl)ethoxy(μ-ethoxy)titanium]. Red-brown liq.

Lappert, M.F. *et al*, *J. Chem. Soc., Dalton Trans.*, 1971, 1314 (*synth, nmr*)

$C_9H_{17}N_2Ti$ Ti-00022

Cyclopentadienylbis(dimethylamido)titanium

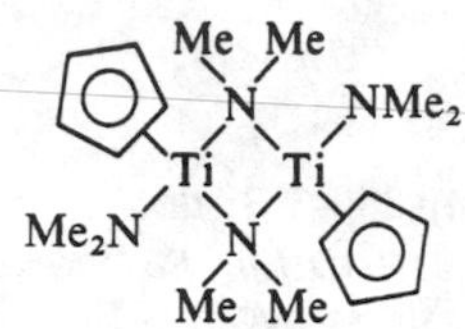

M 201.127
Dimeric.
Dimer: [11067-83-7]. *Di-π-cyclopentadienylbis[μ(dimethylaminato)]bis(dimethylaminato)dititanium,* 8CI. *Dicyclopentadienyltetrakis(dimethylaminato)dititanium. Bis[(η-cyclopentadienyl)(dimethylamido)(μ-dimethylamido)titanium].* Red-brown oil. Not fully characterised but related compds. were characterised analytically and spectroscopically.

Alyea, E.C. *et al*, *J. Chem. Soc., Chem. Commun.*, 1969, 1064 (*synth, nmr*)
Lappert, M.F. *et al*, *J. Chem. Soc., Dalton Trans.*, 1971, 874, 1314 (*synth, props*)

$C_{10}H_{10}BrTi$ Ti-00023

Bromobis(cyclopentadienyl)titanium
Dicyclopentadienyltitanium bromide

X = Br

M 257.973
Dimeric in C_6H_6.
Dimer: [39333-86-3]. *Di-μ-bromotetrakis(η^5-2,4-cyclopentadien-1-yl)dititanium,* 9CI. *Bis[μ-bromobis(η^5-cyclopentadienyl)titanium].*
$C_{20}H_{20}Br_2Ti_2$ M 515.946
Red-brown solid. Sol. H_2O, org. solvs. Subl. *in vacuo*.

Gubin, S.P. *et al*, *J. Organomet. Chem.*, 1969, **20**, 229.
Coutts, R.S.P. *et al*, *Aust. J. Chem.*, 1973, **26**, 2101.
Coutts, R.S.P. *et al*, *J. Organomet. Chem.*, 1973, **47**, 375 (*synth, uv, ir*)

$C_{10}H_{10}Br_2Ti$ Ti-00024

Dibromobis(η^5-2,4-cyclopentadien-1-yl)titanium, 10CI

Dicyclopentadienyltitanium dibromide. Titanocene dibromide

[1293-73-8]

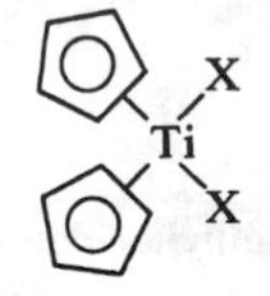

X = Br

M 337.877

Dark-red cryst. Mp 290-313° dec.

Wilkinson, G. *et al*, *J. Am. Chem. Soc.*, 1953, **75**, 1011 (*synth*)
Druce, P.M. *et al*, *J. Chem. Soc. (A)*, 1969, 2106 (*synth, ir, ms, nmr*)
Druce, P.M. *et al*, *J. Chem. Soc. (A)*, 1969, 2814 (*ir*)
Bryukhova, E.V. *et al*, *J. Organomet. Chem.*, 1974, **81**, 195 (*nqr*)
Nesmayanov, A.N. *et al*, *J. Struct. Chem. (Engl. Transl.)*, 1975, **16**, 705 (*cmr*)
McDermott, J.X. *et al*, *J. Am. Chem. Soc.*, 1976, **98**, 6529 (*nmr*)

$C_{10}H_{10}ClTi$ Ti-00025

Chlorobis(cyclopentadienyl)titanium

Dicyclopentadienyltitanium chloride

As Bromobis(cyclopentadienyl)titanium, Ti-00023 with

X = Cl

M 213.522

Dimeric.

Dimer: [1271-18-7]. *Di-μ-chlorotetrakis(η^5-2,4-cyclopentadien-1-yl)dititanium. Bis[bis(η-cyclopentadienyl)-μ-chlorotitanium].*
$C_{20}H_{20}Cl_2Ti_2$ M 427.044
Starting material for Ti(III) cyclopentadienyl complexes. Green cryst. Sol. H_2O, org. solvs. Mp 285-8°. Subl. *in vacuo*. One ref. reports as moisture-sensitive.

Green, M.L.H. *et al*, *J. Chem. Soc., Dalton Trans.*, 1972, 1000 (*synth*)
Chivers, T. *et al*, *Can. J. Chem.*, 1973, **51**, 815.
Coutts, R.S.P. *et al*, *J. Organomet. Chem.*, 1973, **47**, 375 (*synth, ir, uv*)
Wailes, P.C. *et al*, unpublished results cited in *Organic Chemistry of Titanium, Zirconium, and Hafnium*, Academic Press, London, 1974, 207 (*synth*)
Manzer, L.E., *J. Organomet. Chem.*, 1976, **110**, 291 (*synth*)
Jungst, R. *et al*, *Inorg. Chem.*, 1977, **16**, 1645 (*cryst struct*)

$C_{10}H_{10}Cl_2Ti$ Ti-00026

Dichlorobis(η^5-2,4-cyclopentadien-1-yl)titanium, 9CI

Bis(cyclopentadienyl)titanium dichloride. Titanocene dichloride

[1271-19-8]

As Dibromobis(η^5-2,4-cyclopentadien-1-yl)titanium, Ti-00024 with

X = Cl

M 248.975

Cocatalyst for polymerisation. Bright-red acicular cryst. (toluene). Mp 289°.

▷XR2050000.

Wilkinson, G. *et al*, *J. Am. Chem. Soc.*, 1954, **76**, 4281 (*synth*)
Coutts, R.S.P. *et al*, *Adv. Organomet. Chem.*, 1970, **9**, 136 (*rev*)
Siegert, F.W. *et al*, *J. Organomet. Chem.*, 1970, **23**, 177 (*synth, pmr, ms*)
Balutina, O.A. *et al*, *Tr. Khim. Khim. Tekhnol.*, 1974, 119; *CA*, **83**, 170279 (*ir*)
Clearfield, A. *et al*, *Can. J. Chem.*, 1975, **53**, 1622 (*cryst struct*)
Martin, M.L. *et al*, *J. Organomet. Chem.*, 1975, **97**, 261 (*pmr, cmr*)
Fieser, M. *et al*, *Reagents for Organic Synthesis*, Wiley, 1967-83, **6**, 48.

$C_{10}H_{10}Cl_4OTi_2$ Ti-00027

Tetrachlorobis(η^5-2,4-cyclopentadien-1-yl)-μ-oxodititanium, 9CI

Bis[dichloro(η^5-cyclopentadienyl)titanium]oxide. Oxybis[dichlorocyclopentadienyltitanium]

[12146-06-4]

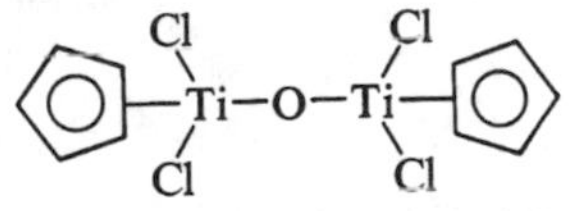

M 383.760

Obt. by hydrol., but is itself hydrol. to monochloro oligomers. Orange cryst. (CCl_4). Spar. sol. org. solvs. Mp 149-51°.

Corradini, P. *et al*, *J. Am. Chem. Soc.*, 1959, **81**, 5510 (*cryst struct*)
Gorsich, R.D., *J. Am. Chem. Soc.*, 1960, **82**, 4211 (*synth*)
Thewalt, U. *et al*, *J. Organomet. Chem.*, 1977, **127**, 169 (*cryst struct*)

$C_{10}H_{10}FTi$ Ti-00028

Bis(cyclopentadienyl)fluorotitanium

Dicyclopentadienyltitanium fluoride

As Bromobis(cyclopentadienyl)titanium, Ti-00023 with

X = F

M 197.067

Dimeric in C_6H_6.

Dimer: [39333-87-4]. *Tetrakis(η^5-2,4-cyclopentadien-1-yl)di-μ-fluorodititanium,* 9CI. *Bis[bis(η^5-cyclopentadienyl)-μ-fluorotitanium].*
$C_{20}H_{20}F_2Ti_2$ M 394.135
Green cryst. by subl. Sol. org. solvs., spar. sol. H_2O.

Gubin, S.P. *et al*, *J. Organomet. Chem.*, 1969, **20**, 229.
Coutts, R.S.P. *et al*, *Aust. J. Chem.*, 1973, **26**, 2101.
Coutts, R.S.P. *et al*, *J. Organomet. Chem.*, 1973, **47**, 375 (*synth, ir, uv*)

$C_{10}H_{10}F_2Ti$ Ti-00029

Bis(η^5-2,4-cyclopentadien-1-yl)difluorotitanium, 10CI

Dicyclopentadienyltitanium difluoride. Titanocene difluoride

[309-89-7]

As Dibromobis(η^5-2,4-cyclopentadien-1-yl)titanium, Ti-00024 with

X = F

M 216.066

Tumour inhibitor. Yellow cryst. Mp 238-80° dec.

Samuel, E., *Bull. Soc. Chim. Fr.*, 1966, 3548 (*synth*)
Druce, P.M. *et al*, *J. Chem. Soc. (A)*, 1969, 2106 (*synth, ir, ms, nmr*)
Dias, A.D., *Rev. Port. Quin.*, 1971, **13**, 222; *CA*, **77**, 120222 (*uv*)

White, D.A., *J. Inorg. Nucl. Chem.*, 1971, **33**, 691 (*pmr*)
Ger. Pat., 2 923 334, (*1979*); *CA*, **94**, 150572 (*use*)

$C_{10}H_{10}ITi$ Ti-00030

Bis(cyclopentadienyl)iodotitanium

Dicyclopentadienyltitanium iodide

As Bromobis(cyclopentadienyl)titanium, Ti-00023 with

X = I

M 304.974

Dimeric in C_6H_6.

Dimer: [39333-90-9]. *Tetrakis(η^5-2,4-cyclopentadien-1-yl)di-μ-iododititanium*, 9CI. *Bis[bis(η^5-cyclopentadienyl)-μ-iodotitanium]*.

$C_{20}H_{20}I_2Ti_2$ M 609.947

Red-brown solid. Sol. H_2O, org. solvs. Shows anomalous magnetic behaviour.

Gubin, S.P. *et al*, *J. Organomet. Chem.*, 1969, **20**, 229.
Coutts, R.S.P. *et al*, *Aust. J. Chem.*, 1973, **26**, 2101.
Coutts, R.S.P. *et al*, *J. Organomet. Chem.*, 1973, **47**, 375 (*synth, uv, ir*)

$C_{10}H_{10}I_2Ti$ Ti-00031

Bis(η^5-2,4-cyclopentadien-1-yl)diiodotitanium, 10CI

Dicyclopentadienyltitanium diiodide. Titanocene diiodide

[12152-92-0]

As Dibromobis(η^5-2,4-cyclopentadien-1-yl)titanium, Ti-00024 with

X = I

M 431.878

Black cryst. Mp 315-7°.

Chien, J.C.W., *J. Phys. Chem.*, 1963, **67**, 2477 (*uv*)
Druce, P.M. *et al*, *J. Chem. Soc.* (*A*), 1969, 2106, 2814 (*synth, ir, ms, nmr*)
Gubin, S.P. *et al*, *J. Organomet. Chem.*, 1969, **20**, 229.
Bryukhova, E.V. *et al*, *J. Organomet. Chem.*, 1974, **81**, 195 (*nqr*)
Nesmayanov, A.N. *et al*, *J. Struct. Chem.* (*Engl. Transl.*), 1975, **16**, 705 (*pmr, cmr*)

$C_{10}H_{10}N_6Ti$ Ti-00032

Diazidobis(η^5-2,4-cyclopentadien-1-yl)titanium, 10CI

Titanocene diazide

[1298-37-9]

As Dibromobis(η^5-2,4-cyclopentadien-1-yl)titanium, Ti-00024 with

$X = N_3$

M 262.109

Orange cryst. Mp 145-6°. $Bp_{0.001}$ 200° subl.

Langford, C.H. *et al*, *J. Organomet. Chem.*, 1965, **4**, 271 (*synth, uv*)
Coutts, R.S.P. *et al*, *Aust. J. Chem.*, 1971, **24**, 1075 (*synth, ir, ms, nmr*)
de Gil, E.R. *et al*, *Acta Crystallogr., Sect. B*, 1976, **32**, 909 (*cryst struct*)

$C_{10}H_{10}S_5Ti$ Ti-00033

Bis(η^5-2,4-cyclopentadien-1-yl)(pentathio)titanium, 9CI

Bis(η^5-cyclopentadienyl)cyclo(pentasulfur-1,5-diyl)titanium

[12116-82-4]

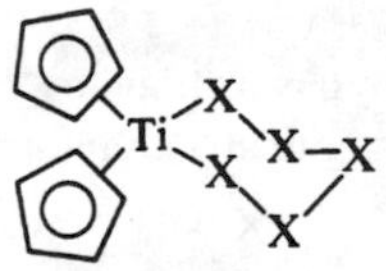

X = S

M 338.369

Intermed. in synth. of S_n (n = 7, 9, 20, etc.) and other ring-systems. Dark purple cryst. Mp 194-8°. Below 120°, C_5H_5 rings nonequivalent owing to S_5 ring conformation. E_{act} (ring inversion) 69.1 kJ mol.$^{-1}$.

Köpf, H. *et al*, *Chem. Ber.*, 1969, **102**, 1504 (*synth*)
Epstein, E.F. *et al*, *J. Organomet. Chem.*, 1971, **26**, 229 (*cryst struct*)
Muller, E.G. *et al*, *J. Organomet. Chem.*, 1976, **111**, 91 (*ir, nmr, cryst struct*)
Abel, E.W. *et al*, *J. Organomet. Chem.*, 1978, **160**, 75 (*nmr*)

$C_{10}H_{10}Se_5Ti$ Ti-00034

Bis(η^5-2,4-cyclopentadien-1-yl)[pentaselenidato(2−)]titanium, 9CI

Bis(η^5-cyclopentadienyl)cyclo-1,5-pentaselenium-1,5-diyl)titanium

[12307-22-1]

As Bis(η^5-2,4-cyclopentadien-1-yl)(pentathio)titanium, Ti-00033 with

X = Se

M 572.869

Monomeric in C_6H_6. Air-stable violet leaflets (C_6H_6). Insol. H_2O, sol. Et_2O, C_6H_6, sl. sol. CS_2. Mp 211°. C_5H_5 rings nonequivalent magnetically (Se_5 ring conformation).

Köpf, H. *et al*, *Chem. Ber.*, 1968, **101**, 272 (*synth, ir, uv, nmr*)

$C_{10}H_{10}Ti$ Ti-00035

Titanocene

Bis(η^5-2,4-cyclopentadien-1-yl)titanium, 9CI. *Dicyclopentadienyltitanium*

[1271-29-0]

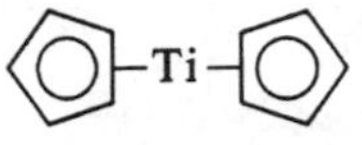

M 178.069

Identity doubtful, except possibly as a reactive intermediate. Black solid. Struct. not established.

▷XR2075000.

Rausch, M.D. *et al*, *J. Organomet. Chem.*, 1978, **160**, 81.
Pez, G.P. *et al*, *Adv. Organomet. Chem.*, 1981, **21**, 1 (*rev*)

$C_{10}H_{11}Cl_3Ti$ Ti-00036

Trichloro(2-methyl-1-phenyl-1-propenyl)titanium, 10CI

Trichloro(1-phenyl-2,2-dimethylethenyl)titanium

[72348-00-6]

$(H_3C)_2C{=}CPhTiCl_3$

M 285.436
Material characterized spectroscopically.

Cardin, D.J. *et al*, *J. Chem. Soc.*, *Chem. Commun.*, 1979, 513 (*synth*, *nmr*)

$C_{10}H_{11}Ti$ Ti-00037

Bis(η^5-cyclopentadienyl)hydrotitanium

As Bromobis(cyclopentadienyl)titanium, Ti-00023 with

X = H

M 179.077
Dimeric.

Dimer: [11078-10-7]. *Tetra-π-cyclopentadienyldi-μ-hydrodititanium*, *8CI*. *Di-π-cyclopentadienyldi-2,4-cyclopentadien-1-yldi-μ-hydrodititanium*, *8CI*. *Bis*[*bis*(η^5-*cyclopentadienyl*)-*μ-hydrotitanium*]. $C_{20}H_{22}Ti_2$ M 358.154
Violet solid. On standing isomerises to polymer. With H_2 gives a fulvalene hydrido dimer.

Bercaw, J.E. *et al*, *J. Am. Chem. Soc.*, 1969, **91**, 7301 (*synth*, *ir*, *esr*)

$C_{10}H_{14}BTi$ Ti-00038

Bis(η^5-2,4-cyclopentadien-1-yl)[tetrahydroborato(1−)*H*,*H'*-]titanium, 10CI

Bis(η^5-*cyclopentadienyl*)*titanium tetrahydridoborate*. *Titanocene borohydride*

[12772-20-2]

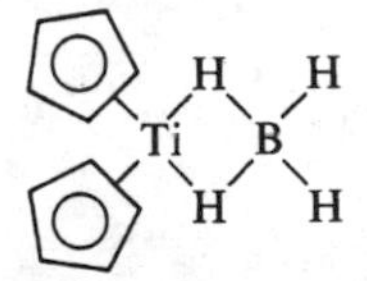

M 192.911
Purple needles. Subl. *in vacuo*.

Nöth, H. *et al*, *Naturwissenschaften*, 1960, **47**, 57 (*esr*)
Davies, N. *et al*, *J. Chem. Soc.* (*A*), 1969, 2601 (*ir*)
Marks, T.J. *et al*, *Inorg. Chem.*, 1972, **11**, 2540 (*synth*, *ir*)
Melmed, K.M. *et al*, *Inorg. Chem.*, 1973, **12**, 232 (*cryst struct*)
Inorg. Synth., 1977, **17**, 91 (*synth*)

$C_{10}H_{15}Cl_2NS_2Ti$ Ti-00039

Dichloro(η^5-2,4-cyclopentadien-1-yl)(diethylcarbamodithioato-*S*,*S'*)titanium, 9CI

Dichloro(*η-cyclopentadienyl*)N,N-*diethyldithiocarbamatotitanium*

[53079-53-1]

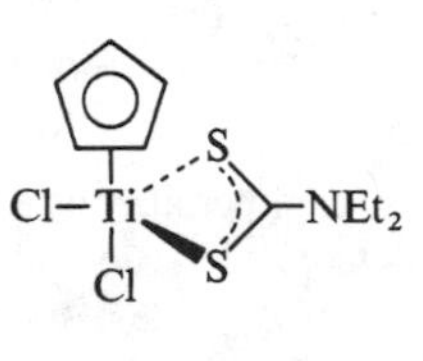

M 332.141
Red solid. Mp 181-4°. Stable to hydrolysis unlike most $(C_5H_5)Ti(IV)$ dihalides.

Coutts, R.S.P. *et al*, *J. Organomet. Chem.*, 1975, **84**, 47 (*synth*, *ir*, *nmr*)

$C_{10}H_{15}Cl_3Ti$ Ti-00040

Trichloro[(1,2,3,4,5-η)-1,2,3,4,5-pentamethyl-2,4-cyclopentadien-1-yl]titanium, 10CI

(*Pentamethylcyclopentadienyl*)*titanium trichloride*

[70756-51-3]

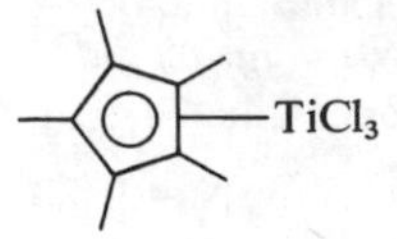

M 289.468
Red solid. Mp 225-7°.

King, R.B. *et al*, *J. Organomet. Chem.*, 1967, **8**, 287 (*synth*, *ir*, *pmr*)
Nesmeyanov, A.N. *et al*, *Izv. Akad. Nauk SSSR*, *Ser. Khim.*, 1968, 514 (*ir*)
Gubin, S.P. *et al*, *J. Organomet. Chem.*, 1969, **20**, 229 (*polarog*)
Nesmayanov, A.N. *et al*, *Dokl. Chem.* (*Engl. Transl.*), 1972, **205**, 632 (*cmr*)

$C_{10}H_{18}O_2Ti$ Ti-00041

(η^5-2,4-Cyclopentadien-1-yl)diethoxymethyltitanium, 10CI

[53695-79-7]

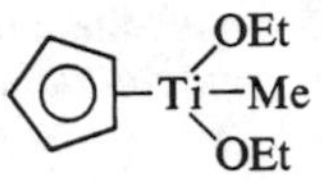

M 218.131
Yellow liq. d_4^{20} 1.059. $Bp_{0.05}$ 32°. Forms complex with Me_2AlOEt.

Blandy, C. *et al*, *J. Organomet. Chem.*, 1977, **128**, 415 (*synth*, *ir*, *pmr*, *cmr*, *props*)

$C_{10}H_{24}O_3Ti$ Ti-00042

Methyltris(2-propanolato)titanium, 10CI

Methyltriisopropoxytitanium. *Methyltris*(*isopropoxy*)-*titanium*

[18006-13-8]

$MeTi[OCH(CH_3)_2]_3$

M 240.178
Yellow liq. Mp 10°. $Bp_{0.018}$ 50°.

Clauss, K., *Justus Liebigs Ann. Chem.*, 1968, **711**, 19 (*synth*)
Dijkgraaf, C. *et al*, *Spectrochem. Acta*, *Part A*, 1969, **25**, 1455 (*synth*)
Blandy, C. *et al*, *C.R. Hebd. Seances Acad. Sci.*, 1974, **278**, 1323 (*nmr*)
Rausch, M.D. *et al*, *J. Organomet. Chem.*, 1974, **74**, 85 (*synth*, *nmr*)
Blandy, C. *et al*, *Synth. React. Inorg. Metal-Org. Chem.*, 1978, **8**, 331 (*nmr*)

$C_{10}H_{29}N_3SiTi$ Ti-00043

Tris(dimethylamido)(trimethylsilylmethyl)titanium

Tris(N-*methylmethanaminato*)[(*trimethylsilyl*)*methyl*]-*titanium*, *9CI*

[59512-75-3]

$Me_3SiCH_2Ti(NMe_2)_3$

M 267.325
Liq. Mp <0°. $Bp_{0.001}$ 50°.

Bürger, H. *et al*, *J. Organomet. Chem.*, 1976, **108**, 69 (*synth*, *ir*, *nmr*)

$C_{11}H_{10}Cl_2N_2Ti$ Ti-00044

Dichloro(η^5-2,4-cyclopentadien-1-yl)(phenylazo)titanium, 10CI

Dichloro(η-cyclopentadienyl)phenyldiazenidotitanium

[68182-14-9]

M 288.999

Orange cryst. (Et_2O).

Dilworth, J.R. *et al*, *J. Organomet. Chem.*, 1978, **159**, 47 (*synth, ir, nmr*)

$C_{11}H_{13}ClTi$ Ti-00045

Chlorobis(η^5-2,4-cyclopentadien-1-yl)methyltitanium, 10CI

Chloro(methyl)titanocene

[1278-83-7]

X = Me

M 228.557

Orange-yellow solid. Mp 168-70°. Cryst. phase transition at 91.6°.

Beachell, H.C. *et al*, *Inorg. Chem.*, 1965, **4**, 1133 (*synth, mr*)
Gubin, S.P. *et al*, *J. Organomet. Chem.*, 1969, **20**, 241 (*polarog*)
Waters, J.A. *et al*, *J. Organomet. Chem.*, 1970, **22**, 417 (*synth, ir, uv, nmr*)
Waters, J.A. *et al*, *J. Organomet. Chem.*, 1971, **33**, 41.

$C_{11}H_{13}Ti$ Ti-00046

Bis(η^5-2,4-cyclopentadien-1-yl)methyltitanium

[12288-70-9]

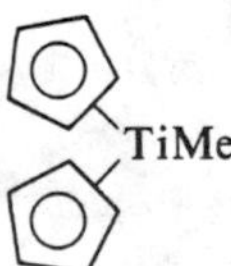

M 193.104

Complex prepd. *in situ*. Effects cyclometalation of pyridines and quinolines.

Klei, B. *et al*, *J. Chem. Soc., Chem. Commun.*, 1978, 659 (*synth, use*)
Klei, B. *et al*, *J. Organomet. Chem.*, 1980, **188**, 97 (*use*)
Fieser, M. *et al*, *Reagents for Organic Synthesis*, Wiley, 1967-83, **8**, 40 (*use*)

$C_{11}H_{14}O_6Ti$ Ti-00047

Tris(acetato-*O*)(η^5-2,4-cyclopentadien-1-yl)titanium, 9CI

Triacetato(η^5-cyclopentadienyl)titanium. (η^5-Cyclopentadienyl)titanium triacetate

[1282-42-4]

As Tribromo(η^5-2,4-cyclopentadien-1-yl)titanium, Ti-00007 with

X = OAc

M 290.108

Mp 115-7°.

Dvoryantseva, G.G. *et al*, *Dokl. Akad. Nauk SSSR, Ser. Sci. Khim.*, 1965, **161**, 603 (*synth, ir*)
Nesmeyanov, A.N. *et al*, *Tetrahedron* Suppl., 1966, **8**, pt. 2, 389 (*synth, nmr*)
Nesmayanov, A.N. *et al*, *Izv. Akad. Nauk SSSR, Ser. Khim.*, 1967, 808 (*synth*)
Nesmayanov, A.N. *et al*, *Izv. Akad. Nauk SSSR, Ser. Khim.*, 1971, 2729 (*nmr*)
Nesmayanov, A.N. *et al*, *J. Struct. Chem.* (*Engl. Transl.*), 1975, 705 (*cmr*)

$C_{11}H_{20}O_3Ti$ Ti-00048

(π-Cyclopentadienyl)triethoxytitanium, 8CI

[1282-41-3]

As Tribromo(η^5-2,4-cyclopentadien-1-yl)titanium, Ti-00007 with

X = OEt

M 248.157

Monomeric in C_6H_6. Colourless liq. Bp_3 106-7°.

Nesmayanov, A.N. *et al*, *Izv. Akad. Nauk SSSR, Ser. Khim.*, 1961, 743 (*synth*)
Nesmayanov, A.N. *et al*, *Dokl. Chem.* (*Engl. Transl.*), 1968, **182**, 874 (*ir*)
Nesmayanov, A.N. *et al*, *J. Struct. Chem.* (*Engl. Transl.*), 1972, **13**, 964 (*nmr*)

$C_{11}H_{23}N_3Ti$ Ti-00049

(η^5-Cyclopentadienyl)tris(dimethylamido)titanium

*(η^5-Cyclopentadien-1-yl)tris(*N*-methylmethanaminato)-titanium,* 9CI

[58057-99-1]

As Tribromo(η^5-2,4-cyclopentadien-1-yl)titanium, Ti-00007 with

X = NMe_2

M 245.203

Red oil or orange cryst. Mp 100°. $Bp_{0.01}$ 95°, $Bp_{0.001}$ 50°. Dec. ~140°.

Chandra, G. *et al*, *Inorg. Nucl. Chem. Lett.*, 1965, **1**, 83 (*synth*)
Chandra, G. *et al*, *J. Chem. Soc.* (*A*), 1968, 1940 (*synth*)
Bürger, H. *et al*, *J. Organomet. Chem.*, 1969, **20**, 129 (*synth, nmr*)
Bürger, H. *et al*, *J. Organomet. Chem.*, 1975, **101**, 295 (*ir, raman*)

$C_{12}H_{10}N_2O_2Ti$ Ti-00050

Dicyanatobis(η^5-cyclopentadienyl)titanium

*Bis(cyanato-*N*)bis(η^5-cyclopentadien-1-yl)titanium,* 9CI. *Biscyclopentadienyldicyanatotitanium*

[12109-61-4]

As Dibromobis(η^5-2,4-cyclopentadien-1-yl)titanium, Ti-00024 with

X = NCO

M 262.103

The ligating atom of the NCO group is uncertain (cf. 9CI name). Orange-red solid. Mp 275-7°. $Bp_{0.001}$ 160° subl.

Coutts, R.S.P. *et al*, *Aust. J. Chem.*, 1966, **19**, 2069 (*synth, nmr*)
Samuel, E., *Bull. Soc. Chim. Fr.*, 1966, 3548 (*synth, ir*)
Burmeister, J.L. *et al*, *Inorg. Chem.*, 1969, **8**, 170; 1970, **9**, 58 (*synth, ir, uv, ms*)
Anderson, S.J. *et al*, *J. Chem. Soc., Chem. Commun.*, 1974, 996 (*cryst struct*)

$C_{12}H_{10}O_2Ti$ Ti-00051

Dicarbonylbis(η^5-2,4-cyclopentadien-1-yl)titanium, 10CI
Dicarbonyldi-π-cyclopentadienyltitanium, 8CI. *Titanocene dicarbonyl*
[12129-51-0]

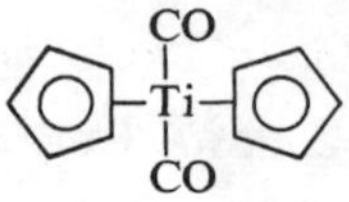

M 234.090
Monomeric in C_6H_6. Dark-red cryst. (toluene or by subl.). Sol. org. solvs. Reacts with CH_2Cl_2 and $CHCl_3$. Mp 90° dec. $Bp_{0.001}$ 40-80° subl. Readily oxidized.

Calderazzo, F. *et al*, *Inorg. Chim. Acta*, 1967, **1**, 65 (*ir*)
Fischer, E.O. *et al*, *J. Organomet. Chem.*, 1967, **9**, P15 (*uv*)
Alt, H. *et al*, *J. Am. Chem. Soc.*, 1974, **96**, 5936 (*synth*)
Demerseman, B. *et al*, *J. Organomet. Chem.*, 1975, **101**, C24 (*synth*)
Fachinetti, G. *et al*, *J. Chem. Soc., Chem. Commun.*, 1976, 230 (*synth, pmr, ir*)
Atwood, J.L. *et al*, *J. Organomet. Chem.*, 1977, **132**, 367 (*cryst struct*)

$C_{12}H_{10}Ti$ Ti-00052

Diphenyltitanium, 10CI
Titanium diphenyl
[14724-88-0]

$TiPh_2$

M 202.091
Not well characterised. Polymeric. Black solid. Sol. C_6H_6. Forms adducts with THF and NH_3. μ_B = 0.59 BM. Dec. slowly at 200-50°.
▷Pyrophoric

Latyaeva, V.N. *et al*, *J. Organomet. Chem.*, 1964, **2**, 388; 1969, **16**, 103 (*synth*)
Thiele, K.H. *et al*, *Z. Anorg. Chem.*, 1978, **441**, 13 (*synth*)

$C_{12}H_{12}S_2Ti$ Ti-00053

Bis(η^5-2,4-cyclopentadien-1-yl)[1,2-ethenedithiolato(2−)-*S,S'*]titanium, 9CI
[12307-60-7]

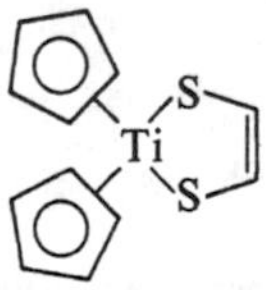

M 268.227
Green-black cryst. (CH_2Cl_2/hexane or by subl.). Mp 177-9°, 202-4°. $Bp_{0.1}$ 110-20° subl.

King, R.B. *et al*, *Inorg. Chem.*, 1968, **7**, 340 (*synth, ir, nmr*)
Köpf, H. *et al*, *Z. Naturforsch., B*, 1968, **23**, 1531 (*synth, ir, nmr*)
Kutoglu, A., *Acta Crystallogr., Sect. B*, 1973, **29**, 2891 (*cryst struct*)

$C_{12}H_{12}Ti$ Ti-00054

Bis(η^6-benzene)titanium, 9CI
[52462-43-8]

M 204.107
Reducing agent in org. synth. Cocatalyst for diene oligomerisations. Burgundy-red cryst.

Akhmedov, V.M. *et al*, *J. Chem. Soc., Dalton Trans.*, 1975, 1412 (*use*)
Anthony, M.T. *et al*, *J. Chem. Soc., Dalton Trans.*, 1975, 1419 (*synth, spectra*)
Hawker, P.N. *et al*, *J. Chem. Soc., Chem. Commun.*, 1978, 730 (*synth*)

$C_{12}H_{12}Ti$ Ti-00055

(η^7-Cycloheptatrienylium)(η^5-2,4-cyclopentadien-1-yl)titanium
π-Cycloheptatrienyl-π-cyclopentadienyltitanium
[51203-49-7]

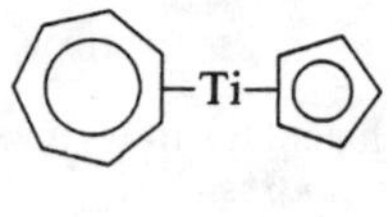

M 204.107
Sky-blue cryst. Sublimes at 125°/1 mm. Diamagnetic.

v. Oven, H.O. *et al*, *J. Organomet. Chem.*, 1970, **23**, 159 (*synth, ir*)
Zeinstra, J.D. *et al*, *J. Organomet. Chem.*, 1973, **54**, 207 (*struct*)
Verkouw, H.T. *et al*, *J. Organomet. Chem.*, 1973, **59**, 259 (*ms, ir, pmr*)
Evans, S. *et al*, *J. Chem. Soc., Dalton Trans.*, 1974, 304.
Groenenboom, C.J. *et al*, *J. Organomet. Chem.*, 1974, **80**, 229 (*cmr, pmr*)
Demerseman, B. *et al*, *J. Organomet. Chem.*, 1975, **101**, C24 (*synth*)

$C_{12}H_{13}ClOTi$ Ti-00056

[(*C,O*-η)-Acetyl]chlorobis(η^5-2,4-cyclopentadien-1-yl)titanium, 10CI
Acetylchlorotitanocene
[66320-88-5]

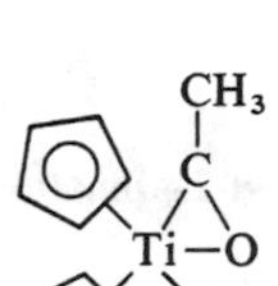

M 256.567
Orange air-stable cryst. (toluene). Sol. $CHCl_3$, polar solvs., insol. hydrocarbons.

Floriani, C. *et al*, *J. Chem. Soc., Chem. Commun.*, 1972, 790 (*synth, ir, ms*)
Fachinetti, G. *et al*, *J. Organomet. Chem.*, 1974, **71**, C5 (*synth, ir, nmr*)
Fachinetti, G. *et al*, *J. Chem. Soc., Dalton Trans.*, 1977, 2297 (*synth, cryst struct*)

$C_{12}H_{13}ClTi$ Ti-00057

Chlorobis(η^5-2,4-cyclopentadien-1-yl)ethenyltitanium
Chloro-π-cyclopentadienylvinyltitanium, 8CI. *Chloro(vinyl)titanocene*
As Chlorobis(η^5-2,4-cyclopentadien-1-yl)methyltitanium, Ti-00045 with

X = —CH=CH_2

M 240.568
Red-brown solid. Mp 153° dec.

Waters, J.A. *et al*, *J. Organomet. Chem.*, 1970, **22**, 417 (*synth, ir, nmr*)

$C_{12}H_{16}NTi$ **Ti-00058**

[Bis(η^5-cyclopentadienyl)dimethylamidotitanium

M 222.145

Dimeric.

Dimer: [11057-67-3]. *Tetra-π-cyclopentadienylbis[μ-(dimethylaminato)]dititanium, 8CI. Bis[bis(η^5-cyclopentadienyl)dimethylamidotitanium].*

$C_{24}H_{32}N_2Ti_2$ M 444.290

Useful starting material for Ti(III) complexes. Brown solid. Highly sensitive to air and moisture.

Lappert, M.F. *et al*, *J. Chem. Soc. (A)*, 1971, 874 (*synth, ir*)
Lappert, M.F. *et al*, *J. Chem. Soc. (A)*, 1971, 1314 (*props*)

$C_{12}H_{16}S_2Ti$ **Ti-00059**

Bis(η^5-2,4-cyclopentadien-1-yl)bis(methanethiolato)titanium, 10CI

Bis(η-cyclopentadienyl)bis(thiomethoxy)titanium. Bis(thiomethoxy)titanocene

[12089-78-0]

As Dibromobis(η^5-2,4-cyclopentadien-1-yl)titanium, Ti-00024 with

X = SMe

M 272.258

Red-violet solid or purple cryst. Mp 184-5°.

Coutts, R.S.P. *et al*, *Aust. J. Chem.*, 1966, **19**, 1377 (*synth, nmr*)
Giddings, S.A., *Inorg. Chem.*, 1967, **6**, 849 (*synth*)
Abel, E.W. *et al*, *J. Organomet. Chem.*, 1968, **14**, 285 (*synth, nmr*)
Köpf, H. *et al*, *Z. Naturforsch., B*, 1968, **23**, 1536 (*synth, ir*)

$C_{12}H_{16}Ti$ **Ti-00060**

Bis(η^5-2,4-cyclopentadien-1-yl)dimethyltitanium, 9CI

[1271-66-5]

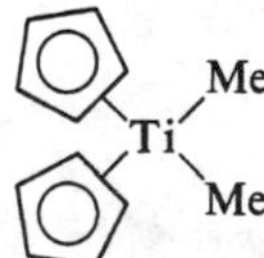

M 208.138

Orange-yellow needles (pentane). Mp 97° dec.

Piper, T.S. *et al*, *J. Inorg. Nucl. Chem.*, 1956, **3**, 104 (*synth*)
Ger. Pat., 1 037 446, (*1958*); *CA*, **54**, 18546 (*synth*)
Clauss, K. *et al*, *Justus Liebigs Ann. Chem.*, 1962, **654**, 8 (*synth*)
Hanlan, J.F. *et al*, *Can. J. Chem.*, 1972, **50**, 747 (*pmr, ir*)
Erskine, G.J. *et al*, *J. Organomet. Chem.*, 1976, **114**, 119 (*synth, props*)
Cardoso, A.M. *et al*, *J. Chem. Soc., Dalton Trans.*, 1980, 1156 (*synth*)

$C_{12}H_{18}Al_2Cl_8Ti$ **Ti-00061**

Tetra-μ-chlorobis(dichloroaluminium)[(1,2,3,4,5,6-η)hexamethylbenzene)titanium, 10CI

Hexamethylbenzenebis(di-μ-dichlorodichloroaluminum)titanium

[39450-72-1]

M 547.741

Catalyst for trimerization of butadiene. Violet cryst. (C_6H_6). Sol. aromatic hydrocarbons, insol. aliphatic hydrocarbons. Br and I analogues also known.

Pasynkiewicz, S. *et al*, *J. Organomet. Chem.*, 1973, **54**, 203 (*synth*)
Dzierzgowski, S. *et al*, *J. Mol. Catal.*, 1977, **2**, 253 (*use*)
Thewalt, U. *et al*, *J. Organomet. Chem.*, 1979, **172**, 317 (*cryst struct*)

$C_{12}H_{34}N_2Si_2Ti$ **Ti-00062**

Bis(dimethylamido)bis(trimethylsilylmethyl)titanium

Bis(N-methylmethanaminato)bis[(trimethylsilyl)methyl]titanium, 9CI

[59512-73-1]

$(Me_3SiCH_2)_2Ti(NMe_2)_2$

M 310.465

Yellow solid (below 0°).

Bürger, H. *et al*, *J. Organomet. Chem.*, 1976, **108**, 69 (*synth, nmr*)

$C_{13}H_{10}ClF_6NTi$ **Ti-00063**

Chlorodi-π-cyclopentadienyl[2,2,2-trifluoro-1-(trifluoromethyl)ethylideniminato]titanium

Chloro[bis(trifluoromethyl)ketimido]titanocene. Chlorobis(η-cyclopentadienyl)[bis(trifluoromethyl)ketimido]titanium

[11077-55-7]

As Chlorobis(η^5-2,4-cyclopentadien-1-yl)methyltitanium, Ti-00045 with

X = —N=C(CF_3)$_2$

M 377.552

Yellow solid. Mp 194-5° dec. $Bp_{0.001}$ <84° subl.

Cetinkaya, B. *et al*, *J. Chem. Soc., Chem. Commun.*, 1971, 215 (*synth, ir, nmr, ms*)

$C_{13}H_{13}Ti$ **Ti-00064**

(η^8-1,3,5,7-Cyclooctatetraene)(η^5-2,4-cyclopentadien-1-yl)titanium, 10CI

[11065-40-0]

M 217.126

Green cryst. $Bp_{0.1}$ 120° subl. Dec. at ~160° without melting (under Ar). Thermally stable but v. sensitive to oxygen. Paramagnetic (1 unpaired electron).

Van Oven, H.O. *et al*, *J. Organomet. Chem.*, 1969, **19**, 373 (*synth, ir, ms*)
Lehmkuhl, H. *et al*, *J. Organomet. Chem.*, 1970, **25**, C44 (*synth*)
Kroon, P.A. *et al*, *J. Organomet. Chem.*, 1970, **25**, 451 (*cryst struct*)
Thomas, J.L. *et al*, *Inorg. Chem.*, 1972, **11**, 348 (*esr*)
Veldman, M.E.E. *et al*, *J. Organomet. Chem.*, 1975, **84**, 247 (*synth, ms, ir, props*)

$C_{13}H_{14}Cl_2Ti$ Ti-00065

Dichloro[1,3-propanediylbis[(1,2,3,4,5-η)-2,4-cyclopentadien-1-ylidene]]titanium, 9CI
Dichloro(1,1'-trimethylenedicyclopentadienyl)titanium
[38117-97-4]

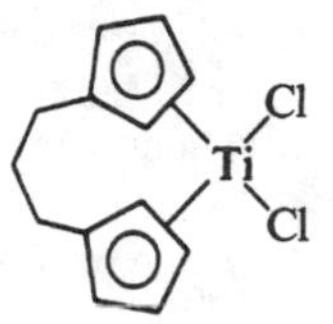

M 289.040
Deep red-brown cryst.

Japan. Pat., 63 5 991, (*1963*) (*synth, ir, uv, nmr*)
Davis, B.R. *et al*, *J. Organomet. Chem.*, 1971, **30**, 75 (*cryst struct*)
Hillman, M. *et al*, *J. Organomet. Chem.*, 1972, **42**, 123 (*synth, props*)
Epstein, E.F. *et al*, *Inorg. Chim. Acta*, 1973, **7**, 211 (*neutron diffraction*)
Smith, J.A. *et al*, *J. Organomet. Chem.*, 1979, **173**, 175 (*nmr*)

$C_{13}H_{18}AlClTi$ Ti-00066

μ-Chlorobis(η⁵-cyclopentadienyl)(dimethylaluminum)-μ-methylenetitanium
[67719-69-1]

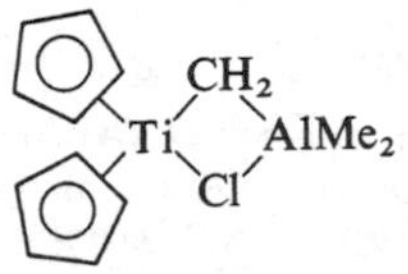

M 284.600
Reagent for olefin homologation and redn. of certain ketones to alkenes and esters to enol ethers. Red-orange cryst. Air and moisture sensitive.

Tebbe, F.N. *et al*, *J. Am. Chem. Soc.*, 1978, **100**, 3611 (*synth, use*)
Pine, S.H. *et al*, *J. Am. Chem. Soc.*, 1980, **102**, 3270 (*use*)
Fieser, M. *et al*, *Reagents for Organic Synthesis*, Wiley, 1967-83, **8**, 83 (*use*)

$C_{13}H_{19}ClSiTi$ Ti-00067

Chlorobis(η⁵-2,4-cyclopentadien-1-yl)(trimethylsilyl)titanium
Chloro(trimethylsilyl)titanocene
[75799-47-2]

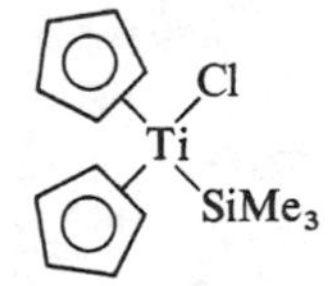

M 286.712
Dark-green cryst. Mp 119-21°.

Rösch, L. *et al*, *J. Organomet. Chem.*, 1980, **197**, 51 (*synth, ir, cryst struct*)

$C_{13}H_{33}N_3Ti$ Ti-00068

Tris(diethylaminato)methyltitanium, 8CI
Tris(diethylamido)methyltitanium
[25483-56-1]

$MeTi(NEt_2)_3$

M 279.304
Yellow liq. Mp 4-5°. Dec. at 120-30°. $TiCD_3$ analogue also prepd.

Bürger, H. *et al*, *J. Organomet. Chem.*, 1969, **20**, 129 (*synth, ir, raman, pmr*)
Bürger, H. *et al*, *J. Organomet. Chem.*, 1970, **21**, 381 (*uv*)

$C_{14}H_{10}F_6O_4Ti$ Ti-00069

Bis(η⁵-2,4-cyclopentadien-1-yl)bis(trifluoroacetato-*O*)titanium, 10CI
Titanocene bis(trifluoroacetate)
[1282-45-7]

As Dibromobis(η⁵-2,4-cyclopentadien-1-yl)titanium, Ti-00024 with

$X = —OOCCF_3$

M 404.101
Orange cryst. (toluene). Mp 175-9°.

Drozdov, G.V. *et al*, *Zh. Obshch. Khim.*, 1962, **32**, 2360 (*synth, uv*)
Beachell, H.C. *et al*, *Inorg. Chem.*, 1965, **4**, 1133 (*synth, nmr*)
Saunders, L. *et al*, *Polymer*, 1965, **6**, 635 (*synth, ir*)
Nesmayanov, A.N. *et al*, *Izv. Akad. Nauk SSSR, Ser. Khim.*, 1971, 2729 (*nmr*)

$C_{14}H_{10}N_2S_2Ti$ Ti-00070

Bis(η⁵-2,4-cyclopentadien-1-yl)[2,3-dimercapto-2-butenedinitrilato(2−)-*S,S'*]titanium, 9CI
Bis(η⁵-cyclopentadienyl)(1,2-dicyanoethylenedithiolato)titanium
[12087-16-0]

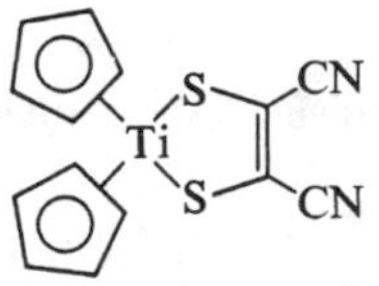

M 318.246
Dark-green cryst. Mp 260°.

Köpf, H. *et al*, *J. Organomet. Chem.*, 1965, **4**, 426 (*synth*)
Chauduri, M.A. *et al*, *J. Chem. Soc.* (*A*), 1966, 838 (*synth, ir*)
Locke, J. *et al*, *Inorg. Chem.*, 1966, **5**, 1187 (*synth, ir, nmr*)
McCleverty, J.A. *et al*, *Inorg. Chem.*, 1969, **8**, 1340.
Chiesi-Villa, A. *et al*, *Acta Crystallogr., Sect. B*, 1976, **33**, 909 (*cryst struct*)

$C_{14}H_{16}O_4Ti$ Ti-00071

Bis(acetato-*O*)bis(η⁵-2,4-cyclopentadien-1-yl)titanium, 10CI
Dicyclopentadienyltitanium diacetate. Titanocene diacetate. Diacetatobis(η⁵-cyclopentadienyl)titanium
[1282-51-5]

As Dibromobis(η⁵-2,4-cyclopentadien-1-yl)titanium, Ti-00024 with

X = OAc

M 296.158
Orange cryst. Mp 126-30°. Hydrolytically sensitive and thermally labile.

Nesmayanov, A.N. *et al*, *Izv. Akad. Nauk SSSR, Ser. Khim.*, 1961, 2146 (*synth*)
Razuvaev, G.A. *et al*, *Dokl. Akad. Nauk SSSR, Ser. Sci. Khim.*, 1961, **138**, 1126 (*synth*)
Nesmayanov, A.N. *et al*, *Tetrahedron* Suppl., 1966, **8**, pt. 2, 382 (*nmr*)

$C_{14}H_{16}Ti$ Ti-00072

Bis[(1,2,3,4,5,6-η)-methylbenzene]titanium, 10CI
Ditoluenetitanium
[55527-82-7]

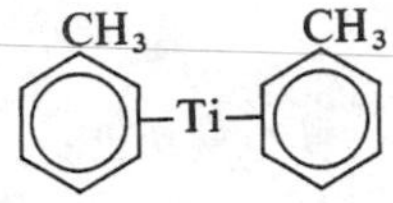

M 232.160
Obt. by metal vapour synth. technique. Burgandy-red cryst.

Anthony, M.T. *et al*, *J. Chem. Soc., Dalton Trans.*, 1975, 1419 (*synth, ir, ms, nmr, pe*)
Hawker, P.N. *et al*, *J. Chem. Soc., Chem. Commun.*, 1978, 730 (*synth*)

$C_{14}H_{18}Ti$ Ti-00073

(1,4-Butanediyl)bis(η⁵-2,4-cyclopentadien-1-yl)titanium, 9CI
1,4-Tetramethylenebis(η⁵-cyclopentadienyl)titanium.
1,1-Bis(η⁵-cyclopentadienyl)titanacyclopentane
[52124-67-1]

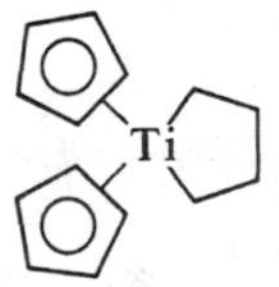

M 234.176
Orange needles. Dec. >–30°. Relatively stable compared with acyclic relatives.

McDermott, J.X. *et al*, *J. Am. Chem. Soc.*, 1974, **96**, 947; 1976, **98**, 6529 (*synth, nmr*)

$C_{14}H_{19}OPTi$ Ti-00074

Carbonylbis(η⁵-2,4-cyclopentadien-1-yl)(trimethylphosphine)-titanium, 10CI
[56770-61-7]

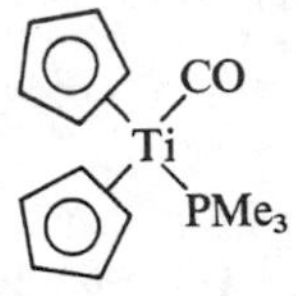

M 282.157
Maroon cryst. (hexane).

Demerseman, B. *et al*, *J. Organomet. Chem.*, 1975, **93**, 199 (*synth, ir*)

$C_{14}H_{20}AlCl_2Ti$ Ti-00075

Di-μ-chlorobis(η⁵-cyclopentadien-1-yl)(diethylaluminum)titanium, 10CI
[39459-00-2]

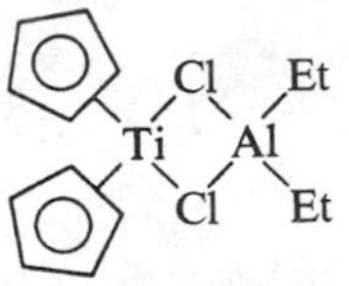

M 334.080
Catalyses olefin polymerization. Blue cryst. (heptane). Mp 80-90°. V. sensitive to O_2.

Natta, G. *et al*, *J. Am. Chem. Soc.*, 1958, **80**, 755 (*cryst struct*)
Breslow, D.S. *et al*, *J. Am. Chem. Soc.*, 1959, **81**, 81 (*synth, use*)
Maki, A.H. *et al*, *J. Am. Chem. Soc.*, 1960, **82**, 4109 (*esr*)
Henrici-Olivé, G. *et al*, *Angew. Chem., Int. Ed. Engl.*, 1967, **6**, 790 (*esr, synth, rev*)
Henrici-Olivé, G. *et al*, *J. Organomet. Chem.*, 1969, **17**, 83 (*esr*)
Bulychev, B.M. *et al*, *Transition Met. Chem.*, 1981, **6**, 32 (*esr, props*)

$C_{14}H_{22}N_2Ti$ Ti-00076

Bis(η⁵-cyclopentadienyl)bis(dimethylamido)titanium
Bis(η⁵-2,4-cyclopentadien-1-yl)bis(N-methylmethanaminato)titanium, 9CI. Bis(dimethylamido)titanocene
[12086-52-1]
As Dibromobis(η⁵-2,4-cyclopentadien-1-yl)titanium, Ti-00024 with

$$X = NMe_2$$

M 266.221
Air-sensitive brown oil. $Bp_{0.2}$ 120-8°.

Chandra, G. *et al*, *J. Chem. Soc. (A)*, 1968, 1940 (*synth, ms*)

$C_{14}H_{23}N_3S_6Ti$ Ti-00077

(η⁵-2,4-Cyclopentadien-1-yl)tris(dimethylcarbamodithioato-*S,S'*)titanium, 10CI
(η-Cyclopentadienyl)tris(N,N-dimethyldithiocarbamato)titanium
[67891-23-0]

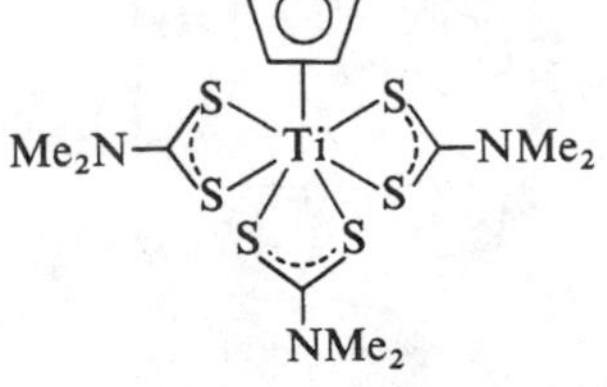

M 473.596
Orange-yellow cryst. + C_6H_6 (CH_2Cl_2/C_6H_6).

Steffen, W.L. *et al*, *Inorg. Chem.*, 1978, **17**, 3498 (*cryst struct*)

$C_{14}H_{28}P_2Ti$ Ti-00078

Bis(η^4-1,3-butadiene)[1,2-ethanediylbis[dimethylphosphine]-P,P′]titanium, 10CI

[1,2-Bis(dimethylphosphino)ethane]bis(η^4-1,3-butadiene)titanium

[75061-57-3]

M 306.203

Catalytically dimerizes alkenes. Blue cryst. (hexane).

Datta, S. *et al*, *J. Organomet. Chem.*, 1980, **188**, 353 (*synth, ms, cmr, nmr*)

Wreford, S.S. *et al*, *J. Chem. Soc., Chem. Commun.*, 1981, 458 (*props*)

$C_{15}H_{15}Ti$ Ti-00079

[(1,2-η)-2,4-cyclopentadien-1-yl]bis(η^5-2,4-cyclopentadien-1-yl)titanium, 9CI

Tris(η-cyclopentadienyl)titanium

[52700-41-1]

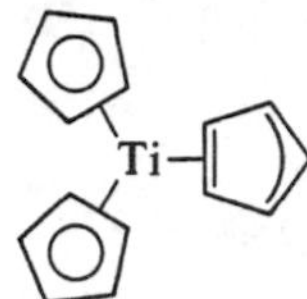

M 243.164

Green cryst. Mp 138-40°. $Bp_{0.001}$ 125° subl. μ_{eff} close to spin-only value.

Canty, A.J. *et al*, *Aust. J. Chem.*, 1968, **21**, 807 (*synth*)

Siegert, F.W. *et al*, *J. Organomet. Chem.*, 1969, **20**, 141 (*synth, ir, nmr*)

Lucas, C.R. *et al*, *J. Chem. Soc., Chem. Commun.*, 1973, 97 (*cryst struct*)

$C_{15}H_{17}O_2Ti^{\oplus}$ Ti-00080

Di-π-cyclopentadienyl(2,4-pentanedionato)titanium(1+)

Acetylacetonatotitanocene

M 277.178 (ion)

Perchlorate: [12267-13-9].

$C_{15}H_{17}ClO_6Ti$ M 376.629

Violet cryst. (MeOH). Other salts also prepd.

Doyle, G. *et al*, *Inorg. Chem.*, 1967, **6**, 1111 (*synth, ir, nmr*)

Doyle, G. *et al*, *Inorg. Chem.*, 1968, **7**, 2484.

$C_{15}H_{19}Ti$ Ti-00081

Bis(η^5-2,4-cyclopentadien-1-yl)[(1,2,3-η)-2-methyl-2-butenyl]titanium, 10CI

Bis(cyclopentadienyl)(1,2-dimethylallyl)titanium

[12111-85-2]

M 247.195

Monomeric in C_6H_6. Sol. org. solvs. Mp 70.5-71° dec. V. sensitive to O_2 and H_2O even as cryst.

Helmholdt, R.B. *et al*, *Recl. Trav. Chim. Pays-Bas*, 1967, **86**, 1263 (*ir, cryst struct*)

Martin, H.A. *et al*, *J. Organomet. Chem.*, 1967, **12**, 149 (*synth, uv*)

$C_{15}H_{21}ClTi$ Ti-00082

Chlorobis(η^5-2,4-cyclopentadien-1-yl)(2,2-dimethylpropyl)titanium

Chlorobis(η^5-cyclopentadienyl)neopentyltitanium, 8CI. *Chloro(neopentyl)titanocene*

[11082-41-0]

As Chlorobis(η^5-2,4-cyclopentadien-1-yl)methyltitanium, Ti-00045 with

$$X = —CH_2C(CH_3)_3$$

M 284.664

Red-brown solid. Mp 95°.

Waters, J.A. *et al*, *J. Organomet. Chem.*, 1970, **22**, 417 (*synth, ir, nmr, uv*)

$C_{15}H_{26}O_3Ti$ Ti-00083

Phenyltris(2-propanolato)titanium, 9CI

Triisopropoxyphenyltitanium. Phenyltris(isopropoxy)titanium

[16635-23-7]

$$PhTi[OCH(CH_3)_2]_3$$

M 302.249

White or pale-yellow cryst. Mp 88-90°.

Herman, D.F. *et al*, *J. Am. Chem. Soc.*, 1952, **74**, 2693; 1953, **75**, 3877 (*synth*)

Herman, D.F., *Adv. Chem. Ser.*, 1959, **23**, 265 (*ir*)

Holloway, H., *Chem. Ind. (London)*, 1962, 214 (*synth*)

Rausch, M.D. *et al*, *J. Organomet. Chem.*, 1974, **74**, 85 (*nmr*)

$C_{16}H_{10}F_5Ti$ Ti-00084

Bis(η^5-2,4-cyclopentadien-1-yl)pentafluorophenyltitanium, 10CI

[39333-59-0]

M 345.127

Monomeric. Brown-purple leaflets (pentane). μ_{eff} 1.92 BM (20°). Dec. at 106°.

Teuben, J.H. *et al*, *J. Organomet. Chem.*, 1972, **46**, 313 (*synth, ir, props*)

De Boer, E.J.M. *et al*, *J. Organomet. Chem.*, 1978, **153**, 53; 1979, **166**, 193; **181**, 61 (*props*)

$C_{16}H_{10}F_6Ti$ Ti-00085

Bis-π-cyclopentadienylbis(3,3,3-trifluoro-1-propynyl)titanium, 8CI

Bis(trifluoromethylethynyl)titanocene

As Dibromobis(η^5-2,4-cyclopentadien-1-yl)titanium, Ti-00024 with

X = —C≡CCF_3

M 364.125

Orange solid. Mp ca. 125° dec. $Bp_{0.001}$ 100° subl.

▷Explosion hazard on heating

Bruce, M.I. *et al*, *J. Chem. Soc.* (*A*), 1968, 356 (*synth, ir, pmr, nmr*)

$C_{16}H_{14}O_2Ti$ Ti-00086

[1,2-Benzenediolato(2−)-*O,O'*]bis(η^5-2,4-cyclopentadien-1-yl)titanium

Di-π-cyclopentadienyl[pyrocatecholato(2−)]titanium, 8CI

[12308-48-4]

X = O

M 286.165

Monomeric in C_6H_6. Uses claimed in paints, varnishes and related materials. Brown-red cryst. Sol. C_6H_6, toluene, ethers, spar. sol. hexane, MeOH, EtOH. Mp 100-3°.

Andrä, K., *J. Organomet. Chem.*, 1968, **11**, 567 (*synth, ir*)

$C_{16}H_{14}S_2Ti$ Ti-00087

[1,2-Benzenedithiolato(2−)-*S,S'*]bis(η^5-2,4-cyclopentadien-1-yl)titanium, 9CI

[12155-21-4]

As [1,2-Benzenediolato(2−)-*O,O'*]bis(η^5-2,4-cyclopentadien-1-yl)titanium, Ti-00086 with

X = S

M 318.287

Monomeric in C_6H_6. Blue-black needles (C_6H_6/propanol). Sol. nonpolar solvs., spar. sol. Et_2O, alcohols. Mp 222-4°.

Köpf, H. *et al*, *J. Organomet. Chem.*, 1965, **4**, 426 (*synth*)
Kutoglu, A., *Z. Anorg. Allg. Chem.*, 1972, **390**, 195 (*cryst struct*)

$C_{16}H_{15}ClTi$ Ti-00088

Chlorobis(η^5-2,4-cyclopentadien-1-yl)phenyltitanium, 9CI

Chloro(phenyl)titanocene

[12663-63-7]

As Chlorobis(η^5-2,4-cyclopentadien-1-yl)methyltitanium, Ti-00045 with

X = Ph

M 290.628

Orange platelets (pentane). Mp 121° dec. Can be handled briefly in air, even in soln.

Waters, J.A. *et al*, *J. Organomet. Chem.*, 1970, **21**, 417 (*synth, nmr, uv*)
Waters, J.A. *et al*, *J. Organomet. Chem.*, 1971, **33**, 41.
Rausch, M.D. *et al*, *J. Organomet. Chem.*, 1974, **74**, 85 (*synth, nmr*)

$C_{16}H_{15}Ti$ Ti-00089

Bis(η^5-2,4-cyclopentadien-1-yl)phenyltitanium, 10CI

[11136-26-8]

As Bis(η^5-2,4-cyclopentadien-1-yl)pentafluorophenyltitanium, Ti-00084 with

R = Ph

M 255.175

Thermally unstable monomer. Green cryst. (pentane). η_{eff} 1.58 BM (20°). Dec. at 29°.

Teuben, J.H. *et al*, *Recl. Trav. Chim. Pays-Bas*, 1971, **90**, 360 (*synth*)
Teuben, J.H. *et al*, *J. Organomet. Chem.*, 1972, **46**, 313 (*synth, ir, props*)
De Boer, E.J.M. *et al*, *J. Organomet. Chem.*, 1978, **15**, 53; 1979, **166**, 193 (*props*)
Middleton, A.R. *et al*, *J. Chem. Soc., Dalton Trans.*, 1980, 1888 (*props*)

$C_{16}H_{16}Ti$ Ti-00090

Bis(1,3,5,7-cyclooctatetraene)titanium, 9CI

[η-Cyclooctatetraene(2−)](η^4-cyclooctatetraene)titanium

[12112-27-5]

M 256.182

Structurally has η^8- and η^4-COT ligands but fluxional in soln. Violet-red cryst. (toluene). Spar. sol. aromatic hydrocarbons.

Breil, H. *et al*, *Angew. Chem., Int. Ed. Engl.*, 1966, **5**, 898 (*synth*)
Dietrich, H. *et al*, *Angew. Chem., Int. Ed. Engl.*, 1969, **8**, 765 (*cryst struct*)
Lehmkuhl, H. *et al*, *J. Organomet. Chem.*, 1970, **25**, C44 (*synth*)
Schwartz, J. *et al*, *J. Chem. Soc., Chem. Commun.*, 1973, 172 (*nmr*)

$C_{16}H_{22}Cl_4O_2Ti_2$ Ti-00091

Tetrachlorobis(η^5-2,4-cyclopentadien-1-yl)[μ-[2,3-dimethyl-2,3-butanediolato(2−)-*O*:*O'*]dititanium, 10CI

1,1,2,2-Tetramethylethanediolatobis[dichloro(η-cyclopentadienyl)titanium]

[42178-05-2]

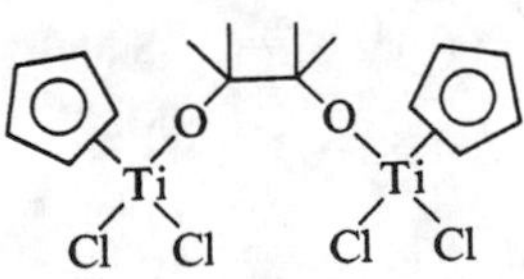

M 483.921

Struct. data indicate some Ti-*O* multiple bonding. Yellow cryst. (THF/Et_2O). V. spar. sol. C_6H_6.

Coutts, R.S.P. *et al*, *J. Organomet. Chem.*, 1973, **50**, 145 (*synth, ir, uv*)
Huffman, J.C. *et al*, *J. Am. Chem. Soc.*, 1980, **102**, 3009 (*cryst struct*)

Marsella, J.A. *et al*, *J. Organomet. Chem.*, 1980, **201**, 389 (*synth*, *ir*, *nmr*)

$C_{16}H_{28}O_6P_2Ti$ Ti-00092

Bis(η^5-2,4-cyclopentadien-1-yl)bis(trimethylphosphite)titanium, 10CI

[76083-35-7]

As Dibromobis(η^5-2,4-cyclopentadien-1-yl)titanium, Ti-00024 with

$$X = P(OMe)_3$$

M 426.221

V. air-sensitive and less stable than Zr analogue. Red-brown cryst. (pentane).

Chang, M. *et al*, *J. Organomet. Chem.*, 1980, **199**, C3 (*synth*, *nmr*, *ms*, *props*)

$C_{16}H_{44}Si_4Ti$ Ti-00093

Tetrakis(trimethylsilylmethyl)titanium, 9CI

[33948-28-6]

$$Ti(CH_2SiMe_3)_4$$

M 396.746

Olefin polymerisation catalyst. Pale-yellow-green liq. Mp 0-1°. $Bp_{0.001}$ 25°. Thermally robust compared with simple Ti alkyls.

Ballard, D.G.H., *Proc. 23rd IUPAC Congress*, Boston, 1971, **6**, 213 (*use*)
Collier, M.R. *et al*, *J. Chem. Soc., Dalton Trans.*, 1973, 445 (*synth*, *ir*, *nmr*)
Lappert, M.F. *et al*, *J. Organomet. Chem.*, 1974, **66**, 271 (*pe*)
Lappert, M.F. *et al*, *J. Chem. Soc., Chem. Commun.*, 1975, 830.

$C_{17}H_{17}Ti$ Ti-00094

Benzylbis(η^5-cyclopentadienyl)titanium

Bis(η^5-2,4-cyclopentadien-1-yl)(phenylmethyl)titanium, 10CI

[39333-72-7]

As Bis(η^5-2,4-cyclopentadien-1-yl)-pentafluorophenyltitanium, Ti-00084 with

$$R = CH_2Ph$$

M 269.201

Brown cryst. (Et_2O/dioxan). Dec. >43°.

Teuben, J.H. *et al*, *J. Organomet. Chem.*, 1972, **46**, 313 (*synth*, *esr*, *ir*)

$C_{17}H_{38}Si_3Ti$ Ti-00095

(η^5-2,4-Cyclopentadien-1-yl)tris[(trimethylsilyl)methyl]titanium, 9CI

[54056-33-6]

$CpTi(CH_2SiMe_3)_3$

M 374.624

Yellow cryst. Mp 60-2°.

Chivers, T. *et al*, *J. Organomet. Chem.*, 1974, **77**, 241 (*synth*, *ir*, *nmr*)
Green, M.L.H. *et al*, *J. Organomet. Chem.*, 1974, **73**, 259 (*synth*, *nmr*)

$C_{18}H_{14}Cl_2Ti$ Ti-00096

Dichlorobis[(1,2,3,3*a*,7*a*-η)-1*H*-inden-1-yl]titanium, 9CI

Bisindenyltitanium dichloride.

Dichlorodiindenyltitanium

[12113-02-9]

M 349.095

Dark-brown cryst. Mp 210°.

▷XR2056000.

Marconi, W. *et al*, *Chem. Ind. (London)*, 1962, **44**, 229 (*synth*)
Samuel, E. *et al*, *J. Organomet. Chem.*, 1965, **4**, 156 (*synth*, *ir*)
Samuel, E., *Bull. Soc. Chim. Fr.*, 1966, 3548 (*ir*)
Joseph, P.T. *et al*, *Indian J. Chem.*, 1971, **9**, 175 (*synth*)

$C_{18}H_{18}N_2Ti$ Ti-00097

Bis(η^5-2,4-cyclopentadien-1-yl)di-1*H*-pyrrol-1-yltitanium, 10CI

Di(η^1-pyrrolyl)titanocene

[11077-90-0]

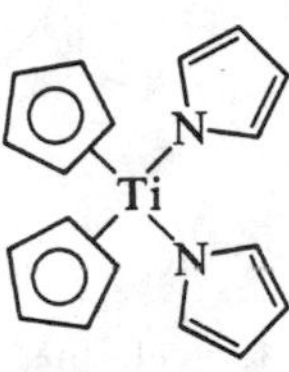

M 310.234

Red-brown cryst. Mp 175-7° dec. Cryst. struct. shows pyrrolyl rings to be essentially σ-bonded.

Issleib, K. *et al*, *Z. Anorg. Allg. Chem.*, 1969, **369**, 83 (*synth*)
Tille, D., *Z. Naturforsch., B*, 1970, **25**, 1538 (*ir*, *nmr*)
Van Bynum, R. *et al*, *Inorg. Chem.*, 1980, **19**, 2368 (*cryst struct*)

$C_{18}H_{19}Ti$ Ti-00098

Bis(η^5-2,4-cyclopentadien-1-yl)(2,6-dimethylphenyl)titanium, 10CI

[39333-76-1]

As Bis(η^5-2,4-cyclopentadien-1-yl)pentafluorophenyltitanium, Ti-00084 with

$$R = 2,6\text{-Dimethylphenyl}$$

M 283.228

Green cryst. (pentane/Et_2O). Mp 180°. $Bp_{0.1}$ 160° subl. η_{eff} 1.66 BM. V. air-sensitive.

Teuben, J.H. *et al*, *J. Organomet. Chem.*, 1972, **46**, 313 (*synth*, *ir*, *props*)
Olthof, G.J. *et al*, *J. Organomet. Chem.*, 1976, **122**, 47 (*cryst struct*)

$C_{18}H_{24}N_2Ti$ Ti-00099

Bis(dimethylamido)(η^5-methylcyclopentadienyl)phenylethynyltitanium

$(H_3CC_5H_4)Ti(NMe_2)_2(C{\equiv}CPh)$

M 316.281

Dark viscous oil.

Jenkins, A.D. *et al*, *J. Organomet. Chem.*, 1970, **23**, 165 (*synth, ir, ms, nmr*)

$C_{18}H_{24}Ti$ Ti-00100

Bis[(1,2,3,4,5,6-η)-1,3,5-trimethylbenzene]titanium, 10CI
Dimesitylenetitanium
[57347-27-0]

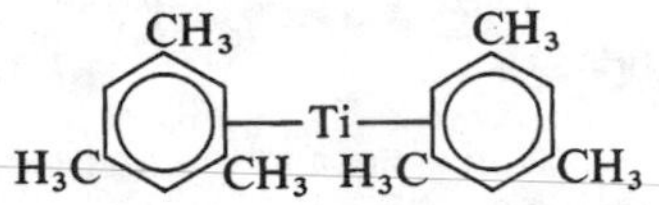

M 288.268
Burgundy-red solid.

Anthony, M.T. *et al*, *J. Chem. Soc., Dalton Trans.*, 1975, 1419 (*synth, ir, nmr*)

$C_{18}H_{32}Si_2Ti$ Ti-00101

Bis(η⁵-2,4-cyclopentadien-1-yl)bis[(trimethylsilyl)methyl]titanium, 10CI
Bis(trimethylsilylmethyl)titanocene
[11077-96-6]

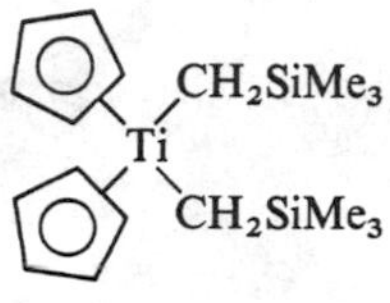

M 352.502
Orange cryst. Mp 185°. Bp 60-80° subl. *in vacuo*. Also said to dec. without melting.

Yagupsky, G. *et al*, *J. Chem. Soc., Chem. Commun.*, 1970, 1369 (*synth, ms*)
Wozniak, B. *et al*, *J. Chem. Soc. (A)*, 1971, 3116 (*synth, nmr, ir*)
Collier, M.R. *et al*, *J. Chem. Soc., Dalton Trans.*, 1973, 445 (*synth, ir, ms, nmr*)

$C_{19}H_{15}O_4Ti$ Ti-00102

Bis(μ-benzoato)(η⁵-cyclopentadienyl)titanium

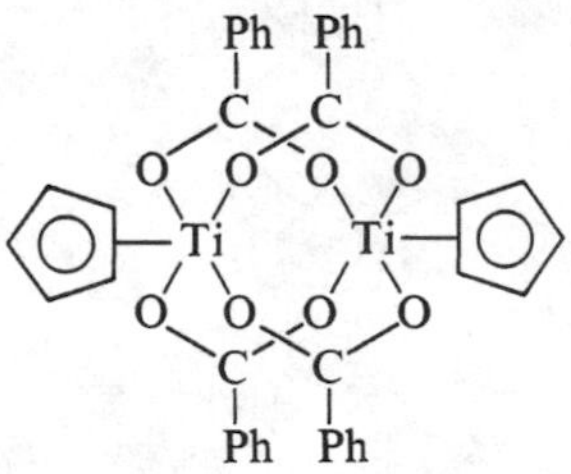

M 355.205
Dimeric.

Dimer: [11088-05-4]. *Tetrakis[μ-benzoato-O:O′]-bis(η⁵-2,4-cyclopentadien-1-yl)dititanium, 10CI.*
$C_{38}H_{30}O_8Ti_2$ M 710.410
Diamagnetic but no Ti-Ti bonding (l = 363 pm). Brown cryst. Spar. sol. hexane, THF.

Razuvaev, G.A. *et al*, *Dokl. Akad. Nauk SSSR, Ser. Sci. Khim.*, 1969, **187**, 340 (*synth*)
Coutts, R.S.P. *et al*, *Aust. J. Chem.*, 1973, **26**, 941 (*synth, esr, uv, ir*)
Tarkhova, T.N. *et al*, *Zh. Strukt. Khim.*, 1976, **17**, 1052 (*cryst struct*)

$C_{19}H_{28}ClNTi$ Ti-00103

(1-*tert*-Butyl-2,2-dimethylpropylideniminato)chlorodi-π-cyclopentadienyltitanium, 8CI
*Chlorobis(η⁵-cyclopentadienyl)di-*tert*-butylketimidotitanium*
[11105-88-7]

As Chlorobis(η⁵-2,4-cyclopentadien-1-yl)methyltitanium, Ti-00045 with

$$X = —N{=}C[C(CH_3)_3]_2$$

M 353.770
Yellow-orange solid.

Collier, M.R. *et al*, *Inorg. Nucl. Chem. Lett.*, 1971, **7**, 689 (*synth, ir, nmr, ms*)

$C_{20}H_{19}Ti_2$ Ti-00104

μ(η¹:η⁵-Cyclopentadienyl)tris(η-cyclopentadienyl)dititanium-(Ti-Ti)
Tris(η⁵-2,4-cyclopentadien-1-yl)[μ-[(1-η:1,2,3,4,5-η)-2,4-cyclopentadien-1-ylidene)]dititanium(Ti-Ti), 9CI
[61529-60-0]

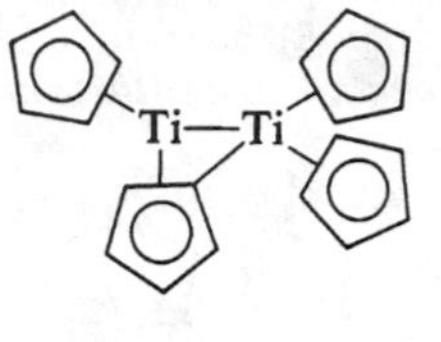

M 355.130
Grey-black solid. Forms cryst. adduct with two moles THF possibly identical with "black titanocene".

▷Pyrophoric

Pez, G.P., *J. Am. Chem. Soc.*, 1976, **98**, 8072 (*synth, struct*)
Pez, G.P. *et al*, *Adv. Organomet. Chem.*, 1981, **191** (*rev*)

$C_{20}H_{20}Cl_4O_4Ti_4$ Ti-00105

Tetrachlorotetrakis(η⁵-2,4-cyclopentadien-1-yl)tetra-μ-oxo-tetratitanium, 10CI
Cyclotetra[μ-oxochloro-η-cyclopentadienyltitanium]
[12171-57-2]

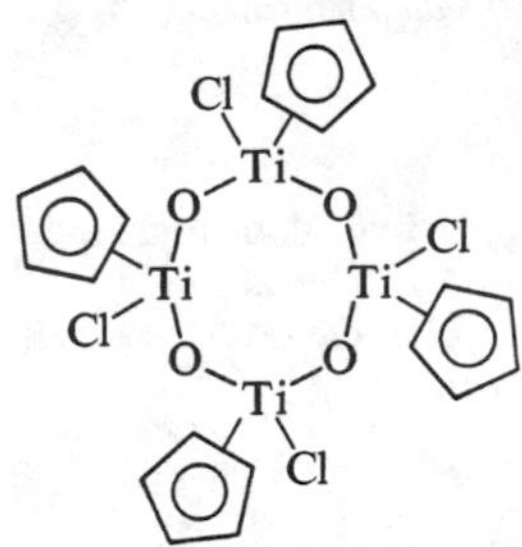

M 657.708
Yellow cryst. V. spar. sol. org. solvs. Mp 260°. Ring nonplanar. Tetrahydroindenyl and other analogues known.

Monomer:
C_5H_5ClOTi M 164.427
Unknown.

Gorsich, R.D. *et al*, *J. Am. Chem. Soc.*, 1960, **82**, 4211 (*synth*)
Saunders, L. *et al*, *Polymer*, 1965, **6**, 635 (*synth, ir, struct*)
Skapski, A.C. *et al*, *Acta Crystallogr., Sect. B*, 1970, **26**, 716 (*cryst struct*)
Coutts, R.S.P. *et al*, *Inorg. Nucl. Chem. Lett.*, 1973, **9**, 49 (*ir, synth*)
Marsella, J.A. *et al*, *J. Organomet. Chem.*, 1980, **201**, 389 (*nmr*)

$C_{20}H_{20}O_2Ti_2$ **Ti-00106**

[μ-[(1,2,3,4,5-η:1′,2′,3′,4′,5′-η)[Bi-2,4-cyclopentadien-1-yl]-1,1′-diyl]bis(η^5-2,4-cyclopentadien-1-yl)di-μ-hydroxodititanium, 9CI

μ(η^5:η^5-Fulvalene)di-μ-hydroxobis(η^5-cyclopentadienyltitanium)

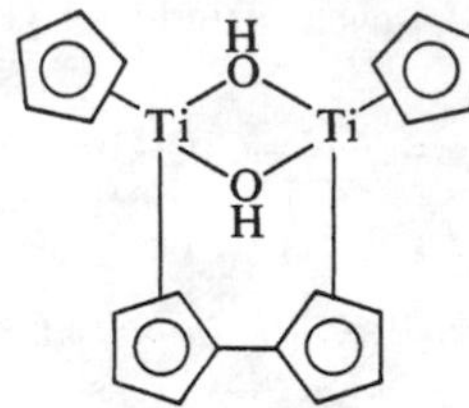

M 388.137

Reddish-purple cryst. (THF). V. oxygen sensitive, stable under inert atm. at r.t.

Guggenberger, L.J. *et al*, *J. Am. Chem. Soc.*, 1976, **98**, 4137 (*synth, ir, ms, cryst struct*)

$C_{20}H_{20}Ti$ **Ti-00107**

Di-2,4-cyclopentadien-1-ylbis(η^5-2,4-cyclopentadien-1-yl)-titanium, 10CI

Tetra(cyclopentadienyl)titanium. Bis(η^1-cyclopentadienyl)bis(η-cyclopentadienyl)titanium

[63726-15-8]

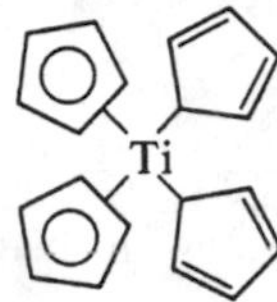

M 308.258

Fluxional. Shows both ring rotation and η^1:η^5 interchange. Violet-black solid or green-black cryst. Mp 128°. Oxygen-sensitive.

Siegert, F.W. *et al*, *J. Organomet. Chem.*, 1969, **20**, 141 (*synth, ir, nmr*)
Calderon, J.L. *et al*, *J. Am. Chem. Soc.*, 1971, **93**, 3587 (*synth, nmr*)
Calderon, J.L. *et al*, *J. Am. Chem. Soc.*, 1971, **93**, 3592 (*cryst struct*)
Lee, J.G. *et al*, *J. Organomet. Chem.*, 1977, **135**, 115 (*synth, derivs*)

$C_{20}H_{20}Ti_2$ **Ti-00108**

μ(η^5:η^5-Fulvalene)di-μ-hydridobis(cyclopentadienyltitanium)

[μ-[(1,2,3,4,5-η:1′,2′,3′,4′,5′-η)[Bis-2,4-cyclopentadien-1-yl]-1,1′-diyl]]bis(η^5-2,4-cyclopentadien-1-yl)di-μ-hydrodititanium, 9CI

[52676-23-0]

M 356.138

Green solid.

Watt, G.W. *et al*, *J. Am. Chem. Soc.*, 1966, **88**, 1138 (*synth*)
Salzmann, J.J. *et al*, *Helv. Chim. Acta*, 1967, **50**, 1831 (*synth*)
Brintzinger, H.H. *et al*, *J. Am. Chem. Soc.*, 1970, **92**, 6182 (*spectra, struct*)
Bercaw, J.E. *et al*, *J. Am. Chem. Soc.*, 1972, **94**, 1219.
Davison, A. *et al*, *J. Am. Chem. Soc.*, 1974, **96**, 3017 (*cmr, struct*)
Guggenberger, L.J. *et al*, *J. Am. Chem. Soc.*, 1976, **98**, 4137.
Pez, G.P. *et al*, *Adv. Organomet. Chem.*, 1981, **19**, 1 (*rev*)

$C_{20}H_{21}ClTi$ **Ti-00109**

Chlorobis(η^5-cyclopentadienyl)(2,2-dimethyl-1-phenylethenyl)titanium

As Chlorobis(η^5-2,4-cyclopentadien-1-yl)methyltitanium, Ti-00045 with

$$X = \text{—CPh=C(CH}_3)_2$$

M 344.719

Characterised by full x-ray analysis and spectroscopically. Dark-red cryst.

Cardin, D.J. *et al*, unpublished (*synth, nmr, ir, cryst struct*)

$C_{20}H_{24}Si_2Ti_2$ **Ti-00110**

Tetrakis(η^5-2,4-cyclopentadien-1-yl)di-μ-silylenedititanium, 9CI

Bis[bis(η-cyclopentadienyl)(μ-silylene)titanium]

[50923-29-0]

M 416.341

Olive-green cryst. Sol. C_6H_6.

Hencken, G. *et al*, *Chem. Ber.*, 1973, **106**, 1747 (*synth, cryst struct*)

$C_{20}H_{30}Cl_2Ti$ **Ti-00111**

Dichlorobis[(1,2,3,4,5-η)-1,2,3,4,5-pentamethyl-2,4-cyclopentadien-1-yl]titanium, 10CI

Bis(pentamethylcyclopentadienyl)titanium dichloride. Decamethyltitanocene dichloride

[11136-36-0]

X = Cl

M 389.243

Red-brown cryst. Mp 273° dec.

Bercaw, J.E. *et al*, *J. Am. Chem. Soc.*, 1972, **94**, 1219 (*synth, nmr*)
Bercaw, J.E., *J. Am. Chem. Soc.*, 1974, **96**, 5087 (*synth, ir*)
Harrigan, R.W. *et al*, *J. Organomet. Chem.*, 1974, **81**, 79 (*uv*)
McKenzie, T.C. *et al*, *J. Organomet. Chem.*, 1975, **102**, 457 (*cryst struct*)

$C_{20}H_{30}Ti$ **Ti-00112**

Bis[(1,2,3,4,5-η)-1,2,3,4,5-pentamethyl-2,4-cyclopentadien-1-yl]titanium, 10CI

Decamethyltitanocene. Permethyltitanocene

[11136-37-1]

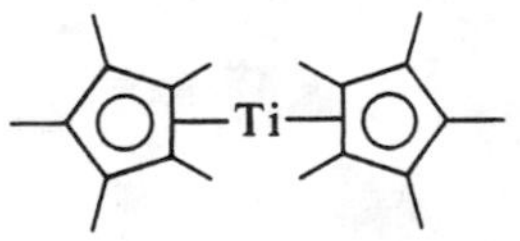

M 318.337

Exists as tautomeric mixt. with H transfer $CH_3 \rightarrow Ti$. Monomeric in soln. Orange cryst. Sol. Et_2O, hydrocarbons. μ_{eff} 2.6 BM.

Bercaw, J.E. *et al*, *J. Am. Chem. Soc.*, 1971, **93**, 2046; 1972, **94**, 1219; 1974, **96**, 5087 (*synth, nmr, ir, props*)

$C_{20}H_{44}Ti$ **Ti-00113**

Tetrakis[2,2-dimethylpropyl)titanium

Tetraneopentyltitanium

[13356-21-3]

$$Ti[CH_2C(CH_3)_3]_4$$

M 332.448

Yellow cryst. Mp ~105° dec. Bp ~50° subl. *in vacuo*. Photosensitive in soln. Thermally unstable at r.t. over long periods.

Davidson, P.J. *et al*, *J. Organomet. Chem.*, 1973, **57**, 269 (*synth, nmr, ir, raman*)
Mowat, W. *et al*, *J. Chem. Soc., Dalton Trans.*, 1973, 1120 (*synth, nmr, ir*)
Lappert, M.F. *et al*, *J. Organomet. Chem.*, 1974, **66**, 271 (*pe*)

$C_{21}H_{57}Si_6Ti$ **Ti-00114**

Tris[bis(trimethylsilyl)methyl]titanium, 10CI

[53668-82-9]

$$Ti[CH(SiMe_3)_2]_3$$

M 526.074

Blue-green cryst. Sol. hydrocarbons. Extremely air- and water-sensitive.

Barker, G.K. *et al*, *J. Chem. Soc., Dalton Trans.*, 1978, 734 (*synth, esr*)

$C_{22}H_{10}F_{10}Ti$ **Ti-00115**

Bis(η^5-2,4-cyclopentadien-1-yl)bis(pentafluorophenyl)titanium, 10CI

Bis(pentafluorophenyl)titanocene

[12155-89-4]

As Dibromobis(η^5-2,4-cyclopentadien-1-yl)titanium, Ti-00024 with

$$X = C_6F_5$$

M 512.185

Orange solid. Mp 228-30°.

Bourn, A.J.R. *et al*, *Proc. Chem. Soc., London*, 1963, 200 (*nmr*)
Chaudari, M.A. *et al*, *J. Organomet. Chem.*, 1964, **2**, 206 (*synth, ir, nmr*)
Tamborski, C. *et al*, *J. Organomet. Chem.*, 1965, **4**, 446 (*synth*)
Bruce, M.I. *et al*, *Org. Mass Spectrom.*, 1968, **1**, 835 (*ms*)

$C_{22}H_{20}O_2Ti$ **Ti-00116**

Bis(η^5-2,4-cyclopentadien-1-yl)bis(phenolato-*O*)titanium

Bis(η^5-cyclopentadienyl)diphenoxytitanium

[12246-19-4]

As Dibromobis(η^5-2,4-cyclopentadien-1-yl)titanium, Ti-00024 with

$$X = OPh$$

M 364.279

Monomeric in C_6H_6. Deep-yellow cryst. (C_6H_6). Mp 142°. $E_{1/2} = -1174$ mV.

Andrä, K., *Z. Chem.*, 1967, **7**, 318 (*synth*)
Andrä, K., *J. Organomet. Chem.*, 1968, **11**, 567 (*synth, ir, props*)

$C_{22}H_{20}S_2Ti$ **Ti-00117**

Bis(benzenethiolato)bis(η^5-2,4-cyclopentadien-1-yl)titanium, 10CI

Bis(η-cyclopentadienyl)bis(thiophenoxy)titanium. Bis(thiophenoxy)titanocene

[1292-47-3]

As Dibromobis(η^5-2,4-cyclopentadien-1-yl)titanium, Ti-00024 with

$$X = —SPh$$

M 396.400

Deep-red cryst. Mp 199-206°.

Köpf, H. *et al*, *Z. Anorg. Allg. Chem.*, 1965, **340**, 139 (*synth, ir, nmr*)
Coutts, R.S.P. *et al*, *Aust. J. Chem.*, 1966, **19**, 1377 (*synth, nmr*)
Dessy, R.E. *et al*, *J. Am. Chem. Soc.*, 1966, **88**, 5112 (*polarog*)
Abel, E.W. *et al*, *J. Organomet. Chem.*, 1968, **14**, 285 (*synth*)
Fachinetti, G. *et al*, *J. Chem. Soc., Dalton Trans.*, 1974, 2433 (*synth*)
Muller, E.G. *et al*, *J. Organomet. Chem.*, 1976, **111**, 73 (*cryst struct*)

$C_{22}H_{20}Ti$ **Ti-00118**

Bis(η^5-cyclopentadien-1-yl)diphenyltitanium, 10CI

Diphenyltitanocene

[1273-09-2]

As Ti-00024 with

$$X = Ph$$

M 332.280

Orange-yellow cryst. Mp 142-8°.

Summers, L. *et al*, *J. Am. Chem. Soc.*, 1954, **76**, 2278 (*synth*)
Beachell, H.C. *et al*, *Inorg. Chem.*, 1965, **4**, 1133 (*synth, nmr*)
Tel'noi, V.I. *et al*, *Dokl. Akad. Nauk SSSR, Ser. Sci. Khim.*, 1967, **174**, 1374.
Kocman, V. *et al*, *J. Chem. Soc., Chem. Commun.*, 1971, 1340 (*cryst struct*)
Boekel, C.P. *et al*, *J. Organomet. Chem.*, 1974, **81**, 371 (*synth*)
Boekel, C.P. *et al*, *J. Organomet. Chem.*, 1975, **102**, 161.

$C_{22}H_{36}Ti$ **Ti-00119**

Dimethylbis[(1,2,3,4,5-η)-1,2,3,4,5-pentamethyl-2,4-cyclopentadien-1-yl]titanium, 9CI

[11136-41-7]

As Dichlorobis[(1,2,3,4,5-η)-1,2,3,4,5-pentamethyl-2,4-cyclopentadien-1-yl]titanium, Ti-00111 with

$$X = Me$$

M 348.406

Yellow solid. Thermal stability greatly enhanced over C_5H_5 analogue.

Bercaw, J.E. *et al*, *J. Am. Chem. Soc.*, 1971, **93**, 2045 (*synth, nmr*)
Bercaw, J.E. *et al*, *J. Am. Chem. Soc.*, 1972, **94**, 1219 (*synth, ms*)
Pez, G.I. *et al*, *Adv. Organomet. Chem.*, 1981, **19**, 1 (*rev*)

$C_{22}H_{39}FeN_3Ti$ Ti-00120

Tris(diethylaminato)ferrocenyltitanium

Tris(N-*ethylethanaminato*)*ferrocenyltitanium*, 9CI. [*Tris*(*diethylamino*)*titanium*]*ferrocene*

[42993-54-4]

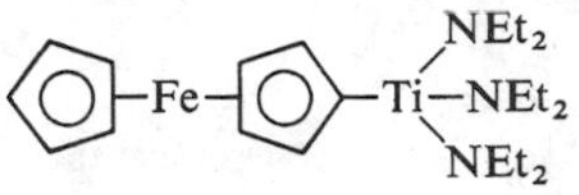

Fe–Ti(NEt_2)$_3$

M 449.297

Dark-red cryst. (Et_2O/pet. ether). Mp 31.5-32.5°.

Bürger, H. *et al*, *J. Organomet. Chem.*, 1973, **56**, 269 (*synth, pmr, cmr, ir*)

$C_{23}H_{20}ClNTi$ Ti-00121

Chlorodi-π-cyclopentadienyl(1,1-diphenylmethyleniminato)titanium, 8CI

Chlorobis(η^5-*cyclopentadienyl*)(*diphenylketimido*)*titanium. Chloro*(*diphenylketimido*)*titanocene*

[11106-00-6]

As Chlorobis(η^5-2,4-cyclopentadien-1-yl)methyltitanium, Ti-00045 with

X = —N=CPh_2

M 393.751

Yellow-orange solid.

Collier, M.R. *et al*, *Inorg. Nucl. Chem. Lett.*, 1971, **7**, 689 (*synth, ir, nmr, ms*)

$C_{23}H_{20}O_3Ti$ Ti-00122

(η^5-2,4-Cyclopentadien-1-yl)triphenoxytitanium, 9CI

[12149-47-2]

As Tribromo(η^5-2,4-cyclopentadien-1-yl)titanium, Ti-00007 with

R = OPh

M 392.289

Yellow solid. Mp 102-4°.

Nesmayanov, A.N. *et al*, *Izv. Akad. Nauk SSSR, Ser. Khim.*, 1971, 2354, 2588 (*synth, ir, pmr*)
Nesmayanov, A.N. *et al*, *J. Struct. Chem.* (*Engl. Transl.*), 1975, **16**, 705 (*cmr*)

$C_{24}H_{18}N_2O_8Ti$ Ti-00123

Bis(η^5-2,4-cyclopentadien-1-yl)bis(4-nitrobenzoato-*O'*)titanium, 9CI

Bis(*4-nitrobenzoato*)*titanocene*

[11082-60-3]

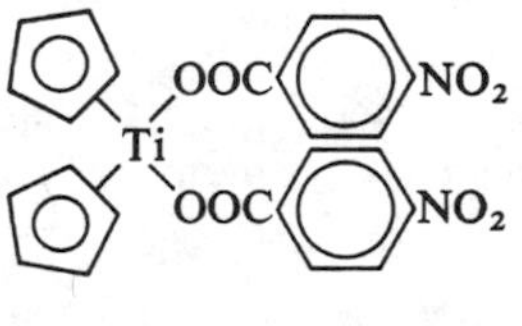

Ti(OOC–C_6H_4–NO_2)$_2$

M 510.295

Orange cryst. Mp 180° dec.

Razuvaev, G.A. *et al*, *J. Gen. Chem. USSR* (*Engl. Transl.*), 1971, **41**, 1560 (*synth, ir*)
Kuntsevich, T.S. *et al*, *Kristallografiya*, 1976, **21**, 80 (*cryst struct*)

$C_{24}H_{20}Ti$ Ti-00124

Tetraphenyltitanium, 9CI

Titanium tetraphenyl

[2371-72-4]

$TiPh_4$

M 356.302

Monomeric in C_6H_6. Yellow cryst. Mp ca. 45° dec. Stable when pure in soln., but deteriorates in the solid state. Forms adducts in soln. with *N*-donors.

Boustany, K.S. *et al*, *Helv. Chim. Acta*, 1967, **50**, 1305 (*synth*)
Jacot-Guillarmod, A. *et al*, *Chimia*, 1969, **23**, 188 (*synth*)
Tabacchi, R. *et al*, *Chimia*, 1970, **24**, 271 (*derivs*)
Tabacchi, R. *et al*, *Helv. Chim. Acta*, 1970, **53**, 1971; 1977 (*synth, nmr*)

$C_{24}H_{24}Ti$ Ti-00125

Dibenzylbis(η^5-2,4-cyclopentadien-1-yl)titanium

Bis(η^5-*2,4-cyclopentadien-1-yl*)*bis*(*phenylmethyl*)*titanium*, 10CI. *Dibenzyltitanocene*

[37299-11-9]

As Dibromobis(η^5-2,4-cyclopentadien-1-yl)titanium, Ti-00024 with

X = CH_2Ph

M 360.334

Orange-yellow solid. Mp ca. 100° dec.

Fachinetti, G. *et al*, *J. Chem. Soc., Chem. Commun.*, 1972, 654 (*nmr*)
Boekel, C.P. *et al*, *J. Organomet. Chem.*, 1975, **102**, 317 (*synth*)
Clark, R.J.H. *et al*, *J. Chem. Soc., Dalton Trans.*, 1976, 120 (*synth*)

$C_{24}H_{24}Ti_2$ Ti-00126

Tris(1,3,5,7-cyclooctatetraene)dititanium, 10CI

Bis[η-*cyclooctatetraene*(*2−*)*titanium*](μ-η^4,η^4-*cyclooctatetraene*)

[12149-66-5]

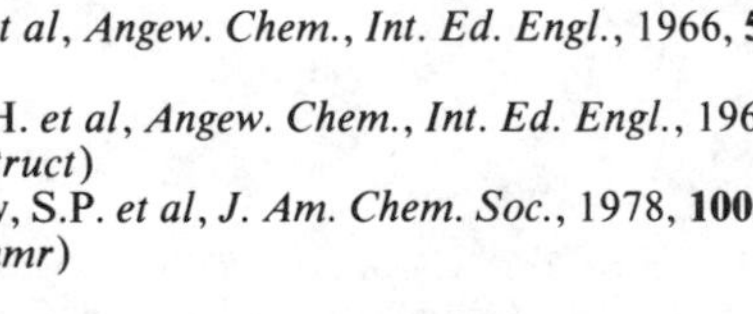

Ti–Ti

M 408.214

Extremely air-sensitive yellow cryst. Sl. sol. aromatic solvs. Gives a chain ion with K in THF. Paramagnetic, μ = 2.56 BM.

Breil, H. *et al*, *Angew. Chem., Int. Ed. Engl.*, 1966, **5**, 898 (*synth*)
Dietrich, H. *et al*, *Angew. Chem., Int. Ed. Engl.*, 1966, **5**, 899 (*cryst struct*)
Kolesnikov, S.P. *et al*, *J. Am. Chem. Soc.*, 1978, **100**, 999 (*esr, props, nmr*)

$C_{24}H_{44}Ti$ Ti-00127

Tetracyclohexyltitanium, 10CI

[30169-70-1]

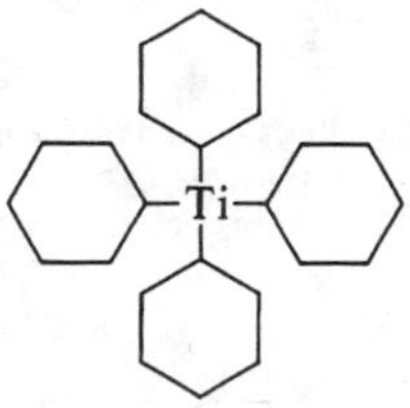

M 380.492

Yellow cryst. (pentane). Stable at −80° for weeks and for days at −50°. Dec. at low temp.

▷Inflames in air

Weber, J.B. *et al*, *Helv. Chim. Acta*, 1978, **61**, 2949 (*synth*, *props*)

$C_{25}H_{33}OTi$ Ti-00128

Bis(η^5-cyclopentadienyl)(2,6-di-*tert*-butyl-4-methylphenoxy)-titanium

[2,6-Bis(1,1-dimethylethyl)-4-methylphenolato]bis(η^5-2,4-cyclopentadien-1-yl)titanium, 10CI

[74481-90-6]

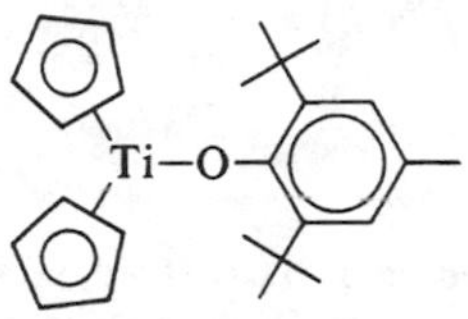

M 397.415

Monomeric, unlike analogues with less bulky groups. Deep-purple cryst. Mp 167-9°.

Cetinkaya, B. *et al*, *J. Organomet. Chem.*, 1980, **188**, C31 (*synth*, *cryst struct*, *ir*)

$C_{26}H_{20}Ti$ Ti-00129

Bis(η^5-2,4-cyclopentadien-1-yl)bis(phenylethynyl)titanium, 10CI

Di(phenylethynyl)titanocene

[12303-93-4]

As Dibromobis(η^5-2,4-cyclopentadien-1-yl)titanium, Ti-00024 with

X = —C≡CPh

M 380.324

Orange cryst.

Köpf, H. *et al*, *J. Organomet. Chem.*, 1967, **10**, 383 (*synth*, *ir*, *nmr*, *uv*)

Teuben, J.H. *et al*, *J. Organomet. Chem.*, 1969, **17**, 87 (*synth*, *ir*, *ms*, *uv*)

Jenkins, A.D. *et al*, *J. Organomet. Chem.*, 1970, **23**, 165 (*ir*)

$C_{26}H_{31}NTi$ Ti-00130

Tribenzyl(piperidinyl)titanium

Tris(phenylmethyl)-1-piperidinyltitanium, 9CI

[59512-63-9]

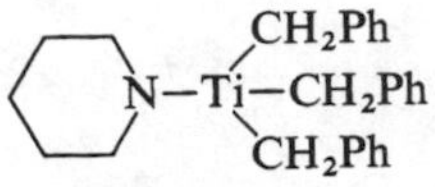

M 405.418

Red solid below 5°. Mp 5°. Dec. at 30°.

Bürger, H., *J. Organomet. Chem.*, 1976, **108**, 69 (*synth*, *ir*, *nmr*)

$C_{28}H_{25}ClGeTi$ Ti-00131

Chlorobis(η^5-cyclopentadienyl)(triphenylgermyl)titanium, 8CI

Chloro(triphenylgermyl)titanocene

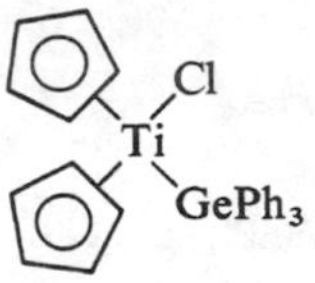

M 517.429

Green solid. Mp 193-6° dec. Unusual colour for Ti(IV) complex.

Coutts, R.S.P. *et al*, *J. Chem. Soc., Chem. Commun.*, 1968, 260 (*synth*, *nmr*)

$C_{28}H_{25}ClSnTi$ Ti-00132

Chlorobis(η^5-cyclopentadienyl)(triphenylstannyl)titanium, 8CI

Chloro(triphenylstannyl)titanocene

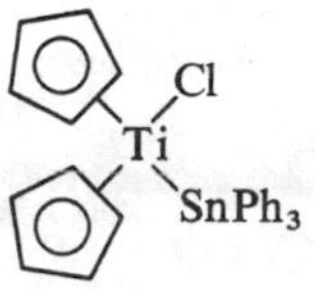

M 563.529

Green solid. Mp 177-80° dec. Unusual colour for Ti(IV) complex.

Coutts, R.S.P. *et al*, *J. Chem. Soc., Chem. Commun.*, 1968, 260 (*synth*, *mr*)

$C_{28}H_{25}GeTi$ Ti-00133

Bis(η^5-cyclopentadienyl)triphenylgermyltitanium

Di-π-cyclopentadienyl(triphenylgermyl)titanium, 8CI

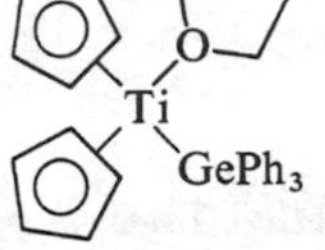

THF complex

M 481.976

THF complex:

$C_{32}H_{33}GeOTi$ M 554.082

Green solid. Paramagnetic with 1 unpaired electron.

Coutts, R.S.P. *et al*, *J. Chem. Soc., Chem. Commun.*, 1968, 260 (*synth*, *ir*, *nmr*)

$C_{28}H_{25}P_3Ti$ Ti-00134

Bis(η^5-2,4-cyclopentadien-1-yl)[1,2,3-triphenyltriphosphinato(2−)-P^1,P^3]titanium, 10CI

[37299-20-0]

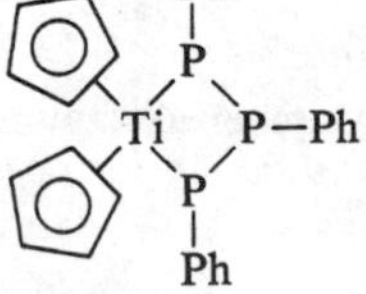

M 502.307

Monomeric in dioxan. Mod. air-stable, dark-violet solid. Mp 243-5°.

Issleib, K. *et al*, *Angew. Chem., Int. Ed. Engl.*, 1972, **11**, 527 (*synth*, *ms*, *nmr*)

$C_{28}H_{25}SnTi$ Ti-00135

Bis(η^5-cyclopentadienyl)triphenylstannyltitanium

Di-π-cyclopentadienyl(triphenylstannyl)titanium, 8CI

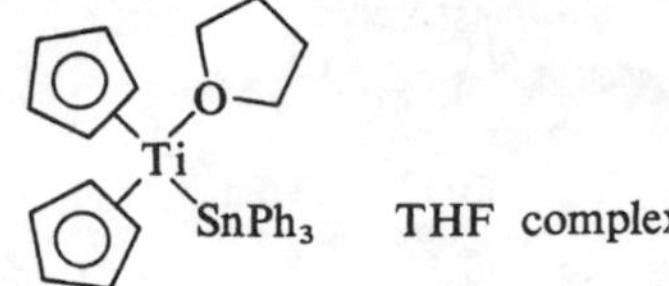

M 528.076

THF complex:

$C_{32}H_{33}SnOTi$ M 600.182

Green solid. Mp 177-8°. Paramagnetic with 1 unpaired electron.

Coutts, R.S.P. *et al*, *J. Chem. Soc., Chem. Commun.*, 1968, 260 (*synth, ir, nmr*)

$C_{28}H_{28}Ti$ Ti-00136

Tetrabenzyltitanium

Tetrakis(phenylmethyl)titanium, 9CI

[17520-19-3]

$Ti(CH_2Ph)_4$

M 412.409

Olefin polymerisation catalyst. Red cryst. (pentane). Spar. sol. aliphatic hydrocarbons, sol. aromatic hydrocarbons. Mp ca. 70°. Dec. after several hours at 100° in heptane. Nonequiv. Ti—C—Ph angles in solid state.

Giannini, U. *et al*, *J. Chem. Soc., Chem. Commun.*, 1968, 940 (*synth, use, nmr, ir*)
Tabacchi, R., *et al*, *Chimia*, 1970, **24**, 221 (*props*)
Bassi, I.W. *et al*, *J. Am. Chem. Soc.*, 1971, **93**, 3788 (*cryst struct*)
Bröser, W. *et al*, *J. Organomet. Chem.*, 1971, **32**, 385 (*ir*)
Davies, G.R. *et al*, *J. Chem. Soc., Chem. Commun.*, 1971, 1511 (*cryst struct*)
Jacot-Guillarmod, A. *et al*, *Chimia*, 1974, **28**, 15 (*props*)

$C_{28}H_{30}O_3Ti$ Ti-00137

[(1,2,3,4,5-η)-1,2,3,4,5-Pentamethyl-2,4-cyclopentadien-1-yl]triphenoxytitanium, 9CI

[36523-12-3]

$Ti(OPh)_3$

M 462.423

Liq. Bp_1 214-6°.

Nesmayanov, A.N. *et al*, *Izv. Akad. Nauk SSSR, Ser. Khim.*, 1971, 2354, 2588 (*synth, ir, pmr*)
Nesmayanov, A.N. *et al*, *J. Struct. Chem. (Engl. Transl.)*, 1975, **16**, 705 (*cmr*)

$C_{28}H_{38}Fe_2N_2Ti$ Ti-00138

Diferrocenylbis(diethylaminato)titanium

*Bis(*N*-ethylethanaminato)diferrocenyltitanium*, 9CI

[39470-19-4]

NEt_2 — Fe — Ti — Fe — NEt_2

M 562.196

Cherry-red cryst. (Et_2O/pet. ether). Mp 72° dec.

Bürger, H. *et al*, *J. Organomet. Chem.*, 1973, **56**, 269 (*synth, pmr, cmr, ir*)

$C_{28}H_{44}Ti$ Ti-00139

Tetrakis(bicyclo[2.2.1]-hept-1-yl)titanium, 9CI

Tetrakis(1-norbornyl)titanium

[36333-76-3]

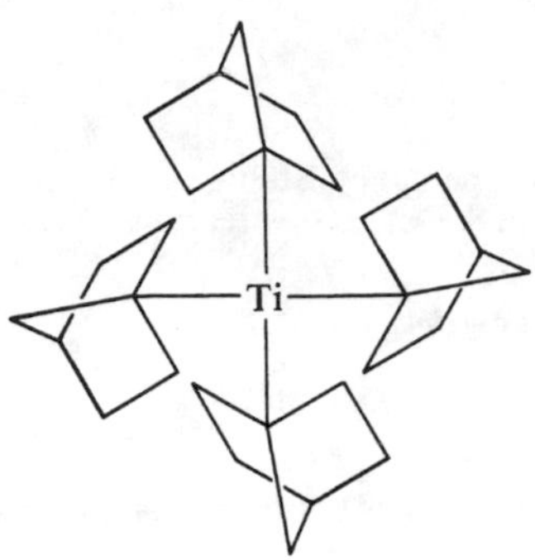

M 428.536

Yellow cryst. Air stable in isooctane soln. No pyridine adducts detected.

Bower, B.K. *et al*, *J. Am. Chem. Soc.*, 1972, **94**, 2512 (*synth, props, ms*)

$C_{29}H_5Co_7O_{24}Ti$ Ti-00140

(η^5-2,4-Cyclopentadien-1-yl)[μ_4-[methanolato(4−)-*C*:*C*:*C*:*O*]](nonacarbonyltricobalt)(tetracarbonylcobalt)titanium, 10CI

(η-Cyclopentadienyl)bis[μ_3-(oxymethylidyne)cyclotris(tricarbonylcobalt)](tetracarbonylcobaltio)titanium

[67421-95-8]

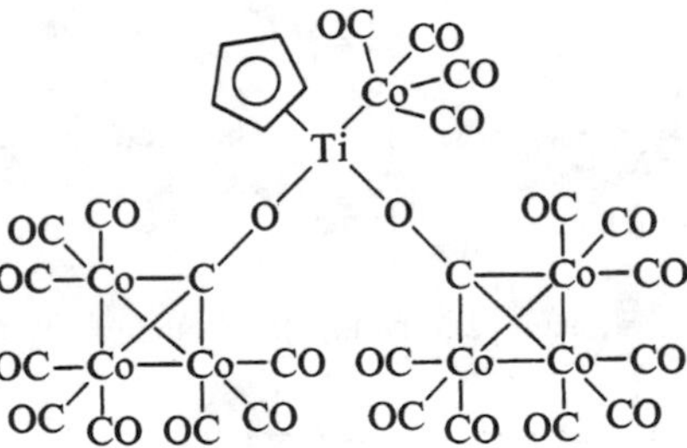

M 1197.757

Air-sensitive cryst. Black cryst. (toluene). Spar. sol. aromatic solvs., almost insol. aliphatic hydrocarbons. Mp 125° dec.

Schmid, G. *et al*, *Chem. Ber.*, 1978, **111**, 1239 (*synth, ir, nmr, cryst struct*)

$C_{30}H_{24}Ti$ Ti-00141

Bis[(1,2,3,3*a*,7*a*-η)-1*H*-inden-1-yl]diphenyltitanium, 9CI

[49596-07-8]

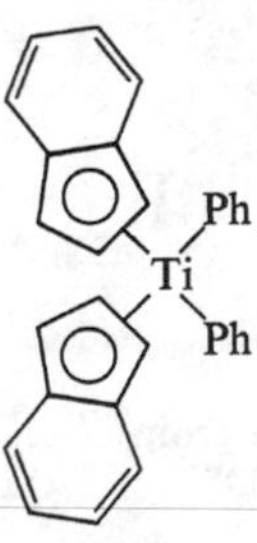

M 432.400

Orange-red air-stable cryst. Mp 105° dec.

Rausch, M.D., *Pure Appl. Chem.*, 1970, **30**, 523 (*synth, nmr*)
Samuel, E. *et al, J. Am. Chem. Soc.*, 1973, **95**, 6263 (*synth, nmr*)

$C_{30}H_{28}Fe_2Ti$ Ti-00142

Bis(η[5]-2,4-cyclopentadien-1-yl)diferrocenyltitanium

Diferrocenyldicyclopentadienyltitanium. Di-π-cyclopentadienyl-di-σ-ferrocenyltitanium

[65274-19-3]

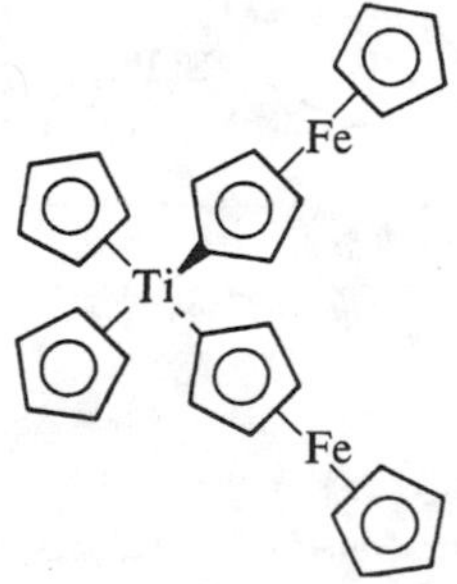

M 548.125

Deep-green cryst. Solns. are air-sensitive.

Razuvaev, G.A. *et al, J. Organomet. Chem.*, 1977, **141**, 313 (*synth*)
Osborne, A.G. *et al, J. Organomet. Chem.*, 1979, **181**, 425.
Kotz, J.C. *et al, Organometallics*, 1983, **2**, 68 (*props*)

$C_{34}H_{68}FeN_6Ti_2$ Ti-00143

Hexakis(diethylaminato)-μ-1,1′-ferrocenediyldititanium

Hexakis(N*-ethylethanaminato)-μ-1,1′-ferrocenediyldititanium,* 9CI. *1,1′-Bis[tris(diethylamino)titanium]-ferrocene*

[42920-73-0]

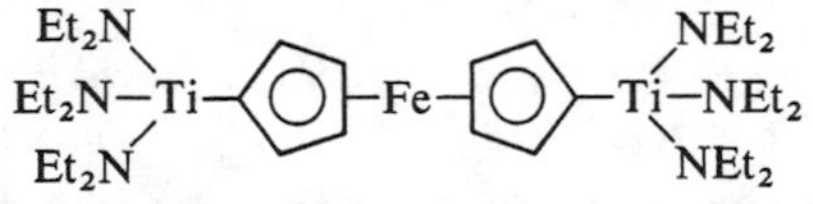

M 712.558

Red-brown, water-sensitive cryst. (Et_2O/pet. ether). Mp 40-50° dec.

Bürger, H. *et al, J. Organomet. Chem.*, 1973, **56**, 269 (*synth, pmr, cmr, ir*)
Thewalt, U. *et al, Z. Naturforsch, B*, 1975, **30**, 636 (*struct*)
Bürger, H. *et al, Z. Anorg. Allg. Chem.*, 1976, **423**, 112 (*pmr, cmr*)

$C_{34}H_{68}N_6RuTi_2$ Ti-00144

Hexakis(diethylaminato)-μ-1,1′-ruthenocenediyldititanium

Hexakis(N*-ethylethanaminato)-μ-1,1′-ruthenocenediyldititanium,* 9CI. *1,1′-Bis[tris(diethylamino)titanium]-ruthenocene*

[60898-12-6]

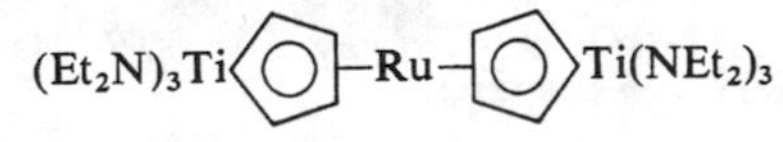

M 757.781

Yellow cryst. (Et_2O). Mp 90° dec.

Bürger, H. *et al, Z. Anorg. Allg. Chem.*, 1976, **423**, 112 (*synth, pmr, cmr*)

$C_{36}H_{30}Ti_2$ Ti-00145

Tetra-π-cyclopentadienylbis(phenylethynyl)dititanium, 8CI

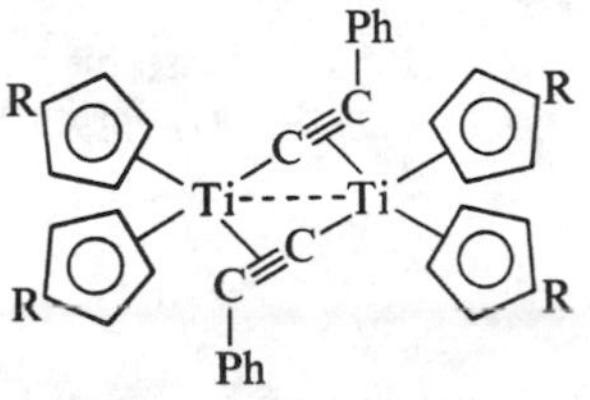

R = H

M 558.393

Dark-green cryst. Mp >200°.

Monomer: Bis(η[5]-2,4-cyclopentadien-1-yl)-(phenylethynyl)titanium.
$C_{18}H_{15}Ti$ M 279.197
Unknown.

Teuben, J.H. *et al, J. Organomet. Chem.*, 1969, **17**, 87 (*synth, ir, uv, ms*)

$C_{38}H_{30}Ti$ Ti-00146

Bis(η[5]-2,4-cyclopentadien-1-yl)(1,2,3,4-tetraphenyl-1,3-butadiene-1,4-diyl)titanium, 10CI

Bis(η[5]-cyclopentadienyl)-2,3,4,5-tetraphenyltitanole. Bis(η[5]-cyclopentadienyl)-2,3,4,5-tetraphenyltitanacyclopentadiene

[1317-21-1]

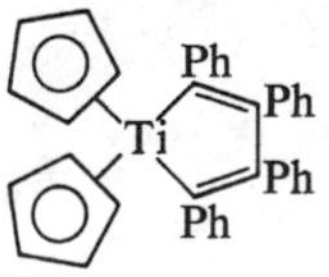

M 534.535

Accessible by a variety of routes. Green cryst., air-stable as solid. Mp 157-9° (150°).

Vol'pin, M.E. *et al, Dokl. Akad. Nauk SSSR, Ser. Sci. Khim.*, 1963, **151**, 1100 (*synth*)
Sonogashira, K. *et al, Bull. Chem. Soc. Jpn.*, 1966, **39**, 1178 (*synth, ir, nmr, uv*)
Siegert, F.W. *et al, Recl. Trav. Chim. Pays-Bas*, 1970, **89**, 764 (*synth, ir, nmr, uv*)
Watt, G.W. *et al, J. Am. Chem. Soc.*, 1970, **92**, 826 (*synth*)
Atwood, J.L. *et al, J. Am. Chem. Soc.*, 1976, **98**, 2454 (*cryst struct*)

$C_{40}H_{38}Ti_2$ Ti-00147

[μ-(1,4-Diphenyl-1,3-butadiene-1,2,3,4-tetrayl-*C*[1],*C*[3]:*C*[2],*C*[4])]tetrakis[(1,2,3,4,5-η)-1-methyl-2,4-cyclopentadien-1-yl]dititanium, 10CI

Bis[bis(η[5]-methylcyclopentadienyl)titanium](μ-η[2]:η[2]-1,4-diphenyl-1,3-butadiene)

[59645-02-2]

As Tetra-π-cyclopentadienylbis(phenylethynyl)dititanium, Ti-00145 with

R = CH_3

M 614.500

Green cryst. Sol. toluene, spar. sol. other common solvs.

Sekutowski, D.G. *et al*, *J. Am. Chem. Soc.*, 1976, **98**, 1376 (*synth, cryst struct, ir, ms*)

$C_{40}H_{44}Ti$ — Ti-00148

Tetrakis(2-methyl-1-phenyl-1-propenyl)titanium, 10CI

Tetrakis(2,2-dimethyl-1-phenylethenyl)titanium

[63422-29-7]

$[(H_3C)_2C{=}CPh]_4Ti$

M 572.668

Yellow solid. Mp >−20° dec. Unstable above −20°.

Cardin, C.J. *et al*, *J. Organomet. Chem.*, 1977, **132**, C23 (*synth, nmr*)

$C_{40}H_{60}N_2Ti_2$ — Ti-00149

[η-(Dinitrogen-*N,N'*)]tetrakis[(1,2,3,4,5-η)-1,2,3,4,5-pentamethyl-2,4-cyclopentadien-1-yl]dititanium, 10CI

Bis[bis(η⁵-pentamethylcyclopentadienyl)titanium]-μ-η¹,η¹'-dinitrogen

[11136-46-2]

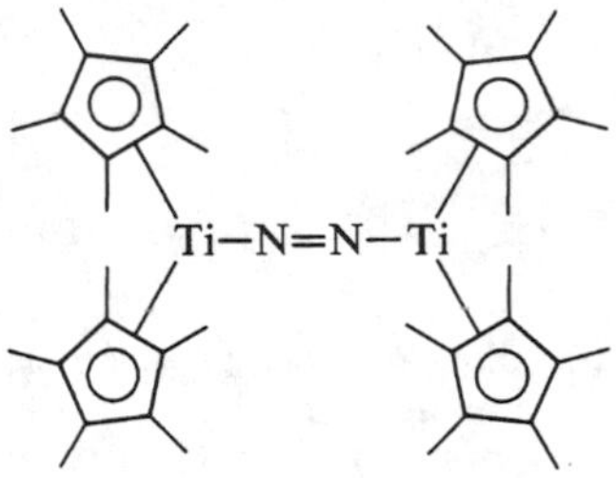
Ti—N=N—Ti

M 664.687

Blue-black cryst. Below −42° coordinates 2 further moles of N_2 end-on, one on each Ti atom.

Sanner, R.D. *et al*, *J. Am. Chem. Soc.*, 1976, **98**, 8358 (*synth, cryst struct, props*)

Bercaw, J., *Kinet. Katal*, 1977, **18**, 549 (*synth, cryst struct, props*)

$C_{46}H_{40}Ge_2Ti$ — Ti-00150

Bis(η⁵-2,4-cyclopentadien-1-yl)bis(triphenylgermyl)titanium, 9CI

Bis(triphenylgermyl)titanocene

[39330-96-6]

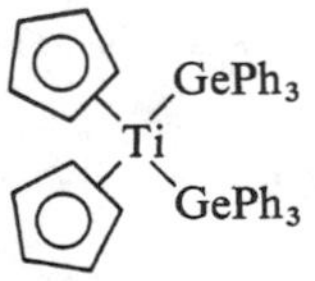
GePh₃
Ti
GePh₃

M 785.882

Burgundy solid. Mp 171-3°.

Coutts, R.S.P. *et al*, *J. Chem. Soc., Chem. Commun.*, 1968, 260 (*synth, nmr*)

Razuvaev, G.A. *et al*, *Izv. Akad. Nauk SSSR, Ser. Khim.*, 1972, 1605 (*synth*)

$C_{46}H_{40}O_2Si_2Ti$ — Ti-00151

Bis(η⁵-2,4-cyclopentadien-1-yl)bis(triphenylsilyloxy)titanium

Di-π-cyclopentadienylbis(triphenylsilanolato)titanium, 8CI

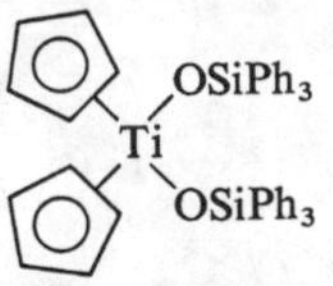
OSiPh₃
Ti
OSiPh₃

M 728.872

Bright-yellow cryst. (ligroin). Sol. most org. solvs., insol. MeOH. Mp 199-200°, 210°. Air-stable but dec. by acid or alkali.

Noltes, J.G. *et al*, *Recl. Trav. Chim. Pays-Bas*, 1962, **81**, 39 (*synth, ir*)

U.S.P., 3 030 394, (*1962*); *CA*, **57**, 3589 (*synth*)

Takiguchi, T. *et al*, *Bull. Chem. Soc. Jpn.*, 1968, **41**, 2810 (*synth, struct*)

$C_{46}H_{40}Sn_2Ti$ — Ti-00152

Bis(η-cyclopentadienyl)bis(triphenylstannyl)titanium, 8CI

Bis(triphenylstannyl)titanocene

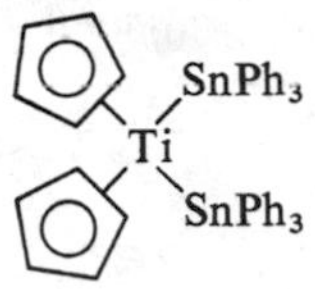
SnPh₃
Ti
SnPh₃

M 878.082

Green solid. Mp 80° dec.

Coutts, R.S.P. *et al*, *J. Chem. Soc., Chem. Commun.*, 1968, 260 (*synth, nmr*)

$C_{58}H_{50}Si_4Ti$ — Ti-00153

Bis(η⁵-2,4-cyclopentadien-1-yl)(1,1,2,2,3,3,4,4-octaphenyl-1,4-tetrasilanediyl)titanium, 10CI

Bis(η⁵-cyclopentadienyl)cyclo(1,1,2,2,3,3,4,4-octaphenyltetrasilicon-1,4-diyl)titanium

[71776-21-1]

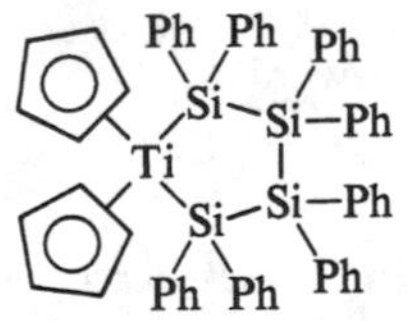
Ph Ph Ph
Si
Si—Ph
Ti
Si—Ph
Si
Ph Ph Ph

M 907.255

Monomeric in C_6H_6. Small green cryst. (Me_2CO/pet. ether). Sol. aromatics, ethers, hydrocarbons, dec. in $CHCl_3$. Mp 140° dec. Air-stable solid. Cryst. give powder *in vacuo*.

Holtman, M.S. *et al*, *J. Organomet. Chem.*, 1980, **187**, 147 (*synth, ir, nmr, uv*)

Suggestions for new Entries are welcomed. Please write to the Editor, Dictionary of Organometallic Compounds, Chapman and Hall Ltd, 11 New Fetter Lane, London EC4P 4EE

Tm Thulium

S. A. Cotton

Thulium (Fr., Ger.), Tulio (Sp.), Tullio (Ital.), Тулий (Tuulii) (Russ.), ツリウム (Japan.)

Atomic Number. 69

Atomic Weight. 168.9342

Electronic Configuration. [Xe] $4f^{13}\,6s^2$

Oxidation State. +3

Coordination Number. 6 or greater.

Colour. Green or yellow-green.

Availability. Starting materials are Tm_2O_3 and anhydrous $TmCl_3$ which are expensive. See also under Lanthanum.

Handling. See under Lanthanum.

Toxicity. See under Lanthanum.

Isotopic Abundance. ^{169}Tm, 100%.

Spectroscopy. ^1H nmr spectra of paramagnetic thulium compounds show large paramagnetic shifts with broad lines. See also under Lanthanum.

Analysis. See under Lanthanum.

References. See under Lanthanum.

$C_6H_{18}Tm^{\ominus\ominus\ominus}$ Tm-00001

Hexamethylthulate(3−)

$TmMe_6^{\ominus\ominus\ominus}$

M 259.142 (ion)

Tris[(tetramethylethylenediamine)lithium] salt: [76206-92-3].
$C_{24}H_{66}Li_3N_6Tm$ M 628.583
White cryst. Sol. Et_2O. Mp 109-14°. Extremely air- and moisture-sensitive.

Schumann, H. *et al*, *Angew. Chem., Int. Ed. Engl.*, 1981, **20**, 120 (*synth*)

$C_{10}H_{10}ClTm$ Tm-00002

Chlorobis(η^5-2,4-cyclopentadien-1-yl)thulium

Chlorodicyclopentadienylthulium. Dicyclopentadienylthulium chloride

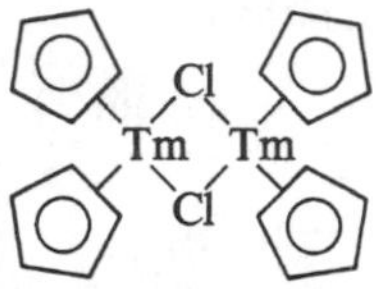

M 334.576
Dimeric.

Dimer: [60482-72-6]. *Di-μ-chlorotetrakis(η^5-2,4-cyclopentadien-1-yl)dithulium*, 9CI.
$C_{20}H_{20}Cl_2Tm$ M 500.218
Yellow-green cryst. Sol. THF. Subl. at 200° *in vacuo*.

Holton, J. *et al*, *J. Chem. Soc., Dalton Trans.*, 1979, 45 (*synth*)

$C_{11}H_{13}Tm$ Tm-00003

Bis(η^5-2,4-cyclopentadien-1-yl)methylthulium

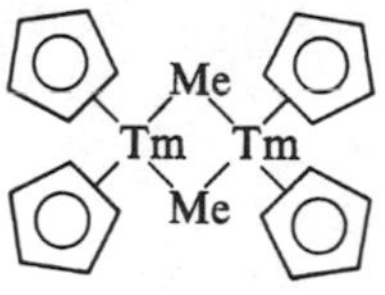

M 314.158
Dimeric.

Dimer: [70844-68-7]. *Tetrakis(η^5-2,4-cyclopentadien-1-yl)di-μ-methyldithulium*, 9CI.
$C_{22}H_{26}Tm_2$ M 628.316
Pale-green cryst. Sol. CH_2Cl_2, C_6H_6, toluene. Dec. >160°. Air-sensitive. $\mu_{eff} = 7.5\mu_B$.

Holton, J. *et al*, *J. Chem. Soc., Dalton Trans.*, 1979, 54 (*synth, ir*)

$C_{15}H_{15}Tm$ Tm-00004

Tris(η^5-2,4-cyclopentadien-1-yl)thulium

Tricyclopentadienylthulium

[1272-26-0]

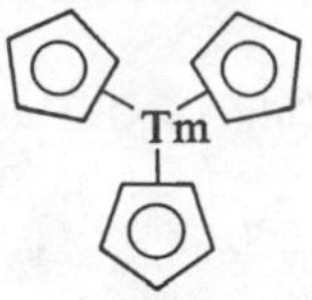

M 364.218
Yellow-green cryst. Sol. THF. Mp 278°. Hydrol. by H_2O. Subl. at 220° *in vacuo*. $\mu_{eff} = 7.1\mu_B$.

Isocyanocyclohexane complex: [37299-05-1]. *Tris(2,4-cyclopentadien-1-yl)isocyanocyclohexanethulium.*
$C_{22}H_{26}NTm$ M 473.388
Sol. C_6H_6. Mp 162°.

Fischer, E.O. *et al*, *J. Organomet. Chem.*, 1965, **3**, 181 (*synth*)
Pappalardo, R., *J. Mol. Spectrosc.*, 1969, **29**, 13 (*uv*)
v. Ammon, R. *et al*, *Inorg. Nucl. Chem. Lett.*, 1969, **5**, 315 (*nmr*)
v. Ammon, R. *et al*, *Ber. Bunsenges. Phys. Chem.*, 1972, **76**, 995.
Aleksanyan, V.T. *et al*, *J. Raman Spectrosc.*, 1974, **2**, 345 (*raman*)

U Uranium

S. A. Cotton

Uranium (Fr.), Uran (Ger.), Uranio (Sp., Ital.), Уран (Uuran) (Russ.), ウラン (Japan.)

Atomic Number. 92

Atomic Weight. 238.029. Individual samples may vary in atomic weight since commercial materials are always depleted in the fissionable isotope (see below).

Electronic Configuration. [Rn] $5f^3\, 6d^1\, 7s^2$

Oxidation States. +3, +4 are the most common.

Coordination Number. Generally 6 or higher with conventional ligands.

Colour. Complexes of U(IV) are usually green or red-brown. The small number of U(III) complexes are brown, green or blue-black.

Availability. The most common starting material is anhydrous UCl_4 which is commercially available and relatively expensive but which may be synthesized from the reaction between U_3O_8 and hexachloropropene. See *Inorg. Synth.*, 1957, **5,** 143; 1974, **15,** 243 and Karraker, D. G. *et al.*, *J. Am. Chem. Soc.*, 1974, **96,** 6885.

Handling. All organouranium compounds and those of other actinides should be assumed to be destroyed by traces of air and/or water, unless specifically stated otherwise and should therefore be handled in an inert atmosphere. In general they are soluble in organic solvents such as THF and may also be soluble in less polar solvents such as toluene. Some compounds are pyrophoric.

Toxicity. Although not as toxic as other actinides, the radioactivity of α-emitting uranium compounds means that they should be treated with respect. The main hazard is oral ingestion.

Isotopic Abundance. Natural uranium consists of ^{235}U, 0.72% ($t_{1/2} = 7 \times 10^8$ y) and ^{238}U, 99.2% ($t_{1/2} = 4.51 \times 10^9$ y) but commercial materials are always depleted in the fissionable ^{235}U.

Spectroscopy. Uranium itself is not commonly used as an nmr probe although nmr parameters have been obtained for UF_6 in solution. Compounds of U(IV) are paramagnetic ($5f^2$) but their ^1H nmr spectra display reasonably sharp resonances with large isotropic shifts. The linewidths in ^1H nmr spectra of U(III) ($5f^3$) are broader than those of U(IV).

Analysis. Microanalytical results for carbon content of organometallic compounds of uranium and other actinides are frequently low owing to carbide formation.

Analysis for uranium can be carried out by gravimetric or volumetric methods, both first requiring destruction of the compound (eg. by combustion in a furnace) and destruction of any remaining organic material with acid. The hydroxide can be precipitated from solution, ignited, and weighed as U_3O_8; alternatively the 8-hydroxyquinoline complex has been used.

Titrimetric methods involve reduction in solution to U^{4+} and titration with Ce^{4+} or $Cr_2O_7^{2-}$.

References. In addition to reviews listed in the introduction to the *Sourcebook*, the following provide further reading:

General

Marks, T. J., *Prog. Inorg. Chem.*, 1979, **25,** 224.
Cotton, S. A., *J. Organomet. Chem. Lib.*, 1977, **3,** 189.
Marks, T. J., *Acc. Chem. Res.*, 1976, **9,** 223.
Bagnall, K. W., *The Actinide Elements*, Elsevier, 1972.
Kannellakopulos, B. and Bagnall, K. W., *MTP International Review of Science, Inorganic Chemistry Series 1*, Emeléus, H. J. and Bagnall, K. W. Eds, University Book Press, 1971, **7,** 229.
Organometallics of the f-Elements, Marks, T. J. and Fischer, R. D. Eds, Reidel, Dortrecht, 1979.

Nmr

Von Ammon, R. *et al.*, *Inorg. Nucl. Chem. Lett.*, 1969, **5,** 219.
Marks, T. J. *et al.*, *J. Am. Chem. Soc.*, 1973, **95,** 5529; 1976, **98,** 703.
Fagan, P. J. *et al.*, *J. Am. Chem. Soc.*, 1981, **103,** 6650.
Fagan, P. J. *et al.*, *Organometallics*, 1981, **1,** 170.

Mass spectra

Müller, J., *Chem. Ber.*, 1969, **102,** 52.

Analysis

Halstead, G. W. *et al.*, *J. Am. Chem. Soc.*, 1975, **97,** 3050.
Brown, D., *Halides of the Lanthanides and Actinides*, Wiley, 1968.
Reynolds, L. T. *et al.*, *J. Inorg. Nucl. Chem.*, 1956, **2,** 246.
Bagnall, K. W. *et al.*, *J. Chem. Soc.*, 1961, 1611.
Bibler, J. P. *et al.*, *Inorg. Chem.*, 1968, **7,** 982.

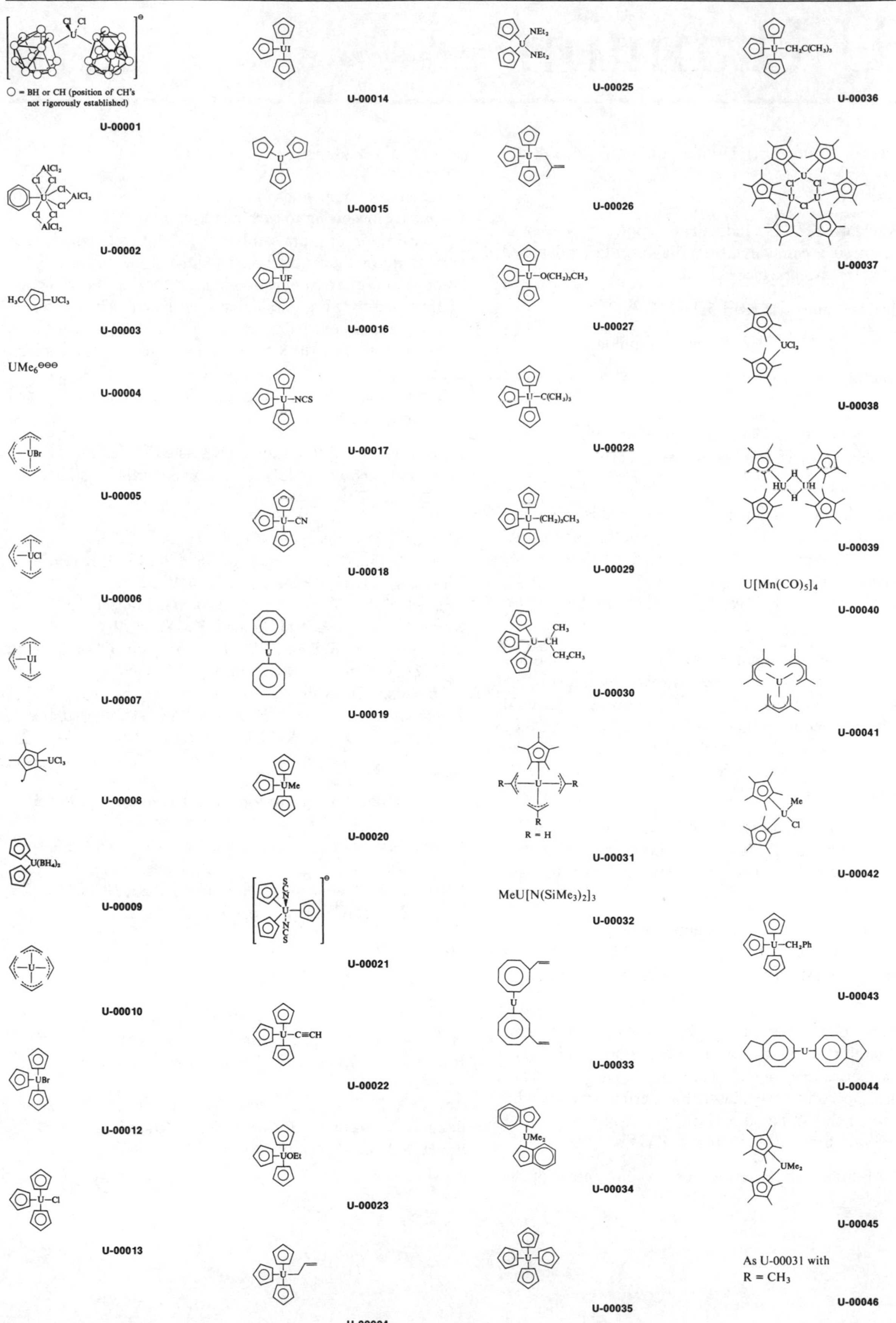

○ = BH or CH (position of CH's not rigorously established)
U-00001
U-00002
U-00003
UMe6⊖⊖⊖
U-00004
U-00005
U-00006
U-00007
U-00008
U-00009
U-00010
U-00012
U-00013
U-00014
U-00015
U-00016
U-00017
U-00018
U-00019
U-00020
U-00021
U-00022
U-00023
U-00024
U-00025
U-00026
U-00027
U-00028
U-00029
U-00030
R = H
U-00031
MeU[N(SiMe3)2]3
U-00032
U-00033
U-00034
U-00035
U-00036
U-00037
U-00038
U-00039
U[Mn(CO)5]4
U-00040
U-00041
U-00042
U-00043
U-00044
U-00045
As U-00031 with
R = CH3
U-00046

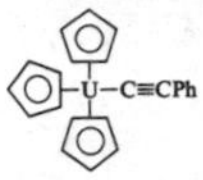

U-00047

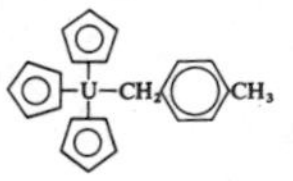

U-00048

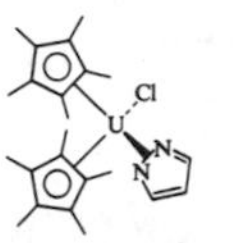

U-00049

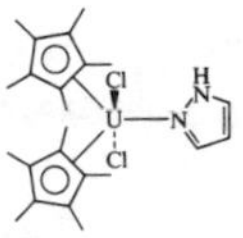

U-00050

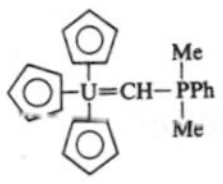

U-00051

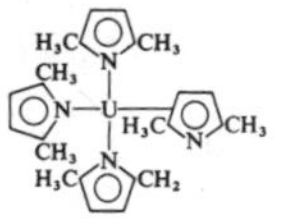

U-00052

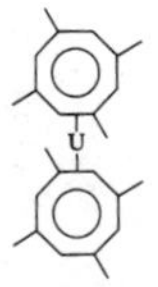

U-00053

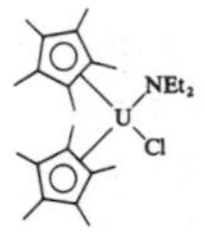

U-00054

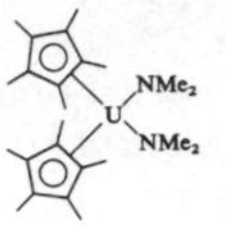

U-00055

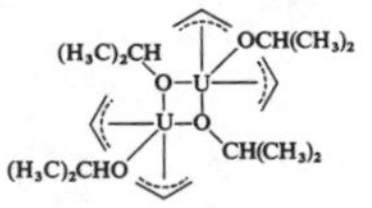

U-00056

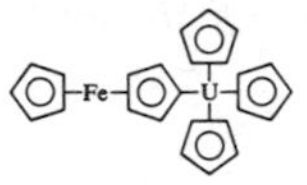

U-00057

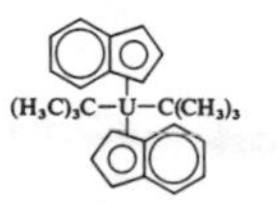

U-00058

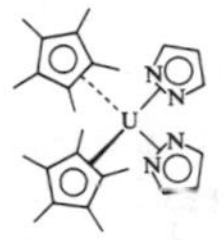

U-00059

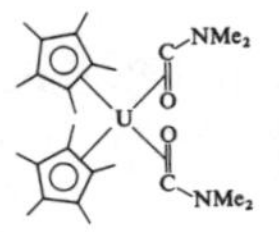

U-00060

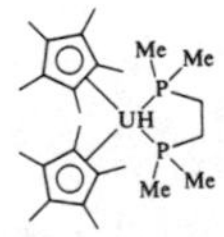

U-00061

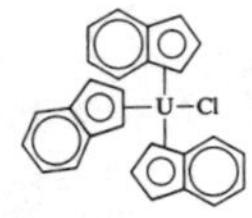

U-00062

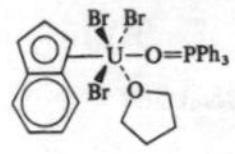

THF adduct

U-00063

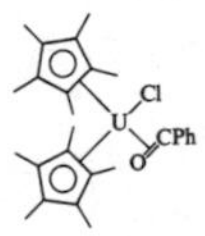

U-00064

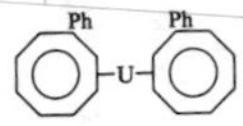

U-00065

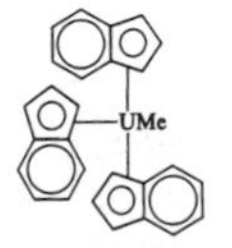

U-00066

$UI_4(CNC_6H_{11})_4$

U-00067

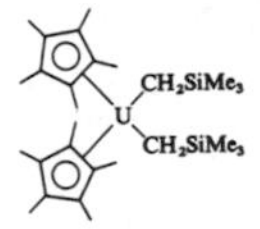

U-00068

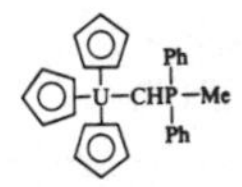

U-00069

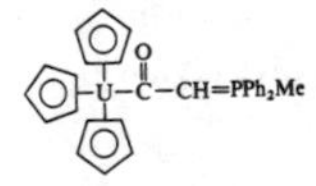

U-00070

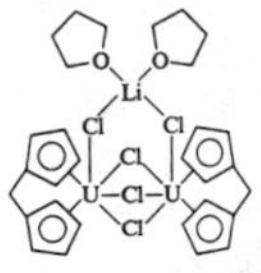

U-00071

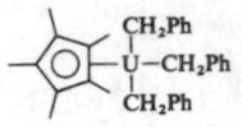

U-00072

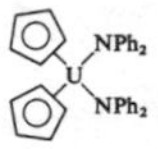

U-00073

$UPh_6^{\ominus\ominus}$

U-00074

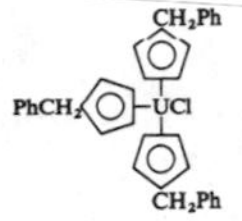

U-00075

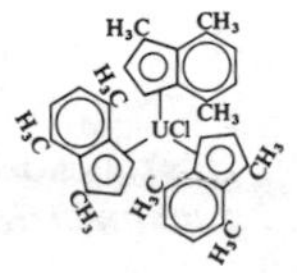

U-00076

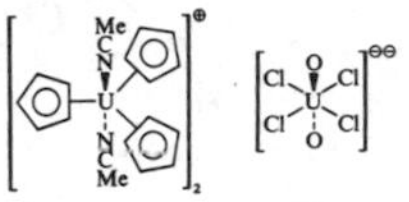

U-00077

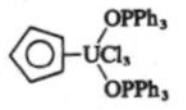

U-00078

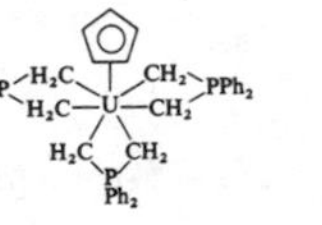

U-00079

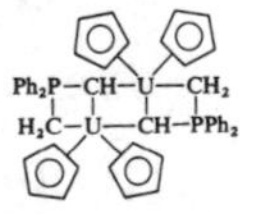

U-00080

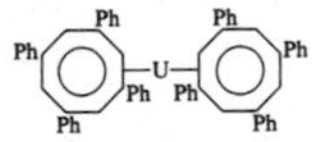

U-00081

$C_4H_{22}B_{18}Cl_2U^{\ominus\ominus}$ U-00001

Dichlorobis[(7,8,9,10,11-η)undecahydro-7,8-dicarbaundecaborato]uranate(2−)

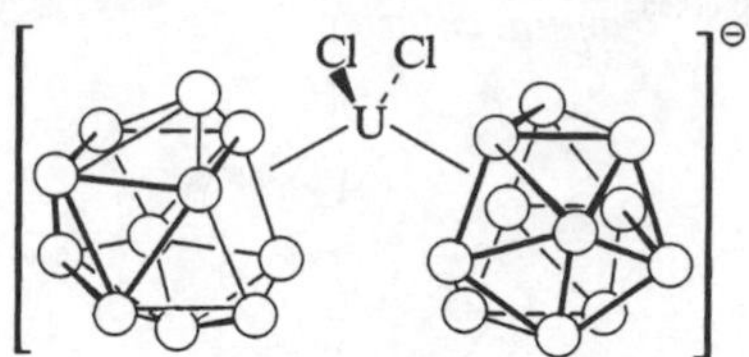

○ = BH or CH (position of CH's not rigorously established)

M 573.733 (ion)

Di-Li salt, octa-THF complex: [62559-95-9].
$C_{36}H_{86}B_{18}Cl_2Li_2O_8U$ M 1164.468
Yellow-orange cryst. Dec. by air and water.

Fronczek, F.R. *et al*, *J. Am. Chem. Soc.*, 1977, **99**, 1769 (*synth, uv, cryst struct*)

$C_6H_6Al_3Cl_{12}U$ U-00002

(Benzene)hexa-μ-chlorohexachlorotrialuminumuranium, 8CI
(Benzene)dodecachlorotrialuminumuranium
[34738-51-7]

$AlCl_2$
Cl Cl
U Cl $AlCl_2$ Cl
Cl Cl
$AlCl_2$

M 822.523

Black cryst. Insol. common solvs. Immediately dec. by air. Dec. by polar. solvs., e.g. THF. Stable to 110°.

Cesari, M. *et al*, *Inorg. Chim. Acta.*, 1971, **5**, 439 (*synth, struct, ir*)

$C_6H_7Cl_3U$ U-00003

Trichloro[(1,2,3,4,5-η)-1-methyl-2,4-cyclopentadien-1-yl]uranium

H_3C UCl_3

M 423.509

Bis-THF complex: [67507-18-0].
$C_{14}H_{23}Cl_3O_2U$ M 567.722
Bright-green cryst. Air-sensitive.

Ernst, R.D. *et al*, *J. Am. Chem. Soc.*, 1979, **101**, 2656 (*synth, struct, ir, nmr*)

$C_6H_{18}U^{\ominus\ominus\ominus}$ U-00004

Hexamethyluranate(3−)

$UMe_6^{\ominus\ominus\ominus}$

M 328.237 (ion)

Tri-Li salt: [63816-93-3].
$C_6H_{18}Li_3U$ M 349.060
Pale-green solid + 3 dioxan. Sol. THF. Mp 265-8° dec. Air- and moisture-sensitive.

Sigurdson, E.R. *et al*, *J. Chem. Soc., Dalton Trans.*, 1977, 812 (*synth, nmr, ir*)

$C_9H_{15}BrU$ U-00005

Bromotris(η^3-2-propenyl)uranium, 9CI
Tris(π-allyl)uranium bromide. Triallyluranium bromide
[42801-93-4]

M 441.150

Catalyses sterospecific polymerisation of butadiene.

Lugli, G. *et al*, *Makromol. Chem.*, 1974, **175**, 2021 (*synth*)

$C_9H_{15}ClU$ U-00006

Chlorotris(η^3-2-propenyl)uranium, 9CI
Tris(π-allyl)uranium chloride. Triallyluranium chloride
[41556-44-9]

M 396.699

Catalyses stereospecific polymerisation of butadiene. Dark-violet cryst.

Lugli, G. *et al*, *Makromol. Chem.*, 1974, **175**, 2021 (*synth*)

$C_9H_{15}IU$ U-00007

Iodotris(η^3-2-propenyl)uranium, 9CI
Triallyliodouranium. Triallyluranium iodide
[42801-92-3]

M 488.151

Catalyses stereospecific polymerisation of butadiene.

Lugli, G. *et al*, *Makromol. Chem.*, 1974, **175**, 2021.

$C_{10}H_{15}Cl_3U$ U-00008

[(1,2,3,4,5-η)-1,2,3,4,5-Pentamethyl-2,4-cyclopentadienyl-]trichlorouranium
Pentamethylcyclopentadienyluranium trichloride

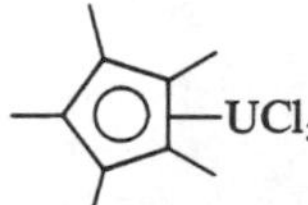

M 479.616

Bis-THF complex: [82511-72-6].
$C_{18}H_{31}Cl_3O_2U$ M 623.830
Dark-red needles. Sol. THF, dioxan.

Mintz, E.A. *et al*, *J. Am. Chem. Soc.*, 1982, **104**, 4692 (*synth, nmr, ir*)

$C_{10}H_{18}B_2U$ U-00009

Bis(η^5-2,4-cyclopentadien-1-yl)bis[tetrahydroborato(1−)-*H,H'*]uranium, 10CI

Bis(cyclopentadienyl)uranium bis(tetrahydroborate)

[65888-45-1]

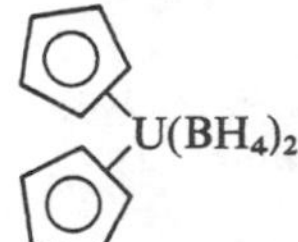

M 397.901

Red-brown solid. Sol. THF, dimethoxyethane. Subl. at 60° *in vacuo*. V. oxygen- and moisture-sensitive.

Zanella, P. *et al*, *J. Organomet. Chem.*, 1977, **142**, C21 (*synth, struct, ir, nmr*)

$C_{12}H_{20}U$ U-00010

Tetrakis(η^3-2-propenyl)uranium

Tetraallyluranium

[12701-96-1]

M 402.319

Reacts with alcohols to yield mixed complexes $U(C_3H_5)_2(OR)_2$. Reacts with HX to form $U(C_3H_5)$-X(X = Cl, Br, I). Dark-red solid. Sol. pentane. Mp −20° dec. Burns spontaneously in air. $\mu_{eff} = 2.6\mu_B$.

Lugli, G. *et al*, *Inorg. Chim. Acta*, 1969, **3**, 253 (*synth, ir*)
Brunelli, M. *et al*, *J. Magn. Reson.*, 1973, **9**, 247 (*nmr*)
Ger. Pat., 2 257 787, (*1973*); *CA*, **79**, 53579
Brunelli, M. *et al*, *J. Chem. Soc., Dalton Trans.*, 1979, 861 (*props*)

$C_{15}H_{15}BrU$ U-00012

Bromotris(η^5-2,4-cyclopentadien-1-yl)uranium

Tricyclopentadienyluranium bromide

[67507-08-8]

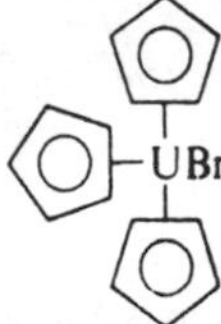

M 513.216

Dark-brown cryst. Sl. sol. pentane, sol. C_6H_6. Subl. at 160° *in vacuo*.

Fischer, R.D. *et al*, *J. Organomet. Chem.*, 1970, **25**, 123 (*synth, uv, ir, nmr*)
Sung-Yu, N.K. *et al*, *Inorg. Chem.*, 1981, **20**, 2727 (*synth, ir, ms, nmr*)

$C_{15}H_{15}ClU$ U-00013

Chlorotris(η^5-2,4-cyclopentadien-1-yl)uranium, 9CI

Chlorotri-π-cyclopentadienyluranium, 8CI. *Tris(cyclopentadienyl)uranium chloride*

[1284-81-7]

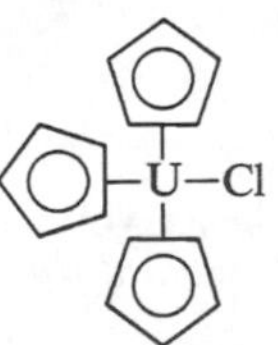

M 468.765

Air-sensitive red-black cryst. Mp 260-5°. Gives green aq. solns.

Reynolds, L.T. *et al*, *J. Inorg. Nucl. Chem.*, 1956, **2**, 246 (*synth*)
Wong, C.-H. *et al*, *Acta Crystallogr.*, 1965, **18**, 340 (*cryst struct*)
Anderson, M.L. *et al*, *J. Organomet. Chem.*, 1969, **17**, 345 (*ir, uv, ms*)
Fischer, R.D. *et al*, *J. Organomet. Chem.*, 1970, **25**, 123 (*pmr*)
Fukushima, E. *et al*, *Chem. Phys. Lett.*, 1976, **44**, 285 (*cmr*)
Inorg. Synth., 1976, **16**, 147 (*synth*)

$C_{15}H_{15}IU$ U-00014

Iodotris(η^5-2,4-cyclopentadien-1-yl)uranium

Tricyclopentadienyluranium iodide

[69030-40-6]

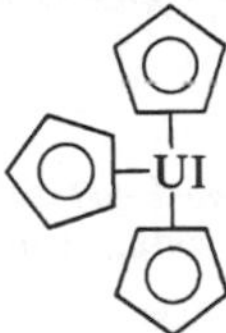

M 560.217

Brown solid. Sol. C_6H_6. Subl. at 170° *in vacuo*.

Fischer, R.D. *et al*, *J. Organomet. Chem.*, 1970, **25**, 123.
Aderhold, C. *et al*, *Z. Naturforsch., A*, 1978, **33**, 1268.
Sung-Yu, N.K. *et al*, *Inorg. Chem.*, 1981, **20**, 2727 (*synth, nmr, ms*)

$C_{15}H_{15}U$ U-00015

Tris(η^5-2,4-cyclopentadien-1-yl)uranium, 9CI

Tricyclopentadienyluranium

[54007-00-0]

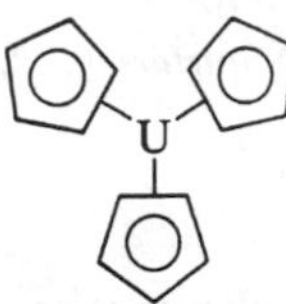

M 433.312

Forms adducts with Lewis bases e.g. THF. Bronze solid. Sol. THF, Et_2O, sl. sol. C_6H_6. Mp >200°. $\mu_{eff} = 2.33\mu_B$. Subl. at 120-50° *in vacuo*.

Nicotine complex: Tris(η^5-2,4-cyclopentadienyl)-(nicotine)uranium.
$C_{25}H_{29}N_2U$ M 595.546
Brown. Sl. sol. pentane.

Isocyanocyclohexane complex: Tris(η^5-2,4-cyclopentadien-1-yl)isocyanocyclohexaneuranium.
$C_{22}H_{26}NU$ M 542.483
Pale-brown solid. Sol. pentane, C_6H_6. $\mu_{eff} = 2.33\mu_B$.

THF complex:
$C_{19}H_{23}OU$ M 505.419

Brown solid (THF/C_6H_6).

Kanellakopulos, B. *et al*, *J. Organomet. Chem.*, 1970, **24**, 507 (*synth, ir, uv*)
Karraker, D.G. *et al*, *Inorg. Chem.*, 1972, **11**, 1742 (*synth*)
Zanella, P. *et al*, *Inorg. Chim. Acta.*, 1980, **44**, L155 (*synth*)
Wasserman, H.J. *et al*, *J. Organomet. Chem.*, 1983, **254**, 305 (*cryst struct*)

$C_{15}H_{15}UF$ U-00016

Tris(η^5-2,4-cyclopentadien-1-yl)fluorouranium, 9CI
Tricyclopentadienyluranium fluoride
[12715-68-3]

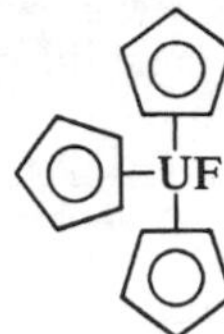

M 452.311
Green cryst. Sol. C_6H_6. Air- and moisture-sensitive. Subl. at 170° *in vacuo*. μ_{eff} 3.35μ_B.

Fischer, R.D. *et al*, *J. Organomet. Chem.*, 1970, **25**, 123 (*synth, ir*)
Kanellakopulos, B. *et al*, *Angew. Chem., Int. Ed. Engl.*, 1970, **9**, 957 (*ir*)
Ryan, R.R. *et al*, *J. Am. Chem. Soc.*, 1975, **97**, 4258 (*cryst struct*)
Aderhold, C. *et al*, *Z. Naturforsch., A*, 1978, **33**, 1268.

$C_{16}H_{15}NSU$ U-00017

Tris(η^5-2,4-cyclopentadien-1-yl)(thiocyanato-*N*)uranium
Tricyclopentadienyluranium thiocyanate
[69526-48-3]

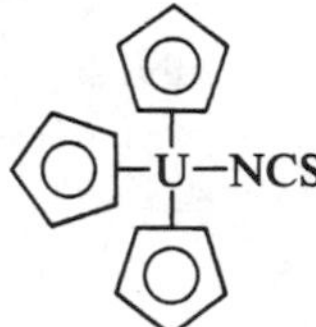

M 491.390
Brown cryst. Sol. most org. solvs. Mp 200° dec.
MeCN adduct: [69377-30-6].
$C_{18}H_{18}N_2SU$ M 532.443
Green cryst. Dec. >90°.

Fischer, R.D. *et al*, *Z. Naturforsch., B*, 1978, **33**, 1393 (*synth, ir, uv, struct*)

$C_{16}H_{15}NU$ U-00018

(Cyano-*C*)tris(η^5-2,4-cyclopentadien-1-yl)uranium
Tris(cyclopentadienyl)cyanouranium
[54006-99-4]

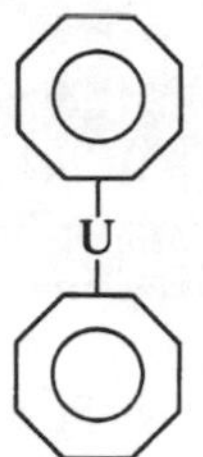

M 459.330
Green solid. Insol. all common org. solvs. Mp >200°. Involatile, ms. obt. >150°. Dec. above 200° leaving black pyrophoric residue.

Kanellakopulos, B. *et al*, *J. Organomet. Chem.*, 1974, **76**, C42 (*synth, ir*)
Bagnall, K.W. *et al*, *J. Chem. Soc., Dalton Trans.*, 1982, 1999 (*synth, ir, uv, ms*)

$C_{16}H_{16}U$ U-00019

Uranocene
Bis(η^8-1,3,5,7-cyclooctatetraene)uranium, 9CI. *Di-π-cyclooctatrienyleneuranium*, 8CI
[11079-26-8]

M 446.331
Green plates. Flammable in air but stable to H_2O. Subl. at 180°/0.03mm.

Streitwieser, A. *et al*, *J. Am. Chem. Soc.*, 1968, **90**, 7364 (*synth*)
Zalkin, A. *et al*, *J. Am. Chem. Soc.*, 1969, **91**, 5667 (*struct*)
Hocks, L. *et al*, *Spectrochim. Acta, Part. A*, 1974, **30**, 904 (*ir*)
Inorg. Synth., 1978, **19**, 149 (*synth*)

$C_{16}H_{18}U$ U-00020

Tris(η^5-2,4-cyclopentadien-1-yl)methyluranium, 9CI
Tricyclopentadienylmethyluranium
[37205-28-0]

M 448.347
Pale-yellow powder. Sol. C_6H_6, THF, Et_2O. Pyrophoric in air.

Brandi, G. *et al*, *Inorg. Chem. Acta.*, 1973, **7**, 319 (*synth, ir, nmr*)
Marks, T.J. *et al*, *J. Am. Chem. Soc.*, 1973, **95**, 5529 (*synth, ir, nmr*)

$C_{17}H_{15}N_2S_2U^{\ominus}$ U-00021

Tris(η^5-cyclopentadienyl)diisothiocyanatouranate(IV)

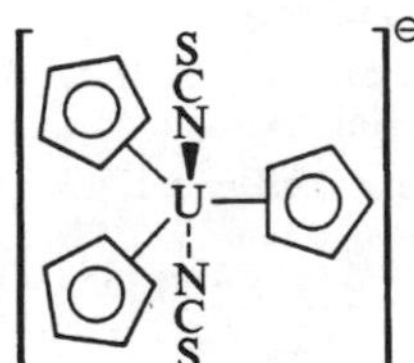

M 549.468 (ion)
Tetraphenylarsonium salt:
$C_{41}H_{35}AsN_2S_2U$ M 932.811
Bright-green cryst. (MeCN). Sol. MeCN, THF.

Bagnall, K.W. *et al*, *J. Chem. Soc., Dalton Trans.*, 1982, 1999 (*synth, ir, uv, nmr*)
Borubieri, G. *et al*, *J. Chem. Soc., Dalton Trans.*, 1983, 45 (*cryst struct*)

$C_{17}H_{16}U$ **U-00022**

Tris(η^5-2,4-cyclopentadien-1-yl)ethynyluranium, 9CI

Tricyclopentadienyluranium acetylide

[52827-35-7]

M 458.342

Yellow-green solid. Sol. THF, toluene. Sensitive to oxygen and moisture.

Tsutsui, M. *et al*, *Inorg. Chem.*, 1975, **14**, 78 (*synth*, *ir*, *nmr*)
Atwood, J.L. *et al*, *J. Coord. Chem.*, 1976, **5**, 209 (*struct*)

$C_{17}H_{20}OU$ **U-00023**

Ethoxytris(η^5-2,4-cyclopentadien-1-yl)uranium

[63415-86-1]

M 478.373

Can be synthesized by reaction of $(C_5H_5)_3UCl$ with $LiBHEt_3$. Exchanges alkyl groups with AlR_3. Green cryst. Sol. THF, C_6H_6, sl. sol. pentane. Mp 210-3°.

v. Ammon, R. *et al*, *Radiochim. Acta.*, 1969, **11**, 162 (*synth*, *nmr*, *uv*)
Marques-Ellis, H. *et al*, *J. Organomet. Chem.*, 1977, **131**, 257 (*synth*, *nmr*)
Vasil'ev, V.K., *J. Organomet. Chem.*, 1977, **142**, C7.

$C_{18}H_{20}U$ **U-00024**

Tris(η^5-2,4-cyclopentadien-1-yl)-2-propenyluranium, 9CI

Tricyclopentadienylallyluranium. Allyltricyclopentadienyluranium

[37298-76-3]

M 474.385

Dark-brown cryst. Sol. toluene. Mp 155° dec. Inflames in air.

Marks, T.J. *et al*, *J. Am. Chem. Soc.*, 1973, **95**, 5529 (*synth*, *ir*, *nmr*)

$C_{18}H_{30}N_2U$ **U-00025**

Bis(η^5-2,4-cyclopentadien-1-yl)bis(*N*-ethylethanaminato)-uranium

Dicyclopentadienylbis(diethylamino)uranium

[54068-37-0]

M 512.477

Inserts CO_2, CS_2, COS. Reacts with carboxylic and thiocarboxylic acids. Gold flakes. Sol. hexane. Air- and moisture-sensitive.

Arduini, A.L. *et al*, *Inorg. Chem.*, 1981, **20**, 2470, 2474, 2480 (*synth*, *nmr*, *ir*, *ms*, *props*)

$C_{19}H_{22}U$ **U-00026**

Tris(η^5-2,4-cyclopentadien-1-yl)(2-methyl-1-propenyl)uranium, 9CI

Tricyclopentadienyl-η^1-2-methylallyluranium(IV)

[50643-51-1]

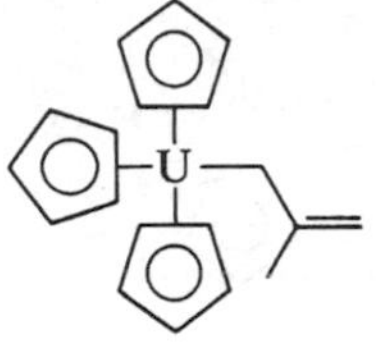

M 488.412

Deep reddish-brown cryst. Sol. toluene. Air-sensitive.

Halstead, G.W. *et al*, *J. Am. Chem. Soc.*, 1975, **97**, 3049 (*synth*, *struct*, *ir*, *ms*)

$C_{19}H_{24}OU$ **U-00027**

Butoxytris(η^5-2,4-cyclopentadien-1-yl)uranium

[1284-95-3]

M 506.427

Green cryst. Sol. dimethoxymethane, cyclohexane, THF, sl. sol. pentane. Mp 149-51°. Extremely air-sensitive. Subl. at 120° *in vacuo*. Liberates butanol quantitatively with HCl. Can be prepared by redox reaction from UCl_3 and C_5H_5Na in THF.

Ter Haar, G.L., *Inorg. Chem.*, 1964, **3**, 1648 (*synth*, *ir*)
v. Ammon, R. *et al*, *Radiochim, Acta.*, 1969, **11**, 162 (*synth*, *nmr*, *uv*)
McLaren, A.B., *Inorg. Nucl. Chem. Lett.*, 1980, **16**, 223 (*synth*, *ms*)

$C_{19}H_{24}U$ U-00028

tert-Butyltricyclopentadienyluranium

Tris(η^5-2,4-cyclopentadien-1-yl)(1,1-dimethylethyl)-uranium, 9CI. *Tricyclopentadienyl*-tert-*butyluranium(IV)*

[50647-46-6]

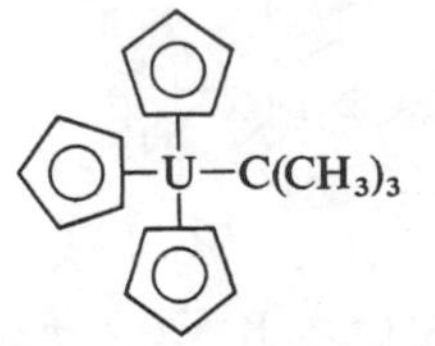

$(C_5H_5)_3U-C(CH_3)_3$

M 490.428

Red solid. Sol. C_6H_6, Et_2O. Pyrophoric in air.

Marks, T.J. *et al*, *J. Am. Chem. Soc.*, 1973, **95**, 5529 (*synth, ir, nmr*)

$C_{19}H_{24}U$ U-00029

Butyltris(η^5-2,4-cyclopentadien-1-yl)uranium

Tricyclopentadienyl(butyl)uranium(IV)

[37298-84-3]

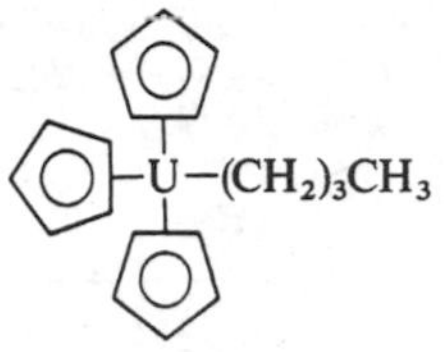

$(C_5H_5)_3U-(CH_2)_3CH_3$

M 490.428

Dark-red cryst. Sol. C_6H_6, toluene, Et_2O, hot hexane. Mp 130° dec. Smokes on exposure to air. μ_{eff} = 3.36μ_B.

Brandi, G. *et al*, *Inorg. Chim. Acta.*, 1973, **7**, 319 (*synth, ir, nmr*)

Marks, T.J. *et al*, *J. Am. Chem. Soc.*, 1973, **95**, 5529 (*synth, ir, nmr*)

Perego, G. *et al*, *Acta Crystallogr., Sect. B*, 1976, **32**, 3034 (*struct*)

$C_{19}H_{24}U$ U-00030

Tris(η^5-2,4-cyclopentadien-1-yl)(1-methylpropyl)uranium

*Tricyclopentadienyl(*sec-*butyl)uranium.* sec-*Butyltricyclopentadienyluranium*

[80410-09-9]

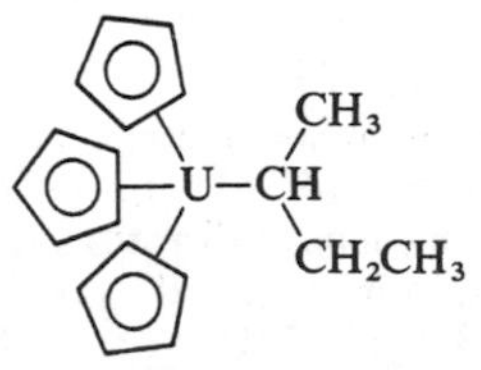

$(C_5H_5)_3U-CH(CH_3)CH_2CH_3$

M 490.428

Yields chiefly butane and $(C_5H_5)_3U$ on photolysis. Dark black-green microcryst. Sol. toluene, C_6H_6.

Bruno, J.W. *et al*, *J. Am. Chem. Soc.*, 1982, **104**, 1860 (*synth, ir, nmr*)

$C_{19}H_{30}U$ U-00031

[(1,2,3,4,5-η)-1,2,3,4,5-Pentamethyl-2,4-cyclopentadien-1-yl)]tris(2-propenyl)uranium

(Pentamethylcyclopentadienyl)tris(allyl)uranium. Triallyl(pentamethylcyclopentadienyl)uranium

[84895-69-2]

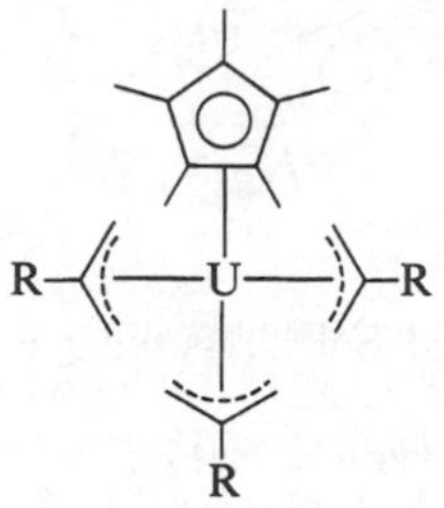

R = H

M 496.475

Dark cryst. Sol. THF, pentane. Slowly dec. at r.t. V. air-sensitive.

Cymbaluk, T.H. *et al*, *Organometallics*, 1983, **2**, 963 (*synth, ir, nmr*)

$C_{19}H_{57}N_3Si_6U$ U-00032

Methyltris[(hexamethyldisilyl)amido]uranium

*Methyltris[1,1,1-trimethyl-*N*-(trimethylsilyl)silanaminato]uranium*

[69517-44-8]

$MeU[N(SiMe_3)_2]_3$

M 734.221

Light-brown needles (pentane). Sol. Et_2O, pentane. Mp 130-2°. Air- and moisture-sensitive.

Turner, H.W. *et al*, *Inorg. Chem.*, 1979, **18**, 1221 (*synth, ir, nmr*)

$C_{20}H_{20}U$ U-00033

Bis[(1,2,3,4,5,6,7,8-η)-1-ethenyl-1,3,5,7-cyclooctatetraene]uranium, 9CI

1,1'-Divinyluranocene

[37274-09-2]

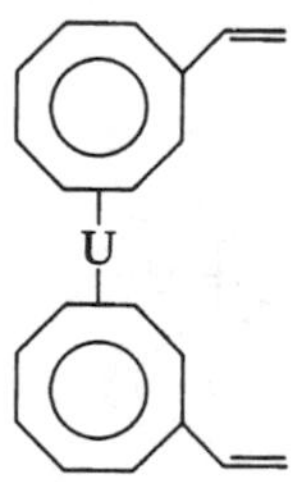

M 498.407

Yields 1,1'-diethyluranocene (94%) on hydrogenation using Pd/C catalyst. Green solid. Sol. cyclohexane. Oxygen- and moisture-sensitive.

Streitweiser, A. *et al*, *Inorg. Chem.*, 1973, **12**, 1102 (*synth, uv, ms, ir, nmr*)

$C_{20}H_{20}U$ **U-00034**

Bis[(1,2,3,3*a*,7*a*-*η*)-1*H*-inden-1-yl]dimethyluranium

[63720-03-6]

UMe_2

M 498.407

Yellow-brown solid. Sol. THF, toluene. Extremely air- and moisture-sensitive. Pyrophoric in air.

Seyam, A.M. *et al*, *Inorg. Nucl. Chem. Lett.*, 1977, **13**, 115 (*synth*, *ir*)

$C_{20}H_{20}U$ **U-00035**

Tetrakis(η^5-2,4-cyclopentadien-1-yl)uranium, 9CI

[1298-76-6]

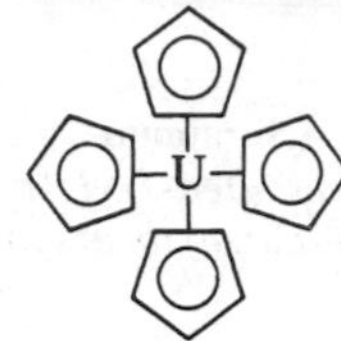

M 498.407

Air-sensitive red cryst. (pentane). Dec. at 250° under N_2.

Fischer, E.O. *et al*, *Z. Naturforsch.*, *B*, 1962, **17**, 275 (*synth*, *ir*)
Anderson, M.L. *et al*, *J. Organomet. Chem.*, 1969, **17**, 345 (*ir*, *uv*, *ms*)
v. Ammon, R. *et al*, *Chem. Phys. Lett.*, 1970, **4**, 553 (*pmr*)
Burns, J.H. *et al*, *J. Organomet. Chem.*, 1974, **69**, 225 (*cryst struct*)

$C_{20}H_{26}U$ **U-00036**

Tris(η^5-2,4-cyclopentadien-1-yl)(2,2-dimethylpropyl)uranium, 9CI

Tricyclopentadienylneopentyluranium(IV)

[37298-90-1]

U–$CH_2C(CH_3)_3$

M 504.454

Dark-red cryst. Sol. toluene, Et_2O. Mp 148° dec. Smokes in air.

Marks, T.J. *et al*, *J. Am. Chem. Soc.*, 1973, **95**, 5529 (*synth*, *ir*, *nmr*)

$C_{20}H_{30}ClU$ **U-00037**

Chlorobis(pentamethylcyclopentadienyl)uranium

[71794-95-1]

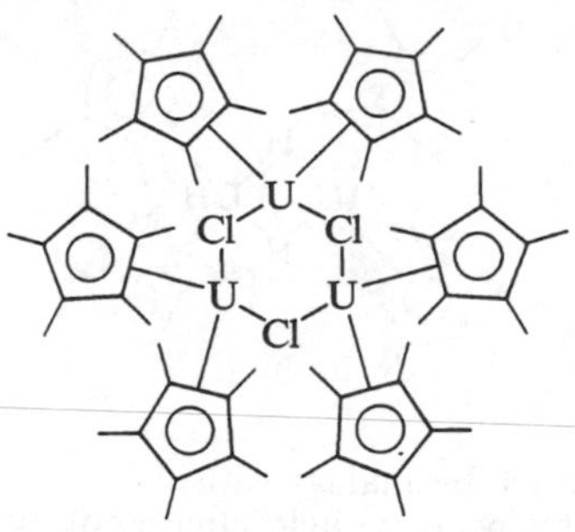

M 543.939

Trimeric.

Trimer: Tri-μ-chlorohexakis[(1,2,3,4,5-η)-1,2,3,4,5-pentamethyl-2,4-cyclopentadien-1-yl]triuranium, 9CI.

$C_{60}H_{90}Cl_3U_3$ M 1631.817

Forms adducts with Lewis bases Py, PMe_3, Et_2O, THF. Unusual reactions with unsaturated organic molecules. Forms alkyls, amides, and hydrides. Green cryst. Sol. THF. Air- and moisture-sensitive.

Finke, R.G. *et al*, *J. Chem. Soc., Chem. Commun.*, 1981, 232 (*props*)
Fagan, P.J. *et al*, *Organometallics*, 1982, **1**, 170 (*synth*, *struct*, *ir*, *nmr*)

$C_{20}H_{30}Cl_2U$ **U-00038**

Dichlorobis[(1,2,3,4,5-*η*)-1,2,3,4,5-pentamethyl-2,4-cyclopentadien-1-yl]uranium

Bis(pentamethylcyclopentadienyl)uranium dichloride

[67506-89-2]

UCl_2

M 579.392

Starting material for a wide range of organouranium compounds. Maroon cryst. Sol. toluene, Et_2O, THF. Air-sensitive.

Manriquez, J.M. *et al*, *J. Am. Chem. Soc.*, 1978, **100**, 3939 (*synth*, *ir*)
Fagan, P.J. *et al*, *J. Am. Chem. Soc.*, 1981, **103**, 6650 (*synth*, *ir*, *pmr*)

$C_{20}H_{32}U$ U-00039

Dihydrobis(pentamethylcyclopentadienyl)uranium

Bis(pentamethylcyclopentadienyl)uranium dihydride

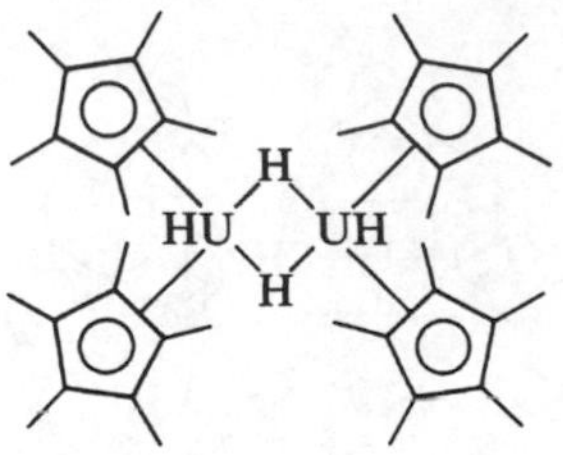

M 510.502

Dimeric, struct. by analogy with Dihydrobis(pentamethylcyclopentadienyl)thorium.

Dimer: [67588-76-5]. *Di-μ-hydrodihydrotetrakis[(1,2,3,4,5-η)-1,2,3,4,5-pentamethyl-2,4-cyclopentadien-1-yl]diuranium.*

$C_{40}H_{64}U_2$ M 1021.003

Catalyst for homogeneous hydrogenation of hexene. Black-green cryst. Sol. toluene. Sensitive to N_2 and air, must be kept under Ar at −20°.

Fagan, P.J. *et al*, *J. Am. Chem. Soc.*, 1981, **103**, 6650 (*synth, ir, nmr, use*)

$C_{20}Mn_4O_{20}U$ U-00040

Tetrakis(pentacarbonylmanganio)uranium, 8CI

[31135-22-5]

$U[Mn(CO)_5]_4$

M 1017.989

Struct. not yet established. Bright-orange solid. Sol. Me_2CO, THF. Mp 180° dec. Extremely air-sensitive; air oxidn. affords $Mn_2(CO)_{10}$, reaction with bromine yields UBr_4 + $Mn(CO)_5Br$. Reacts quickly with cold chloroform forming $Mn(CO)_5Cl$.

Bennett, R.L. *et al*, *J. Organomet. Chem.*, 1971, **26**, 355 (*synth, ir, ms*)

$C_{21}H_{27}U$ U-00041

Tris(2,4-dimethylpentadienyl)uranium

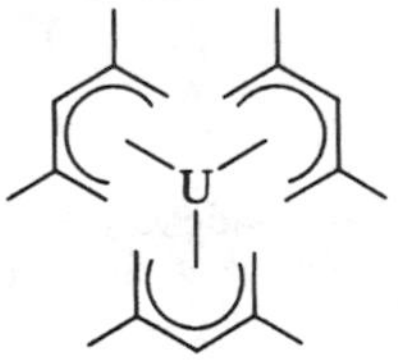

M 517.473

Dark cryst. giving red-brown soln. Sol. pentane, THF, toluene. μ_{eff} = 2.4 μ_B. Dec. at once by halogenated solvs.

▷Pyrophoric in air

Cymbaluk, T.H. *et al*, *J. Organomet. Chem.*, 1983, **255**, 311 (*synth, ir, nmr*)

$C_{21}H_{33}ClU$ U-00042

Chloro(methyl)bis[(1,2,3,4,5-η)-1,2,3,4,5-pentamethyl-2,4-cyclopentadien-1-yl]uranium

[67506-91-6]

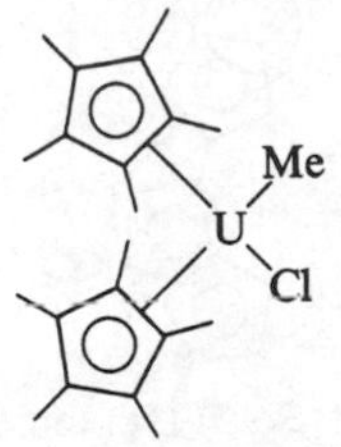

M 558.974

Yields Chlorobis(pentamethylcyclopentadienyl)uranium, U-00037 on hydrogenolysis. Reacts additively with carbonyl compds., alcohol. Maroon cryst. Sol. C_6H_6, toluene. Air- and moisture-sensitive.

Fagan, P.J. *et al*, *J. Am. Chem. Soc.*, 1981, **103**, 6650 (*synth, nmr*)

Fagan, P.J. *et al*, *Organometallics*, 1982, **1**, 170.

$C_{22}H_{22}U$ U-00043

Benzyltricyclopentadienyluranium

Tris(η⁵-2,4-cyclopentadien-1-yl)(phenylmethyl)uranium, 9CI. Tricyclopentadienylbenzyluranium

[37206-36-3]

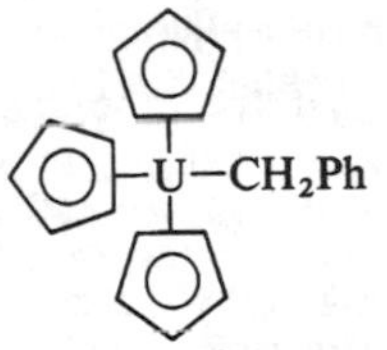

M 524.445

Dark-violet needles. Sol. C_6H_6, THF. Mp 200° dec. Air-sensitive.

Brandi, G. *et al*, *Inorg. Chim. Acta.*, 1973, **7**, 319 (*synth, ir, nmr*)

$C_{22}H_{24}U$ U-00044

Dicyclopentenouranocene

Bis(η⁸-bicyclo[6.3.0]undeca-2,4,6-triene-1,8-diide)uranium

[80287-59-8]

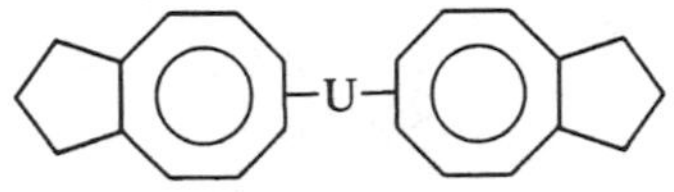

M 526.461

Green cryst. Sol. THF, toluene, dioxan. μ_{eff} = 2.4μ_B. Air-sensitive.

Zalkin, A. *et al*, *Organometallics*, 1982, **1**, 619 (*synth, struct, ir, nmr, cmr*)

$C_{22}H_{36}U$ U-00045

Dimethylbis[(1,2,3,4,5-η)-1,2,3,4,5-pentamethyl-2,4-cyclopentadien-1-yl]uranium

[67605-92-9]

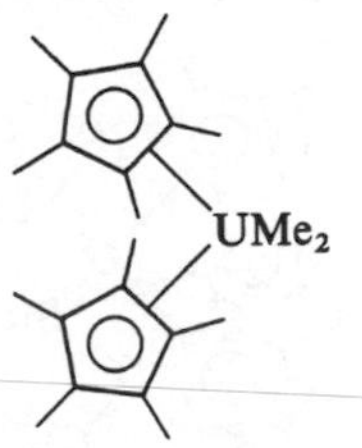

M 538.555

When supported on alumina, very active catalyst for hydrogenation of propene and polymerization of ethylene. Inserts CO. U-C bond undergoes hydrogenolysis to form Dihydrobis(pentamethylcyclopentadienyl)uranium, U-00039 . Orange needles. Sol. toluene, pentane. V. air-sensitive.

Fagan, P.J. *et al*, *J. Am. Chem. Soc.*, 1978, **100**, 3939 (*synth*)
Manriquez, J.M. *et al*, *J. Am. Chem. Soc.*, 1978, **100**, 7112.
Bowman, R.G. *et al*, *J. Chem. Soc., Chem. Commun.*, 1981, 257.
Fagan, P.J. *et al*, *J. Am. Chem. Soc.*, 1981, **103**, 6650 (*synth, ir, pmr*)

$C_{22}H_{36}U$ U-00046

Tris(2-methyl-1-propenyl)[(1,2,3,4,5-η)-1,2,3,4,5-pentamethyl-2,4-cyclopentadien-1-yl]uranium

(Pentamethylcyclopentadienyl)tris(2-methylallyl)uranium

[84895-68-1]

As [(1,2,3,4,5-η)-1,2,3,4,5-Pentamethyl-2,4-cyclopentadien-1-yl)]tris(2-propenyl)uranium, U-00031 with

$$R = CH_3$$

M 538.555

Dark cryst. Sol. THF, pentane, Et_2O. Slowly dec. at r.t. V. air-sensitive.

Cymbaluk, T.H. *et al*, *Organometallics*, 1983, **2**, 963 (*synth, ir, struct, nmr*)

$C_{23}H_{20}U$ U-00047

Tris(η^5-2,4-cyclopentadien-1-yl)(phenylethynyl)uranium

Tricyclopentadienylphenylethynyluranium(IV)

[39311-60-9]

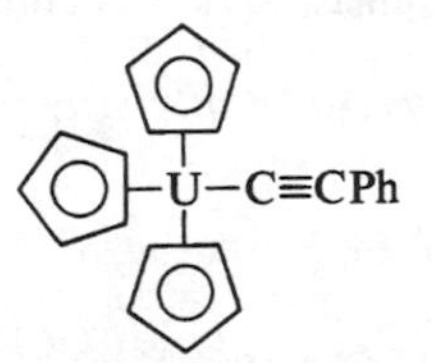

M 534.440

Yellow-green cryst. Sol. hexane, C_6H_6, THF. Mp 183-5° dec. Water- and oxygen-sensitive. $\mu_{eff} = 2.88\mu_B$.

Gebala, A.E. *et al*, *J. Am. Chem. Soc.*, 1973, **95**, 91 (*synth, ir, nmr*)
Atwood, J.L. *et al*, *J. Chem. Soc., Chem. Commun.*, 1973, 453 (*struct*)

$C_{23}H_{24}U$ U-00048

[Tris(η^5-2,4-cyclopentadien-1-yl)][(4-methylphenyl)methyl]-uranium, 9CI

Tricyclopentadienyl-4-methylbenzyluranium

[39448-14-1]

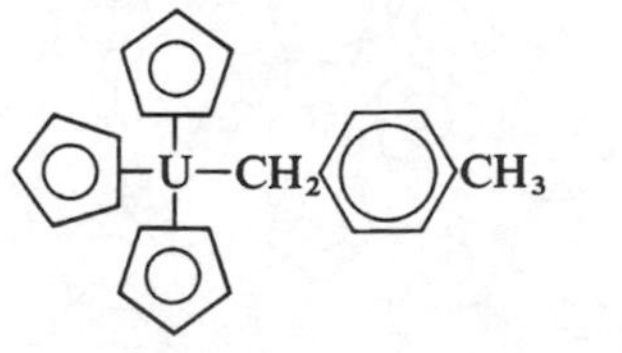

M 538.472

Dark-violet cryst. Sol. C_6H_6, toluene. Air-sensitive.

Brandi, G. *et al*, *Inorg. Chim. Acta.*, 1973, **7**, 319 (*synth, ir, nmr*)
Perego, G. *et al*, *Acta Crystallogr., Sect. B*, 1976, **32**, 3034 (*struct*)

$C_{23}H_{33}ClN_2U$ U-00049

Chlorobis[(1,2,3,4,5-η)-1,2,3,4,5-pentamethyl-2,4-cyclopentadien-1-yl)(1*H*-pyrazolato-N^1,N^2)uranium

[81277-18-1]

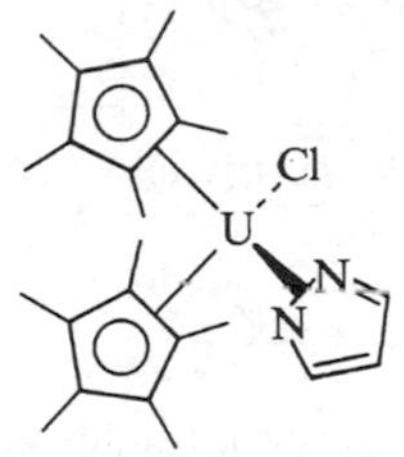

M 611.009

Red-brown cryst. Sol. hexane, toluene, THF. $\mu_{eff} = 2.42\mu_B$.

Eigenbrot, C.W. *et al*, *Inorg. Chem.*, 1982, **21**, 2653 (*synth, struct*)

$C_{23}H_{34}Cl_2N_2U$ U-00050

Dichlorobis[(1,2,3,4,5-η)-1,2,3,4,5-pentamethyl-2,4-cyclopentadien-1-yl](1*H*-pyrazole-N^2)uranium

[81277-17-0]

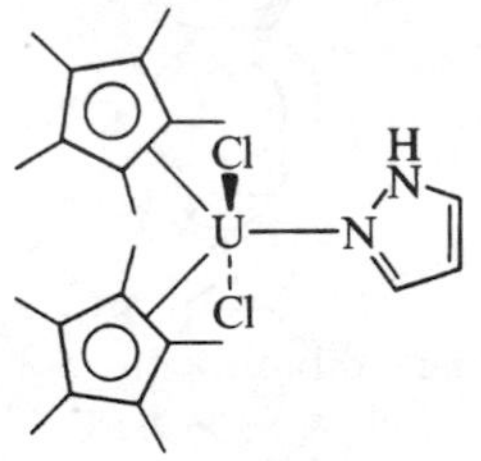

M 647.470

Red-brown cryst. Sol. toluene, THF. $\mu_{eff} = 3.42\mu_B$.

Eigenbrot, C.W. *et al*, *Inorg. Chem.*, 1982, **21**, 2653 (*synth, cryst struct, ir, ms, nmr*)

$C_{24}H_{27}PU$ **U-00051**

Tris(η^5-2,4-cyclopentadien-1-yl)[(dimethylphenylphosphoranylidene)methyl)]uranium

[77357-85-8]

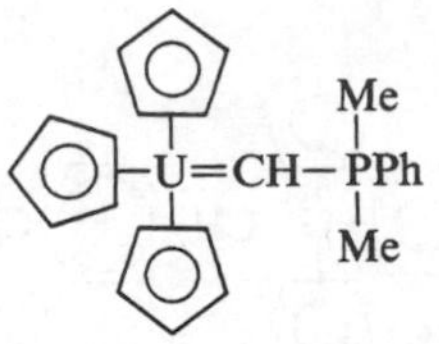

M 584.480

Dark-green cryst. Sol. THF, toluene. Pyrophoric in air.

Cramer, R.E. *et al*, *Inorg. Chem.*, 1981, **20**, 2466 (*synth*, *struct*, *ir*, *uv*, *nmr*)

Cramer, R.E. *et al*, *Organometallics*, 1983, **2**, 1336.

$C_{24}H_{32}N_4U$ **U-00052**

[(1,2,3,4,5-η)-2,5-Dimethyl-1*H*-pyrrol-1-yl]tris(2,5-dimethyl-1*H*-pyrrol-1-yl)uranium, 9CI

Tetrakis(2,5-dimethylpyrrolyl)uranium (IV)

[55224-20-9]

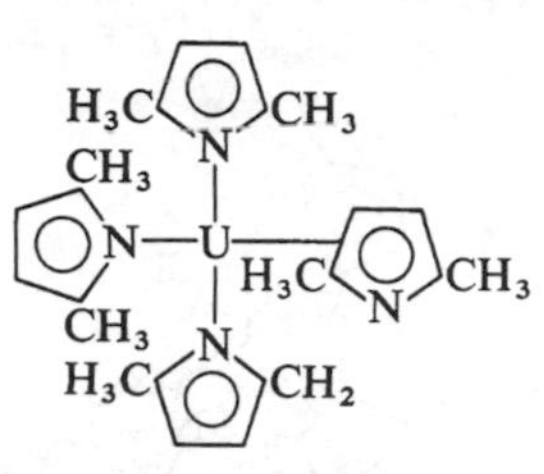

M 614.573

nmr shows nonequivalent dimethylpyrrole moieties. Red-brown microcryst. solid. Sol. THF, Et_2O. Mp 98-102° dec. $\mu_{eff} = 2.20\mu_B$.

Marks, T.J. *et al*, *J. Organomet. Chem.*, 1974, **82**, C35 (*synth*, *ms*, *ir*, *nmr*)

$C_{24}H_{32}U$ **U-00053**

Bis[(1,2,3,4,5,6,7,8-η)-1,3,5,7-tetramethyl-1,3,5,7-cyclooctatetraene]uranium, 9CI

1,1′,3,3′,5,5′,7,7′-Octamethyluranocene

[12715-87-6]

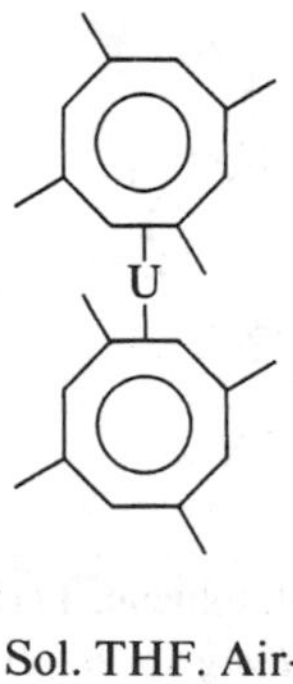

M 558.546

Burgundy-red cryst. Sol. THF. Air- and moisture-sensitive. $\mu_{eff} = 2.2\mu_B$.

Streitweiser, A. *et al*, *J. Am. Chem. Soc.*, 1971, **93**, 7343 (*synth*, *nmr*)

Hodgson, K.O. *et al*, *Inorg. Chem.*, 1973, **12**, 458 (*struct*)

Streitweiser, A., *Inorg. Chem.*, 1973, **12**, 1102 (*synth*, *ir*, *uv*, *ms*, *nmr*)

Edelstein, N. *et al*, *Inorg. Chem.*, 1976, **15**, 1397.

$C_{24}H_{40}ClNU$ **U-00054**

Chloro(diethylamino)bis(pentamethylcyclopentadienyl)uranium

Chloro(N-*ethylethanaminato*)*bis*[(*1,2,3,4,5-η*)-*1,2,3,4,5-pentamethyl-2,4-cyclopentadien-1-yl*]*uranium*

[77260-83-4]

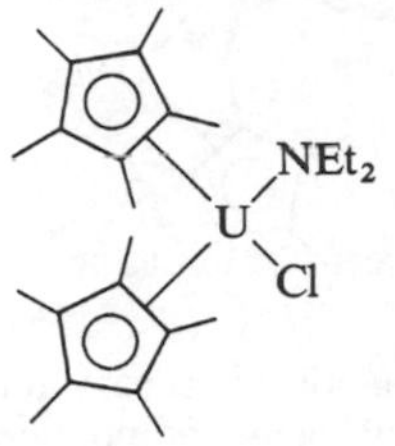

M 616.069

Inserts CO into U-N bond. Dark-red cryst. Sol. toluene. Air- and moisture-sensitive.

Fagan, P.J. *et al*, *J. Am. Chem. Soc.*, 1981, **103**, 2206 (*synth*, *ir*, *nmr*)

$C_{24}H_{42}N_2U$ **U-00055**

Bis(dimethylamino)bis(pentamethylcyclopentadienyl)uranium

Bis(N-*methylmethanaminato*)*bis*(*1,2,3,4,5-η*)-*1,2,3,4,5-pentamethyl-2,4-cyclopentadien-1-yl*]*uranium*

[77260-85-6]

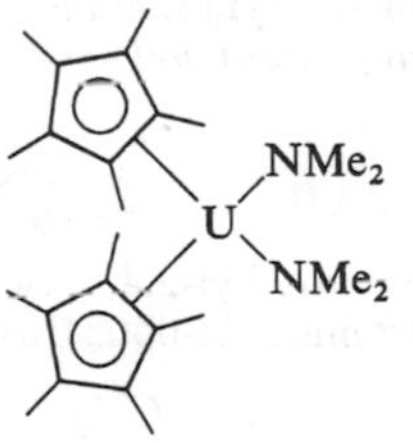

M 596.638

Inserts CO into both U-N bonds at 60° to form Bis[(dimethylamino)carbonyl-*C*,*O*]bis[(1,2,3,4,5-η)-1,2,3,4,5-pentamethyl-2,4-cyclopentadien-1-yl]uranium, U-00060 . Orange cryst. Sol. pentane, Et_2O. Air-sensitive.

Fagan, P.J. *et al*, *J. Am. Chem. Soc.*, 1981, **103**, 2206 (*synth*, *ir*, *nmr*)

$C_{24}H_{48}O_4U_2$ **U-00056**

Bis[μ-(2-propanolato)]bis(2-propanolato)tetrakis(η^3-2-propenyl)diuranium

Di-μ-isopropoxobis[di(η-allyl)isopropoxouranium(IV-)]

[71427-99-1]

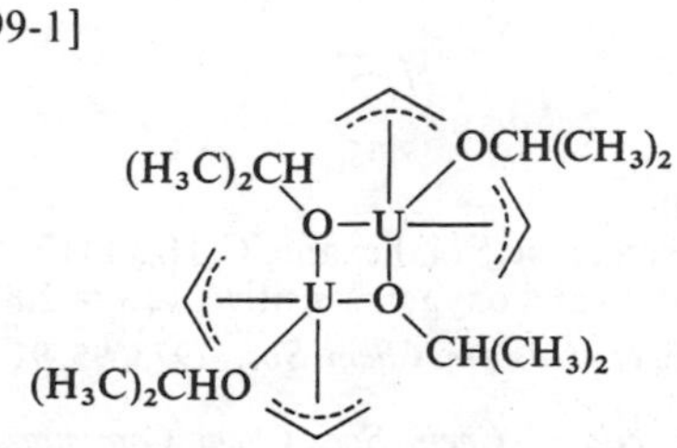

M 876.699

Light-brown cryst. Sol. pentane, Et_2O. Air-sensitive, dec. by moisture. Dec. in a few days at r.t.

Brunelli, M. *et al*, *J. Chem. Soc.*, *Dalton Trans.*, 1979, 861 (*synth*, *ir*, *pmr*, *struct*)

$C_{25}H_{24}FeU$ **U-00057**

Tris(η^5-2,4-cyclopentadien-1-yl)ferrocenyluranium, 9CI

[52827-36-8]

M 618.341

Brown solid (hexane); stable under Ar, but sensitive to O_2 and H_2O. Sl. sol. hydrocarbons. Mp >190° dec. Bp 180° *in vacuo* subl.

Tsutsui, M. *et al*, *J. Am. Chem. Soc.*, 1974, **96**, 3650.
Tsutsui, M. *et al*, *Inorg. Chem.*, 1975, **14**, 78 (*synth, pmr, ir*)

$C_{26}H_{32}U$ **U-00058**

Di-*tert*-butyldiindenyluranium

Bis(1,1-dimethylethyl)bis[(1,2,3,3a,7a-η)-1H-inden-1-yl]uranium, 9CI

[63720-04-7]

M 582.568

Brown powder. Sol. THF, toluene. Extremely oxygen- and moisture-sensitive. Pyrophoric in air.

Seyam, A.M. *et al*, *Inorg. Nucl. Chem. Lett.*, 1977, **13**, 115 (*synth, ir*)

$C_{26}H_{36}N_4U$ **U-00059**

Bis[(1,2,3,4,5-η)-1,2,3,4,5-pentamethyl-2,4-cyclopentadien-1-yl]bis(1*H*-pyrazolato-*N*[1],*N*[2])uranium

[81293-74-5]

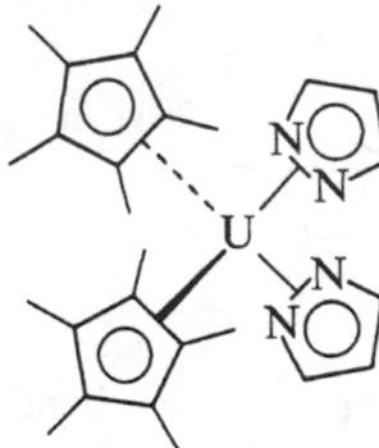

M 642.626

Red-brown cryst. Sol. hexane, THF, toluene.

Eigenbrot, C.W. *et al*, *Inorg. Chem.*, 1982, **21**, 2653 (*synth, cryst struct, ir, ms, pmr*)

$C_{26}H_{42}N_2O_2U$ **U-00060**

Bis[(dimethylamino)carbonyl-*C,O*]bis[(1,2,3,4,5-η)-1,2,3,4,5-pentamethyl-2,4-cyclopentadien-1-yl]uranium

[77260-96-9]

M 652.659

Prepd. by CO insertion of Bis(dimethylamino)bis(pentamethylcyclopentadienyl)uranium, U-00055 . Emerald-green prisms. Sol. pentane, toluene. Air-sensitive. Partially decarboxylates at 110°.

Fagan, P.J. *et al*, *J. Am. Chem. Soc.*, 1981, **103**, 2206 (*synth, ir, nmr, cryst struct*)

$C_{26}H_{47}P_2U$ **U-00061**

[1,2-Ethanediylbis(dimethylphosphine)-*P,P'*]hydrobis[(1,2,3,4,5-η)-1,2,3,4,5-pentamethyl-2,4-cyclopentadien-1-yl]uranium

[80602-96-6]

M 659.634

Formed by reaction of Dimethylbis[(1,2,3,4,5-η)-1,2,3,4,5-pentamethyl-2,4-cyclopentadien-1-yl]uranium, U-00045 with H_2 in presence of excess bis(1,2-dimethylphosphino)ethane. Reacts with N_2, THF and CO to yield complex mixt. of prods. Black microcryst. Sol. C_6H_6, toluene. $\mu_{eff} = 3.47\mu_B$.

Duttera, M.R. *et al*, *J. Am. Chem. Soc.*, 1982, **104**, 865 (*synth, struct, nmr, ir*)

$C_{27}H_{21}ClU$ **U-00062**

Chlorotris[(1,2,3,3*a*,7*a*-η)-1*H*-inden-1-yl]uranium

Chlorotri-π-indenyluranium, 8CI. *Triindenyluranium chloride*

[11082-70-5]

M 618.945

Red-brown cryst. Sol. THF, sl. sol. pentane. Mp 203° dec., >300°. V. reactive to moisture and air. Subl. *in vacuo* at 160°.

Laubereau, P.G. *et al*, *Inorg. Chem.*, 1971, **10**, 2274 (*synth, ir, ms, nmr, uv*)

Burns, J.H. *et al*, *Inorg. Chem.*, 1971, **10**, 2789 (*struct*)
Fragala, I.L. *et al*, *Inorg. Chem.*, 1983, **22**, 216 (*pe*)

$C_{27}H_{22}Br_3OPU$ U-00063

Tribromo[(1,2,3,3*a*,7*a*-η)-1*H*-inden-1-yl](triphenylphosphine oxide-*O*)uranium

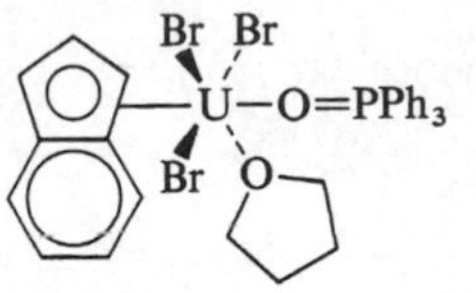

THF adduct

M 871.185

THF adduct: [74191-17-6].

$C_{31}H_{30}Br_3O_2PU$ M 943.291

Red solid. Oxygen- and moisture-sensitive.

Meunier-Piret, J. *et al*, *Bull. Soc. Chim. Belg.*, 1980, **89**, 241 (*struct*)
Goffart, J. *et al*, *Inorg. Nucl. Chem. Lett.*, 1980, **16**, 233 (*synth, ms, ir*)

$C_{27}H_{35}ClOU$ U-00064

(Benzoyl-*C*,*O*)chlorobis[(1,2,3,4,5-η)-1,2,3,4,5-pentamethyl-2,4-cyclopentadien-1-yl]uranium

[77841-33-9]

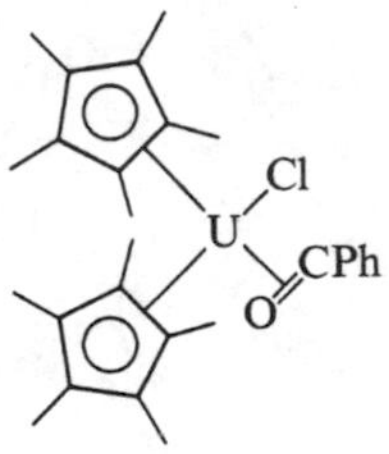

M 649.055

Coordinated PhCO reduced to $PhCH_2O$ by H_2 at 25° using catalyst of Dihydrobis(pentamethylcyclopentadienyl)thorium. Sol. C_6H_6. Air-sensitive.

Maatta, E.A. *et al*, *J. Am. Chem. Soc.*, 1981, **103**, 3576.

$C_{28}H_{24}U$ U-00065

Bis[(1,2,3,4,5,6,7,8-η)-1-phenyl-1,3,5,7-cyclooctatetraene]uranium

1,1'-Diphenyluranocene

[37274-13-8]

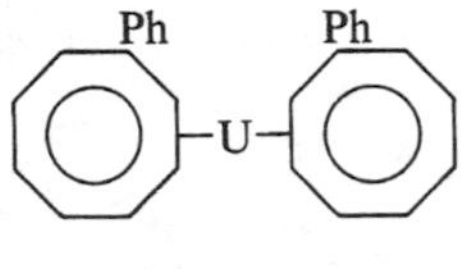

M 598.527

Green solid. Sol. hexane. Stable to air for several minutes.

Streitweiser, A., *et al*, *Inorg. Chem.*, 1973, **12**, 1102 (*synth, uv, ms, ir, nmr*)

$C_{28}H_{24}U$ U-00066

Tris[(1,2,3,3*a*,7*a*-η)-1*H*-inden-1-yl]methyluranium, 9CI

[63643-55-0]

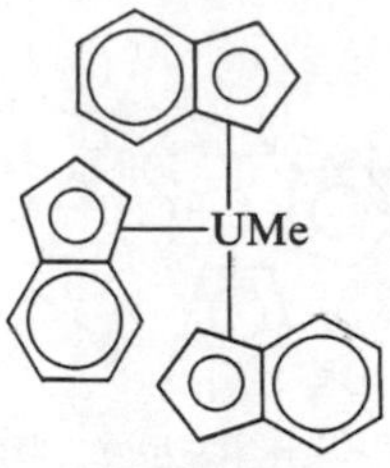

M 598.527

Red-brown solid. Sol. THF, C_6H_6, sl. sol. pentane. Mp 211-3°. Subl. at 185-90° *in vacuo*. Dec. rapidly in air. Undergoes methanolysis nearly quantitatively to afford $(C_9H_7)_3UOMe$.

Goffart, J. *et al*, *Inorg. Nucl. Chem. Lett.*, 1977, **13**, 189 (*synth, ir, ms, uv*)

$C_{28}H_{44}I_4N_4U$ U-00067

Tetrakis(cyclohexylisocyanide)tetraiodouranium

Tetraiodotetrakis(isocyanocyclohexane)uranium

[32518-07-3]

$UI_4(CNC_6H_{11})_4$

M 1182.329

Off-white cryst. Sol. C_6H_6, $CHCl_3$, Me_2CO, MeOH. Extremely sensitive to moisture, gradually dec. at 0°.

Lux, F. *et al*, *Angew. Chem., Int. Ed. Engl.*, 1971, **10**, 274 (*synth, ir*)

$C_{28}H_{52}Si_2U$ U-00068

Bis[(1,2,3,4,5-η)-1,2,3,4,5-pentamethyl-2,4-cyclopentadien-1-yl]bis[(trimethylsilyl)methyl]uranium

[68963-85-9]

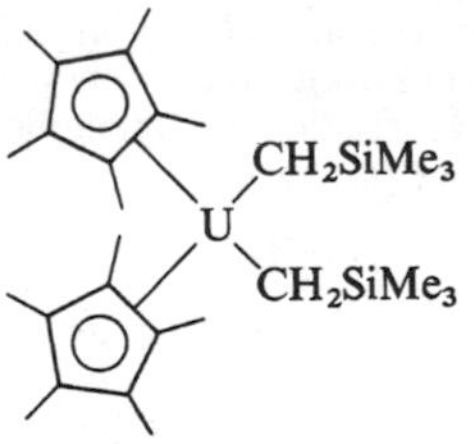

M 682.919

Inserts two CO groups into U-C bonds at 25°. Orange needles. Sol. pentane. Air- and moisture-sensitive.

Manriquez, J.M. *et al*, *J. Am. Chem. Soc.*, 1978, **100**, 7112 (*props*)
Fagan, P.J. *et al*, *J. Am. Chem. Soc.*, 1981, **103**, 6650 (*synth, ir, nmr*)

$C_{29}H_{29}PU$ U-00069

Tris(η^5-2,4-cyclopentadien-1-yl)(methyldiphenylphosphoranylidenemethyl)uranium

[77357-86-9]

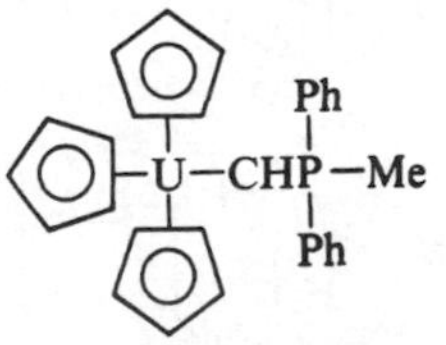

M 646.551

Dark-green microcryst. Sol. toluene, THF. Inserts CO to give Tris(η^5-2,4-cyclopentadien-1-yl)[(methyldiphenylphosphoranylidene)acetyl]uranium, U-00070 V. reactive to traces of O_2/H_2O.

▷Inflames in air

Cramer, R.E. *et al*, *Inorg. Chem.*, 1981, **20**, 2466 (*synth, struct, ir, uv*)

Cramer, R.E. *et al*, *Organometallics*, 1982, **1**, 869 (*props*)

$C_{30}H_{29}OPU$ U-00070

Tris(η^5-2,4-cyclopentadien-1-yl)[(methyldiphenylphosphoranylidene)acetyl]uranium

[81388-97-8]

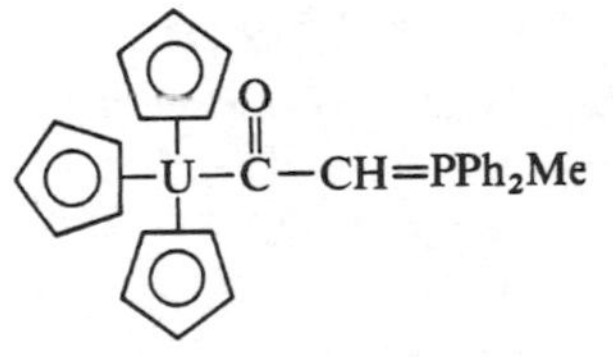

M 674.561

Formed by CO insertion into $(C_5H_5)_3U{=}CHPMePh_2$. Red cryst. Sol. toluene.

Cramer, R.E. *et al*, *Organometallics*, 1982, **1**, 869 (*synth, struct, nmr*)

$C_{30}H_{36}Cl_5LiO_2U_2$ U-00071

Di-μ-chlorodi-μ_3-chloro(μ-chlorobis[methylenebis(η^5-2,4-cyclopentadien-1-ylidene)])diuraniumbis(tetrahydrofuran)lithium

[60109-96-8]

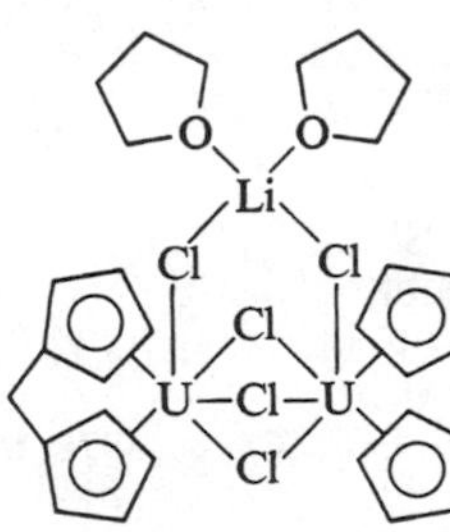

M 1088.877

Dimer cleaved by various reagents including Lewis bases. Readily alkylated. Dark-red cryst. Sol. toluene, hot pentane. Air-sensitive.

Secaur, C.A. *et al*, *J. Am. Chem. Soc.*, 1976, **98**, 3713 (*synth, struct, ms, nmr*)

$C_{31}H_{36}U$ U-00072

Tribenzyl(pentamethylcyclopentadienyl)uranium

[(1,2,3,4,5-η)-1,2,3,4,5-Pentamethyl-2,4-cyclopentadien-1-yl]tris(phenylmethyl)uranium. Pentamethylcyclopentadienyltribenzyluranium

[82511-74-8]

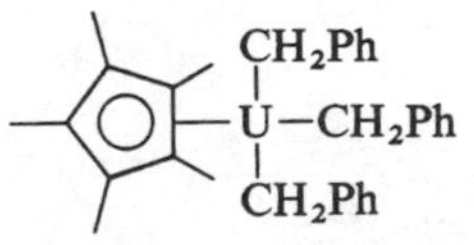

M 646.654

Black cryst. Sol. toluene.

Mintz, E.A. *et al*, *J. Am. Chem. Soc.*, 1982, **104**, 4692 (*synth, nmr, cmr*)

$C_{34}H_{30}N_2U$ U-00073

Bis(η^5-2,4-cyclopentadien-1-yl)bis(*N*-phenylbenzenaminato)uranium

Dicyclopentadienylbis(diphenylamino)uranium

[77507-88-1]

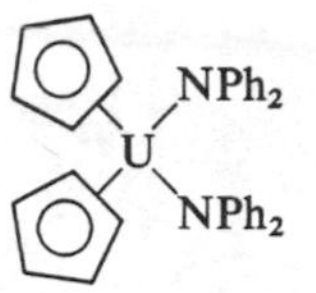

M 704.653

Dark-red solid. Sol. THF. Air- and moisture-sensitive.

Arduini, A.L. *et al*, *Inorg. Chem.*, 1981, **20**, 2470 (*synth, nmr, ms*)

$C_{36}H_{30}U^{\ominus\ominus}$ U-00074

Hexaphenyluranate(2−)

$UPh_6^{\ominus\ominus}$

M 700.662 (ion)

Di-Li salt: [63816-92-2].

$C_{36}H_{30}Li_2U$ M 714.544

Red solid + $8Et_2O$ (Et_2O). Dec. >5° in solid state, rapidly in soln. Inflames in air.

Sigurdson, E.R. *et al*, *J. Chem. Soc., Dalton Trans.*, 1977, 812 (*synth, nmr, ir*)

$C_{36}H_{33}ClU$ U-00075

Chlorotris(benzylcyclopentadienyl)uranium

Chlorotris[(1,2,3,4,5-η)-1-phenylmethyl-2,4-cyclopentadien-1-yl]uranium. Tris(benzylcyclopentadienyl)uranium(IV) chloride

[37325-10-3]

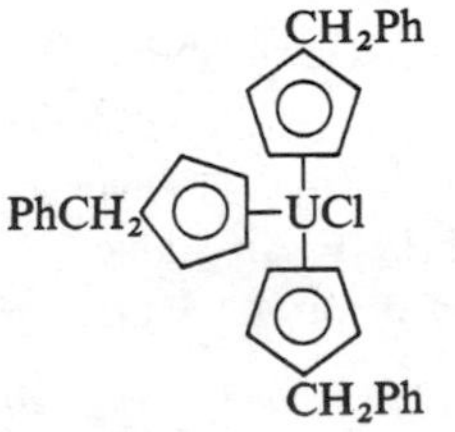

M 739.139

Dark-red cryst. Sol. dimethoxyethane, cyclohexane. Oxygen-stable but water-sensitive.

Leong, J. *et al*, *Inorg. Chem.*, 1973, **12**, 1329 (*synth, struct*)

$C_{36}H_{39}ClU$ U-00076

Chlorotris[(1,2,3,3*a*,7*a*-η)-1,4,7-trimethyl-1*H*-inden-1-yl]uranium

Tris(1,4,7-trimethylindenyl)uranium chloride

[78599-85-6]

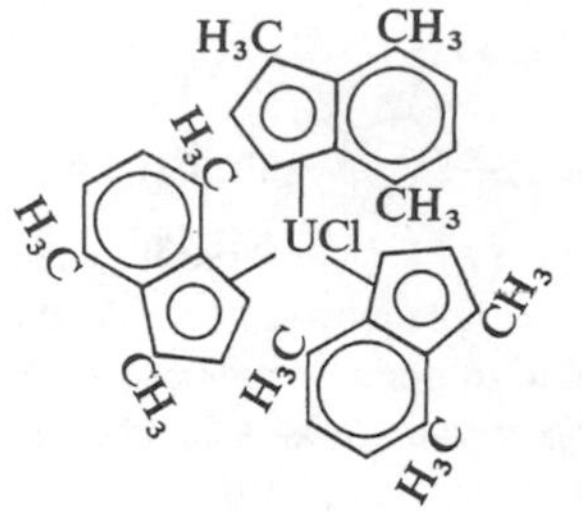

M 745.186

Dark-brown cryst. Sol. THF, sl. sol. pentane. Air- and moisture-sensitive.

Goffart, J. *et al*, *J. Organomet. Chem.*, 1981, **209**, 281 (*synth, struct, ir, ms, uv*)

$C_{38}H_{42}Cl_4N_4O_2U_3$ U-00077

Bis[bis(acetonitrile)tris(η⁵-cyclopentadienyl)uranium(1+)]tetrachlorodioxouranate(2−)

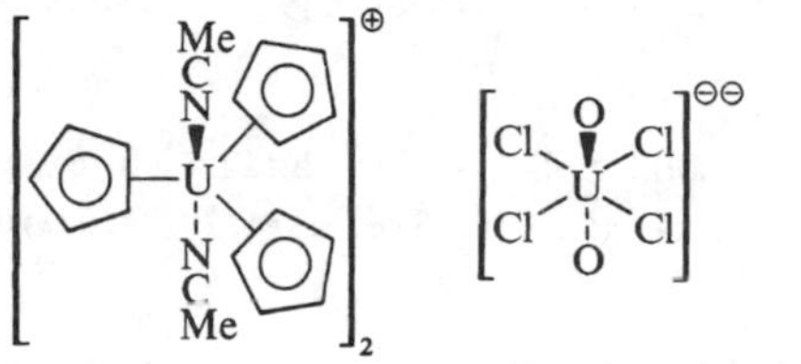

M 1442.674

Bis(1,3-butadiene) adduct:

$C_{46}H_{54}Cl_4N_4O_2U_3$ M 1550.857

Prepd. by reaction of butadiene with $(C_5H_5)_3UCl$ in MeCN. Dark-green needles.

Cyclopentadiene adduct:

$C_{43}H_{48}Cl_4N_4O_2U_3$ M 1508.777

Formed by reaction of $(C_5H_5)_3UCl$ with lab. grade MeCN over 3-4 months. Green-brown amorph. solid.

Bombieri, G. *et al*, *J. Chem. Soc., Dalton Trans.*, 1983, 1115 (*synth, cryst struct, ir, nmr*)

$C_{41}H_{35}Cl_3O_2P_2U$ U-00078

Trichloro(η⁵-2,4-cyclopentadien-1-yl)bis(triphenylphosphine oxide-*O*)uranium

[67507-20-4]

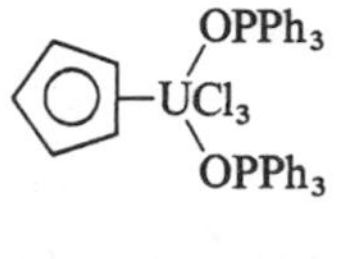

M 966.062

Green cryst. Sol. THF. Air-stable.

Bagnall, K.W. *et al*, *J. Chem. Soc., Dalton Trans.*, 1978, 295 (*synth, ir*)

Bombieri, G. *et al*, *Inorg. Nucl. Chem. Lett.*, 1978, **14**, 359 (*struct*)

$C_{47}H_{47}P_3U$ U-00079

(η⁵-2,4-Cyclopentadien-1-yl)tris[(diphenylphosphinidenio)bis(methylene)]uranium

[77357-87-0]

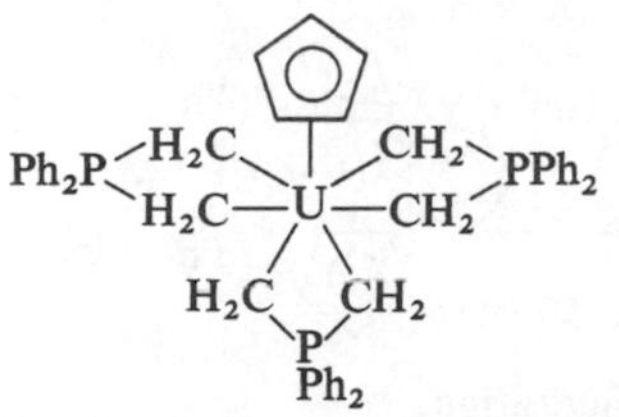

M 942.838

Gold rod-shaped cryst. Sol. Et_2O. Air- and moisture-sensitive.

Cramer, R.E. *et al*, *Inorg. Chem.*, 1981, **20**, 2466 (*synth, struct, ir, uv, pmr*)

$C_{48}H_{46}P_2U_2$ U-00080

Tetrakis(η⁵-2,4-cyclopentadien-1-yl)bis[μ-[methylene(diphenylphosphinidenio)methylidyne]]diuranium

[67771-71-5]

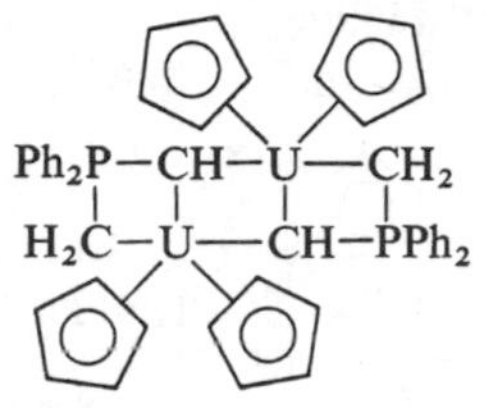

M 1160.897

Deep-red cryst. + $0.5Et_2O$. Sol. Et_2O, C_6H_6. Also exists as pentane solvate.

Cramer, R.E. *et al*, *Inorg. Chem.*, 1980, **19**, 2565 (*synth, ms, struct*)

$C_{64}H_{48}U$ U-00081

Bis[(1,2,3,4,5,6,7,8-η)-1,3,5,7-tetraphenyl-1,3,5,7-cyclooctatetraene]uranium, 9CI

1,1′,3,3′,5,5′,7,7′-Octaphenyluranocene

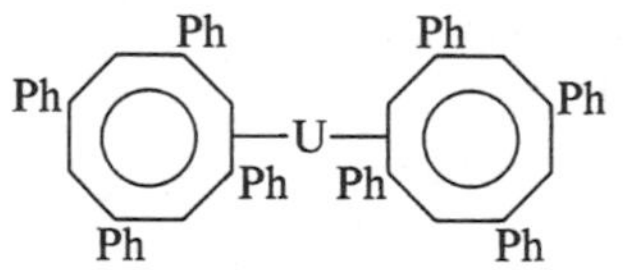

M 1055.112

Dark-brown cryst. Sl. sol. THF, insol. most solvs. $Bp_{0.00001}$ 400° subl.

Streitweiser, A. *et al*, *J. Organomet. Chem.*, 1975, **97**, C41 (*synth, ms, uv*)

Templeton, L.K. *et al*, *Inorg. Chem.*, 1976, **15**, 3000 (*struct*)

V Vanadium

J. A. Labinger

Vanadium (Fr., Ger.), Vanadio (Sp., Ital.), Ванадий (Vanadii) (Russ.), バナジウム (Japan.)

Atomic Number. 23

Atomic Weight. 50.9414(3)

Electronic Configuration. [Ar] $3d^3\,4s^2$

Oxidation States. These range from −3 (for $V(CO)_5{}^{3-}$) to +5. The more common values are −1 to +4; relatively few pentavalent organovanadium compounds are known.

Coordination Number. 3–8 for conventional ligands.

Colour. Organovanadium(+IV) compounds tend to be strongly coloured, from dark brown to blue or black. This is also the case for the other odd-electron states. Colours of compounds of the remaining oxidation states typically range from colourless (fairly rare) through yellow to red or brown.

Availability. Convenient starting materials include VCl_4 (a liquid), VCl_3 and $V(acac)_3$. Cp_2VCl_2 is commercially available but expensive.

Handling. Vanadium complexes are air-sensitive and require standard inert atmosphere techniques for their manipulation.

Toxicity. Vanadium compounds are not especially toxic although the carbonyls should be considered highly toxic as are those of other metals.

Isotopic Abundance. ^{51}V, 99.76%.

Spectroscopy. ^{51}V ($I = \frac{7}{2}$) has relative sensitivity 0.381, a relatively low quadrupole moment and has therefore been used extensively as an nmr probe, primarily in work by Rehder *et al.* on carbonyl derivatives. Whilst many vanadium compounds are paramagnetic, nmr has still proven useful for these. A large amount of ^{1}H and ^{13}C nmr spectroscopy has been carried out on substituted vanadocenes and related compounds by Kohler and others.

Mass spectroscopy is straightforward since vanadium has essentially only one naturally occurring isotope.

Esr is a very powerful tool, especially for the +4 oxidation state which has electronic configuration d^1 and hence is highly suitable. A vast amount of work has been done on compounds of the type Cp_2VX_2. Esr has also been used to some extent for vanadium compounds of oxidation states 0 and +2.

Analysis. Vanadium may be determined volumetrically by conversion to standard oxidation state (+2 or +4) and redox titration, or colorimetrically as a peroxo complex or heteropoly acid.

References. General reviews are listed in the introduction to the *Sourcebook*.

$NH_3V(CO)_5^{\ominus}$

V-00001

V-00002

X = Br

V-00003

V-00004

As V-00003 with
X = Cl

V-00005

V-00006

V-00007

$V(CO)_5^{\ominus\ominus\ominus}$

V-00008

$PhV(O)Cl_2$

V-00009

V-00010

$V(CO)_5CN^{\ominus\ominus}$

V-00011

$V(CO)_6^{\ominus}$

V-00012

$V(CO)_6$

V-00013

$EtHgV(CO)_6$

V-00014

V-00015

V-00016

R = H

V-00017

V-00018

L = CS

V-00019

V-00020

As V-00017 with
R = Me

V-00021

V-00022

V-00023

V-00024

$[(H_3C)_3CNC]V(CO)_5^{\ominus}$

V-00025

X = Br

V-00026

X = Y = Br

V-00027

V-00028

As V-00026 with
X = Cl

V-00029

As V-00027 with
X = Y = Cl

V-00030

As V-00026 with
X = I

V-00031

V-00032

V-00033

V-00034

V-00035

$[(H_3C)_3CO]_2VMe_2$

V-00036

V-00037

V-00038

As V-00027 with
X = CO, Y = I

V-00039

As V-00026 with
X = CO

V-00040

V-00041

V-00042

V-00043

V-00044

V-00045

As V-00027 with
X = Me, Y = Cl

V-00046

As V-00026 with
X = Me

V-00047

V-00048

As V-00027 with
X = Y = CN

V-00049

V-00050

V-00051

V-00052

V-00053

V-00054

As V-00026 with
X = Et

V-00055

As V-00027 with
X = Y = Me

V-00056

$OV(CH_2SiMe_3)_3$

V-00057

As V-00027 with
X = Ac, Y = CO

V-00058

As V-00026 with
X = $-CH_2CH{=}CH_2$

V-00059

V-00060

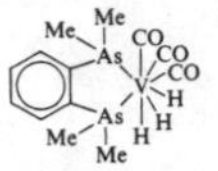

V-00061

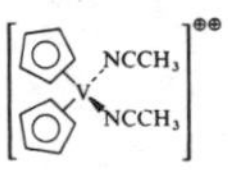

V-00062

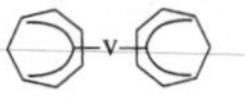

V-00063

As V-00027 with
X = Y = Et

V-00064

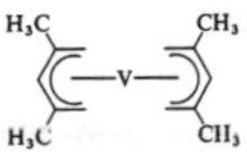

V-00065

$EtV(NEt_2)_3$

V-00066

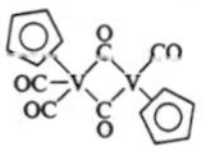

V-00067

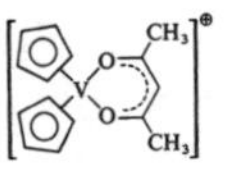

V-00068

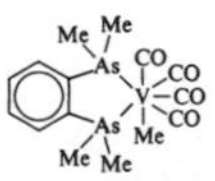

V-00069

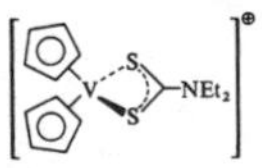

V-00070

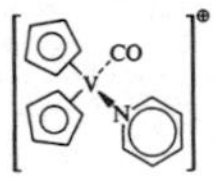

V-00071

As V-00026 with
X = SPh

V-00072

As V-00026 with
X = Ph

V-00073

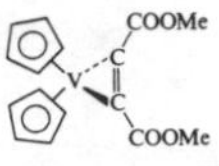

V-00074

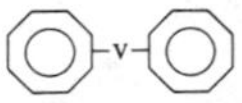

V-00075

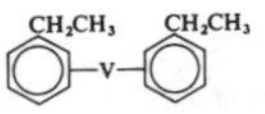

V-00076

$V(CH_2SiMe_3)_4$

V-00077

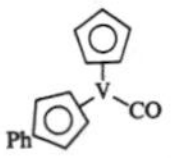

V-00078

As V-00026 with
$X = CH_2Ph$

V-00079

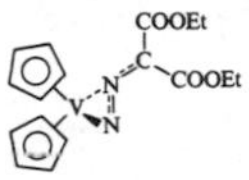

V-00080

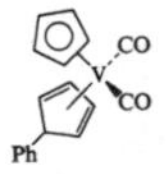

V-00081

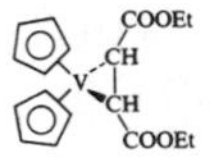

V-00082

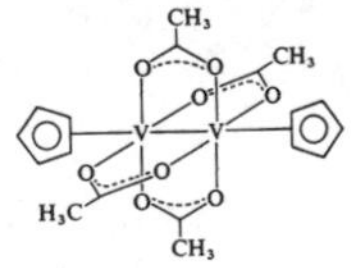

V-00083

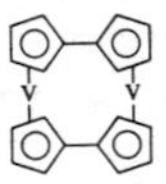

V-00084

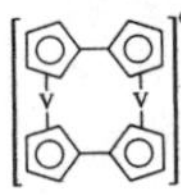

V-00085

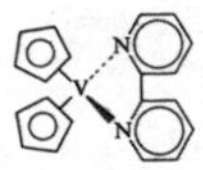

V-00086

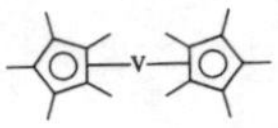

V-00087

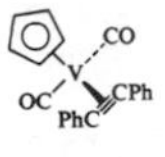

V-00088

$V[CH(SiMe_3)_2]_3$

V-00089

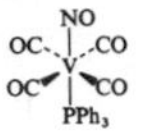

V-00090

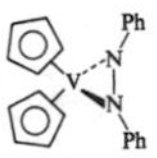

V-00091

As V-00027 with
X = Y = SPh

V-00092

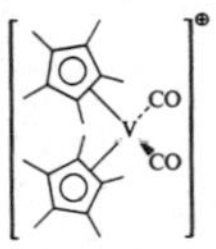

V-00093

$Ph_3PV(CO)_5^{\ominus}$

V-00094

$(OC)_5VPPh_3$

V-00095

$Ph_3SnV(CO)_6$

V-00096

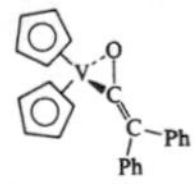

V-00097

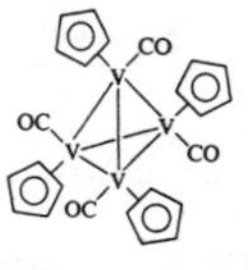

V-00098

As V-00027 with
$X = Y = CH_2Ph$

V-00099

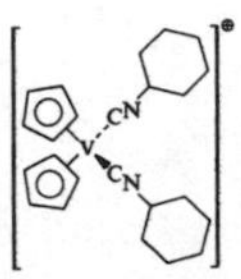

V-00100

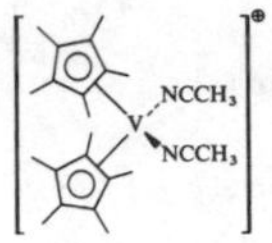

V-00101

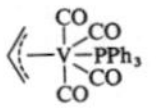

V-00102

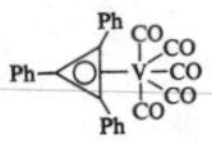

V-00103

As V-00019 with
$L = PPh_3$

V-00104

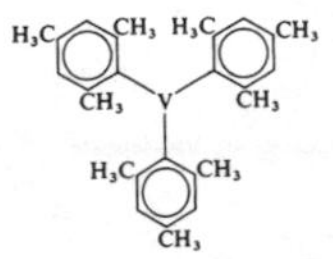

V-00105

$V(CH_2Ph)_4$

V-00106

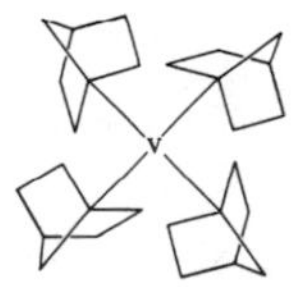

V-00107

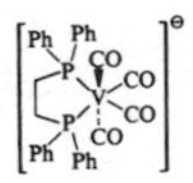

V-00108

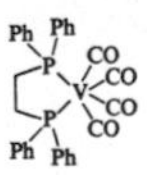

V-00109

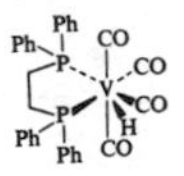

V-00110

$[Ph_2PCH_2CH_2PPh_2V(CO)_5]^{\ominus}$

V-00111

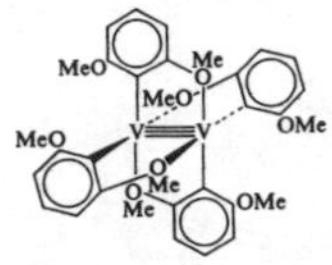

V-00112

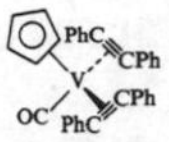

V-00113

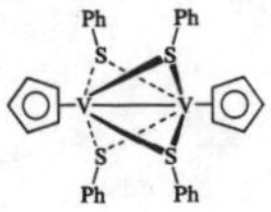

V-00114

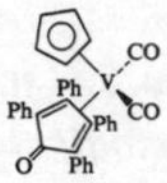

V-00115

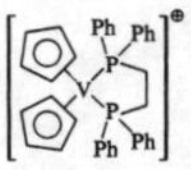

V-00116

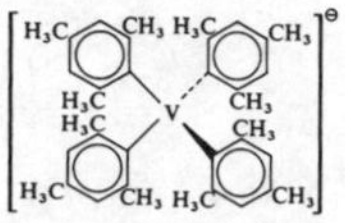

V-00117

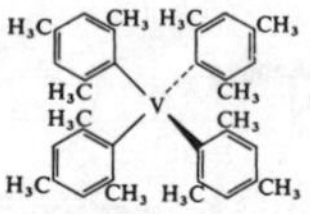

V-00118

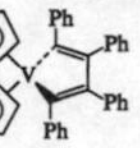

V-00119

$C_5H_3NO_5V^{\ominus}$ **V-00001**

Amminepentacarbonylvanadate(1−), 9CI

$NH_3V(CO)_5^{\ominus}$

M 208.024 (ion)

Na salt: [81831-31-4].
$C_5H_3NNaO_5V$ M 231.014
Red-purple soln. in liq. NH_3.

Tetraphenylphosphonium salt: [36580-51-5].
$C_{29}H_{23}NO_5PV$ M 547.420
Dark-purple cryst. Insol. Et_2O, alkanes, EtOH, sl. sol. liq. NH_3, v. sol. MeCN, THF when dec. is rapid. Turns black in air after several hrs. Dec. >130°.

Tetraphenylarsonium salt: [78937-16-3].
$C_{29}H_{23}AsNO_5V$ M 591.368
Magenta solid (liq. NH_3). Dec. in solvs. in which it is sol. Air-stable. Dec. >110°.

Bis(triphenylphosphine)iminium salt: [81831-32-5].
$C_{41}H_{33}N_2O_5P_2V$ M 746.611
Dark-purple cryst., unstable at r.t.

Rehder, D., *J. Organomet. Chem.*, 1972, **37**, 303 (*synth, ir, nmr, uv*)
Ellis, J.E. *et al*, *J. Am. Chem. Soc.*, 1981, **103**, 6100 (*synth*)
Ellis, J.E. *et al*, *Organometallics*, 1982, **1**, 898 (*synth*)

$C_5H_5Br_2OV$ **V-00002**

Dibromo(η^5-2,4-cyclopentadien-1-yl)oxovanadium
Cyclopentadienyloxodibromovanadium

(Cp)–V(=O)(Br)(Br)

M 291.843
Green-black cryst. Sol. aromatic hydrocarbons, $CHCl_3$, CCl_4. Air-stable, diamagnetic.

Fischer, E.O. *et al*, *Chem. Ber.*, 1960, **93**, 701 (*synth*)
B.P., 857 138, (*1960*); *CA*, **55**, 17538
Skachillova, S.Ya. *et al*, *J. Gen. Chem. USSR* (*Engl. Transl.*), 1966, **36**, 1073

$C_5H_5Br_3V$ **V-00003**

Tribromo(η^5-2,4-cyclopentadien-1-yl)vanadium, 9CI
Cyclopentadienyltribromovanadium. Cyclopentadienylvanadium tribromide
[60530-27-0]

(Cp)–VX_3

X = Br

M 355.748
Green cryst. (toluene). Sol. aromatic hydrocarbons, $CHCl_3$, CCl_4. Paramagnetic.

Fischer, E.O. *et al*, *Chem. Ber.*, 1960, **93**, 701 (*synth*)

$C_5H_5Cl_2OV$ **V-00004**

Dichloro(η^5-2,4-cyclopentadien-1-yl)oxovanadium, 9CI
Cyclopentadienyloxodichlorovanadium. Cyclopentadienylvanadium oxydichloride
[1293-64-7]

(Cp)–V(=O)(Cl)(Cl)

M 202.941
Blue-black cryst. (toluene). Sol. aromatic hydrocarbons, $CHCl_3$, CCl_4. $Bp_{0.1}$ 100° subl. Diamagnetic.

Fischer, E.O. *et al*, *Chem. Ber.*, 1960, **93**, 701 (*synth*)
de Liefde Meijer, H.J. *et al*, *Recl. Trav. Chim. Pays-Bas*, 1965, **84**, 1418 (*synth, ir*)

$C_5H_5Cl_3V$ **V-00005**

Trichloro(η^5-2,4-cyclopentadien-1-yl)vanadium, 9CI
Cyclopentadienyltrichlorovanadium. Cyclopentadienylvanadium trichloride
[34767-30-1]

As Tribromo(η^5-2,4-cyclopentadien-1-yl)vanadium, V-00003 with

X = Cl

M 222.395
Alkene polymerisation catalyst. Violet cryst. (toluene). Sol. aromatic hydrocarbons, $CHCl_3$, CCl_4. Bp 130-60° subl. *in vacuo*. Paramagnetic.

Fischer, E.O. *et al*, *Chem. Ber.*, 1960, **93**, 701 (*synth*)
U.S.P., 3 080 305, (*1963*); *CA*, **59**, 3957 (*use*)
Ger. Pat., 1 959 322, (*1971*); *CA*, **75**, 88768 (*use*)
Thiele, K.-H. *et al*, *Z. Anorg. Allg. Chem.*, 1976, **423**, 231 (*synth*)

$C_5H_5N_3O_3V^{\oplus}$ **V-00006**

Cyclopentadienyltrinitrosylvanadium(1+)

[(Cp)–V(NO)(NO)(NO)]$^{\oplus}$

M 206.054 (ion)
Diamagnetic.

Hexafluorophosphate: [69439-84-5].
$C_5H_5F_6N_3O_3PV$ M 351.018
Green cryst. ($MeNO_2$). Sol. $MeNO_2$. Dec. at 160°.

Herberhold, M. *et al*, *Angew. Chem., Int. Ed. Engl.*, 1979, **18**, 220 (*synth, ir, pmr, ms*)

$C_5H_5O_2V$ **V-00007**

(η^5-2,4-Cyclopentadien-1-yl)dioxovanadium, 9CI
[57088-86-5]

(Cp)–VO_2

M 148.035

Latyaeva, V.N. *et al*, *CA*, 1975, **83**, 179261t.

$C_5O_5V^{\ominus\ominus\ominus}$ **V-00008**

Pentacarbonylvanadate(3−)

$V(CO)_5^{\ominus\ominus\ominus}$

M 190.994 (ion)

Tri-K salt: [78937-14-1].

$C_5K_3O_5V$ M 308.288

Dark red-brown powder (NH_3). Insol. most solvs., sl. sol. liq. NH_3. Dec. >60°.

▷Pyrophoric. Shock-sensitive, solid deflagrates on scratching

Ellis, J.E. *et al*, *J. Am. Chem. Soc.*, 1976, **98**, 8264 (*synth*)
Ellis, J.E. *et al*, *J. Am. Chem. Soc.*, 1981, **103**, 6100 (*synth*, *haz*)

$C_6H_5Cl_2OV$ V-00009

Dichlorooxophenylvanadium

Phenylvanadium oxydichloride

[28597-01-5]

$PhV(O)Cl_2$

M 214.952

Ethylene polymerisation catalyst. Deep-red cryst. (pentane). Diamagnetic, highly unstable.

▷Dec. explosively above 10°

Carrick, W.L. *et al*, *J. Am. Chem. Soc.*, 1960, **82**, 3887 (*use*)
Reichle, W.T. *et al*, *J. Organomet. Chem.*, 1970, **24**, 419 (*synth*)
Thiele, K.-H. *et al*, *Z. Anorg. Allg. Chem.*, 1972, **390**, 280 (*synth*, *uv*, *ir*, *haz*)

$C_6H_5N_2O_3V$ V-00010

Carbonyl(η^5-2,4-cyclopentadien-1-yl)dinitrosylvanadium

Cyclopentadienylcarbonyldinitrosylvanadium

[31811-51-5]

M 204.059

Dark-brown powder by subl. Sol. aromatic, aliphatic hydrocarbons, Et_2O. Mp 50°. Bp 40° subl. *in vacuo*. Diamagnetic, light-sensitive.

Fischer, E.O. *et al*, *J. Organomet. Chem.*, 1968, **14**, P4 (*synth*, *ir*, *pmr*, *ms*)
Herberhold, M. *et al*, *Angew. Chem., Int. Ed. Engl.*, 1979, **18**, 220.

$C_6NO_5V^{\ominus\ominus}$ V-00011

Pentacarbonyl(cyano-*C*)vanadate(2−), 9CI

$V(CO)_5CN^{\ominus\ominus}$

M 217.011 (ion)

Bis(tetraphenylphosphonium) salt: [36580-53-7].

$C_{54}H_{40}NO_5P_2V$ M 895.803

Black cryst. (NH_3). Sol. MeCN, DMF, NH_3.

▷Pyrophoric

Rehder, D., *J. Organomet. Chem.*, 1972, **37**, 303 (*synth*, *ir*, *nmr*, *uv*)

$C_6O_6V^{\ominus}$ V-00012

Hexacarbonylvanadate(1−)

$V(CO)_6^{\ominus}$

M 219.004 (ion)

Na salt, bis(diglyme) complex: [15531-13-2].

$C_{18}H_{28}NaO_{12}V$ M 510.343

Yellow solid. Sol. H_2O. Mp 173-6° dec. (under N_2).

Werner, R.P.M. *et al*, *Chem. Ind.* (*London*), 1961, 144 (*synth*, *ir*)

C_6O_6V V-00013

Vanadium hexacarbonyl

Vanadium carbonyl, 9CI. *Hexacarbonylvanadium*

[14024-00-1]

$V(CO)_6$

M 219.004

Dark-blue cryst. by vac. subl. Dec. at 60-70°; air-sensitive, paramagnetic with one unpaired electron.

▷Toxic

Ercoli, R. *et al*, *J. Am. Chem. Soc.*, 1960, **82**, 2966 (*synth*)
Werner, R.P.M. *et al*, *Chem. Ind.* (*London*), 1961, 144 (*synth*)
Silvestri, G. *et al*, *J. Chem. Soc., Dalton Trans.*, 1972, 2558 (*synth*)
Rietz, R.R. *et al*, *Inorg. Chem.*, 1975, **14**, 2818.
Schmidling, D.G., *J. Mol. Struct.*, 1975, **24**, 1 (*struct*)
Rubinson, K.A., *J. Am. Chem. Soc.*, 1976, **98**, 5188 (*epr*, *uv*)
Sax, N.I., *Dangerous Properties of Industrial Materials*, 5th Ed., Van Nostrand-Reinhold, 1979, 1082.

$C_8H_5HgO_6V$ V-00014

Hexacarbonyl(ethylmercury)vanadium, 9CI

[36571-13-8]

$EtHgV(CO)_6$

M 448.655

Orange platelets (pentane). Sol. CH_2Cl_2, mod. sol. aliphatic hydrocarbons. Mp 47-8° dec. Subl. at r.t. *in vacuo*, with part. dec.

Davison, A. *et al*, *J. Organomet. Chem.*, 1972, **36**, 113 (*synth*, *ir*)

$C_8H_5O_3V^{\ominus\ominus}$ V-00015

Tricarbonyl(η^5-2,4-cyclopentadien-1-yl)vanadate(2−)

M 200.067 (ion)

Di-Na salt: [68688-11-9].

$C_8H_5Na_2O_3V$ M 246.047

Yellow powder + 1THF (THF). Sol. THF, HMPA.

Fischer, E.O. *et al*, *Chem. Ber.*, 1970, **103**, 3684 (*synth*)
Ellis, J.E. *et al*, *J. Organomet. Chem.*, 1976, **120**, 389 (*ir*)
Kinney, R.J. *et al*, *J. Am. Chem. Soc.*, 1978, **100**, 7902 (*synth*, *ir*, *pmr*)

$C_8H_5O_5V$ V-00016

Pentacarbonyl(η^3-2-propenyl)vanadium, 9CI

π-Allylpentacarbonylvanadium

[53593-70-7]

M 232.066

Dark-red needles (Et_2O at −78°). Mp 93° dec. Unstable.

Schneider, M. *et al*, *J. Organomet. Chem.*, 1976, **121**, 345 (*synth*, *spectra*)

$C_8H_6O_3V^\ominus$ **V-00017**

Tricarbonyl(η^5-2,4-cyclopentadien-1-yl)hydrovanadate(1−)

Cyclopentadienyltricarbonylhydrovanadate(1−)

R = H

M 201.075 (ion)

Diamagnetic.

Tetraethylammonium salt:

$C_{16}H_{26}NO_3V$ M 331.328

Relatively air-stable orange-yellow microcryst. Sl. sol. THF. Dec. in MeCN over 2 days.

Bis(triphenylphosphine)iminium salt: [68738-02-3].

$C_{44}H_{36}NO_3P_2V$ M 739.662

Yellow solid (THF). Sol. THF, HMPA, insol. aliphatic hydrocarbons. Mp 201° dec.

Kinney, R.J. *et al*, *J. Am. Chem. Soc.*, 1973, **100**, 635, 7902 (*synth, ir, pmr*)

Puttfarcken, U. *et al*, *J. Organomet. Chem.*, 1980, **185**, 219 (*synth, ir, nmr, pmr*)

$C_9H_5NO_3V^\ominus$ **V-00018**

Tricarbonyl(cyano-*C*)(η^5-2,4-cyclopentadien-1-yl)vanadate(1−), 9CI

Cyclopentadienyltricarbonylcyanovanadate(1−)

M 226.085 (ion)

Na salt: [46238-54-4].

$C_9H_5NNaO_3V$ M 249.075

Orange-red cryst. (H_2O). Sol. H_2O.

Tetramethylammonium salt:

$C_{13}H_{17}N_2O_3V$ M 300.230

Orange-red needles (H_2O). Dec. at 130°.

Fischer, E.O. *et al*, *Chem. Ber.*, 1970, **103**, 3684 (*synth, ir, pmr*)

Rehder, D., *Z. Naturforsch., B*, 1976, **31**, 273 (*ir, nmr*)

$C_9H_5O_3SV$ **V-00019**

Tricarbonyl(η^5-2,4-cyclopentadien-1-yl)(thiocarbonyl)vanadium

[65892-02-6]

L = CS

M 244.138

Yellow solid (hexane). Mp 69-72°.

Rajan, S., *Indian J. Chem., Sect. A*, 1977, **15**, 920 (*synth, ir*)

$C_9H_5O_4V$ **V-00020**

Tetracarbonyl(η^5-2,4-cyclopentadien-1-yl)vanadium, 9CI

Cyclopentadienyltetracarbonylvanadium

[12108-04-2]

M 228.078

Orange cryst. by subl. Mp 139°. Sublimes at 80-100°/0.5 mm.

Fischer, E.O. *et al*, *Chem. Ber.*, 1960, **93**, 1006 (*props*)

Werner, R.P.M. *et al*, *Inorg. Chem.*, 1964, **3**, 298 (*synth, ir*)

Wilford, J.B. *et al*, *J. Organomet. Chem.*, 1967, **8**, 495 (*struct*)

Durig, J.R. *et al*, *J. Organomet. Chem.*, 1969, **16**, 425 (*ir, raman*)

Rietz, R.R. *et al*, *Inorg. Chem.*, 1975, **14**, 2818.

Fachinetti, G. *et al*, *J. Chem. Soc., Dalton Trans.*, 1976, 1046 (*synth*)

$C_9H_8O_3V^\ominus$ **V-00021**

Tricarbonyl(η^5-2,4-cyclopentadien-1-yl)methylvanadate(1−)

Cyclopentadienyltricarbonylmethylvanadate(1−)

As Tricarbonyl(η^5-2,4-cyclopentadien-1-yl)-hydrovanadate(1−), V-00017 with

R = Me

M 215.102 (ion)

Diamagnetic.

Bis(triphenylphosphine)iminium salt: [68688-15-3].

$C_{45}H_{38}NO_3P_2V$ M 753.689

Brick-red cryst. (THF). Sol. THF, insol. aliphatic hydrocarbons, Et_2O. Air-sensitive.

Kinney, R.J. *et al*, *J. Am. Chem. Soc.*, 1973, **100**, 635, 7902 (*synth, ir, pmr*)

$C_9H_{15}V$ **V-00022**

Tris(η^3-2-propenyl)vanadium, 9CI

Triallylvanadium

[12170-02-4]

M 174.159

Highly unstable. Ethylene polymerisation catalyst. Brown cryst.

▷Deflagrates above −30°

Wilke, G. *et al*, *Angew. Chem., Int. Ed. Engl.*, 1966, **5**, 151 (*synth*)

Ballard, D.G.H. *et al*, *Makromol. Chem.*, 1971, **148**, 175 (*use*)

$C_{10}H_7O_3V$ **V-00023**

Tricarbonyl(η^7-cycloheptatrienylium)vanadium, 9CI

Tricarbonyl(tropylium)vanadium. Cycloheptatrienyltricarbonylvanadium

[12083-16-8]

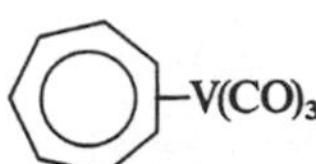

M 226.105

Dark-green cryst. (hexane). Mp 134-7° dec.

Werner, R.P.M. *et al*, *J. Am. Chem. Soc.*, 1961, **83**, 2023 (*synth*)
Fritz, H.P. *et al*, *Chem. Ber.*, 1964, **97**, 1398 (*ir*, *pmr*)
Whitesides, G.M. *et al*, *J. Am. Chem. Soc.*, 1969, **91**, 2245 (*pmr*)
Mueller, J. *et al*, *J. Organomet. Chem.*, 1972, **34**, 165 (*synth*, *ms*)
Burkert, P.K. *et al*, *CA*, 1974, **80**, 139152 (*nqr*)
Rietz, R.R. *et al*, *Inorg. Chem.*, 1975, **14**, 2818.

$C_{10}H_7O_4V$ V-00024

Tetracarbonyl(η^5-methyl-2,4-cyclopentadien-1-yl)vanadium
[63339-27-5]

M 242.104
Yellow solid by subl. Sol. aliphatic hydrocarbons, THF. $Bp_{0.01}$ 50° subl.

Jones, W.D. *et al*, *J. Am. Chem. Soc.*, 1981, **103**, 4415 (*synth*, *ms*, *pmr*)

$C_{10}H_9NO_5V^{\ominus}$ V-00025

Pentacarbonyl(2-isocyano-2-methylpropane)vanadate(1−)
Pentacarbonyl(tert-*butyl isocyanide*)*vanadate*(*1*−)

$[(H_3C)_3CNC]V(CO)_5^{\ominus}$

M 274.126 (ion)
Tetraethylammonium salt: [78954-02-6].
$C_{18}H_{29}N_2O_5V$ M 404.379
Orange cryst. (THF), dec. over several minutes in air. Sol. THF, insol. Et_2O. Mp 102-5° dec.

Ellis, J.E. *et al*, *Organometallics*, 1982, **1**, 898 (*synth*, *ir*, *pmr*)

$C_{10}H_{10}BrV$ V-00026

Bromobis(η^5-2,4-cyclopentadien-1-yl)vanadium
Bis(*cyclopentadienyl*)*bromovanadium*
[64815-29-8]

X = Br

M 261.035
Deep-blue cryst. by subl. Sl. sol. Et_2O, hydrocarbons. Mp 221-2°. $Bp_{0.2}$ 165° subl. Air-sensitive.

de Liefde Meijer, H.J. *et al*, *Recl. Trav. Chim. Pays-Bas*, 1961, **80**, 831 (*synth*)

$C_{10}H_{10}Br_2V$ V-00027

Dibromobis(η^5-2,4-cyclopentadien-1-yl)vanadium
Bis(*cyclopentadienyl*)*dibromovanadium*
[69109-78-0]

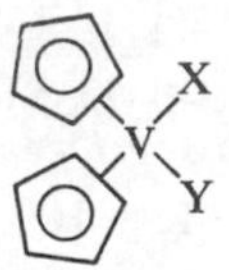

X = Y = Br

M 340.939
Dark-green solid, unstable in solid state. Solns. in $CHCl_3$, CCl_4 are stable. Sol. CCl_4, $CHCl_3$. Paramagnetic.

Wilkinson, G. *et al*, *J. Am. Chem. Soc.*, 1954, **76**, 4281 (*synth*, *ir*)

$C_{10}H_{10}ClOV$ V-00028

Chlorobis(η^5-2,4-cyclopentadien-1-yl)oxovanadium
Vanadocene oxychloride
[71191-42-9]

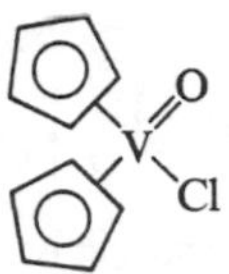

M 232.583
Blue cryst. by subl. Bp 170-90° subl. *in vacuo*.

Holliday, A.K. *et al*, *J. Chem. Soc., Dalton Trans.*, 1979, 228 (*synth*, *pmr*)

$C_{10}H_{10}ClV$ V-00029

Chlorobis(η^5-2,4-cyclopentadien-1-yl)vanadium, 9CI
Bis(*cyclopentadienyl*)*chlorovanadium. Vanadocene chloride*
[12701-79-0]
As Bromobis(η^5-2,4-cyclopentadien-1-yl)vanadium, V-00026 with

X = Cl

M 216.584
Indigo cryst. (CH_2Cl_2). Sol. CH_2Cl_2, THF. Mp 194-8°, 206-7°. $Bp_{0.0001}$ 100° subl. Paramagnetic.

de Liefde Meijer, H.J. *et al*, *Inorg. Chem. Acta*, 1970, **4**, 651 (*synth*)
Manzer, L.E., *J. Organomet. Chem.*, 1976, **110**, 291 (*synth*)
Fieselman, B.F. *et al*, *J. Organomet. Chem.*, 1977, **137**, 43 (*cryst struct*, *synth*, *props*)

$C_{10}H_{10}Cl_2V$ V-00030

Dichlorobis(η^5-2,4-cyclopentadien-1-yl)vanadium, 9CI
Bis(*cyclopentadienyl*)*dichlorovanadium. Vanadocene dichloride*
[12083-48-6]
As Dibromobis(η^5-2,4-cyclopentadien-1-yl)vanadium, V-00027 with

X = Y = Cl

M 252.037
Dark-green cryst. ($CHCl_3$). Sol. $CHCl_3$, H_2O, alcohols. Paramagnetic. Dec. at 250°.

Wilkinson, G. *et al*, *J. Am. Chem. Soc.*, 1954, **76**, 4281 (*synth, ir*)
Organomet. Synth., 1965, **1**, 75 (*synth*)
Holloway, J.D.L. *et al*, *J. Am. Chem. Soc.*, 1979, **101**, 2038.

$C_{10}H_{10}IV$ V-00031

Bis(η^5-2,4-cyclopentadien-1-yl)iodovanadium, 9CI

Vanadocene iodide. Bis(cyclopentadienyl)vanadium iodide

[53291-02-4]

As Bromobis(η^5-2,4-cyclopentadien-1-yl)vanadium, V-00026 with

X = I

M 308.035

Dark-green cryst. by subl. Insol. Et_2O, aliphatic hydrocarbons. Mp 214-5°. $Bp_{0.2}$ 165° subl.

de Liefde Meijer, H.J. *et al*, *Recl. Trav. Chim. Pays-Bas*, 1961, **80**, 831 (*synth*)
Köhler, F.H. *et al*, *Chem. Ber.*, 1978, **111**, 3464 (*synth*)

$C_{10}H_{10}S_5V$ V-00032

Bis(η^5-2,4-cyclopentadienyl)pentathiovanadium, 9CI

Bis(cyclopentadienyl)vanadium pentasulfide. Vanadocene pentasulfide

[11077-28-4]

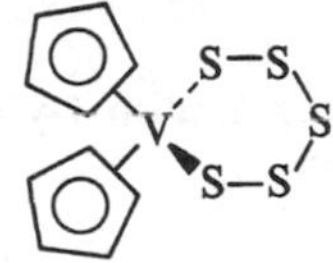

M 341.431

Air-stable dark-red cryst. (THF aq.). Sol. THF. Dec. >150°.

Köpf, H. *et al*, *Angew. Chem., Int. Ed. Engl.*, 1971, **10**, 137 (*synth*)
Muller, E.G., *J. Organomet. Chem.*, 1976, **111**, 91 (*synth, esr, ir, cryst struct, props*)

$C_{10}H_{10}V$ V-00033

Bis(η^5-2,4-cyclopentadien-1-yl)vanadium

Vanadocene, 9CI

[1277-47-0]

M 181.131

Polymerisation catalyst for acetylenes. Air-sensitive violet cryst. by subl. Sol. C_6H_6, THF. Mp 167-8°. Paramagnetic.

Fischer, E.O. *et al*, *Chem. Ber.*, 1958, **91**, 2205 (*synth, ir*)
de Liefde Meijer, H.J. *et al*, *Recl. Trav. Chim. Pays-Bas*, 1966, **85**, 1007 (*synth*)
Müller, J. *et al*, *J. Organomet. Chem.*, 1967, **10**, 313 (*ms*)
Köhler, F.H. *et al*, *J. Organomet. Chem.*, 1976, **110**, 235 (*cmr, pmr*)
Aleksanyan, V.T. *et al*, *J. Organomet. Chem.*, 1977, **124**, 293 (*ir, raman*)
Handlir, K. *et al*, *Z. Chem.*, 1979, **19**, 265 (*synth*)

$C_{10}H_{14}BV$ V-00034

Bis(η^5-2,4-cyclopentadien-1-yl)[tetrahydroborato(1−)-*H,H'*-]vanadium, 9CI

Bis(cyclopentadienyl)vanadium borohydride

[55744-43-9]

M 195.972

Deep-violet cryst. by subl. Sol. aromatic hydrocarbons. $Bp_{0.001}$ 55° subl. Dec. slowly at r.t. under N_2. Dec. >80°. Diamagnetic.

▷Pyrophoric

Marks, T.J. *et al*, *J. Am. Chem. Soc.*, 1975, **97**, 1439 (*synth, ir, pmr*)

$C_{10}H_{17}O_4P_2V$ V-00035

Tetracarbonyl[1,2-ethanediylbis[dimethylphosphine]-*P,P'*]-hydrovanadium, 9CI

Hydrotetracarbonyl[bis(dimethylphosphino)ethane]-vanadium

[56408-47-0]

M 314.131

Yellow cryst. (CH_2Cl_2/heptane). Sol. CH_2Cl_2. Mp 123° dec.

Ellis, J.E. *et al*, *J. Organomet. Chem.*, 1975, **93**, 205 (*synth, ir, pmr*)

$C_{10}H_{24}O_2V$ V-00036

Di-*tert*-butoxydimethylvanadium

*Dimethylbis(2-methyl-2-propanolato-O)vanadium. Dimethyldi-*tert*-butoxyvanadium*

[63243-40-3]

$[(H_3C)_3CO]_2VMe_2$

M 227.240

Ethylene polymerisation catalyst. Blue-violet liq. Sol. hydrocarbons, ethers. Bp_1 35-40°. Paramagnetic. Dec. >77°.

Razuvaev, G.A. *et al*, *J. Organomet. Chem.*, 1977, **131**, 43 (*synth, esr*)
Razuvaev, G.A., *CA*, 1978, **89**, 180433 (*use*)

$C_{11}H_7O_5V$ V-00037

(η^5-1-Acetyl-2,4-cyclopentadien-1-yl)tetracarbonylvanadium

M 270.115

Red cryst. by subl. at 80-90° *in vacuo*. Sol. CH_2Cl_2. Mp 98°.

Fischer, E.O. *et al*, *Chem. Ber.*, 1960, **93**, 1006 (*synth, ir*)
Fischer, E.O. *et al*, *J. Organomet. Chem.*, 1967, **7**, 113 (*synth, pmr*)
Palyi, G. *et al*, *Inorg. Chim. Acta*, 1975, **15**, L23 (*synth*)

$C_{11}H_9O_2V$ **V-00038**

Dicarbonyl(η^4-1,3-cyclobutadiene)(η^5-2,4-cyclopentadien-1-yl)vanadium

[68033-40-9]

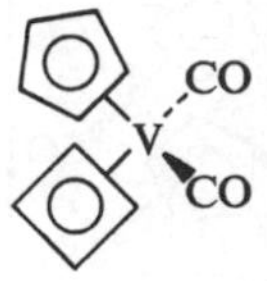

M 224.132

Bright-orange cryst. by subl. Sol. common org. solvs. Subl. at r.t. *in vacuo*.

Rausch, M.D. *et al*, *J. Chem. Soc., Chem. Commun.*, 1978, 401 (*synth, ms, nmr, ir*)

$C_{11}H_{10}IOV$ **V-00039**

Carbonylbis(η^5-2,4-cyclopentadien-1-yl)iodovanadium, 9CI

Bis(cyclopentadienyl)iodocarbonylvanadium

[53291-19-3]

As Dibromobis(η^5-2,4-cyclopentadien-1-yl)vanadium, V-00027 with

X = CO, Y = I

M 336.045

Dec. *in vacuo* to Bis(η^5-2,4-cyclopentadien-1-yl)-iodovanadium, V-00031 Diamagnetic.

Calderazzo, F. *et al*, *J. Am. Chem. Soc.*, 1974, **96**, 3695 (*synth, ir*)

$C_{11}H_{10}OV$ **V-00040**

Carbonylbis(η^5-2,4-cyclopentadien-1-yl)vanadium, 9CI

Bis(cyclopentadienyl)carbonylvanadium. Vanadocene carbonyl. Bis(cyclopentadienyl)vanadium carbonyl

[53339-41-6]

As Bromobis(η^5-2,4-cyclopentadien-1-yl)vanadium, V-00026 with

X = CO

M 209.141

Dark-brown solid (heptane). Sol. hydrocarbons. Paramagnetic.

Calderazzo, F. *et al*, *J. Am. Chem. Soc.*, 1974, **96**, 3695 (*synth, ir*)

$C_{11}H_{10}S_2V$ **V-00041**

[(*C,S*-η)-Carbon disulfide]bis(η^5-2,4-cyclopentadien-1-yl)-vanadium, 9CI

Bis(cyclopentadienyl)vanadium(carbon disulfide)

[60674-62-6]

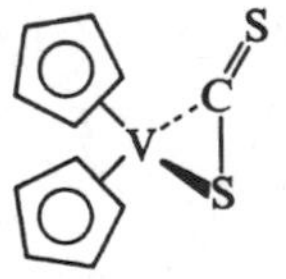

M 257.262

Maroon cryst. (toluene). Sol. toluene, THF. Paramagnetic.

Fachinetti, G. *et al*, *J. Chem. Soc., Dalton Trans.*, 1976, 1046 (*synth, ir*)

Fachinetti, G. *et al*, *J. Chem. Soc., Dalton Trans.*, 1979, 1612 (*cryst struct*)

$C_{11}H_{11}O_2V$ **V-00042**

(η^4-1,3-Butadiene)dicarbonyl(η^5-2,4-cyclopentadien-1-yl)-vanadium

Cyclopentadienyldicarbonyl(η^4-1,3-butadiene)vanadium

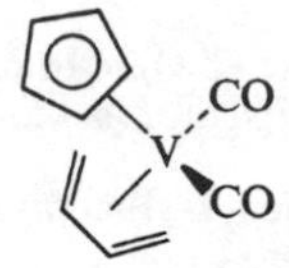

M 226.148

Dark-red needles by subl. Mp 135-40° dec. Bp 60-80° subl. *in vacuo*.

Fischer, E.O. *et al*, *Chem. Ber.*, 1960, **93**, 3006 (*synth, ir*)

$C_{11}H_{11}V$ **V-00043**

(η^6-Benzene)(η^5-2,4-cyclopentadien-1-yl)vanadium, 9CI

Cyclopentadienyl(benzene)vanadium

[39358-56-0]

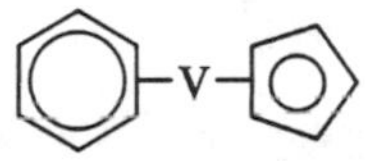

M 194.149

Not well characterised.

Müller, J. *et al*, *J. Organomet. Chem.*, 1974, **71**, 257.

$C_{11}H_{12}NO_3V$ **V-00044**

Tricarbonyl(η^5-2,4-cyclopentadien-1-yl)(dimethylaminomethylene)vanadium

[67407-28-7]

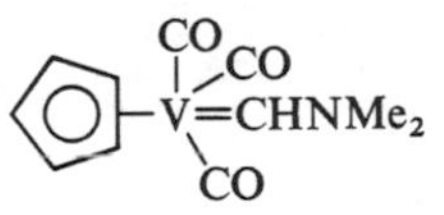

M 257.162

Yellow needles (hexane). Sol. THF, hydrocarbons. Mp 120°. Diamagnetic.

Hartshorn, A.J. *et al*, *J. Chem. Soc., Dalton Trans.*, 1978, 348 (*synth, ir, pmr*)

$C_{11}H_{12}OV$ **V-00045**

Bis(η^5-2,4-cyclopentadien-1-yl)(η^2-*C,O*-formaldehyde)vanadium

[80737-39-9]

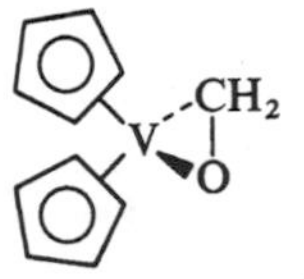

M 211.157

Deep-green cryst. (toluene), thermally labile. Sol. toluene.

Gambarotta, S. *et al*, *J. Am. Chem. Soc.*, 1982, **104**, 2019 (*synth, ir, cryst struct*)

$C_{11}H_{13}ClV$ **V-00046**

Chlorobis(η^5-2,4-cyclopentadien-1-yl)methylvanadium

Bis(cyclopentadienyl)chloro(methyl)vanadium

[66908-84-7]

As Dibromobis(η^5-2,4-cyclopentadien-1-yl)vanadium, V-00027 with

X = Me, Y = Cl

M 231.618
Brown cryst. (toluene). Sol. THF, toluene. Paramagnetic.

Razuvaev, G.A. *et al*, *Bull. Acad. Sci. USSR, Div. Chem. Sci.*, 1978, **27**, 605.
Razuvaev, G.A. *et al*, *Inorg. Chim. Acta*, 1980, **44**, L103 (*synth, esr*)

$C_{11}H_{13}V$ V-00047

Bis(η^5-2,4-cyclopentadien-1-yl)methylvanadium, 9CI
[54111-39-6]
As Bromobis(η^5-2,4-cyclopentadien-1-yl)vanadium, V-00026 with

X = Me

M 196.165
Green-black cryst. (pentane). Sol. aromatic, aliphatic hydrocarbons, Et_2O. Mp 52°. $Bp_{0.001}$ 40-60° subl. Paramagnetic. Dec. >138°.

Razuvaev, G.A. *et al*, *Dokl. Chem.* (*Engl. Transl.*), 1976, **231**, 690
Bourman, H. *et al*, *J. Organomet. Chem.*, 1976, **110**, 327 (*synth, ir, uv*)
Foust, D.F. *et al*, *J. Organomet. Chem.*, 1980, **193**, 209 (*synth, ms, props*)

$C_{11}H_{23}Cl_2P_2V$ V-00048

Dichloro(η^5-2,4-cyclopentadien-1-yl)bis(trimethylphosphine)-vanadium
Cyclopentadienyldichlorobis(trimethylphosphine)vanadium

M 339.098
Dark-blue needles (pentane). Sol. aromatic, aliphatic hydrocarbons. Air-sensitive, paramagnetic.

Nieman, J. *et al*, *J. Organomet. Chem.*, 1980, **186**, C12 (*synth, ir, uv, props*)

$C_{12}H_{10}N_2V$ V-00049

Bis(cyano-*C*)bis(η^5-2,4-cyclopentadien-1-yl)vanadium, 9CI
Bis(cyclopentadienyl)dicyanovanadium. Bis(cyclopentadienyl)vanadium dicyanide. Vanadocene dicyanide
[39333-50-1]
As Dibromobis(η^5-2,4-cyclopentadien-1-yl)vanadium, V-00027 with

X = Y = CN

M 233.166
Green cryst. ($CHCl_3$). Sl. sol. polar org. solvs. Air-stable, paramagnetic.

Doyle, G. *et al*, *Inorg. Chem.*, 1968, **7**, 2479, 2484 (*synth, ir, esr*)

$C_{12}H_{10}O_2V^{\oplus}$ V-00050

Dicarbonylbis(2,4-cyclopentadien-1-yl)vanadium(1+)
Bis(cyclopentadienyl)dicarbonylvanadium(1+)

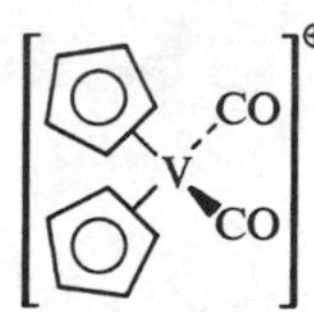

M 237.151 (ion)
Diamagnetic.
Triiodide:
$C_{12}H_{10}I_3O_2V$ M 617.865
Orange ppt.
Hexacarbonylvanadate:
$C_{18}H_{10}O_8V_2$ M 456.155
Orange solid (Me_2CO/pentane). Insol. H_2O, hydrocarbons, spar. sol. CS_2. Dec. at 140° (sealed tube under N_2) without melting. Mod. air-stable as solid, air-sensitive in soln. ir ν_{CO} 2050, 2010, 1860 cm^{-1}.
Trichlorostannate:
$C_{12}H_{10}Cl_3O_2SnV$ M 462.200
Yellow-orange. Sol. Me_2CO.
Tetraphenylborate: [57088-91-2].
$C_{36}H_{30}BO_2V$ M 556.383
Air-stable, yellow-orange solid (Me_2CO). Sol. Me_2CO, THF, v. sl. sol. MeOH, insol. hydrocarbons.

Calderazzo, F. *et al*, *Inorg. Chem.*, 1963, **2**, 721 (*synth, ir, pmr*)
Fachinetti, G. *et al*, *J. Chem. Soc., Dalton Trans.*, 1976, 203, 1046 (*synth, pmr*)
Atwood, J.L. *et al*, *Inorg. Chem.*, 1980, **19**, 3812 (*cryst struct*)

$C_{12}H_{12}V$ V-00051

Bis(η^6-benzene)vanadium, 9CI
Dibenzenevanadium
[12129-72-5]

M 207.168
Butadiene polymerisation catalyst. Red-black cryst. by subl. Sol. most org. solvs., insol. CCl_4, MeOH. Mp 277-8°. Bp 120-50° subl. *in vacuo*. Paramagnetic. Dec. >300°.

Fischer, E.O. *et al*, *Chem. Ber.*, 1957, **90**, 250 (*synth*)
Brauer, G., *Handbook of Preparative Inorg. Chem.*, Academic Press, N.Y., 1965, **2**, 1289.
Akhmadev, V.M. *et al*, *J. Chem. Soc., Dalton Trans.*, 1975, 1412 (*use*)

$C_{12}H_{12}V$ V-00052

(η^7-Cycloheptatrienylium)(η^5-2,4-cyclopentadien-1-yl)vanadium, 9CI
[12636-68-9]

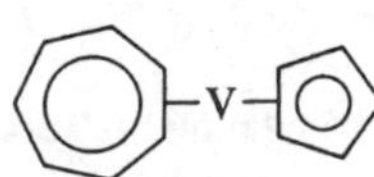

M 207.168
Purple cryst. by subl. Sol. aliphatic hydrocarbons. $Bp_{0.1}$ 100° subl. Paramagnetic.

Engebretson, G. *et al*, *J. Am. Chem. Soc.*, 1963, **85**, 481 (*cryst struct*)

Organomet. Synth., 1965, **1**, 140 (*synth*)
Gulick, W.M. *et al*, *Inorg. Chem.*, 1967, **6**, 1320 (*uv, props*)
Rettig, M.S. *et al*, *J. Am. Chem. Soc.*, 1970, **92**, 5100 (*pmr, ms, esr*)
Evans, S. *et al*, *J. Chem. Soc., Dalton Trans.*, 1974, 304 (*pe*)
Müller, J. *et al*, *J. Organomet. Chem.*, 1974, **71**, 257 (*synth, ms*)

$C_{12}H_{13}O_3V$ V-00053

Tricarbonylhydro(η^6-1,3,5-trimethylbenzene)vanadium

Tricarbonylhydro(η^6-mesitylene)vanadium

[31868-99-2]

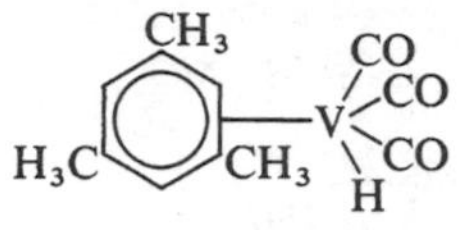

M 256.174

Orange-yellow cryst. (pentane). Sol. aliphatic hydrocarbons. Dec. at 69-72°.

Davison, A. *et al*, *J. Organomet. Chem.*, 1970, **23**, 491 (*synth, ir, pmr, ms*)

$C_{12}H_{14}V$ V-00054

Bis(η^5-1-methyl-2,4-cyclopentadien-1-yl)vanadium

1,1′-Dimethylvanadocene

[12146-93-9]

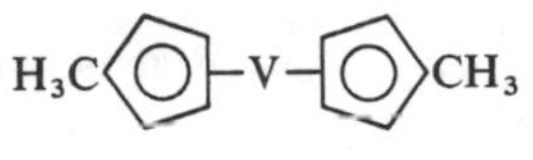

M 209.184

Deep-violet liq. or deep-violet cryst. Sol. aromatic, aliphatic hydrocarbons, THF. Mp 26-8°.

Eberl, K. *et al*, *Angew. Chem.*, 1976, **88**, 575.
Köhler, F.H., *J. Organomet. Chem.*, 1976, **110**, 235 (*synth, nmr*)
Köhler, F.H. *et al*, *Z. Naturforsch., B*, 1977, **32**, 1026 (*synth, nmr*)

$C_{12}H_{15}V$ V-00055

Bis(η^5-2,4-cyclopentadien-1-yl)ethylvanadium, 9CI

[59424-00-9]

As Bromobis(η^5-2,4-cyclopentadien-1-yl)vanadium, V-00026 with

X = Et

M 210.192

Green-black cryst. (pentane). Sol. aliphatic hydrocarbons, Et_2O. Mp 27°. Paramagnetic. Dec. >94°.

Bouman, H. *et al*, *J. Organomet. Chem.*, 1976, **110**, 327 (*synth, ir, uv*)
Razuvaev, G.A. *et al*, *J. Organomet. Chem.*, 1977, **129**, 169 (*synth*)

$C_{12}H_{16}V$ V-00056

Bis(2,4-cyclopentadien-1-yl)dimethylvanadium

[62363-03-5]

As Dibromobis(η^5-2,4-cyclopentadien-1-yl)vanadium, V-00027 with

X = Y = Me

M 211.200

Black cryst. (toluene/hexane). Sol. aromatic hydrocarbons. Mp 130-2° dec. $Bp_{0.001}$ 65-100° subl. Paramagnetic.

Razuvaev, G.A. *et al*, *J. Organomet. Chem.*, 1977, **129**, 169 (*synth, esr*)
Faust, D.F. *et al*, *J. Organomet. Chem.*, 1980, **193**, 209 (*synth, ms*)

$C_{12}H_{33}OSi_3V$ V-00057

Oxotris(trimethylsilylmethyl)vanadium

[31011-82-2]

$OV(CH_2SiMe_3)_3$

M 328.590

Pale-yellow cryst. (pet. ether). Sol. hydrocarbons, ethers. Mp 75°. Diamagnetic, light-sensitive, mod. air-stable.

Mowat, W. *et al*, *J. Chem. Soc., Dalton Trans.*, 1972, 533 (*synth, ir, pmr, uv*)

$C_{13}H_{13}O_2V$ V-00058

Acetylcarbonylbis(η^5-2,4-cyclopentadien-1-yl)vanadium

[54111-41-0]

As Dibromobis(η^5-2,4-cyclopentadien-1-yl)vanadium, V-00027 with

X = Ac, Y = CO

M 252.186

Green-maroon cryst. (toluene). Sol. aromatic hydrocarbons. Diamagnetic.

Fachinetti, G. *et al*, *J. Chem. Soc., Dalton Trans.*, 1976, 203 (*synth, ir*)

$C_{13}H_{15}V$ V-00059

Bis(2,4-cyclopentadien-1-yl)(η^1-2-propenyl)vanadium

(η^1-Allyl)bis(cyclopentadienyl)vanadium

As Bromobis(η^5-2,4-cyclopentadien-1-yl)vanadium, V-00026 with

$X = -CH_2CH{=}CH_2$

M 222.203

σ-bonded allyl. Black cryst. (THF/pentane), solns. unstable >0°. Mp 49°. Air-sensitive, paramagnetic.

Martin, H.A. *et al*, *Angew. Chem., Int. Ed. Engl.*, 1964, **3**, 311.
Siegert, F.W. *et al*, *J. Organomet. Chem.*, 1968, **15**, 131 (*synth, pmr, ir*)
de Liefde Meijer, H.J. *et al*, *Inorg. Chim. Acta*, 1970, **4**, 651.

$C_{13}H_{16}OV^{\oplus}$ V-00060

Bis(η^5-2,4-cyclopentadien-1-yl)(2-propanone)vanadium(1+), 9CI

Acetonebis(cyclopentadienyl)vanadium(1+)

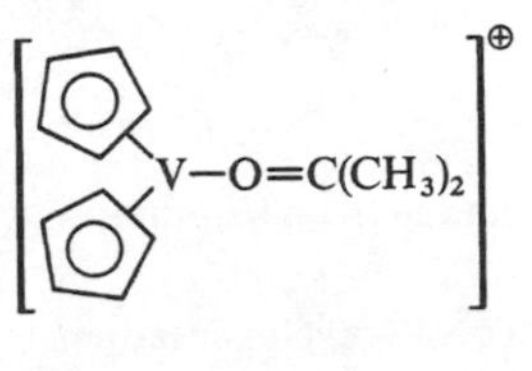

$[Cp_2V{-}O{=}C(CH_3)_2]^{\oplus}$

M 239.210 (ion)

Tetraphenylborate: [60686-18-2].

$C_{37}H_{36}BOV$ M 558.442
Deep-blue cryst. (Me_2CO). Sol. THF, Me_2CO, insol. H_2O. Paramagnetic.

Fachinetti, G. *et al*, *J. Chem. Soc., Dalton Trans.*, 1976, 1046 (*synth, ir*)
Gambarotta, S. *et al*, *Inorg. Chem.*, 1981, **20**, 1173 (*cryst struct*)

$C_{13}H_{19}As_2O_3V$ V-00061

Tricarbonyltrihydro[1,2-phenylenebis[dimethylarsine]-*As,A-s'*]vanadium

[1,2-Bis(dimethylarsino)benzene]tricarbonyltrihydrovanadium

[61917-87-1]

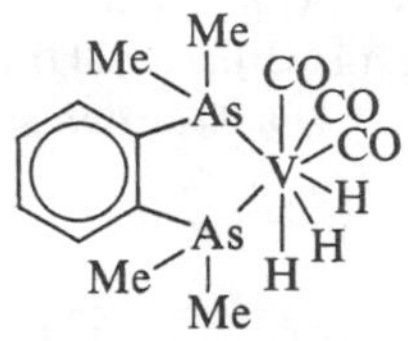

M 424.076
Red cryst. (CH_2Cl_2). Sol. CH_2Cl_2, THF. Air-stable, diamagnetic. Dec. >130°.

Ellis, J.E. *et al*, *J. Am. Chem. Soc.*, 1977, **99**, 626 (*synth, pmr, ir*)

$C_{14}H_{16}N_2V^{\oplus\oplus}$ V-00062

Bis(acetonitrile)bis(η^5-2,4-cyclopentadien-1-yl)vanadium(2+)

Bis(cyclopentadienyl)bis(acetonitrile)vanadium(2+)

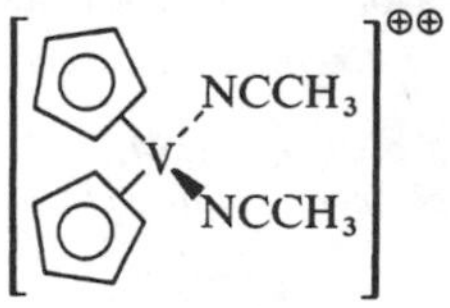

M 263.235 (ion)

Bis(hexafluorophosphate): [37205-22-4].
$C_{14}H_{16}F_{12}N_2P_2V$ M 553.164
Dark-green needles (MeCN). Dec. slowly at r.t. Sol. polar org. solvs.

Connelly, N.G. *et al*, *J. Organomet. Chem.*, 1972, **38**, 385 (*synth, ir*)

$C_{14}H_{16}V$ V-00063

Bis[(1,2,3,4,5,6-η)-1,3,5-cycloheptatriene]vanadium, 9CI

[12671-05-5]

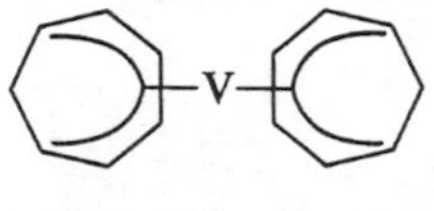

M 235.222
Green-brown solid (hexane). Sol. common org. solvs. Mp 132° dec. Paramagnetic.

Müller, J. *et al*, *Chem. Ber.*, 1972, **105**, 3346 (*synth, esr, ms*)

$C_{14}H_{20}V$ V-00064

Bis(η^5-2,4-cyclopentadien-1-yl)diethylvanadium

[63118-93-4]

As Dibromobis(η^5-2,4-cyclopentadien-1-yl)vanadium, V-00027 with

X = Y = Et

M 239.254
Thermally unstable black soln. in toluene. Dec. in soln. >0° to Bis(η^5-2,4-cyclopentadien-1-yl)ethylvanadium, V-00055 .

Razuvaev, G.A. *et al*, *J. Organomet. Chem.*, 1977, **129**, 169 (*synth, esr*)
Evans, A.G. *et al*, *J. Chem. Soc., Dalton Trans.*, 1978, 57.

$C_{14}H_{22}V$ V-00065

Bis(η^5-2,4-dimethyl-2,4-pentadien-1-yl)vanadium

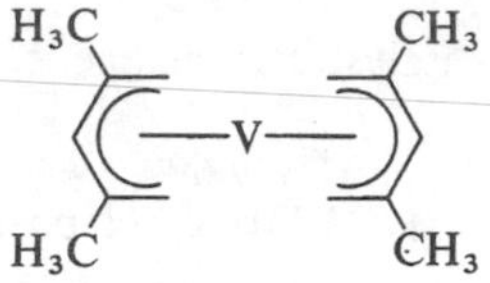

M 241.269
Sandwich struct. Deep-green cryst. (pentane). V. sol. aliphatic hydrocarbons. Subl. *in vacuo*. Paramagnetic.

Wilson, D.R. *et al*, *J. Am. Chem. Soc.*, 1982, **104**, 1120 (*synth, struct*)

$C_{14}H_{35}N_3V$ V-00066

Tris(diethylamido)ethylvanadium

Ethyltris(N-ethylethanaminato-N)vanadium. Ethyltris(diethylamino)vanadium

[56765-58-3]

$EtV(NEt_2)_3$

M 296.392
Ethylene polymerisation catalyst. Deep-green liq. Stable to 115°. Sol. hydrocarbons. $Bp_{0.001}$ 71-3°. Paramagnetic.

Fröhlich, H.-O. *et al*, *Z. Chem.*, 1975, **15**, 233 (*synth*)
U.S.P., 4 042 610, (*1977*); *CA*, **87**, 184686 (*use*)
Fröhlich, H.-O. *et al*, *Z. Anorg. Allg. Chem.*, 1979, **458**, 89 (*synth, props*)

$C_{15}H_{10}O_5V_2$ V-00067

Di-μ-carbonyltricarbonylbis(η^5-2,4-cyclopentadien-1-yl)divanadium, 9CI

Pentacarbonylbis(cyclopentadienyl)divanadium. Bis(cyclopentadienyl)divanadium pentacarbonyl

[41699-43-8]

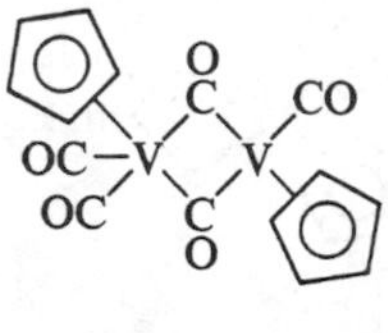

M 372.124
Contains two semibridging CO groups. Dark-green cryst. (C_6H_6). Sol. hydrocarbons, THF. Mp 99° dec. Thermally unstable >40°.

Cotton, F.A. *et al*, *J. Organomet. Chem.*, 1978, **160**, 93 (*ir, cryst struct*)
Herrmann, W.A. *et al*, *Chem. Ber.*, 1979, **112**, 392 (*synth, ms*)
Lewis, L.N. *et al*, *Inorg. Chem.*, 1980, **19**, 1840 (*synth, ir, pmr*)
Huffman, J.C. *et al*, *Inorg. Chem.*, 1980, **19**, 2755 (*cryst struct*)

$C_{15}H_{17}O_2V^{\oplus}$ V-00068

Bis(η^5-2,4-cyclopentadien-1-yl)(pentanedionato)vanadium(1+), 8CI
(Acetylacetonato)bis(cyclopentadienyl)vanadium(1+)

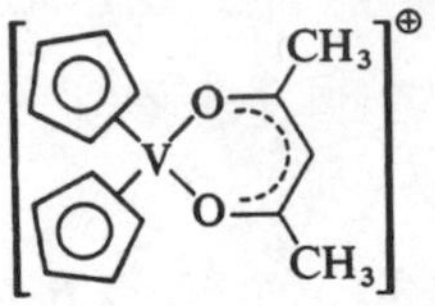

M 280.240 (ion)
Styrene polym. catalyst. Paramagnetic.
Perchlorate:
$C_{15}H_{17}ClO_6V$ M 379.690
Olive-green cryst. (H_2O). Sl. sol. polar org. solvs.
▷Explosive

Doyle, G. *et al*, *Inorg. Chem.*, 1968, **7**, 2479, 2484 (*synth, ir, esr*)
U.S.P., 3 577 448, (*1971*); *CA*, **75**, 37306 (*use*)

$C_{15}H_{19}As_2O_4V$ V-00069

Tetracarbonylmethyl[1,2-phenylenebis(dimethylarsine)-*As,As'*]vanadium, 9CI
Tetracarbonyl[bis(dimethylarsino)benzene]methylvanadium
[56408-48-1]

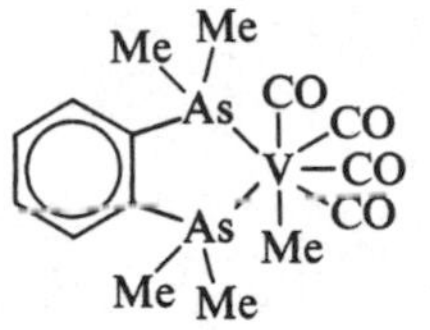

M 464.097
Violet cryst. (CH_2Cl_2/heptane). Sol. CH_2Cl_2, insol. aliphatic hydrocarbons. Mp 112° dec. Mod. air-stable, diamagnetic.

Ellis, J.E. *et al*, *J. Organomet. Chem.*, 1975, **93**, 205 (*synth, ir, pmr*)

$C_{15}H_{20}NS_2V^{\oplus}$ V-00070

Bis(η^5-2,4-cyclopentadien-1-yl)(diethylcarbamodithioato-*S,S'*)vanadium(1+), 9CI

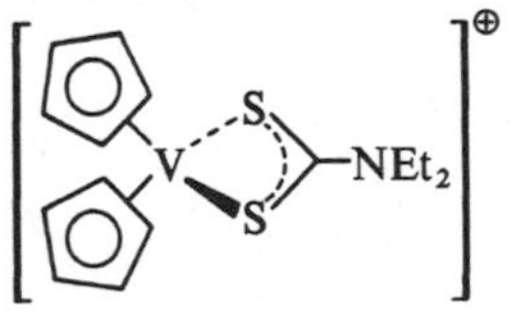

M 329.391 (ion)
Paramagnetic.
Triiodide:
$C_{15}H_{20}I_3NS_2V$ M 710.105
Air-stable, violet-maroon cryst. (THF/C_6H_6). Sol. Me_2CO.
Tetrafluoroborate: [37215-22-8].
$C_{15}H_{20}BF_4NS_2V$ M 416.195
Styrene polymerisation catalyst. Green cryst. (H_2O). Sl. sol. common org. solvs.

U.S.P., 3 577 448, (*1971*); *CA*, **75**, 37306 (*use*)
Bond, A.M. *et al*, *Inorg. Chem.*, 1973, **12**, 887 (*props*)
Casey, A.T. *et al*, *Aust. J. Chem.*, 1974, **27**, 757 (*synth, ir, esr, uv*)
Fachinetti, G. *et al*, *J. Chem. Soc., Dalton Trans.*, 1976, 203 (*synth*)

$C_{16}H_{15}NOV^{\oplus}$ V-00071

Carbonylbis(η^5-2,4-cyclopentadien-1-yl)(pyridine)vanadium(1+), 9CI
Bis(cyclopentadienyl)carbonyl(pyridine)vanadium(1+)

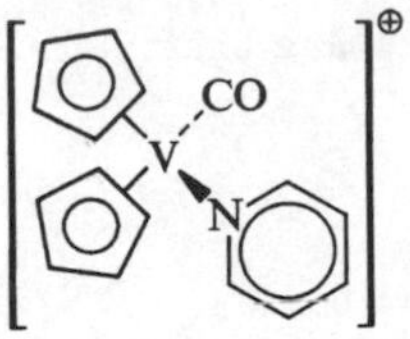

M 288.242 (ion)
Diamagnetic.
Tetraphenylborate: [57088-95-6].
$C_{40}H_{35}BNOV$ M 607.474
Red-violet cryst. (THF). Sol. THF.

Fachinetti, G. *et al*, *J. Chem. Soc., Dalton Trans.*, 1976, 1046 (*synth, ir*)

$C_{16}H_{15}SV$ V-00072

(Benzenethiolato)bis(η^5-2,4-cyclopentadien-1-yl)vanadium, 9CI
Bis(cyclopentadienyl)vanadium phenylsulfide
[55500-02-2]
As Bromobis(η^5-2,4-cyclopentadien-1-yl)vanadium, V-00026 with

X = SPh

M 290.296
Olefin polymerisation catalyst. Green-black cryst. (toluene). Sol. aromatic hydrocarbons. Paramagnetic.

Fachinetti, G. *et al*, *J. Chem. Soc., Dalton Trans.*, 1974, 2433 (*synth*)
Ger. Pat., 2 521 973, (*1975*); *CA*, **84**, 744356 (*use*)
Razuvaev, G.A. *et al*, *Inorg. Chim. Acta*, 1978, **31**, L357.

$C_{16}H_{15}V$ V-00073

Bis(η^5-2,4-cyclopentadien-1-yl)phenylvanadium, 9CI
[12212-56-5]
As Bromobis(η^5-2,4-cyclopentadien-1-yl)vanadium, V-00026 with

X = Ph

M 258.236
Green-black cryst. (toluene). Sol. aromatic hydrocarbons. Mp 90-2°. Paramagnetic.

de Liefde Meijer, H.J. *et al*, *Chem. Ind. (London)*, 1960, **79**, 119 (*synth*)
Razuvaev, G.A. *et al*, *J. Organomet. Chem.*, 1977, **129**, 169 (*synth, esr*)

$C_{16}H_{16}O_4V$ V-00074

Bis(η^5-2,4-cyclopentadien-1-yl)[(2,3-η)-dimethyl 2-butynedioate]vanadium, 9CI
Bis(cyclopentadienyl)(dimethylacetylenedicarboxylate)vanadium
[12155-24-7]

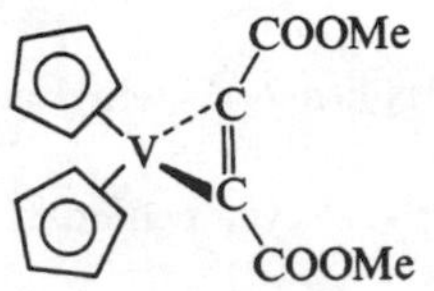

M 323.242

Dark-green cryst. (toluene). Mp 146-7° dec.

Tsumura, R. *et al*, *Bull. Chem. Soc. Jpn.*, 1965, **38**, 861 (*synth, esr*)
Fachinetti, G. *et al*, *Inorg. Chem.*, 1979, **18**, 2282 (*synth, ir, cryst struct*)
Peterson, J.L. *et al*, *Inorg. Chem.*, 1980, **19**, 1852 (*synth, esr, cryst struct*)

$C_{16}H_{16}V$ V-00075

Bis(η^8-1,3,5,7-cyclooctatetraene)vanadium, 9CI

[35796-61-3]

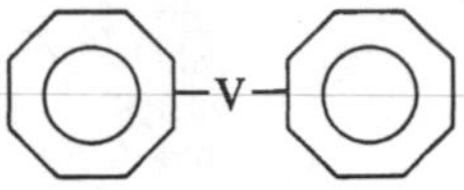

M 259.244

Purple solid. Paramagnetic.

Breil, H. *et al*, *Angew. Chem., Int. Ed. Engl.*, 1966, **5**, 898 (*synth*)
Thomas, J.L. *et al*, *Inorg. Chem.*, 1972, **11**, 348 (*esr*)
Hocks, L. *et al*, *Spectrochim Acta, Part A*, 1974, **30**, 907 (*ir*)
Hocks, L. *et al*, *J. Organomet. Chem.*, 1976, **104**, 199 (*esr, uv*)

$C_{16}H_{20}V$ V-00076

Bis[(1,2,3,4,5,6-η)ethylbenzene]vanadium, 9CI

[36955-48-3]

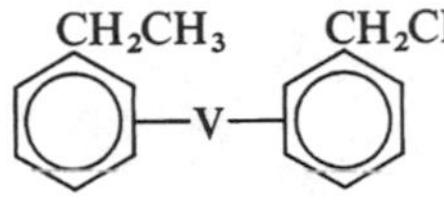

CH₂CH₃ CH₂CH₃

M 263.276

Metal deposition reagent. Red liq. Mp 130-40°. Stable to gas chromatography.

Petukhov, G.G. *et al*, *J. Gen. Chem. USSR*, (*Engl. Transl.*), 1974, **44**, 2184 (*use*)
Zborovskaya, L.S. *et al*, *CA*, 1975, **83**, 179260 (*synth*)
Dozorov, V.A. *et al*, *CA*, 1979, **91**, 159914 (*purifn*)

$C_{16}H_{44}Si_4V$ V-00077

Tetrakis[(trimethylsilyl)methyl]vanadium, 9CI

[41572-08-1]

$V(CH_2SiMe_3)_4$

M 399.807

Deep-green cryst. (pet. ether). V. sol. hydrocarbons. Mp 43°. Subl. at r.t. *in vacuo*, stable indef. below −30°.

▷Pyrophoric

Mowat, W. *et al*, *J. Chem. Soc., Dalton Trans.*, 1972, 533 (*synth, ir, uv*)

$C_{17}H_{14}OV$ V-00078

Carbonyl(η^5-2,4-cyclopentadien-1-yl)(η^5-1-phenyl-2,4-cyclopentadien-1-yl)vanadium

[54111-38-5]

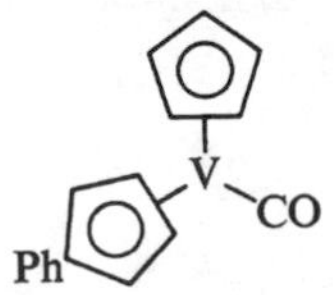

M 285.239

Green cryst. Mp 46°. Paramagnetic.

Fachinetti, G. *et al*, *J. Chem. Soc., Dalton Trans.*, 1976, 203 (*synth, ir*)

$C_{17}H_{17}V$ V-00079

Benzylbis(cyclopentadienyl)vanadium

Bis(η^5-2,4-cyclopentadien-1-yl)(phenylmethyl)vanadium, 9CI

[54111-40-9]

As Bromobis(η^5-2,4-cyclopentadien-1-yl)vanadium, V-00026 with

$X = CH_2Ph$

M 272.263

Black cryst. (hexane/toluene). Sol. aromatic hydrocarbons. Mp 107-9°. Paramagnetic.

de Liefde Meijer, H.J. *et al*, *Recl. Trav. Chim. Pays-Bas*, 1961, **80**, 831 (*synth*)
Razuvaev, G.A. *et al*, *J. Organomet. Chem.*, 1977, **129**, 169 (*synth, esr*)

$C_{17}H_{20}N_2O_4V$ V-00080

Bis(η^5-2,4-cyclopentadien-1-yl)(η^2-diethyldiazomalonate-*N,N'*)vanadium

[80679-14-7]

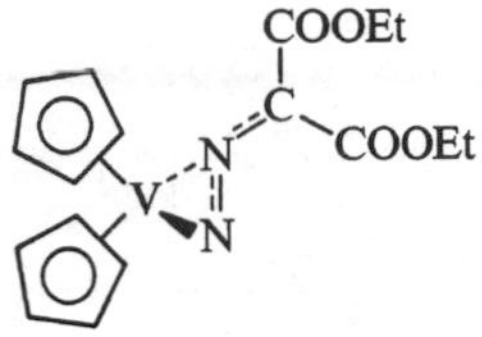

COOEt COOEt

M 367.298

Black cryst. (toluene), v. air-sensitive. Sol. aromatic hydrocarbons, insol. aliphatic hydrocarbons. $\mu_{eff} = 1.77$ BM (293K).

Gambarotta, S. *et al*, *J. Am. Chem. Soc.*, 1982, **104**, 1918 (*synth, ir*)

$C_{18}H_{15}O_2V$ V-00081

Dicarbonyl(η^5-2,4-cyclopentadien-1-yl)(phenylcyclopentadiene)vanadium

Dicarbonyl(η^5-2,4-cyclopentadien-1-yl)[[(2,3,4,5-η)-2,4-cyclopentadien-1-yl]benzene]vanadium, 9CI. (*Cyclopentadienyl)(phenylcyclopentadiene)dicarbonylvanadium*

[54162-08-2]

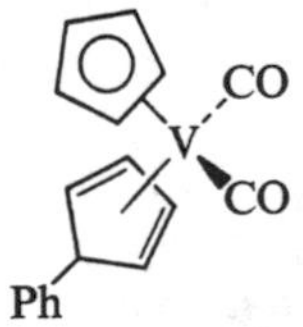

M 314.257

Red cryst. Mp 46°. Diamagnetic. Thermally dec. to green $(C_5H_5)V(C_5H_4Ph)(CO)$.

Fachinetti, G. *et al*, *J. Chem. Soc., Dalton Trans.*, 1976, 203 (*synth, ir*)

$C_{18}H_{22}O_4V$ V-00082

Bis(η^5-2,4-cyclopentadien-1-yl)[(2,3-η)-diethyl 2-butene-dioate]vanadium, 10CI

Bis(cyclopentadienyl)(η^2-diethyl fumarate)vanadium

[60674-63-7]

M 353.311

Green-maroon cryst. (C_6H_6). Sol. aromatic hydrocarbons. Paramagnetic.

Fachinetti, G. *et al*, *J. Chem. Soc., Dalton Trans.*, 1976, 1046 (*synth*)

Fachinetti, G. *et al*, *Inorg. Chem.*, 1979, **18**, 2282 (*cryst struct*)

$C_{18}H_{22}O_8V_2$ V-00083

Tetrakis[μ-(acetato-*O,O'*)]bis(η^5-2,4-cyclopentadien-1-yl)-divanadium, 9CI

[33221-79-3]

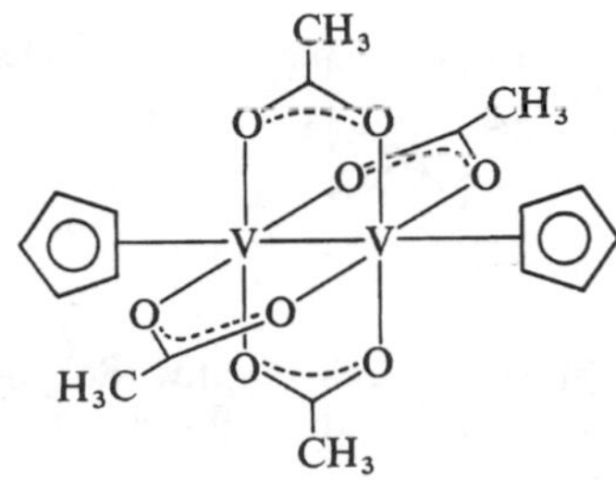

M 468.250

Mediator in Pd catalysed carbonylation of aromatic nitro compds. Violet cryst. by subl. Spar. sol. common org. solvs. Mp 239-40° dec. $Bp_{0.1}$ 125° subl. Antiferromagnetic.

Monomer: Bis(acetato)cyclopentadienylvanadium. Diacetoxycyclopentadienylvanadium. Cyclopentadienylvanadium diacetate.

$C_9H_{11}O_4V$ M 234.125

Unknown.

King, R.B., *Inorg. Chem.*, 1966, **5**, 2231 (*synth, ms, ir*)

Razuvaev, G.A., *Dokl. Chem. (Engl. Transl.)*, 1974, **216**, 355

McGregor, K.T. *et al*, *J. Organomet. Chem.*, 1975, **101**, 321 (*esr, props*)

$C_{20}H_{16}V_2$ V-00084

1,1″;1,1‴-Bivanadocene

Bis[μ-(1,2,3,4,5-η:1′,2′,3′,4′,5′-η)[bi-2,4-cyclopentadien-1-yl]-1,1′-diyl]divanadium, 10CI. Bis(fulvalene)divanadium

[71743-66-3]

M 358.229

Dark-purple cryst. by subl., stable to >300° under N_2. Spar. sol. common org. solvs. $Bp_{0.1}$ 230° subl. Diamagnetic.

▷Pyrophoric

Smart, J.C. *et al*, *J. Am. Chem. Soc.*, 1979, **101**, 4371; 1980, **102**, 1009 (*synth, ir, ms*)

$C_{20}H_{16}V_2^{\oplus\oplus}$ V-00085

1,1″:1′,1‴-Bivanadocenium(2+), 10CI

Bis(fulvalene)divanadium(2+)

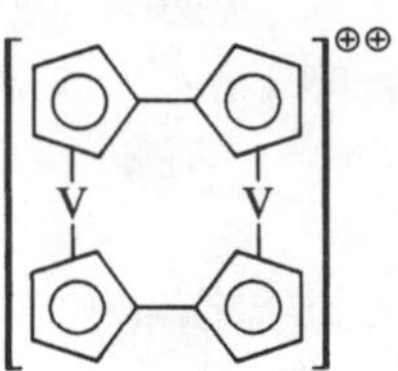

M 358.229 (ion)

Bis(hexafluorophosphate): [73153-05-6].

$C_{20}H_{16}F_{12}P_2V_2$ M 648.158

Brown air-sensitive cryst. (MeCN). Sol. MeCN. Paramagnetic, hygroscopic.

Smart, J.C. *et al*, *J. Am. Chem. Soc.*, 1979, **101**, 4371; 1980, **102**, 1009 (*synth, ir, uv, props, struct*)

$C_{20}H_{18}N_2V$ V-00086

(2,2′-Bipyridine-*N,N′*)bis(η^5-2,4-cyclopentadien-1-yl)vanadium, 9CI

Bis(cyclopentadienyl)(bipyridyl)vanadium

[60818-71-5]

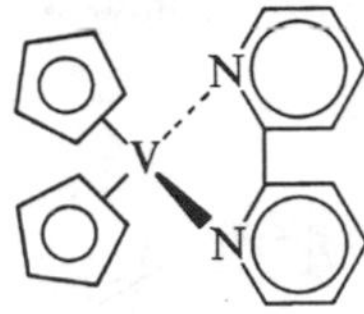

M 337.317

Black cryst. (heptane). Sol. hydrocarbons. Paramagnetic.

Fachinetti, G. *et al*, *J. Chem. Soc., Dalton Trans.*, 1976, 1046 (*synth*)

$C_{20}H_{30}V$ V-00087

Bis[(1,2,3,4,5-η)-1,2,3,4,5-pentamethyl-2,4-cyclopentadien-1-yl]vanadium

Decamethylvanodocene

[74507-60-1]

M 321.399

Dark-red prisms (pentane). Sol. THF, pentane, Et_2O, CH_2Cl_2, aromatic hydrocarbons, sl. sol. MeCN, Me_2CO. Mp 299-300°. $Bp_{0.00001}$ 100° subl. Paramagnetic.

Robbins, J.L. *et al*, *J. Am. Chem. Soc.*, 1982, **104**, 1882 (*synth, ir, ms, esr, uv*)

$C_{21}H_{15}O_2V$ V-00088

Dicarbonyl(η^5-2,4-cyclopentadien-1-yl)(diphenylacetylene)-vanadium

M 350.290

Green cryst. (C_6H_6). Sol. common org. solvs. Mp 58-60°. Bp_1 90° subl. dec. Dec. at 122-5°.

▷Pyrophoric

Nesmeyanov, A.N. *et al*, *Dokl. Chem.* (*Engl. Transl.*), 1968, **182**, 789 (*synth*, *ir*, *pmr*)

$C_{21}H_{57}Si_6V$ V-00089

Tris[bis(trimethylsilyl)methyl]vanadium, 9CI

[53726-77-5]

$V[CH(SiMe_3)_2]_3$

M 529.136

Air-sensitive blue-green cryst. (hexane). Sol. hexane, Et_2O. Paramagnetic.

Barker, G.K. *et al*, *J. Chem. Soc., Dalton Trans.*, 1978, 734 (*synth*)

$C_{22}H_{15}NO_5PV$ V-00090

Tetracarbonylnitrosyl(triphenylphosphine)vanadium

M 455.279

trans-form

Orange cryst. Mp 88-90° dec.

Werner, R.P.M., *Z. Naturforsch., B*, 1961, **16**, 478.

$C_{22}H_{20}N_2V$ V-00091

(Azobenzene)bis(cyclopentadienyl)vanadium

*Bis(η^5-2,4-cyclopentadien-1-yl)(diphenyldiazene-*N,N′*-)vanadium*, 9CI

[51159-66-1]

M 363.355

Dark-maroon cryst. (toluene). Sol. aromatic hydrocarbons. Paramagnetic.

Fachinetti, G., *J. Organomet. Chem.*, 1973, **57**, C51 (*synth*)
Peterson, J.L. *et al*, *Inorg. Chem.*, 1980, **19**, 1852 (*synth*, *esr*)

$C_{22}H_{20}S_2V$ V-00092

Bis(benzenethiolato)bis(η^5-2,4-cyclopentadien-1-yl)vanadium, 9CI

Bis(cyclopentadienyl)vanadium bis(phenylsulfide)

[56009-67-7]

As Dibromobis(η^5-2,4-cyclopentadien-1-yl)vanadium, V-00027 with

X = Y = SPh

M 399.462

Dark-violet cryst. (C_6H_6), green-black cryst. (toluene/hexane). Sol. aromatic hydrocarbons, THF. Mp 140° dec. Paramagnetic. Dec. at *ca* 100° liberating Ph_2S_2.

Fachinetti, G. *et al*, *J. Chem. Soc.*, 1974, 2433 (*synth*)
Muller, E.G. *et al*, *J. Organomet. Chem.*, 1976, **111**, 73 (*synth*, *ir*, *props*, *esr*, *cryst struct*)

$C_{22}H_{30}O_2V^{\oplus}$ V-00093

Dicarbonylbis[(1,2,3,4,5-η)-1,2,3,4,5-pentamethyl-2,4-cyclopentadien-1-yl]vanadium(1+)

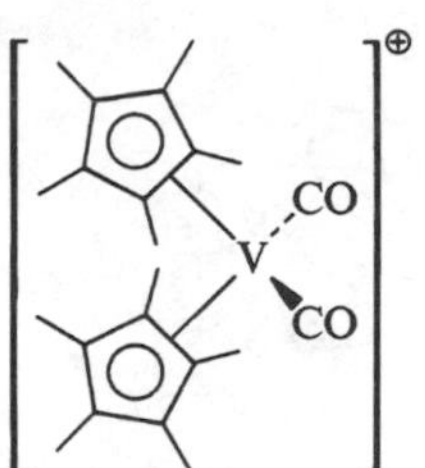

M 377.419 (ion)

Diamagnetic.

Hexafluorophosphate: [80679-49-8].
$C_{22}H_{30}F_6O_2PV$ M 522.383
Bright-yellow prisms (Me_2CO). Sol. Me_2CO, insol. aliphatic hydrocarbons.

Robbins, J.L. *et al*, *J. Am. Chem. Soc.*, 1982, **104**, 1882 (*synth*, *cmr*, *pmr*, *ir*)

$C_{23}H_{15}O_5PV^{\ominus}$ V-00094

Pentacarbonyl(triphenylphosphine)vanadate(1−), 9CI

$Ph_3PV(CO)_5^{\ominus}$

M 453.284 (ion)

Tetraethylammonium salt: [56328-27-9].
$C_{31}H_{35}NO_5PV$ M 583.536
Bright yellow-orange cryst. (Me_2CO). Sol. Me_2CO. Mp 191-3°.

Tetraphenylphosphonium salt:
$C_{47}H_{35}O_5P_2V$ M 792.680
Bright-orange cryst. (Me_2CO/Et_2O). Sol. THF. Dec. >150°.

Rehder, D., *J. Organomet. Chem.*, 1972, **37**, 303 (*ir*, *uv*, *nmr*)
Darensbourg, M.Y. *et al*, *J. Am. Chem. Soc.*, 1981, **103**, 6100 (*synth*, *ir*)
Fjare, K.L. *et al*, *Organometallics*, 1982, **1**, 1373 (*synth*)

$C_{23}H_{15}O_5PV$ V-00095

Pentacarbonyl(triphenylphosphine)vanadium

[72622-82-3]

$(OC)_5VPPh_3$

M 453.284

Olive-green solid, unstable at r.t. Sol. CH_2Cl_2, toluene, alkanes. Mp 104° dec. Paramagnetic. Dec. in polar org. solvs.

Ellis, J.E. *et al*, *Inorg. Chem.*, 1980, **19**, 1082 (*synth*, *ir*, *props*)

$C_{24}H_{15}O_6SnV$ V-00096

Hexacarbonyl(triphenylstannyl)vanadium, 9CI

Triphenylstannylvanadium hexacarbonyl

[27323-98-4]

$Ph_3SnV(CO)_6$

M 569.010

Bright orange-yellow cryst. (CH_2Cl_2). Sol. CH_2Cl_2, Et_2O, hydrocarbons. Mp 81-2° dec.

Davison, A. *et al*, *J. Organomet. Chem.*, 1972, **36**, 113 (*synth*, *ir*)

$C_{24}H_{20}OV$ V-00097

Bis(2,4-cyclopentadien-1-yl)(η^2-*C,O*-diphenylketene)vanadium

[76173-79-0]

M 375.363

Deep-green cryst. (toluene). Sol. toluene, insol. THF. Paramagnetic.

Gambarotta, S. *et al*, *Inorg. Chem.*, 1981, **20**, 1173 (*cryst struct, synth, ir*)

$C_{24}H_{20}O_4V_4$ V-00098

Tetracarbonyltetrakis(cyclopentadienyl)tetravanadium

[69276-81-9]

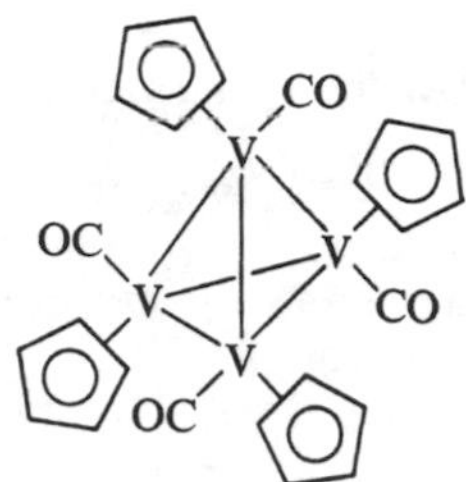

M 576.186

Proposed struct. shown. Black cryst. with metallic lustre (CH_2Cl_2/Et_2O). Sol. CH_2Cl_2, Et_2O, C_6H_6. Air-sensitive, paramagnetic.

Herrmann, W.A. *et al*, *J. Organomet. Chem.*, 1979, **164**, C25 (*synth, ir, ms*)

$C_{24}H_{24}V$ V-00099

Dibenzylbis(cyclopentadienyl)vanadium

Bis(η^5-2,4-cyclopentadien-1-yl)bis(phenylmethyl)vanadium, 9CI

[63118-94-5]

As Dibromobis(η^5-2,4-cyclopentadien-1-yl)vanadium, V-00027 with

$$X = Y = CH_2Ph$$

M 363.395

Not isol. in pure form. Soln. half-life ~30 min. at r.t.

Razuvaev, G.A. *et al*, *J. Organomet. Chem.*, 1977, **129**, 169 (*synth, esr*)

Having problems with locating a compound? Have you checked the indexes?

$C_{24}H_{32}N_2V^{\oplus}$ V-00100

Bis(η^5-2,4-cyclopentadien-1-yl)bis(isocyanocyclohexane)vanadium(1+), 9CI

Bis(cyclopentadienyl)bis(cyclohexyl isocyanide)-vanadium(1+)

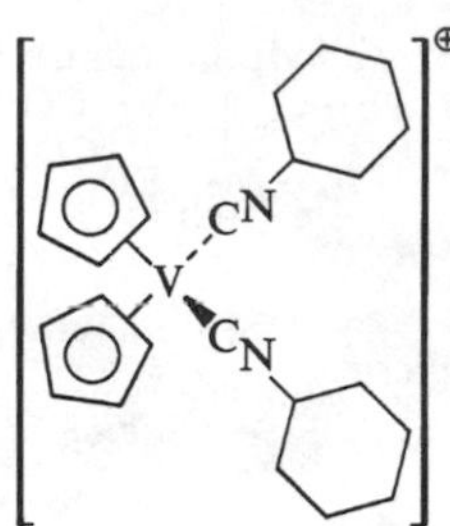

M 399.472 (ion)

Diamagnetic.

Chloride: [60686-15-9].
$C_{24}H_{32}ClN_2V$ M 434.925
Maroon cryst. (THF). Sol. Me_2CO, insol. aromatic hydrocarbons.

Fachinetti, G. *et al*, *J. Chem. Soc., Dalton Trans.*, 1976, 1046 (*synth, ir*)

$C_{24}H_{36}N_2V^{\oplus}$ V-00101

Bis(acetonitrile)bis(η^5-pentamethyl-2,4-cyclopentadien-1-yl)vanadium(1+)

M 403.503 (ion)

Hexafluorophosphate: [80679-48-7].
$C_{24}H_{36}F_6PV$ M 520.454
Olive-green needles (MeCN/toluene). Sol. MeCN, Me_2CO.

Robbins, J.L. *et al*, *J. Am. Chem. Soc.*, 1982, **104**, 1882 (*synth, ir*)

$C_{25}H_{20}O_4V$ V-00102

Tetracarbonyl(η^3-propenyl)(triphenylphosphine)vanadium

Allyltetracarbonyl(triphenylphosphine)vanadium

[61443-37-6]

M 435.372

Red-brown needles (C_6H_6/hexane). Sol. aromatic hydrocarbons, Et_2O, THF, spar. sol. aliphatic hydrocarbons.

Schneider, M. *et al*, *J. Organomet. Chem.*, 1974, **73**, C7; 1976, **121**, 345 (*synth*)

Schneider, M. *et al*, *J. Organomet. Chem.*, 1976, **121**, 189 (*cryst struct*)

$C_{26}H_{15}O_5V$ V-00103

Pentacarbonyl[(1,2,3-η)-1,2,3-triphenyl-2-cyclopropen-1-yl-]vanadium, 9CI

[62050-66-2]

M 458.343

Red needles (Et_2O). Sol. org. solvs. Mp 230° dec.

Schneider, M. *et al*, *J. Organomet. Chem.*, 1976, **121**, 345 (*synth, ir, nmr, ms*)

$C_{26}H_{20}O_3PV$ V-00104

Tricarbonyl(η^5-2,4-cyclopentadien-1-yl)(triphenylphosphine)-vanadium, 9CI

Cyclopentadienyltricarbonyl(triphenylphosphine)vanadium

[12213-09-1]

As Tricarbonyl(η^5-2,4-cyclopentadien-1-yl)-(thiocarbonyl)vanadium, V-00019 with

$$L = PPh_3$$

M 462.357

Orange cryst. ($CHCl_3$). Sol. CH_2Cl_2, $CHCl_3$, toluene, EtOH. Mp 158-62°. Air-stable as solid.

Connelly, N.G. *et al*, *J. Chem. Soc., Dalton Trans.*, 1976, 2165 (*ir, synth*)
Rehder, D., *J. Magn. Reson.*, 1977, **25**, 177 (*nmr*)
Alway, D.G. *et al*, *Inorg. Chem.*, 1980, **19**, 779 (*synth, ir, pmr, uv*)

$C_{27}H_{33}V$ V-00105

Tris(2,4,6-trimethylphenyl)vanadium

Trimesitylvanadium

[59774-49-1]

M 408.499

THF complex:
$C_{31}H_{41}OV$ M 480.606
Butadiene polymerisation catalyst. Deep-blue cryst. + 1THF (THF). Sol. THF, poorly sol. Et_2O. Paramagnetic. Dec. at 142°.

Seidel, W. *et al*, *Z. Chem.*, 1974, **14**, 25; 1977, **17**, 73 (*synth*)
Seidel, W. *et al*, *Z. Anorg. Allg. Chem.*, 1977, **435**, 146 (*synth, uv, ir*)
East Ger. Pat., 126 998, (*1977*); *CA*, **88**, 106023 (*use*)

$C_{28}H_{28}V$ V-00106

Tetrabenzylvanadium

Tetrakis(phenylmethyl)vanadium, 9CI. *Vanadium tetrabenzyl*

[41328-40-9]

$$V(CH_2Ph)_4$$

M 415.471

Green-black cryst. (toluene). Sol. hydrocarbons. Light-sensitive, paramagnetic. Solns. stable to 60°.

Ibekwe, S.D. *et al*, *J. Organomet. Chem.*, 1971, **31**, C65 (*synth, esr*)
Razuvaev, G.A. *et al*, *J. Organomet. Chem.*, 1975, **93**, 113 (*synth, esr*)
Köhler, E. *et al*, *Z. Anorg. Allg. Chem.*, 1976, **421**, 129 (*esr, nmr*)

$C_{28}H_{44}V$ V-00107

Tetrakis(bicyclo[2.2.1]hept-1-yl)vanadium, 9CI

Tetrakis(norbornyl)vanadium. Tetranorbornylvanadium

[36333-77-4]

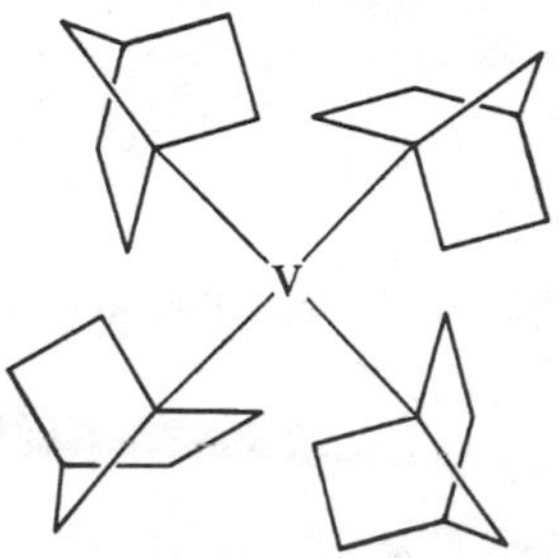

M 431.597

Alkene oligomerisation catalyst. Purple cryst. (toluene). Sol. hydrocarbons. Dec. >100°, solns. mod. stable to 0.1M H_2SO_4. Paramagnetic.

Bower, B.K. *et al*, *J. Am. Chem. Soc.*, 1972, **94**, 2512 (*synth*)
Ger. Pat., 2 204 885, (*1972*); *CA*, **77**, 165313 (*use*)
Ger. Pat., 2 204 920, (*1972*); *CA*, **77**, 165307 (*use*)
Bower, B.K. *et al*, *Inorg. Chem.*, 1974, **13**, 759 (*esr*)

$C_{30}H_{24}O_4P_2V^{\ominus}$ V-00108

Tetracarbonyl[1,2-ethanediylbis[diphenylphosphine]-*P,P'*]-vanadate(1–)

[1,2-Bis(diphenylphosphino)ethane]tetracarbonylvanadate(1–)

M 561.406 (ion)

Tetraethylammonium salt: [67204-23-3].
$C_{38}H_{44}NO_4P_2V$ M 691.659
Orange-red cryst. (THF/C_6H_6). Sol. THF, C_6H_6, insol. aliphatic hydrocarbons. Mp 256-60° dec.

Davison, A. *et al*, *J. Organomet. Chem.*, 1971, **31**, 239 (*synth, ir*)
Rehder, D. *et al*, *J. Organomet. Chem.*, 1976, **122**, 53 (*synth, ir, nmr*)

$C_{30}H_{24}O_4P_2V$ V-00109

Tetracarbonyl[1,2-ethanediylbis[diphenylphosphine]-*P,P'*]-vanadium

[Bis(diphenylphosphino)ethane]tetracarbonylvanadium

[18347-09-6]

M 561.406

Green cryst. (pentane). Sl. sol. aliphatic hydrocarbons. Paramagnetic. Dec. in polar solvs.

Behrens, H. *et al*, *Z. Anorg. Allg. Chem.*, 1968, **356**, 225 (*synth, ir, props*)

$C_{30}H_{25}O_4P_2V$ V-00110

Tetracarbonyl[1,2-ethanediylbis[diphenylphosphine]-*P,P'*]-hydrovanadium, 10CI

Hydrotetracarbonyl[bis(diphenylphosphino)ethane]vanadium

[36681-92-2]

M 562.414

Yellow-brown cryst. (CH_2Cl_2/heptane). Sol. CH_2Cl_2, C_6H_6. Dec. to $V(CO)_4(Ph_2PCH_2CH_2PPh_2)$ above 50°.

Davison, A. *et al*, *J. Organomet. Chem.*, 1972, **36**, 131 (*ir, synth*)

Greiser, T. *et al*, *Transition Met. Chem.*, 1979, **4**, 168 (*cryst struct*)

$C_{31}H_{24}O_5P_2V^{\ominus}$ V-00111

Pentacarbonyl[η^1-1,2-ethanediylbis[diphenylphosphine]-*P*]-vanadate(1−)

Pentacarbonyl[bis(diphenylphosphino)ethane]-vanadate(1−)

$[Ph_2PCH_2CH_2PPh_2V(CO)_5]^{\ominus}$

M 589.417 (ion)

One P not coordinated.

Tetraethylammonium salt: [34089-19-5].
$C_{39}H_{44}NO_5P_2V$ M 719.669
Yellow platelets (Me_2CO). Solns. are light-sensitive. Sol. Me_2CO, THF, MeCN, sl. sol. toluene, insol. Et_2O, EtOH, aliphatic hydrocarbons. Mp 123°.

Fjare, K.L. *et al*, *Organometallics*, 1982, **1**, 1373 (*synth, ir, pmr, nmr*)

$C_{32}H_{36}O_8V_2$ V-00112

Tetrakis[μ-(2,6-dimethoxyphenyl-*C:O*)]divanadium, 10CI

[62613-18-7]

M 650.515

V—V triple bond proposed. Black cryst. (THF). Diamagnetic.

Seidel, W. *et al*, *Z. Chem.*, 1976, **16**, 492 (*synth*)

Cotton, F.A. *et al*, *J. Am. Chem. Soc.*, 1977, **99**, 7886 (*struct*)

$C_{34}H_{25}OV$ V-00113

Carbonyl-η^5-2,4-cyclopentadien-1-ylbis(diphenylacetylene)vanadium

M 500.512

Orange cryst. (C_6H_6). Sol. common org. solvs. Mp 105-10° dec. Light-sensitive, dec. in air yielding tetraphenylcyclopentadienone.

Nesmeyanov, A.N. *et al*, *Dokl. Chem. (Engl. Transl.)*, 1968, **182**, 789 (*synth, ir, pmr*)

$C_{34}H_{30}S_4V_2$ V-00114

Tetrakis[μ-(benzenethiolato)]bis(η^5-2,4-cyclopentadien-1-yl)divanadium, 9CI

[59487-44-4]

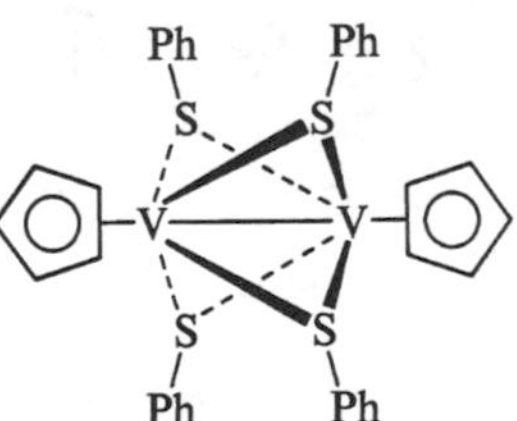

M 668.734

Brown cryst. Insol. common org. solvs. Paramagnetic. Volatile under high vacuum. Dec. at 300°.

Muller, E.G. *et al*, *J. Organomet. Chem.*, 1976, **111**, 73 (*synth, props*)

Pasynskii, A.A. *et al*, *Inorg. Chim. Acta*, 1980, **39**, 91 (*synth, ir, ms*)

$C_{36}H_{25}O_3V$ V-00115

Dicarbonyl-η^5-2,4-cyclopentadien-1-yl(η^4-tetraphenylcyclopentadienone)vanadium

Dicarbonylcyclopentadienyltetracyclonevanadium

M 556.533

Red-orange cryst. (C_6H_6). Sol. aromatic hydrocarbons, Et_2O. Mp 123-6° dec. Dec. in air yielding tetraphenylcyclopentadienone.

Pasynskii, A.A. *et al*, *Dokl. Chem. (Engl. Transl.)*, 1969, **185**, 253 (*synth, ir*)

$C_{36}H_{34}P_2V^{\oplus}$ V-00116

Bis(η^5-2,4-cyclopentadien-1-yl)[1,2-ethanediylbis[diphenylphosphine]-*P,P'*]vanadium(1+), 9CI

[*Bis(diphenylphosphino)ethane*]*bis(cyclopentadienyl)-vanadium(1+)*

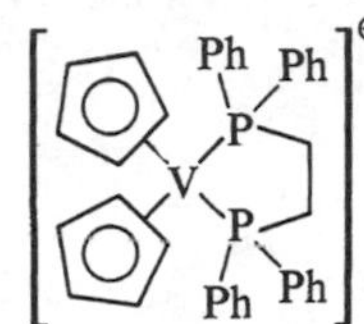

M 579.554 (ion)

Tetraphenylborate: [57088-89-8].
$C_{60}H_{54}BP_2V$ M 898.786
Green solid (THF). Sol. THF. Diamagnetic.

Fachinetti, G. *et al*, *J. Chem. Soc., Dalton Trans.*, 1976, 1046.

$C_{36}H_{44}V^{\ominus}$ V-00117

Tetrakis(2,4,6-trimethylphenyl)vanadate(1−)

Tetramesitylvanadate(1−)

M 527.685 (ion)

Paramagnetic.

Li salt, tetrakis(THF) complex: [59834-80-9].
$C_{52}H_{76}LiO_4V$ M 823.053
Deep-violet cryst. (THF/Et_2O). Sol. THF, spar. sol. Et_2O, C_6H_6. Dec. at 161°.

Seidel, W. *et al*, *Z. Anorg. Allg. Chem.*, 1976, **426**, 150 (*synth*)
Seidel, W. *et al*, *Z. Chem.*, 1977, **17**, 73 (*synth, props*)
Kreisel, G. *et al*, *Z. Anorg. Allg. Chem.*, 1980, **460**, 51 (*uv*)

$C_{36}H_{44}V$ V-00118

Tetrakis(2,4,6-trimethylphenyl)vanadium

Tetramesitylvanadium

[59774-48-0]

M 527.685

Air-stable red cryst. (C_6H_6). Sol. aromatic hydrocarbons. Paramagnetic.

Seidel, W. *et al*, *Z. Chem.*, 1976, **16**, 115 (*synth*)
Glowiak, T. *et al*, *J. Organomet. Chem.*, 1978, **155**, 39 (*struct, uv, props*)
Kirmse, R. *et al*, *Z. Chem.*, 1980, **20**, 385 (*esr*)

$C_{38}H_{30}V$ V-00119

Bis(η^5-2,4-cyclopentadien-1-yl)(1,2,3,4-tetraphenyl-1,3-butadiene-1,4-diyl)vanadium, 9CI

[59547-67-0]

M 537.597

Green cryst. (Et_2O).

Vogelaar-Van der Huizen, T.M. *et al*, *J. Organomet. Chem.*, 1976, **105**, 321 (*synth, ir, esr*)

W Tungsten

M. Green

Tungstène (Fr.), Wolfram (Ger.), Tungsteno (Sp., Ital.), Вольфрам (Vol'fram) (Russ.), タングステン (Japan.)

Atomic Number. 74

Atomic Weight. 183.85

Electronic Configuration. [Xe] $5d^4\ 6s^2$

Oxidation States. −4, −2, −1, 0, +2, +4, +6

Coordination Number. The most common coordination number with conventional ligands is 6, eg. $[W_2(CO)_{10}]^{2-}$, $[W(CS)(CO)_5]$, WMe_6 but 4, 5 and 7 are also known.

Colour. Variable, including colourless, yellow, red, green and blue.

Availability. Readily available starting materials are $W(CO)_6$ and anhydrous WCl_6 which are moderately expensive.

Handling. See under Chromium.

Toxicity. Tungsten compounds, with the exception of the carbonyl complexes, are not regarded as highly toxic.

Isotopic Abundance. ^{180}W, 0.135%; ^{182}W, 26.4%; ^{183}W, 14.4%; ^{184}W, 30.6%; ^{186}W, 28.4%. Fifteen radioisotopes of tungsten have been detected. ^{185}W (β-emitter, $t_{1/2}$ = 75d) and ^{187}W (β-emitter, $t_{1/2} \sim$ 23.8h) are most often used for tracer work.

Spectroscopy. ^{183}W, $I = \frac{1}{2}$, has a low relative receptivity, but even so, a range of ^{183}W nmr chemical shift data is available for organometallics. See for example McFarlane, H. C. E. *et al.*, *J. Chem. Soc.*, *Dalton Trans.*, 1976, 1616.

Analysis. For standard methods see *Scott's Standard methods of Chemical Analysis*, Furman, N. H. Ed., 6th Edn, Van Nostrand, 1962.

References. General reviews are listed in the introduction to the *Sourcebook*.

$WCl_5CH_2SiMe_3$
W-00001

$[W(CO)_4]^{\ominus\ominus\ominus\ominus}$
W-00002

W-00003

W-00004

X = Br
W-00005

As W-00005 with
X = Cl
W-00006

W-00007

As W-00005 with
X = I
W-00008

$[MeW(CO)_5]^{\ominus}$
W-00009

W-00010

WMe_6 (O_h)
W-00011

$(OC)_5WCN^{\ominus}$
W-00012

$(OC)_5WCS$
W-00013

$W(CO)_6$ (O_h)
W-00014

W-00015

W-00016

W-00017

L = Br
W-00018

As W-00018 with
L = Cl
W-00019

W-00020

As W-00018 with
L = I
W-00021

W-00022

W-00023

As W-00018 with
L = H
W-00024

$(OC)_5W{=}C(SMe)_2$
W-00025

W-00026

$[WMe_8]^{\ominus\ominus}$
W-00027

W-00028

W-00029

W-00030

R = H
W-00031

W-00032

W-00033

As W-00018 with
L = Me
W-00034

As W-00018 with
L = SO_2Me
W-00035

$(OC)_5W{=}C(CH_3)OEt$
W-00036

W-00037

W-00038

W-00039

$[W_2H(CO)_{10}]^{\ominus}$
W-00040

W-00041

W-00042

As W-00031 with
R = CH_3
W-00043

W-00044

W-00045

X = Br
W-00046

$W(O)Cl_2(\eta\text{-}C_5H_5)_2$
W-00047

As W-00046 with
X = Cl
W-00048

W-00049

W-00050

As W-00018 with
L = $AsMe_2$
W-00051

W-00052

As W-00018 with
L = $SbMe_2$
W-00053

W-00054

W-00055

W-00056

X = Br
W-00057

As W-00057 with
X = Cl
W-00058

As W-00057 with
X = I
W-00059

W-00060

W-00061

W-00062

As W-00018 with
L = $GeMe_3$

W-00063

As W-00018 with
L = $SiMe_3$

W-00064

As W-00018 with
L = $SnMe_3$

W-00065

W-00066

$W{\equiv}CC(CH_3)_3(O{=}PEt_3)Cl_3$

W-00067

W-00068

W-00069

W-00070

W-00071

W-00072

W-00073

W-00074

W-00075

W-00076

W-00077

W-00078

W-00079

W-00080

W-00081

W-00082

$(OC)_5W{=}CPhOMe$

W-00083

W-00084

W-00085

W-00086

$(OC)_4WP[CH(CH_3)_2]_3^{\ominus\ominus}$

W-00087

W-00088

$[WCl(CO)_2(C_5H_5)]_2$

W-00089

W-00090

W-00091

W-00092

W-00093

W-00094

W-00095

trans-form

W-00096

W-00097

W-00098

W-00099

X = Br

W-00100

As W-00100 with
X = Cl

W-00101

W-00102

W-00103

W-00104

W-00105

$W_2(\mu\text{-}SCF_3)_2(CO)_4(\eta\text{-}C_5H_5)_2$

W-00106

W-00107

W-00108

W-00109

$[(H_3C)_3CO]_4WO$

W-00110

$(OC)_5W{=}CPh_2$

W-00111

W-00112

W-00113

W-00114

W-00115

W-00116

$(C_5H_5)_2ErW(C_5H_5)(CO)_3$

W-00117

$(C_5H_5)_2YbW(C_5H_5)(CO)_3$

W-00118

R = CH_2CH_3

W-00119

W-00120

W-00121

W-00122

W-00123

W-00124

W-00125

W-00126

W-00127

$(H_3CC_5H_4)_2DyW(C_5H_5)(CO)_3$

W-00128

$(H_3CC_5H_4)_2HoW(C_5H_5)(CO)_3$

W-00129

W-00130

W-00131

W-00132

$(OC)_3WCl(NO)PPh_3$

W-00133

W-00134

W-00135

$[(OC)_5W-PPh_2-W(CO)_5]^{\ominus}$

W-00136

W-00137

W-00138

W-00139

W-00140

W-00141

W-00142

W-00143

W-00144

$(Me_3SiCH_2)_3W{\equiv}W(CH_2SiMe_3)_3$

W-00145

$(OC)_5W{=}C(OEt)SiPh_3$

W-00146

W-00147

W-00148

$W(CH_2Ph)_4$

W-00149

W-00150

W-00151

$[(H_3C)_2CH]_3PW(CO)_4(SnPh_3)^{\ominus}$

W-00152

W-00153

W-00154

As W-00124 with R = Ph

W-00155

W-00156

$C_4H_{11}Cl_5SiW$ W-00001

Pentachloro[(trimethylsilyl)methyl]tungsten

[57595-19-4]

$WCl_5CH_2SiMe_3$

M 448.331

Dolgoplosk, B.A. *et al*, *Dokl. Chem.* (*Engl. Transl.*), 1975, **223**, 501 (*synth*)

$C_4O_4W^{\ominus\ominus\ominus\ominus}$ W-00002

Tetracarbonyltungstate(4−)

$[W(CO)_4]^{\ominus\ominus\ominus\ominus}$

M 295.892 (ion)

Tetra-Na salt: [67202-46-4].

$C_4Na_4O_4W$ M 387.851

▷Pyrophoric

Ellis, J.E. *et al*, *J. Am. Chem. Soc.*, 1983, **105**, 2296 (*synth*, *props*)

$C_5H_5ClN_2O_2W$ W-00003

Chloro(η^5-cyclopentadien-1-yl)dinitrosyltungsten, 10CI

[53419-14-0]

(Cp)–W(NO)(NO)Cl

M 344.410

Green cryst. $Bp_{0.005}$ 40-50° subl.

Hoyano, J.K. *et al*, *J. Chem. Soc., Dalton Trans.*, 1975, 1022.

Inorg. Synth., 1978, **18**, 126 (*synth*)

Botto, R.E. *et al*, *Inorg. Chem.*, 1979, **18**, 2049 (*synth*, *nmr*, *props*)

Legzdins, P. *et al*, *Inorg. Chem.*, 1979, **18**, 1250.

$C_5H_9Cl_4W^{\ominus}$ W-00004

Tetrachloro(1,1-dimethylpropylidyne)tungsten(1−), 10CI

$[Cl_4W{\equiv}CC(CH_3)_3]^{\ominus}$

M 394.788 (ion)

Tetraethylammonium salt: [78251-20-4].

$C_{13}H_{29}Cl_4NW$ M 525.041

Wengrovius, J.H. *et al*, *J. Am. Chem. Soc.*, 1981, **103**, 3932 (*synth*, *cmr*, *props*)

$C_6H_3BrO_4W$ W-00005

Bromotetracarbonylethylidynetungsten, 10CI

[50726-31-3]

$X-W(CO)_4{\equiv}CCH_3$

X = Br

M 402.841

Colourless cryst.

Fischer, E.O. *et al*, *Chem. Ber.*, 1976, **109**, 1673 (*synth*)

Fischer, E.O. *et al*, *Angew. Chem., Int. Ed. Engl.*, 1977, **16**, 465 (*props*)

Neugebauer, D. *et al*, *J. Organomet. Chem.*, 1978, **153**, C41 (*cryst struct*)

$C_6H_3ClO_4W$ W-00006

Tetracarbonylchloroethylidynetungsten, 9CI

[60872-91-5]

As Bromotetracarbonylethylidynetungsten, W-00005 with

X = Cl

M 358.390

Brown cryst. (pentane).

Fischer, E.O. *et al*, *Chem. Ber.*, 1976, **109**, 1673 (*synth*, *ir*, *nmr*, *mr*)

$C_6H_3IO_4SW$ W-00007

Tetracarbonyliodo[(methylthiomethylidyne)]tungsten, 9CI

[59831-50-4]

$I-W(CO)_4{\equiv}CSMe$

M 481.902

Mod. air-sensitive yellow cryst.

Angelici, R.J. *et al*, *J. Am. Chem. Soc.*, 1973, **95**, 7516; 1974, **96**, 7586; 1975, **97**, 1261 (*synth*, *nmr*)

$C_6H_3IO_4W$ W-00008

Tetracarbonylethylidyneiodotungsten, 10CI

[50726-32-4]

As Bromotetracarbonylethylidynetungsten, W-00005 with

X = I

M 449.842

Yellow cryst. stable below 30°.

Fischer, E.O. *et al*, *Chem. Ber.*, 1976, **109**, 1673 (*synth*, *ir*, *nmr*)

Neugebauer, D. *et al*, *J. Organomet. Chem.*, 1978, **153**, C41 (*cryst struct*)

$C_6H_3O_5W^{\ominus}$ W-00009

Pentacarbonylmethyltungstate(1−), 9CI

Methylpentacarbonyltungsten(1−)

$[MeW(CO)_5]^{\ominus}$

M 338.937 (ion)

$(Ph_3P)_2N^{\oplus}$ salt: [62197-79-9]. *(T-4)Triphenyl(*P,P,P-*triphenylphosphineimidato-*N*)phosphorus(1+) pentacarbonylmethyltungstate(1−)*, 9CI. *Bis(triphenylphosphine)iminium methylpentacarbonyltungstate.*

$C_{42}H_{33}NO_5P_2W$ M 877.524

Yellow cryst.

$Me_4N^{\oplus}$ salt: [68866-90-0]. N,N,N-*Trimethylmethanaminium(1+) pentacarbonylmethyltungstate(1−)*, 9CI. *Tetramethylammonium methylpentacarbonyltungstate.*

$C_{10}H_{15}NO_5W$ M 413.082

Yellow cryst. (CH_2Cl_2/Et_2O).

$Et_4N^{\oplus}$ salt: [57574-51-3]. N,N,N-*Triethylethanaminium(1+) pentacarbonylmethyltungstate(1−)*, 9CI. *Tetraethylammonium methylpentacarbonyltungstate.*

$C_{14}H_{23}NO_5W$ M 469.189

Yellow cryst. (CH_2Cl_2/hexane).

Ellis, J.E. *et al*, *Inorg. Chem.*, 1977, **16**, 1357 (*synth*)
Casey, C.P. *et al*, *J. Am. Chem. Soc.*, 1978, **100**, 7565 (*synth*, *nmr*)

$C_6H_5N_2O_3W^{\oplus}$ W-00010

Carbonyl(η^5-2,4-cyclopentadien-1-yl)dinitrosyltungsten(1+), 9CI

[53419-02-6]

M 336.967 (ion)

Hexafluorophosphate: [53419-03-7].
$C_6H_5F_6N_2O_3PW$ M 481.931
Green powder.

Stewart, R.P. *et al*, *Inorg. Chem.*, 1975, **14**, 2699 (*synth*)

$C_6H_{18}W$ W-00011

Hexamethyltungsten, 9CI

Tungsten hexamethyl

[36133-73-0]

WMe_6 (O_h)

M 274.058

Dark-red cryst.; stable when stored at −40°. Sol. hydrocarbons. Dec. at r.t.

▷Reacts explosively with atmos. oxygen and may also detonate *in vacuo* or under N_2 or Ar

Shortland, A.J. *et al*, *J. Chem. Soc., Dalton Trans.*, 1973, 872 (*ir*)
Smith, J. *et al*, *J. Chem. Soc., Dalton Trans.*, 1974, 1742.
Gayler, A.L. *et al*, *J. Chem. Soc., Dalton Trans.*, 1976, 2235 (*synth*, *nmr*, *ms*)
Chiu, K.W. *et al*, *J. Am. Chem. Soc.*, 1980, **102**, 7978.

$C_6NO_5W^{\ominus}$ W-00012

Pentacarbonyl(cyano-*C*)tungstate(1−), 10CI

[68927-82-2]

$(OC)_5WCN^{\ominus}$

M 349.920 (ion)

Tetraethylammonium salt: [77827-07-7].
$C_{14}H_{20}N_2O_5W$ M 480.172
White cryst. Sol. H_2O.

Bis(triphenylphosphine)iminium salt: [22043-88-5].
$C_{42}H_{30}N_2O_5P_2W$ M 888.507
White cryst.

King, R.B., *Inorg. Chem.*, 1967, **6**, 25.
Ruff, J.K. *et al*, *Synth. React. Inorg. Metal-Org. Chem.*, 1971, **1**, 215 (*synth*)
Fehlhammer, W.P. *et al*, *Angew. Chem., Int. Ed. Engl.*, 1979, **18**, 75 (*props*)
Angelici, R.J. *et al*, *Inorg. Chem.*, 1981, **20**, 2977 (*synth*)

C_6O_5SW W-00013

(Carbonothioyl)pentacarbonyltungsten, 9CI

Thiocarbonylpentacarbonyltungsten

[50358-92-4]

$(OC)_5WCS$

M 367.973

Yellow cryst. (hexane).

Angelici, R.J. *et al*, *J. Am. Chem. Soc.*, 1973, **95**, 7516 (*props*)
Angelici, R.J. *et al*, *Inorg. Chem.*, 1976, **15**, 1089 (*synth*)
Fenske, R.F. *et al*, *Inorg. Chem.*, 1976, **15**, 2015 (*struct*)
Angelici, R.J. *et al*, *Inorg. Chem.*, 1976, **15**, 2403 (*props*)
Poliakoff, M., *Inorg. Chem.*, 1976, **15**, 2892 (*ir*)
Butler, I.S., *Acc. Chem. Res.*, 1977, **10**, 359 (*synth*, *props*)

C_6O_6W W-00014

Tungsten carbonyl, 9CI

Hexacarbonyltungsten. Tungsten hexacarbonyl

[14040-11-0]

$W(CO)_6$ (O_h)

M 351.912

Important starting material. Air-stable, colourless, odourless cryst. Sl. sol. hexane. Mp 150° dec.

Arnesen, S.P. *et al*, *Acta Chem. Scand.*, 1966, **20**, 2711 (*ed*)
Jones, L.H. *et al*, *Inorg. Chem.*, 1969, **11**, 2349 (*ir*)
Beneze, L., *J. Organomet. Chem.*, 1972, **37**, C37 (*struct*)
Braterman, P.S. *et al*, *J. Chem. Soc., Dalton Trans.*, 1973, 1027 (*nmr*)
Dellien, I. *et al*, *Chem. Rev.*, 1976, **76**, 283.
Hagen, A.P. *et al*, *Inorg. Chem.*, 1978, **17**, 1369 (*synth*, *bibl*)

$C_7H_5Cl_3O_2W$ W-00015

Dicarbonyltrichloro(η^5-2,4-cyclopentadien-1-yl)tungsten, 8CI

[12107-08-3]

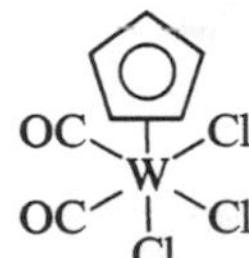

M 411.324

Yellow cryst. Dec. at 150°.

Green, M.L.H. *et al*, *J. Chem. Soc. (A)*, 1967, 686 (*synth*, *nmr*, *ir*)
King, R.B. *et al*, *J. Organomet. Chem.*, 1973, **60**, 125 (*synth*)

$C_7H_5NO_3W$ W-00016

Dicarbonyl(η^5-2,4-cyclopentadien-1-yl)nitrosyltungsten, 10CI

[12128-14-2]

M 334.971

Orange-red solid. $Bp_{0.005}$ 50-60° subl.

Crease, A.E. *et al*, *J. Chem. Soc., Dalton Trans.*, 1973, 1501.
Rosan, A.M. *et al*, *Synth. React. Inorg. Metal-Org. Chem.*, 1976, **6**, 357 (*synth*)
Inorg. Synth., 1978, **18**, 126 (*synth*)
Botto, R.E. *et al*, *Inorg. Chem.*, 1979, **18**, 2049 (*struct*)
Ginzberg, A.E. *et al*, *J. Organomet. Chem.*, 1979, **182**, C1.

$C_8H_2O_8W_2^{\ominus\ominus}$ W-00017

Octacarbonyldi-μ-hydroditungstate(2−), 9CI

M 593.799 (ion)

Bis(tetraethylammonium) salt:
$C_{24}H_{42}N_2O_8W_2$ M 854.304
Red cryst.

Davison, A. *et al*, *J. Chem. Soc., Chem. Commun.*, 1973, 691 (*synth*)
Churchill, M.R. *et al*, *Inorg. Chem.*, 1974, **13**, 2413 (*cryst struct*)

$C_8H_5BrO_3W$ W-00018

Bromotricarbonyl(η^5-2,4-cyclopentadien-1-yl)tungsten, 9CI
[37131-50-3]

L = Br

M 412.880
Red cryst. ($CHCl_3$/pentane).

Sloan, T.E. *et al*, *Inorg. Chem.*, 1968, **7**, 1268 (*synth, ir, nmr*)

$C_8H_5ClO_3W$ W-00019

Tricarbonylchloro(η^5-2,4-cyclopentadien-1-yl)tungsten, 9CI
Chlorotricarbonyl(η^5-cyclopentadienyl)tungsten
[12128-24-4]

As Bromotricarbonyl(η^5-2,4-cyclopentadien-1-yl)-tungsten, W-00018 with

L = Cl

M 368.429
Deep-red cryst. (CCl_4). Mp 160° dec.

Coffey, C.E., *J. Inorg. Nucl. Chem.*, 1963, **25**, 179 (*synth, ir*)
Green, M.L.H. *et al*, *J. Chem. Soc. (A)*, 1967, 687 (*pmr*)
King, R.B., *Org. Mass Spectrom.*, 1969, **2**, 401 (*ms*)
Laine, R.M. *et al*, *Inorg. Chem.*, 1977, **16**, 388 (*synth, nmr*)

$C_8H_5IO_2SW$ W-00020

(Carbonothioyl)dicarbonyl(η^5-2,4-cyclopentadien-1-yl)iodotungsten, 10CI
Thiocarbonyldicarbonyl(η^5-cyclopentadienyl)iodotungsten
[72303-50-5]

M 475.941
Air-stable purple cryst. (CH_2Cl_2/hexane at −20°).

Greaves, W.W. *et al*, *J. Organomet. Chem.*, 1980, **191**, 49 (*synth, ir, nmr*)

$C_8H_5IO_3W$ W-00021

Tricarbonyl(η^5-2,4-cyclopentadien-1-yl)iodotungsten, 10CI
Iodotricarbonyl(η^5-cyclopentadienyl)tungsten
[31870-69-6]

As Bromotricarbonyl(η^5-2,4-cyclopentadien-1-yl)-tungsten, W-00018 with

X = I

M 459.880
Red cryst. ($CHCl_3$/pentane).

Sloan, T.E. *et al*, *Inorg. Chem.*, 1968, **7**, 1268 (*synth, ir, nmr*)

$C_8H_5O_2SW^{\ominus}$ W-00022

(Carbonothioyl)dicarbonyl(η^5-2,4-cyclopentadien-1-yl)tungstate(1−), 10CI
Thiocarbonyldicarbonyl(η^5-cyclopentadienyl)-tungsten(1−)

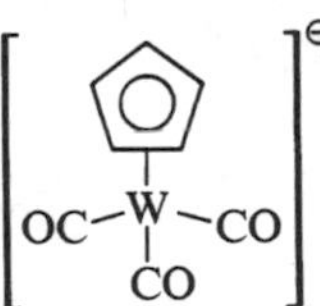

M 349.036 (ion)

Tetrabutylammonium salt: [74989-73-4].
$C_{24}H_{41}O_2NSW$ M 591.503
Characterised spectroscopically, not isol. ir ν_{CO} 1890, 1804 cm^{-1} (CH_2Cl_2).

Greaves, W.W. *et al*, *J. Organomet. Chem.*, 1980, **191**, 49 (*synth, ir, nmr*)

$C_8H_5O_3W^{\ominus}$ W-00023

Tricarbonyl(η^5-2,4-cyclopentadien-1-yl)tungstate(1−), 10CI
[12126-17-9]

M 332.976 (ion)

Na salt: [12107-36-7].
$C_8H_5NaO_3W$ M 355.965
Reactive, air-sensitive.

K salt: [62866-03-9].
$C_8H_5KO_3W$ M 372.074
Reactive, air-sensitive.

Brunner, H. *et al*, *J. Organomet. Chem.*, 1976, **107**, 307 (*props*)
Tsutsui, M. *et al*, *J. Organomet. Chem.*, 1976, **116**, 91 (*props*)
Inorg. Synth., 1977, **17**, 104 (*synth*)
Nesmeyanov, A.N. *et al*, *J. Organomet. Chem.*, 1979, **166**, 217 (*props*)
Malisch, W. *et al*, *Chem. Ber.*, 1980, **113**, 3318 (*props*)
Olgemueller, B. *et al*, *Chem. Ber.*, 1981, **114**, 867 (*props*)

$C_8H_6O_3W$ W-00024

Tricarbonyl(η^5-2,4-cyclopentadien-1-yl)hydrotungsten, 9CI
Hydridotricarbonyl(η^5-cyclopentadienyl)tungsten. Cyclopentadienyltungsten tricarbonyl hydride
[12128-26-6]

As Bromotricarbonyl(η^5-2,4-cyclopentadien-1-yl)-tungsten, W-00018 with

L = H

M 333.984
Yellow cryst. by subl. *in vacuo*. Mp 68.5-69.5°.

Inorg. Synth., 1963, **7**, 136 (*synth*)
King, R.B. *et al*, *Inorg. Chem.*, 1966, **5**, 1837 (*synth*)

Faller, J.W. *et al*, *J. Chem. Soc., Chem. Commun.*, 1969, 719 (*nmr*)
Johnson, B.F.G. *et al*, *J. Chem. Soc. (A)*, 1970, 1684 (*ms*)
Kalck, P. *et al*, *J. Organomet. Chem.*, 1970, **24**, 445.
Keppie, S.A. *et al*, *J. Chem. Soc. (A)*, 1971, 3216 (*synth, nmr*)
Davidson, G. *et al*, *J. Organomet. Chem.*, 1976, **120**, 229 (*ir, raman*)
McFarlane, H.C. *et al*, *J. Chem. Soc., Dalton Trans.*, 1976, 1616 (*nmr*)

$C_8H_6O_5S_2W$ W-00025

[Bis(methylthio)methylene]pentacarbonyltungsten, 10CI
[76262-42-5]

$(OC)_5W{=}C(SMe)_2$

M 430.102
Red cryst. (pentane at −20°). Mp 63-5°.

Angelici, R.J. *et al*, *J. Am. Chem. Soc.*, 1981, **103**, 817 (*synth, props*)

$C_8H_{10}INOW$ W-00026

(η^5-2,4-Cyclopentadien-1-yl)iodonitrosyl(η^3-2-propenyl)tungsten, 10CI
η^3-Allylcyclopentadienyliodonitrosyltungsten
[71341-42-9]

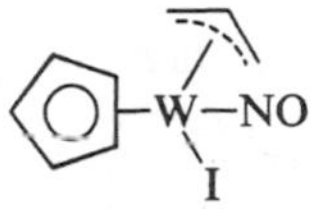

M 446.928
Orange-brown cryst. Mp 165° dec.

Legzdins, P. *et al*, *Inorg. Chem.*, 1979, **18**, 3268 (*synth, cryst struct, nmr*)

$C_8H_{24}W^{\ominus\ominus}$ W-00027

Octamethyltungstate(2−), 9CI

$[WMe_8]^{\ominus\ominus}$

M 304.128 (ion)
Di-Li salt: [62337-31-9].
$C_8H_{24}Li_2W$ M 318.010
Yellow powder.

Gayler, A.L. *et al*, *J. Organomet. Chem.*, 1975, **85**, C37.
Gayler, A.L. *et al*, *J. Chem. Soc., Dalton Trans.*, 1976, 2235 (*synth, props*)

$C_8I_2O_8W_2$ W-00028

Octacarbonyldi-μ-iododitungsten, 9CI
[55405-97-5]

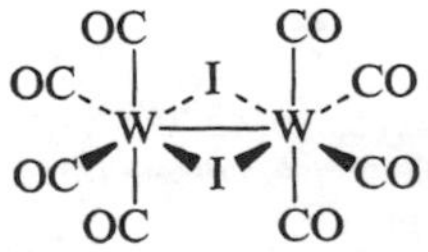

M 845.592
W—W bond length 2.155 Å.

Schmid, G. *et al*, *Chem. Ber.*, 1975, **108**, 260 (*synth*)
Schmid, G. *et al*, *Chem. Ber.*, 1976, **109**, 2148 (*props, cryst struct*)

$C_9HNO_{10}W_2$ W-00029

Nonacarbonyl-μ-hydronitrosylditungsten, 9CI
[40270-21-1]

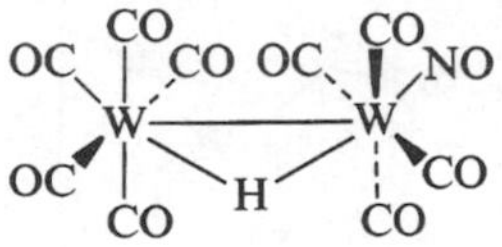

M 650.808
NO group axial, bent staggered struct. Orange cryst. (CH_2Cl_2). Sol. common org. solvs. Mp 128-129.5°. Air-stable as solid but mod. air-sensitive in soln.

Bau, R. *et al*, *J. Am. Chem. Soc.*, 1974, **96**, 6621 (*synth, cryst struct*)
King, R.B. *et al*, *Inorg. Chem.*, 1974, **13**, 3038 (*props*)
Angelici, R.J. *et al*, *Inorg. Chem.*, 1976, **15**, 2397 (*synth*)

$C_9H_5O_4W^{\oplus}$ W-00030

Tetracarbonyl(η^5-2,4-cyclopentadien-1-yl)tungsten(1+)

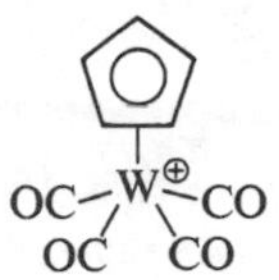

M 360.986 (ion)
Hexafluorophosphate: [37668-41-0].
$C_9H_5F_6O_4PW$ M 505.950
Yellow powder (Me_2CO/Et_2O).

Fischer, E.O. *et al*, *Chem. Ber.*, 1962, **95**, 249 (*synth*)
Dub, M. *Organometallic Compounds*, Springer-Verlag, Berlin, 2nd Ed., 1966, Vol. 1 (*synth*)
Treichel, P.M. *et al*, *Inorg. Chem.*, 1967, **6**, 1328.
Treichel, P.M. *et al*, *J. Organomet. Chem.*, 1967, **7**, 449 (*synth*)
Nolte, M.J. *et al*, *J. Chem. Soc., Dalton Trans.*, 1978, 932.

$C_9H_6O_3W$ W-00031

(η^6-Benzene)tricarbonyltungsten, 9CI
Tricarbonylbenzenetungsten
[12128-53-9]

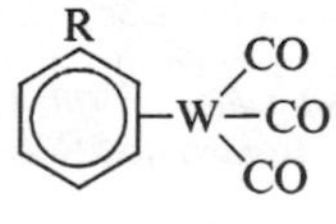

R = H

M 345.995
Yellow air-stable cryst. Sol. org. solvs. Mp 140-5°.

King, R.B. *et al*, *Inorg. Chem.*, 1966, **5**, 1837 (*synth*)
Brown, D.A. *et al*, *J. Chem. Soc. (A)*, 1969, 1534 (*struct*)
Pidcock, A. *et al*, *J. Chem. Soc. (A)*, 1969, 1604 (*props*)
Pettit, R. *et al*, *J. Am. Chem. Soc.*, 1971, **93**, 7087 (*props*)
Mann, B.E., *J. Chem. Soc., Dalton Trans.*, 1973, 2012 (*nmr*)
Lokshin, B.V. *et al*, *J. Organomet. Chem.*, 1975, **102**, 535 (*props*)

$C_9H_7ClO_2W$ W-00032

Dicarbonylchloro(η^7-cycloheptatrienylium)tungsten, 9CI

Dicarbonylchlorotropyliumtungsten

[59589-12-7]

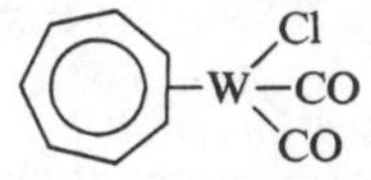

M 366.456

King, R.B. *et al*, *Inorg. Chem.*, 1966, **5**, 1837 (*synth*)
Hoch, G. *et al*, *Z. Naturforsch., B*, 1976, **31**, 294 (*props*)
Powell, P. *et al*, *J. Organomet. Chem.*, 1978, **149**, C1 (*props*)

$C_9H_7IO_2W$ W-00033

Dicarbonyl(η^7-cycloheptatrienylium)iodotunsten, 9CI

Dicarbonyliodotropyliumtungsten

[12082-10-9]

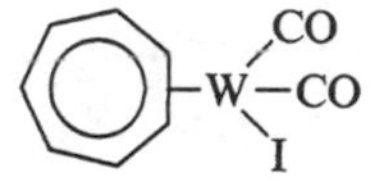

M 457.908

Greenish-black cryst. Mp 185-91° dec.

King, R.B. *et al*, *Inorg. Chem.*, 1966, **5**, 1837 (*synth, ir, nmr*)
Powell, P. *et al*, *J. Organomet. Chem.*, 1978, **149**, C1 (*props*)

$C_9H_8O_3W$ W-00034

Tricarbonyl(η^5-2,4-cyclopentadien-1-yl)methyltungsten, 10CI

Methyltricarbonyl(η^5-cyclopentadienyl)tungsten

[12082-27-8]

As Bromotricarbonyl(η^5-2,4-cyclopentadien-1-yl)-tungsten, W-00018 with

L = Me

M 348.010

Lemon-yellow cryst. Mp 144.7-145.3°.

Piper, T.S. *et al*, *J. Inorg. Nucl. Chem.*, 1956, **3**, 104 (*synth, ir, pmr*)
Davidson, G. *et al*, *J. Organomet. Chem.*, 1976, **120**, 229 (*ir, raman*)
McFarlane, H.C. *et al*, *J. Chem. Soc., Dalton Trans.*, 1976, 1616 (*nmr*)
Alt, H.G., *J. Organomet. Chem.*, 1977, **124**, 167 (*synth*)
Severson, R.G. *et al*, *J. Organomet. Chem.*, 1978, **157**, 173 (*synth, uv*)

$C_9H_8O_5SW$ W-00035

Tricarbonyl(η^5-2,4-cyclopentadien-1-yl)methanesulfonyltungsten, 10CI

Methanesulfonyltricarbonyl(η^5-cyclopentadienyl)tungsten

[52409-63-9]

As Bromotricarbonyl(η^5-2,4-cyclopentadien-1-yl)-tungsten, W-00018 with

L = SO_2Me

M 412.069

Yellow cryst. Mp 176-9° dec.

Kroll, J.O. *et al*, *J. Organomet. Chem.*, 1974, **66**, 95 (*synth, nmr, ir*)
Lorenz, I.P. *et al*, *Z. Naturforsch., B*, 1976, **31**, 888 (*synth*)
Severson, R.G. *et al*, *J. Am. Chem. Soc.*, 1979, **101**, 877 (*synth, nmr*)

$C_9H_8O_6W$ W-00036

Pentacarbonyl(1-ethoxyethylidene)tungsten, 8CI

[25879-47-4]

$(OC)_5W{=}C(CH_3)OEt$

M 396.009

Yellow cryst. (pentane).

Aumann, R. *et al*, *Angew. Chem., Int. Ed. Engl.*, 1967, **6**, 878 (*synth*)
Darensbourg, M.Y. *et al*, *Inorg. Chem.*, 1970, **9**, 32 (*synth*)

$C_9H_{10}BrNO_4W$ W-00037

Bromotetracarbonyl[(diethylamino)methylidyne]tungsten

[60872-94-8]

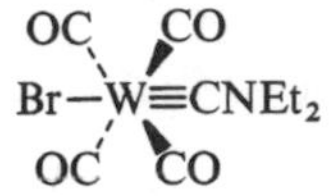

M 459.936

Yellow cryst. Mp 47°.

Fischer, E.O. *et al*, *J. Organomet. Chem.*, 1974, **65**, C53 (*synth, ir, nmr*)

$C_9H_{10}OW$ W-00038

Carbonyl(η^5-2,4-cyclopentadien-1-yl)(η^2-ethyne)methyltungsten, 10CI

(Acetylene)carbonyl(η^5-cyclopentadienyl)methyltungsten

[62798-73-6]

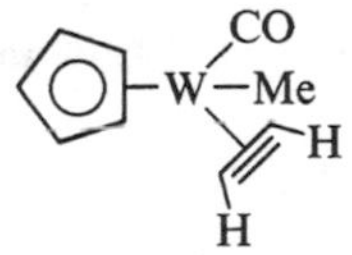

M 318.027

Cryst. Mp 91°.

Alt, H.G., *J. Organomet. Chem.*, 1977, **127**, 349 (*synth, ir, nmr*)
Alt, H.G. *et al*, *Z. Naturforsch., B*, 1977, **32**, 144 (*synth*)

$C_9H_{11}ClN_2O_2W$ W-00039

Bis(acetonitrile)dicarbonylchloro(η^3-2-propenyl)tungsten, 10CI

Bis(acetonitrile)(η^3-allyl)dicarbonylchlorotungsten

[62680-77-7]

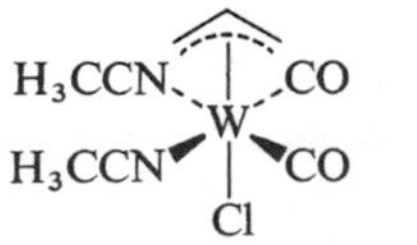

M 398.501

Yellow cryst.

Hayter, R.E., *J. Organomet. Chem.*, 1968, **13**, P1 (*synth*)
Hohmann, F., *J. Organomet. Chem.*, 1977, **137**, 315 (*synth*)
Doyle, G., *J. Organomet. Chem.*, 1978, **150**, 67 (*props*)
Brisdon, B.J. *et al*, *Inorg. Chim. Acta*, 1980, **40**, 191 (*props*)
Brisdon, B.J. *et al*, *Transition Met. Chem.*, 1981, **6**, 83 (*props*)

$C_{10}HO_{10}W_2^{\ominus}$ W-00040

Decacarbonyl-μ-hydroditungstate(1−)

[19773-10-5]

$[W_2H(CO)_{10}]^{\ominus}$

M 648.812 (ion)
$Et_4N^{\oplus}$ salt adopts linear eclipsed struct. $PPN^{\oplus}$ salt adopts bent staggered struct.
Tetraethylammonium salt: [12083-01-1].
$C_{18}H_{21}NO_{10}W_2$ M 779.065
Yellow cryst.
Bis(triphenylphosphine)iminium salt: [56172-01-1].
$C_{46}H_{31}NO_{10}P_2W_2$ M 1187.399
Yellow prisms (CH_2Cl_2/Et_2O).

Hayter, R.G., *J. Am. Chem. Soc.*, 1966, **88**, 4376 (*synth*)
Gray, H.B. *et al*, *J. Am. Chem. Soc.*, 1975, **97**, 3073 (*synth, ir*)
Wilson, R.D. *et al*, *J. Organomet. Chem.*, 1975, **91**, C49 (*cryst struct*)
Bau, R. *et al*, *Acc. Chem. Res.*, 1979, **12**, 176 (*cryst struct*)

$C_{10}H_7O_3W^{\oplus}$ W-00041

Tricarbonyl(η^7-cycloheptatrienylium)tunsten(1+), 9CI
Tricarbonyl(η^7-tropylium)tunsten(1+)
[46238-46-4]

M 359.014 (ion)
Tetrafluoroborate: [12083-17-9].
$C_{10}H_7BF_4O_3W$ M 445.817
Orange cryst.

King, R.B. *et al*, *Inorg. Chem.*, 1966, **5**, 1837 (*synth, props*)
Kane-Maguire, L.A.P. *et al*, *J. Chem. Soc., Dalton Trans.*, 1973, 1683 (*props*)
Kane-Maguire, L.A.P. *et al*, *J. Organomet. Chem.*, 1975, **102**, C4 (*props*)
John, G.R. *et al*, *J. Organomet. Chem.*, 1976, **120**, C47 (*props*)
Ziegler, M.L. *et al*, *Z. Naturforsch., B*, 1976, **31**, 294 (*props*)

$C_{10}H_8O_3W$ W-00042

Tricarbonyl[(1,2,3,4,5,6-η)-1,3,5-cycloheptatriene]tungsten, 9CI
[12128-81-3]

M 360.021
Air-stable red cryst. Sol. hydrocarbons. Sublimes.

King, R.B. *et al*, *Inorg. Chem.*, 1966, **5**, 1837 (*synth*)
King, R.B. *et al*, *Inorg. Chem.*, 1971, **10**, 1841 (*props*)
Kreiter, C.G. *et al*, *Chem. Ber.*, 1975, **108**, 1502 (*nmr*)
Brown, D.L.S. *et al*, *J. Organomet. Chem.*, 1977, **142**, 321 (*props*)
Fragala, I. *et al*, *J. Organomet. Chem.*, 1979, **182**, 511 (*pe*)
Gower, M. *et al*, *Inorg. Chim. Acta*, 1979, **37**, 79 (*props*)

$C_{10}H_8O_3W$ W-00043

Tricarbonyl[(1,2,3,4,5,6-η)-methylbenzene]tungsten, 9CI
Tricarbonyltoluenetungsten
[12128-82-4]
As (η^6-Benzene)tricarbonyltungsten, W-00031 with

$R = CH_3$

M 360.021
Yellow cryst. Mp 139-42° dec. $Bp_{0.2}$ 120° subl.

King, R.B. *et al*, *Inorg. Chem.*, 1966, **5**, 1837 (*synth*)
Kobayashi, H. *et al*, *Bull. Chem. Soc. Jpn.*, 1975, **48**, 1222 (*ir*)

$C_{10}H_8O_4W$ W-00044

Acetyltricarbonyl(η^5-2,4-cyclopentadien-1-yl)tungsten, 10CI
Tricarbonyl(η^5-2,4-cyclopentadien-1-yl)acetyltungsten.
Cyclopentadienyltricarbonylacetyltungsten
[64666-36-0]

M 376.021
Yellow cryst.

Adams, R.D. *et al*, *Inorg. Chem.*, 1978, **17**, 266 (*synth, props*)

$C_{10}H_9O_3W^{\oplus}$ W-00045

Tricarbonyl(η^5-2,4-cyclopentadien-1-yl)(η^2-ethene)tungsten(1+), 10CI
Tricarbonyl(η^5-cyclopentadienyl)(ethylene)-tungsten(1+)

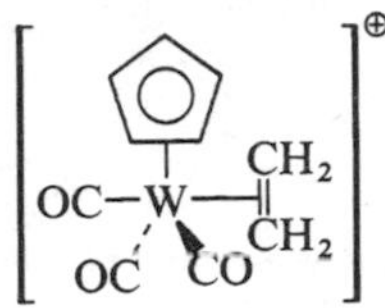

M 361.029 (ion)
Tetrafluoroborate: [63079-59-4].
$C_{10}H_9BF_4O_3W$ M 447.833
Green cryst.

Fischer, E.O. *et al*, *Chem. Ber.*, 1961, **94**, 1200 (*synth*)
Knoth, W.H., *Inorg. Chem.*, 1975, **14**, 1566 (*synth, props*)

$C_{10}H_{10}Br_2W$ W-00046

Dibromobis(η^5-2,4-cyclopentadien-1-yl)tungsten, 10CI
[12184-19-9]

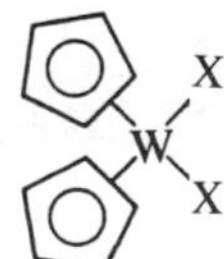

X = Br

M 473.847
Green cryst. Dec. at 250°.

Cooper, R.L. *et al*, *J. Chem. Soc. (A)*, 1967, 1155 (*synth*)

$C_{10}H_{10}Cl_2OW$ W-00047

Dichlorobis(η^5-2,4-cyclopentadien-1-yl)oxotungsten, 10CI
[53504-76-0]

$W(O)Cl_2(\eta\text{-}C_5H_5)_2$

M 400.944
Green cryst. (pet. ether).

Anand, S.P. *et al*, *J. Organomet. Chem.*, 1969, **17**, 423 (*synth*)
Anand, S.P. *et al*, *J. Organomet. Chem.*, 1969, **19**, 387; 1971, **28**, 265.

$C_{10}H_{10}Cl_2W$ W-00048

Dichlorobis(η^5-2,4-cyclopentadien-1-yl)tungsten, 10CI

[12184-26-8]

As Dibromobis(η^5-2,4-cyclopentadien-1-yl)tungsten, W-00046 with

X = Cl

M 384.945

Green cryst. Dec. at 250°.

Cooper, R.L. *et al*, *J. Chem. Soc. (A)*, 1967, 1155 (*synth*)
Crabtree, R.H. *et al*, *J. Chem. Soc. (A)*, 1971, 1350.
Green, M.L.H. *et al*, *J. Chem. Soc., Dalton Trans.*, 1972, 1445.
Lindsell, W.E., *J. Chem. Soc., Dalton Trans.*, 1975, 2548.
Bell, L.G. *et al*, *J. Organomet. Chem.*, 1977, **135**, 173.
Dias, A.R. *et al*, *J. Organomet. Chem.*, 1979, **175**, 193.

$C_{10}H_{10}N_2O_5W$ W-00049

Pentacarbonyl(1,3-dimethyl-2-imidazolidinylidene)tungsten, 10CI

[64515-01-1]

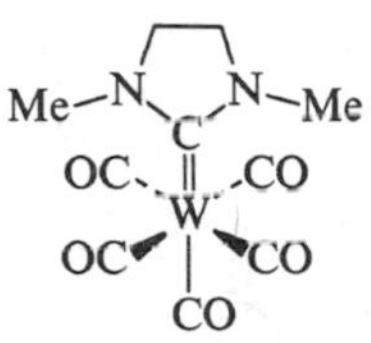

M 422.049

White cryst. Mp 118°.

Lappert, M.F. *et al*, *J. Chem. Soc., Dalton Trans.*, 1977, 1283 (*synth, props, ir, nmr*)

$C_{10}H_{10}O_2W$ W-00050

Dicarbonyl(η^5-2,4-cyclopentadien-1-yl)(η^3-2-propenyl)tungsten

π-Allyldicarbonyl-π-cyclopentadienyltungsten, 8CI

[31811-41-3]

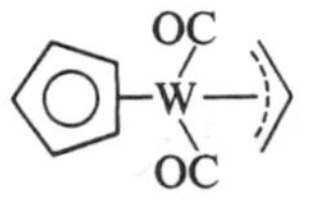

M 346.038

Yellow air-stable cryst.

Dub, M., *Organometallic Compounds*, Springer-Verlag, Berlin, 2nd Ed., 1966, **1**.
Davison, A. *et al*, *Inorg. Chem.*, 1967, **6**, 2124 (*synth, nmr, ir*)

$C_{10}H_{11}AsO_3W$ W-00051

Tricarbonyl(η^5-2,4-cyclopentadien-1-yl)(dimethylarsino)tungsten, 10CI

[52456-36-7]

As Bromotricarbonyl(η^5-2,4-cyclopentadien-1-yl)tungsten, W-00018 with

L = $AsMe_2$

M 437.967

Light-orange needles. Mp 69-70° (sealed tube).

Malisch, W. *et al*, *Angew. Chem., Int. Ed. Engl.*, 1974, **13**, 84 (*synth, ir, nmr*)
Panster, P. *et al*, *J. Organomet. Chem.*, 1977, **134**, C32 (*synth, props*)
Müller, R. *et al*, *Chem. Ber.*, 1977, **110**, 3910 (*synth*)

$C_{10}H_{11}LiW$ W-00052

Bis(η^5-2,4-cyclopentadien-1-yl)hydro(tungsten)lithium

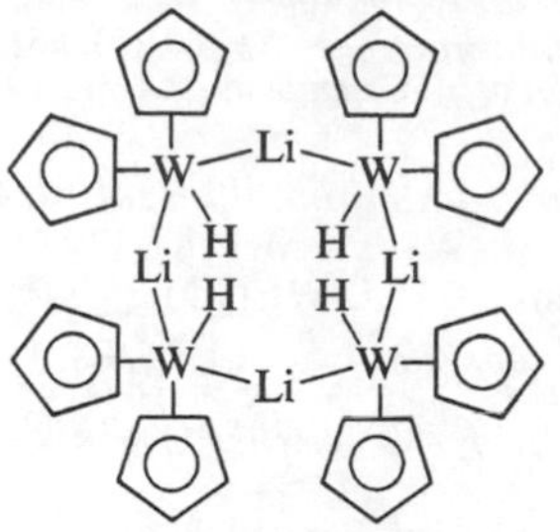

M 321.988

Tetrameric. Important synthetic intermediate.

Tetramer: [53322-18-2].

$C_{40}H_{44}Li_4W_4$ M 1287.952

Yellow-brown cryst. Insol. aliphatic and aromatic hydrocarbons. Reacts with other solvs. inc. THF. Dec. in air or H_2O.

Prout, K. *et al*, *Acta Crystallogr., Sect. B*, 1974, **30**, 2318 (*cryst struct*)
Francis, B.R. *et al*, *J. Chem. Soc., Dalton Trans.*, 1976, 1339 (*synth, props*)

$C_{10}H_{11}O_3SbW$ W-00053

Tricarbonyl(η^5-2,4-cyclopentadien-1-yl)(dimethylstilbino)tungsten, 10CI

[52571-36-5]

As Bromotricarbonyl(η^5-2,4-cyclopentadien-1-yl)tungsten, W-00018 with

L = $SbMe_2$

M 484.795

Cryst. Mp 64-6°.

Malisch, W. *et al*, *J. Organomet. Chem.*, 1975, **99**, 421 (*synth, ir, nmr, ms*)

$C_{10}H_{11}O_3TlW$ W-00054

Tricarbonyl(η^5-2,4-cyclopentadien-1-yl)(dimethylthallium)tungsten, 10CI

[66125-45-9]

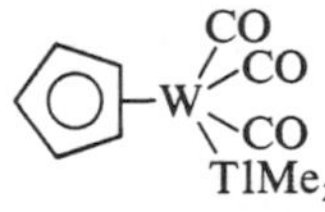

M 567.428

Mp 76-8°.

▷Toxic

Walther, B. *et al*, *J. Organomet. Chem.*, 1978, **145**, 285 (*synth, ir, pmr*)

$C_{10}H_{12}W$ W-00055

Bis(η^5-2,4-cyclopentadien-1-yl)dihydrotungsten, 10CI

Bis(η^5-cyclopentadienyl)dihydridotungsten

[1271-33-6]

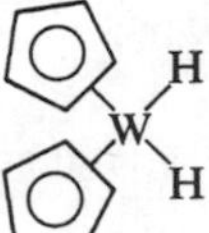

M 316.055

Yellow cryst. Mp 163-5°.

Green, M.L.H. *et al*, *J. Chem. Soc.*, 1961, 4854.
Johnson, M.P. *et al*, *J. Am. Chem. Soc.*, 1966, **88**, 301.
Otsuka, S. *et al*, *J. Am. Chem. Soc.*, 1973, **95**, 5091.
Barefield, E.U. *et al*, *J. Organomet. Chem.*, 1974, **76**, C50.
Hoxmeier, R.J. *et al*, *Inorg. Chem.*, 1979, **18**, 3453.
Kochi, J.K. *et al*, *J. Am. Chem. Soc.*, 1980, **102**, 208.

$C_{10}H_{13}W^{\oplus}$ W-00056

Bis(η^5-2,4-cyclopentadien-1-yl)trihydrotungsten(1+), 10CI
Bis(η^5-cyclopentadienyl)trihydridotungsten(1+)
[11108-17-1]

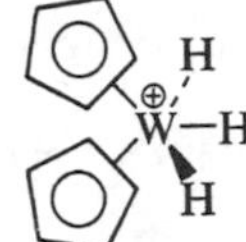

M 317.063 (ion)
Hexafluorophosphate: [11108-17-1].
$C_{10}H_{13}F_6PW$ M 462.027
Characterised spectroscopically.

Green, M.L.H. *et al*, *J. Chem. Soc.*, 1961, 4854 (*synth*)
Bau, R. *et al*, *J. Am. Chem. Soc.*, 1977, **99**, 1775 (*struct*)

$C_{11}H_5BrO_4W$ W-00057

Bromotetracarbonyl(phenylmethylidyne)tungsten, 9CI
Bromotetracarbonylphenylcarbynetungsten
[60872-93-7]

OC CO
X–W≡CPh
OC CO
X = Br

M 464.912
Cryst. (CH_2Cl_2/pentane).

Fischer, E.O. *et al*, *Chem. Ber.*, 1976, **109**, 1673 (*synth*)
Fischer, E.O. *et al*, *J. Organomet. Chem.*, 1977, **134**, C37 (*synth*)

$C_{11}H_5ClO_4W$ W-00058

Tetracarbonylchloro(phenylmethylidyne)tungsten, 9CI
Chlorotetracarbonylphenylcarbynetungsten
[60873-00-9]
As Bromotetracarbonyl(phenylmethylidyne)tungsten, W-00057 with

X = Cl

M 420.461
Light-yellow needle-like cryst.

Huttner, G. *et al*, *Angew. Chem., Int. Ed. Engl.*, 1974, **13**, 609 (*struct*)
Fischer, E.O. *et al*, *Chem. Ber.*, 1976, **109**, 1673 (*synth, nmr, ir, ms*)

$C_{11}H_5IO_4W$ W-00059

Tetracarbonyliodo(phenylmethylidyne)tungsten, 9CI
Iodotetracarbonylphenylcarbynetungsten
[60872-92-6]
As Bromotetracarbonyl(phenylmethylidyne)tungsten, W-00057 with

X = I

M 511.913
Cryst. (pentane).

Huttner, G. *et al*, *Angew. Chem., Int. Ed. Engl.*, 1974, **13**, 609 (*struct*)
Fischer, E.O. *et al*, *Chem. Ber.*, 1976, **109**, 1673 (*synth*)
Fischer, E.O. *et al*, *J. Organomet. Chem.*, 1977, **134**, C37 (*synth*)

$C_{11}H_8O_3W$ W-00060

Tricarbonyl[(1,2,3,4,5,6-η)-1,3,5,7-cyclooctatetraene]tungsten, 10CI
[12108-95-1]

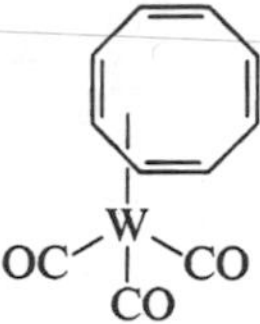

M 372.032
Red cryst.

King, R.B. *et al*, *Inorg. Chem.*, 1966, **5**, 1837 (*synth*)
Holloway, C.E. *et al*, *J. Chem. Soc. (A)*, 1969, 931 (*nmr*)
Mann, B.E. *et al*, *J. Chem. Soc., Dalton Trans.*, 1979, 1021 (*nmr*)
Laine, R.M., *Transition Met. Chem.*, 1980, **5**, 158 (*synth*)

$C_{11}H_8O_4W$ W-00061

[(2,3,5,6-η)-Bicyclo[2.2.1]hepta-2,5-diene]tetracarbonyltungsten, 9CI
Tetracarbonyl(2,5-norbornadiene)tungsten, 8CI. *Norbornadienetetracarbonyltungsten*
[12129-25-8]

M 388.032
Yellow cryst. Mp 90-2°. Sublimes at 80°/0.1 mm.

Bennett, M.A. *et al*, *J. Chem. Soc.*, 1961, 2037 (*synth, ir*)
Fronzaglia, A. *et al*, *Inorg. Chem.*, 1966, **5**, 1837 (*synth, pmr, ir*)
Dobson, G.R., *Inorg. Chem.*, 1969, **8**, 90 (*synth, ir*)
Mann, B.E., *J. Chem. Soc., Dalton Trans.*, 1973, 2012 (*cmr, ir*)

$C_{11}H_{10}OW$ W-00062

Carbonylbis(η^5-2,4-cyclopentadien-1-yl)tungsten, 10CI
[39333-44-3]

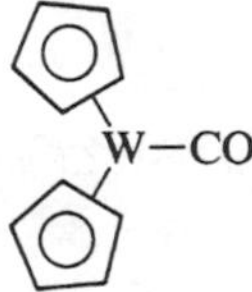

M 342.049
Green solid. $Bp_{0.001}$ 60° subl.

Thomas, J.L. *et al*, *J. Am. Chem. Soc.*, 1973, **95**, 1838 (*synth*)
Brintzinger, H.H. *et al*, *J. Am. Chem. Soc.*, 1974, **96**, 3694.
Perutz, R.N. *et al*, *J. Chem. Soc., Chem. Commun.*, 1979, 742.

$C_{11}H_{14}GeO_3W$ W-00063

Tricarbonyl(η^5-2,4-cyclopentadien-1-yl)(trimethylgermyl)tungsten, 9CI
[33306-93-3]

As Bromotricarbonyl(η^5-2,4-cyclopentadien-1-yl)tungsten, W-00018 with

L = $GeMe_3$

M 450.670

Yellow cryst. Mp 113-4°. $Bp_{0.001}$ 90° subl.

Carrick, A. *et al*, *J. Chem. Soc. (A)*, 1968, 913.
Cardin, D.J. *et al*, *J. Chem. Soc. (A)*, 1970, 2594 (*synth*)
Keppie, S.A. *et al*, *J. Chem. Soc. (A)*, 1971, 3216 (*synth*)

$C_{11}H_{14}O_3SiW$ W-00064

Tricarbonyl(η^5-2,4-cyclopentadien-1-yl)(trimethylsilyl)tungsten, 9CI

[12282-35-8]

As Bromotricarbonyl(η^5-2,4-cyclopentadien-1-yl)tungsten, W-00018 with

L = $SiMe_3$

M 406.165

Yellow-orange needles, air- and light-sensitive. $Bp_{0.001}$ 100° subl.

Cardin, D.J. *et al*, *J. Chem. Soc. (A)*, 1970, 2594 (*synth*)
Hagen, A.P. *et al*, *Inorg. Chem.*, 1971, **10**, 1657 (*synth*)
Malisch, W. *et al*, *Chem. Ber.*, 1974, **107**, 979 (*synth*)

$C_{11}H_{14}O_3SnW$ W-00065

Tricarbonyl(η^5-2,4-cyclopentadien-1-yl)(trimethylstannyl)-tungsten, 9CI

[12093-29-7]

As Bromotricarbonyl(η^5-2,4-cyclopentadien-1-yl)tungsten, W-00018 with

L = $SnMe_3$

M 496.770

Pale-yellow cryst. Mp 120°.

Patil, H.R.H. *et al*, *Inorg. Chem.*, 1966, **5**, 1401 (*synth*)
Carrick, A. *et al*, *J. Chem. Soc. (A)*, 1968, 913 (*synth*)
Cardin, D.J. *et al*, *J. Chem. Soc. (A)*, 1970, 2594 (*synth*)
Keppie, S.A. *et al*, *J. Chem. Soc. (A)*, 1971, 3216 (*synth*)

$C_{11}H_{22}O_3P_2W$ W-00066

Tricarbonyl(η^2-ethene)bis(trimethylphosphine)tungsten, 10CI

[66496-24-0]

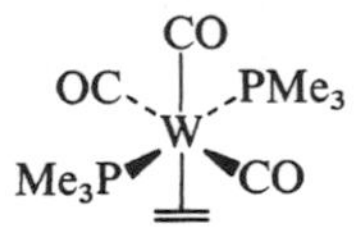

CO, OC, PMe_3, W, Me_3P, CO

M 448.091

mer-trans-form

Yellow cryst.

Koemm, U. *et al*, *J. Organomet. Chem.*, 1978, **148**, 179 (*synth*, *nmr*)

$C_{11}H_{24}Cl_3OPW$ W-00067

Trichloro(1,1-dimethylpropylidyne)(triethylphosphine oxide-*O*)tungsten, 10CI

[78251-21-5]

$W{\equiv}CC(CH_3)_3(O{=}PEt_3)Cl_3$

M 493.493

Blue cryst.

Wengrovius, J.H. *et al*, *J. Am. Chem. Soc.*, 1981, **103**, 3932 (*synth*, *cmr*, *nmr*)

$C_{11}H_{28}Cl_2OP_2W$ W-00068

Dichloro(2,2-dimethylpropylidene)oxobis(trimethylphosphine)tungsten, 10CI

[76603-92-4]

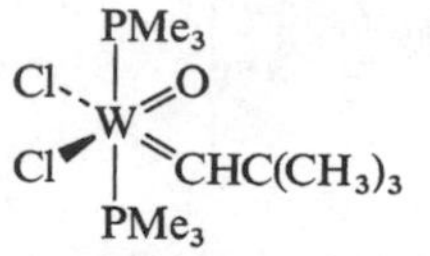

PMe_3, Cl, O, W, Cl, $CHC(CH_3)_3$, PMe_3

M 493.045

Olefin metathesis catalyst.

Wengrovius, J.H. *et al*, *Organometallics*, 1982, **1**, 148 (*synth*)

$C_{12}H_8O_4W$ W-00069

Tetracarbonyl[(1,2,5,6-η)-1,3,5,7-cyclooctatetraene]tungsten, 10CI

[74468-66-9]

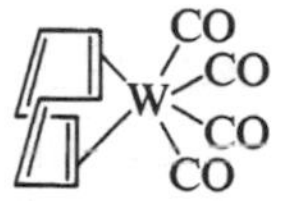

CO, W, CO, CO, CO

M 400.043

Red-orange cryst. Sol. hydrocarbons.

Winstein, S., *Chem. Soc., Spec. Publ.*, 1967, **21**, 5.
Laine, R.M., *Transition Met. Chem.*, 1980, **5**, 158 (*synth*)

$C_{12}H_{10}O_3W$ W-00070

[(2,3,4,5,6,7-η)Bicyclo[6.1.0]nona-2,4,6-triene]tricarbonyltungsten, 9CI

[59714-56-6]

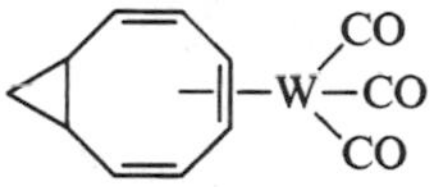

W, CO, CO, CO

M 386.059

Salzer, A., *J. Organomet. Chem.*, 1976, **117**, 245 (*synth*, *props*)

$C_{12}H_{12}O_3W$ W-00071

Tricarbonyl[(1,2,3,4,5,6-η)-1,3,5-trimethylbenzene]tungsten, 9CI

Tricarbonylmesitylenetungsten

[12129-69-0]

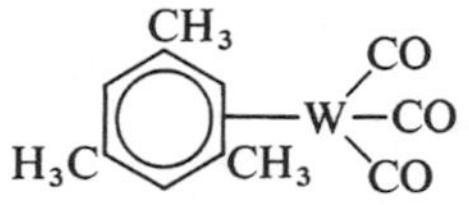

CH_3, CO, W, CO, H_3C, CH_3, CO

M 388.075

Yellow cryst. Bp_1 140° subl. Dec. >180°.

Fischer, E.O. *et al*, *Chem. Ber.*, 1958, **91**, 2763 (*synth*)
King, R.B. *et al*, *Inorg. Chem.*, 1966, **5**, 1837.
Kobayashi, H. *et al*, *Bull. Chem. Soc. Jpn.*, 1975, **48**, 1222 (*ir*)
Brown, D.L.S. *et al*, *J. Organomet. Chem.*, 1977, **142**, 321 (*props*)
Sennikov, P.E. *et al*, *J. Organomet. Chem.*, 1980, **190**, 167.

$C_{12}H_{12}O_4W$ **W-00072**

Tetracarbonyl[(1,2,5,6-η)-1,5-cyclooctadiene]tungsten, 9CI

[12129-70-3]

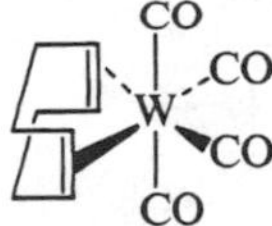

M 404.074

Fischer, E.O. *et al*, *Chem. Ber.*, 1959, **92**, 2995 (*synth*)
Leigh, G.J. *et al*, *J. Organomet. Chem.*, 1965, **4**, 461 (*synth*)
King, R.B. *et al*, *Inorg. Chem.*, 1966, **5**, 1837 (*synth*)

$C_{12}H_{12}W$ **W-00073**

Bis(η^6-benzene)tungsten, 9CI

[12089-23-5]

M 340.077

Green cryst.

Fischer, E.O. *et al*, *Chem. Ber.*, 1960, **93**, 2065 (*synt*)
Skell, P.S. *et al*, *J. Am. Chem. Soc.*, 1974, **96**, 1945 (*synth*)
Cloke, F.G.W. *et al*, *J. Chem. Soc., Chem. Commun.*, 1978, 72 (*synth*)
Green, M.L.H., *J. Organomet. Chem.*, 1980, **200**, 119 (*synth*)

$C_{12}H_{14}W$ **W-00074**

Bis(η^5-2,4-cyclopentadien-1-yl)(η^2-ethene)tungsten, 9CI

Dicyclopentadienyl(ethylene)tungsten

[37343-06-9]

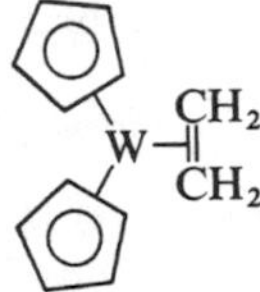

M 342.093

Orange cryst. Mp 176° dec.

Thomas, J.L., *J. Am. Chem. Soc.*, 1973, **95**, 1838 (*synth*)
Green, M.L.H. *et al*, *J. Chem. Soc., Dalton Trans.*, 1974, 1324 (*synth*)
Green, M.L.H. *et al*, *J. Organomet. Chem.*, 1974, **76**, 49 (*props*)
Green, J.C. *et al*, *J. Chem. Soc., Dalton Trans.*, 1975, 403 (*pe*)

$C_{12}H_{15}O_3TlW$ **W-00075**

Tricarbonyl(η^5-2,4-cyclopentadien-1-yl)(diethylthallium)tungsten

[66125-46-0]

M 595.482

Used in the synth. of tungsten carbonyls. Mp 119-21°.

▷Toxic

Walther, B. *et al*, *J. Organomet. Chem.*, 1978, **145**, 285 (*synth*, *ir*, *nmr*)

$C_{12}H_{15}W^{\oplus}$ **W-00076**

Bis(η^5-2,4-cyclopentadien-1-yl)(η^2-ethene)hydrotungsten(1+)

[37343-09-2]

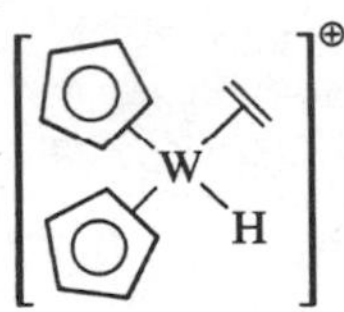

M 343.101 (ion)

Hexafluorophosphate: [53770-66-4].
$C_{12}H_{15}F_6PW$ M 488.065
White cryst.

Benfield, F.W.S., *J. Chem. Soc., Dalton Trans.*, 1974, 1324 (*synth*)

$C_{12}H_{16}O_2W$ **W-00077**

(Acetato-*O*)[(1,2,3,4,5-η)-methylbenzene](η^3-2-propenyl)tungsten, 10CI

η^3-Allyl(acetato)-η^6-toluenetungsten

[78607-84-8]

M 376.107

Green, M.L.H., *J. Organomet. Chem.*, 1980, **200**, 119 (*synth*)

$C_{12}H_{18}O_3W$ **W-00078**

Tris[(3,4-η)-3-buten-2-one-*O*]tungsten, 8CI

Tris(methyl vinyl ketone)tungsten

[12178-34-6]

M 394.122

Yellow cryst. (EtOAc); air-stable. Mp 137-40°. $Bp_{0.1}$ 90° subl.

King, R.B. *et al*, *Inorg. Chem.*, 1966, **5**, 1837 (*synth*, *ir*, *ms*, *nmr*)
Moriarty, R.E. *et al*, *J. Chem. Soc., Chem. Commun.*, 1972, 1242 (*cryst struct*)

$C_{12}H_{18}W$ **W-00079**

Tris(η^4-1,3-butadiene)tungsten

[51733-16-5]

M 346.124

Catalyses polym. of ethylene and butadiene. Yellow cryst. Darkens after 1 week in air. Sol. common org. solvs. Dec. at 130°. Trigonal prismatic struct. by analogy with corresponding Me deriv.

Skell, P.S. *et al*, *J. Am. Chem. Soc.*, 1974, **96**, 626 (*synth*, *nmr*, *ms*)

Gansing, W. *et al*, *Angew. Chem.*, 1981, **93**, 201.

$C_{13}H_5ClF_{12}W$ W-00080

Chloro(η^5-2,4-cyclopentadien-1-yl)bis[(2,3-η)-1,1,1,4,4,4-hexafluoro-2-butyne]tungsten, 10CI

Bis[bis(trifluoromethyl)acetylene]chloro(cyclopentadienyl)tungsten

[54720-97-7]

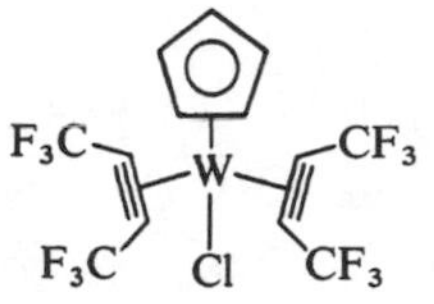

M 608.466

Pale-yellow cryst. Mp 147°.

Davidson, J.L. *et al*, *J. Chem. Soc., Dalton Trans.*, 1976, 738 (*synth, cryst struct*)
Davidson, J.L. *et al*, *J. Chem. Soc., Dalton Trans.*, 1977, 287, 2246 (*props*)
Davidson, J.L., *J. Organomet. Chem.*, 1980, **186**, C19 (*props*)

$C_{13}H_5ClO_4W$ W-00081

Tetracarbonylchloro(3-phenyl-2-propynylidyne)tungsten, 9CI

[55336-79-3]

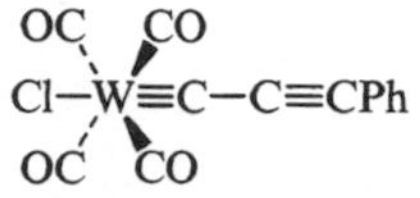

M 444.483

Cryst.

Fischer, E.O. *et al*, *J. Organomet. Chem.*, 1974, **81**, C23 (*synth, ir, nmr*)

$C_{13}H_8O_5W$ W-00082

Pentacarbonyl[(η^2-ethenyl)benzene]tungsten, 9CI

Pentacarbonyl(η^2-styrene)tunsten

[60635-81-6]

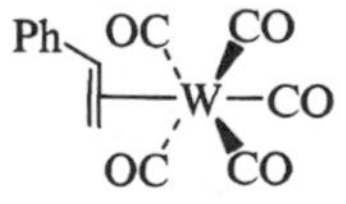

M 428.053

Fischer, E.O. *et al*, *J. Organomet. Chem.*, 1976, **112**, C59 (*synth*)

$C_{13}H_8O_6W$ W-00083

Pentacarbonyl(methoxyphenylmethylene)tungsten, 10CI

[37823-96-4]

$(OC)_5W{=}CPhOMe$

M 444.053

Orange-red cryst. Mp 59°.

Fischer, E.O. *et al*, *Chem. Ber.*, 1967, **100**, 2445 (*synth*)
Casey, C.P. *et al*, *J. Organomet. Chem.*, 1974, **77**, 345 (*props*)
Fischer, E.O. *et al*, *Chem. Ber.*, 1976, **109**, 1673 (*props*)
Casey, C.P. *et al*, *J. Am. Chem. Soc.*, 1977, **99**, 1651 (*props*)
Fischer, E.O. *et al*, *Chem. Ber.*, 1977, **110**, 1651 (*props*)

$C_{13}H_{15}W^{\oplus}$ W-00084

Bis(η^5-2,4-cyclopentadien-1-yl)(η^3-2-propenyl)tungsten(1+), 9CI

(η^3-Allyl)bis(η^5-cyclopentadienyl)tungsten(1+)

[55650-02-7]

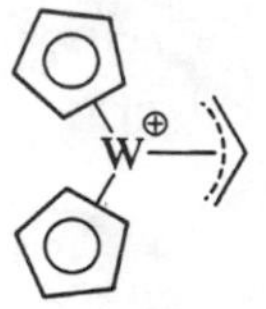

M 355.112 (ion)

Hexafluorophosphate:

$C_{13}H_{15}F_6PW$ M 500.076

Orange cryst.

Ephritikhine, M. *et al*, *J. Chem. Soc., Dalton Trans.*, 1977, 1131 (*synth*)

$C_{13}H_{17}W^{\oplus}$ W-00085

Bis(η^5-2,4-cyclopentadien-1-yl)(η^2-ethene)methyltungsten(1+), 10CI

[53504-75-9]

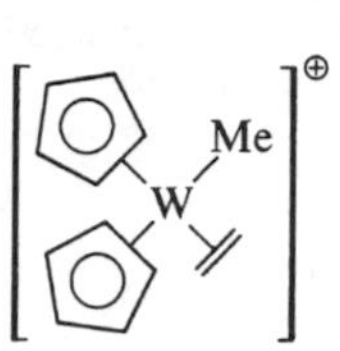

M 357.127 (ion)

Hexafluorophosphate: [53504-75-9].

$C_{13}H_{17}F_6PW$ M 502.091

Pale-yellow cryst.

Benfield, F.W.S. *et al*, *J. Organomet. Chem.*, 1974, **76**, 49 (*props*)
Benfield, F.W.S. *et al*, *J. Chem. Soc., Dalton Trans.*, 1974, 1324 (*synth*)
Green, M.L.H. *et al*, *J. Chem. Soc., Dalton Trans.*, 1979, 262 (*props*)
Cooper, N.J. *et al*, *J. Chem. Soc., Dalton Trans.*, 1979, 1121 (*props*)

$C_{13}H_{20}N_2O_3S_4W$ W-00086

Tricarbonylbis(diethylcarbamodithioato-*S,S'*)tungsten, 10CI

[72827-54-4]

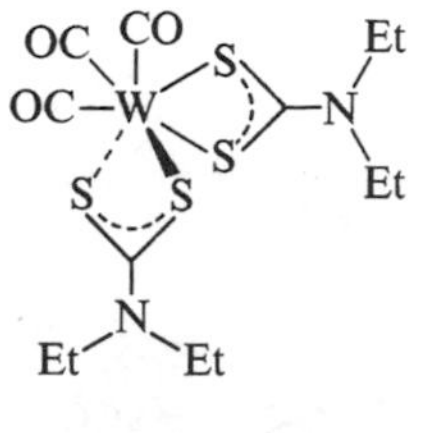

M 564.403

7-Coordinate tungsten. Orange-red cryst. turning green on exp. to air for several days. Loses CO reversibly *in vacuo*.

Templeton, J.L. *et al*, *Inorg. Chem.*, 1980, **9**, 1753 (*synth, ir, pmr, cmr, cryst struct*)

$C_{13}H_{21}O_4PW^{\ominus\ominus}$ W-00087

Tetracarbonyl(triisopropylphosphine)tungstate(2−)

Tetracarbonyl[tris(1-methylethyl)phosphine]-tungsten(2−)

[79135-08-3]

$(OC)_4WP[CH(CH_3)_2]_3^{\ominus\ominus}$

M 456.130 (ion)

Di-Na salt:

$C_{13}H_{21}NaO_4PW$ M 479.120

Thermally unstable orange soln. in THF.

Cooper, N.J. *et al*, *Organometallics*, 1982, **1**, 215 (*synth*)

$C_{13}H_{37}ClP_4W$ W-00088

Chloromethylidynetetrakis(trimethylphosphine)tungsten, 10CI

[76642-46-1]

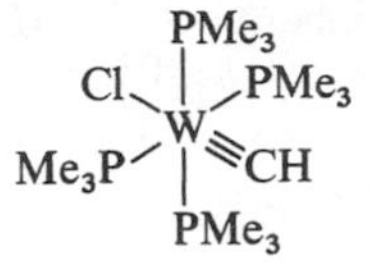

M 536.633

Orange-yellow cryst.

Sharp, P.R. *et al*, *J. Am. Chem. Soc.*, 1981, **103**, 965 (*synth, nmr, props*)

$C_{14}H_{10}Cl_2O_4W_2$ W-00089

Tetracarbonyldichlorobis(η^5-2,4-cyclopentadien-1-yl)ditungsten, 8CI

[51185-44-5]

$[WCl(CO)_2(C_5H_5)]_2$

M 680.837

Obt. by photochemical dec. of $(C_5H_5)W(CO)_3Cl$. Not isol.; identified spectroscopically in soln.

Ali, L.H. *et al*, *J. Chem. Soc., Dalton Trans.*, 1973, 1475 (*synth, props, uv*)

Allen, D.M. *et al*, *J. Chem. Soc., Dalton Trans.*, 1973, 1899.

Allen, D.M. *et al*, *Inorg. Chim. Acta*, 1977, **21**, 191 (*ir*)

$C_{14}H_{10}O_4W_2$ W-00090

Tetracarbonylbis(η^5-2,4-cyclopentadien-1-yl)ditungsten, 10CI

[62853-03-6]

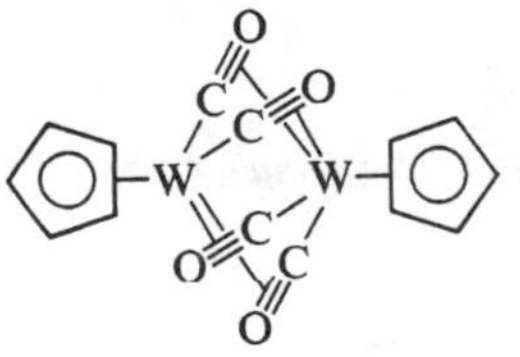

M 609.931

Sol. hydrocarbons, $CHCl_3$. Not obt. pure. ir ν_{CO} 1885, 1830 cm^{-1} (isooctane).

Ginley, D.S. *et al*, *Inorg. Chim. Acta*, 1977, **23**, 85 (*synth, pmr, ir*)

$C_{14}H_{12}O_2W$ W-00091

Dicarbonyl(η^5-2,4-cyclopentadien-1-yl)(η^3-6-methylene-2,4-cyclohexadien-1-yl)tungsten, 10CI

[68349-04-2]

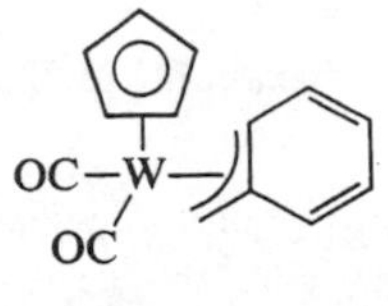

M 396.098

Orange-brown cryst. Mp 103-5°.

Cotton, F.A. *et al*, *J. Am. Chem. Soc.*, 1969, **91**, 1339 (*synth, nmr*)

Severson, R.G. *et al*, *J. Organomet. Chem.*, 1978, **157**, 173.

$C_{14}H_{16}NW^{\oplus}$ W-00092

(Cyanomethyl)bis(η^5-2,4-cyclopentadien-1-yl)(η^2-ethene)tungsten(1+)

[71191-62-3]

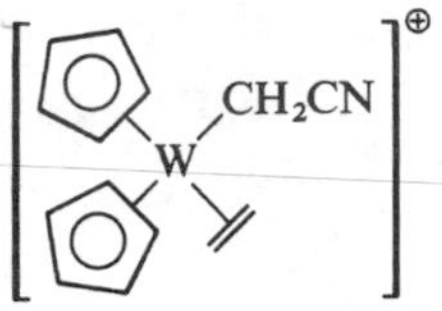

M 382.137 (ion)

Hexafluorophosphate: [71191-63-4].

$C_{14}H_{16}F_6NPW$ M 527.101

Yellow-brown cryst.

Benfield, F.W.S. *et al*, *J. Organomet. Chem.*, 1974, **76**, 49 (*props*)

Green, M.L.H. *et al*, *J. Chem. Soc., Dalton Trans.*, 1979, 262 (*synth*)

$C_{14}H_{16}O_2W$ W-00093

Dicarbonylbis[(1,2,3,4-η)-1,3-cyclohexadiene]tungsten, 9CI

[12131-26-9]

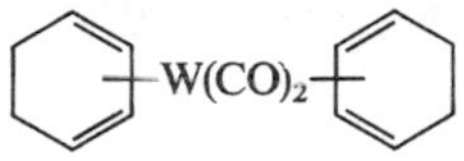

M 400.129

Red cryst.

King, R.B. *et al*, *Inorg. Chem.*, 1966, **5**, 1837 (*synth*)

King, R.B. *et al*, *Inorg. Chem.*, 1974, **13**, 74 (*props*)

$C_{14}H_{16}W$ W-00094

Bis[(1,2,3,4,5,6-η)methylbenzene]tungsten, 10CI

Bis(η^6-toluene)tungsten

[52346-44-8]

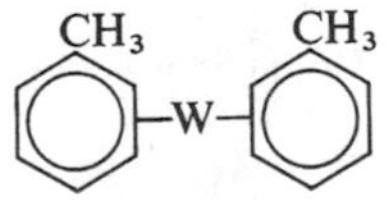

M 368.130

Green cryst.

Cloke, F.G.W. *et al*, *J. Chem. Soc., Chem. Commun.*, 1978, 72 (*synth*)

Green, M.L.H., *J. Organomet. Chem.*, 1980, **200**, 119 (*synth*)

$C_{14}H_{17}OW^{\oplus}$ W-00095

Bis[(2,3-η)-2-butyne]carbonyl(η^5-2,4-cyclopentadien-1-yl)tungsten(1+), 10CI

Carbonyl(η^5-cyclopentadienyl)(dimethylacetylene)tungsten

[74380-93-1]

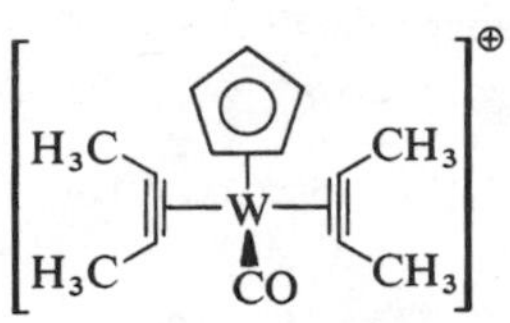

M 385.138 (ion)

Hexafluorophosphate: [74380-94-2].

$C_{14}H_{17}F_6OPW$ M 530.102
Yellow cryst. (CH_2Cl_2/Et_2O).

Beck, W. *et al*, *Z. Naturforsch.*, *B*, 1978, **33**, 1214 (*synth*)
Watson, P.L. *et al*, *J. Am. Chem. Soc.*, 1980, **102**, 2698 (*synth*)

$C_{14}H_{20}N_4O_4W$ W-00096

Tetracarbonylbis(1,3-dimethyl-2-imidazolidinylidene)tungsten, 10CI
[64514-96-1]

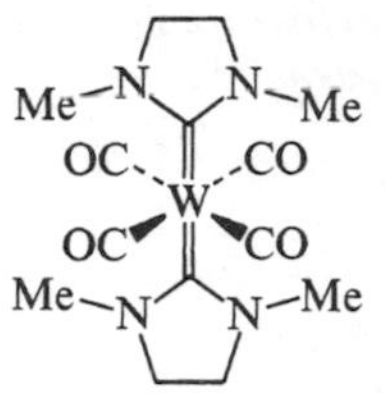

trans-form

M 492.186

cis-form
Yellow cryst. Sol. $CHCl_3$, Me_2CO. Mp 278° dec.

trans-form
Orange-red cryst. Sol. $CHCl_3$. Isom. to *cis*-form at 160-80°.

Lappert, M.F. *et al*, *J. Chem. Soc., Dalton Trans.*, 1977, 1283 (*synth, pmr*)

$C_{14}H_{22}IO_4PW$ W-00097

Carbonyl(η^7-cycloheptatrienylium)iodo(triethyl phosphite)-tungsten, 9CI
Carbonyliodo(triethyl phosphite)(tropylium)tungsten
[36580-33-3]

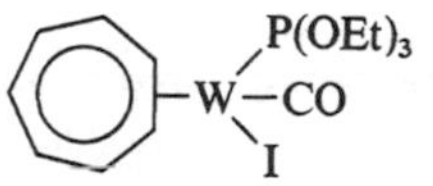

M 596.054
Light-green cryst. Mp 78-80°.

Beall, T.W. *et al*, *Inorg. Chem.*, 1972, **11**, 915 (*synth, ir, nmr*)

$C_{15}H_8N_3O_4W^{\oplus}$ W-00098

Tricarbonylnitrosyl(1,10-phenanthroline-N^1,N^{10})tungsten(1+), 10CI
[63890-73-3]

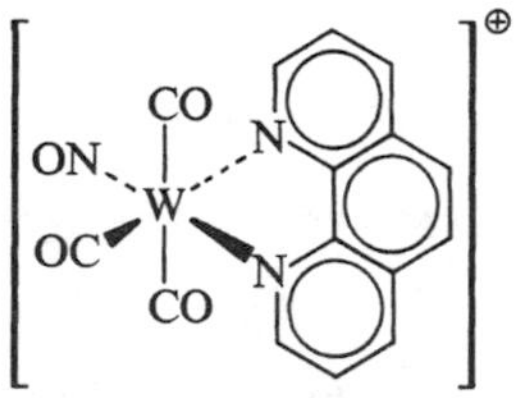

M 478.096 (ion)

mer-form
Hexafluorophosphate: [63890-74-4].
$C_{15}H_8F_6N_3PO_4W$ M 623.060
Air-stable orange solid. Sol. Me_2CO, CH_2Cl_2.

Connelly, N.G. *et al*, *J. Chem. Soc., Dalton Trans.*, 1977, 925 (*synth, props*)

$C_{15}H_9BrFeO_4W$ W-00099

Bromotetracarbonyl(ferroceniumylmethylidyne)tungsten, 9CI
Bromotetracarbonyl(ferrocenylmethylidyne)tungsten
[68830-04-6]

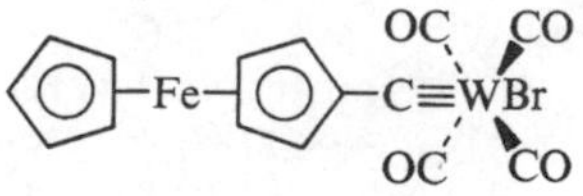

M 572.835

trans-form [65230-41-3]
Raspberry-red cryst. Stable at r.t. in the solid state for several hours even on exp. to air. In soln. dec. slowly >−10°. Spar. sol. hexane, sol. CH_2Cl_2, Me_2CO, other polar solvs. Mp >70° dec. Diamagnetic.

Fischer, E.O. *et al*, *Chem. Ber.*, 1977, **110**, 3397; 1978, **111**, 3530 (*synth, ir, pmr*)

$C_{15}H_9BrO_4RuW$ W-00100

Bromotetracarbonyl(ruthenocenylmethylidyne)tungsten, 10CI

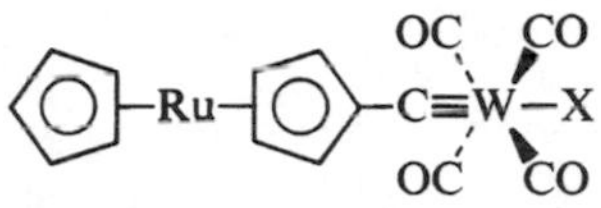

X = Br

M 618.058

trans-form [75293-20-8]
Bright-yellow powder. Mp >−5° dec.

Fischer, E.O. *et al*, *J. Organomet. Chem.*, 1980, **191**, 261 (*synth, ir, ms, cmr, pmr*)

$C_{15}H_9ClO_4RuW$ W-00101

Tetracarbonylchloro(ruthenocenomethylidyne)tungsten, 10CI
As Bromotetracarbonyl(ruthenocenylmethylidyne)tungsten, W-00100 with

X = Cl

M 573.607

trans-form
Bright-yellow.

Fischer, E.O. *et al*, *J. Organomet. Chem.*, 1980, **191**, 261.

$C_{15}H_{10}FeO_5W$ W-00102

Tricarbonyl(η^5-2,4-cyclopentdien-1-yl)[dicarbonyl(η^5-2,4-cyclopentadien-1-yl)iron]tungsten, 10CI
Pentacarbonylbis(η^5-2,4-cyclopentadien-1-yl)irontungsten
[64041-01-6]

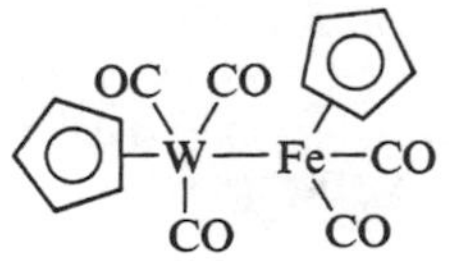

M 509.938
Red cryst. (CH_2Cl_2/hexane). Mp 220° dec. (softens ca. 150°).

Pandey, V.N., *Inorg. Chim. Acta*, 1977, **23**, L26 (*synth, ir*)
Abrahamson, H.B. *et al*, *Inorg. Chem.*, 1978, **17**, 1003 (*synth, ir, pmr, uv*)
Madach, T. *et al*, *Chem. Ber.*, 1980, **113**, 2675 (*ir, pmr*)

$C_{15}H_{13}ClN_2O_2W$ W-00103

(2,2′-Bipyridine-*N,N′*)dicarbonylchloro(η^3-2-propenyl)tungsten

π-Allyl(2,2′-bipyridine)dicarbonylchlorotungsten, 8CI

[12245-58-8]

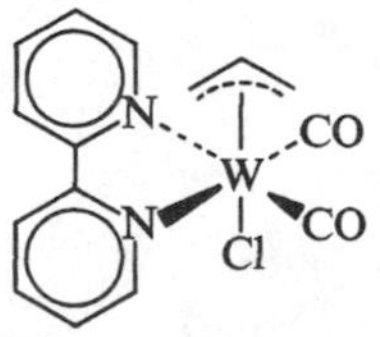

M 472.583

Air-stable maroon cryst. Sl. sol. polar solvs., insol. nonpolar solvs.

Hull, C.G. *et al*, *J. Organomet. Chem.*, 1967, **9**, 519 (*synth, pmr, ir*)

Brisdon, B.J. *et al*, *J. Chem. Soc., Dalton Trans.*, 1975, 1999 (*synth*)

$C_{15}H_{17}BN_6O_2W$ W-00104

Dicarbonyl[hydrotris(pyrazolato)borato(1−)](η^3-2-methylpropenyl)tungsten

Dicarbonyl[hydrotris(pyrazolato)borato(1−)](2-methyl-π-allyl)tungsten, 8CI

[12242-63-6]

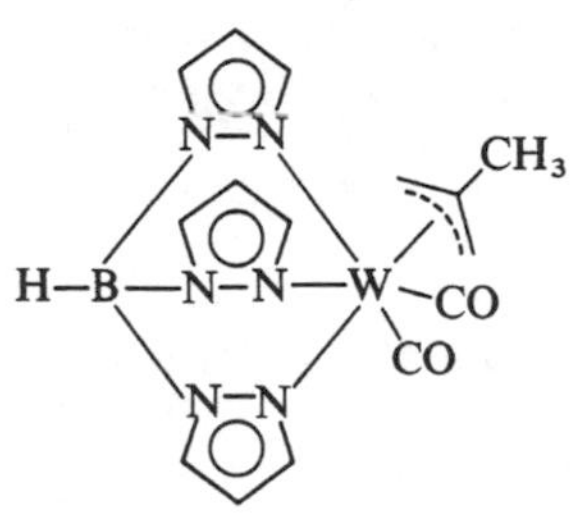

M 507.998

Yellow air-stable cryst.

Trofimenko, S., *J. Am. Chem. Soc.*, 1967, **89**, 3904; 1969, **91**, 558 (*synth, props*)

$C_{15}H_{18}O_3W$ W-00105

Tricarbonyl[(1,2,3,4,5,6-η)-hexamethylbenzene]tungsten, 9CI

[33505-53-2]

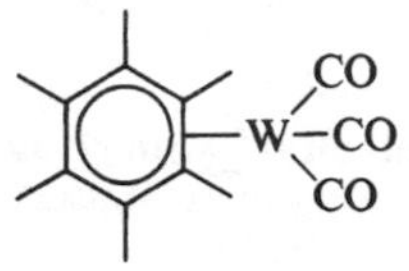

M 430.155

Pale-yellow cryst.

Manuel, T.A. *et al*, *Chem. Ind.* (*London*), 1960, 231

Stiddard, M.H.B. *et al*, *J. Chem. Soc.* (*A*), 1969, 2355 (*synth*)

Muetterties, E.L. *et al*, *J. Organomet. Chem.*, 1979, **178**, 197 (*synth*)

$C_{16}H_{10}F_6O_4S_2W_2$ W-00106

Tetracarbonylbis(η^5-2,4-cyclopentadien-1-yl)bis[μ-(trifluoromethanethiolato)]ditungsten, 8CI

[12771-97-0]

$W_2(\mu\text{-}SCF_3)_2(CO)_4(\eta\text{-}C_5H_5)_2$

M 812.063

Apparently not well characterised.

Havlin, R. *et al*, *Z. Naturforsch., B*, 1966, **21**, 1108 (*synth*)

Davidson, J.L. *et al*, *J. Chem. Soc., Dalton Trans.*, 1972, 107 (*synth*)

Watkins, D.D. *et al*, *J. Organomet. Chem.*, 1975, **102**, 71.

$C_{16}H_{10}O_6W_2$ W-00107

Hexacarbonylbis(η^5-2,4-cyclopentadien-1-yl)ditungsten, 9CI

Hexacarbonylbis(η^5-2,4-cyclopentadien-1-yl)ditungsten, 9CI. *Bis(cyclopentadienyltricarbonyltungsten)*. *Bis(cyclopentadienyl)hexacarbonylditungsten*

[12566-66-4]

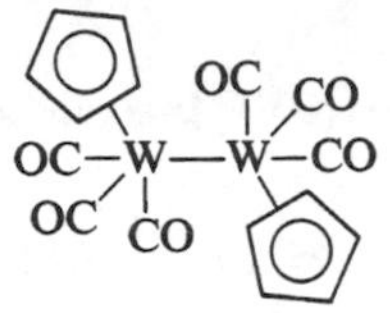

M 665.951

Purple-red cryst. Mp 240-2° dec.

Wilkinson, G., *J. Am. Chem. Soc.*, 1954, **76**, 209 (*synth*)

Davison, A. *et al*, *J. Chem. Soc.*, 1962, 3653 (*ir, pmr*)

Parker, D.J., *J. Chem. Soc.* (*A*), 1970, 1382 (*ir*)

Barnett, R.W. *et al*, *J. Organomet. Chem.*, 1972, **44**, 1 (*rev*)

Adams, R.D. *et al*, *Inorg. Chem.*, 1974, **13**, 1086 (*cryst struct, pmr*)

Wrighton, M.S. *et al*, *Pure Appl. Chem.*, 1975, **41**, 671 (*uv*)

Ginley, D.S. *et al*, *Inorg. Chim. Acta*, 1977, **23**, 85 (*synth, props*)

Birdwhistell, R. *et al*, *J. Organomet. Chem.*, 1978, **157**, 239 (*synth*)

$C_{16}H_{16}W$ W-00108

Bis(η^5-2,4-cyclopentadien-1-yl)hydrophenyltungsten, 10CI

Bis(η^5-cyclopentadienyl)hydrido(phenyl)tungsten

[11077-71-7]

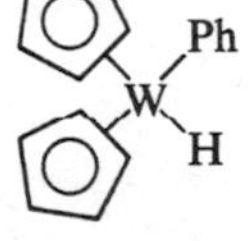

M 392.152

Orange needles.

Green, M.L.H. *Pure Appl. Chem.*, 1978, **50**, 27 (*synth*)

Berry, M. *et al*, *J. Chem. Soc., Dalton Trans.*, 1979, 1950 (*synth*)

Cooper, N.J. *et al*, *J. Chem. Soc., Dalton Trans.*, 1979, 1557 (*synth*)

$C_{16}H_{23}NO_2W$ W-00109

Dicarbonyl(η^5-2,4-cyclopentadien-1-yl)(2,2,4,4-tetramethyl-3-pentaniminato)tungsten, 9CI

[32610-90-5]

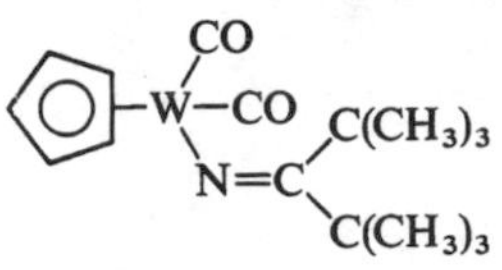

M 445.213

Deep-blue cryst. Mp 120-1° dec. $Bp_{0.001}$ 100° subl.

Kilner, M. *et al*, *J. Chem. Soc.* (*A*), 1971, 292 (*synth, ir, ms*)

Inglis, T. *et al*, *J. Chem. Soc., Dalton Trans.*, 1976, 562.

$C_{16}H_{36}O_5W$ W-00110

Tetrakis(2-methyl-2-propanolato)oxotungsten, 10CI

Tetra-tert-*butoxyoxotungsten*

[58832-09-0]

$[(H_3C)_3CO]_4WO$

M 492.307

Wengrovius, J.H. *et al*, *Organometallics*, 1982, **1**, 148 (*synth, props*)

$C_{18}H_{10}O_5W$ W-00111

Pentacarbonyl(diphenylmethylene)tungsten, 9CI

[50276-12-5]

$(OC)_5W{=}CPh_2$

M 490.124

Black cryst. (pentane). Mp 65-6°.

Casey, C.P. *et al*, *J. Am. Chem. Soc.*, 1973, **95**, 5833; 1974, **96**, 7808; 1977, **99**, 2127, 2533 (*synth, props, cryst struct*)

Fischer, E.O. *et al*, *Chem. Ber.*, 1977, **110**, 3842 (*props*)

$C_{18}H_{14}FeO_2W$ W-00112

Dicarbonyl(η^5-2,4-cyclopentadien-1-yl)(ferrocenylmethylidyne)tungsten, 10CI

[65230-44-6]

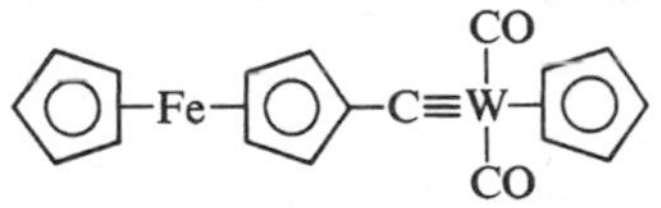

M 502.004

Bright-red cryst. (pentane). Spar. sol. nonpolar solvs. Mp 112°.

Fischer, E.O. *et al*, *Chem. Ber.*, 1977, **110**, 3397; 1978, **111**, 3530 (*synth, ir, pmr, cmr, ms*)

$C_{18}H_{14}FeO_3W$ W-00113

Tricarbonyl(η^5-2,4-cyclopentadien-1-yl)ferrocenyltungsten, 9CI

π-Cyclopentadienyl-σ-ferrocenyltungsten tricarbonyl

[52472-04-5]

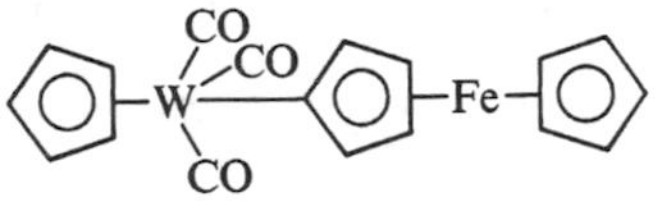

M 518.004

Bright-orange air stable cryst. (CH_2Cl_2/heptane). Mp 173-5° (170-3°).

Nesmeyanov, A.N. *et al*, *Izv. Akad. Nauk SSSR, Ser. Khim.*, 1973, 2796 (*synth, ir, pmr*)

Pannell, K.H. *et al*, *Inorg. Chem.*, 1976, **15**, 2671 (*synth, ir*)

$C_{18}H_{14}FeO_6W$ W-00114

Pentacarbonyl(ethoxyferrocenylmethylene)tungsten, 9CI

[Ethoxyferrocenylcarbene]pentacarbonyltungsten

[34691-66-2]

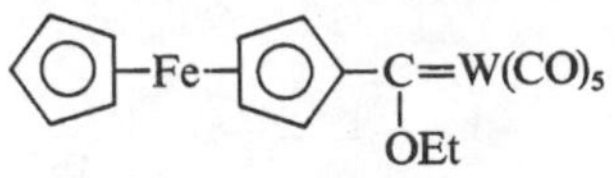

M 566.002

Dark-brown needles (Et_2O/hexane). Mp 155-6°.

Connor, J.A. *et al*, *J. Chem. Soc., Dalton Trans.*, 1972, 1470 (*synth, pmr, ir, uv, ms*)

$C_{18}H_{14}O_3W$ W-00115

Tricarbonyl[(1,2,3,4,5,6-η)-7-(1-phenylethylidene)-1,3,5-cycloheptatriene]tungsten, 9CI

Tricarbonyl(8-phenyl-8-methylheptafulvene)tungsten

[52471-95-1]

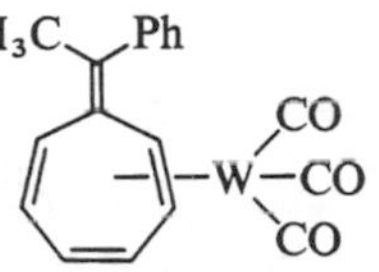

M 462.157

Not isol.; observed spectroscopically *in situ*.

Howell, J.A.S. *et al*, *J. Chem. Soc., Dalton Trans.*, 1974, 293 (*synth*)

$C_{18}H_{14}O_6RuW$ W-00116

Pentacarbonyl(ethoxyruthenocenylmethylene)tungsten, 10CI

[75280-67-0]

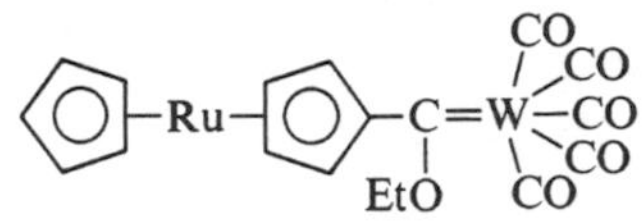

M 611.225

Orange-red cryst. (CH_2Cl_2/hexane). Mp 142°.

Fischer, E.O. *et al*, *J. Organomet. Chem.*, 1980, **191**, 261 (*synth, ir, ms, pmr, cmr, struct*)

$C_{18}H_{15}ErO_3W$ W-00117

[μ-(Carbonyl-*C,O*)]bis(η^5-2,4-cyclopentadien-1-yl)[dicarbonyl(η^5-2,4-cyclopentadien-1-yl)tungsten]erbium

[39477-06-0]

$(C_5H_5)_2ErW(C_5H_5)(CO)_3$

M 630.425

Nitrosylated by *N*-methyl-*N*-nitroso-*p*-toluenesulfonamide to afford $(C_5H_5)W(CO)_2NO$. Sol. THF, DMSO. Air- and moisture-sensitive. Dec. >220°.

Crease, A.E. *et al*, *J. Chem. Soc., Dalton Trans.*, 1973, 1501 (*synth, ir*)

$C_{18}H_{15}O_3WYb$ W-00118

[μ-(Carbonyl-*C,O*)]bis(η^5-2,4-cyclopentadien-1-yl)[dicarbonyl(η^5-2,4-cyclopentadien-1-yl)tungsten]ytterbium

[39477-07-1]

$(C_5H_5)_2YbW(C_5H_5)(CO)_3$

M 636.205
Nitrosylated by *N*-methyl-*N*-nitroso-*p*-toluenesulfonamide to afford $(C_5H_5)W(CO)_2NO$. Golden cryst. Sol. THF, DMSO. Mp >220° dec. Air- and moisture-sensitive.

Crease, A.E. *et al*, *J. Chem. Soc., Dalton Trans.*, 1973, 1501 (*synth, ir*)

$C_{18}H_{21}O_2PW$ W-00119

Carbonyl(η^5-2,4-cyclopentadien-1-yl)[(1,2-η)(4-methylphenyl)oxoethenyl](trimethylphosphine)tungsten, 10CI
[61202-59-3]

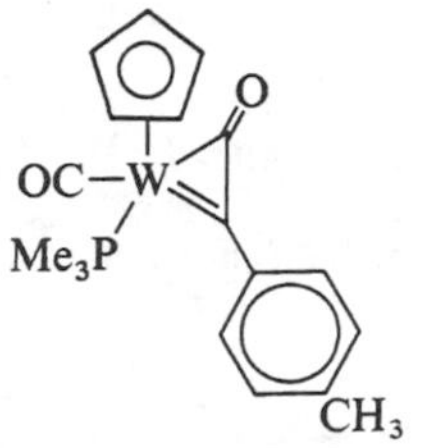

M 484.186
η^2-Ketenyl complex. Brick-red cryst.

Kreissl, F.R. *et al*, *Chem. Ber.*, 1977, **110**, 3782; 1979, **112**, 3376 (*synth, ir, nmr*)

$C_{18}H_{46}N_4W_2$ W-00120

Dimethyltetrakis(*N*-dimethylamino)ditungsten
*Bis(methyl)tetrakis(*N*-methylmethanaminato)ditungsten, 10CI*
[72286-69-2]

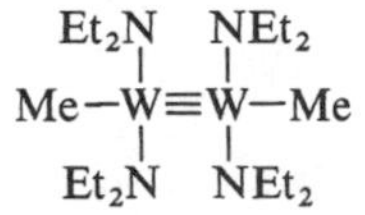

M 686.288
Red cryst.

Chisholm, M.H. *et al*, *J. Am. Chem. Soc.*, 1976, **98**, 6393 (*synth*)
Chisholm, M.H. *et al*, *Inorg. Chem.*, 1977, **16**, 320, 603 (*ms, cryst struct, props*)
Chisholm, M.H. *et al*, *J. Am. Chem. Soc.*, 1979, **101**, 6784.

$C_{19}H_{14}FeO_4W$ W-00121

Tricarbonyl(η^5-2,4-cyclopentadien-1-yl)(ferrocenylcarbonyl)-tungsten, 9CI
π-Cyclopentadienyl-σ-ferrocenoyltungsten tricarbonyl
[52680-31-6]

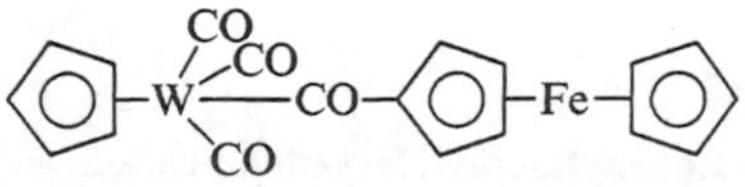

M 546.014
Bright-orange air-stable cryst. (heptane or CH_2Cl_2/hexane). Mp 147° dec. Decarbonylates at 130° to Tricarbonyl(η^5-2,4-cyclopentadien-1-yl)ferrocenyltungsten, W-00113 .

Nesmeyanov, A.N. *et al*, *Izv. Akad. Nauk SSSR, Ser. Khim.*, 1973, 2796 (*synth, ir, pmr*)
Pannell, K.H. *et al*, *Inorg. Chem.*, 1976, **15**, 2671 (*synth, ms, ir, pmr*)

$C_{19}H_{15}N_3O_2W$ W-00122

Dicarbonyl(η^5-2,4-cyclopentadien-1-yl)(1,3-diphenyltriazenato-N^1,N^3)tungsten, 10CI
*Dicarbonyl(1,3-diphenyl-1-triazenato-*N^1,N^3*)(η^5-cyclopentadienyl)tungsten*
[63641-12-3]

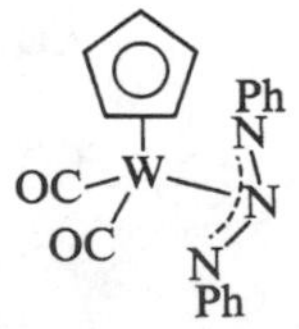

M 501.196
Red-orange cryst. (CH_2Cl_2/hexane). Mp 107-9° dec.

Pfeiffer, E. *et al*, *J. Organomet. Chem.*, 1976, **105**, 371 (*synth*)
King, R.B. *et al*, *Inorg. Chem.*, 1977, **16**, 1164, 2648 (*synth, props*)

$C_{19}H_{24}O_2PW^{\oplus}$ W-00123

Dicarbonyl(dimethylphenylphosphine)hydro[(1,2,3,4,5,6-η)-1,3,5-trimethylbenzene]tunsten(1+), 10CI
Dicarbonyl(dimethylphenylphosphine)hydrido(mesitylene)tungsten(1+)
[64426-76-2]

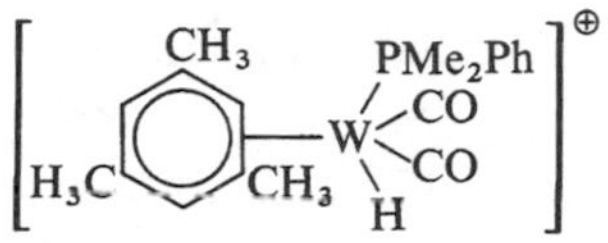

M 499.221 (ion)
Trifluoromethanesulfonate: [64426-77-3].
$C_{20}H_{24}F_3O_5PSW$ M 648.286
Observed in soln. by nmr.

Flood, T.C. *et al*, *J. Am. Chem. Soc.*, 1977, **99**, 4334 (*synth, nmr*)

$C_{19}H_{30}OW$ W-00124

Carbonyltris(3-hexyne)tungsten, 8CI
[12131-66-7]

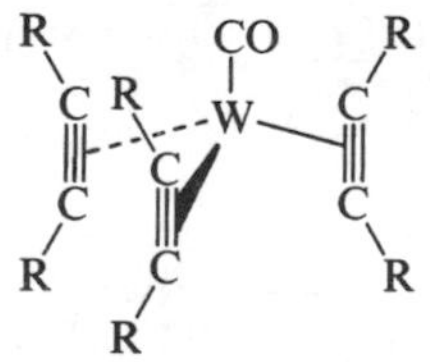

R = CH_2CH_3

M 458.295
Pale-yellow cryst. Mp 55-6°.

Tate, D.P. *et al*, *J. Am. Chem. Soc.*, 1963, **85**, 2174; 1964, **86**, 3261 (*synth, ir, nmr*)
King, R.B., *Inorg. Chem.*, 1968, **7**, 1044 (*struct*)

$C_{20}H_{15}NO_2W$ W-00125

Dicarbonyl(η^5-2,4-cyclopentadien-1-yl)(α-phenylbenzenemethanimінato)tungsten, 9CI

[31833-59-7]

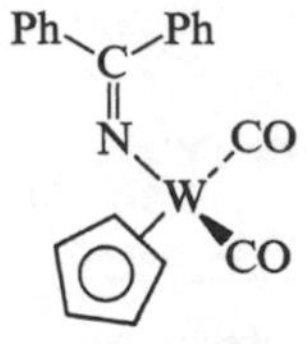

M 485.194

Green-black dichroic cryst. (toluene/hexane). Mp 110-2°.

Farmery, K. *et al*, *J. Chem. Soc. (A)*, 1970, 2279 (*synth, ir, ms*)
Keable, H.R. *et al*, *J. Chem. Soc., Dalton Trans.*, 1972, 153; 1974, 639 (*synth, props*)

$C_{20}H_{15}O_3TlW$ W-00126

Tricarbonyl(η^5-2,4-cyclopentadien-1-yl)(diphenylthallium)-tungsten

[66125-42-6]

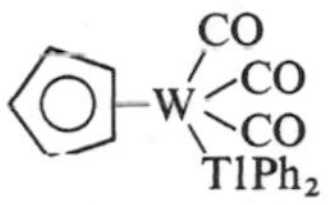

M 691.570

Mp 38-40°.

▷Toxic

Walther, B. *et al*, *J. Organomet. Chem.*, 1978, **145**, 285 (*synth, ir*)

$C_{20}H_{16}O_8W_2$ W-00127

Tetracarbonylbis(η^5-2,4-cyclopentadien-1-yl)[μ-[(2,3-η:2,3-η)-dimethyl-2-butynedioate]]ditungsten, 10CI

[62303-78-0]

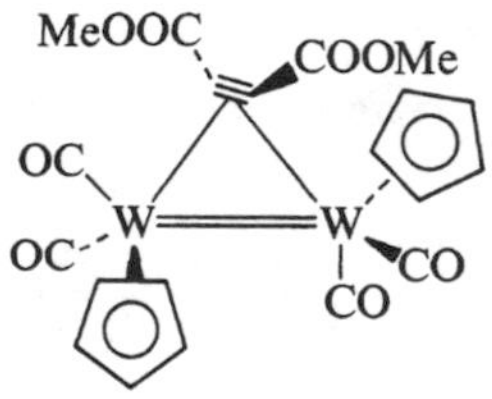

M 752.042

Air-stable red solid. Mp 188-90° dec. (under N_2). Struct. by analogy with unsubst. acetylene analogue.

Maine, R.M. *et al*, *J. Organomet. Chem.*, 1977, **124**, 29 (*synth, ir, cmr, pmr, uv, ms*)
Finnimore, S.R. *et al*, *J. Chem. Soc., Chem. Commun.*, 1980, 411 (*synth*)

$C_{20}H_{19}DyO_3W$ W-00128

[μ-(Carbonyl-*C,O*)]bis[(1,2,3,4,5-η)-1-methyl-2,4-cyclopentadien-1-yl][dicarbonyl(η^5-2,4-cyclopentadien-1-yl)tungsten]dysprosium

$(H_3CC_5H_4)_2DyW(C_5H_5)(CO)_3$

M 653.718

Nitrosylated by *N*-methyl-*N*-nitroso-*p*-toluenesulfonamide to give $(C_5H_5)W(CO)_2NO$. Sol. THF, DMSO. Mp >220° dec. Air- and moisture-sensitive.

Crease, A.E. *et al*, *J. Chem. Soc., Dalton Trans.*, 1973, 1501 (*synth, ir*)

$C_{20}H_{19}HoO_3W$ W-00129

[μ-(Carbonyl-*C,O*)]bis[(1,2,3,4,5-η)-1-methyl-2,4-cyclopentadien-1-yl][dicarbonyl(η^5-2,4-cyclopentadien-1-yl)tungsten]holmium

[39477-08-2]

$(H_3CC_5H_4)_2HoW(C_5H_5)(CO)_3$

M 656.149

Nitrosylated by *N*-methyl-*N*-nitroso-*p*-toluenesulfonamide to give $(C_5H_5)_2W(CO)_2NO$. Sol. THF, DMSO. Mp >220° dec. Air- and moisture-sensitive.

Crease, A.E. *et al*, *J. Chem. Soc., Dalton Trans.*, 1973, 1501 (*synth, ir*)

$C_{20}H_{23}W_2^{\oplus}$ W-00130

Tetrakis(η^5-2,4-cyclopentadien-1-yl)-μ-hydrodihydroditungsten(1+), 10CI

Tetrakis(η^5-cyclopentadienyl)-μ-hydridodihydridoditungsten(1+)

[72932-64-0]

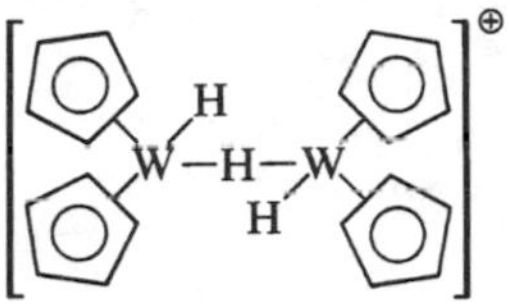

M 631.102 (ion)

Perchlorate: [72932-65-1].
$C_{20}H_{23}ClO_4W_2$ M 730.552
Dark-purple cryst.

Klinger, R.J. *et al*, *J. Am. Chem. Soc.*, 1980, **102**, 208 (*synth, cryst struct*)

$C_{20}H_{48}N_4W_2$ W-00131

Bis[μ-[2-methyl-2-propanaminato(2−)]]bis[2-methyl-2-propanaminato(2−)]tetramethylditungsten, 10CI

Tetrakis(tert-butylaminato)tetramethylditungsten

[73448-07-4]

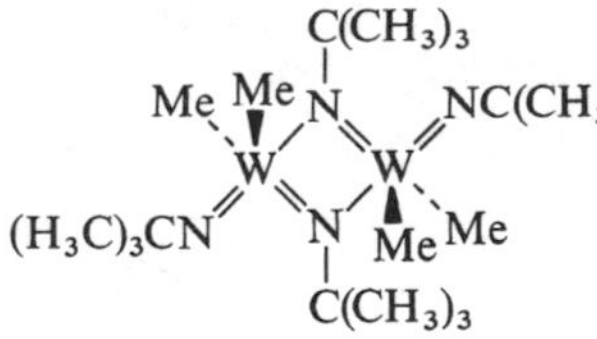

M 712.326

Air-sensitive, red-orange cryst. Sol. hexane.

Nugent, W.A. *et al*, *J. Am. Chem. Soc.*, 1980, **102**, 1759 (*synth*)
Thorn, D.L. *et al*, *J. Am. Chem. Soc.*, 1981, **103**, 357 (*struct*)

$C_{21}H_{14}O_3W$ W-00132

Tricarbonyl(diphenylmethylidenecyclopentadiene)tungsten

Tricarbonyl[(η^6-2,4-cyclopentadien-1-ylidenephenylmethyl)benzene]tungsten, 10CI. Tricarbonyldiphenylfulvenetungsten

[63882-22-4]

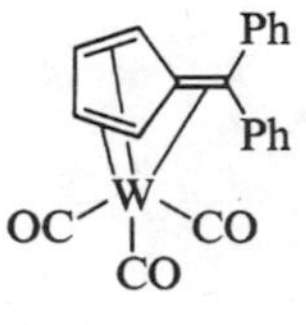

M 498.190

Violet needles. Mp 215°.

Edelmann, F. *et al*, *J. Organomet. Chem.*, 1977, **134**, 31 (*synth*)

$C_{21}H_{15}ClNO_4PW$ W-00133

Tricarbonylchloronitrosyl(triphenylphosphine)tungsten, 9CI

[42536-19-6]

$(OC)_3WCl(NO)PPh_3$

M 595.631

Meridonal, but complete stereochemistry not known. Yellow solid. Sol. common org. solvs.

Colton, R. *et al*, *Aust. J. Chem.*, 1973, **26**, 1087 (*synth, ir, pmr*)

$C_{21}H_{24}O_3W$ W-00134

Tricarbonyl(η^6-octamethylnaphthalene)tungsten

[79391-58-5]

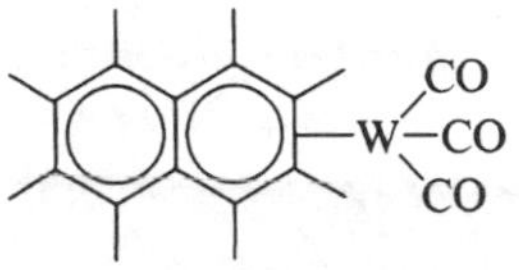

M 508.269

Bright-orange cryst. (CH_2Cl_2/hexane).

Hull, J.W. *et al*, *Organometallics*, 1982, **1**, 264 (*synth, struct*)

$C_{21}H_{28}O_2P_2W$ W-00135

Carbonyl(η^5-2,4-cyclopentadien-1-yl)[1,2-ethanediylbis(dimethylphosphine)]-*P*,*P'*][(4-methylphenyl)oxoethenyl]tungsten, 10CI

[75899-98-8]

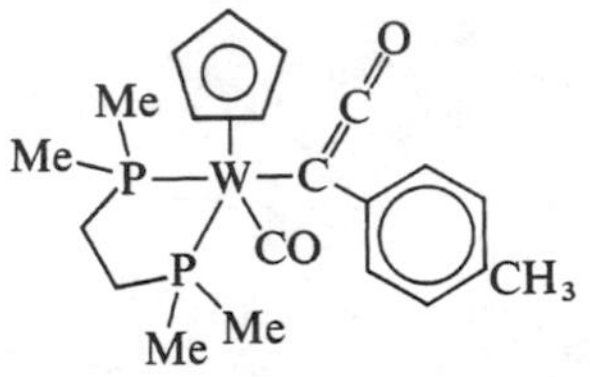

M 558.249

η'-Ketenyl complex. Yellow cryst.

Kreissl, F.R. *et al*, *Chem. Ber.*, 1980, **113**, 3377 (*synth*)

$C_{22}H_{10}O_{10}PW^{\ominus}$ W-00136

Decacarbonyl(μ-diphenylphosphido)ditungstate(1−)

[80049-78-1]

$[(OC)_5W{-}PPh_2{-}W(CO)_5]^{\ominus}$

M 649.139 (ion)

K salt:

$C_{22}H_{10}KO_{10}PW$ M 688.237

No phys. props. reported.

Keiter, R.L., *Organometallics*, 1982, **1**, 409 (*synth*)

$C_{23}H_{15}BrO_4SiW$ W-00137

Bromotetracarbonyl[(triphenylsilyl)methylidyne]tungsten, 10CI

Bromotetracarbonyltriphenylsilylcarbynetungsten

[62319-82-8]

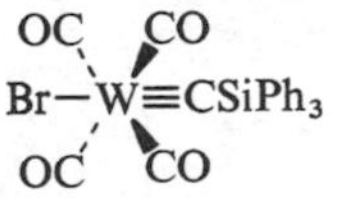

M 647.209

Ivory cryst. (CH_2Cl_2/pentane at −78°). Dec. >ca. 65°.

Fischer, E.O. *et al*, *Angew. Chem., Int. Ed. Engl.*, 1977, **16**, 401 (*synth, ir, nmr*)

$C_{24}H_{14}O_6W_2$ W-00138

Hexacarbonylbis[(1,2,3,3*a*,7*a*-η)-1*H*-inden-1-yl]ditungsten, 10CI

Hexacarbonylbis(η^5-indenyl)tungsten

[68586-58-3]

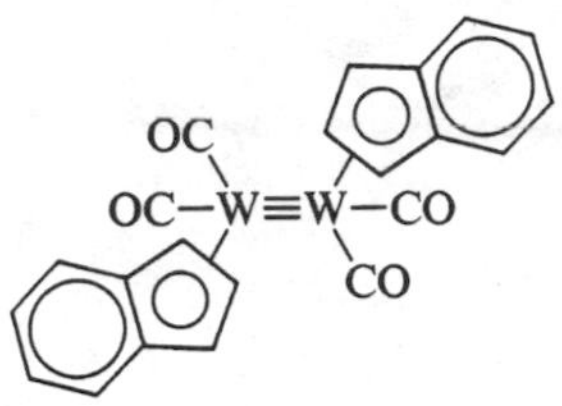

M 766.071

Cryst. (CH_2Cl_2/heptane). Mp 204-5° dec.

Birdwhistell, R. *et al*, *J. Organomet. Chem.*, 1978, **157**, 239 (*synth, props*)

Nesmeyanov, A.N. *et al*, *J. Organomet. Chem.*, 1978, **159**, 189 (*synth, props*)

$C_{24}H_{17}PO_4W$ W-00139

Tetracarbonyl[2-(η^2-ethenyl)phenyldiphenylphosphine-*P*]tungsten, 9CI

[12290-83-4]

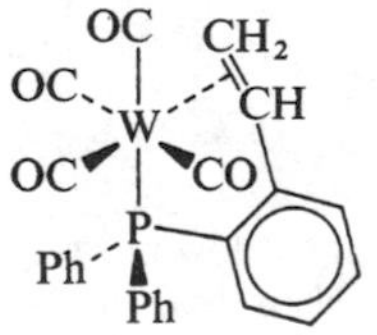

M 584.220

Yellow cryst. Mp 160-75° dec.

Bennett, M.A. *et al*, *J. Organomet. Chem.*, 1967, **10**, 301; 1973, **51**, 289 (*synth*)

$C_{24}H_{20}NO_2PW$ W-00140

Carbonyl(η^5-2,4-cyclopentadien-1-yl)nitrosyl(triphenylphosphine)tungsten, 8CI

[33113-68-7]

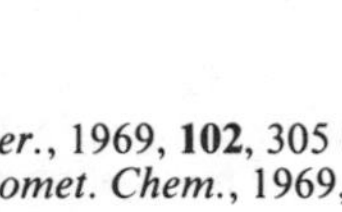

M 569.251

Solid. Mp 222-4°.

Brunner, H., *Chem. Ber.*, 1969, **102**, 305 (*synth, props*)

Brunner, H., *J. Organomet. Chem.*, 1969, **16**, 119 (*synth, nmr, ir*)

$C_{24}H_{24}W$ W-00141

Dibenzylbis(η^5-2,4-cyclopentadien-1-yl)tungsten

Bis(η^5-2,4-cyclopentadien-1-yl)bis(phenylmethyl)tungsten. Bis(benzyl)bis(η^5-cyclopentadienyl)tungsten

M 496.304

Orange needles (C_6H_6/pet. ether). Mp 200° dec.

Francis, B.R. *et al*, *J. Chem. Soc., Dalton Trans.*, 1976, 1339 (*synth, pmr, ms*)

$C_{24}H_{24}W_2$ W-00142

[μ-[(1,2,3,4-η:5,6,7,8-η)-1,3,5,7-cyclooctatetraene-]]bis[(1,2,3,4-η)-1,3,5,7-cyclooctatetraene]ditungsten, 10CI

[66719-23-1]

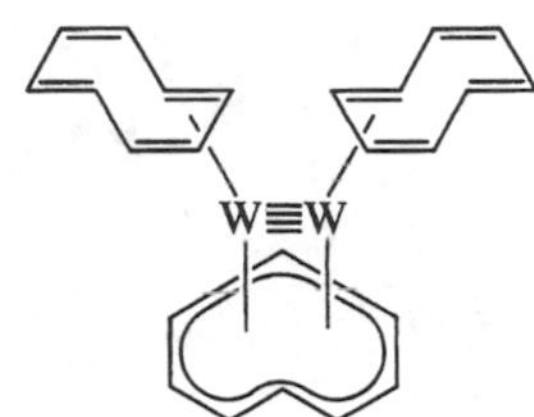

M 680.154

Green-black cryst. (toluene). Mp >360°.

Cotton, F.A. *et al*, *Inorg. Chem.*, 1978, **17**, 2093 (*synth, cryst struct*)

$C_{24}H_{30}O_4W_2$ W-00143

Tetracarbonylbis[(1,2,3,4,5-η)-1,2,3,4,5-pentamethyl-2,4-cyclopentadien-1-yl]ditungsten, 10CI

[70634-77-4]

M 750.199

Deep-red cryst. (Et_2O at low temp.). Mp 237° dec. Adds 2 mol. CO reversibly across the W-W triple bond.

King, R.B. *et al*, *J. Organomet. Chem.*, 1979, **171**, 53 (*synth, nmr, ir, props*)

$C_{24}H_{62}Si_6W_2$ W-00144

Bis[μ-[(trimethylsilyl)methylidyne]tetrakis[(trimethylsilyl)-methyl]ditungsten, 10CI

[59654-41-0]

M 886.967

Brown-red polycrystalline, glassy solid. $Bp_{0.001}$ 140-50° subl.

Andersen, R.A. *et al*, *Angew. Chem., Int. Ed. Engl.*, 1976, **15**, 609 (*synth, nmr*)

Chisholm, M.H. *et al*, *Inorg. Chem.*, 1976, **15**, 2250; 1978, **17**, 696 (*cryst struct*)

$C_{24}H_{66}Si_6W_2$ W-00145

Hexakis[(trimethylsilyl)methyl]ditungsten, 9CI

[36643-37-5]

$(Me_3SiCH_2)_3W{\equiv}W(CH_2SiMe_3)_3$

M 890.998

D_{3d} ethane-like struct. Yellow plates. Mp 99°.

Huq, F. *et al*, *J. Chem. Soc., Chem. Commun.*, 1971, 1079 (*cryst struct*)

Mowat, W. *et al*, *J. Chem. Soc., Dalton Trans.*, 1972, 533 (*synth*)

Chisholm, M.H. *et al*, *J. Am. Chem. Soc.*, 1976, **98**, 4683 (*struct*)

Chisholm, M.H. *et al*, *Inorg. Chem.*, 1976, **15**, 2252 (*struct*)

$C_{26}H_{20}O_6SiW$ W-00146

Pentacarbonyl[ethoxy(triphenylsilyl)methylene]tungsten, 10CI

[65573-55-9]

$(OC)_5W{=}C(OEt)SiPh_3$

M 640.376

Dark-red cryst. (pentane). Mp 93°.

Fischer, E.O. *et al*, *Chem. Ber.*, 1977, **110**, 3467 (*synth, cryst struct*)

$C_{27}H_{26}O_2PTlW$ W-00147

Dicarbonyl(η^5-2,4-cyclopentadien-1-yl)(dimethylthallium)-(triphenylphosphine)tungsten

[66125-49-3]

M 801.708

Mp 160-2°.

▷Toxic

Walther, B. *et al*, *J. Organomet. Chem.*, 1978, **145**, 285.

$C_{28}H_{26}GeW$ W-00148

Bis(η^5-2,4-cyclopentadien-1-yl)hydro(triphenylgermyl)tungsten, 10CI

Bis(η^5-cyclopentadienyl)hydrido(triphenylgermyl)tungsten

[70675-84-2]

M 618.953

Francis, B.R. *et al*, *J. Chem. Soc., Dalton Trans.*, 1976, 1339 (*synth, props*)

Perevozchikova, N.V. *et al*, *Zh. Obshch. Khim.*, 1979, **49**, 945; *CA*, **91**, 39621 (*synth*)

$C_{28}H_{28}W$ W-00149

Tetrabenzyltungsten

Tetrakis(phenylmethyl)tungsten, 9CI. *Tetrakis(benzyl)tungsten. Tungsten tetrabenzyl*

[53235-29-3]

$W(CH_2Ph)_4$

M 548.379
Solid, stable at r.t.

Thiele, K.H. *et al*, *Z. Anorg. Allg. Chem.*, 1975, **412**, 11 (*synth, ir, nmr*)

$C_{29}H_{26}FeNO_4PW$ W-00150

Tetracarbonyl[1-[(dimethylamino)methyl]-2-(diphenylphosphino)ferrocene-*N,P*]tungsten, 9CI

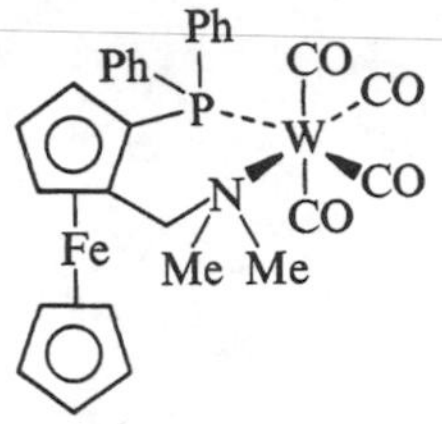

M 723.199
Cryst. (Me_2CO/pet. ether). Mp 203-6° dec.

Kotz, J.C. *et al*, *J. Organomet. Chem.*, 1975, **84**, 255 (*synth, ir, pmr*)

$C_{30}H_{24}O_3P_2SW$ W-00151

(Carbonothioyl)tricarbonyl[1,2-ethanediylbis[diphenylphosphine]-*P,P'*]tungsten, 10CI
[1,2-Bis(diphenylphosphino)ethane](carbonothioyl)tricarbonyltungsten
[54204-88-5]

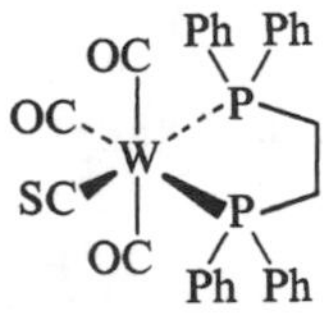

M 710.375
Yellow cryst. (CH_2Cl_2/hexane).

Angelici, R.J. *et al*, *Inorg. Chem.*, 1976, **15**, 1089 (*synth, ir, nmr*)
Jaouen, G. *et al*, *J. Organomet. Chem.*, 1974, **72**, 377 (*synth*)

$C_{31}H_{36}O_4PSnW^{\ominus}$ W-00152

Tetracarbonyl(triphenylstannane)(triisopropylphosphine)-tungstate(1–)

$[(H_3C)_2CH]_3PW(CO)_4(SnPh_3)^{\ominus}$

M 806.137 (ion)
Stereochemistry not reported.
Li salt: [79135-09-4].
$C_{31}H_{36}LiO_4PSnW$ M 813.078
Off-white cryst. + 2THF (Et_2O).
Tetraethylammonium salt: [79135-11-8].
$C_{39}H_{56}NO_4PSnW$ M 936.389
Off-white needles + $\frac{1}{2}CH_2Cl_2$ (CH_2Cl_2).

Maher, J.M. *et al*, *Organometallics*, 1982, **1**, 215 (*synth, ir, pmr*)

$C_{31}H_{42}OWZr$ W-00153

[Bis(η^5-2,4-cyclopentadien-1-yl)hydro[μ-[methanolato(3–)-*C:O*]]bis[(1,2,3,4,5-η)-1,2,3,4,5-pentamethyl-2,4-cyclopentadien-1-yl]zirconium, 10CI
Bis(η^5-cyclopentadienyl)hydrido[bis(η-pentamethylcyclopentadienyl)hydridozirconoxymethylene]tungsten
[69289-49-2]

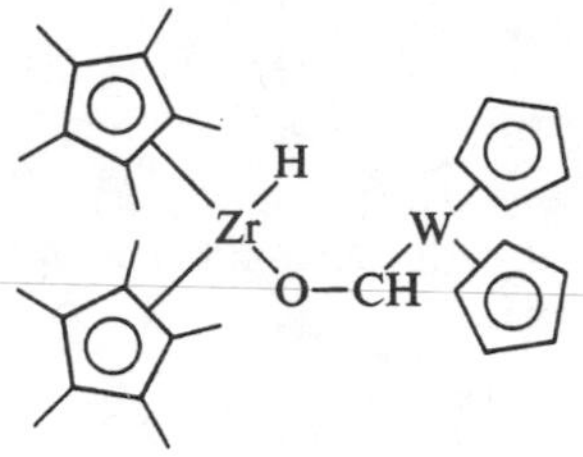

M 705.742
Red-brown cryst. (toluene).

Wolczanski, P.T. *et al*, *J. Am. Chem. Soc.*, 1979, **101**, 218 (*synth, ir, nmr, cryst struct*)

$C_{37}H_{30}O_2PTlW$ W-00154

Dicarbonyl(η^5-2,4-cyclopentadien-1-yl)(diphenylthallium)-(triphenylphosphine)tungsten
[66125-50-6]

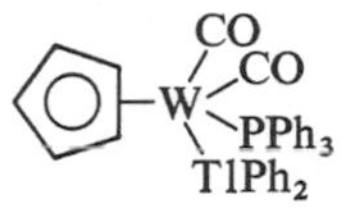

M 925.850
Mp 188-90°.

Walther, B. *et al*, *J. Organomet. Chem.*, 1978, **145**, 285 (*synth, nmr, ir*)

$C_{43}H_{30}OW$ W-00155

Carbonyltris(diphenylacetylene)tungsten
As Carbonyltris(3-hexyne)tungsten, W-00124 with

R = Ph

M 746.559
Yellow prismatic cryst.

Tate, D.P. *et al*, *J. Am. Chem. Soc.*, 1964, **86**, 3261 (*synth*)
Bau, R. *et al*, *J. Am. Chem. Soc.*, 1972, **94**, 1402 (*cryst struct*)

Handle all chemicals with care

$C_{52}H_{48}N_4P_4W$ **W-00156**

Bis(dinitrogen)bis[1,2-bis(diphenylphosphino)ethane]tungsten

Bis(dinitrogen)bis[1,2-ethanediylbis(diphenylphosphine)-P,P′]tungsten, 9CI. *Bis(dinitrogen)bis[ethylenebis(diphenylphosphine)]tungsten*, 8CI

[55954-53-5]

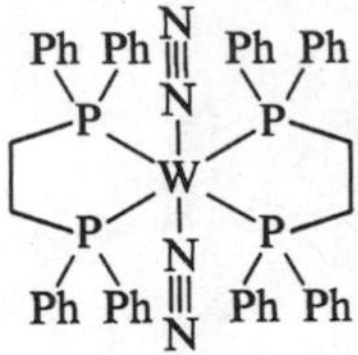

M 1036.723

Orange cryst.

Bell, B. *et al*, *J. Chem. Soc., Chem. Commun.*, 1970, 842 (*synth*, *ir*)

Y Yttrium

S. A. Cotton

Yttrium (Fr., Ger.), Ytrio or Itrio (Sp.), Ittrio (Ital.), Иттрий (Ittrii) (Russ.), イットリウム (Japan.)

Atomic Number. 39

Atomic Weight. 88.9059

Electronic Configuration. [Kr] $4d^1$ $5s^2$

Oxidation State. +3

Coordination Number. Usually 6 or greater, unless bulky ligands are present.

Colour. Generally colourless but may be yellow or brown.

Availability. Usual starting materials are Y_2O_3 and anhydrous YCl_3. See also under Lanthanum.

Handling. See under Lanthanum.

Toxicity. Yttrium is not generally regarded as toxic but little is known about this element.

Isotopic Abundance. ^{89}Y, 100%.

Spectroscopy. ^{89}Y has $I = \frac{1}{2}$, but has not been studied as an nmr probe.

Analysis. See under Lanthanum.

References. See under Lanthanum.

$C_5H_5Cl_2Y$ Y-00001

Dichloro(η^5-2,4-cyclopentadien-1-yl)yttrium, 9CI
Cyclopentadienylyttrium dichloride

$(C_5H_5)YCl_2$

M 224.906

Tris-THF adduct: [52810-84-1].
$C_{17}H_{29}Cl_2O_3Y$ M 441.226
Dec. by air and moisture. Struct. probably resembles Dichloro(η^5-2,4-cyclopentadien-1-yl)erbium.

Jamerson, J.D. *et al*, *J. Organomet. Chem.*, 1974, **65**, C33 (*synth*)

$C_{10}H_{10}ClY$ Y-00002

Chlorobis(η^5-2,4-cyclopentadien-1-yl)yttrium
Chlorodicyclopentadienylyttrium. Dicyclopentadienylyttrium chloride

M 254.548
Dimeric.

Dimer: Dichlorotetrakis(η^5-2,4-cyclopentadien-1-yl)-diyttrium, 9CI. Colourless cryst. Sol. THF. Subl. *in vacuo* at 250°.

Okamato, J. *et al*, *CA*, 1966, **64**, 9762a (*synth*)
Holton, J. *et al*, *J. Chem. Soc., Dalton Trans.*, 1979, 45 (*synth*)
Green, J.C. *et al*, *J. Organomet. Chem.*, 1981, **212**, 329 (*pe*)

$C_{11}H_{13}Y$ Y-00003

Bis(η^5-2,4-cyclopentadien-1-yl)methylyttrium

M 234.130
Dimeric.

Dimer: [60997-40-2]. *Tetrakis(η^5-2,4-cyclopentadien-1-yl)di-μ-methyldiyttrium.*
$C_{22}H_{26}Y_2$ M 468.259
Colourless cryst. Sol. C_6H_6, toluene, CH_2Cl_2. Air-sensitive solid. Dec. >158°.

Holton, J. *et al*, *J. Chem. Soc., Dalton Trans.*, 1979, 54 (*synth, struct, ir, nmr*)

$C_{13}H_{13}Y$ Y-00004

(η^8-1,3,5,7-Cyclooctatetraene)(η^5-2,4-cyclopentadien-1-yl)-yttrium, 9CI
[52668-25-4]

M 258.152
Cryst. Sol. THF. Extremely air-sensitive. Forms labile THF adduct.

Jamerson, J.D. *et al*, *J. Organomet. Chem.*, 1974, **65**, C33 (*synth, nmr*)

$C_{14}H_{22}AlY$ Y-00005

Bis(η^5-2,4-cyclopentadien-1-yl)(dimethylaluminum)di-μ-methylyttrium
[57398-65-9]

M 306.215
Homogeneous ethylene polymerization catalyst. Colourless cryst. Sol. toluene, CH_2Cl_2, C_6H_6. Mp 143°. Air-sensitive.

Ballard, D.G.H. *et al*, *J. Chem. Soc., Chem. Commun.*, 1978, 994 (*use*)
Scollary, G.R., *Aust. J. Chem.*, 1978, **31**, 411 (*struct*)
Holton, J. *et al*, *J. Chem. Soc., Dalton Trans.*, 1979, 45 (*synth, nmr, ir, cmr*)

$C_{15}H_{15}Y$ Y-00006

Tris(η^5-2,4-cyclopentadien-1-yl)yttrium, 9CI
Tri-π-cyclopentadienylyttrium, 8CI
[1294-07-1]

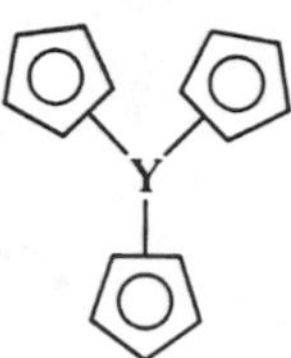

M 284.189
Pale-yellow cryst. Sol. THF. Mp 295°. Subl. at 200-50° *in vacuo*, hydrol. by H_2O.
Cyclohexylisocyanide adduct: [12098-35-0]. Colourless solid. Sol. C_6H_6. Mp 165°.
THF adduct: White cryst. Sol. THF. Air-sensitive. Subl. at 160-80° *in vacuo*.

Wilkinson, G. *et al*, *J. Am. Chem. Soc.*, 1954, **76**, 6210 (*synth*)
Fischer, E.O. *et al*, *J. Organomet. Chem.*, 1966, **6**, 141 (*synth*)
Okamato, Y. *et al*, *CA*, 1966, **64**, 9762a (*synth, rev*)
Von Ammon, R. *et al*, *Ber. Bunsenges. Phys. Chem.*, 1972, **76**, 995 (*ir, nmr*)
Devyatykh, G.G. *et al*, *Dokl. Akad. Nauk. SSSR*, 1973, **208**, 1094 (*synth, ms*)
Aleksanyan, V.T. *et al*, *J. Organomet. Chem.*, 1977, **131**, 251 (*raman*)
Rogers, R.D. *et al*, *J. Organomet. Chem.*, 1981, **216**, 383 (*struct*)

$C_{16}H_{16}Y^{\ominus}$ Y-00007

Bis(η^8-1,3,5,7-cyclooctatetraene)yttrate(1−), 9CI
Dicyclooctatetraeneyttrate(1−)

M 297.208 (ion)
K salt: [51187-41-8].
$C_{16}H_{16}KY$ M 336.307
Yellow solid. Sol. dioxan, THF. Mp >345°. Extremely air- and moisture-sensitive.

Hodgson, K.O. *et al*, *J. Am. Chem. Soc.*, 1973, **95**, 8650 (*synth, ir*)

$C_{16}H_{23}Y$ Y-00008

tert-Butylbis(methylcyclopentadienyl)yttrium

Bis[(1,2,3,4,5-η)-1-methyl-2,4-cyclopentadien-1-yl-](1,1-dimethylethyl)yttrium, 9CI

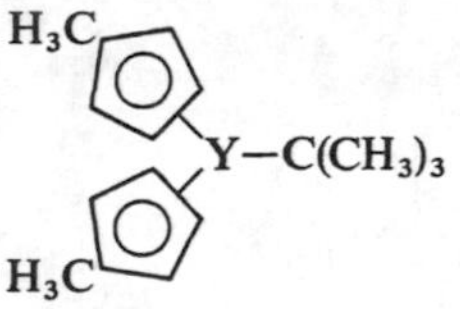

M 304.264

THF complex: [80642-67-7].

$C_{20}H_{31}OY$ M 376.370

Light-yellow powder. Reacts with H_2 to give $[(C_5H_4CH_3)_2YH(C_4H_8O)]_2$. Extremely air- and moisture-sensitive.

Evans, W.J. *et al*, *J. Am. Chem. Soc.*, 1982, **104**, 2008 (*synth, ir, nmr*)

$C_{16}H_{28}AlClNY$ Y-00009

Chlorobis(cyclopentadienyl)[trihydro(triethylamino)aluminum]yttrium

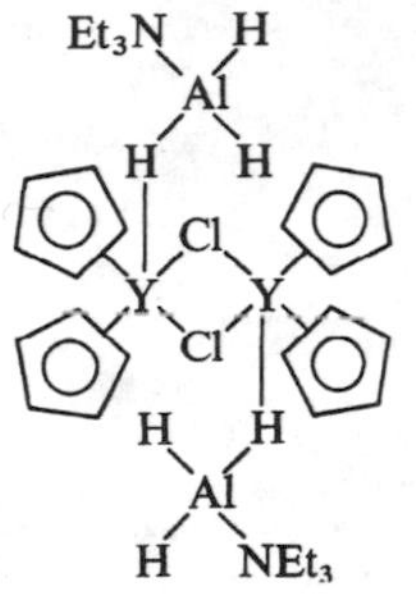

M 385.744

Dimeric.

Dimer: Di-μ-chlorotetrakis(η⁵-2,4-cyclopentadien-1-yl-)bis[trihydro(triethylalumino)aluminum]diyttrium.

$C_{32}H_{56}Al_2Cl_2N_2Y_2$ M 771.489

Colourless cryst. Sol. C_6H_6. Air- and moisture-sensitive.

Lobkorsky, E.B. *et al*, *J. Organomet. Chem.*, 1983, **254**, 167 (*synth, struct*)

$C_{16}H_{44}Si_4Y^{\ominus}$ Y-00010

Tetrakis[(trimethylsilyl)methyl]yttrate(1−)

$[Y(CH_2SiMe_3)_4]^{\ominus}$

M 437.772 (ion)

Li salt, tetrakis-THF complex: [67483-85-6].

$C_{32}H_{76}LiO_4Si_4Y$ M 733.139

Colourless cryst. Sol. Et_2O, THF. Mp 85-9°.

Atwood, J.L. *et al*, *J. Chem. Soc., Chem. Commun.*, 1978, 140 (*synth, cmr, nmr, ir*)

$C_{18}H_{15}Y$ Y-00011

Triphenylyttrium, 8CI

Yttrium triphenyl

[25292-08-4]

Ph_3Y

M 320.222

Brown solid. Sol. THF. Stable to 210° *in vacuo*. Pyrophoric in air.

Hart, F.A. *et al*, *J. Organomet. Chem.*, 1970, **21**, 147 (*synth, ir*)

$C_{21}H_{33}Y$ Y-00012

Methylbis[(1,2,3,4,5-η)-1,2,3,4,5-pentamethyl-2,4-cyclopentadien-1-yl]yttrium

[87136-56-9]

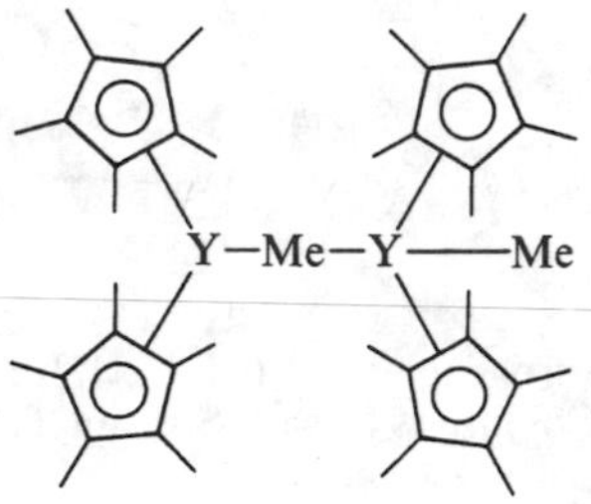

M 374.398

Asymmetric dimeric struct. in solid state, dissociates in soln. (by analogy with Lu analogue, Methylbis[(1,2,3,4,5-η)-1,2,3,4,5-pentamethyl-2,4-cyclopentadien-1-yl]lutetium. Activates methane in soln.

Dimer:

$C_{42}H_{66}Y_2$ M 748.795

White solid. Sol. toluene, cyclohexane.

Watson, P.L., *J. Am. Chem. Soc.*, 1983, **105**, 6491 (*nmr, cmr, props*)

$C_{21}H_{57}Si_6Y$ Y-00013

Tris[bis(trimethylsilyl)methyl]yttrium

[53668-81-8]

$Y[CH(SiMe_3)_2]_3$

M 567.100

White cryst. Sol. Et_2O.

Bis-THF complex: [53659-75-9].

$C_{29}H_{73}O_2Si_6Y$ M 711.313

White cryst. Sol. THF. Air-sensitive.

Barker, G.K. *et al*, *J. Organomet. Chem.*, 1974, **76**, C45 (*synth, pmr*)

$C_{24}H_{33}AlCl_2OY_2$ Y-00014

μ-Chloro-μ₃-chlorotetrakis(η⁵-2,4-cyclopentadien-1-yl-)trihydro[1,1'-oxybis[ethane]]aluminumdiyttrium, 10CI

[83243-22-5]

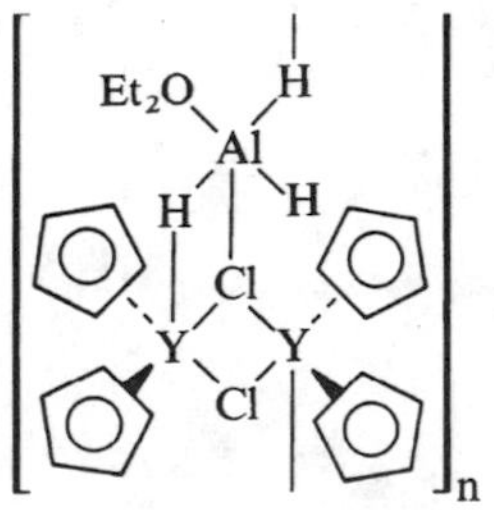

M 613.223

Polymeric chain-struct. with Y—H—Al chain linkages. Formed by reaction of AlH_3 with $(C_5H_5)_2YCl$. Colourless cryst. Sol. Et_2O. Sensitive to air and moisture.

Lobkovskii, E.B. *et al*, *J. Organomet. Chem.*, 1982, **235**, 151 (*synth, cryst struct*)

$C_{32}H_{46}O_2Y_2$ **Y-00015**

Di-μ-hydrotetrakis[(1,2,3,4,5-η)-1-methyl-2,4-cyclopentadien-1-yl]bis(tetrahydrofuran)diyttrium

[80658-44-2]

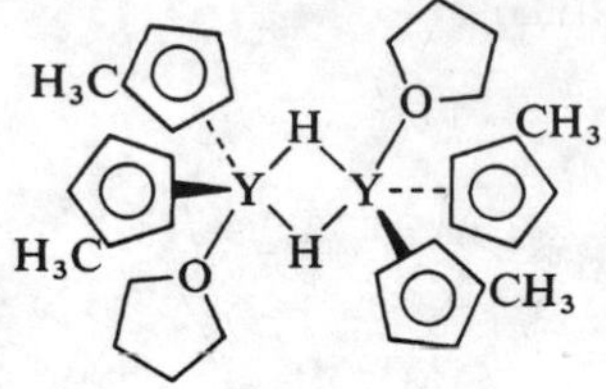

M 640.526

Formed by hydrogenolysis of $(MeC_5H_4)_2Y(C_4H_9)$-(OC_4H_8). Colourless prisms. Sol. hot THF. Extremely air- and moisture-sensitive.

Evans, W.J. *et al*, *J. Am. Chem. Soc.*, 1982, **104**, 2008 (*synth, cryst struct, ir*)

Yb Ytterbium

S. A. Cotton

Ytterbium (Fr., Ger.), Yterbio or Iterbio (Sp.), Itterbio (Ital.), Иттербий (Itterbii) (Russ.), イッテルビウム (Japan.)

Atomic Number. 70

Atomic Weight. 173.04

Electronic Configuration. [Xe] $4f^{14}\,6s^2$

Oxidation States. +2, +3, the latter being more common.

Coordination Number. 6 or greater, but less with bulky ligands such as $—CH_2SiMe_3$, $—C(CH_3)_3$.

Colour. Yb(II) complexes are usually red or green. Those of Yb(III) are variable in colour.

Availability. Starting materials are Yb_2O_3 and anhydrous $YbCl_3$, which are relatively expensive. See also under Lanthanum.

Handling. See under Lanthanum.

Toxicity See under Lanthanum.

Isotopic Abundance. ^{170}Yb, 3.03%; ^{171}Yb, 14.31%; ^{172}Yb, 21.82%; ^{173}Yb, 16.13%; ^{174}Yb, 31.84%; ^{176}Yb, 12.73%.

Spectroscopy. Yb(III) compounds give ^{1}H nmr spectra with very broad, shifted lines but a considerable amount of data is becoming available. Yb(II) compounds are diamagnetic.

Analysis. See under Lanthanum.

References. See under Lanthanum.

$(C_5H_5)YbCl_2$

Yb-00001

$YbMe_6^{\ominus\ominus\ominus}$

Yb-00002

$(C_8H_8)Yb$

Yb-00003

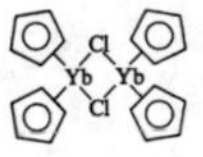

Yb-00004

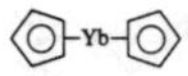

Yb-00005

$(C_6F_5)_2Yb$

Yb-00007

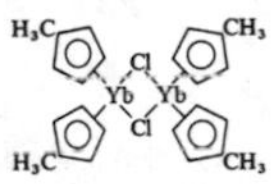

Yb-00008

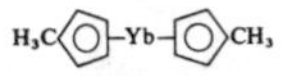

Yb-00009

$Yb(CH_2SiMe_3)_3$

Yb-00010

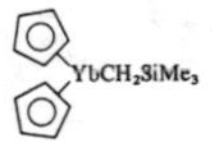

Yb-00011

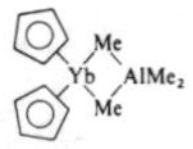

Yb-00012

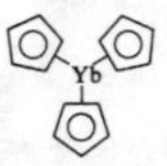

Yb-00013

$Yb(C{\equiv}CPh)_2$

Yb-00014

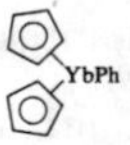

Yb-00015

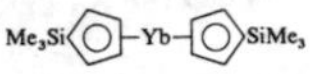

Yb-00016

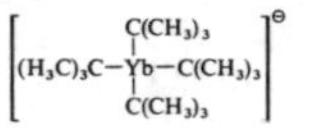

Yb-00017

$(C_5H_5)_2YbW(C_5H_5)(CO)_3$

Yb-00018

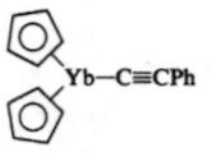

Yb-00019

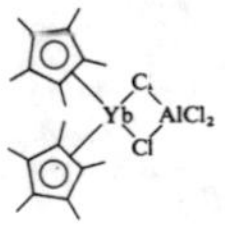

Yb-00020

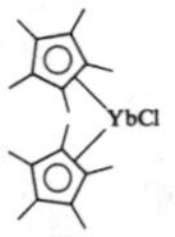

Yb-00021

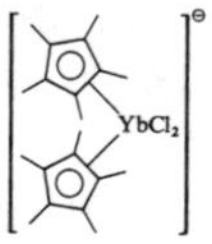

Yb-00022

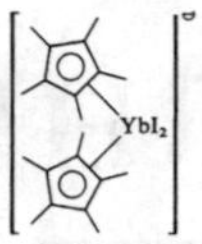

Yb-00023

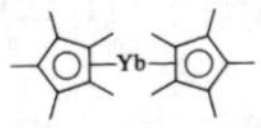

Yb-00024

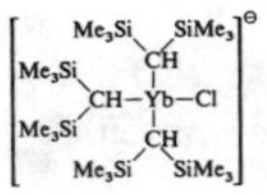

Yb-00025

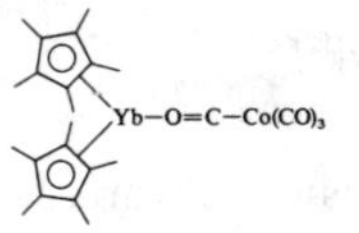

Yb-00026

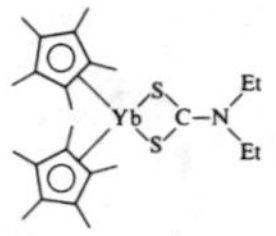

Yb-00027

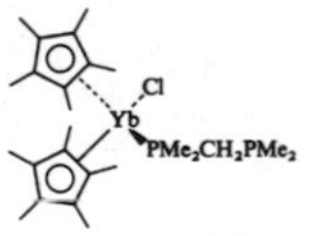

Yb-00028

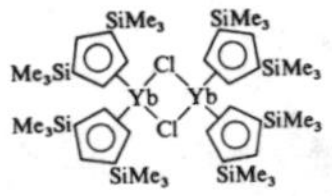

Yb-00029

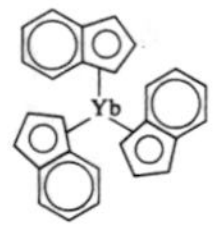

Yb-00030

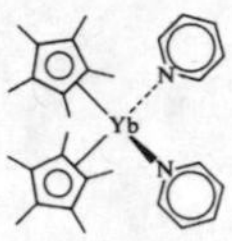

Yb-00031

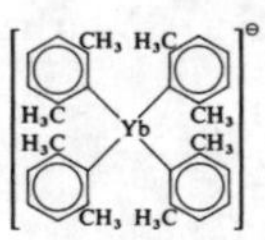

Yb-00032

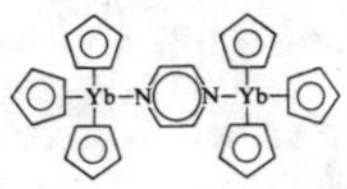

Yb-00033

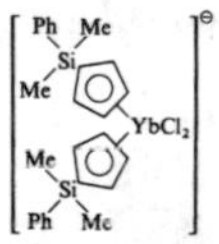

Yb-00034

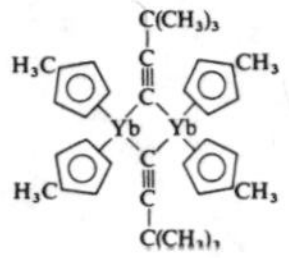

Yb-00035

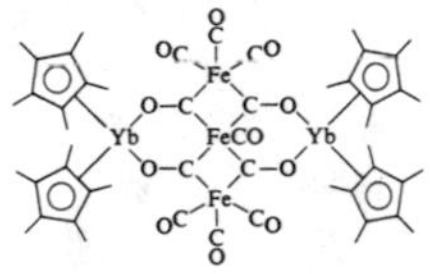

Yb-00036

$C_5H_5Cl_2Yb$ **Yb-00001**

Dichloro(η^5-2,4-cyclopentadien-1-yl)ytterbium

Cyclopentadienylytterbium dichloride

$(C_5H_5)YbCl_2$

M 309.041

Tris-THF adduct:

$C_{17}H_{29}Cl_2O_3Yb$ M 525.360

Orange cryst. Mp 78-81°. Extremely air-sensitive. μ = 4.33μ_B. Struct. probably resembles Dichloro(η^5-2,4-cyclopentadien-1-yl)erbium.

Manastyrskyj, S. *et al*, *Inorg. Chem.*, 1963, **2**, 904 (*synth*)

$C_6H_{18}Yb^{\ominus\ominus\ominus}$ **Yb-00002**

Hexamethylytterbate(3−)

$YbMe_6^{\ominus\ominus\ominus}$

M 263.248 (ion)

Tris[(tetramethylethylenediamine)lithium] salt: [76206-94-5].

$C_{24}H_{66}Li_3N_6Yb$ M 632.689

White cryst. Sol. Et_2O. Mp 141-2°. Extremely air- and moisture-sensitive.

Schumann, H. *et al*, *Angew. Chem., Int. Ed. Engl.*, 1981, **20**, 120 (*synth*)

C_8H_8Yb **Yb-00003**

(η^8-1,3,5,7-Cyclooctatetraene)ytterbium

[65036-19-3]

$(C_8H_8)Yb$

M 277.191

Pink solid. Sol. Py, DMF. Mp >500°. Extremely oxygen-sensitive. Dec. by water.

Hayes, R.G. *et al*, *J. Am. Chem. Soc.*, 1969, **91**, 6876 (*synth*)
de Kock, C.W. *et al*, *Inorg. Chem.*, 1978, **17**, 625 (*synth*, *ir*)

$C_{10}H_{10}ClYb$ **Yb-00004**

Chlorobis(η^5-2,4-cyclopentadien-1-yl)ytterbium

Chlorodicyclopentadienylytterbium. Dicyclopentadienylytterbium chloride

M 338.682

Dimeric.

Dimer: [42612-73-7]. *Di-μ-chlorotetrakis(η^5-2,4-cyclopentadien-1-yl)diytterbium*, 9CI.

$C_{20}H_{20}Cl_2Yb_2$ M 677.364

Orange-red cryst. Sol. THF. Mp >200° dec. Dec. in air. Subl. *in vacuo*. μ_{eff} = 4.81μ_B.

Maginn, R.E. *et al*, *J. Am. Chem. Soc.*, 1963, **85**, 672 (*synth*, *ir*)
Marks, T.J. *et al*, *Inorg. Chem.*, 1976, **15**, 1302 (*synth*)

$C_{10}H_{10}Yb$ **Yb-00005**

Bis(η^5-2,4-cyclopentadien-1-yl)ytterbium

Dicyclopentadienylytterbium. Ytterbocene

[1271-31-4]

M 303.229

Red solid. Sol. THF. Subl. at 400° *in vacuo*.

Fischer, E.O. *et al*, *J. Organomet. Chem.*, 1965, **3**, 181 (*synth*, *ir*)
Aleksanyan, V.T. *et al*, *J. Organomet. Chem.*, 1977, **131**, 113 (*ir*)
Evans, W.J. *et al*, *J. Chem. Soc., Chem. Commun.*, 1981, 292 (*synth*)

$C_{11}H_{13}Yb$ **Yb-00006**

Bis(η^5-2,4-cyclopentadien-1-yl)methylytterbium

[55672-19-0]

M 318.264

Dimeric.

Dimer: [60997-43-5]. *Tetrakis(η^5-2,4-cyclopentadien-1-yl)di-μ-methyldiytterbium*, 9CI.

$C_{22}H_{26}Yb_2$ M 636.527

Orange cryst. Sol. C_6H_6, toluene, CH_2Cl_2. Air- and moisture-sensitive. μ_{eff} = 4.0μ_B (295K). Dec. >165°.

Ely, N. *et al*, *Inorg. Chem.*, 1975, **14**, 2680 (*synth*, *ir*, *uv*)
Holton, J. *et al*, *J. Chem. Soc., Dalton Trans.*, 1979, 54 (*synth*, *ir*, *uv*, *ms*, *struct*)

$C_{12}F_{10}Yb$ **Yb-00007**

Bis(pentafluorophenyl)ytterbium

$(C_6F_5)_2Yb$

M 507.156

Tetrakis-THF complex: [59975-65-4].

$C_{28}H_{32}F_{10}O_4Yb$ M 795.582

Orange-red cryst. Mp 78° dec. (under N_2).

Deacon, G.B. *et al*, *J. Organomet. Chem.*, 1977, **135**, 103 (*synth*, *ir*, *nmr*, *uv*)

$C_{12}H_{14}ClYb$ **Yb-00008**

Chlorobis(η^5-methylcyclopentadienyl)ytterbium

Di(methylcyclopentadienyl)ytterbium chloride

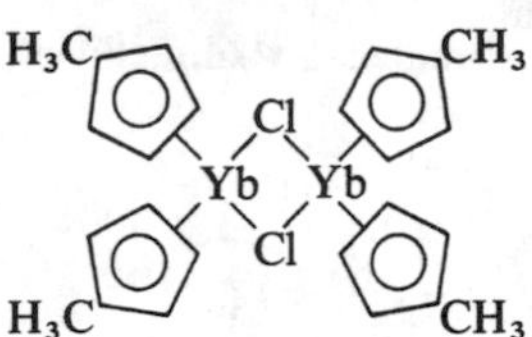

M 366.736

Dimeric.

Dimer: [54365-60-5]. *Di-μ-chlorotetrakis[(1,2,3,4,5-η)-1-methyl-2,4-cyclopentadien-1-yl-]diytterbium*, 9CI.

$C_{24}H_{28}Cl_2Yb_2$ M 733.471

Red cryst. Sol. C_6H_6, THF. Mp 115-20°. Subl. *in vacuo*. Air-sensitive.

Maginn, R.E. *et al*, *J. Am. Chem. Soc.*, 1963, **85**, 672 (*synth*)
Baker, G.C. *et al*, *Inorg. Chem.*, 1975, **14**, 1376 (*cryst struct*)

$C_{12}H_{14}Yb$ **Yb-00009**

Bis[(1,2,3,4,5-η)-1-methyl-2,4-cyclopentadien-1-yl]ytterbium

H_3C–Cp–Yb–Cp–CH_3

M 331.283

THF complex: [76137-73-0].

$C_{16}H_{22}OYb$ M 403.389

Yellow cryst. V. air- and moisture-sensitive.

Zinnen, H.A. *et al*, *J. Chem. Soc., Chem. Commun.*, 1980, 810 (*synth, ir, uv, pmr, cmr, struct*)

$C_{12}H_{33}Yb$ **Yb-00010**

Tris[(trimethylsilyl)methyl]ytterbium

$Yb(CH_2SiMe_3)_3$

M 350.433

Bis-THF complex: [67624-22-0]. *Bis(tetrahydrofuran)-tris[(trimethylsilyl)methyl]ytterbium.*

$C_{20}H_{49}O_2Si_3Yb$ M 578.902

Red cryst. Mp 53-4°. Air-sensitive in soln., stable as large cryst. $\mu_{eff} = 4.64\mu_B$.

Atwood, J.L. *et al*, *J. Chem. Soc., Chem. Commun.*, 1978, 140 (*synth, ir*)

$C_{14}H_{21}SiYb$ **Yb-00011**

Bis(η[5]-2,4-cyclopentadien-1-yl)[(trimethylsilyl)methyl]ytterbium

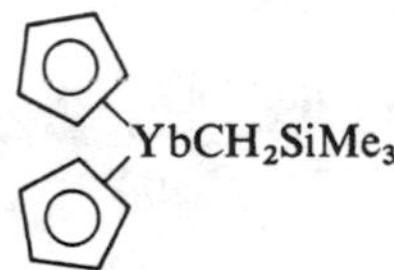

M 390.445

THF complex: [82293-72-9].

$C_{18}H_{29}OSiYb$ M 462.552

Red-brown solid. Sol. THF, C_6H_6, toluene. Mp 40-50° dec. Extremely sensitive to moisture and oxygen.

Schumann, H. *et al*, *Organometallics*, 1982, **1**, 1194 (*synth, ir, nmr*)

$C_{14}H_{22}AlYb$ **Yb-00012**

Bis(η[5]-2,4-cyclopentadien-1-yl)(dimethylaluminum)di-μ-methylytterbium

[60475-14-1]

Me
Yb AlMe$_2$
Me

M 390.349

Orange-red cryst. Sol. C_6H_6, toluene, CH_2Cl_2. Mp 133-5°. Air-sensitive.

Holton, J. *et al*, *J. Chem. Soc., Dalton Trans.*, 1979, 45 (*synth, struct, ir*)

$C_{15}H_{15}Yb$ **Yb-00013**

Tris(η[5]-2,4-cyclopentadien-1-yl)ytterbium, 9CI

Tricyclopentadienylytterbium

[1295-20-1]

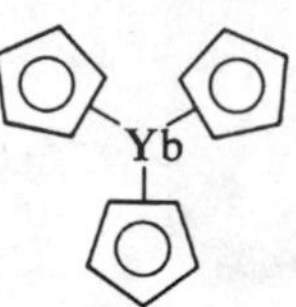

M 368.324

Dark-green cryst. Sol. THF. Mp 273° dec. Hydrol. by H_2O. Subl. at 150° *in vacuo*. $\mu_{eff} = 4.0\mu_B$.

Isocyanocyclohexane adduct: [12098-36-1]. *Tris(η[5]-2,4-cyclopentadien-1-yl)isocyanocyclohexaneytterbium.*

$C_{22}H_{26}NYb$ M 477.494

Dark-green cryst. Sol. C_6H_6. Mp 167°. $\mu_{eff} = 4.4\mu_B$. Subl. at 150° *in vacuo*.

THF complex: Tris(η[5]-2,4-cyclopentadien-1-yl)-tetrahydrofuranytterbium.

$C_{19}H_{23}OYb$ M 440.430

Dark-green cryst. Sol. THF. Mp 223-6°. Subl. *in vacuo* at 100-20°.

Triphenylphosphine adduct: Tris(η[5]-2,4-cyclopentadien-1-yl)triphenylphosphineytterbium.

$C_{33}H_{30}PYb$ M 630.614

V. dark-green cryst. Sol. C_6H_6, sl. sol. pentane. Mp 185°.

NH_3 adduct: Tris(2,4-cyclopentadien-1-yl)-ammineytterbium.

$C_{15}H_{18}NYb$ M 385.354

Green cryst. Sol. C_6H_6. Forms $(C_5H_5)_2YbNH_2$ at 200°. Subl. at 150° in high vacuum.

Birmingham, J.M. *et al*, *J. Am. Chem. Soc.*, 1956, **78**, 42 (*synth*)
Fischer, R.D. *et al*, *J. Organomet. Chem.*, 1967, **6**, 141; **8**, 155 (*uv*)
Pappalardo, R. *et al*, *J. Chem. Phys.*, 1967, **46**, 632 (*uv*)
Thomas, J.L. *et al*, *J. Organomet. Chem.*, 1967, **23**, 486 (*ms*)
v. Ammon, R. *et al*, *Inorg. Nucl. Chem. Lett.*, 1969, **5**, 315 (*nmr*)
v. Ammon, R. *et al*, *Ber. Bunsenges. Phys. Chem.*, 1972, **76**, 995 (*ir, nmr*)

$C_{16}H_{10}Yb$ **Yb-00014**

Bis(phenylethynyl)ytterbium

[66080-21-5]

$Yb(C{\equiv}CPh)_2$

M 375.295

Purple-black cryst. Sol. THF. Mp 200° dec. V. air- and water-sensitive.

Deacon, G.B. *et al*, *J. Organomet. Chem.*, 1978, **146**, C43 (*synth, ir, pmr*)

$C_{16}H_{15}Yb$ **Yb-00015**

Bis(η[5]-2,4-cyclopentadien-1-yl)phenylytterbium, 9CI

[55672-22-5]

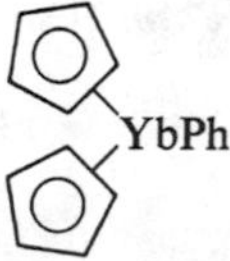

M 380.335

Orange cryst. Sol. C_6H_6, THF. Air- and moisture-sensitive. $\mu_{eff} = 3.86\mu_B$ (295K).

Ely, N.M. *et al*, *Inorg. Chem.*, 1975, **14**, 2680 (*synth, ir*)

$C_{16}H_{26}Si_2Yb$ Yb-00016

Bis[(1,2,3,4,5-η)-1-(trimethylsilyl)-2,4-cyclopentadien-1-yl]ytterbium

Me_3Si–Cp–Yb–Cp–$SiMe_3$

M 447.592

Bis-THF complex:
$C_{24}H_{42}O_2Si_2Yb$ M 591.806
Purple cryst. Mp 120°. Loses THF *in vacuo* at 30°.

Lappert, M.F. *et al*, *J. Chem. Soc., Chem. Commun.*, 1980, 987 (*synth, ms*)

$C_{16}H_{36}Yb^{\ominus}$ Yb-00017

Tetra-*tert*-butylytterbate(1–)
Tetrakis(1,1-dimethylethyl)ytterbate(1–)

$[(H_3C)_3C-Yb(C(CH_3)_3)_3]^{\ominus}$

M 401.500 (ion)

Li salt, tri-THF complex: [68868-91-7].
$C_{28}H_{60}LiO_3Yb$ M 624.761
Red-purple solid. Sol. Et_2O, THF. V. air- and moisture-sensitive. $\mu_{eff} = 4.6\mu_B$.

Wayda, A.L. *et al*, *J. Am. Chem. Soc.*, 1978, **100**, 7119 (*synth, ir*)

$C_{18}H_{15}O_3WYb$ Yb-00018

[μ-(Carbonyl-*C,O*)]bis(η⁵-2,4-cyclopentadien-1-yl)[dicarbonyl(η⁵-2,4-cyclopentadien-1-yl)tungsten]ytterbium
[39477-07-1]

$(C_5H_5)_2YbW(C_5H_5)(CO)_3$

M 636.205
Nitrosylated by *N*-methyl-*N*-nitroso-*p*-toluenesulfonamide to afford $(C_5H_5)W(CO)_2NO$. Golden cryst. Sol. THF, DMSO. Mp >220° dec. Air- and moisture-sensitive.

Crease, A.E. *et al*, *J. Chem. Soc., Dalton Trans.*, 1973, 1501 (*synth, ir*)

$C_{18}H_{15}Yb$ Yb-00019

Bis(η⁵-2,4-cyclopentadien-1-yl)(phenylethynyl)ytterbium
[53224-34-3]

$Cp_2Yb-C{\equiv}CPh$

M 404.357
Orange cryst. Sol. C_6H_6, THF. Air- and moisture-sensitive. $\mu_{eff} = 4.31\mu_B$.

Ely, N.M. *et al*, *Inorg. Chem.*, 1975, **14**, 2680 (*synth, ir*)

$C_{20}H_{30}AlCl_4Yb$ Yb-00020

Di-μ-chloro(dichloroaluminum-)bis[(1,2,3,4,5-η)-1,2,3,4,5-pentamethyl-2,4-cyclopentadienyl]ytterbium
[78064-54-7]

$Cp^*_2Yb(\mu-Cl)_2AlCl_2$

M 612.291
Deep-blue cryst. Sol. pentane. Mp 175° (part. melts). Air- and moisture-sensitive.

Watson, P.L. *et al*, *Inorg. Chem.*, 1981, **20**, 3271 (*synth, struct, ir, uv, nmr*)

$C_{20}H_{30}ClYb$ Yb-00021

Chlorobis[(1,2,3,4,5-η)-1,2,3,4,5-pentamethyl-2,4-cyclopentadien-1-yl]ytterbium

Cp^*_2YbCl

M 478.950
THF complex: [78064-47-8].
$C_{24}H_{38}ClOYb$ M 551.057
Violet cryst. Sol. toluene, THF. Mp 221-3°. μ_{eff} $4.20\mu_B$.

Tilley, T.D. *et al*, *Inorg. Chem.*, 1981, **20**, 3267 (*synth, ir, ms, nmr*)

$C_{20}H_{30}Cl_2Yb^{\ominus}$ Yb-00022

Dichlorobis[(1,2,3,4,5-η)-1,2,3,4,5-pentamethyl-2,4-cyclopentadien-1-yl]ytterbate(1–)

$[Cp^*_2YbCl_2]^{\ominus}$

M 514.403 (ion)
Li salt, tetramethylethylenediamine complex: [78128-20-8].
$C_{26}H_{46}Cl_2LiN_2Yb$ M 637.550
Violet needles. Sol. Et_2O. Mp 255° dec.
Li salt, bis (Et_2O) complex: [78128-04-8].
$C_{28}H_{50}Cl_2LiO_2Yb$ M 669.589
Dark-purple cryst. Sol. Et_2O. Mp >250°. Loses Et_2O *in vacuo* or on stirring with pentane or toluene.

Tilley, T.D. *et al*, *Inorg. Chem.*, 1981, **20**, 3267 (*synth, struct, ir*)
Watson, P.L. *et al*, *Inorg. Chem.*, 1981, **20**, 3271 (*synth, struct, ir, uv, nmr*)

$C_{20}H_{30}I_2Yb^{\ominus}$ **Yb-00023**

Diiodobis[(1,2,3,4,5-η)-1,2,3,4,5-pentamethyl-2,4-cyclopentadien-1-yl]ytterbate(1−)

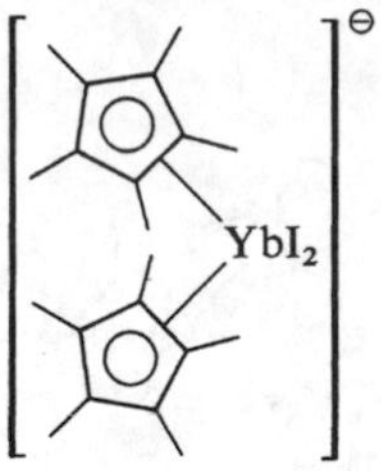

M 697.306 (ion)

Li salt: [78128-00-4].
$C_{28}H_{50}I_2LiO_2Yb$ M 852.492
Purple cryst. Sol. toluene, Et_2O. Mp >250°. Air- and moisture-sensitive.

Watson, P.L., *J. Chem. Soc., Chem. Commun.*, 1980, 652 (*synth*)

Watson, P.L. *et al*, *Inorg. Chem.*, 1981, **20**, 3271 (*synth, struct, ir, uv, nmr*)

$C_{20}H_{30}Yb$ **Yb-00024**

Bis[(1,2,3,4,5-η)-1,2,3,4,5-pentamethyl-2,4-cyclopentadien-1-yl]ytterbium

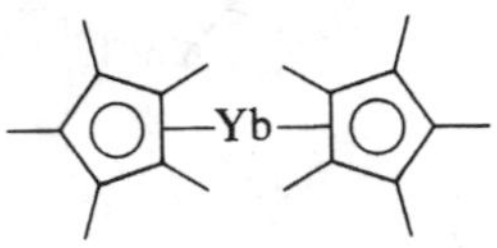

M 443.497

THF complex: [74282-46-5].
$C_{24}H_{38}OYb$ M 515.604
Red prisms. Sol. hydrocarbons, ethers. Mp 206-9°. Diamagnetic.

Complex + 1THF + ½toluene: [74282-48-7]. Brown-red prisms. Mp 204-6°. V. air- and moisture-sensitive.

Tilley, T.D. *et al*, *Inorg. Chem.*, 1980, **19**, 2999 (*synth, struct, ir, nmr, cmr*)

$C_{21}H_{57}ClSi_6Yb^{\ominus}$ **Yb-00025**

Tris[bis(triethylsilyl)methyl]chloroytterbate(1−)

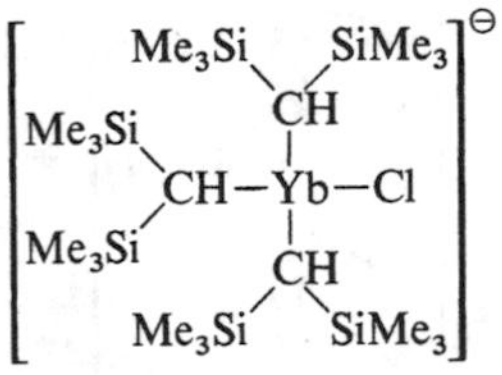

M 686.687 (ion)

Li salt, tetrakis-THF complex: [67483-92-5].
$C_{37}H_{89}ClLiO_4Si_6Yb$ M 982.055
Red cryst. Sol. hexane, THF. Mp 123-4° dec.

Atwood, J.L. *et al*, *J. Chem. Soc., Chem. Commun.*, 1978, 140 (*synth, struct*)

$C_{24}H_{30}CoO_4Yb$ **Yb-00026**

[μ-(Carbonyl-*C,O*)]bis[(1,2,3,4,5-η)-1,2,3,4,5-pentamethyl-2,4-cyclopentadien-1-yl](tricarbonylcobalt)ytterbium

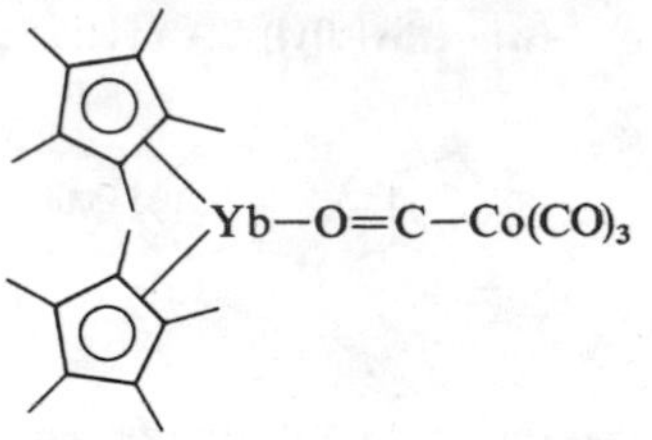

M 614.472
Blue ppt.

THF adduct: [80758-23-2].
$C_{28}H_{38}CoO_5Yb$ M 686.578
Blue prisms. $\mu_{eff} = 4.1\mu_B$.

Tilley, T.D. *et al*, *J. Chem. Soc., Chem. Commun.*, 1981, 985 (*synth, struct, ir, nmr*)

$C_{25}H_{40}NS_2Yb$ **Yb-00027**

Bis[(1,2,3,4,5-η)-1,2,3,4,5-pentamethyl-2,4-cyclopentadien-1-yl]diethylcarbamodithioato-*S,S'*-ytterbium, 10CI

[81276-69-9]

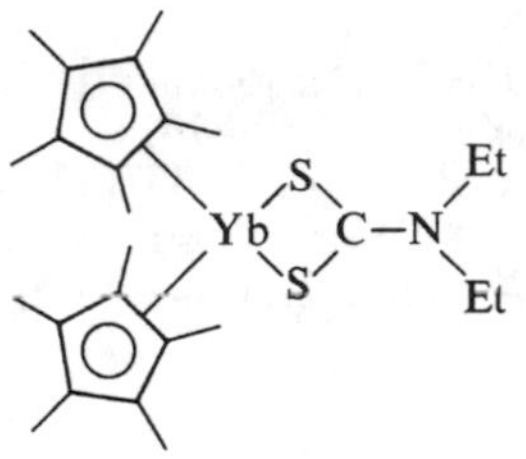

M 591.758
Purple prisms. Sol. pentane, Et_2O. Mp 226-7°. Air-sensitive. $\mu_{eff} = 3.39\mu_B$.

Tilley, T.D. *et al*, *Inorg. Chem.*, 1982, **21**, 2644 (*synth, ir, pmr, ms, struct*)

$C_{25}H_{44}ClP_2Yb$ **Yb-00028**

[Bis(1,2,3,4,5-η)-1,2,3,4,5-pentamethyl-2,4-cyclopentadien-1-yl][bis(dimethylphosphino)methane]chloroytterbium

[84254-54-6]

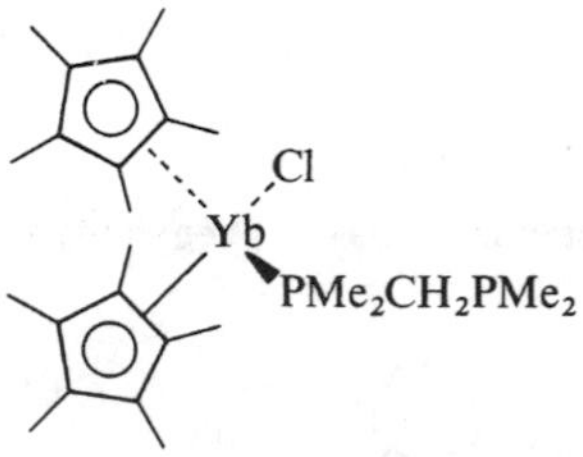

M 615.063
Purple prisms. Sol. C_6H_6, toluene. Mp 208° dec. $\mu_{eff} = 4.4\mu_B$. The diphosphine is monodentate in this complex.

Tilley, T.D. *et al*, *Inorg. Chem.*, 1983, **22**, 856 (*synth, cryst struct, ir*)

$C_{26}H_{50}ClSi_4Yb$ Yb-00029

Bis[1,3-bis(trimethylsilylmethyl)cyclopentadienyl]chloroytterbium

M 683.516

Bridged dimer.

Dimer: [81536-99-4]. *Tetrakis[(1,2,3,4,5-η)-1,3-bis(trimethylsilylmethyl)-2,4-cyclopentadien-1-yl]di-μ-chlorodiytterbium.*

$C_{52}H_{100}Cl_2Si_8Yb_2$ M 1367.032

Red cryst. Sol. toluene, THF.

Lappert, M.F. *et al*, *J. Chem. Soc., Chem. Commun.*, 1981, 1190 (*synth, struct*)

$C_{27}H_{21}Yb$ Yb-00030

Triindenylytterbium, 8CI

Tris(1,2,3,3a,7a-η)-1H-inden-1-ylytterbium

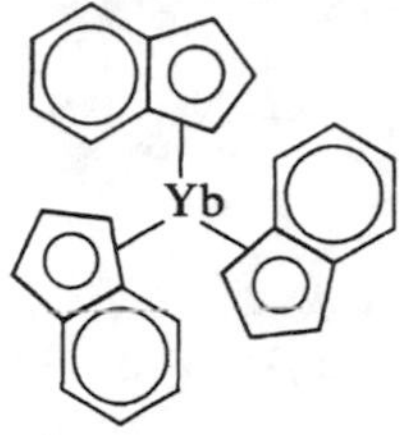

M 518.503

THF complex: [22696-76-0].

$C_{31}H_{29}OYb$ M 590.610

Dark green. $\mu = 4.10\mu_B$.

Tsutsui, M. *et al*, *J. Am. Chem. Soc.*, 1969, **91**, 3175 (*synth*)

$C_{30}H_{40}N_2Yb$ Yb-00031

Bis[(1,2,3,4,5-η)-1,2,3,4,5-pentamethyl-2,4-cyclopentadien-yl]bis(pyridine)ytterbium

[81276-72-4]

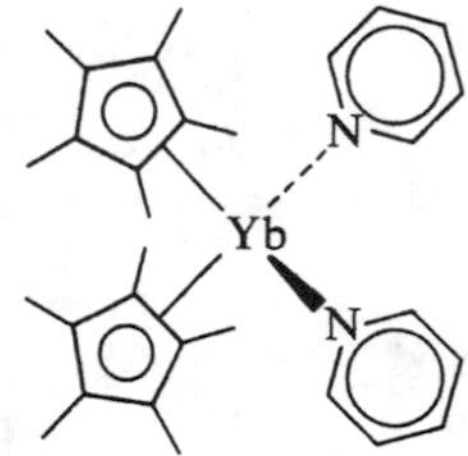

M 601.699

Dark-green prisms. Sol. pentane, toluene. Mp 208-10° (darkens at 190°).

Tilley, T.D. *et al*, *Inorg. Chem.*, 1982, **21**, 2647 (*synth, struct, pmr, ir*)

$C_{32}H_{36}Yb^{\ominus}$ Yb-00032

Tetrakis(2,6-dimethylphenyl)ytterbate(1−)

M 593.676 (ion)

Li salt, tetra-THF complex:

$C_{48}H_{68}LiO_4Yb$ M 889.044

Yellow cryst. Sol. THF. Dec. by H_2O and air.

Cotton, S.A. *et al*, *J. Chem. Soc., Chem. Commun.*, 1972, 1225 (*synth, struct*)

$C_{34}H_{34}N_2Yb_2$ Yb-00033

Hexakis(η⁵-2,4-cyclopentadien-1-yl)[μ-(pyrazine-N¹:N⁴)]-diytterbium

[63950-86-7]

M 816.736

Green-brown cryst. Sol. dimethoxyethane, sl. sol. toluene.

Baker, E.C. *et al*, *Inorg. Chem.*, 1977, **16**, 2710 (*synth, struct, ms, ir*)

$C_{36}H_{34}Cl_2Si_2Yb$ Yb-00034

Dichlorobis[(1,2,3,4,5-η)-1-(methyldiphenylsilyl)-2,4-cyclopentadien-1-yl]ytterbate(1−)

M 766.782

Li salt, bis(diethyl ether) complex: [78128-08-2].

$C_{44}H_{54}Cl_2O_2Si_2Yb$ M 915.026

Orange cryst. Air- and moisture-sensitive.

Watson, P.L. *et al*, *Inorg. Chem.*, 1981, **20**, 3271 (*synth, struct, ir, uv*)

$C_{36}H_{42}Yb_2$ **Yb-00035**

Tetrakis(1,2,3,4,5-η-1-methyl-2,4-cyclopentadien-1-yl-)bis[μ-3,3-dimethyl-1-butyn-1-yl)diytterbium

[76565-31-6]

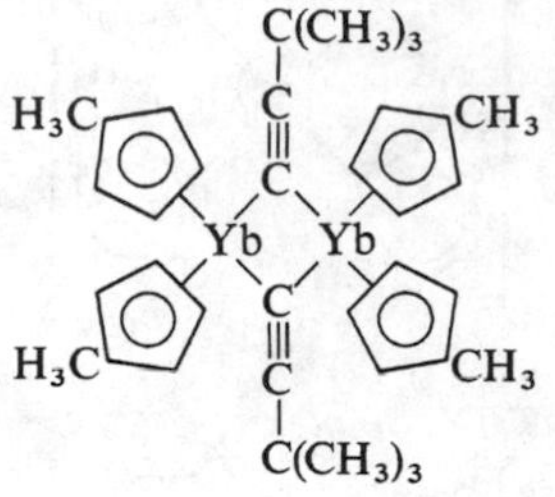

M 820.808

Struct. by analogy with Bis(2,4-cyclopentadien-1-yl)-(3,3-dimethylbutynyl)erbium. Bright-orange cryst. Sol. THF. Extremely air- and moisture-sensitive.

Atwood, J.L. *et al*, *Inorg. Chem.*, 1981, **20**, 4115 (*synth, ir, uv*)

$C_{51}H_{60}Fe_3Yb_2$ **Yb-00036**

Tetrakis[μ₃-(carbonyl-*C,C,O*)](heptacarbonyltriiron)tetrakis[(1,2,3,4,5-η)-1,2,3,4,5-pentamethyl-2,4-cyclopentadien-1-yl]diytterbium

[80878-91-7]

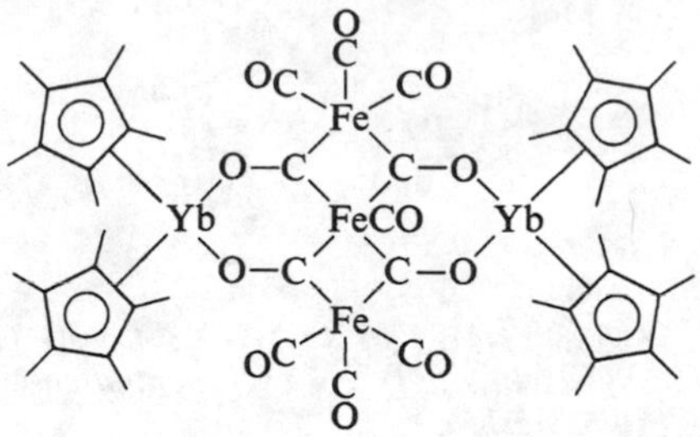

M 1186.656

Violet prisms. Sol. toluene. Mp 307-10°. $\mu_{eff} = 3.91\mu_B$.

Tilley, T.D. *et al*, *J. Am. Chem. Soc.*, 1982, **104**, 1772 (*synth, struct, ir, nmr*)

Zr Zirconium

D. J. Cardin

Zirconium (Fr.), Zirkon (Ger.), Zirconio (Sp., Ital.), Цирконий (Tsirkonii) (Russ.), ジルコニウム (Japan.)

Atomic Number. 40

Atomic Weight. 91.22

Electronic Configuration. [Kr] $4d^2\ 5s^2$

Oxidation States. The oxidation state (IV) predominates for the organic derivatives of zirconium, and to a significantly greater extent than for its congener titanium. Such Zr(III) derivatives as are known include the species $ZrR_2(\eta^5\text{-}C_5H_5)$ (R = alkyl), which are kinetically labile, and some bridged bimetallic compounds, eg. $[Zr(\mu\text{-}PR_2)(\eta^5\text{-}C_5H_5)]_2$ which are diamagnetic. Monomeric compounds of the type $ZrX(\eta^5\text{-}C_5H_5)_2$ (X = halide) have not yet been obtained. The best characterised Zr(II) compounds are the metallocene adducts of neutral donors (eg. CO, PR_3) and compounds having the ligand pentamethylcyclopentadienyl. The diene metallocene adducts can be formulated as Zr(IV) species, having a metallocyclic structure. Zr(0) is represented by arene and conjugated diene adducts, generally together with phosphine, eg. $Zr(PMe_3)(\eta^6\text{-}C_6H_6)_2$, $Zr(PMe_2CH_2CH_2PMe_2)(\eta^4\text{-}C_4H_6)_2$.

Coordination Number. (Note: all ligands in this section are treated as monohapto, including cyclic hydrocarbons, which may, or more frequently, may not be). For Zr(IV), and particularly the metallocene derivatives, tetrahedral four coordinate geometry predominates. Chelating ligands may however confer higher coordination numbers, eg. 7 in $Zr(diket)_3(\eta^5\text{-}C_5H_5)$ (diket = β-diketonate) or 5 in $ZrR(\eta^2\text{-}RCO)(\eta^5\text{-}C_5H_5)_2$ (R = alkyl). The coordination numbers of Zr(III) compounds are frequently not known, structural information being unavailable. Metal(II) complexes are four coordinate for the type $ZrL_2(\eta^5\text{-}C_5H_5)_2$, while the zero-valent Zr(P-P)-$(diene)_2$ type are believed to be pseudo-octahedral.

Colour. The remarks concerning colour for titanium generally apply here, with the difference that the intense charge transfer bands present for the lighter metal are frequently absent from, or of reduced intensity, in the visible spectrum of analogous compounds of the heavier element, Titanocene dichloride, for example, is deep red, whereas zirconocene dichloride is colourless.

Availability. Analogous starting materials to those for titanium are generally employed with $ZrCl_2$ and $ZrCl_2(\eta^5\text{-}C_5H_5)_2$ being the most important. The range of organometallic derivatives commercially available is smaller than for titanium.

Handling. Standard inert atmosphere techniques are required for those compounds of zirconium which are air-sensitive.

Toxicity. Organozirconium compounds are not associated with any known specific toxicity.

Isotopic Abundance. ^{90}Zr, 51.46%; ^{91}Zr, 11.23%; ^{92}Zr, 17.11%; ^{94}Zr, 17.40%; ^{96}Zr, 2.8%.

Spectroscopy. ^{91}Zr, $I = \frac{5}{2}$, is the only isotope with nuclear spin and has receptivity relative to ^{13}C of 6.04. There is only one report of nmr of zirconium to date. Nmr of other nuclei is not generally useful for paramagnetic materials.

Esr is a valuable tool for d^1 compounds where ^{91}Zr hyperfine couplings are generally observed.

Analysis. For precise analysis compounds are generally decomposed by ashing followed by fusion, or by wet oxidation, eg. with $HClO_4/HNO_3$. Aqueous solutions can be estimated titrimetrically, colorimetrically, or gravimetrically. Detailed procedures for particular uses are given in standard analytical works. X-ray fluorescence and polarographic methods have also been recommended.

Spot tests for elemental detection are available with a range of organic reagents. *p*-Dimethylaminophenylarsonic acid gives a red-brown precipitate, and pyrocatechol violet a blue colouration. Several ions (including organic ions) interfere. Recommended procedures are given in standard works.

References. General reviews are listed in the introduction to the *Sourcebook*.

Wailes, P. C. *et al.*, *Organometallic Chemistry of Titanium, Zirconium, and Hafnium*, Academic Press, London, 1974.

Larsen, E. M., *Adv. Inorg. Chem. Radiochem.*, 1970, **13**, 1.

Schwartz, J., *Organic Synthesis-Today-Tomorrow*, Trost, B. M. and Hutchinson, C. R. Eds, Pergamon, Oxford, 1981 (*hydrozirconation*)

Schwartz, J. and Labinger, J. A., *Angew. Chem., Int. Ed. Engl.*, 1976, **15**, 333 (*hydrozirconation*)

Cardin, D. J. *et al.*, *The Organometallic Chemistry of Zirconium and Hafnium*, Ellis Horwood, in press (1984)

ZrX_3
X = Br

Zr-00001

As Zr-00001 with
X = Cl

Zr-00002

As Zr-00001 with
X = I

Zr-00003

$ZrCl_2$

Zr-00004

ZrH_2

Zr-00005

X = Y = Br

Zr-00006

As Zr-00006 with
X = Y = Cl

Zr-00007

As Zr-00006 with
X = Y = F

Zr-00008

As Zr-00006 with
X = Y = I

Zr-00009

Zr-00010

Zr-00011

Zr-00012

Zr-00013

$ZrCl_3$

Zr-00014

Zr-00015

As Zr-00006 with
X = OMe, Y = Cl

Zr-00016

X = OMe

Zr-00017

As Zr-00001 with
X = OAc

Zr-00018

As Zr-00006 with
X = Y = NCO or CNO

Zr-00019

As Zr-00006 with
X = Y = NCS

Zr-00020

As Zr-00006 with
X = Y = CO

Zr-00021

X = Cl

Zr-00022

As Zr-00006 with
X = Cl, Y = OAc

Zr-00023

As Zr-00006 with
X = Y = OMe

Zr-00024

As Zr-00006 with
X = Y = Me

Zr-00025

Zr-00026

Zr-00027

Zr-00028

As Zr-00022 with
X = Me

Zr-00029

As Zr-00006 with
X = Cl, Y = $SiMe_3$

Zr-00030

Zr-00031

As Zr-00006 with
X = Y = $-OOCCF_3$

Zr-00032

Zr-00033

Zr-00034

Zr-00035

As Zr-00006 with
X = Y = NMe_2

Zr-00036

As Zr-00017 with
X = NMe_2

Zr-00037

$Me_2Zr[N(SiMe_3)_2]_2$

Zr-00038

Zr-00039

Zr-00040

Zr-00041

Zr-00042

S-cis-form

Zr-00043

Zr-00044

$Zr(CH_2SiMe_3)_4$

Zr-00045

Zr-00046

Zr-00047

X = Cl

Zr-00048

Zr-00049

Zr-00050

Zr-00051

Zr-00052

As Zr-00006 with
X = Y = NEt_2

Zr-00053

Zr-00054

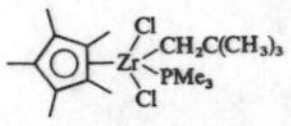

Zr-00055

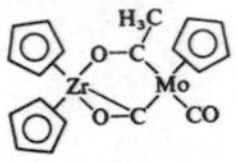

Zr-00056

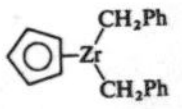

Zr-00057

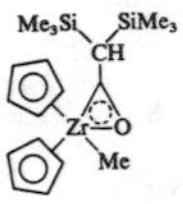

Zr-00058

$ZrMePh(Et_2O)_3$

Zr-00059

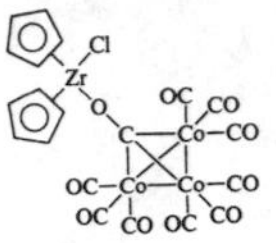

Zr-00060

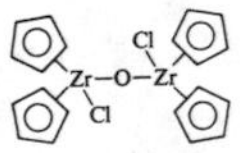

Zr-00061

As Zr-00048 with
X = Me

Zr-00062

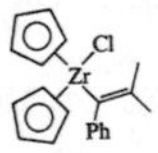

Zr-00063

As Zr-00006 with
X = Y = —$CH_2C(CH_3)_3$

Zr-00064

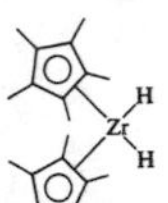

Zr-00065

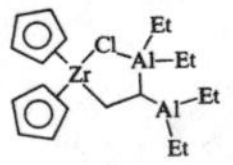

Zr-00066

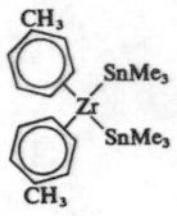

Zr-00067

$Zr[CH_2C(CH_3)_3]_4$

Zr-00068

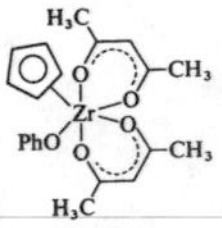

Zr-00069

$ClZr[CH(SiMe_3)_2]_3$

Zr-00070

As Zr-00006 with
X = Y = C_6F_5

Zr-00071

As Zr-00006 with
X = Y = SPh

Zr-00072

As Zr-00006 with
X = Y = Ph

Zr-00073

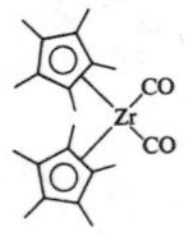

Zr-00074

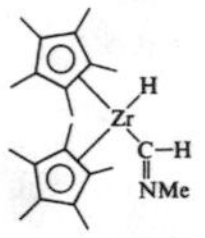

Zr-00075

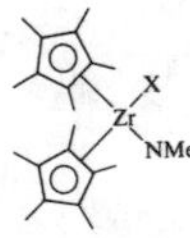

X = I

Zr-00076

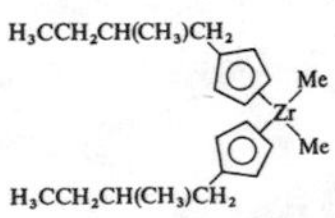

Zr-00077

As Zr-00076 with
X = H

Zr-00078

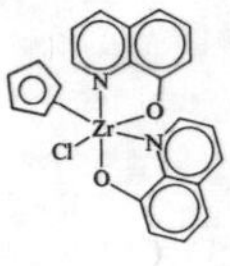

Zr-00079

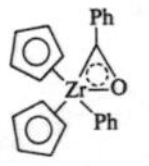

Zr-00080

As Zr-00006 with
X = Cl, Y = —CH_2PPh_2

Zr-00081

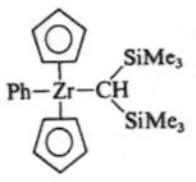

Zr-00082

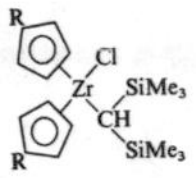

R = $SiMe_3$

Zr-00083

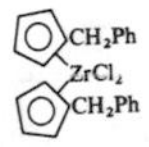

Zr-00084

As Zr-00006 with
X = Y = —CH_2Ph

Zr-00085

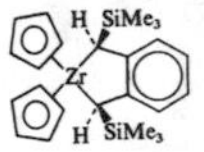

Zr-00086

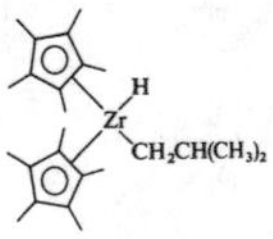

Zr-00087

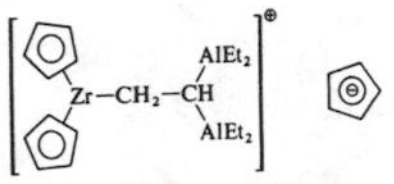

Zr-00088

As Zr-00083 with
R = $(H_3C)_3C$—

Zr-00089

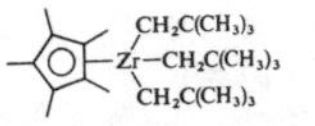

Zr-00090

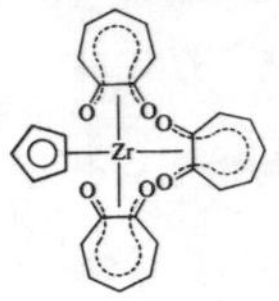

Zr-00091

As Zr-00006 with
X = Y = —C≡CPh

Zr-00092

As Zr-00006 with
X = Y = —CH=CHPh

Zr-00093

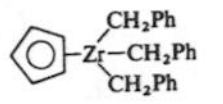

Zr-00094

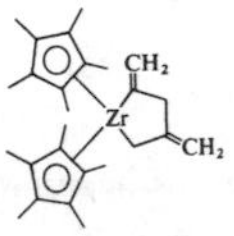

Zr-00095

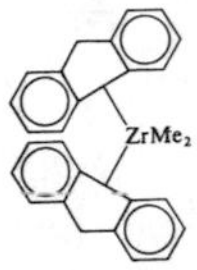

Zr-00096

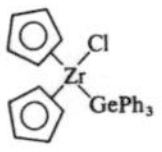

Zr-00097

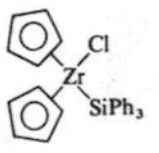

Zr-00098

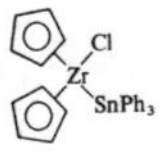

Zr-00099

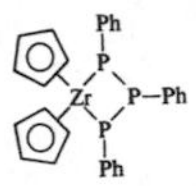

Zr-00100

$Zr(CH_2Ph)_4$

Zr-00101

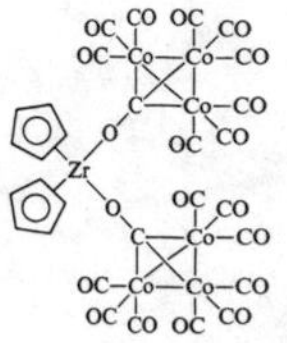

Zr-00102

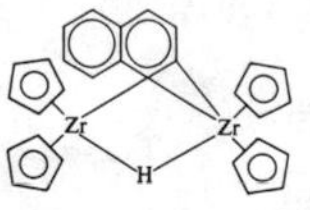

Zr-00103

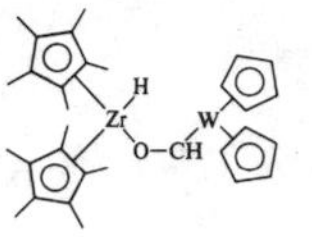

Zr-00104

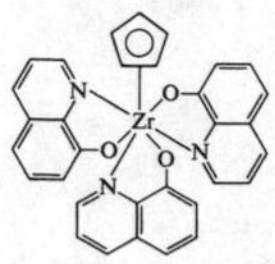

Zr-00105

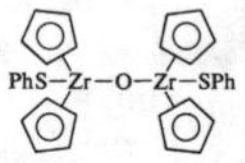

Zr-00106

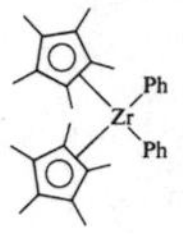

Zr-00107

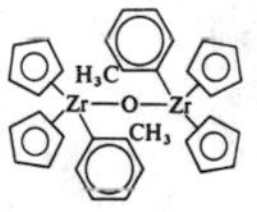

Zr-00108

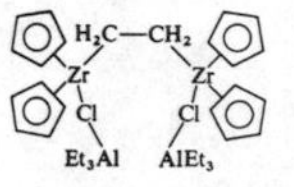

Zr-00109

As Zr-00006 with
X = Y = —N=CPh$_2$

Zr-00110

As Zr-00006 with
X = Y = —CHPh$_2$

Zr-00111

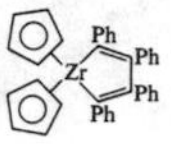

Zr-00112

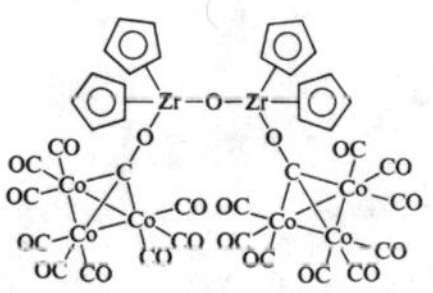

Zr-00113

$[(H_3C)_2C=CPh]_4Zr$

Zr-00114

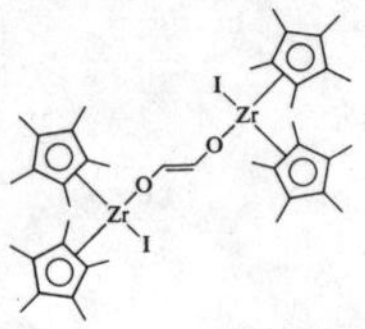

Zr-00115

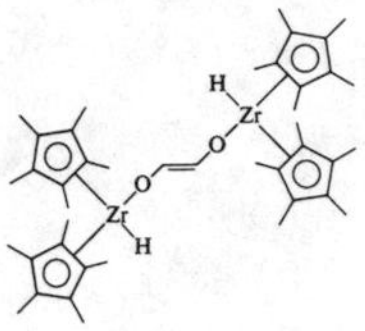

Zr-00116

$C_5H_5Br_3Zr$ **Zr-00001**
Tribromo(η^5-2,4-cyclopentadien-1-yl)zirconium

–ZrX_3

X = Br

M 396.027
Pale-yellow cryst. Mp 197-8°.

Reid, A.F. *et al*, *J. Organomet. Chem.*, 1964, **2**, 329 (*ir*, *nmr*, *uv*)

$C_5H_5Cl_3Zr$ **Zr-00002**
Trichloro(η^5-2,4-cyclopentadien-1-yl)zirconium, 9CI
[34767-44-7]

As Tribromo(η^5-2,4-cyclopentadien-1-yl)zirconium, Zr-00001 with

X = Cl

M 262.674
Cream solid. Mp 237° dec. Forms adducts with dimethoxyethane and TMEDA.

Reid, A.F. *et al*, *J. Organomet. Chem.*, 1964, **2**, 329 (*ir*, *uv*, *nmr*)
Wells, N.J. *et al*, *J. Organomet. Chem.*, 1981, **213**, C17 (*synth*)

$C_5H_5I_3Zr$ **Zr-00003**
(η^5-2,4-Cyclopentadien-1-yl)triiodozirconium
As Tribromo(η^5-2,4-cyclopentadien-1-yl)zirconium, Zr-00001 with

X = I

M 537.028
Red solid. Mp 133-133.5°.

Reid, A.F. *et al*, *J. Organomet. Chem.*, 1964, **2**, 329 (*ir*, *nmr*, *uv*)

$C_8H_8Cl_2Zr$ **Zr-00004**
Dichloro(η^8-1,3,5,7-cyclooctatetraene)zirconium
[42423-66-5]

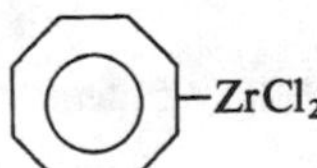

M 266.277
THF complex:
$C_{12}H_{16}Cl_2OZr$ M 338.384
Orange-yellow cryst. (THF). Spar. sol. Loses THF *in vacuo*.

Lehmkuhl, H. *et al*, *J. Organomet. Chem.*, 1972, **46**, C1 (*synth*)
Kablitz, H.J. *et al*, *J. Organomet. Chem.*, 1973, **51**, 241 (*synth*, *ir*, *ms*)
Brauer, D.J. *et al*, *Inorg. Chem.*, 1975, **14**, 3052 (*cryst struct*)

$C_8H_{10}Zr$ **Zr-00005**
(η^8-1,3,5,7-cyclooctatetraene)dihydrozirconium, 9CI
(η-Cyclooctatetraene(2−))dihydridozirconium
[42077-14-5]

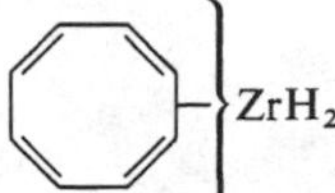

M 197.387

Detailed struct. unknown. Green-black cryst. (toluene).

Kablitz, H.J. *et al*, *J. Organomet. Chem.*, 1973, **51**, 241 (*synth*, *ir*)

$C_{10}H_{10}Br_2Zr$ **Zr-00006**
Dibromobis(η^5-2,4-cyclopentadien-1-yl)zirconium, 9CI
Zirconocene dibromide
[1294-67-3]

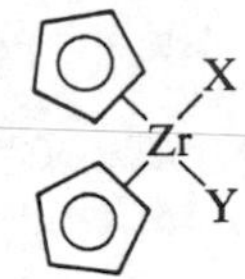

X = Y = Br

M 381.217
Colourless solid. Mp 259-60°.

Druce, P.M. *et al*, *J. Chem. Soc. (A)*, 1969, 2106, 2814 (*ir*, *ms*, *nmr*)
Samuel, E. *et al*, *Inorg. Chem.*, 1973, **12**, 881 (*ir*)
Sayer, B.G. *et al*, *Inorg. Chim. Acta*, 1981, **48**, 53 (*nmr*)

$C_{10}H_{10}Cl_2Zr$ **Zr-00007**
Dichlorobis(η^5-2,4-cyclopentadien-1-yl)zirconium, 9CI
Zirconocene dichloride. Bis(cyclopentadienyl)zirconium chloride
[1291-32-3]

As Dibromobis(η^5-2,4-cyclopentadien-1-yl)zirconium, Zr-00006 with

X = Y = Cl

M 292.315
Colourless cryst. Mp 248° (242-5°).
▷ZH7525000.

Samuel, E. *et al*, *C.R. Hebd. Seances Acad. Sci.*, 1962, **254**, 308 (*synth*)
Reid, A.F. *et al*, *J. Organomet. Chem.*, 1964, **2**, 329 (*ir*, *nmr*, *uv*)
Dillard, J.G. *et al*, *J. Organomet. Chem.*, 1969, **16**, 265 (*ms*)
Druce, P.M. *et al*, *J. Chem. Soc. (A)*, 1969, 2106 (*ir*, *ms*, *nmr*)
Samuel, E. *et al*, *Inorg. Chem.*, 1973, **12**, 881 (*ir*)
Prout, K. *et al*, *Acta Crystallogr., Sect. B*, 1974, **30**, 2290 (*cryst struct*)
Cauletti, C. *et al*, *J. Electron Spectrosc. Relat. Phenom.*, 1980, **18**, 61 (*pe*)

$C_{10}H_{10}F_2Zr$ **Zr-00008**
Bis(η^5-2,4-cyclopentadien-1-yl)difluorozirconium, 9CI
Zirconocene difluoride
[11090-85-0]

As Dibromobis(η^5-2,4-cyclopentadien-1-yl)zirconium, Zr-00006 with

X = Y = F

M 259.406
Colourless solid. Mp >250°.

Druce, P.M. *et al*, *J. Chem. Soc. (A)*, 1969, 2106, 2814 (*ir*, *ms*, *nmr*)
Bush, M.A. *et al*, *J. Chem. Soc. (A)*, 1971, 2225 (*cryst struct*)
Samuel, E. *et al*, *Inorg. Chem.*, 1973, **12**, 881 (*ir*)

Cauletti, C. *et al*, *J. Electron Spectrosc. Relat. Phenom.*, 1980, **18**, 61 (*pe*)

$C_{10}H_{10}I_2Zr$ Zr-00009

Bis(η^5-2,4-cyclopentadien-1-yl)diiodozirconium, 9CI

Zirconocene diiodide

[1298-41-5]

As Dibromobis(η^5-2,4-cyclopentadien-1-yl)zirconium, Zr-00006 with

X = Y = I

M 475.218

Yellow cryst. (C_6H_6). Mp 299-300°.

Druce, P.M. *et al*, *J. Chem. Soc. (A)*, 1969, 2106, 2814 (*ir*, *ms*, *nmr*)
Bush, M.A. *et al*, *J. Chem. Soc. (A)*, 1971, 2225 (*cryst struct*)
Samuel, E. *et al*, *Inorg. Chem.*, 1973, **12**, 881 (*ir*)
Sayer, B.G. *et al*, *Inorg. Chim. Acta*, 1981, **48**, 53 (*nmr*)

$C_{10}H_{10}S_5Zr$ Zr-00010

Bis(η^5-2,4-cyclopentadien-1-yl)(pentathio)zirconium, 10CI

Bis(η^5-cyclopentadienyl)cyclo(pentasulfur-1,5-diyl)zirconium

[75213-09-1]

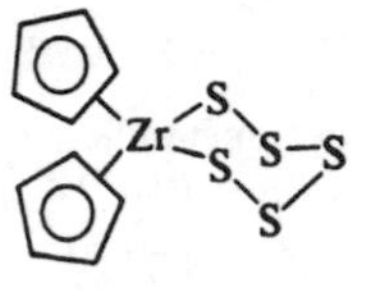

M 381.709

Air-stable orange cryst. (CH_2Cl_2). Mp 160-70° dec. E_{act} (ring inversion) 48.6 kJ $mol.^{-1}$ (CD_2Cl_2) chain conformation at r.t.

McCall, J.M. *et al*, *J. Organomet. Chem.*, 1980, **193**, C37 (*synth*, *ir*, *nmr*, *cryst struct*)

$C_{10}H_{11}ClZr$ Zr-00011

Chlorobis(η^5-2,4-cyclopentadien-1-yl)hydrozirconium, 10CI

Chlorobis(η-cyclopentadienyl)hydridozirconium. Zirconocene hydridochloride. Schwartz's reagent

[37342-97-5]

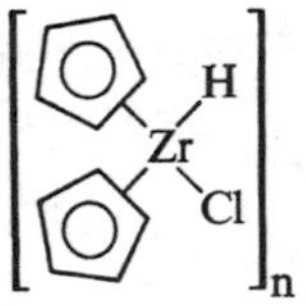

M 257.870

Polymeric. Reagent for hydrozirconation in org. synth., crosscoupling of alkynes etc. White solid. Insol. Light-sensitive, involatile.

Kautzner, B. *et al*, *J. Chem. Soc., Chem. Commun.*, 1969, 1105 (*synth*, *ir*)
Wailes, P.C. *et al*, *J. Organomet. Chem.*, 1970, **24**, 405 (*synth*, *ir*)
Schwartz, J. *et al*, *Angew. Chem., Int. Ed. Engl.*, 1976, **15**, 333 (*use*)
Sorrell, T.N., *Tetrahedron Lett.*, 1978, 4985 (*use*)
Laycock, D.E. *et al*, *J. Org. Chem.*, 1981, **46**, 289 (*use*)
Fieser, M. *et al*, *Reagents for Organic Synthesis*, Wiley, 1967-83, **8**, 84 (*use*)

$C_{10}H_{12}Zr$ Zr-00012

Bis(η^5-2,4-cyclopentadien-1-yl)dihydrozirconium, 10CI

Zirconocene dihydride

[37342-98-6]

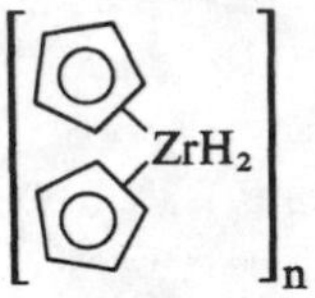

M 223.425

Polymeric. Hydrogenation catalyst for alkenes and alkynes. White solid. Insol. common solvs.

Couturier, S. *et al*, *J. Organomet. Chem.*, 1978, **157**, C61 (*synth*, *ms*, *use*)
Kautzner, B. *et al*, *J. Chem. Soc., Chem. Commun.*, 1969, 1105 (*synth*, *ir*)
Wailes, P.C. *et al*, *J. Organomet. Chem.*, 1972, **43**, C29 (*ir*)
Labinger, J.A. *et al*, *J. Organomet. Chem.*, 1978, **155**, C25.
Inorg. Synth., 1979, **19**, 223.

$C_{10}H_{15}BZr$ Zr-00013

Bis(η^5-2,4-cyclopentadien-1-yl)hydro[tetrahydroborato(1−)-*H,H'*]zirconium

Bis(η-cyclopentadienyl)hydrido(tetrahydridoborato)-zirconium

[12116-93-7]

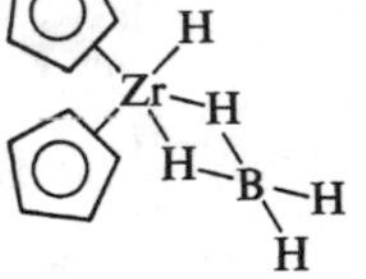

M 237.259

White solid. Subl. *in vacuo*.

James, B.D. *et al*, *Inorg. Chem.*, 1967, **6**, 1979 (*synth*, *ir*, *ms*, *nmr*)
Marks, T.J. *et al*, *Inorg. Chem.*, 1972, **11**, 2540 (*ir*, *raman*)
Marks, T.J. *et al*, *J. Am. Chem. Soc.*, 1975, **97**, 3397 (*nmr*)

$C_{10}H_{15}Cl_3Zr$ Zr-00014

Trichloro[(1,2,3,4,5-η)-1,2,3,4,5-pentamethyl-2,4-cyclopentadien-1-yl]zirconium, 10CI

[75181-07-6]

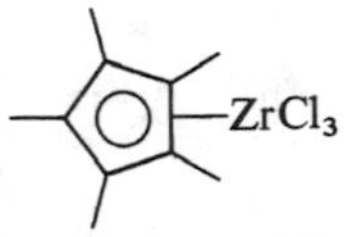

M 332.808

Yellow or white solid. $Bp_{0.001}$ 160° subl.

Wengrovius, J.H. *et al*, *J. Organomet. Chem.*, 1981, **205**, 319 (*synth*, *nmr*)
Blenkers, J. *et al*, *J. Organomet. Chem.*, 1981, **218**, 383 (*nmr*, *synth*)

$C_{10}H_{18}B_2Zr$ Zr-00015

Bis(η^5-2,4-cyclopentadien-1-yl)bis[tetrahydroborato(1−)-*H,H'*]zirconium, 10CI
Zirconocene bis(borohydride)
[12083-77-1]

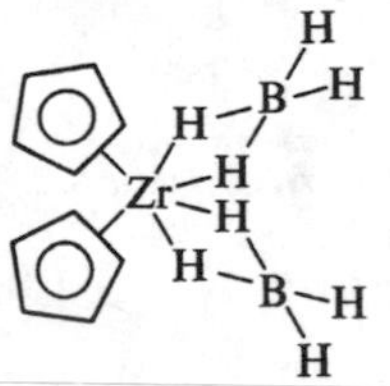

M 251.092
Pale-yellow solid by subl. Mp 155° dec. Bp 110-5° subl. *in vacuo*. Hydrolysed only slowly by water. BD_4 analogue also made.

Nanda, R.K. *et al*, *Inorg. Chem.*, 1964, **3**, 1798 (*synth, ir*)
Davies, N. *et al*, *J. Chem. Soc. (A)*, 1969, 2601 (*ir, nmr*)
Marks, T.J. *et al*, *J. Am. Chem. Soc.*, 1975, **97**, 3397 (*ir, nmr*)
Smith, B.E. *et al*, *J. Inorg. Nucl. Chem.*, 1976, **38**, 1973 (*synth, ir, raman*)

$C_{11}H_{13}ClOZr$ Zr-00016

Chlorobis(η-cyclopentadienyl)methoxyzirconium
Chlorodi-π-cyclopentadienylmethoxyzirconium, 8CI
[11087-26-6]
As Dibromobis(η^5-2,4-cyclopentadien-1-yl)zirconium, Zr-00006 with

X = OMe, Y = Cl

M 287.896
White cryst. by subl. Mp 111-114.5°.

Gray, D.R. *et al*, *Inorg. Chem.*, 1971, **10**, 2143 (*synth, ir, nmr*)
Fachinetti, G. *et al*, *J. Chem. Soc., Chem. Commun.*, 1978, 269 (*synth*)

$C_{11}H_{14}O_3S_6Zr$ Zr-00017

(η^5-2,4-Cyclopentadien-1-yl)tris(*O*-methylcarbonodithioato-*S,S'*)zirconium, 10CI
(η-Cyclopentadienyl)tris(methylxanthato)zirconium
[66199-99-3]

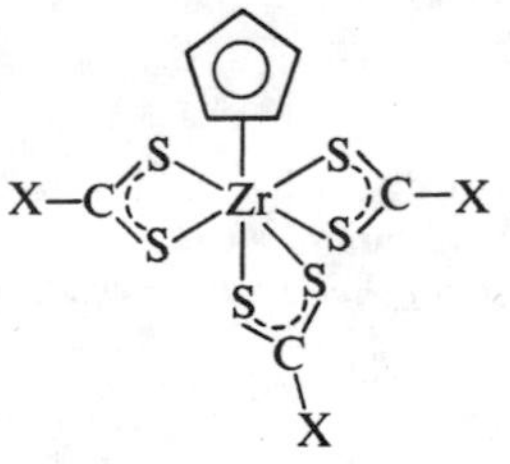

X = OMe

M 477.810
Colourless cryst. (CH_2Cl_2/pet. ether). Mp 116-21° dec.

Jain, V.K. *et al*, *Inorg. Chim. Acta*, 1978, **26**, 51 (*synth, ir, nmr*)

$C_{11}H_{14}O_6Zr$ Zr-00018

Tris(acetato)(η-cyclopentadienyl)zirconium
Triacetoxycyclopentadienylzirconium
[31760-74-4]
As Tribromo(η^5-2,4-cyclopentadien-1-yl)zirconium, Zr-00001 with

X = OAc

M 333.448
White cryst. (C_6H_6/pet. ether). Mp ~170°, 178° dec.

Brainina, E.M. *et al*, *Izv. Akad. Nauk SSSR, Ser. Khim.*, 1963, 835 (*synth*)
Wailes, P.C. *et al*, *J. Organomet. Chem.*, 1970, **24**, 413 (*synth, nmr*)

$C_{12}H_{10}N_2O_2Zr$ Zr-00019

Bis(cyanato-*N*)bis(η^5-2,4-cyclopentadien-1-yl)zirconium, 9CI
Bis(cyanato-O)bis(η^5-2,4-cyclopentadien-1-yl)zirconium, 9CI. *Bis(η-cyclopentadienyl)dicyanatozirconium. Zirconocene dicyanate*
[12109-63-6]
As Dibromobis(η^5-2,4-cyclopentadien-1-yl)zirconium, Zr-00006 with

X = Y = NCO or CNO

M 305.443
Bonding of NCO group (cyanate or isocyanate) uncertain. Colourless solid. Mp 218°. Readily hydrol. to μ-oxo compd.

Coutts, R.S.P. *et al*, *Aust. J. Chem.*, 1966, **19**, 2069 (*synth, ir, nmr*)
Burmeister, J.L. *et al*, *Inorg. Chem.*, 1970, **9**, 58 (*ir, ms, uv*)
Jensen, A. *et al*, *Acta Chem. Scand.*, 1972, **26**, 2898 (*ir, props*)

$C_{12}H_{10}N_2S_2Zr$ Zr-00020

Bis(η^5-2,4-cyclopentadien-1-yl)bis(thiocyanato-*N*)zirconium, 9CI
Bis(η-cyclopentadienyl)dithiocyanatozirconium. Zirconocene dithiocyanate
[12109-67-0]
As Dibromobis(η^5-2,4-cyclopentadien-1-yl)zirconium, Zr-00006 with

X = Y = NCS

M 337.564
Yellow solid. Mp 279-81°. $Bp_{0.0004}$ 200° subl.

Coutts, R.S.P. *et al*, *Aust. J. Chem.*, 1966, **19**, 2069 (*synth, ir, nmr*)
Samuel, E., *Bull. Soc. Chim. Fr.*, 1966, 3548 (*synth, ir, uv*)
Burmeister, J.L. *et al*, *Inorg. Chem.*, 1970, **9**, 58 (*ir, uv*)
Jensen, A. *et al*, *Acta Chem. Scand.*, 1972, **26**, 2898 (*ir*)

$C_{12}H_{10}O_2Zr$ Zr-00021

Dicarbonylbis(η^5-2,4-cyclopentadien-1-yl)zirconium, 10CI
Zirconocene dicarbonyl
[59487-85-3]
As Dibromobis(η^5-2,4-cyclopentadien-1-yl)zirconium, Zr-00006 with

X = Y = CO

M 277.430
Violet or black cryst. (heptane). $Bp_{0.01}$ 70° subl.

Demerseman, B. *et al*, *J. Organomet. Chem.*, 1976, **107**, C19 (*synth, ir*)
Fachinetti, G. *et al*, *J. Chem. Soc., Chem. Commun.*, 1976, 230 (*synth, ms, nmr*)
Thomas, J.L. *et al*, *J. Organomet. Chem.*, 1976, **111**, 297 (*synth, ir*)
Atwood, J.L. *et al*, *Inorg. Chem.*, 1980, **19**, 3812 (*cryst struct*)
Gell, K.I. *et al*, *J. Am. Chem. Soc.*, 1981, **103**, 2687 (*synth, ir, nmr*)

$C_{12}H_{13}ClOZr$ Zr-00022

(Acetyl-*C,O*)chlorobis(η^5-2,4-cyclopentadien-1-yl)zirconium

[77001-15-1]

X = Cl

M 299.907

Characterized spectroscopically.

Marsella, J.A. *et al*, *J. Organomet. Chem.*, 1980, **201**, 389 (*synth, ir, nmr*)

$C_{12}H_{13}ClO_2Zr$ Zr-00023

(Acetato-*O*)chlorobis(η^5-2,4-cyclopentadien-1-yl)zirconium, 9CI

(Acetato)chlorodi-π-cyclopentadienylzirconium, 8CI. *Zirconocene acetate chloride*

[60876-22-4]

As Zr-00006 with

X = Cl, Y = OAc

M 315.907

Solid. Mp 137-9°.

Brainina, E.M. *et al*, *Izv. Akad. Nauk SSSR, Ser. Khim.*, 1969, 2492 (*synth, ir, nmr*)
Brainina, E.M. *et al*, *Izv. Akad. Nauk SSSR, Ser. Khim.*, 1976, 1611 (*synth*)
Suzuki, H. *et al*, *Bull. Chem. Soc. Jpn.*, 1978, **51**, 1764 (*synth, ir, nmr, ms*)

$C_{12}H_{16}O_2Zr$ Zr-00024

Bis(η^5-2,4-cyclopentadien-1-yl)dimethoxyzirconium

Di-π-cyclopentadienyldimethoxyzirconium, 8CI. *Zirconocene dimethoxide*

[11087-29-9]

As Dibromobis(η^5-2,4-cyclopentadien-1-yl)zirconium, Zr-00006 with

X = Y = OMe

M 283.477

Pale-yellow cryst. by subl. Mp 63-8°. $Bp_{0.1}$ 85° subl.

Gray, D.R. *et al*, *Inorg. Chem.*, 1971, **10**, 2143 (*synth, ir, nmr*)
Etievant, P. *et al*, *Bull. Soc. Chim. Fr.*, 1978, 292 (*synth, nmr*)

$C_{12}H_{16}Zr$ Zr-00025

Bis(η^5-2,4-cyclopentadien-1-yl)dimethylzirconium, 10CI

Zirconocene dimethyl

[12636-72-5]

As Dibromobis(η^5-2,4-cyclopentadien-1-yl)zirconium, Zr-00006 with

X = Y = Me

M 251.478

Cocatalyst in metathesis reactions and α-olefin polymerisations. White cryst. by subl. $Bp_{0.0001}$ 60-80° subl.

Samuel, E. *et al*, *J. Am. Chem. Soc.*, 1973, **95**, 6263 (*synth, nmr*)
Alt, H. *et al*, *J. Am. Chem. Soc.*, 1974, **96**, 5936.
Jordan, R.F. *et al*, *J. Am. Chem. Soc.*, 1979, **101**, 4853 (*cmr*)
Ger. Pat., 3 007 725, (*1981*); *CA*, **95**, 187927j (*use*)
Van Leeuwen, P.W.N.M. *et al*, *J. Organomet. Chem.*, 1981, **209**, 169 (*cidnp*)

$C_{12}H_{20}Zr$ Zr-00026

Tetrakis(η^3-2-propenyl)zirconium, 10CI

Tetraallylzirconium

[12090-34-5]

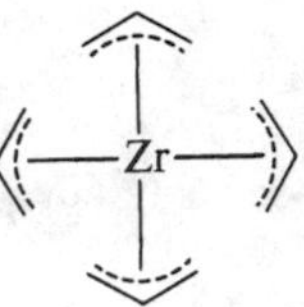

M 255.510

Dynamic tetra-π-allyl struct. in soln. Used in the catalytic polymerisation of vinyl monomers. Red cryst. (pentane). $Bp_{0.0005}$ ~20° subl. Dec. from 0°.

Wilke, G. *et al*, *Angew. Chem., Int. Ed. Engl.*, 1966, **5**, 151 (*synth*)
Becconsall, J.K. *et al*, *J. Chem. Soc. (A)*, 1967, 423 (*synth, nmr, ms*)
Ballard, D.G.H. *et al*, *J. Chem. Soc. (B)*, 1968, 1168 (*use*)

$C_{13}H_{13}ClZr$ Zr-00027

Chloro[η^8-1,3,5,7-cyclooctatetraene](η^5-2,4-cyclopentadien-1-yl)zirconium, 9CI

[39413-64-4]

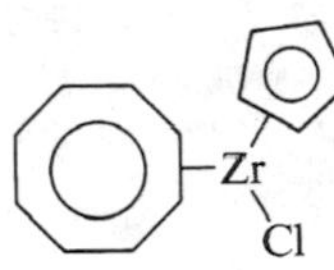

M 295.919

Red cryst. (toluene).

Kablitz, H.J. *et al*, *J. Organomet. Chem.*, 1973, **51**, 241 (*synth, nmr, ms*)

$C_{13}H_{14}Cl_2Zr$ Zr-00028

Dichloro[1,3-propanediylbis[(1,2,3,4,5-η)-2,4-cyclopentadien-1-ylidene]]zirconium, 9CI

Dichloro(1,1′-trimethylenedicyclopentadienyl)zirconium

[38117-95-2]

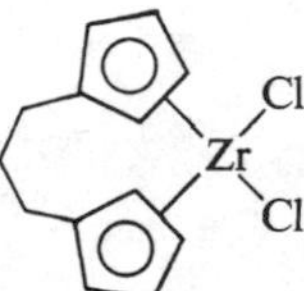

M 332.380

Cryst. by subl. $Bp_{0.0001}$ 170° subl.

Hillman, M. *et al*, *J. Organomet. Chem.*, 1972, **42**, 123 (*synth, ir, nmr, uv*)
Saldarriaga-Molina, C.H. *et al*, *J. Organomet. Chem.*, 1974, **80**, 79 (*cryst struct*)

$C_{13}H_{16}OZr$ Zr-00029

(η^2-Acetyl)bis(η^5-2,4-cyclopentadien-1-yl)methylzirconium, 10CI

[60970-97-0]

As (Acetyl-*C,O*)chlorobis(η^5-2,4-cyclopentadien-1-yl)-zirconium, Zr-00022 with

X = Me

M 279.489

Yellow cryst. In soln. loses CO under vacuum.

Fachinetti, G. *et al*, *J. Chem. Soc., Dalton Trans.*, 1977, 1946 (*synth, ir, nmr, cryst struct*)
Marsella, J.A. *et al*, *J. Am. Chem. Soc.*, 1981, **103**, 5596 (*props*)

$C_{13}H_{19}ClSiZr$ Zr-00030

Chlorobis(η^5-2,4-cyclopentadien-1-yl)(trimethylsilyl)zirconium, 10CI

[76772-61-7]

As Dibromobis(η^5-2,4-cyclopentadien-1-yl)zirconium, Zr-00006 with

X = Cl, Y = $SiMe_3$

M 330.052

White solid by subl. Mp 98-100°. $Bp_{0.0002}$ 60-70° subl.

Blakeney, A.J. *et al*, *J. Organomet. Chem.*, 1980, **202**, 263 (*synth, nmr*)

$C_{13}H_{27}Si_2Zr$ Zr-00031

(η^5-Cyclopentadienyl)bis(trimethylsilylmethyl)zirconium

CH_2SiMe_3 / Zr / CH_2SiMe_3

M 330.747

Unstable, not isol. from chemical or electrochemical reduction. Characterised spectroscopically.

Lappert, M.F. *et al*, *J. Chem. Soc., Dalton Trans.*, 1981, 805 (*esr*)

$C_{14}H_{10}F_6O_4Zr$ Zr-00032

Bis(η^5-2,4-cyclopentadien-1-yl)bis(trifluoroacetato-*O*)zirconium, 9CI

Di-π-cyclopentadienylbis(trifluoroacetato)zirconium, 8CI. *Zirconocene bis(trifluoroacetate)*

[37205-20-2]

As Dibromobis(η^5-2,4-cyclopentadien-1-yl)zirconium, Zr-00006 with

X = Y = $-OOCCF_3$

M 447.441

Colourless cryst. Mp 117°. $Bp_{0.003}$ 150° subl.

Brainina, E.M. *et al*, *Izv. Akad. Nauk SSSR, Ser. Khim.*, 1963, 835 (*synth*)
King, R.B. *et al*, *J. Organomet. Chem.*, 1968, **15**, 457 (*ir, synth*)
Wakes, P.C. *et al*, *J. Organomet. Chem.*, 1970, **24**, 413 (*synth, ir*)
Minacheva, M.K. *et al*, *Izv. Akad. Nauk SSSR, Ser. Khim.*, 1972, 139 (*synth*)

$C_{14}H_{16}Zr$ Zr-00033

(η^4-1,3-Butadiene)bis(η^5-2,4-cyclopentadien-1-yl)zirconium, 10CI

Bis(η^5-cyclopentadienyl)(η^4-1,3-butadiene)zirconium

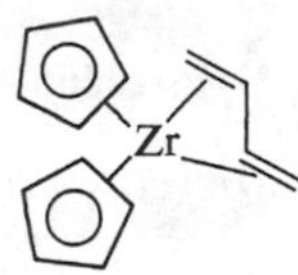

M 275.500

(*s-trans*)-*form* [75374-50-4]

Red cryst. Formed as mixt. with *s-cis*-isomer but photochemical isom. to *s-trans*-form occurs.

Erker, G. *et al*, *J. Am. Chem. Soc.*, 1980, **102**, 6344 (*synth, cryst struct, ir, nmr*)
Yasuda, H. *et al*, *Chem. Lett.*, 1981, 519 (*synth*)

$C_{14}H_{18}Zr$ Zr-00034

(η^8-1,3,5,7-Cyclooctatetraene)bis(η^3-2-propenyl)zirconium, 9CI

Bis(η-allyl)[η-cyclooctatetraene(2−)]zirconium. Diallylcyclooctateraenezirconium

[39413-75-7]

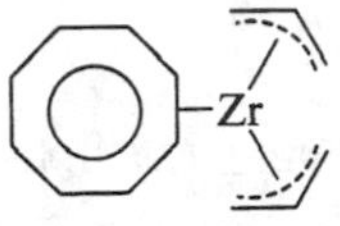

M 277.516

Catalytically oligomerises butadiene. Red-brown cryst. (toluene/Et_2O).

Kablitz, H.J. *et al*, *J. Organomet. Chem.*, 1973, **51**, 241 (*synth, nmr, ir, ms*)
Hoffmann, E.G. *et al*, *J. Organomet. Chem.*, 1975, **97**, 183 (*ir, nmr*)

$C_{14}H_{21}ClSiZr$ Zr-00035

Chlorobis(η^5-2,4-cyclopentadien-1-yl)[(trimethylsilyl)methyl]-zirconium, 10CI

[11082-39-6]

Cl / Zr / CH_2SiMe_3

M 344.078

Pale-yellow cryst. (hexane/toluene). Mp 118-21°.

Collier, M.R., *J. Chem. Soc., Dalton Trans.*, 1973, 445 (*synth, ir, ms, nmr*)

$C_{14}H_{22}N_2Zr$ Zr-00036

Bis(η-cyclopentadienyl)bis(dimethylamido)zirconium

Di-π-cyclopentadienylbis(dimethylaminato)zirconium, 8CI. *Zirconocene bis(dimethylamide)*

[11108-30-8]

As Dibromobis(η^5-2,4-cyclopentadien-1-yl)zirconium, Zr-00006 with

X = Y = NMe_2

M 309.561

Polymerisation catalyst for acrylonitrile. Yellow cryst. $Bp_{0.05}$ 110-20° subl.

Chandra, G. *et al*, *J. Chem. Soc. (A)*, 1968, 1940 (*synth*)
Jenkins, A.D. *et al*, *J. Polym. Sci., Polym. Lett., Part B*, 1968, **6**, 865 (*use*)

$C_{14}H_{23}N_3S_6Zr$ **Zr-00037**

(η^5-2,4-Cyclopentadien-1-yl)tris(dimethylcarbamodithioato-*S,S'*)zirconium, 9CI

(η-Cyclopentadienyl)tris(N,N*-dimethyldithiocarbamato)zirconium*

[61113-31-3]

As (η^5-2,4-Cyclopentadien-1-yl)tris(*O*-methylcarbonodithioato-*S,S'*)zirconium, Zr-00017 with

X = NMe_2

M 516.936

Colourless cryst. + CH_2Cl_2 (CH_2Cl_2/hexane). Mp 270-5° dec.

Bruder, A.H. *et al*, *J. Am. Chem. Soc.*, 1976, **98**, 6932 (*synth, cryst struct, ir, nmr*)

$C_{14}H_{42}N_2Si_4Zr$ **Zr-00038**

Dimethylbis[bis(trimethylsilyl)amido]zirconium

*Dimethylbis[1,1,1-trimethyl-*N*-(trimethylsilyl)silanaminato]zirconium, 10CI*

[70969-30-1]

$Me_2Zr[N(SiMe_3)_2]_2$

M 442.061

Colourless liq. at r.t.; cryst. (pentane). Mp ca. 20°.

Andersen, R.A., *Inorg. Chem.*, 1979, **18**, 2928 (*synth, ir, pmr, cmr, ms*)

$C_{15}H_{15}ClZr$ **Zr-00039**

Chlorotris(η-cyclopentadienyl)zirconium

Chloro-2,4-cyclopentadien-1-ylbis(η^5-2,4-cyclopentadien-1-yl)zirconium, 10CI

[69005-93-2]

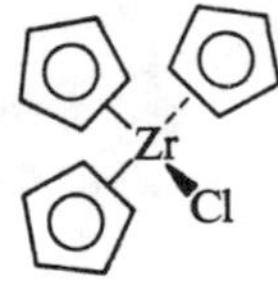

M 321.957

Struct. postulated to be different from Chlorotris(η-methylcyclopentadienyl)zirconium, Zr-00051 (3 η^5-rings). Pale-yellow cryst. Mp 234°.

Etievant, P. *et al*, *Bull. Soc. Chim. Fr., Part II*, 1978, 292 (*synth, ms, nmr*)

$C_{15}H_{19}ClO_4Zr$ **Zr-00040**

Chloro(η^5-2,4-cyclopentadien-1-yl)bis(2,4-pentanedionato-*O,O'*)zirconium, 9CI

Bis(acetylacetonato)chloro(η-cyclopentadienyl)zirconium

[12216-18-1]

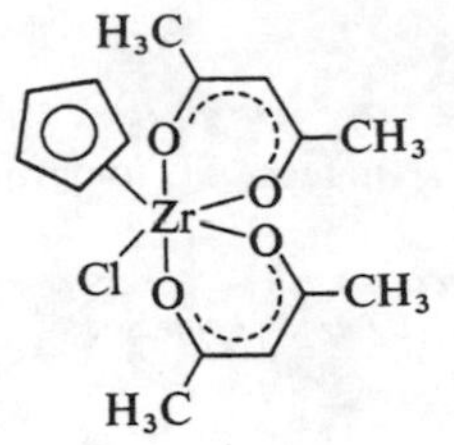

M 389.986

Colourless cryst. Mp 189-90°.

Brainina, E.M. *et al*, *Izv. Akad. Nauk SSSR, Ser. Khim.*, 1964, 1421 (*synth*)
Pinnavaia, T.I. *et al*, *J. Am. Chem. Soc.*, 1968, **90**, 5288 (*nmr*)
Stezowski, J.J. *et al*, *J. Am. Chem. Soc.*, 1969, **91**, 2890 (*cryst struct*)
Fedorov, L.A. *et al*, *Izv. Akad. Nauk SSSR, Ser. Khim.*, 1971, 1844 (*nmr*)
Frazer, M.J. *et al*, *Inorg. Chem.*, 1971, **10**, 2137 (*synth, ir, nmr*)
Minacheva, M.K., *Koord. Khim.*, 1975, **1**, 831 (*cmr*)

$C_{15}H_{22}Zr$ **Zr-00041**

Bis(η^5-cyclopentadienyl)(2,2-dimethylpropyl)hydrozirconium

Bis(η-cyclopentadienyl)hydridoneopentylzirconium. Dicyclopentadienylhydridoneopentylzirconium

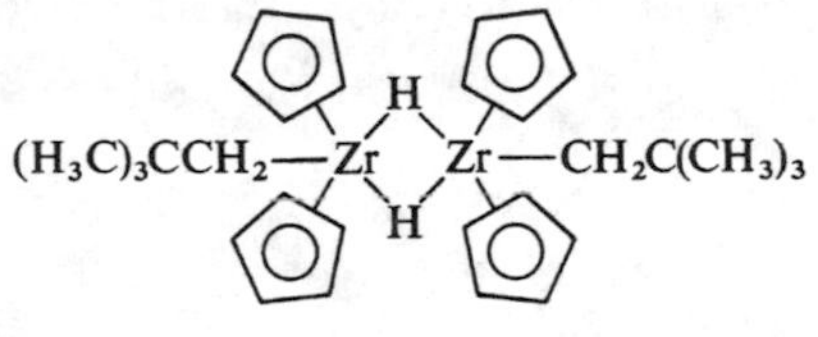

M 293.559

Dimeric.

Dimer: [67063-43-8]. *Tetrakis(η^5-2,4-cyclopentadien-1-yl)bis(2,2-dimethylpropyl)di-μ-hydrodizirconium, 10CI. Bis[bis(cyclopentadienyl)-hydridoneopentylzirconium].*
$C_{30}H_{44}Zr_2$ M 587.118
Not isol., characterised spectroscopically. Several related dimeric hydrides also reported.

Gell, K.I. *et al*, *J. Am. Chem. Soc.*, 1978, **100**, 3246 (*synth, nmr*)

$C_{15}H_{27}Zr$ **Zr-00042**

(η^5-2,4-Cyclopentadien-1-yl)bis(2,2-dimethylpropyl)zirconium

(η-Cyclopentadienyl)dineopentylzirconium

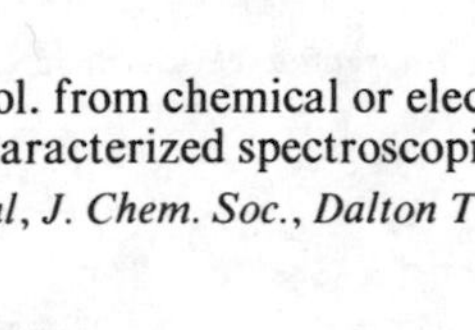

M 298.598

Unstable, not isol. from chemical or electrochemical reduction. Characterized spectroscopically.

Lappert, M.F. *et al*, *J. Chem. Soc., Dalton Trans.*, 1981, 805 (*esr*)

$C_{16}H_{20}Zr$ **Zr-00043**

Bis(η^5-2,4-cyclopentadien-1-yl)[(1,2,3,4-η)-2,3-dimethyl-1,3-butadiene]zirconium, 10CI

[75361-74-9]

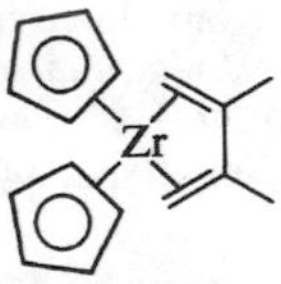

S-cis-form

M 303.554

s-cis and *s-trans* isomers known. Red cryst. solid. *s-trans* obt. from *s-cis* by photochem. isom.

Erker, G. *et al*, *J. Am. Chem. Soc.*, 1980, **102**, 6344 (*synth, nmr, cryst struct*)

$C_{16}H_{20}Zr$ **Zr-00044**

Bis(η^5-2,4-cyclopentadien-1-yl)-η^1-2-propenyl-η^3-2-propenylzirconium

Allyl-π-allyl-π-cyclopentadienylzirconium, 8CI. *Diallylbis(η-cyclopentadienyl)zirconium. Zirconocene diallyl*

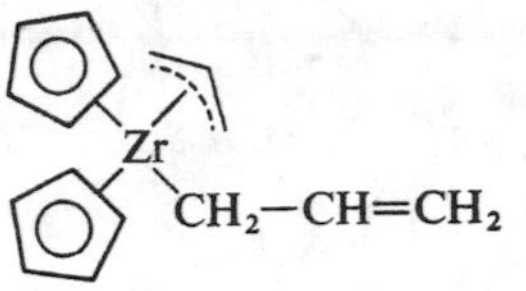

M 303.554

Cream cryst. (pentane). Mp >−18° dec. Dec. at r.t. under N_2.

Martin, H.A. *et al*, *J. Organomet. Chem.*, 1968, **14**, 149 (*synth, ir, nmr*)

$C_{16}H_{44}Si_4Zr$ **Zr-00045**

Tetrakis(trimethylsilylmethyl)zirconium, 9CI

[32665-18-2]

$Zr(CH_2SiMe_3)_4$

M 440.086

Polymerizes 1-alkenes. Colourless cryst. (pentane, <10°); colourless liq. Mp 10-1°. $Bp_{0.001}$ ca. 25°. Shows enhanced thermal stability over sample alkyls.

▷Pyrophoric

Collier, M.R. *et al*, *J. Chem. Soc., Dalton Trans.*, 1973, 445 (*synth, ir, nmr*)
Lappert, M.F. *et al*, *J. Organomet. Chem.*, 1974, **66**, 271 (*pe*)
Lappert, M.F. *et al*, *J. Chem. Soc., Chem. Commun.*, 1975, 830.
Ballard, D.G.H. *et al*, *J. Catal.*, 1976, **44**, 116 (*use*)

$C_{17}H_{25}PZr$ **Zr-00046**

Bis[(1,2,3,4,5,6-η)methylbenzene](trimethylphosphine)zirconium, 10CI

Ditoluene(trimethylphosphine)zirconium

[71361-41-6]

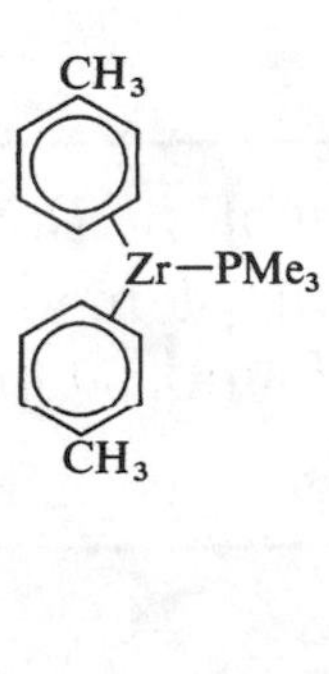

M 351.578

Believed to have (bent) sandwich struct. Dark-green cryst. Sol. aromatic hydrocarbons, spar. sol. pet. ether. Solns. stable under Ar but dec. under N_2.

Cloke, F.G.N. *et al*, *J. Chem. Soc., Dalton Trans.*, 1981, 1938 (*synth, ms, nmr*)

$C_{17}H_{29}ClSi_2Zr$ **Zr-00047**

[Bis(trimethylsilyl)methyl]chlorobis(η^5-2,4-cyclopentadien-1-yl)zirconium, 10CI

[64815-30-1]

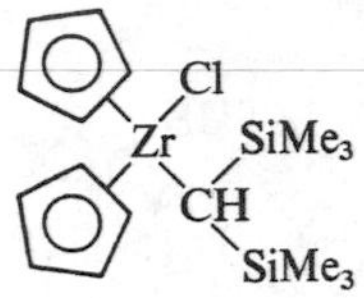

M 416.260

Gives first 'sideways on' N_2 complex on reduction under N_2. Yellow air-stable cryst. (toluene). Mp 221-3°.

Atwood, J.L. *et al*, *J. Am. Chem. Soc.*, 1977, **99**, 6645 (*synth, nmr*)
Gynane, M.J.S. *et al*, *J. Chem. Soc., Chem. Commun.*, 1978, 34.
Jeffery, J. *et al*, *J. Organomet. Chem.*, 1979, **181**, 25.
Lappert, M.F. *et al*, *J. Chem. Soc., Dalton Trans.*, 1981, 814 (*nmr*)

$C_{18}H_{14}Cl_2Zr$ **Zr-00048**

Dichlorobis[(1,2,3,3a,7a-η)-1*H*-inden-1-yl]zirconium, 9CI

[12148-49-1]

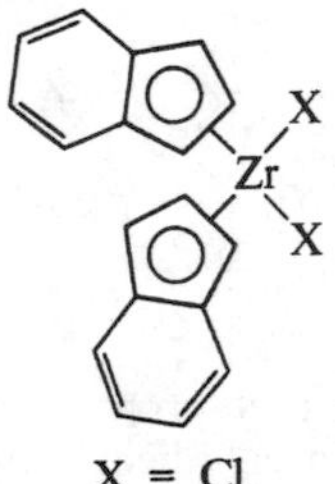

X = Cl

M 392.435

6-membered ring not coordinated. Yellow cryst. Mp 264°. Reported to dec. in org. solvs. 6 membered ring catalytically hydrogenated over Pt. Sensitive to air in soln.

Samuel, E. *et al*, *J. Organomet. Chem.*, 1965, **4**, 156 (*synth, ir*)
Samuel, E., *Bull. Soc. Chim. Fr.*, 1966, 3548 (*synth, nmr*)

$C_{18}H_{18}N_2Zr$ **Zr-00049**

Bis(η^5-cyclopentadien-1-yl)di-1*H*-pyrrol-1-ylzirconium, 10CI

Di-π-cyclopentadienyldipyrrol-1-ylzirconium, 8CI.
Bis(η-cyclopentadienyl)bis(pyrrolyl)zirconium

[73587-39-0]

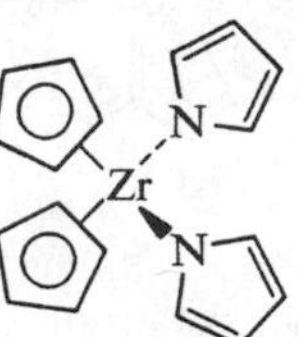

M 353.574

Yellow cryst. Mp 195°. Cyclopentadienyl ligands removed by pyridine-sodium at reflux temp. in THF to give the $ZrPy_6^{\ominus\ominus}$ anion.

Issleib, K. *et al*, *Z. Anorg. Chem.*, 1969, **369**, 83 (*synth, ir*)
Vann Bynum, R. *et al*, *Inorg. Chem.*, 1980, **19**, 2360 (*synth, cryst struct, ms*)

$C_{18}H_{18}Zr$ Zr-00050

Bis(η^5-2,4-cyclopentadien-1-yl)[1,2-phenylenebis[methylene]]-zirconium, 10CI
2,2-Bis(η^5-cyclopentadienyl)-2-zirconaindane
[75070-70-1]

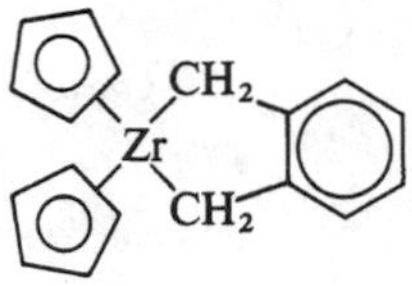

M 325.560
Cryst. Mp 177°.

Lappert, M.F. *et al*, *J. Chem. Soc., Chem. Commun.*, 1980, 476 (*synth, nmr, cryst struct*)
Lappert, M.F. *et al*, *J. Organomet. Chem.*, 1980, **192**, C35 (*props*)

$C_{18}H_{21}ClZr$ Zr-00051

Chlorotris(η-methylcyclopentadienyl)zirconium
Chloro(1-methyl-2,4-cyclopentadien-1-yl)-bis[(1,2,3,4,5-η)-1-methyl-2,4-cyclopentadien-1-yl]zirconium, 10CI
[69005-92-1]

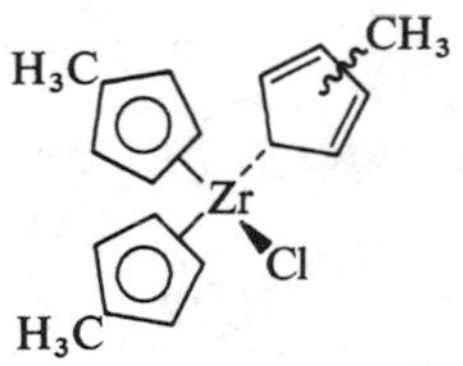

M 364.037
Fluxional η^1-$(\eta^5)_2$ struct. proposed, different from Chlorotris(η-cyclopentadienyl)zirconium, Zr-00039 . Pale-yellow cryst. (toluene). Mp 155°, 166-7°.

Brainina, E.H. *et al*, *Dokl. Akad. Nauk SSSR, Ser. Sci. Khim.*, 1966, **169**, 335 (*synth*)
Etievant, P. *et al*, *Bull. Soc. Chim. Fr., Part II*, 1978, 292 (*synth, ms, nmr*)

$C_{18}H_{26}AlClZr$ Zr-00052

Chlorobis(η^5-2,4-cyclopentadien-1-yl)[1-(dimethylalumino)-2-methyl-1-pentenyl]zirconium, 10CI
Chlorobis(η-cyclopentadienyl)(1-dimethylalumino-2-methyl-2-propylethenyl)zirconium
[77089-97-5]

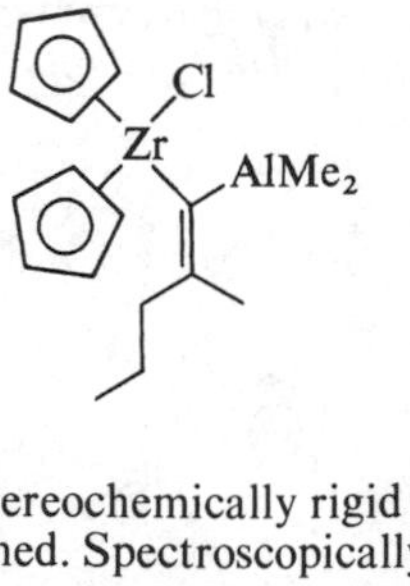

M 396.058
Prod. as a single stereochemically rigid stereoisomer. Synthetic intermed. Spectroscopically and analytically characterised.

Yoshida, T. *et al*, *J. Am. Chem. Soc.*, 1981, **103**, 1276 (*synth, ir, nmr, use*)

$C_{18}H_{30}N_2Zr$ Zr-00053

Bis(η^5-cyclopentadienyl)bis(diethylamido)zirconium
Di-π-cyclopentadienylbis(diethylaminato)zirconium, 8CI. Zirconocene bis(diethylamide)
As Dibromobis(η^5-2,4-cyclopentadien-1-yl)zirconium, Zr-00006 with

$$X = Y = NEt_2$$

M 365.668
Orange-yellow solid. $Bp_{0.03}$ 120-30° subl.

Chandra, G. *et al*, *J. Chem. Soc. (A)*, 1968, 1940 (*synth, ms*)

$C_{18}H_{32}Si_2Zr$ Zr-00054

Bis(η^5-2,4-cyclopentadien-1-yl)bis[(trimethylsilyl)methyl]zirconium, 10CI
[11077-97-7]

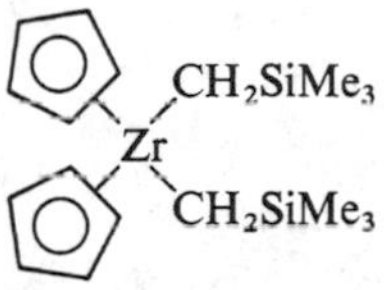

M 395.842
White needles. Mp 96-7°. Sensitive to air but thermally more stable than the analogous alkyls.

Collier, M.R. *et al*, *J. Organomet. Chem.*, 1970, **25**, C36 (*synth*)
Collier, M.R. *et al*, *J. Chem. Soc., Dalton Trans.*, 1973, 445 (*synth, ir, nmr*)
Jeffery, J. *et al*, *J. Chem. Soc., Chem. Commun.*, 1978, 1081 (*nmr*)
Jeffery, J. *et al*, *J. Chem. Soc., Dalton Trans.*, 1981, 1593 (*cryst struct*)

$C_{18}H_{35}Cl_2PZr$ Zr-00055

Dichloro(2,2-dimethylpropyl)[(1,2,3,4,5-η)-1,2,3,4,5-pentamethyl-2,4-cyclopentadien-1-yl](trimethylphosphine)zirconium, 10CI
Dichloro(neopentyl)(trimethylphosphine)(η-pentamethylcyclopentadienyl)zirconium
[78067-94-4]

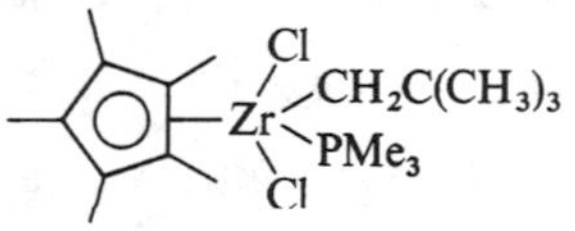

M 444.574
Orange cryst. at −30°. Mp >−30°.

Wengrovius, J.H. *et al*, *J. Organomet. Chem.*, 1981, **205**, 319 (*nmr*)

The first digit of the Entry number defines the Supplement in which the Entry is found. 0 indicates the Main Work

$C_{19}H_{18}MoO_3Zr$ **Zr-00056**

[μ-(Acetyl-*C*:*O*)][μ-(η^2-carbonyl)][carbonyl(η^5-2,4-cyclopentadien-1-yl)molydenum]bis(η^5-2,4-cyclopentadien-1-yl)zirconium

*μ-[Acetyl-*C(Mo):O(Zr)*]μ-[carbonyl-*C-(Mo):O(Zr)*]bis-η-cyclopentadienylzirconium(carbonyl)η-cyclopentadienylmolybdenum*

[76671-79-9]

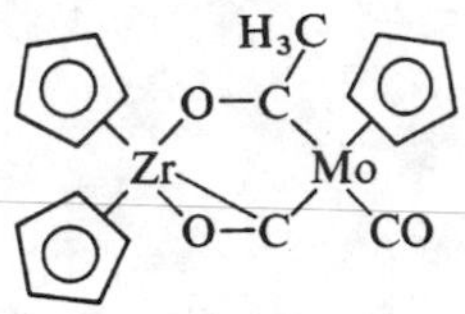

M 481.509

Air-stable solid. Yellow cryst. Unreactive to most donor ligands.

Longato, B. *et al*, *J. Am. Chem. Soc.*, 1981, **103**, 209 (*synth, ir, nmr, cryst struct*)

$C_{19}H_{19}Zr$ **Zr-00057**

Bis(benzyl)(η-cyclopentadienyl)zirconium

(η^5-2,4-Cyclopentadien-1-yl)bis(phenylmethyl)zirconium. Dibenzyl(cyclopentadienyl)zirconium

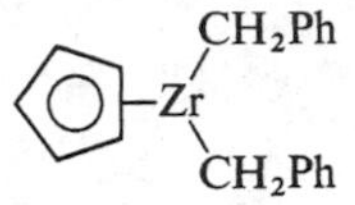

M 338.579

Unstable, not isol.

Lappert, M.F. *et al*, *J. Chem. Soc., Dalton Trans.*, 1981, 805 (*esr*)

$C_{19}H_{32}OSi_2Zr$ **Zr-00058**

Bis(trimethylsilyl)acetyl-*C*,*O*)bis(η^5-2,4-cyclopentadien-1-yl)methylzirconium, 10CI

*Methyl[η^2-bis(trimethylsilyl)acetyl-*C,O*]bis(η-cyclopentadienyl)zirconium*

[71520-96-2]

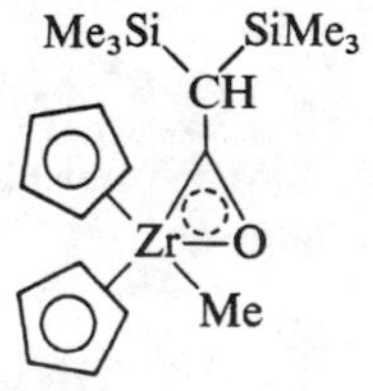

M 423.852

Pale-yellow cryst. (CH_2Cl_2/Et_2O). Mp 123-5°. Stable to CO loss in soln. at 25°.

Lappert, M.F. *et al*, *J. Organomet. Chem.*, 1979, **174**, C35 (*synth, ir*)

Jeffery, J. *et al*, *J. Chem. Soc., Dalton Trans.*, 1981, 1593 (*synth, ir, pmr*)

$C_{19}H_{38}O_3Zr$ **Zr-00059**

Tris(diethyl ether)methylphenylzirconium

Methyltris[(1,1′-oxybis[ethane]]phenylzirconium, 10CI

[69593-37-9]

$ZrMePh(Et_2O)_3$

M 405.727

Associated. Brown cryst. (Et_2O). Spar. sol. org. solvs. Mp ca. 90° dec.

Razuvaev, G.A. *et al*, *Dokl. Akad. Nauk SSSR, Ser. Sci. Khim.*, 1978, **243**, 1212 (*synth, props*)

Razuvaev, G.A. *et al*, *Inorg. Chim. Acta*, 1980, **44**, L285 (*synth, ir*)

$C_{20}H_{10}ClCo_3O_{10}Zr$ **Zr-00060**

Chlorobis(η^5-2,4-cyclopentadien-1-yl)[μ_4-[methanolato(4−)-*C*:*C*:*C*:*O*]](nonacarbonyltricobalt)zirconium, 10CI

Chlorobis(η^5-cyclopentadienyl)[μ_3-(oxymethylidyne)-cyclotris(tricarbonylcobalt)]zirconium

[67422-57-5]

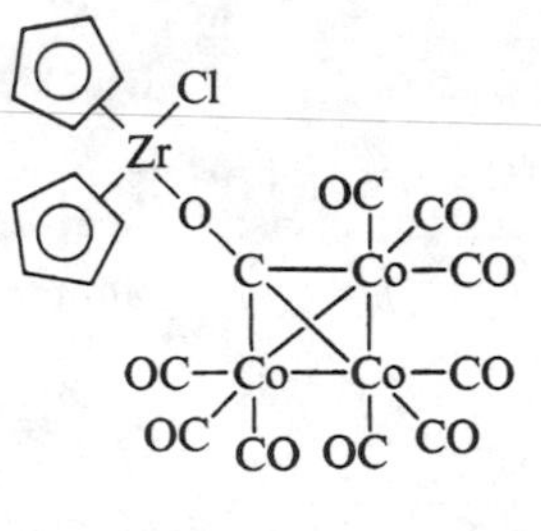

M 713.766

Dark-red cryst. (toluene). Spar. sol. aromatic, v. spar. sol. aliphatic solvs. Mp 138-41°.

Stutte, B. *et al*, *Chem. Ber.*, 1978, **111**, 1603 (*synth,ir, nmr, cryst struct*)

Ishii, M. *et al*, *Ber. Bunsenges. Phys. Chem.*, 1979, **83**, 1026 (*synth, ir, nmr*)

$C_{20}H_{20}Cl_2OZr_2$ **Zr-00061**

Dichlorotetrakis(η^5-2,4-cyclopentadien-1-yl)-μ-oxodizirconium, 10CI

Dichlorotetra-π-cyclopentadienyloxodizirconium. μ-Oxobis[chlorobis(η-cyclopentadienyl)zirconium]

[12097-04-0]

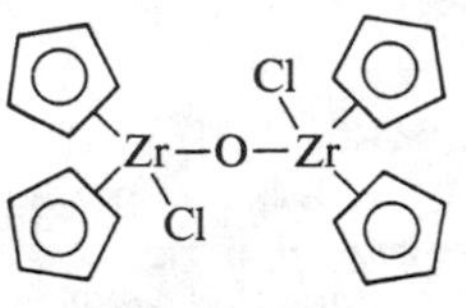

M 529.723

Colourless cryst. Mp 300-9°. Bp 200° subl. *in vacuo*.

Reid, A.F. *et al*, *Aust. J. Chem.*, 1965, **18**, 173 (*synth, ir, nmr, ms*)

Samuel, E., *Bull. Soc. Chim. Fr.*, 1966, 3548 (*synth, ir, uv, nmr*)

Etievant, P. *et al*, *Bull. Soc. Chim. Fr.*, 1978, 292 (*synth*)

Inorg. Synth., 1979, **19**, 223 (*synth*)

$C_{20}H_{20}Zr$ **Zr-00062**

Bis[(1,2,3,3*a*,7*a*-η)-1*H*-inden-1-yl]dimethylzirconium, 9CI

[49596-04-5]

As Dichlorobis[(1,2,3,3*a*,7*a*-η)-1*H*-inden-1-yl]zirconium, Zr-00048 with

X = Me

M 351.598

Straw-coloured solid by subl. Mp 107° dec. Subl. *in vacuo*.

Samuel, E. *et al*, *J. Am. Chem. Soc.*, 1973, **95**, 6263 (*synth, nmr*)

Alt, H.G. *et al*, *J. Am. Chem. Soc.*, 1974, **96**, 5936.

Atwood, J.L. *et al*, *Inorg. Chem.*, 1975, **14**, 1757 (*cryst struct*)

$C_{20}H_{21}ClZr$ Zr-00063

Chlorobis(η^5-2,4-cyclopentadien-1-yl)(2-methyl-1-phenyl-1-propenyl)zirconium, 10CI

Chlorobis(η-cyclopentadienyl)(2,2-dimethyl-1-phenylethenyl)zirconium

[63422-41-3]

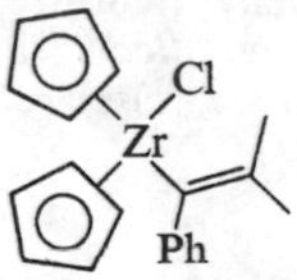

M 388.059

Yellow cryst. (Et_2O). Mp 170° dec. Also characterized by cryst. struct. and elemental analysis.

Cardin, C.J. *et al*, *J. Organomet. Chem.*, 1977, **132**, C23 (*synth, ir, nmr*)

$C_{20}H_{32}Zr$ Zr-00064

Bis(η^5-2,4-cyclopentadien-1-yl)bis(2,2-dimethylpropyl)zirconium, 10CI

Bis(η^5-cyclopentadienyl)bis(neopentyl)zirconium. Zirconocene dineopentyl

[69793-83-5]

As Dibromobis(η^5-2,4-cyclopentadien-1-yl)zirconium, Zr-00006 with

$$X = Y = —CH_2C(CH_3)_3$$

M 363.693

Pale-yellow.

Jeffery, J. *et al*, *J. Chem. Soc., Chem. Commun.*, 1978, 1081 (*nmr*)

Lappert, M.F. *et al*, *J. Chem. Soc., Chem. Commun.*, 1979, 305 (*props*)

Jeffery, J. *et al*, *J. Chem. Soc., Dalton Trans.*, 1981, 1593 (*cryst struct, synth, nmr*)

$C_{20}H_{32}Zr$ Zr-00065

Dihydrobis[(1,2,3,4,5-η)-1,2,3,4,5-pentamethyl-2,4-cyclopentadien-1-yl]zirconium, 10CI

Dihydridobis(η-pentamethylcyclopentadienyl)zirconium

[61396-34-7]

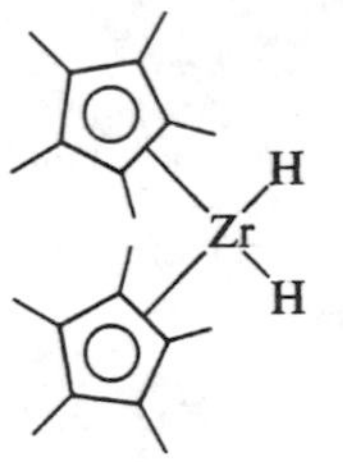

M 363.693

Monomeric. Pale-yellow cryst. (pet. ether). Freely sol. hydrocarbons and ethers. Deuterium analogue also prepd.

Manriquez, J.M. *et al*, *J. Am. Chem. Soc.*, 1976, **98**, 6733 (*synth, mw, ir, nmr*)

Manriquez, J.M. *et al*, *J. Am. Chem. Soc.*, 1978, **100**, 2716 (*synth, nmr, ir*)

Threlkel, R.S. *et al*, *J. Am. Chem. Soc.*, 1981, **103**, 2650 (*use*)

$C_{20}H_{33}Al_2ClZr$ Zr-00066

μ-Chlorobis(η^5-2,4-cyclopentadien-1-yl)bis(diethylaluminum)-μ-1-ethanyl-2-ylidenezirconium, 10CI

μ-Chloro-1-[bis(η-cyclopentadienylzirconio)]-2,2-bis(diethylalumino)ethane

[55466-98-3]

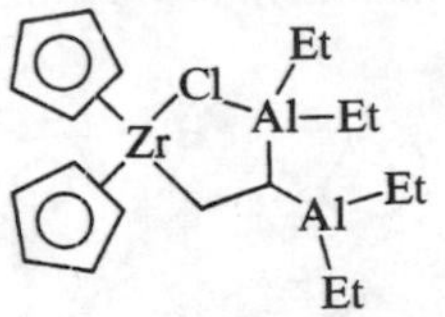

M 454.117

Cryst. (heptane). Mp 128° dec.

Heins, E. *et al*, *Makromol. Chem.*, 1970, **134**, 1 (*synth*)

Kaminsky, W. *et al*, *Justus Liebigs Ann. Chem.*, 1975, 438 (*synth, nmr, bibl*)

Kaminsky, W. *et al*, *Angew. Chem., Int. Ed. Engl.*, 1976, **15**, 629 (*synth, cryst struct*)

Kopf, J. *et al*, *Cryst. Struct. Commun.*, 1980, **9**, 197 (*cryst struct*)

$C_{20}H_{34}Sn_2Zr$ Zr-00067

Bis[(1,2,3,4,5,6-η)-methylbenzene]bis(trimethylstannyl)zirconium, 10CI

Ditoluenebis(trimethylstannyl)zirconium

[78379-41-6]

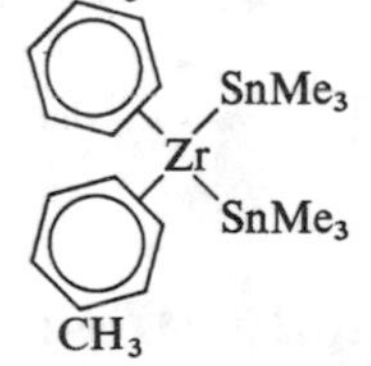

M 603.089

Black cryst. Solid only slowly dec. by air.

Cloke, F.G.N. *et al*, *J. Chem. Soc., Chem. Commun.*, 1981, 117 (*synth, nmr*)

$C_{20}H_{44}Zr$ Zr-00068

Tetrakis(2,2-dimethylpropyl)zirconium, 9CI

Tetraneopentylzirconium

[38010-72-9]

$$Zr[CH_2C(CH_3)_3]_4$$

M 375.788

Polymerizes olefins. Colourless cryst. by subl. Mp [illegible]-11° dec. $Bp_{0.01}$ 50° subl. Shows enhanced thermal stability over straight chain alkyls.

Davidson, P.J. *et al*, *J. Organomet. Chem.*, 1973, **57**, 269 (*synth, nmr, ms, ir, raman*)

Mowat, W. *et al*, *J. Chem. Soc., Dalton Trans.*, 1973, 1120 (*synth, nmr, ir*)

Lappert, M.F. *et al*, *J. Organomet. Chem.*, 1974, **66**, 271 (*pe*)

Lappert, M.F. *et al*, *J. Chem. Soc., Chem. Commun.*, 1975, 830.

Chiu, K.W. *et al*, *J. Chem. Soc., Dalton Trans.*, 1981, 2088.

Wengrovius, J.H. *et al*, *J. Organomet. Chem.*, 1981, **205**, 319 (*synth*)

$C_{21}H_{24}O_5Zr$ **Zr-00069**

π-Cyclopentadienylbis(2,4-pentanedionato)phenoxyzirconium, 8CI

Bis(acetylacetonato)(η-cyclopentadienyl)phenoxyzirconium

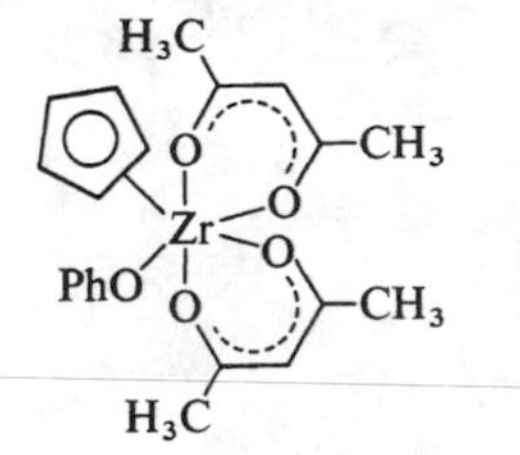

M 447.638

Mp 62-63.5°.

Brainina, E.M. *et al*, *Dokl. Akad. Nauk SSSR, Ser. Sci. Khim.*, 1971, **196**, 1085 (*synth, nmr*)

$C_{21}H_{57}ClSi_6Zr$ **Zr-00070**

Tris[bis(trimethylsilyl)methyl]chlorozirconium, 10CI

Chlorotris[bis(trimethylsilyl)methyl]zirconium

[53668-84-1]

$ClZr[CH(SiMe_3)_2]_3$

M 604.867

White cryst.

Barker, G.K. *et al*, *J. Organomet. Chem.*, 1974, **76**, C45 (*synth, nmr, ir*)

$C_{22}H_{10}F_{10}Zr$ **Zr-00071**

Bis(η⁵-cyclopentadienyl)bis(pentafluorophenyl)zirconium

Dicyclopentadienylbis(pentafluorophenyl)zirconium

As Dibromobis(η^5-2,4-cyclopentadien-1-yl)zirconium, Zr-00006 with

$X = Y = C_6F_5$

M 555.525

White cryst. by subl. Mp 218-9°. $Bp_{0.01}$ 120° subl.

▷Explodes in air above Mp

Chaudari, M. *et al*, *J. Chem. Soc. (A)*, 1966, 838 (*synth, ir*)
Samuel, E. *et al*, *J. Am. Chem. Soc.*, 1973, **95**, 6263 (*nmr*)

$C_{22}H_{20}S_2Zr$ **Zr-00072**

Bis(benzenethiolato)bis(η⁵-2,4-cyclopentadien-1-yl)zirconium, 10CI

Bis(η-cyclopentadienyl)bis(phenylthiolato)zirconium

[37206-34-1]

As Dibromobis(η^5-2,4-cyclopentadien-1-yl)zirconium, Zr-00006 with

X = Y = SPh

M 439.740

Yellow cryst. Mp 147-54° dec.

Köpf, H., *J. Organomet. Chem.*, 1968, **14**, 353 (*synth, ir, nmr*)
Minacheva, M.K. *et al*, *Izv. Akad. Nauk SSSR, Ser. Khim.*, 1972, 139 (*synth*)
Petersen, J.L., *J. Organomet. Chem.*, 1979, **166**, 179 (*synth*)
Brindley, P.B. *et al*, *J. Organomet. Chem.*, 1981, **222**, 89 (*synth, ir, nmr*)

$C_{22}H_{20}Zr$ **Zr-00073**

Bis(η⁵-2,4-cyclopentadien-1-yl)diphenylzirconium, 10CI

Zirconocene diphenyl

[51177-89-0]

As Dibromobis(η^5-2,4-cyclopentadien-1-yl)zirconium, Zr-00006 with

X = Y = Ph

M 375.620

White cryst. (Et_2O). Mp 140° dec.

Samuel, E. *et al*, *J. Am. Chem. Soc.*, 1973, **95**, 6263 (*synth, ir, nmr*)
Razuvaev, G.A. *et al*, *Dokl. Akad. Nauk SSSR, Ser. Sci. Khim.*, 1978, **243**, 1212 (*props*)
Razuvaev, G.A. *et al*, *J. Organomet. Chem.*, 1979, **174**, 67 (*props*)
Erker, G. *et al*, *J. Organomet. Chem.*, 1980, **188**, C1 (*props*)
Hsueh-Sung Tung, *et al*, *Inorg. Chim. Acta*, 1981, **52**, 197.

$C_{22}H_{30}O_2Zr$ **Zr-00074**

Dicarbonylbis[(1,2,3,4,5-η)-1,2,3,4,5-pentamethyl-2,4-cyclopentadien-1-yl)zirconium, 10CI

[61396-31-4]

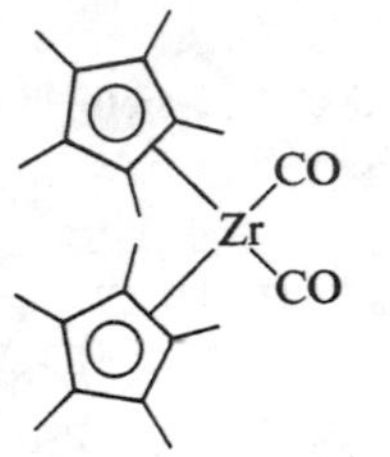

M 417.698

Purple-brown or black cryst. (toluene).

Manriquez, J.M. *et al*, *J. Am. Chem. Soc.*, 1976, **98**, 6733 (*synth, ir, nmr*)
Sikora, D.J. *et al*, *J. Am. Chem. Soc.*, 1981, **103**, 1265 (*cryst struct, synth, ir, ms*)

$C_{22}H_{35}NZr$ **Zr-00075**

Hydro[(methylimino)methyl]bis[(1,2,3,4,5-η)-1,2,3,4,5-pentamethyl-2,4-cyclopentadien-1-yl]zirconium, 10CI

(N-Methylformimidoyl)hydridobis(η⁵-pentamethylcyclopentadienyl)zirconium

[72108-52-2]

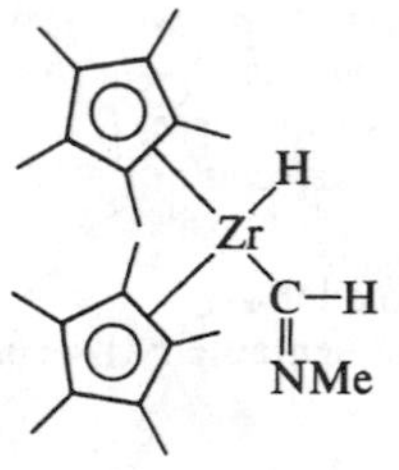

M 404.745

Unstable; characterized spectroscopically.

Wolczanski, P.T. *et al*, *J. Am. Chem. Soc.*, 1979, **101**, 6450 (*synth, ir, nmr*)

$C_{22}H_{36}INZr$ Zr-00076

Dimethylamidoiodobis(η^5-pentamethylcyclopentadienyl)zirconium

Iodo(N-*methylmethanaminato*)*bis*[(*1,2,3,4,5-η*)-*1,2,3,4,5-pentamethyl-2,4-cyclopentadien-1-yl*]*zirconium*, 10CI. *Bis*(η-*pentamethylcyclopentadienyl*)-*dimethylamido*(*iodo*)*zirconium*

[72125-82-7]

X = I

M 532.658

Characterization includes elemental analyses.

Wolczanski, P.T. *et al*, *J. Am. Chem. Soc.*, 1979, **101**, 6450 (*synth*, *nmr*)

$C_{22}H_{36}Zr$ Zr-00077

Dimethylbis[(1,2,3,4,5-η)-1-(2-methylbutyl)-2,4-cyclopentadien-1-yl]zirconium, 10CI

[75663-08-0]

M 391.746

(S,S)-form [75663-08-0]

Chiral hydrogenation catalyst. Oil. $[\alpha]_D$ +14° (c, 2.5 in C_6H_6).

Couturier, S. *et al*, *J. Organomet. Chem.*, 1980, **195**, 291 (*synth*, *nmr*)

$C_{22}H_{37}NZr$ Zr-00078

Dimethylamidohydro(pentamethylcyclopentadienyl)zirconium

Hydro(N-*methylmethanaminato*)*bis*[(*1,2,3,4,5-η*)-*1,2,3,4,5-pentamethyl-2,4-cyclopentadien-1-yl*]*zirconium*, 10CI. *Bis*(η-*pentamethylcyclopentadienyl*)*dimethylamidohydridozirconium*

[72108-53-3]

As Dimethylamidoiodobis(η^5-pentamethylcyclopentadienyl)zirconium, Zr-00076 with

X = H

M 406.761

Solid by subl. Characterization includes elemental analysis.

Wolczanski, P.T. *et al*, *J. Am. Chem. Soc.*, 1979, **101**, 6450 (*synth*, *ir*, *nmr*)

$C_{23}H_{17}ClN_2O_2Zr$ Zr-00079

Chloro(η^5-cyclopentadienyl)bis(8-quinolinolato)zirconium

[12114-15-7]

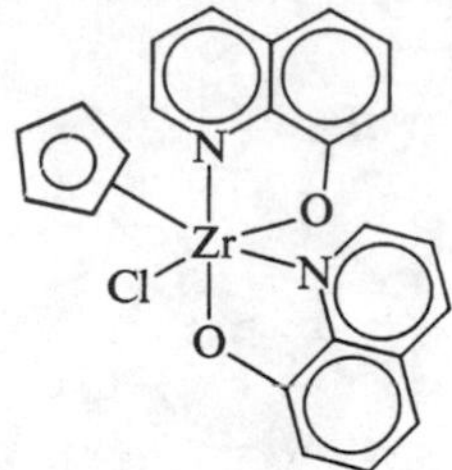

M 480.073

Yellow solid (Et_2O). Mp 260-3°.

Freidlina, R.K.H. *et al*, *Izv. Akad. Nauk SSSR*, *Ser. Khim.*, 1966, 1396 (*synth*)

Charalambous, J. *et al*, *J. Chem. Soc.* (*A*), 1971, 2487 (*synth*, *ir*, *nmr*, *ms*)

$C_{23}H_{20}OZr$ Zr-00080

(η^2-Benzoyl)bis(η^5-2,4-cyclopentadien-1-yl)phenylzirconium, 10CI

[66152-53-2]

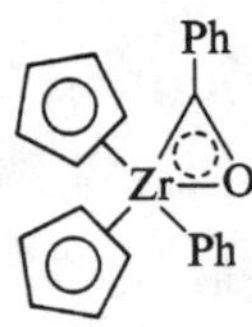

M 403.630

Characterized spectroscopically. Yellow-orange solid (heptane/toluene). Mp 23°. Stable to loss of CO both in solid state and in soln.

Fachinetti, G. *et al*, *J. Chem. Soc.*, *Dalton Trans.*, 1977, 1946 (*synth*, *ir*, *nmr*)

Erker, G. *et al*, *Angew. Chem.*, *Int. Ed. Engl.*, 1978, **17**, 605 (*isom*)

Erker, G. *et al*, *J. Organomet. Chem.*, 1980, **188**, C1.

Marsella, J.A. *et al*, *J. Am. Chem. Soc.*, 1981, **103**, 5596.

$C_{23}H_{22}ClPZr$ Zr-00081

Chlorobis(η^5-2,4-cyclopentadien-1-yl)[(diphenylphosphino)-methyl]zirconium, 10CI

[74380-49-7]

As Dibromobis(η^5-2,4-cyclopentadien-1-yl)zirconium, Zr-00006 with

X = Cl, Y = $-CH_2PPh_2$

M 456.074

Pale-yellow cryst.

Schore, N.E. *et al*, *J. Am. Chem. Soc.*, 1980, **102**, 4251 (*synth*, *cryst struct*, *nmr*, *ir*)

$C_{23}H_{34}Si_2Zr$ Zr-00082

[Bis(trimethylsilyl)methyl]bis(η^5-2,4-cyclopentadien-1-yl)-phenylzirconium, 10CI

Bis(η-cyclopentadienyl)phenyl[bis(trimethylsilyl)methyl]zirconium

[79194-27-7]

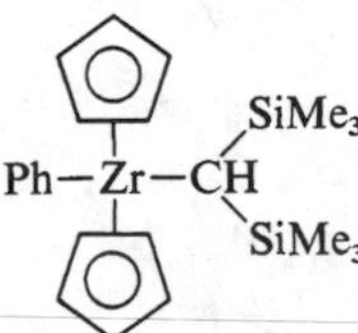

M 457.913

White cryst. Mp 116-8°. Shows unusually high barrier to rotation about Zr—C bonds.

Jeffery, J. *et al*, *J. Chem. Soc., Chem. Commun.*, 1978, 1081 (*nmr, cryst struct*)
Jeffery, J. *et al*, *J. Chem. Soc., Dalton Trans.*, 1981, 1593 (*synth, ir, nmr*)

$C_{23}H_{45}ClSi_4Zr$ Zr-00083

[Bis(trimethylsilyl)methyl]chlorobis[(1,2,3,4,5-η)-1-(trimethylsilyl)-2,4-cyclopentadien-1-yl]zirconium, 10CI

Chlorobis[η-(trimethylsilyl)cyclopentadienyl]bis(trimethylsilyl)methylzirconium

[74042-81-2]

R = $SiMe_3$

M 560.624

Pale-yellow cryst. (hexane). Mp 154-6°.

Lappert, M.F. *et al*, *J. Chem. Soc., Dalton Trans.*, 1981, 805, 814 (*synth, ir, nmr, cryst struct*)

$C_{24}H_{22}Cl_2Zr$ Zr-00084

Bis(η^5-benzylcyclopentadienyl)dichlorozirconium

Dichlorobis[(1,2,3,4,5-η)-1-(phenylmethyl)-2,4-cyclopentadien-1-yl]zirconium, 10CI

[58689-69-3]

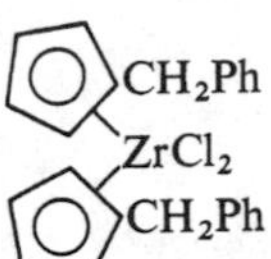

M 472.564

Mp 118°. Triboluminescent and highly piezoelectric.

Renaut, P. *et al*, *J. Organomet. Chem.*, 1978, **148**, 35 (*synth*)
Dusausoy, Y. *et al*, *J. Organomet. Chem.*, 1978, **157**, 167 (*cryst struct, nmr, ir, uv, raman*)

$C_{24}H_{24}Zr$ Zr-00085

Bis(benzyl)bis(η^5-cyclopentadienyl)zirconium

Bis(η^5-2,4-cyclopentadien-1-yl)bis(phenylmethyl)zirconium, 10CI. Zirconocene dibenzyl

[37206-41-0]

As Dibromobis(η^5-2,4-cyclopentadien-1-yl)zirconium, Zr-00006 with

$$X = Y = -CH_2Ph$$

M 403.674

Yellow needles (toluene). Shows markedly higher thermal stability than Ti analogue.

Fachinetti, G. *et al*, *J. Chem. Soc., Chem. Commun.*, 1972, 654 (*synth, ir, nmr*)
Fachinetti, G. *et al*, *J. Chem. Soc., Dalton Trans.*, 1977, 1946 (*ir, nmr*)
Lappert, M.F. *et al*, *J. Chem. Soc., Chem. Commun.*, 1979, 305.
Brindley, P.B. *et al*, *J. Chem. Soc., Perkin Trans. 2*, 1981, 419 (*nmr*)

$C_{24}H_{34}Si_2Zr$ Zr-00086

Bis(η^5-2,4-cyclopentadien-1-yl)[1,2-phenylenebis[(trimethylsilyl)methylene]]zirconium, 10CI

2,2-Bis(η^5-cyclopentadienyl)-1,3-bis(trimethylsilyl)-2-zirconaindane

[76933-95-4]

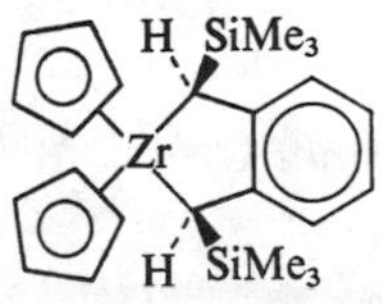

M 469.924

meso-form

Red cryst. (pentane). Sol. org. solvs. Mp 141-2°. Limited air stability but thermally robust and sublimable.

Lappert, M.F. *et al*, *J. Chem. Soc., Chem. Commun.*, 1980, 1284 (*synth, ir, ms, nmr*)

$C_{24}H_{40}Zr$ Zr-00087

Hydro(2-methylpropyl)bis[(1,2,3,4,5-η)-1,2,3,4,5-pentamethyl-2,4-cyclopentadien-1-yl)zirconium, 10CI

Isobutyl(hydrido)bis(η-pentamethylcyclopentadienyl)-zirconium

[67108-86-5]

M 419.800

Yellow, microcryst. solid (pet. ether). Isotopomers also made.

Manriquez, J.M. *et al*, *J. Am. Chem. Soc.*, 1978, **100**, 2716 (*synth, nmr*)
McAlister, D.R. *et al*, *J. Am. Chem. Soc.*, 1978, **100**, 5966 (*synth, nmr*)

$C_{25}H_{38}Al_2Zr$ **Zr-00088**

[2,2-Bis(diethylalumino)ethyl]tris(η^5-2,4-cyclopentadien-1-yl)zirconium, 10CI

1,1-Bis(diethylalumino)-2-[bis(η-cyclopentadienyl)zirconio]ethane cyclopentadienide

[64539-44-2]

M 483.758

Ionic material. Polymerises olefins. Cryst. Extremely air- and water-sensitive.

Kaminsky, W. *et al*, *Makromol. Chem.*, 1974, **175**, 443 (*synth, use*)

Kaminsky, W. *et al*, *Angew. Chem., Int. Ed. Engl.*, 1976, **15**, 629 (*nmr, struct*)

Kopf, J. *et al*, *Cryst. Struct. Commun.*, 1980, **9**, 271 (*cryst struct*)

$C_{25}H_{45}ClSi_2Zr$ **Zr-00089**

Bis(η^5-*tert*-butylcyclopentadienyl)[bis(trimethylsilyl)methyl]-chlorozirconium

[Bis(trimethylsilyl)methyl]chlorobis[(1,2,3,4,5-η)-1-(1,1-dimethylethyl)-2,4-cyclopentadien-1-yl]zirconium, 10CI

[74042-79-8]

As [Bis(trimethylsilyl)methyl]chlorobis[(1,2,3,4,5-η)-1-(trimethylsilyl)-2,4-cyclopentadien-1-yl]zirconium, Zr-00083 with

$$R = (H_3C)_3C—$$

M 528.475

Pale-yellow cryst. (hexane). Mp 159-62°.

Lappert, M.F. *et al*, *J. Chem. Soc., Dalton Trans.*, 1981, 805, 814 (*synth, ir, nmr, cryst struct*)

$C_{25}H_{48}Zr$ **Zr-00090**

Tris(2,2-dimethylpropyl)[(1,2,3,4,5-η)-1,2,3,4,5-pentamethyl-2,4-cyclopentadien-1-yl]zirconium, 10CI

Trineopentyl(η-pentamethylcyclopentadienyl)zirconium

[78067-96-6]

M 439.874

Off-white cryst. at −30°. Mp >−30°.

Wengrovius, J.H. *et al*, *J. Organomet. Chem.*, 1981, **205**, 319 (*nmr*)

$C_{26}H_{20}O_6Zr$ **Zr-00091**

(η^5-2,4-Cyclopentadien-1-yl)tris(2-hydroxy-2,4,6-cycloheptatrien-1-onato-*O,O'*)zirconium, 9CI

(η-Cyclopentadienyl)tris(tropolonato)zirconium

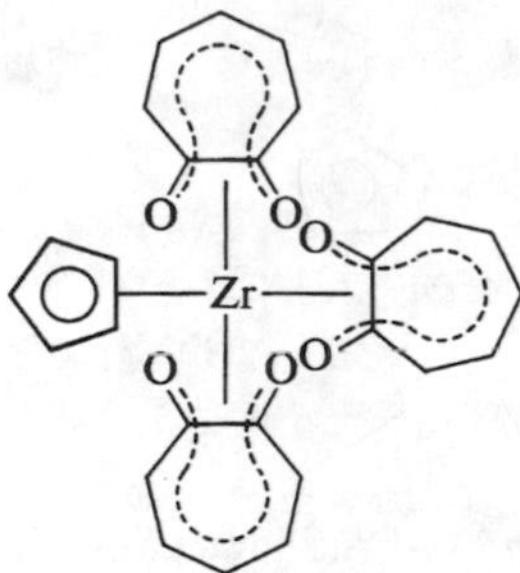

M 519.660

Pale-yellow solid.

Frazer, M.J. *et al*, *Inorg. Chem.*, 1971, **10**, 2137 (*ir, nmr*)

McPartlin, M. *et al*, *J. Organomet. Chem.*, 1976, **104**, C20 (*cryst struct*)

$C_{26}H_{20}Zr$ **Zr-00092**

Bis(η^5-2,4-cyclopentadien-1-yl)bis(phenylethynyl)zirconium, 10CI

Zirconocene bis(phenylacetylide)

[72982-57-1]

As Dibromobis(η^5-2,4-cyclopentadien-1-yl)zirconium, Zr-00006 with

$$X = Y = —C{\equiv}CPh$$

M 423.664

Light-brown solid.

Jenkins, A.D. *et al*, *J. Organomet. Chem.*, 1970, **23**, 165 (*synth, ir, nmr, ms*)

Jimenez, R. *et al*, *J. Organomet. Chem.*, 1979, **182**, 353 (*synth, ir, nmr, uv*)

$C_{26}H_{24}Zr$ **Zr-00093**

Bis(η-cyclopentadienyl)bis(2-phenylethenyl)zirconium

Di-π-cyclopentadienyldistyrylzirconium, 8CI. Bis(η-cyclopentadienyl)bis(β-styryl)zirconium

[11058-18-7]

As Dibromobis(η^5-2,4-cyclopentadien-1-yl)zirconium, Zr-00006 with

$$X = Y = —CH{=}CHPh$$

M 427.696

Black solid.

Wailes, P.C. *et al*, *J. Organomet. Chem.*, 1971, **27**, 373 (*synth, nmr*)

$C_{26}H_{26}Zr$ **Zr-00094**

Tribenzyl(η-cyclopentadienyl)zirconium

(η^5-2,4-Cyclopentadien-1-yl)tris(phenylmethyl)zirconium, 10CI

[78163-63-0]

M 429.711

Yellow-orange cryst. (toluene/hexane).

Brindley, P.B. *et al*, *J. Chem. Soc., Perkin Trans. 2*, 1981, 419.

$C_{26}H_{38}Zr$ **Zr-00095**

[1,3-Bis(methylene)-1,4-butanediyl]bis(1,2,3,4,5-η)-1,2,3,4,5-pentamethyl-2,4-cyclopentadien-1-yl]zirconium, 10CI

1,1-Bis(η⁵-pentamethylcyclopentadienyl)-2,4-dimethylene-1-zirconacyclopentane

[75125-14-3]

M 441.806

Orange cryst.

Schmidt, J.R. *et al*, *Inorg. Chem.*, 1981, **20**, 318 (*synth, cryst struct, ir, nmr*)

$C_{28}H_{24}Zr$ **Zr-00096**

Bis[(4a,4b,8a,9a-η)-9H-fluoren-9-yl]dimethylzirconium, 9CI

[60373-20-8]

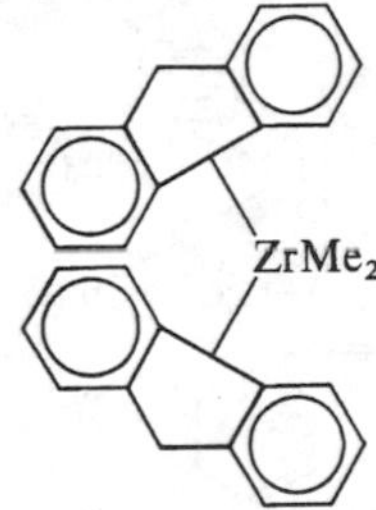

M 451.718

Light-yellow cryst. (pentane). Bp 200° subl. *in vacuo*. Deuterated derivs. also obt.

Samuel, E. *et al*, *J. Organomet. Chem.*, 1976, **113**, 331 (*synth, nmr, ms, ir*)

$C_{28}H_{25}ClGeZr$ **Zr-00097**

Chlorobis(η⁵-2,4-cyclopentadien-1-yl)(triphenylgermyl)zirconium, 9CI

[12636-98-5]

M 560.769

Orange solid. $Bp_{0.001}$ 190° subl.

Coutts, R.S.P. *et al*, *J. Chem. Soc., Chem. Commun.*, 1968, 260 (*synth, nmr*)
Kingston, B.M. *et al*, *J. Chem. Soc., Dalton Trans.*, 1972, 69 (*synth, ms, nmr*)

$C_{28}H_{25}ClSiZr$ **Zr-00098**

Chlorobis(η⁵-2,4-cyclopentadien-1-yl)(triphenylsilyl)zirconium, 9CI

[12283-86-2]

M 516.264

Orange cryst. (THF). Mp 175-8° dec. Bp 180° subl.

Cardin, D.J. *et al*, *J. Chem. Soc., Chem. Commun.*, 1967, 1035 (*synth, ms, nmr*)
Muir, K.W. *et al*, *J. Chem. Soc. (A)*, 1971, 2663 (*cryst struct*)
Kingston, B.M. *et al*, *J. Chem. Soc., Dalton Trans.*, 1972, 69 (*synth, nmr, ms*)

$C_{28}H_{25}ClSnZr$ **Zr-00099**

Chlorobis(η⁵-2,4-cyclopentadien-1-yl)(triphenylstannyl)zirconium, 9CI

[12637-01-3]

M 606.869

Orange solid. Dec. below 220°.

Coutts, R.S.P. *et al*, *J. Chem. Soc., Chem. Commun.*, 1968, 260 (*synth, nmr*)
Kingston, B.M. *et al*, *J. Chem. Soc., Dalton Trans.*, 1972, 69 (*synth, nmr*)

$C_{28}H_{25}P_3Zr$ **Zr-00100**

Bis(η⁵-2,4-cyclopentadien-1-yl)[1,2,3-triphenyltriphosphinato(2−)-P^1,P^3]zirconium, 10CI

[37299-22-2]

M 545.647

Monomeric. Orange solid. Sol. org. solvs. except aliphatic hydrocarbons and alcohols. Mp 233-5°. Solid stable in air, solns. air-sensitive.

Issleib, K. *et al*, *Angew. Chem., Int. Ed. Engl.*, 1972, **11**, 527 (*synth, ms, nmr*)

$C_{28}H_{28}Zr$ **Zr-00101**

Tetrabenzylzirconium

Tetrakis(phenylmethyl)zirconium, 9CI. *Zirconium tetrabenzyl*

[24356-01-2]

$Zr(CH_2Ph)_4$

M 455.749

Olefin polymerisation catalyst and isomerization catalyst. Yellow cryst. (hydrocarbons). Mp 133-4°. Sensitive to light.

Zucchini, U. *et al*, *J. Chem. Soc., Chem. Commun.*, 1969, 1174 (*synth, nmr, use*)

Davies, G.R. *et al*, *J. Chem. Soc., Chem. Commun.*, 1971, 677, 1511 (*cryst struct*)
Zucchini, U. *et al*, *J. Organomet. Chem.*, 1971, **26**, 357 (*synth, nmr*)
Felten, J.J. *et al*, *J. Organomet. Chem.*, 1972, **36**, 87 (*derivs*)
Clarke, J.F. *et al*, *J. Organomet. Chem.*, 1974, **74**, 417 (*derivs*)
Lappert, M.F. *et al*, *J. Chem. Soc., Chem. Commun.*, 1975, 830.

$C_{30}H_{10}Co_6O_{20}Zr$ Zr-00102

Bis(η^5-2,4-cyclopentadien-1-yl)bis[μ_4-[methanolato(4−)-*C*:*C*:*C*:*O*]]bis(nonacarbonyltricobalt)zirconium
Bis(η-cyclopentadienyl)di[μ_3-(oxymethylidyne)cyclotris(tricarbonylcobalt)]zirconium
[67422-53-1]

M 1135.216
Red cryst. (toluene). Mp 145° dec.

Ishii, M. *et al*, *Ber. Bunsenges. Phys. Chem.*, 1978, **83**, 1026 (*cryst struct*)
Stutte, B. *et al*, *Chem. Ber.*, 1978, **111**, 1603 (*synth, ir, nmr, cryst struct*)

$C_{30}H_{28}Zr_2$ Zr-00103

Tetrakis(η^5-2,4-cyclopentadien-1-yl)-μ-hydro[μ-[(1,2-η:2-η)-2-naphthalenyl]]dizirconium, 10CI
μ-Hydrido-μ-[2-η^1:1,2-η^2-naphthyl]bis[bis(η-cyclopentadienyl)zirconium]
[72894-62-3]

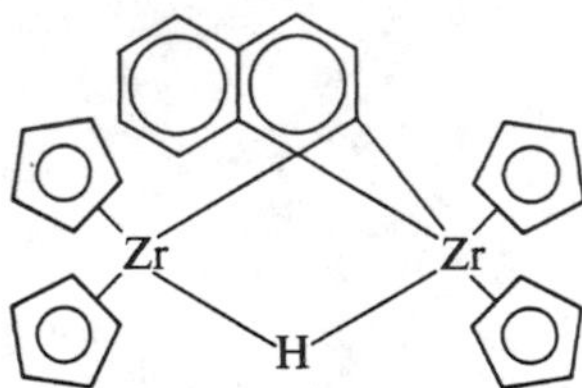

M 570.991
Dark-green cryst. (THF). Sol. THF.

Pez, G.P. *et al*, *J. Am. Chem. Soc.*, 1979, **101**, 6933 (*synth, ir, nmr, cryst struct*)

$C_{31}H_{42}OWZr$ Zr-00104

[Bis(η^5-2,4-cyclopentadien-1-yl)tungsten]hydro[μ-[methanolato(3−)-*C*:*O*]]bis[(1,2,3,4,5-η)-1,2,3,4,5-pentamethyl-2,4-cyclopentadien-1-yl]zirconium, 10CI
Bis(η^5-cyclopentadienyl)hydrido[bis(η-pentamethylcyclopentadienyl)hydridozirconoxymethylene]tungsten
[69289-49-2]

M 705.742
Red-brown cryst. (toluene).

Wolczanski, P.T. *et al*, *J. Am. Chem. Soc.*, 1979, **101**, 218 (*synth, ir, nmr, cryst struct*)

$C_{32}H_{23}N_3O_3Zr$ Zr-00105

π-Cyclopentadienyltris(8-quinolinolato)zirconium
Tris(8-quinolinolato)cyclopentadienylzirconium

M 588.772
Solid + $CHCl_3$. Mp 357-60° dec.

Brainina, E.M. *et al*, *Dokl. Akad. Nauk SSSR, Ser. Sci. Khim.*, 1966, **169**, 335 (*synth*)

$C_{32}H_{30}OS_2Zr_2$ Zr-00106

Bis(benzenethiolato)tetrakis(η^5-2,4-cyclopentadien-1-yl)-μ-oxodizirconium, 10CI
μ-Oxobis[bis(η-cyclopentadienyl)phenylthiolatozirconium]
[69662-01-7]

M 677.148
Pale-yellow cryst. (C_6H_6).

Petersen, J.L., *J. Organomet. Chem.*, 1979, **166**, 179 (*synth, ir, cryst struct*)

$C_{32}H_{40}Zr$ **Zr-00107**

Bis[(1,2,3,4,5-η)-1,2,3,4,5-pentamethyl-2,4-cyclopentadien-1-yl]diphenylzirconium, 10CI

[79847-76-0]

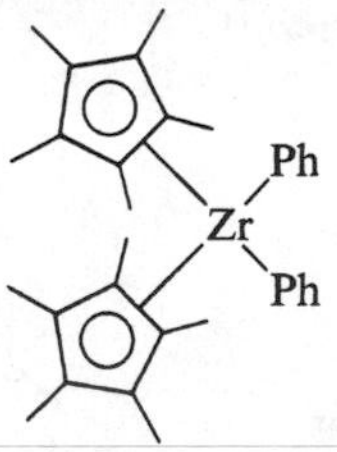

M 515.888

Characterised spectroscopically and by photochemical decomposition. Pale-yellow cryst. (Et_2O).

Tung, H.S. *et al, Inorg. Chim. Acta*, 1981, **52**, 197 (*synth, nmr*)

$C_{34}H_{34}OZr_2$ **Zr-00108**

Tetrakis(η^5-2,4-cyclopentadien-1-yl)bis(2-methylphenyl)-μ-oxodizirconium

μ-Oxobis[bis(η-cyclopentadienyl)-o-tolylzirconium]

[77674-18-1]

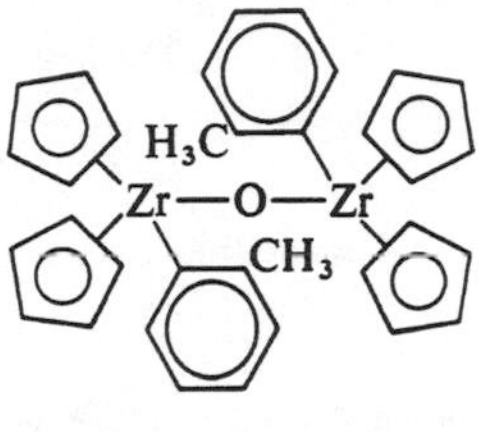

M 641.082

Cryst. *m*-Isomer and other analogues also made.

Chen. S.S. *et al, K'o Hsueh Tung Pao*, 1980, **25**, 270; *CA*, **94**, 208952k (*synth, cryst struct*)

$C_{34}H_{54}Al_2Cl_2Zr_2$ **Zr-00109**

Di-μ-chlorotetrakis(η^5-2,4-cyclopentadien-1-yl-)-μ-1,2-ethanediylbis(triethylaluminum)dizirconium, 9CI

1,2-Bis[Al,Zr-μ-chloro(triethylaluminum)(bis(η-cyclopentadienyl)zirconio]ethane

[56379-20-5]

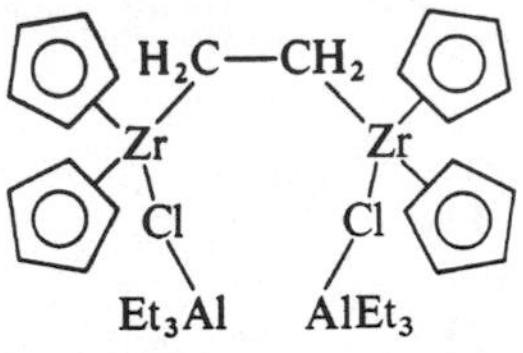

M 770.110

Catalyses olefin polymerisation. Cryst. Extremely air- and water-sensitive.

Kaminsky, W. *et al, Makromol. Chem.*, 1974, **175**, 443 (*synth, use*)

Kaminsky, W. *et al, Angew. Chem., Int. Ed. Engl.*, 1976, **15**, 629 (*synth, nmr, cryst struct*)

$C_{36}H_{30}N_2Zr$ **Zr-00110**

Bis(η^5-cyclopentadienyl)bis(diphenylketimido)zirconium

Di-π-cyclopentadienylbis(1,1-diphenylmethylenimina-to)zirconium, 8CI. *Zirconocene bis(diphenylketimide)*

[11106-18-6]

As Dibromobis(η^5-2,4-cyclopentadien-1-yl)zirconium, Zr-00006 with

X = Y = —N=CPh_2

M 581.866

Red cryst. $Bp_{0.001}$ 150° subl.

Collier, M.R. *et al, Inorg. Nucl. Chem Lett.*, 1971, **7**, 689 (*synth, ir, nmr, ms*)

$C_{36}H_{32}Zr$ **Zr-00111**

Bis(η^5-2,4-cyclopentadien-1-yl)bis(diphenylmethyl)zirconium, 10CI

[60175-22-6]

As Dibromobis(η^5-2,4-cyclopentadien-1-yl)zirconium, Zr-00006 with

X = Y = —$CHPh_2$

M 555.869

Orange-red cryst. (C_6H_6 or toluene). Mp >130° dec. Sl. air-sensitive in the solid state, dec. rapidly in soln.

Atwood, J.L. *et al, J. Am. Chem. Soc.*, 1977, **99**, 6645 (*synth, ir, nmr, cryst struct*)

Lappert, M.F. *et al, J. Chem. Soc., Chem. Commun.*, 1979, 305 (*props*)

Lappert, M.F. *et al, J. Chem. Soc., Dalton Trans.*, 1981, 805 (*props*)

$C_{38}H_{30}Zr$ **Zr-00112**

Bis(η^5-2,4-cyclopentadien-1-yl)(1,2,3,4-tetraphenyl-1,3-butadiene-1,4-diyl)zirconium, 10CI

1,1-Bis(η^5-cyclopentadienyl)-2,3,4,5-tetraphenylzirconole

[53433-58-2]

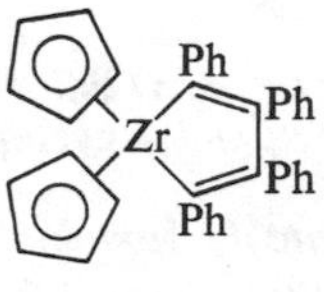

M 577.875

Orange cryst. (CH_2Cl_2/pentane). Sol. CH_2Cl_2, insol. Et_2O, pet. ether. Mp 140°.

Braye, E.H. *et al, J. Am. Chem. Soc.*, 1961, **83**, 4406 (*synth*)

Watt, G.W. *et al, J. Am. Chem. Soc.*, 1970, **92**, 826 (*synth*)

Alt, H.G. *et al, J. Am. Chem. Soc.*, 1974, **96**, 5936 (*synth, nmr*)

Hunter, W.E. *et al, J. Organomet. Chem.*, 1981, **204**, 67 (*cryst struct*)

$C_{40}H_{20}Co_6O_{21}Zr_2$ Zr-00113

Tetrakis(η^5-2,4-cyclopentadien-1-yl)bis[μ_4-[methanolato(4–)-*C*:*C*:*C*:*O*]bis(nonacarbonyltricobalt)-μ-oxodizirconium, 10CI

μ-Oxobis[bis(η^5-cyclopentadienyl)[μ_3-(oxymethylidyne)cyclotris(tricarbonylcobalt)]zirconium]

[67422-52-0]

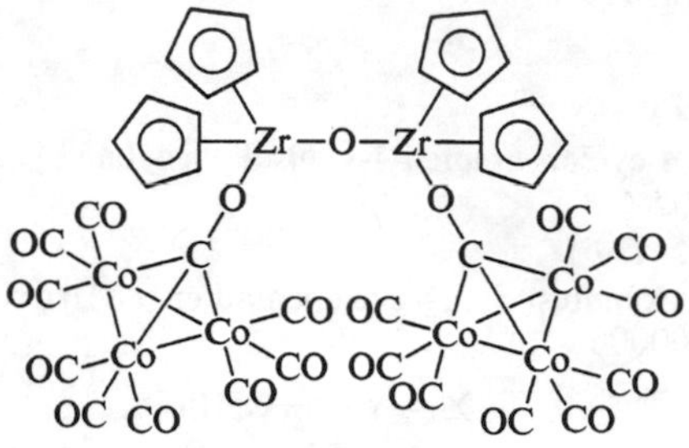

M 1372.625

Mp 160-4° dec.

Stutte, B. *et al*, *Chem. Ber.*, 1978, **111**, 1603 (*synth, ir, nmr*)

$C_{40}H_{44}Zr$ Zr-00114

Tetrakis(2-methyl-1-phenyl-1-propenyl)zirconium, 10CI

Tetrakis(2,2-dimethyl-1-phenylethenyl)zirconium

[63422-30-0]

$[(H_3C)_2C{=}CPh]_4Zr$

M 616.008

Yellow cryst. Mp 70° dec.

Cardin, C.J. *et al*, *J. Organomet. Chem.*, 1977, **132**, C23 (*synth, nmr*)

$C_{42}H_{62}I_2O_2Zr_2$ Zr-00115

[μ-[1,2-Ethenediolato(2–)-*O*:*O'*]]diodotetrakis[(1,2,3,4,5-η)-1,2,3,4,5-pentamethyl-2,4-cyclopentadien-1-yl]dizirconium, 10CI

μ-(1,2-Ethenediolato)bis[iodobis(η^5-pentamethylcyclopentadienyl)zirconium]

[61396-38-1]

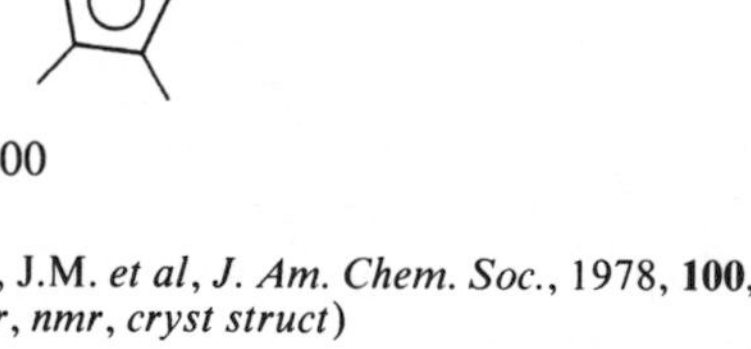

M 1035.200

Cryst.

Manriquez, J.M. *et al*, *J. Am. Chem. Soc.*, 1978, **100**, 2716 (*synth, ir, nmr, cryst struct*)

$C_{42}H_{64}O_2Zr_2$ Zr-00116

[μ-[1,2-Ethenediolato(2–)-*O*:*O'*]]dihydrotetrakis[(1,2,3,4,5-η)-1,2,3,4,5-pentamethyl-2,4-cyclopentadien-1-yl]dizirconium, 10CI

Ethenediolato-1,2-bis[hydridobis(η^5-pentamethylcyclopentadienyl)zirconium]

[61396-37-0]

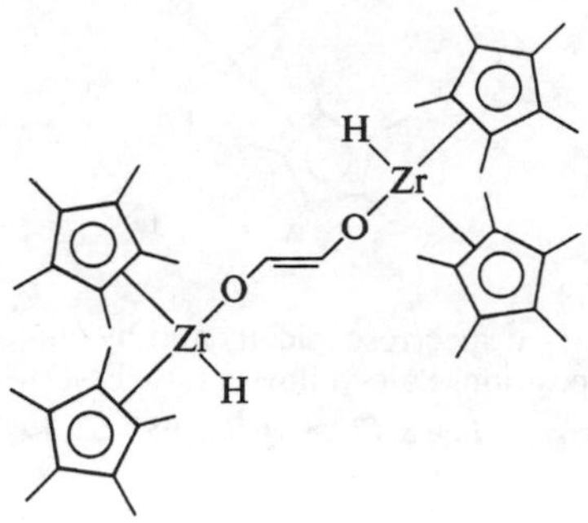

M 783.406

Cryst. Isotopomers also prepd.

Manriquez, J.M. *et al*, *J. Am. Chem. Soc.*, 1978, **100**, 2716 (*synth, ir, nmr, cryst struct*)

Name Index

Bis(η^5-2,4-cyclopentadien-1-yl)[(diphenylphosphinidenio)-bis(methylene)]scandium, Sc-00012
Bis(η^5-cyclopentadien-1-yl)diphenyltitanium, Ti-00118
Bis(η^5-2,4-cyclopentadien-1-yl)diphenylzirconium, Zr-00073
Bis(η^5-2,4-cyclopentadien-1-yl)di-1*H*-pyrrol-1-yltitanium, Ti-00097
Bis(η^5-cyclopentadien-1-yl)di-1*H*-pyrrol-1-ylzirconium, Zr-00049
Bis(η-cyclopentadienyl)dithiocyanatozirconium, *see* Zr-00020
Bis(cyclopentadienyl)divanadium pentacarbonyl, *see* V -00067
[μ-[(1,2,3,4,5-η:1′,2′,3′,4′,5′-η)[Bis-2,4-cyclopentadien-1-yl]-1,1′-diyl]]bis(η^5-2,4-cyclopentadien-1-yl)di-μ-hydroditi-tanium, *see* Ti-00108
Bis(η^5-2,4-cyclopentadien-1-yl)[1,2-ethanediylbis[dimethylphosphine]-*P*,*P′*]tantalum(1+), Ta-00031
Bis(η^5-2,4-cyclopentadien-1-yl)[1,2-ethanediylbis[diphenylphosphine]-*P*,*P′*]vanadium(1+), V -00116
Bis(η^5-2,4-cyclopentadien-1-yl)[1,2-ethenedithiolato(2−)-*S*,*S′*]titanium, Ti-00053
Bis(η^5-2,4-cyclopentadien-1-yl)(η^2-ethene)ethylniobium, Nb-00031
Bis(η^5-2,4-cyclopentadien-1-yl)(η^2-ethene)hydroniobium, Nb-00027
Bis(η^5-2,4-cyclopentadien-1-yl)(η^2-ethene)hydrotungsten(1+), W -00076
Bis(η^5-2,4-cyclopentadien-1-yl)(η^2-ethene)methyltungsten(1+), W -00085
Bis(η^5-2,4-cyclopentadien-1-yl)(η^2-ethene)molybdenum, *see* Mo-00066
Bis(η^5-2,4-cyclopentadien-1-yl)(η^2-ethene)rhenium(1+), Re-00056
Bis(η^5-2,4-cyclopentadien-1-yl)(η^2-ethene)tungsten, W -00074
Bis[(η-cyclopentadienyl)ethoxy(μ-ethoxy)titanium], *in* Ti-00021
Bis(η^5-cyclopentadienyl)(η^2-ethylene)molybdenum, Mo-00066
Bis(η^5-cyclopentadienyl)(ethylene)rhenium(1+), *see* Re-00056
Bis(η^5-2,4-cyclopentadien-1-yl)ethylvanadium, V -00055
Bis(η^5-2,4-cyclopentadien-1-yl)europium, Eu-00003
Bis(cyclopentadienyl)fluorotitanium, Ti-00028
Bis(η^5-2,4-cyclopentadien-1-yl)(η^2-*C*,*O*-formaldehyde)vanadi-um, V -00045
▷Bis(cyclopentadienyl)hafnium dichloride, *see* Hf-00003
Bis(cyclopentadienyl)hexacarbonylditungsten, *see* W -00107
Bis(η^5-cyclopentadienyl)hydrido[bis(η-pentamethylcyclopentadienyl)hydridozirconoxymethylene]-tungsten, *see* W -00153
Bis(η-cyclopentadienyl)hydridoneopentylzirconium, *see* Zr-00041
Bis(η^5-cyclopentadienyl)hydrido(phenyl)tungsten, *see* W -00108
Bis(cyclopentadienyl)hydridorhenium, *see* Re-00042
Bis(cyclopentadienyl)hydridotechnetium, *see* Tc-00006
Bis(η-cyclopentadienyl)hydrido(tetrahydridoborato)zirconium, *see* Zr-00013
Bis(η^5-cyclopentadienyl)hydrido(triphenylgermyl)tungsten, *see* W -00148
Bis(cyclopentadienyl)hydro(carbonyl)niobium, *see* Nb-00018
Bis(cyclopentadienyl)hydro(carbonyl)tantalum, *see* Ta-00014
Bis(η^5-2,4-cyclopentadien-1-yl)-hydro(dimethylphenylphosphine)niobium, Nb-00037
Bis(cyclopentadienyl)hydro(ethylene)niobium, *see* Nb-00027
Bis(η^5-2,4-cyclopentadien-1-yl)hydrolutetium, Lu-00005
[Bis(η^5-2,4-cyclopentadien-1-yl)hydro[μ-[methanolato(3−)-*C*:*O*]]bis[(1,2,3,4,5-η)-1,2,3,4,5-pentamethyl-2,4-cyclopentadien-1-yl]zirconium, W -00153
Bis(η^5-2,4-cyclopentadien-1-yl)hydromolybdenumlithium, Mo-00040
Bis(cyclopentadienyl)hydro(η^2-4-octyne)niobium, Nb-00038
Bis(η^5-2,4-cyclopentadien-1-yl)hydrophenyltungsten, W -00108
Bis(cyclopentadienyl)hydropropenetantalum, Ta-00021
Bis(η^5-2,4-cyclopentadien-1-yl)hydrorhenium, Re-00042
Bis(η^5-2,4-cyclopentadien-1-yl)hydrotechnetium, Tc-00006
Bis(η^5-2,4-cyclopentadien-1-yl)hydro[tetrahydroborato(1−)-*H*,*H′*]zirconium, Zr-00013
Bis(η^5-2,4-cyclopentadien-1-yl)hydro(tetrahydrofuran)lutetium, *in* Lu-00005
Bis(η^5-cyclopentadienyl)hydrotitanium, Ti-00037
Bis(η^5-2,4-cyclopentadien-1-yl)hydro(triethylphosphine)tanta-lum, Ta-00030
Bis(η^5-2,4-cyclopentadien-1-yl)hydro(triphenylgermyl)tungsten, W -00148
Bis(η^5-2,4-cyclopentadien-1-yl)hydro(tungsten)lithium, W -00052
Bis(cyclopentadienyl)iodocarbonylvanadium, *see* V -00039
Bis(η^5-2,4-cyclopentadien-1-yl)iodonitrosylmolybdenum, Mo-00037
Bis(cyclopentadienyl)iodotitanium, Ti-00030
Bis(η^5-2,4-cyclopentadien-1-yl)iodovanadium, V -00031
Bis(η^5-2,4-cyclopentadien-1-yl)(isocyanomethane)propyltanta-lum, Ta-00027
▷Biscyclopentadienylmanganese, *see* Mn-00058
Bis(η^5-2,4-cyclopentadien-1-yl)[(1,2,3-η)-2-methyl-2-butenyl]titanium, Ti-00081
Bis(cyclopentadienyl)methyl(carbonyl)niobium, *see* Nb-00026
Bis(cyclopentadienyl)methylchloroniobium, *see* Nb-00019
Bis(η^5-2,4-cyclopentadien-1-yl)methyldysprosium, Dy-00003
Bis(η^5-2,4-cyclopentadien-1-yl)methylerbium, Er-00005
Bis(η^5-2,4-cyclopentadien-1-yl)methylholmium, Ho-00003
Bis(η^5-2,4-cyclopentadien-1-yl)methyllutetium, Lu-00006
Bis(η^5-2,4-cyclopentadien-1-yl)methyl(methylidene)tantalum, Ta-00017
Bis(cyclopentadienyl)methyloxoniobium, Nb-00020
Bis(η^5-2,4-cyclopentadien-1-yl)(4-methylphenyl)lutetium, Lu-00017
Bis(η^5-2,4-cyclopentadien-1-yl)methylscandium, Sc-00002
Bis(η^5-2,4-cyclopentadien-1-yl)methylthulium, Tm-00003
Bis(η^5-2,4-cyclopentadien-1-yl)methyltitanium, Ti-00046
Bis(cyclopentadienyl)methyl(triethylphosphine)niobium, Nb-00034
Bis(η^5-2,4-cyclopentadien-1-yl)methylvanadium, V -00047
Bis(η^5-2,4-cyclopentadien-1-yl)methylytterbium, Yb-00006
Bis(η^5-2,4-cyclopentadien-1-yl)methylyttrium, Y -00003
Bis(cyclopentadienyl)molybdenum dihydride, *see* Mo-00043
▷Bis(cyclopentadienyl)niobium borohydride, *see* Nb-00015
Bis(η^5-2,4-cyclopentadien-1-yl)(1,1,2,2,3,3,4,4-octaphenyl-1,4-tetrasilanediyl)titanium, Ti-00153
Bis(cyclopentadienyl)oxochloroniobium, *see* Nb-00011
Bis(η^5-2,4-cyclopentadien-1-yl)-μ-oxotetrakis(2,4-pentanedionato-*O*,*O′*)dihafnium, Hf-00047
Bis(η^5-2,4-cyclopentadien-1-yl)pentafluorophenyltitanium, Ti-00084
Bis(η^5-2,4-cyclopentadien-1-yl)(pentanedionato)vanadium(1+), V -00068
Bis(η^5-2,4-cyclopentadien-1-yl)[pentaselenidato(2−)]titanium, Ti-00034
Bis(η^5-2,4-cyclopentadien-1-yl)(pentathio)hafnium, Hf-00006
Bis(η^5-2,4-cyclopentadien-1-yl)(pentathio)titanium, Ti-00033
Bis(η^5-2,4-cyclopentadienyl)pentathiovanadium, V -00032
Bis(η^5-2,4-cyclopentadien-1-yl)(pentathio)zirconium, Zr-00010
Bis(η-cyclopentadienyl)phenyl[bis(trimethylsilyl)methyl]zircon-ium, *see* Zr-00082
Bis(η^5-2,4-cyclopentadien-1-yl)[1,2-phenylenebis[methylene]]zirconium, Zr-00050
Bis(η^5-2,4-cyclopentadien-1-yl)[1,2-phenylenebis[(trimethylsilyl)methylene]]zirconium, Zr-00086
Bis(η^5-2,4-cyclopentadien-1-yl)[1,2-phenylenebis[(trimethylsilyl)methylene]]hafnium, Hf-00040
Bis(η^5-2,4-cyclopentadien-1-yl)(phenylethynyl)gadolinium, Gd-00006
Bis(η^5-2,4-cyclopentadien-1-yl)(phenylethynyl)titanium, *in* Ti-00145
Bis(η^5-2,4-cyclopentadien-1-yl)(phenylethynyl)ytterbium, Yb-00019

Molecular Formula Index

CH_3Br_3Ti
Tribromomethyltitanium, Ti-00001
CH_3Cl_3Ti
Trichloromethyltitanium, Ti-00002
CH_3Cl_4Nb
Tetrachloromethylniobium, Nb-00001
CH_3Cl_4Ta
Tetrachloromethyltantalum, Ta-00001
CH_3O_3Re
Methyltrioxorhenium, Re-00001
$CMnN_3O_4$
Carbonyltrinitrosylmanganese, Mn-00001
$C_2Cl_2NO_3Re$
Dicarbonyldichloronitrosylrhenium, *in* Re-00003
$C_2H_5Cl_3Ti$
Trichloroethyltitanium, Ti-00003
$C_2H_6Cl_3Nb$
Trichlorodimethylniobium, Nb-00002
$C_2H_6Cl_3Ta$
Trichlorodimethyltantalum, Ta-00002
$C_2H_6Cl_4Nb^{\ominus}$
Tetrachlorodimethylniobate(1−), Nb-00003
$C_3Cl_3O_3Re^{\ominus\ominus}$
Tricarbonyltrichlororhenate(2−), Re-00002
$C_3CsCl_3O_3Re$
Tricarbonyltrichlororhenate(2−); Di-Cs salt, *in* Re-00002
$C_3H_9Cl_2Nb$
Dichlorotrimethylniobium, Nb-00004
$C_3H_9Cl_2Ta$
Dichlorotrimethyltantalum, Ta-00003
$C_3H_9Cl_3Nb^{\ominus}$
Trichlorotrimethylniobate(1−), Nb-00005
$C_3H_9CrO_3P_3$
Tricarbonyltris(phosphine)chromium, Cr-00001
$C_3H_9MnN_3O_3^{\oplus}$
Triamminetricarbonylmanganese(1+), Mn-00002
C_4BrMnO_4
Bromotetracarbonylmanganese, *in* Mn-00027
$C_4Br_2MnO_4^{\ominus}$
Dibromotetracarbonylmanganate(1−), Mn-00003
C_4ClMnO_4
Tetracarbonylchloromanganese, *in* Mn-00028
C_4ClO_4Re
Tetracarbonylchlororhenium, *in* Re-00031
$C_4Cl_2MnO_4^{\ominus}$
Tetracarbonyldichloromanganate(1−), Mn-00004
$C_4Cl_2MoO_4$
Tetracarbonyldichloromolybdenum, *in* Mo-00008
$C_4Cl_4N_2O_6Re_2$
Tetracarbonyldi-μ-chlorodichlorodinitrosyldirhenium, Re-00003
$C_4CrNa_4O_4$
▷Tetracarbonylchromate(4−); Tetra-Na salt, *in* Cr-00002
$C_4CrO_4^{\ominus\ominus\ominus\ominus}$
Tetracarbonylchromate(4−), Cr-00002
$C_4H_3MnO_3S$
Tricarbonyl(methanethiolato)manganese, Mn-00005
$C_4H_3O_3ReS$
Tricarbonyl(methanethiolato)rhenium, Re-00004
$C_4H_6CrO_4P_2$
Tetracarbonylbis(phosphine)chromium, Cr-00003
$C_4H_{11}Cl_3SiTi$
Trichloro(trimethylsilylmethyl)titanium, Ti-00004
$C_4H_{11}Cl_5SiW$
Pentachloro[(trimethylsilyl)methyl]tungsten, W -00001
$C_4H_{12}CrO_4Tl_2$
▷Tetramethyl[μ-[chromato(2−)-*O*:*O'*]]dithallium, Cr-00004
$C_4H_{12}ORe$
Tetramethyloxorhenium, Re-00005
$C_4H_{12}O_4Ti$
Tetramethyl titanate, Ti-00005
$C_4H_{12}Ti$
▷Tetramethyltitanium, Ti-00006
$C_4H_{22}B_{18}Cl_2U^{\ominus\ominus}$
Dichlorobis[(7,8,9,10,11-η)undecahydro-7,8-dicarbaundecaborato]uranate(2−), U -00001
C_4IMnO_4
Tetracarbonyliodomanganese, *in* Mn-00040
$C_4I_4MnO_4^{\ominus}$
Tetracarbonyldiiodomanganate(1−), Mn-00006
C_4MnNO_5
Tetracarbonylnitrosylmanganese, Mn-00007
$C_4MnNa_3O_4$
Tetracarbonylmanganate(3−); Tri-Na salt, *in* Mn-00008
$C_4MnO_4^{\ominus\ominus\ominus}$
Tetracarbonylmanganate(3−), Mn-00008
$C_4MoNa_4O_4$
▷Tetracarbonylmolybdate(4−); Tetra-Na salt, *in* Mo-00001
$C_4MoO_4^{\ominus\ominus\ominus\ominus}$
Tetracarbonylmolybdate(4−), Mo-00001
$C_4Na_4O_4W$
▷Tetracarbonyltungstate(4−); Tetra-Na salt, *in* W -00002
$C_4O_4W^{\ominus\ominus\ominus\ominus}$
Tetracarbonyltungstate(4−), W -00002
$C_5AsF_6MnO_7S$
Pentacarbonyl(sulfur dioxide-*S*)manganese(1+); Hexafluoroarsenate, *in* Mn-00015
$C_5AsF_6O_7ReS$
Pentacarbonyl(sulfur dioxide-*S*)rhenium(1+); Hexafluoroarsenate, *in* Re-00013
$C_5BrCrO_5^{\ominus}$
Bromopentacarbonylchromate(1−), Cr-00005
C_5BrMnO_5
Bromopentacarbonylmanganese, Mn-00009
C_5BrO_5Re
Bromopentacarbonylrhenium, Re-00006
C_5BrO_5Tc
Bromopentacarbonyltechnetium, Tc-00001
C_5ClMnO_5
Pentacarbonylchloromanganese, Mn-00010
C_5ClO_5Re
Pentacarbonylchlororhenium, Re-00007
C_5ClO_5Tc
Pentacarbonylchlorotechnetium, Tc-00002
$C_5Cl_3NO_6Re_2$
Pentacarbonyltri-μ-chloronitrosyldirhenium, Re-00008
$C_5CrClO_5^{\ominus}$
Pentacarbonylchlorochromate(1−), Cr-00006

$C_5CrCs_2O_5$
Pentacarbonylchromate(2−); Di-Cs salt, *in* Cr-00008
$C_5CrIO_5^{\ominus}$
Pentacarbonyliodochromate(1−), Cr-00007
$C_5CrNa_2O_5$
Pentacarbonylchromate(2−); Di-Na salt, *in* Cr-00008
$C_5CrO_5^{\ominus\ominus}$
Pentacarbonylchromate(2−), Cr-00008
$C_5F_3O_4ReS$
Tetracarbonyl(trifluoromethanethiolato)rhenium, Re-00009
$C_5HCrO_6^{\ominus}$
Pentacarbonylhydroxychromate(1−), Cr-00009
C_5HMnO_5
▷Pentacarbonylhydromanganese, Mn-00011
C_5HO_5Re
Pentacarbonylhydridorhenium, Re-00010
$C_5H_3AsCrO_5$
(Arsine)pentacarbonylchromium, Cr-00010
$C_5H_3CrNO_5$
Amminepentacarbonylchromium, Cr-00011
$C_5H_3CrO_5P$
Pentacarbonyl(phosphine)chromium, Cr-00012
$C_5H_3CrO_5Sb$
Pentacarbonyl(stibine)chromium, Cr-00013
$C_5H_3MnO_4S$
Tetracarbonyl(methanethiolato)manganese, Mn-00012
$C_5H_3NNaO_5V$
Amminepentacarbonylvanadate(1−); Na salt, *in* V -00001
$C_5H_3NO_5V^{\ominus}$
Amminepentacarbonylvanadate(1−), V -00001
$C_5H_4CrN_2O_5$
Pentacarbonyl(hydrazine-*N*)chromium, Cr-00014
$C_5H_4F_6MnO_7P$
Pentacarbonyl-1,3-dioxalan-2-ylidenemanganese(1+); Hexafluorophosphate, *in* Mn-00030
$C_5H_5Br_2OV$
Dibromo(η^5-2,4-cyclopentadien-1-yl)oxovanadium, V -00002
$C_5H_5Br_3Ti$
Tribromo(η^5-2,4-cyclopentadien-1-yl)titanium, Ti-00007
$C_5H_5Br_3V$
Tribromo(η^5-2,4-cyclopentadien-1-yl)vanadium, V -00003
$C_5H_5Br_3Zr$
Tribromo(η^5-2,4-cyclopentadien-1-yl)zirconium, Zr-00001
$C_5H_5ClCrN_2O_2$
Chloro(η^5-2,4-cyclopentadien-1-yl)dinitrosylchromium, Cr-00015
$C_5H_5ClMoN_2O_2$
Chloro(η^5-2,4-cyclopentadien-1-yl)dinitrosylmolybdenum, Mo-00002
$C_5H_5ClN_2O_2W$
Chloro(η^5-cyclopentadien-1-yl)dinitrosyltungsten, W -00003
C_5H_5ClOTi
Tetrachlorotetrakis(η^5-2,4-cyclopentadien-1-yl)tetra-μ-oxotetratitanium; Monomer, *in* Ti-00105
$C_5H_5Cl_2Dy$
Dichloro(η^5-2,4-cyclopentadien-1-yl)dysprosium, Dy-00001
$C_5H_5Cl_2Er$
Dichloro(η^5-2,4-cyclopentadien-1-yl)erbium, Er-00001
$C_5H_5Cl_2Eu$
Dichloro(η^5-2,4-cyclopentadien-1-yl)europium, Eu-00001
$C_5H_5Cl_2Gd$
Dichloro(η^5-2,4-cyclopentadien-1-yl)gadolinium, Gd-00001
$C_5H_5Cl_2Ho$
Dichloro(η^5-2,4-cyclopentadien-1-yl)holmium, Ho-00001
$C_5H_5Cl_2Lu$
Dichloro(η^5-2,4-cyclopentadien-1-yl)lutetium, Lu-00001
$C_5H_5Cl_2MoNO$
Dichloro(η^5-2,4-cyclopentadien-1-yl)nitrosylmolybdenum, Mo-00003
$C_5H_5Cl_2OV$
Dichloro(η^5-2,4-cyclopentadien-1-yl)oxovanadium, V -00004
$C_5H_5Cl_2Sm$
Dichloro(η^5-2,4-cyclopentadien-1-yl)samarium, Sm-00001
$C_5H_5Cl_2Ti$
Dichloro(η^5-2,4-cyclopentadien-1-yl)titanium, Ti-00008
$C_5H_5Cl_2Y$
Dichloro(η^5-2,4-cyclopentadien-1-yl)yttrium, Y -00001
$C_5H_5Cl_2Yb$
Dichloro(η^5-2,4-cyclopentadien-1-yl)ytterbium, Yb-00001
$C_5H_5Cl_3Hf$
Trichloro(η^5-2,4-cyclopentadien-1-yl)hafnium, Hf-00001
$C_5H_5Cl_3Ti$
Trichloro(η^5-2,4-cyclopentadien-1-yl)titanium, Ti-00009
$C_5H_5Cl_3V$
Trichloro(η^5-2,4-cyclopentadien-1-yl)vanadium, V -00005
$C_5H_5Cl_3Zr$
Trichloro(η^5-2,4-cyclopentadien-1-yl)zirconium, Zr-00002
$C_5H_5Cl_4Nb$
Tetrachloro(η^5-2,4-cyclopentadien-1-yl)niobium, Nb-00006
$C_5H_5Cl_4Ta$
Tetrachloro(η^5-2,4-cyclopentadien-1-yl)tantalum, Ta-00004
$C_5H_5F_3Ti$
(η^5-2,4-Cyclopentadien-1-yl)trifluorotitanium, Ti-00010
$C_5H_5F_6N_3O_3PV$
Cyclopentadienyltrinitrosylvanadium(1+); Hexafluorophosphate, *in* V -00006
$C_5H_5I_3Ti$
(η^5-Cyclopentadien-1-yl)triiodotitanium, Ti-00011
$C_5H_5I_3Zr$
(η^5-2,4-Cyclopentadien-1-yl)triiodozirconium, Zr-00003
$C_5H_5N_3O_3V^{\oplus}$
Cyclopentadienyltrinitrosylvanadium(1+), V -00006
$C_5H_5O_2V$
(η^5-2,4-Cyclopentadien-1-yl)dioxovanadium, V -00007
$C_5H_9Cl_4W^{\ominus}$
Tetrachloro(1,1-dimethylpropylidyne)tungsten(1−), W -00004
$C_5H_{15}Ta$
▷Pentamethyltantalum, Ta-00005
C_5IMnO_5
Pentacarbonyliodomanganese, Mn-00013
$C_5IMoO_5^{\ominus}$
Pentacarbonyliodomolybdate(1−), Mo-00004
C_5IO_5Re
Pentacarbonyliodorhenium, Re-00011
C_5IO_5Tc
Pentacarbonyliodotechnetium, Tc-00003
C_5KMnO_5
Pentacarbonylmanganate(1−); K salt, *in* Mn-00014
C_5KO_5Re
Pentacarbonylrhenate(1−); K salt, *in* Re-00012
$C_5K_3O_5V$
▷Pentacarbonylvanadate(3−); Tri-K salt, *in* V -00008
C_5LiMnO_5
Pentacarbonylmanganate(1−); Li salt, *in* Mn-00014
C_5MnNaO_5
Pentacarbonylmanganate(1−); Na salt, *in* Mn-00014

$C_5MnO_5^{\ominus}$
Pentacarbonylmanganate(1−), Mn-00014
$C_5MnO_7S^{\oplus}$
Pentacarbonyl(sulfur dioxide-*S*)manganese(1+), Mn-00015
C_5NaO_5Re
Pentacarbonylrhenate(1−); Na salt, *in* Re-00012
$C_5O_5Re^{\ominus}$
Pentacarbonylrhenate(1−), Re-00012
$C_5O_5V^{\ominus\ominus\ominus}$
Pentacarbonylvanadate(3−), V -00008
$C_5O_7ReS^{\oplus}$
Pentacarbonyl(sulfur dioxide-*S*)rhenium(1+), Re-00013
$C_6BF_4MnO_6$
Hexacarbonylmanganese(1+); Tetrafluoroborate, *in* Mn-00017
$C_6Cl_3F_5Ti$
Trichloro(pentafluorophenyl)titanium, Ti-00012
$C_6CrNNaO_5$
Pentacarbonyl(cyano-*C*)chromate(1−); Na salt, *in* Cr-00016
$C_6CrNO_5^{\ominus}$
Pentacarbonyl(cyano-*C*)chromate(1−), Cr-00016
$C_6CrN_2O_4^{\ominus\ominus}$
Tetracarbonyldicyanochromate(2−), Cr-00017
C_6CrO_5S
(Carbonothioyl)pentacarbonylchromium, Cr-00018
C_6CrO_5Se
(Carbonoselenoyl)pentacarbonylchromium, Cr-00019
C_6CrO_6
▷Chromium carbonyl, Cr-00020
$C_6F_6O_6PRe$
Hexacarbonylrhenium(1+); Hexafluorophosphate, *in* Re-00017
$C_6H_3BCrNO_5^{\ominus}$
Pentacarbonyl[(cyano-*C*)trihydroborato(1−)-*N*]-chromate(1−), Cr-00021
$C_6H_3BrO_4W$
Bromotetracarbonylethylidynetungsten, W -00005
$C_6H_3ClCrO_4$
Tetracarbonylchloroethylidynechromium, Cr-00022
$C_6H_3ClO_4W$
Tetracarbonylchloroethylidynetungsten, W -00006
$C_6H_3CrIO_4$
Tetracarbonyliodoethylidynechromium, Cr-00023
$C_6H_3CrO_5^{\ominus}$
Pentacarbonylmethylchromate(1−), Cr-00024
$C_6H_3IO_4SW$
Tetracarbonyliodo[(methylthiomethylidyne)]tungsten, W -00007
$C_6H_3IO_4W$
Tetracarbonylethylidyneiodotungsten, W -00008
$C_6H_3MnO_5$
Pentacarbonylmethylmanganese, Mn-00016
$C_6H_3O_5Re$
Pentacarbonylmethylrhenium, Re-00014
$C_6H_3O_5W^{\ominus}$
Pentacarbonylmethyltungstate(1−), W -00009
$C_6H_5Cl_2OV$
▷Dichlorooxophenylvanadium, V -00009
$C_6H_5F_6N_2O_3PW$
Carbonyl(η^5-2,4-cyclopentadien-1-yl)-dinitrosyltungsten(1+); Hexafluorophosphate, *in* W -00010
$C_6H_5N_2O_3V$
Carbonyl(η^5-2,4-cyclopentadien-1-yl)dinitrosylvanadium, V -00010
$C_6H_5N_2O_3W^{\oplus}$
Carbonyl(η^5-2,4-cyclopentadien-1-yl)-dinitrosyltungsten(1+), W -00010
$C_6H_6Al_3Cl_{12}U$
(Benzene)hexa-μ-chlorohexachlorotrialuminumuranium, U -00002
$C_6H_6NO_2Re$
Carbonyl(η^5-2,4-cyclopentadien-1-yl)-hydronitrosylrhenium, Re-00015
$C_6H_7Cl_3U$
Trichloro[(1,2,3,4,5-η)-1-methyl-2,4-cyclopentadien-1-yl]uranium, U -00003
$C_6H_8CrN_2O_2$
(η^5-2,4-Cyclopentadienyl)methyldinitrosylchromium, Cr-00025
$C_6H_{10}CrI$
Iodobis(η^3-2-propenyl)chromium, Cr-00026
$C_6H_{18}Cr^{\ominus\ominus\ominus}$
Hexamethylchromate(3−), Cr-00027
$C_6H_{18}CrLi_3$
Hexamethylchromate(3−); Tri-Li salt, *in* Cr-00027
$C_6H_{18}Er^{\ominus\ominus\ominus}$
Hexamethylerbate(3−), Er-00002
$C_6H_{18}Li_3U$
Hexamethyluranate(3−); Tri-Li salt, *in* U -00004
$C_6H_{18}Lu^{\ominus\ominus\ominus}$
Hexamethyllutetate(3−), Lu-00002
$C_6H_{18}MoO_6P$
Diperoxooxohexamethylphosphoramidomolybdenum, Mo-00005
$C_6H_{18}N_2Ti$
Bis(dimethylamido)dimethyltitanium, Ti-00013
$C_6H_{18}Nd^{\ominus\ominus\ominus}$
Hexamethylneodymate(3−), Nd-00001
$C_6H_{18}Pr^{\ominus\ominus\ominus}$
Hexamethylpraesodymate(3−), Pr-00001
$C_6H_{18}Re$
▷Hexamethylrhenium, Re-00016
$C_6H_{18}Sm^{\ominus\ominus\ominus}$
Hexamethylsamarate(3−), Sm-00002
$C_6H_{18}Tm^{\ominus\ominus\ominus}$
Hexamethylthulate(3−), Tm-00001
$C_6H_{18}U^{\ominus\ominus\ominus}$
Hexamethyluranate(3−), U -00004
$C_6H_{18}W$
▷Hexamethyltungsten, W -00011
$C_6H_{18}Yb^{\ominus\ominus\ominus}$
Hexamethylytterbate(3−), Yb-00002
$C_6MnO_6^{\oplus}$
Hexacarbonylmanganese(1+), Mn-00017
$C_6MoNO_5^{\ominus}$
Pentacarbonyl(cyano-*C*)molybdate(1−), Mo-00006
C_6MoO_6
Molybdenum carbonyl, Mo-00007
$C_6NO_5V^{\ominus\ominus}$
Pentacarbonyl(cyano-*C*)vanadate(2−), V -00011
$C_6NO_5W^{\ominus}$
Pentacarbonyl(cyano-*C*)tungstate(1−), W -00012
$C_6NbO_6^{\ominus}$
Hexacarbonylniobate(1−), Nb-00007
C_6O_5SW
(Carbonothioyl)pentacarbonyltungsten, W -00013
$C_6O_6Re^{\oplus}$
Hexacarbonylrhenium(1+), Re-00017
$C_6O_6Ta^{\ominus}$
Hexacarbonyltantalate(1−), Ta-00006
$C_6O_6V^{\ominus}$
Hexacarbonylvanadate(1−), V -00012
▷Vanadium hexacarbonyl, V -00013

C_6O_6W
Tungsten carbonyl, W -00014

$C_7F_3MnO_7$
Pentacarbonyl(trifluoroacetato-*O*)manganese, Mn-00018

$C_7F_5MnO_5$
Pentacarbonyl(pentafluoroethyl)manganese, Mn-00019

$C_7H_3CrLiO_6$
Acetylpentacarbonylchromate(1−); Li salt, *in* Cr-00029

$C_7H_3CrNO_5S$
Pentacarbonyl(methyl thiocyanate-*N*)chromium, Cr-00028

$C_7H_3CrO_6^{\ominus}$
Acetylpentacarbonylchromate(1−), Cr-00029

$C_7H_3F_6MnNO_5P$
(Acetonitrile)pentacarbonylmanganese(1+); Hexafluorophosphate, *in* Mn-00020
Pentacarbonyl(isocyanomethane)manganese(1+); Hexafluorophosphate, *in* Mn-00021

$C_7H_3F_6NO_5PRe$
(Acetonitrile)pentacarbonylrhenium(1+); Hexafluorophosphate, *in* Re-00018

$C_7H_3MnNO_5^{\oplus}$
(Acetonitrile)pentacarbonylmanganese(1+), Mn-00020
Pentacarbonyl(isocyanomethane)manganese(1+), Mn-00021

$C_7H_3NO_5Re^{\oplus}$
(Acetonitrile)pentacarbonylrhenium(1+), Re-00018

$C_7H_3O_5ReS_3$
Pentacarbonyl(monomethylcarbonotrithioato-*S*)rhenium, Re-00019

$C_7H_3O_6Re$
Acetylpentacarbonylrhenium, Re-00020

$C_7H_4CrO_3S$
Tricarbonylthiophenechromium, Cr-00030

$C_7H_4F_6O_5ReP$
Pentacarbonyl(ethene)rhenium(1+); Hexafluorophosphate, *in* Re-00021

$C_7H_4MnNO_3$
Tricarbonyl-π-pyrrolylmanganese, Mn-00022

$C_7H_4O_5Re$
Pentacarbonyl(ethene)rhenium(1+), Re-00021

$C_7H_5BF_4NO_3Re$
Dicarbonyl(η^5-2,4-cyclopentadien-1-yl)-nitrosylrhenium(1+); Tetrafluoroborate, *in* Re-00023

$C_7H_5Br_2O_2Re$
Dibromodicarbonyl(η^5-2,4-cyclopentadien-1-yl)rhenium, Re-00022

$C_7H_5Cl_3O_2W$
Dicarbonyltrichloro(η^5-2,4-cyclopentadien-1-yl)-tungsten, W -00015

$C_7H_5CrNO_2S$
Dicarbonyl(η^5-2,4-cyclopentadien-1-yl)-thionitrosylchromium, Cr-00031

$C_7H_5CrNO_3$
Dicarbonyl(η^5-2,4-cyclopentadien-1-yl)nitrosylchromium, Cr-00032

$C_7H_5CrNO_5$
(1-Aminoethylidene)pentacarbonylchromium, Cr-00033

$C_7H_5F_6MnNO_3P$
Dicarbonyl(η^5-2,4-cyclopentadien-1-yl)-nitrosylmanganese(1+); Hexafluorophosphate, *in* Mn-00023

$C_7H_5MnNO_3^{\oplus}$
Dicarbonyl(η^5-2,4-cyclopentadien-1-yl)-nitrosylmanganese(1+), Mn-00023

$C_7H_5MnN_2O_2$
Dicarbonyl(η^5-2,4-cyclopentadien-1-yl)(dinitrogen)-manganese, Mn-00024

$C_7H_5MnO_4$
Tetracarbonyl(η^3-2-propenyl)manganese, Mn-00025

$C_7H_5NO_3Re^{\oplus}$
Dicarbonyl(η^5-2,4-cyclopentadien-1-yl)-nitrosylrhenium(1+), Re-00023

$C_7H_5NO_3W$
Dicarbonyl(η^5-2,4-cyclopentadien-1-yl)nitrosyltungsten, W -00016

$C_7H_5O_4Re$
Tetracarbonyl(η^3-2-propenyl)rhenium, Re-00024

$C_7H_6CrO_5S$
Pentacarbonyl(dimethyl sulfide)chromium, Cr-00034

$C_7H_6KO_2Re$
Dicarbonyl(η^5-2,4-cyclopentadien-1-yl)hydrorhenate(1−); K salt, *in* Re-00027

$C_7H_6NO_3Re$
Carbonyl(η^5-2,4-cyclopentadien-1-yl)formyl(nitrosyl)-rhenium, Re-00025

$C_7H_6NO_4ReS_2$
Tetracarbonyl(dimethylcarbamodithioato-*S*,*S′*)rhenium, Re-00026

$C_7H_6O_2Re^{\ominus}$
Dicarbonyl(η^5-2,4-cyclopentadien-1-yl)hydrorhenate(1−), Re-00027

$C_7H_7O_2Re$
Dicarbonyl(η^5-2,4-cyclopentadien-1-yl)dihydrorhenium, Re-00028

$C_7H_8CrF_6N_3O_2P$
Acetonitrile(η^5-2,4-cyclopentadien-1-yl)-dinitrosylchromium(1+); Hexafluorophosphate, *in* Cr-00035

$C_7H_8CrN_3O_2^{\oplus}$
Acetonitrile(η^5-2,4-cyclopentadien-1-yl)-dinitrosylchromium(1+), Cr-00035

$C_7H_8NO_2Re$
Carbonyl(η^5-2,4-cyclopentadien-1-yl)-methylnitrosylrhenium, Re-00029

$C_7H_8NO_3Re$
Carbonyl(η^5-2,4-cyclopentadien-1-yl)(hydroxymethyl)-nitrosylrhenium, Re-00030

$C_7H_9MnN_2O_2$
Dicarbonyl(η^5-2,4-cyclopentadien-1-yl)(hydrazine-*N*)-manganese, Mn-00026

$C_7H_{17}CrO_5^{\oplus\oplus}$
Pentaaquabenzylchromium(2+), Cr-00036

$C_7H_{21}N_3Ti$
Tris(dimethylaminato)methyltitanium, Ti-00014

$C_8Br_2Mn_2O_8$
Di(μ-bromo)octacarbonyldimanganese, Mn-00027

$C_8Cl_2Mn_2O_8$
Di(μ-chloro)octacarbonyldimanganese, Mn-00028

$C_8Cl_2O_8Re_2$
Octacarbonyldi-μ-chlorodirhenium, Re-00031

$C_8Cl_4Mo_2O_8$
Octacarbonyltetrachlorodimolybdenum, Mo-00008

$C_8H_2Cr_2O_8^{\ominus\ominus}$
Octacarbonyl-μ-dihydrodichromate(2−), Cr-00037

$C_8H_2Mo_2O_8^{\ominus\ominus}$
Octacarbonyl-μ-dihydrodimolybdate(2−), Mo-00009

$C_8H_2O_8Re_2$
Octacarbonyldi-μ-hydrodirhenium, Re-00032

$C_8H_2O_8W_2^{\ominus\ominus}$
Octacarbonyldi-μ-hydroditungstate(2−), W -00017

$C_8H_3MnO_7$
Pentacarbonyl(1,2-dioxopropyl)manganese, Mn-00029

$C_8H_4MnO_7^{\oplus}$
Pentacarbonyl-1,3-dioxalan-2-ylidenemanganese(1+), Mn-00030

$C_8H_5AsMoO_3$
(η^6-Arsenin)tricarbonylmolybdenum, Mo-00010
$C_8H_5BrMoO_3$
Bromotricarbonyl(η^5-2,4-cyclopentadien-1-yl)molybdenum, Mo-00011
$C_8H_5BrO_3W$
Bromotricarbonyl(η^5-2,4-cyclopentadien-1-yl)tungsten, W -00018
$C_8H_5ClMoO_3$
Tricarbonylchloro(η^5-2,4-cyclopentadien-1-yl)-molybdenum, Mo-00012
$C_8H_5ClO_3W$
Tricarbonylchloro(η^5-2,4-cyclopentadien-1-yl)tungsten, W -00019
$C_8H_5Cl_2NbO_3$
Tricarbonyldichloro(η^5-2,4-cyclopentadien-1-yl)niobium, Nb-00008
$C_8H_5CrO_3^{\ominus}$
Tricarbonyl(η^5-2,4-cyclopentadien-1-yl)chromate(1−), Cr-00038
$C_8H_5CrO_3Na$
Tricarbonyl(η^5-2,4-cyclopentadien-1-yl)chromate(1−); Na salt, *in* Cr-00038
$C_8H_5HgO_6V$
Hexacarbonyl(ethylmercury)vanadium, V -00014
$C_8H_5IMoO_3$
Tricarbonyl(η^5-2,4-cyclopentadien-1-yl)iodomolybdenum, Mo-00013
$C_8H_5IO_2SW$
(Carbonothioyl)dicarbonyl(η^5-2,4-cyclopentadien-1-yl)-iodotungsten, W -00020
$C_8H_5IO_3W$
Tricarbonyl(η^5-2,4-cyclopentadien-1-yl)iodotungsten, W -00021
$C_8H_5KO_3W$
Tricarbonyl(η^5-2,4-cyclopentadien-1-yl)tungstate(1−); K salt, *in* W -00023
$C_8H_5LiMoO_3$
Tricarbonyl(η^5-2,4-cyclopentadien-1-yl)molybdate(1−); Li salt, *in* Mo-00014
$C_8H_5MnO_2S$
(Carbonothioyl)dicarbonyl(η^5-2,4-cyclopentadien-1-yl)-manganese, Mn-00031
$C_8H_5MnO_2Se$
(Carbonoselenoyl)dicarbonyl(η^5-2,4-cyclopentadien-1-yl)-manganese, Mn-00032
$C_8H_5MnO_3$
▷Cyclopentadienyltricarbonylmanganese, Mn-00033
$C_8H_5MnO_5$
Pentacarbonyl-2-propenylmanganese, Mn-00034
$C_8H_5MnO_7$
Pentacarbonyl(ethoxycarbonyl)manganese, Mn-00035
$C_8H_5MoO_3^{\ominus}$
Tricarbonyl(η^5-2,4-cyclopentadien-1-yl)molybdate(1−), Mo-00014
$C_8H_5MoO_3Tl$
Tricarbonyl(η^5-2,4-cyclopentadien-1-yl)-thalliummolybdenum, Mo-00015
$C_8H_5N_3S_3Ti$
(η^5-Cyclopentadienyl)trithiocyanatotitanium, Ti-00015
$C_8H_5NaO_3W$
Tricarbonyl(η^5-2,4-cyclopentadien-1-yl)tungstate(1−); Na salt, *in* W -00023
$C_8H_5Na_2O_3V$
Tricarbonyl(η^5-2,4-cyclopentadien-1-yl)vanadate(2−); Di-Na salt, *in* V -00015
$C_8H_5O_2SW^{\ominus}$
(Carbonothioyl)dicarbonyl(η^5-2,4-cyclopentadien-1-yl)-tungstate(1−), W -00022
$C_8H_5O_3Re$
Tricarbonyl(η^5-2,4-cyclopentadien-1-yl)rhenium, Re-00033
$C_8H_5O_3Tc$
Cyclopentadienyltricarbonyltechnetium, Tc-00004
$C_8H_5O_3V^{\ominus\ominus}$
Tricarbonyl(η^5-2,4-cyclopentadien-1-yl)vanadate(2−), V -00015
$C_8H_5O_3W^{\ominus}$
Tricarbonyl(η^5-2,4-cyclopentadien-1-yl)tungstate(1−), W -00023
$C_8H_5O_5V$
Pentacarbonyl(η^3-2-propenyl)vanadium, V -00016
$C_8H_5O_7Re$
Pentacarbonyl(ethoxycarbonyl)rhenium, Re-00034
$C_8H_6BCl_4CrNO_5$
Pentacarbonyl[(dimethylamino)methylidyne]chromium(1+); Tetrachloroborate, *in* Cr-00039
$C_8H_6ClMnNO_5^{\oplus}$
Pentacarbonyl[chloro(dimethylamino)methylene]-manganese(1+), Mn-00036
$C_8H_6Cl_2MnNO_9$
Pentacarbonyl[chloro(dimethylamino)methylene]-manganese(1+); Perchlorate, *in* Mn-00036
$C_8H_6CrF_6NO_5P$
Pentacarbonyl[(dimethylamino)methylidyne]chromium(1+); Hexafluorophosphate, *in* Cr-00039
$C_8H_6CrNO_5^{\oplus}$
Pentacarbonyl[(dimethylamino)methylidyne]chromium(1+), Cr-00039
$C_8H_6CrN_2O_2$
(η^6-Benzene)dicarbonyl(dinitrogen)chromium, Cr-00040
$C_8H_6CrO_3$
Tricarbonyl(η^5-2,4-cyclopentadien-1-yl)hydrochromium, Cr-00041
$C_8H_6CrO_6$
Pentacarbonyl(1-methoxyethylidene)chromium, Cr-00042
$C_8H_6LiMnO_6$
Diacetyltetracarbonylmanganate(1−); Li salt, *in* Mn-00037
$C_8H_6MnO_6^{\ominus}$
Diacetyltetracarbonylmanganate(1−), Mn-00037
$C_8H_6MoO_3$
Tricarbonyl(η^5-2,4-cyclopentadien-1-yl)hydromolybdenum, Mo-00016
$C_8H_6MoO_6$
Pentacarbonyl(1-methoxyethylidene)molybdenum, Mo-00017
$C_8H_6O_3V^{\ominus}$
Tricarbonyl(η^5-2,4-cyclopentadien-1-yl)-hydrovanadate(1−), V -00017
$C_8H_6O_3W$
Tricarbonyl(η^5-2,4-cyclopentadien-1-yl)hydrotungsten, W -00024
$C_8H_6O_5S_2W$
[Bis(methylthio)methylene]pentacarbonyltungsten, W -00025
$C_8H_7CrNO_3$
Tricarbonyl(1-methylpyrrole)chromium, Cr-00043
$C_8H_7CrNO_5$
Pentacarbonyl[(dimethylamino)methylene]chromium, Cr-00044
Pentacarbonyl[1-(methylamino)ethylidene]chromium, Cr-00045
C_8H_8CeCl
Chloro(cyclooctatetraene)cerium, Ce-00001
C_8H_8ClEr
(η^8-1,3,5,7-Cyclooctatetraene)chloroerbium, Er-00003

C_8H_8ClLa
(η^8-1,3,5,7-Cyclooctatetraene)chlorolanthanum, La-00001
C_8H_8ClLu
(η^8-1,3,5,7-Cyclooctatetraene)chlorolutetium, Lu-00003
C_8H_8ClNd
Chloro(cyclooctatetraene)neodymium, Nd-00002
C_8H_8ClPr
Chloro(cyclooctatetraene)praesodymium, Pr-00002
C_8H_8ClSm
Chloro(cyclooctatetraene)samarium, Sm-00003
$C_8H_8Cl_2Th$
Dichloro(η^8-1,3,5,7-cyclooctatetraene)thorium, Th-00001
$C_8H_8Cl_2Zr$
Dichloro(η^8-1,3,5,7-cyclooctatetraene)zirconium, Zr-00004
C_8H_8Eu
η^8-1,3,5,7-Cyclooctatetraeneeuropium, Eu-00002
C_8H_8Yb
(η^8-1,3,5,7-Cyclooctatetraene)ytterbium, Yb-00003
$C_8H_9ClMoN_2O_3$
Tricarbonyl(η^5-2,4-cyclopentadien-1-yl)(hydrazine-*N*)-molybdenum(1+); Chloride, *in* Mo-00018
$C_8H_9CrNO_2$
Carbonyl(η^5-2,4-cyclopentadien-1-yl)(η^2-ethene)-nitrosylchromium, Cr-00046
$C_8H_9CrO_5P$
Pentacarbonyl(trimethylphosphine)chromium, Cr-00047
$C_8H_9CrO_5PS$
Pentacarbonyl(trimethylphosphine sulfide-*S*)chromium, Cr-00048
$C_8H_9MnO_5Pb$
Pentacarbonyl(trimethylplumbyl)manganese, Mn-00038
$C_8H_9MnO_5Si$
Pentacarbonyl(trimethylsilyl)manganese, Mn-00039
$C_8H_9Mn_2N_3O_8$
Triamminetricarbonylmanganese(1+); Pentacarbonylmanganate(1−), *in* Mn-00002
$C_8H_9MoN_2O_3^{\oplus}$
Tricarbonyl(η^5-2,4-cyclopentadien-1-yl)(hydrazine-*N*)-molybdenum(1+), Mo-00018
$C_8H_9O_5PbRe$
Pentacarbony(trimethylplumbyl)rhenium, Re-00035
$C_8H_{10}IMoNO$
(η^5-2,4-Cyclopentadien-1-yl)iodonitrosyl(η^3-2-propenyl)-molybdenum, Mo-00019
$C_8H_{10}INOW$
(η^5-2,4-Cyclopentadien-1-yl)iodonitrosyl(η^3-2-propenyl)-tungsten, W -00026
$C_8H_{10}Zr$
(η^8-1,3,5,7-cyclooctatetraene)dihydrozirconium, Zr-00005
$C_8H_{12}CrF_6N_3O_3P$
(η^5-2,4-Cyclopentadien-1-yl)[methoxy(methylamino)-methylene]dinitrosylchromium; Hexafluorophosphate, *in* Cr-00049
$C_8H_{12}CrN_3O_3^{\oplus}$
(η^5-2,4-Cyclopentadien-1-yl)[methoxy(methylamino)-methylene]dinitrosylchromium, Cr-00049
$C_8H_{12}I_2MnNO_4$
Tetracarbonyldiiodomanganate(1−); Tetramethylammonium salt, *in* Mn-00006
$C_8H_{14}ClTa$
Chloro(η^5-2,4-cyclopentadien-1-yl)trimethyltantalum, Ta-00007
$C_8H_{14}O_3Ti$
(η^5-Cyclopentadienyl)trimethoxytitanium, Ti-00016
$C_8H_{14}Ti$
(η^5-2,4-Cyclopentadien-1-yl)trimethyltitanium, Ti-00017
$C_8H_{20}CrO_4Tl_2$
▷(Chromato-*O*)bis(diethylthallium), Cr-00050
$C_8H_{22}Cl_3P_2Ta$
Trichloro(η^2-ethene)bis(trimethylphosphine)tantalum, Ta-00008
$C_8H_{24}Li_2W$
Octamethyltungstate(2−); Di-Li salt, *in* W -00027
$C_8H_{24}Mo_2^{\ominus\ominus\ominus\ominus}$
Octamethyldimolybdenum(4−), Mo-00020
$C_8H_{24}N_4Ti$
Tetrakis(dimethylamino)titanium, Ti-00018
$C_8H_{24}Re_2^{\ominus\ominus}$
Octamethyldirhenate(2−), Re-00036
$C_8H_{24}W^{\ominus\ominus}$
Octamethyltungstate(2−), W -00027
$C_8H_{32}B_{20}Ti^{\ominus\ominus}$
Bis[η^6-decahydro-*C*,*C'*-dimethyldicarbadodecaborato(2−)]-titanate(2−), Ti-00019
$C_8I_2Mn_2O_8$
Di(μ-iodo)octacarbonyldimanganese, Mn-00040
$C_8I_2Mo_2O_8$
Octacarbonyldi-μ-iododimolybdenum, Mo-00021
$C_8I_2O_8W_2$
Octacarbonyldi-μ-iododitungsten, W -00028
$C_9FeMnO_9^{\ominus}$
Nonacarbonylferratemanganate(1−), Mn-00041
$C_9HNO_{10}W_2$
Nonacarbonyl-μ-hydronitrosylditungsten, W -00029
$C_9H_5BF_4MoO_4$
Tetracarbonyl(η^5-2,4-cyclopentadien-1-yl)molybdenum(1+); Tetrafluoroborate, *in* Mo-00022
$C_9H_5ClCrHgO_3$
Tricarbonyl(chloromercury)[μ-[(1-η:1,2,3,4,5,6-η)-phenyl]]chromium, Cr-00051
$C_9H_5ClCrO_3$
Tricarbonyl(η^6-chlorobenzene)chromium, Cr-00052
$C_9H_5ClF_6MnPO_3$
(Chlorobenzene)tricarbonylmanganese(I); Hexafluorophosphate, *in* Mn-00042
$C_9H_5ClMnO_3^{\oplus}$
(Chlorobenzene)tricarbonylmanganese(I), Mn-00042
$C_9H_5CrFO_3$
(η^6-Fluorobenzene)tricarbonylchromium, Cr-00053
$C_9H_5CrLiO_3$
[μ(η:η^6-Phenyl)](tricarbonylchromium)lithium, Cr-00054
$C_9H_5F_6MoO_4P$
Tetracarbonyl(η^5-2,4-cyclopentadien-1-yl)molybdenum(1+); Hexafluorophosphate, *in* Mo-00022
$C_9H_5F_6O_4PW$
Tetracarbonyl(η^5-2,4-cyclopentadien-1-yl)tungsten(1+); Hexafluorophosphate, *in* W -00030
$C_9H_5MnN_3OS_2^{\ominus}$
π-Cyclopentadienyl[dimercaptomaleonitrilato(2−)]-nitrosylmanganate(1−), Mn-00043
$C_9H_5MnO_5$
Tricarbonyl(carboxy-π-cyclopentadienyl)manganese, Mn-00044
$C_9H_5MoO_4^{\oplus}$
Tetracarbonyl(η^5-2,4-cyclopentadien-1-yl)molybdenum(1+), Mo-00022
$C_9H_5NNaO_3V$
Tricarbonyl(cyano-*C*)(η^5-2,4-cyclopentadien-1-yl)-vanadate(1−); Na salt, *in* V -00018
$C_9H_5NO_3V^{\ominus}$
Tricarbonyl(cyano-*C*)(η^5-2,4-cyclopentadien-1-yl)-vanadate(1−), V -00018

$C_9H_5NbO_4$
Tetracarbonyl(η^5-2,4-cyclopentadien-1-yl)niobium, Nb-00009

$C_9H_5O_3SV$
Tricarbonyl(η^5-2,4-cyclopentadien-1-yl)(thiocarbonyl)-vanadium, V -00019

$C_9H_5O_4Ta$
Tetracarbonyl(η^5-2,4-cyclopentadien-1-yl)tantalum, Ta-00009

$C_9H_5O_4V$
Tetracarbonyl(η^5-2,4-cyclopentadien-1-yl)vanadium, V -00020

$C_9H_5O_4W^{\oplus}$
Tetracarbonyl(η^5-2,4-cyclopentadien-1-yl)tungsten(1+), W -00030

$C_9H_6CrN_2O_5$
Pentacarbonyl[cyano(dimethylamino)methylene]chromium, Cr-00055

$C_9H_6CrO_2S$
(η^6-Benzene)(carbonothioyl)dicarbonylchromium, Cr-00056

$C_9H_6CrO_3$
▷(η^6-Benzene)tricarbonylchromium, Cr-00057

$C_9H_6CrO_4$
Tricarbonyl(η^6-phenol)chromium, Cr-00058

$C_9H_6F_6MnO_3P$
(η^6-Benzene)tricarbonylmanganese(1+); Hexafluorophosphate, *in* Mn-00045

$C_9H_6F_6O_3PRe$
(η^6-Benzene)tricarbonylrhenium(1+); Hexafluorophosphate, *in* Re-00037

$C_9H_6MnNO_4$
Tricarbonyl(carboxy-π-cyclopentadienyl)manganese; Amide, *in* Mn-00044

$C_9H_6MnO_3^{\oplus}$
(η^6-Benzene)tricarbonylmanganese(1+), Mn-00045

$C_9H_6MoO_3$
η^6-Benzenetricarbonylmolybdenum, Mo-00023

$C_9H_6O_3Re^{\oplus}$
(η^6-Benzene)tricarbonylrhenium(1+), Re-00037

$C_9H_6O_3W$
(η^6-Benzene)tricarbonyltungsten, W -00031

$C_9H_7ClMoO_2$
Dicarbonylchloro(η^7-cycloheptatrienylium)molybdenum, Mo-00024

$C_9H_7ClO_2W$
Dicarbonylchloro(η^7-cycloheptatrienylium)tungsten, W -00032

$C_9H_7CrNO_3$
(η^6-Aniline)tricarbonylchromium, Cr-00059

$C_9H_7IO_2W$
Dicarbonyl(η^7-cycloheptatrienylium)iodotunsten, W -00033

$C_9H_7MnO_3$
▷(Methylcyclopentadienyl)tricarbonylmanganese, Mn-00046
Tricarbonyl[(1,2,3,4,5-η)-2,4-cyclohexadien-1-yl]-manganese, Mn-00047

$C_9H_8BCl_4MnO_2$
Dicarbonyl(η^5-2,4-cyclopentadien-1-yl)-ethylidynemanganese(1+); Tetrachloroborate, *in* Mn-00048

$C_9H_8CrO_3$
Cyclopentadienyltricarbonylmethylchromium, Cr-00060

$C_9H_8CrO_6$
Pentacarbonyl(tetrahydrofuran)chromium, Cr-00061

$C_9H_8MnO_2^{\oplus}$
Dicarbonyl(η^5-2,4-cyclopentadien-1-yl)-ethylidynemanganese(1+), Mn-00048

$C_9H_8MoO_3$
Tricarbonyl(η^5-2,4-cyclopentadien-1-yl)-methylmolybdenum, Mo-00025

$C_9H_8O_3V^{\ominus}$
Tricarbonyl(η^5-2,4-cyclopentadien-1-yl)-methylvanadate(1−), V -00021

$C_9H_8O_3W$
Tricarbonyl(η^5-2,4-cyclopentadien-1-yl)methyltungsten, W -00034

$C_9H_8O_5SW$
Tricarbonyl(η^5-2,4-cyclopentadien-1-yl)-methanesulfonyltungsten, W -00035

$C_9H_8O_6W$
Pentacarbonyl(1-ethoxyethylidene)tungsten, W -00036

$C_9H_9F_6MnN_3O_3P$
Tricarbonyltris(isocyanomethane)manganese(1+); Hexafluorophosphate, *in* Mn-00049
Tris(acetonitrile)tricarbonylmanganese(1+); Hexafluorophosphate, *in* Mn-00050

$C_9H_9F_6N_3O_3PRe$
Tris(acetonitrile)tricarbonylrhenium(1+); Hexafluorophosphate, *in* Re-00038

$C_9H_9K_2Re_2$
μ-Carbonyloctacarbonyldirhenium(2−); Di-K salt, *in* Re-00040

$C_9H_9MnN_3O_3^{\oplus}$
Tricarbonyltris(isocyanomethane)manganese(1+), Mn-00049
Tris(acetonitrile)tricarbonylmanganese(1+), Mn-00050

$C_9H_9MoN_3O_3$
Tris(acetonitrile)tricarbonylmolybdenum, Mo-00026

$C_9H_9N_3O_3Re^{\oplus}$
Tris(acetonitrile)tricarbonylrhenium(1+), Re-00038

$C_9H_{10}BrCrNO_4$
Bromotetracarbonyl[(diethylamino)methylidyne]chromium, Cr-00062

$C_9H_{10}BrNO_4W$
Bromotetracarbonyl[(diethylamino)methylidyne]tungsten, W -00037

$C_9H_{10}ClCrNO_4$
Tetracarbonylchloro[(diethylamino)methylidyne]chromium, Cr-00063

$C_9H_{10}F_6MoNO_2P$
Carbonyl(η^5-2,4-cyclopentadien-1-yl)nitrosyl(η^3-2-propenyl)molybdenum(1+); Hexafluorophosphate, *in* Mo-00027

$C_9H_{10}MoNO_2^{\oplus}$
Carbonyl(η^5-2,4-cyclopentadien-1-yl)nitrosyl(η^3-2-propenyl)molybdenum(1+), Mo-00027

$C_9H_{10}OW$
Carbonyl(η^5-2,4-cyclopentadien-1-yl)(η^2-ethyne)-methyltungsten, W -00038

$C_9H_{11}ClMo$
(η^6-Benzene)chloro(η^3-2-propenyl)molybdenum, Mo-00028

$C_9H_{11}ClMoN_2O_2$
Bis(acetonitrile)dicarbonylchloro(η^3-2-propenyl)-molybdenum, Mo-00029

$C_9H_{11}ClN_2O_2W$
Bis(acetonitrile)dicarbonylchloro(η^3-2-propenyl)-tungsten, W -00039

$C_9H_{11}Cl_3Ti$
Trichloro(η^5-tetrahydroindenyl)titanium, Ti-00020

$C_9H_{11}IMoN_2O_2$
Bis(acetonitrile)dicarbonyliodo(η^3-2-propenyl)-molybdenum, Mo-00030

$C_9H_{11}MoNO_2$
Carbonyl(η^5-2,4-cyclopentadien-1-yl)nitrosyl(1,2-η-1-propene)molybdenum, Mo-00031

$C_9H_{11}O_2Re$
Dicarbonyl(η^5-2,4-cyclopentadien-1-yl)dimethylrhenium, Re-00039
$C_9H_{11}O_4V$
Bis(acetato)cyclopentadienylvanadium, *in* V -00083
$C_9H_{12}BrCrNO_5$
Bromopentacarbonylchromate(1−); Tetramethylammonium salt, *in* Cr-00005
$C_9H_{12}BrMnN_4O$
Bromocarbonyltetrakis(isocyanomethane)manganese, Mn-00051
$C_9H_{12}MnO$
Bis(η^4-1,3-butadiene)carbonylmanganese, Mn-00052
$C_9H_{13}Cl_2OTi$
Dichloro(η^5-2,4-cyclopentadien-1-yl)titanium; THF complex, *in* Ti-00008
$C_9H_{15}BrU$
Bromotris(η^3-2-propenyl)uranium, U -00005
$C_9H_{15}ClU$
Chlorotris(η^3-2-propenyl)uranium, U -00006
$C_9H_{15}Cr$
Tris(η^3-2-propenyl)chromium, Cr-00064
$C_9H_{15}IU$
Iodotris(η^3-2-propenyl)uranium, U -00007
$C_9H_{15}O_2Ti$
Cyclopentadienyldiethoxytitanium, Ti-00021
$C_9H_{15}V$
▷Tris(η^3-2-propenyl)vanadium, V -00022
$C_9H_{17}N_2Ti$
Cyclopentadienylbis(dimethylamido)titanium, Ti-00022
$C_9H_{18}BrMnO_9P_2$
Bromotricarbonylbis(trimethylphosphite-*P*)manganese, Mn-00053
$C_9O_9Re_2^{\ominus\ominus}$
μ-Carbonyloctacarbonyldirhenium(2−), Re-00040
$C_{10}CrIO_{10}^{\ominus}$
Decacarbonyl-μ-iododichromate(1−), Cr-00065
$C_{10}Cr_2K_2O_{10}$
Decacarbonyldichromate(2−); Di-K salt, *in* Cr-00066
$C_{10}Cr_2Na_2O_{10}$
Decacarbonyldichromate(2−); Di-Na salt, *in* Cr-00066
$C_{10}Cr_2O_{10}^{\ominus\ominus}$
Decacarbonyldichromate(2−), Cr-00066
$C_{10}F_6O_8Re_2S_2$
Octacarbonylbis[μ-(trifluoromethanethiolato)]dirhenium, *in* Re-00009
$C_{10}HCr_2O_{10}^{\ominus}$
Decacarbonyl-μ-hydrodichromate(1−), Cr-00067
$C_{10}HO_{10}W_2^{\ominus}$
Decacarbonyl-μ-hydroditungstate(1−), W -00040
$C_{10}H_4Cr_2O_{10}P_2$
Decacarbonyl[μ-(diphosphine-*P*:*P'*)]dichromium, Cr-00068
$C_{10}H_5O_4ReS$
(Benzenethiolato)tetracarbonylrhenium, Re-00041
$C_{10}H_6CrO_4$
Tricarbonyl[(2,3,4,5,6,7-η)-2,4,6-cycloheptatrien-1-one]chromium, Cr-00069
$C_{10}H_6Mn_2O_8S_2$
Octacarbonylbis(μ-methanethiolato)dimanganese, *in* Mn-00012
$C_{10}H_7BCrF_4O_3$
Tricarbonyl(η^7-cycloheptatrienylium)chromium(1+); Tetrafluoroborate, *in* Cr-00070
$C_{10}H_7BF_4MoO_3$
Tricarbonyl(η^7-cycloheptatrienylium)molybdenum(1+); Tetrafluoroborate, *in* Mo-00032
$C_{10}H_7BF_4O_3W$
Tricarbonyl(η^7-cycloheptatrienylium)tunsten(1+); Tetrafluoroborate, *in* W -00041
$C_{10}H_7ClCrO_7$
▷Tricarbonyl(η^7-cycloheptatrienylium)chromium(1+); Perchlorate, *in* Cr-00070
$C_{10}H_7CrI_3O_3$
Tricarbonyl(η^7-cycloheptatrienylium)chromium(1+); Triiodide, *in* Cr-00070
$C_{10}H_7CrO_3^{\oplus}$
Tricarbonyl(η^7-cycloheptatrienylium)chromium(1+), Cr-00070
$C_{10}H_7MnO_3$
Tricarbonyl[(1,2,3,4,5-η)-2,4,6-cycloheptatrien-1-yl]-manganese, Mn-00054
$C_{10}H_7MnO_4$
(Acetylcyclopentadienyl)tricarbonylmanganese, Mn-00055
$C_{10}H_7MnO_5$
Tricarbonyl(carboxy-π-cyclopentadienyl)manganese; Me ester, *in* Mn-00044
$C_{10}H_7MoO_3^{\oplus}$
Tricarbonyl(η^7-cycloheptatrienylium)molybdenum(1+), Mo-00032
$C_{10}H_7O_3V$
Tricarbonyl(η^7-cycloheptatrienylium)vanadium, V -00023
$C_{10}H_7O_3W^{\oplus}$
Tricarbonyl(η^7-cycloheptatrienylium)tunsten(1+), W -00041
$C_{10}H_7O_4V$
Tetracarbonyl(η^5-methyl-2,4-cyclopentadien-1-yl)-vanadium, V -00024
$C_{10}H_8CrO_3$
Tricarbonyl[(1,2,3,4,5,6-η)-1,3,5-cycloheptatriene]-chromium, Cr-00071
Tricarbonyl[(1,2,3,4,5,6-η)methylbenzene]chromium, Cr-00072
$C_{10}H_8CrO_4$
Tricarbonyl-(1,2,3,4,5-η-methoxybenzene)chromium, Cr-00073
$C_{10}H_8MoO_3$
▷Tricarbonyl[(1,2,3,4,5,6η)-1,3,5-cycloheptatriene]-molybdenum, Mo-00033
Tricarbonyl[(1,2,3,4,5,6-η)methylbenzene]molybdenum, Mo-00034
$C_{10}H_8O_3W$
Tricarbonyl[(1,2,3,4,5,6-η)-1,3,5-cycloheptatriene]-tungsten, W -00042
Tricarbonyl[(1,2,3,4,5,6-η)-methylbenzene]tungsten, W -00043
$C_{10}H_8O_4W$
Acetyltricarbonyl(η^5-2,4-cyclopentadien-1-yl)tungsten, W -00044
$C_{10}H_9BF_4MoO_3$
Tricarbonyl(η^5-2,4-cyclopentadien-1-yl)(η^2-ethene)-molybdenum(1+); Tetrafluoroborate, *in* Mo-00035
$C_{10}H_9BF_4O_3W$
Tricarbonyl(η^5-2,4-cyclopentadien-1-yl)(η^2-ethene)-tungsten(1+); Tetrafluoroborate, *in* W -00045
$C_{10}H_9MnO_3$
Tricarbonyl[(1,2,3,4,5-η)-2,4-cycloheptadien-1-yl]-manganese, Mn-00056
$C_{10}H_9MnO_4$
(η^6-Benzene)dicarbonyl(methoxycarbonyl)manganese, Mn-00057
$C_{10}H_9MoO_3^{\oplus}$
Tricarbonyl(η^5-2,4-cyclopentadien-1-yl)(η^2-ethene)-molybdenum(1+), Mo-00035
$C_{10}H_9NO_5V^{\ominus}$
Pentacarbonyl(2-isocyano-2-methylpropane)vanadate(1−), V -00025

$C_{10}H_9O_3W^{\oplus}$
Tricarbonyl(η^5-2,4-cyclopentadien-1-yl)(η^2-ethene)-tungsten(1+), W -00045

$C_{10}H_{10}BrTi$
Bromobis(cyclopentadienyl)titanium, Ti-00023

$C_{10}H_{10}BrV$
Bromobis(η^5-2,4-cyclopentadien-1-yl)vanadium, V -00026

$C_{10}H_{10}Br_2Hf$
Dibromobis(η^5-2,4-cyclopentadien-1-yl)hafnium, Hf-00002

$C_{10}H_{10}Br_2Ti$
Dibromobis(η^5-2,4-cyclopentadien-1-yl)titanium, Ti-00024

$C_{10}H_{10}Br_2V$
Dibromobis(η^5-2,4-cyclopentadien-1-yl)vanadium, V -00027

$C_{10}H_{10}Br_2W$
Dibromobis(η^5-2,4-cyclopentadien-1-yl)tungsten, W -00046

$C_{10}H_{10}Br_2Zr$
Dibromobis(η^5-2,4-cyclopentadien-1-yl)zirconium, Zr-00006

$C_{10}H_{10}Br_3Nb$
Tribromobis(η^5-2,4-cyclopentadien-1-yl)niobium, Nb-00010

$C_{10}H_{10}ClCrNO_5$
Pentacarbonyl[chloro(diethylamino)methylene]chromium, Cr-00074

$C_{10}H_{10}ClDy$
Chlorobis(η^5-2,4-cyclopentadienyl)dysprosium, Dy-00002

$C_{10}H_{10}ClEr$
Chlorobis(η^5-2,4-cyclopentadien-1-yl)erbium, Er-00004

$C_{10}H_{10}ClGd$
Chlorobis(η^5-2,4-cyclopentadien-1-yl)gadolinium, Gd-00002

$C_{10}H_{10}ClHo$
Chlorobis(η^5-2,4-cyclopentadien-1-yl)holmium, Ho-00002

$C_{10}H_{10}ClLu$
Chlorobis(η^5-2,4-cyclopentadien-1-yl)lutetium, Lu-00004

$C_{10}H_{10}ClNbO$
Chlorobis(η^5-2,4-cyclopentadien-1-yl)oxoniobium, Nb-00011

$C_{10}H_{10}ClOV$
Chlorobis(η^5-2,4-cyclopentadien-1-yl)oxovanadium, V -00028

$C_{10}H_{10}ClSc$
Chlorobis(η^5-2,4-cyclopentadien-1-yl)scandium, Sc-00001

$C_{10}H_{10}ClSm$
Chlorobis(η^5-2,4-cyclopentadien-1-yl)samarium, Sm-00004

$C_{10}H_{10}ClTi$
Chlorobis(cyclopentadienyl)titanium, Ti-00025

$C_{10}H_{10}ClTm$
Chlorobis(η^5-2,4-cyclopentadien-1-yl)thulium, Tm-00002

$C_{10}H_{10}ClV$
Chlorobis(η^5-2,4-cyclopentadien-1-yl)vanadium, V -00029

$C_{10}H_{10}ClY$
Chlorobis(η^5-2,4-cyclopentadien-1-yl)yttrium, Y -00002

$C_{10}H_{10}ClYb$
Chlorobis(η^5-2,4-cyclopentadien-1-yl)ytterbium, Yb-00004

$C_{10}H_{10}Cl_2Hf$
▷Dichlorobis(η^5-2,4-cyclopentadien-1-yl)hafnium, Hf-00003

$C_{10}H_{10}Cl_2Mo$
Dichlorobis(η^5-2,4-cyclopentadien-1-yl)molybdenum, Mo-00036

$C_{10}H_{10}Cl_2Nb$
▷Dichlorobis(η^5-2,4-cyclopentadien-1-yl)niobium, Nb-00012

$C_{10}H_{10}Cl_2OW$
Dichlorobis(η^5-2,4-cyclopentadien-1-yl)oxotungsten, W -00047

$C_{10}H_{10}Cl_2Ta$
Dichlorobis(η^5-2,4-cyclopentadien-1-yl)tantalum, Ta-00010

$C_{10}H_{10}Cl_2Ti$
▷Dichlorobis(η^5-2,4-cyclopentadien-1-yl)titanium, Ti-00026

$C_{10}H_{10}Cl_2V$
Dichlorobis(η^5-2,4-cyclopentadien-1-yl)vanadium, V -00030

$C_{10}H_{10}Cl_2W$
Dichlorobis(η^5-2,4-cyclopentadien-1-yl)tungsten, W -00048

$C_{10}H_{10}Cl_2Zr$
▷Dichlorobis(η^5-2,4-cyclopentadien-1-yl)zirconium, Zr-00007

$C_{10}H_{10}Cl_4Mo_2N_2O_2$
Di-μ-chlorodichlorobis(η^5-2,4-cyclopentadien-1-yl)-dinitrosyldimolybdenum, *in* Mo-00003

$C_{10}H_{10}Cl_4OTi_2$
Tetrachlorobis(η^5-2,4-cyclopentadien-1-yl)-μ-oxodititanium, Ti-00027

$C_{10}H_{10}Cr$
Chromocene, Cr-00075

$C_{10}H_{10}CrN_2O_5$
Pentacarbonyl(diethylcyanamide-*N′*)chromium, Cr-00076

$C_{10}H_{10}CrO_5S_2$
[Bis(ethylthio)methylene]pentacarbonylchromium, Cr-00077

$C_{10}H_{10}Cr_2N_4O_4$
Bis(η^5-2,4-cyclopentadien-1-yl)di-μ-nitrosyldinitrosyldichromium, Cr-00078

$C_{10}H_{10}Eu$
Bis(η^5-2,4-cyclopentadien-1-yl)europium, Eu-00003

$C_{10}H_{10}FTi$
Bis(cyclopentadienyl)fluorotitanium, Ti-00028

$C_{10}H_{10}F_2Hf$
Bis(η^5-2,4-cyclopentadien-1-yl)difluorohafnium, Hf-00004

$C_{10}H_{10}F_2Ti$
Bis(η^5-2,4-cyclopentadien-1-yl)difluorotitanium, Ti-00029

$C_{10}H_{10}F_2Zr$
Bis(η^5-2,4-cyclopentadien-1-yl)difluorozirconium, Zr-00008

$C_{10}H_{10}HfI_2$
Bis(η^5-2,4-cyclopentadien-1-yl)diiodohafnium, Hf-00005

$C_{10}H_{10}HfS_5$
Bis(η^5-2,4-cyclopentadien-1-yl)(pentathio)hafnium, Hf-00006

$C_{10}H_{10}IMoNO$
Bis(η^5-2,4-cyclopentadien-1-yl)iodonitrosylmolybdenum, Mo-00037

$C_{10}H_{10}ITi$
Bis(cyclopentadienyl)iodotitanium, Ti-00030

$C_{10}H_{10}IV$
Bis(η^5-2,4-cyclopentadien-1-yl)iodovanadium, V -00031

$C_{10}H_{10}I_2Nb$
Bis(η^5-2,4-cyclopentadien-1-yl)diiodoniobium, Nb-00013

$C_{10}H_{10}I_2Ti$
Bis(η^5-2,4-cyclopentadien-1-yl)diiodotitanium, Ti-00031

$C_{10}H_{10}I_2Zr$
Bis(η^5-2,4-cyclopentadien-1-yl)diiodozirconium, Zr-00009

$C_{10}H_{10}Mn$
▷Manganocene, Mn-00058

$C_{10}H_{10}MoN_2O_5$
Pentacarbonyl(1,3-dimethyl-2-imidazolidinylidene)-molybdenum, Mo-00038

$C_{10}H_{10}MoO_2$
Dicarbonyl(η^5-2,4-cyclopentadien-1-yl)(η^3-2-propenyl)-molybdenum, Mo-00039

$C_{10}H_{10}N_2O_5W$
Pentacarbonyl(1,3-dimethyl-2-imidazolidinylidene)-tungsten, W -00049

$C_{10}H_{10}N_6Ti$
Diazidobis(η^5-2,4-cyclopentadien-1-yl)titanium, Ti-00032

$C_{10}H_{10}O_2W$
Dicarbonyl(η^5-2,4-cyclopentadien-1-yl)(η^3-2-propenyl)-tungsten, W -00050

$C_{10}H_{10}S_5Ti$
Bis(η^5-2,4-cyclopentadien-1-yl)(pentathio)titanium, Ti-00033
$C_{10}H_{10}S_5V$
Bis(η^5-2,4-cyclopentadienyl)pentathiovanadium, V -00032
$C_{10}H_{10}S_5Zr$
Bis(η^5-2,4-cyclopentadien-1-yl)(pentathio)zirconium, Zr-00010
$C_{10}H_{10}Se_5Ti$
Bis(η^5-2,4-cyclopentadien-1-yl)[pentaselenidato(2−)]-titanium, Ti-00034
$C_{10}H_{10}Sm$
Bis(η^5-2,4-cyclopentadien-1-yl)samarium, Sm-00005
$C_{10}H_{10}Tc_2$
Decacarbonylditechnetium, Tc-00005
$C_{10}H_{10}Ti$
▷Titanocene, Ti-00035
$C_{10}H_{10}V$
Bis(η^5-2,4-cyclopentadien-1-yl)vanadium, V -00033
$C_{10}H_{10}Yb$
Bis(η^5-2,4-cyclopentadien-1-yl)ytterbium, Yb-00005
$C_{10}H_{11}AsO_3W$
Tricarbonyl(η^5-2,4-cyclopentadien-1-yl)(dimethylarsino)-tungsten, W -00051
$C_{10}H_{11}ClHf$
Chlorobis(η^5-2,4-cyclopentadien-1-yl)hydrohafnium, Hf-00007
$C_{10}H_{11}ClZr$
Chlorobis(η^5-2,4-cyclopentadien-1-yl)hydrozirconium, Zr-00011
$C_{10}H_{11}Cl_3Ti$
Trichloro(2-methyl-1-phenyl-1-propenyl)titanium, Ti-00036
$C_{10}H_{11}LiMo$
Bis(η^5-2,4-cyclopentadien-1-yl)hydromolybdenumlithium, Mo-00040
$C_{10}H_{11}LiW$
Bis(η^5-2,4-cyclopentadien-1-yl)hydro(tungsten)lithium, W -00052
$C_{10}H_{11}Lu$
Bis(η^5-2,4-cyclopentadien-1-yl)hydrolutetium, Lu-00005
$C_{10}H_{11}MnO_2$
Dicarbonyl(η^5-2,4-cyclopentadien-1-yl)(1-methylethylidene)manganese, Mn-00059
$C_{10}H_{11}MnO_3$
Dicarbonyl(η^5-2,4-cyclopentadien-1-yl)(1-methoxyethylidene)manganese, Mn-00060
$C_{10}H_{11}MoNO_3$
Dicarbonyl(η^5-2,4-cyclopentadien-1-yl)(2-propanoneoximato-*N*,*O*)molybdenum, Mo-00041
$C_{10}H_{11}MoO_3Tl$
▷Tricarbonyl(η^5-2,4-cyclopentadien-1-yl)-(dimethylthallium)molybdenum, Mo-00042
$C_{10}H_{11}O_3SbW$
Tricarbonyl(η^5-2,4-cyclopentadien-1-yl)-(dimethylstilbino)tungsten, W -00053
$C_{10}H_{11}O_3TlW$
▷Tricarbonyl(η^5-2,4-cyclopentadien-1-yl)-(dimethylthallium)tungsten, W -00054
$C_{10}H_{11}Re$
Bis(η^5-2,4-cyclopentadien-1-yl)hydrorhenium, Re-00042
$C_{10}H_{11}Tc$
Bis(η^5-2,4-cyclopentadien-1-yl)hydrotechnetium, Tc-00006
$C_{10}H_{11}Ti$
Bis(η^5-cyclopentadienyl)hydrotitanium, Ti-00037
$C_{10}H_{12}CrN_2O_5$
[Bis(dimethylamino)methylene]pentacarbonylchromium, Cr-00079
$C_{10}H_{12}CrO_6Si$
Pentacarbonyl[1-(trimethylsiloxy)ethylidene]chromium, Cr-00080
$C_{10}H_{12}Mo$
Bis(η^5-2,4-cyclopentadien-1-yl)dihydromolybdenum, Mo-00043
$C_{10}H_{12}W$
Bis(η^5-2,4-cyclopentadien-1-yl)dihydrotungsten, W -00055
$C_{10}H_{12}Zr$
Bis(η^5-2,4-cyclopentadien-1-yl)dihydrozirconium, Zr-00012
$C_{10}H_{13}F_6PW$
Bis(η^5-2,4-cyclopentadien-1-yl)trihydrotungsten(1+); Hexafluorophosphate, *in* W -00056
$C_{10}H_{13}Nb$
Bis(η^5-2,4-cyclopentadien-1-yl)trihydroniobium, Nb-00014
$C_{10}H_{13}Ta$
Bis(cyclopentadienyl)trihydrotantalum, Ta-00011
$C_{10}H_{13}W^{\oplus}$
Bis(η^5-2,4-cyclopentadien-1-yl)trihydrotungsten(1+), W -00056
$C_{10}H_{14}BNb$
▷Bis(η^5-2,4-cyclopentadien-1-yl)[tetrahydroborato(1−)-*H*,*H′*]niobium, Nb-00015
$C_{10}H_{14}BTi$
Bis(η^5-2,4-cyclopentadien-1-yl)[tetrahydroborato(1−)-*H*,*H′*]titanium, Ti-00038
$C_{10}H_{14}BV$
▷Bis(η^5-2,4-cyclopentadien-1-yl)[tetrahydroborato(1−)-*H*,*H′*]vanadium, V -00034
$C_{10}H_{15}BCrN_2O_5$
Pentacarbonyl[(cyano-*C*)trihydroborato(1−)-*N*]-chromate(1−); Tetramethylammonium salt, *in* Cr-00021
$C_{10}H_{15}BZr$
Bis(η^5-2,4-cyclopentadien-1-yl)-hydro[tetrahydroborato(1−)-*H*,*H′*]zirconium, Zr-00013
$C_{10}H_{15}Cl_2NS_2Ti$
Dichloro(η^5-2,4-cyclopentadien-1-yl)-(diethylcarbamodithioato-*S*,*S′*)titanium, Ti-00039
$C_{10}H_{15}Cl_3Hf$
Trichloro[(1,2,3,4,5-η)-1,2,3,4,5-pentamethyl-2,4-cyclopentadien-1-yl]hafnium, Hf-00008
$C_{10}H_{15}Cl_3Th$
[(1,2,3,4,5-η)-1,2,3,4,5-Pentamethyl-2,4-cyclopentadien-1-yl]trichlorothorium, Th-00002
$C_{10}H_{15}Cl_3Ti$
Trichloro[(1,2,3,4,5-η)-1,2,3,4,5-pentamethyl-2,4-cyclopentadien-1-yl]titanium, Ti-00040
$C_{10}H_{15}Cl_3U$
[(1,2,3,4,5-η)-1,2,3,4,5-Pentamethyl-2,4-cyclopentadienyl]trichlorouranium, U -00008
$C_{10}H_{15}Cl_3Zr$
Trichloro[(1,2,3,4,5-η)-1,2,3,4,5-pentamethyl-2,4-cyclopentadien-1-yl]zirconium, Zr-00014
$C_{10}H_{15}Cl_4Nb$
Tetrachloro[(1,2,3,4,5-η)-1,2,3,4,5-pentamethyl-2,4-cyclopentadien-1-yl]niobium, Nb-00016
$C_{10}H_{15}Cl_4Ta$
Tetrachloro[(1,2,3,4,5-η)-1,2,3,4,5-pentamethyl-2,4-cyclopentadien-1-yl]tantalum, Ta-00012
$C_{10}H_{15}MnN_2OS_2$
π-Cyclopentadienyl(diethyldithiocarbamato)-nitrosylmanganese, Mn-00061
$C_{10}H_{15}NO_5W$
N,*N*,*N*-Trimethylmethanaminium(1+) pentacarbonylmethyltungstate(1−), *in* W -00009
$C_{10}H_{17}O_4P_2V$
Tetracarbonyl[1,2-ethanediylbis[dimethylphosphine]-*P*,*P′*]hydrovanadium, V -00035

$C_{10}H_{18}B_2Hf$
Bis(η^5-2,4-cyclopentadien-1-yl)bis[tetrahydroborato(1−)-*H,H'*]hafnium, Hf-00009
$C_{10}H_{18}B_2U$
Bis(η^5-2,4-cyclopentadien-1-yl)bis[tetrahydroborato(1−)-*H,H'*]uranium, U -00009
$C_{10}H_{18}B_2Zr$
Bis(η^5-2,4-cyclopentadien-1-yl)bis[tetrahydroborato(1−)-*H,H'*]zirconium, Zr-00015
$C_{10}H_{18}CrO_4P_2$
Tetracarbonylbis(trimethylphosphine)chromium, Cr-00081
$C_{10}H_{18}CrO_{10}P_2$
Tetracarbonylbis(trimethyl phosphite-*P*)chromium, Cr-00082
$C_{10}H_{18}O_2Ti$
(η^5-2,4-Cyclopentadien-1-yl)diethoxymethyltitanium, Ti-00041
$C_{10}H_{24}O_2V$
Di-*tert*-butoxydimethylvanadium, V -00036
$C_{10}H_{24}O_3Ti$
Methyltris(2-propanolato)titanium, Ti-00042
$C_{10}H_{26}Cl_4NNb$
Tetrachlorodimethylniobate(1−); Tetraethylammonium salt, *in* Nb-00003
$C_{10}H_{29}N_3SiTi$
Tris(dimethylamido)(trimethylsilylmethyl)titanium, Ti-00043
$C_{10}H_{30}Mo_2N_4$
Dimethyltetrakis(*N*-dimethylamino)dimolybdenum, Mo-00044
$C_{10}K_2Mo_2O_{10}$
Decarbonyldimolybdate(2−); Di-K salt, *in* Mo-00045
$C_{10}MnO_{10}Re$
Pentacarbonyl(pentacarbonylmanganese)rhenium, Re-00043
$C_{10}Mn_2O_{10}$
Decacarbonyldimanganese, Mn-00063
$C_{10}Mo_2O_{10}^{\ominus\ominus}$
Decarbonyldimolybdate(2−), Mo-00045
$C_{10}O_{10}Re_2$
Decacarbonyldirhenium, Re-00044
$C_{11}F_5MnO_5$
Pentacarbonyl(pentafluorophenyl)manganese, Mn-00064
$C_{11}H_3Cr_2O_{10}S^{\ominus}$
Decarbonyl(μ-methanethiolato)dichromate(1−), Cr-00083
$C_{11}H_3Mn_2NO_9$
Nonacarbonyl(isocyanomethane)dimanganese, Mn-00065
$C_{11}H_5BrCrO_4$
Bromotetracarbonyl(phenylmethylidyne)chromium, Cr-00084
$C_{11}H_5BrO_4W$
Bromotetracarbonyl(phenylmethylidyne)tungsten, W -00057
$C_{11}H_5ClO_4W$
Tetracarbonylchloro(phenylmethylidyne)tungsten, W -00058
$C_{11}H_5F_9MoOS$
Carbonyl(η^5-2,4-cyclopentadien-1-yl)[(2,3-η)-1,1,1,4,4,4-hexafluoro-2-butyne]-(trifluoromethanethiolato)molybdenum, Mo-00046
$C_{11}H_5IO_4W$
Tetracarbonyliodo(phenylmethylidyne)tungsten, W -00059
$C_{11}H_5MnO_5$
Pentacarbonylphenylmanganese, Mn-00066
$C_{11}H_7O_5V$
(η^5-1-Acetyl-2,4-cyclopentadien-1-yl)-tetracarbonylvanadium, V -00037
$C_{11}H_8CrO_3$
Tricarbonyl[(1,2,3,4,5,6-η)-1,3,5,7-cyclooctatetraene]-chromium, Cr-00085
Tricarbonyl[(1,2,3,4,5,6-η)-7-methylene-1,3,5-cycloheptatriene]chromium, Cr-00086
$C_{11}H_8CrO_4$
[(2,3,5,6-η)-Bicyclo[2.2.1]hepta-2,5-diene]-tetracarbonylchromium, Cr-00087
Tricarbonyl[1-(η^6-phenyl)ethanone]chromium, Cr-00088
$C_{11}H_8CrO_5$
Tricarbonyl[(1,2,3,4,5,6-η)methyl benzoate]chromium, Cr-00089
$C_{11}H_8MoO_3$
Tricarbonyl[(1,2,3,4,5,6-η)-1,3,5,7-cyclooctatetraene]-molybdenum, Mo-00047
Tricarbonyl[(1,2,3,4,5,6-η)ethenylbenzene]molybdenum, Mo-00048
$C_{11}H_8MoO_4$
[(2,3,5,6-η)-Bicyclo[2.2.1]hepta-2,5-diene]-tetracarbonylmolybdenum, Mo-00049
$C_{11}H_8O_3W$
Tricarbonyl[(1,2,3,4,5,6-η)-1,3,5,7-cyclooctatetraene]-tungsten, W -00060
$C_{11}H_8O_4W$
[(2,3,5,6-η)-Bicyclo[2.2.1]hepta-2,5-diene]-tetracarbonyltungsten, W -00061
$C_{11}H_9O_2V$
Dicarbonyl(η^4-1,3-cyclobutadiene)(η^5-2,4-cyclopentadien-1-yl)vanadium, V -00038
$C_{11}H_{10}ClNbO$
Carbonylchlorobis(η^5-2,4-cyclopentadien-1-yl)niobium, Nb-00017
$C_{11}H_{10}ClOTa$
Carbonylchlorobis(η^5-2,4-cyclopentadien-1-yl)tantalum, Ta-00013
$C_{11}H_{10}Cl_2N_2Ti$
Dichloro(η^5-2,4-cyclopentadien-1-yl)(phenylazo)-titanium, Ti-00044
$C_{11}H_{10}CrO_3$
Tricarbonyl[(1,2,3,4,5,6-η)-1,3,5-cyclooctatriene]-chromium, Cr-00090
$C_{11}H_{10}F_3HfOP$
Carbonylbis(η^5-2,4-cyclopentadien-1-yl)(phosphorous trifluoride)hafnium, Hf-00010
$C_{11}H_{10}IOV$
Carbonylbis(η^5-2,4-cyclopentadien-1-yl)iodovanadium, V -00039
$C_{11}H_{10}MoO$
Carbonylbis(η^5-2,4-cyclopentadien-1-yl)molybdenum, Mo-00050
$C_{11}H_{10}MoO_3$
Tricarbonyl[(1,2,3,4,5,6-η)-1,3,5-cyclooctatriene]-molybdenum, Mo-00051
Tricarbonyl[η^6-5-(1-methylethylidene)-1,3-cyclopentadiene]molybdenum, Mo-00052
$C_{11}H_{10}OV$
Carbonylbis(η^5-2,4-cyclopentadien-1-yl)vanadium, V -00040
$C_{11}H_{10}OW$
Carbonylbis(η^5-2,4-cyclopentadien-1-yl)tungsten, W -00062
$C_{11}H_{10}S_2V$
[(*C,S*-η)-Carbon disulfide]bis(η^5-2,4-cyclopentadien-1-yl)vanadium, V -00041
$C_{11}H_{11}BF_4MoO_2$
(η^4-1,3-Butadiene)dicarbonyl(η^5-2,4-cyclopentadien-1-yl)molybdenum(1+); Tetrafluoroborate, *in* Mo-00055
$C_{11}H_{11}F_6MoN_2O_2P$
Bis(acetonitrile)dicarbonyl(η^5-2,4-cyclopentadien-1-yl)-molybdenum(1+); Hexafluorophosphate, *in* Mo-00054
$C_{11}H_{11}Mn$
(η^6-Benzene)(η^5-2,4-cyclopentadien-1-yl)manganese, Mn-00067
$C_{11}H_{11}Mo$
(η^6-Benzene)(η^5-2,4-cyclopentadien-1-yl)molybdenum, Mo-00053

$C_{11}H_{11}MoN_2O_2^{\oplus}$
Bis(acetonitrile)dicarbonyl(η^5-2,4-cyclopentadien-1-yl)-molybdenum(1+), Mo-00054

$C_{11}H_{11}MoO_2^{\oplus}$
(η^4-1,3-Butadiene)dicarbonyl(η^5-2,4-cyclopentadien-1-yl)molybdenum(1+), Mo-00055

$C_{11}H_{11}NbO$
Carbonylbis(η^5-2,4-cyclopentadien-1-yl)hydroniobium, Nb-00018

$C_{11}H_{11}OTa$
Carbonylbis(η^5-2,4-cyclopentadien-1-yl)hydrotantalum, Ta-00014

$C_{11}H_{11}O_2V$
(η^4-1,3-Butadiene)dicarbonyl(η^5-2,4-cyclopentadien-1-yl)vanadium, V -00042

$C_{11}H_{11}V$
(η^6-Benzene)(η^5-2,4-cyclopentadien-1-yl)vanadium, V -00043

$C_{11}H_{12}NO_3V$
Tricarbonyl(η^5-2,4-cyclopentadien-1-yl)-(dimethylaminomethylene)vanadium, V -00044

$C_{11}H_{12}OV$
Bis(η^5-2,4-cyclopentadien-1-yl)(η^2-*C,O*-formaldehyde)-vanadium, V -00045

$C_{11}H_{13}ClNb$
Chlorobis(η^5-2,4-cyclopentadien-1-yl)methylniobium, Nb-00019

$C_{11}H_{13}ClOZr$
Chlorobis(η-cyclopentadienyl)methoxyzirconium, Zr-00016

$C_{11}H_{13}ClTi$
Chlorobis(η^5-2,4-cyclopentadien-1-yl)methyltitanium, Ti-00045

$C_{11}H_{13}ClV$
Chlorobis(η^5-2,4-cyclopentadien-1-yl)methylvanadium, V -00046

$C_{11}H_{13}Dy$
Bis(η^5-2,4-cyclopentadien-1-yl)methyldysprosium, Dy-00003

$C_{11}H_{13}Er$
Bis(η^5-2,4-cyclopentadien-1-yl)methylerbium, Er-00005

$C_{11}H_{13}Ho$
Bis(η^5-2,4-cyclopentadien-1-yl)methylholmium, Ho-00003

$C_{11}H_{13}Lu$
Bis(η^5-2,4-cyclopentadien-1-yl)methyllutetium, Lu-00006

$C_{11}H_{13}NbO$
Bis(cyclopentadienyl)methyloxoniobium, Nb-00020

$C_{11}H_{13}O_2Re$
Dicarbonyl(η^5-2,4-cyclopentadien-1-yl)-tetramethylenerhenium, Re-00045

$C_{11}H_{13}O_3Re$
Dicarbonyl(η^5-2,4-cyclopentadien-1-yl)(tetrahydrofuran)-rhenium, Re-00046

$C_{11}H_{13}Sc$
Bis(η^5-2,4-cyclopentadien-1-yl)methylscandium, Sc-00002

$C_{11}H_{13}Ti$
Bis(η^5-2,4-cyclopentadien-1-yl)methyltitanium, Ti-00046

$C_{11}H_{13}Tm$
Bis(η^5-2,4-cyclopentadien-1-yl)methylthulium, Tm-00003

$C_{11}H_{13}V$
Bis(η^5-2,4-cyclopentadien-1-yl)methylvanadium, V -00047

$C_{11}H_{13}Y$
Bis(η^5-2,4-cyclopentadien-1-yl)methylyttrium, Y -00003

$C_{11}H_{13}Yb$
Bis(η^5-2,4-cyclopentadien-1-yl)methylytterbium, Yb-00006

$C_{11}H_{14}CrGeO_3$
Tricarbonyl-π-cyclopentadienyl(trimethylgermyl)-chromium, Cr-00091

$C_{11}H_{14}CrO_3Sn$
Tricarbonyl-π-cyclopentadienyl(trimethylstannyl)-chromium, Cr-00092

$C_{11}H_{14}GeMoO_3$
Tricarbonyl(η^5-2,4-cyclopentadien-1-yl)-(trimethylgermyl)molybdenum, Mo-00056

$C_{11}H_{14}GeO_3W$
Tricarbonyl(η^5-2,4-cyclopentadien-1-yl)-(trimethylgermyl)tungsten, W -00063

$C_{11}H_{14}MoO_3Pb$
Tricarbonyl(η^5-2,4-cyclopentadien-1-yl)-(trimethylplumbyl)molybdenum, Mo-00057

$C_{11}H_{14}MoO_3Si$
Tricarbonyl(η^5-2,4-cyclopentadien-1-yl)(trimethylsilyl)-molybdenum, Mo-00058

$C_{11}H_{14}MoO_3Sn$
Tricarbonyl(η^5-2,4-cyclopentadien-1-yl)-(trimethylstannyl)molybdenum, Mo-00059

$C_{11}H_{14}O_3S_6Zr$
(η^5-2,4-Cyclopentadien-1-yl)tris(*O*-methylcarbonodithioato-*S,S'*)zirconium, Zr-00017

$C_{11}H_{14}O_3SiW$
Tricarbonyl(η^5-2,4-cyclopentadien-1-yl)(trimethylsilyl)-tungsten, W -00064

$C_{11}H_{14}O_3SnW$
Tricarbonyl(η^5-2,4-cyclopentadien-1-yl)-(trimethylstannyl)tungsten, W -00065

$C_{11}H_{14}O_6Ti$
Tris(acetato-*O*)(η^5-2,4-cyclopentadien-1-yl)titanium, Ti-00047

$C_{11}H_{14}O_6Zr$
Tris(acetato)(η-cyclopentadienyl)zirconium, Zr-00018

$C_{11}H_{15}CrNO_6$
Acetylpentacarbonylchromate(1−); Tetramethylammonium salt, *in* Cr-00029

$C_{11}H_{15}F_6MnN_5OP$
Carbonylpentakis(isocyanomethane)manganese(1+); Hexafluorophosphate, *in* Mn-00068

$C_{11}H_{15}MnN_5O^{\oplus}$
Carbonylpentakis(isocyanomethane)manganese(1+), Mn-00068

$C_{11}H_{18}CrO_5Sn_2Te$
Pentacarbonyl(hexamethyldistannatellurane)chromium, Cr-00093

$C_{11}H_{20}O_3Ti$
(π-Cyclopentadienyl)triethoxytitanium, Ti-00048

$C_{11}H_{21}MnO_3P^{\ominus}$
Bis(η^4-1,3-butadiene)(trimethyl phosphite)manganate(1−), Mn-00069
Bis(η^4-1,3-butadiene)(trimethyl phosphite)manganese, Mn-00070

$C_{11}H_{22}O_3P_2W$
Tricarbonyl(η^2-ethene)bis(trimethylphosphine)tungsten, W -00066

$C_{11}H_{23}Cl_2P_2V$
Dichloro(η^5-2,4-cyclopentadien-1-yl)-bis(trimethylphosphine)vanadium, V -00048

$C_{11}H_{23}N_3Ti$
(η^5-Cyclopentadienyl)tris(dimethylamido)titanium, Ti-00049

$C_{11}H_{24}Cl_3OPW$
Trichloro(1,1-dimethylpropylidyne)(triethylphosphine oxide-*O*)tungsten, W -00067

$C_{11}H_{27}BrMnO_{11}P_3$
Bromodicarbonyltris(trimethylphosphite-*P*)manganese, Mn-00071

$C_{11}H_{27}BrO_2P_3Re$
Bromodicarbonyltris(trimethylphosphine)rhenium, Re-00047

$C_{11}H_{28}Cl_2OP_2W$
Dichloro(2,2-dimethylpropylidene)-oxobis(trimethylphosphine)tungsten, W -00068

$C_{11}H_{28}Cl_3P_2Ta$
Trichloro(2,2-dimethylpropylidene)-bis(trimethylphosphine)tantalum, Ta-00015

$C_{11}H_{29}Cl_3NNb$
Trichlorotrimethylniobate(1−); Tetraethylammonium salt, *in* Nb-00005

$C_{11}H_{31}NbP_2$
[1,2-Ethanediylbis[dimethylphosphine]-*P*,*P'*]-pentamethylniobium, Nb-00021

$C_{12}Co_3ErO_{12}$
Tris(tetracarbonylcobalto)erbium, Er-00006

$C_{12}F_{10}Yb$
Bis(pentafluorophenyl)ytterbium, Yb-00007

$C_{12}Fe_2MnO_{12}^{\ominus}$
Tetracarbonyl(di-μ-carbonylhexacarbonyldiferrate)-manganate(1−), Mn-00072

$C_{12}HMnO_{12}Os_3$
Dodecacarbonylhydromanganesediosmium, Mn-00073

$C_{12}H_3Mn_3O_{12}$
Dodecacarbonyltrihydrotrimanganese, Mn-00074

$C_{12}H_5CoMoO_7$
Tricarbonyl(η^5-2,4-cyclopentadien-1-yl)-(tetracarbonylcobalt)molybdenum, Mo-00060

$C_{12}H_5CrLiO_6$
Benzoylpentacarbonylchromate(1−); Li salt, *in* Cr-00094

$C_{12}H_5CrO_6^{\ominus}$
Benzoylpentacarbonylchromate(1−), Cr-00094

$C_{12}H_5FeMnO_7$
Pentacarbonyl[dicarbonyl(η^5-2,4-cyclopentadien-1-yl)-iron]manganese, Mn-00075

$C_{12}H_5HgO_6Ta$
Hexacarbonyl(phenylmercury)tantalum, Ta-00016

$C_{12}H_5O_7ReRu$
Dicarbonyl(η^5-2,4-cyclopentadien-1-yl)-(pentacarbonylrhenio)ruthenium, Re-00048

$C_{12}H_5O_{12}Os_3Re$
Tricarbonyltri-μ-hydro(nonacarbonyldi-μ-hydrotriosmium)-rhenium, Re-00049

$C_{12}H_6MnO_{10}Re$
Pentacarbonyl[tetracarbonyl(1-methoxyethylidene)]-manganeserhenium, Re-00050

$C_{12}H_6Mn_2N_2O_8$
Octacarbonylbis(isocyanomethane)dimanganese, Mn-00077

$C_{12}H_6Mn_2O_{10}Pb$
Bis(pentacarbonylmanganese)dimethyllead, Mn-00078

$C_{12}H_6O_{10}PbRe_2$
Bis(pentacarbonylrhenium)dimethyllead, Re-00051

$C_{12}H_6O_{10}Re_2$
Nonacarbonyl(1-methoxyethylidene)dirhenium, Re-00052

$C_{12}H_6O_{10}Tc_2$
Nonacarbonyl(1-methoxyethylidene)ditechnetium, Tc-00007

$C_{12}H_7BrCrO_4$
Bromotetracarbonyl[(4-methylphenyl)methylidyne]-chromium, Cr-00095

$C_{12}H_7CrNO_5$
(Aminophenylmethylene)pentacarbonylchromium, Cr-00096

$C_{12}H_7FeMnO_6$
μ-Carbonylpentacarbonyl(η^5-2,4-cyclopentadienyl)iron-μ-methylenemanganese, Mn-00079

$C_{12}H_7MnO_5$
(2-Acetylphenyl-*C*,*O*)tetracarbonylmanganese, Mn-00080

$C_{12}H_8O_4W$
Tetracarbonyl[(1,2,5,6-η)-1,3,5,7-cyclooctatetraene]-tungsten, W -00069

$C_{12}H_{10}Cl_2Cr$
Bis(η^6-chlorobenzene)chromium, Cr-00097

$C_{12}H_{10}Cl_2Mo$
▷Bis(η^6-chlorobenzene)molybdenum, Mo-00061

$C_{12}H_{10}Cl_3O_2SnV$
Dicarbonylbis(2,4-cyclopentadien-1-yl)vanadium(1+); Trichlorostannate, *in* V -00050

$C_{12}H_{10}CrF_2$
Bis(η^6-fluorobenzene)chromium, Cr-00098

$C_{12}H_{10}CrO_4$
Tricarbonyl[(3*a*,4,5,6,7,7*a*-η)-2,3-dihydro-1*H*-inden-1-ol]chromium, Cr-00099

$C_{12}H_{10}HfN_2O_2$
Bis(cyanato)di-π-cyclopentadienylhafnium, Hf-00011

$C_{12}H_{10}HfO_2$
Dicarbonylbis(η^5-2,4-cyclopentadien-1-yl)hafnium, Hf-00012

$C_{12}H_{10}I_3O_2V$
Dicarbonylbis(2,4-cyclopentadien-1-yl)vanadium(1+); Triiodide, *in* V -00050

$C_{12}H_{10}Mn_2N_2O_4$
μ-Carbonylcarbonylbis(η^5-2,4-cyclopentadien-1-yl)[μ-(nitrosyl-*N*:*N*)]nitrosyldimanganese, Mn-00081

$C_{12}H_{10}MoO_3$
[(2,3,4,5,6,7-η)-Bicyclo[6.1.0]nona-2,4,6-triene]-tricarbonylmolybdenum, Mo-00062

$C_{12}H_{10}N_2Nb$
Bis(cyano-*C*)bis(η^5-2,4-cyclopentadien-1-yl)niobium, Nb-00022

$C_{12}H_{10}N_2O_2Ti$
Dicyanatobis(η^5-cyclopentadienyl)titanium, Ti-00050

$C_{12}H_{10}N_2O_2Zr$
Bis(cyanato-*N*)bis(η^5-2,4-cyclopentadien-1-yl)zirconium, Zr-00019

$C_{12}H_{10}N_2S_2Zr$
Bis(η^5-2,4-cyclopentadien-1-yl)bis(thiocyanato-*N*)-zirconium, Zr-00020

$C_{12}H_{10}N_2V$
Bis(cyano-*C*)bis(η^5-2,4-cyclopentadien-1-yl)vanadium, V -00049

$C_{12}H_{10}O_2Ti$
Dicarbonylbis(η^5-2,4-cyclopentadien-1-yl)titanium, Ti-00051

$C_{12}H_{10}O_2V^{\oplus}$
Dicarbonylbis(2,4-cyclopentadien-1-yl)vanadium(1+), V -00050

$C_{12}H_{10}O_2Zr$
Dicarbonylbis(η^5-2,4-cyclopentadien-1-yl)zirconium, Zr-00021

$C_{12}H_{10}O_3W$
[(2,3,4,5,6,7-η)Bicyclo[6.1.0]nona-2,4,6-triene]-tricarbonyltungsten, W -00070

$C_{12}H_{10}Ti$
▷Diphenyltitanium, Ti-00052

$C_{12}H_{11}BCrF_4O_4$
Tricarbonyl(ethoxy-π-cycloheptatrienylium)chromium(1+); Tetrafluoroborate, *in* Cr-00100

$C_{12}H_{11}CrO_4^{\oplus}$
Tricarbonyl(ethoxy-π-cycloheptatrienylium)chromium(1+), Cr-00100

$C_{12}H_{11}IO_4PRe$
Tetracarbonyl(dimethylphenylphosphine)iodorhenium, Re-00053

$C_{12}H_{11}NbO_2$
Dicarbonyl(cyclopentadiene)-π-cyclopentadienylniobium, Nb-00023

$C_{12}H_{12}BF_4O_4Re$
Tetracarbonyl[(1,2,5,6-η)-1,5-cyclooctadiene)]-rhenium(1+); Tetrafluoroborate, *in* Re-00054

$C_{12}H_{12}Cr$
▷Bis(η⁶-benzene)chromium, Cr-00101
(η⁷-Cycloheptatrienylium)(η⁵-2,4-cyclopentadien-1-yl)-chromium, Cr-00102

$C_{12}H_{12}CrO_3$
Tricarbonyl[(1,2,3,4,5,6-η)-1,3,5-trimethylbenzene]-chromium, Cr-00103

$C_{12}H_{12}CrO_4$
Tetracarbonyl[(1,2,5,6-η)-1,5-cyclooctadiene]chromium, Cr-00104

$C_{12}H_{12}F_6MnP$
(η⁷-Cycloheptatrienylium)(η⁵-2,4-cyclopentadien-1-yl)-manganese(1+); Hexafluorophosphate, *in* Mn-00082

$C_{12}H_{12}F_6PRe$
Bis(η⁶-benzene)rhenium(1+); Hexafluorophosphate, *in* Re-00055

$C_{12}H_{12}F_6PTe$
Bis(η⁶-benzene)technetium(1+); Hexafluorophosphate, *in* Tc-00008

$C_{12}H_{12}Mn^{\oplus}$
(η⁷-Cycloheptatrienylium)(η⁵-2,4-cyclopentadien-1-yl)-manganese(1+), Mn-00082

$C_{12}H_{12}Mo$
Bis(η⁶-benzene)molybdenum, Mo-00063

$C_{12}H_{12}MoO_3$
(η⁶-1,3,5-Trimethylbenzene)tricarbonylmolybdenum, Mo-00064

$C_{12}H_{12}MoO_4$
Tetracarbonyl[(1,2,5,6-η)-1,5-cyclooctadiene]-molybdenum, Mo-00065

$C_{12}H_{12}Nb$
Bis(η⁶-benzene)niobium, Nb-00024
(η⁷-Cycloheptatrienylium)(η⁵-2,4-cyclopentadien-1-yl)-niobium, Nb-00025

$C_{12}H_{12}O_3W$
Tricarbonyl[(1,2,3,4,5,6-η)-1,3,5-trimethylbenzene]-tungsten, W -00071

$C_{12}H_{12}O_4Re^{\oplus}$
Tetracarbonyl[(1,2,5,6-η)-1,5-cyclooctadiene)]-rhenium(1+), Re-00054

$C_{12}H_{12}O_4W$
Tetracarbonyl[(1,2,5,6-η)-1,5-cyclooctadiene]tungsten, W -00072

$C_{12}H_{12}Re^{\oplus}$
Bis(η⁶-benzene)rhenium(1+), Re-00055

$C_{12}H_{12}S_2Ti$
Bis(η⁵-2,4-cyclopentadien-1-yl)[1,2-ethenedithiolato(2−)-*S*,*S′*]titanium, Ti-00053

$C_{12}H_{12}Tc^{\oplus}$
Bis(η⁶-benzene)technetium(1+), Tc-00008

$C_{12}H_{12}Ti$
Bis(η⁶-benzene)titanium, Ti-00054
(η⁷-Cycloheptatrienylium)(η⁵-2,4-cyclopentadien-1-yl)-titanium, Ti-00055

$C_{12}H_{12}V$
Bis(η⁶-benzene)vanadium, V -00051
(η⁷-Cycloheptatrienylium)(η⁵-2,4-cyclopentadien-1-yl)-vanadium, V -00052

$C_{12}H_{12}W$
Bis(η⁶-benzene)tungsten, W -00073

$C_{12}H_{13}ClOTi$
[(*C*,*O*-η)-Acetyl]chlorobis(η⁵-2,4-cyclopentadien-1-yl)-titanium, Ti-00056

$C_{12}H_{13}ClOZr$
(Acetyl-*C*,*O*)chlorobis(η⁵-2,4-cyclopentadien-1-yl)-zirconium, Zr-00022

$C_{12}H_{13}ClO_2Zr$
(Acetato-*O*)chlorobis(η⁵-2,4-cyclopentadien-1-yl)-zirconium, Zr-00023

$C_{12}H_{13}ClTi$
Chlorobis(η⁵-2,4-cyclopentadien-1-yl)ethenyltitanium, Ti-00057

$C_{12}H_{13}Mn$
[(1,2,3,4,5,6-η)-1,3,5-Cycloheptatriene](η⁵-2,4-cyclopentadien-1-yl)manganese, Mn-00083

$C_{12}H_{13}NbO$
Carbonylbis(η⁵-2,4-cyclopentadien-1-yl)methylniobium, Nb-00026

$C_{12}H_{13}O_3V$
Tricarbonylhydro(η⁶-1,3,5-trimethylbenzene)vanadium, V -00053

$C_{12}H_{14}BF_4Re$
Bis(η⁵-2,4-cyclopentadien-1-yl)(η²-ethene)rhenium(1+); Tetrafluoroborate, *in* Re-00056

$C_{12}H_{14}ClEr$
Chlorobis(η⁵-methylcyclopentadienyl)erbium, Er-00007

$C_{12}H_{14}ClGd$
Chlorobis(η⁵-methylcyclopentadienyl)gadolinium, Gd-00003

$C_{12}H_{14}ClSm$
Chlorobis[(1,2,3,4,5-η)-1-methyl-2,4-cyclopentadien-1-yl]samarium, Sm-00006

$C_{12}H_{14}ClYb$
Chlorobis(η⁵-methylcyclopentadienyl)ytterbium, Yb-00008

$C_{12}H_{14}CrO_3Si$
[Trimethyl(η⁶-phenyl)silane]tricarbonylchromium, Cr-00105

$C_{12}H_{14}CrO_3Sn$
Tricarbonyl(trimethylphenylstannane)chromium, Cr-00106

$C_{12}H_{14}Mo$
Bis(η⁵-cyclopentadienyl)(η²-ethylene)molybdenum, Mo-00066

$C_{12}H_{14}MoO_3Sn$
Tricarbonyl(η⁶-trimethylstannylbenzene)molybdenum, Mo-00067

$C_{12}H_{14}Re^{\oplus}$
Bis(η⁵-2,4-cyclopentadien-1-yl)(η²-ethene)rhenium(1+), Re-00056

$C_{12}H_{14}V$
Bis(η⁵-1-methyl-2,4-cyclopentadien-1-yl)vanadium, V -00054

$C_{12}H_{14}W$
Bis(η⁵-2,4-cyclopentadien-1-yl)(η²-ethene)tungsten, W -00074

$C_{12}H_{14}Yb$
Bis[(1,2,3,4,5-η)-1-methyl-2,4-cyclopentadien-1-yl]-ytterbium, Yb-00009

$C_{12}H_{15}F_6PW$
Bis(η⁵-2,4-cyclopentadien-1-yl)(η²-ethene)-hydrotungsten(1+); Hexafluorophosphate, *in* W -00076

$C_{12}H_{15}MoO_2$
Dicarbonyl(pentamethylcyclopentadienyl)molybdenum, *in* Mo-00124

$C_{12}H_{15}MoO_3Tl$
▷Tricarbonyl(η⁵-2,4-cyclopentadien-1-yl)-(diethylthallium)molybdenum, Mo-00068

$C_{12}H_{15}Nb$
Bis(η⁵-2,4-cyclopentadien-1-yl)(η²-ethene)hydroniobium, Nb-00027

$C_{12}H_{15}O_3TlW$
▷Tricarbonyl(η⁵-2,4-cyclopentadien-1-yl)-(diethylthallium)tungsten, W -00075

$C_{12}H_{15}Ta$
Bis(η⁵-2,4-cyclopentadien-1-yl)methyl(methylidene)-tantalum, Ta-00017

$C_{12}H_{15}V$
Bis(η^5-2,4-cyclopentadien-1-yl)ethylvanadium, V -00055

$C_{12}H_{15}W^{\oplus}$
Bis(η^5-2,4-cyclopentadien-1-yl)(η^2-ethene)-hydrotungsten(1+), W -00076

$C_{12}H_{16}ClErO$
(η^8-1,3,5,7-Cyclooctatetraene)chloro(tetrahydrofuran)-erbium, *in* Er-00003

$C_{12}H_{16}ClLuO$
(η^8-1,3,5,7-Cyclooctatetraene)chloro(tetrahydrofuran)-lutetium, *in* Lu-00003

$C_{12}H_{16}Cl_2OZr$
Dichloro(η^8-1,3,5,7-cyclooctatetraene)zirconium; THF complex, *in* Zr-00004

$C_{12}H_{16}F_6PRe$
Bis(η^5-2,4-cyclopentadien-1-yl)dimethylrhenium(1+); Hexafluorophosphate, *in* Re-00057

$C_{12}H_{16}Hf$
Bis(η^5-2,4-cyclopentadien-1-yl)dimethylhafnium, Hf-00013

$C_{12}H_{16}Mo$
(η^6-Benzene)bis(η^3-2-propenyl)molybdenum, Mo-00069

$C_{12}H_{16}NTi$
[Bis(η^5-cyclopentadienyl)dimethylamidotitanium, Ti-00058

$C_{12}H_{16}Nb$
▷Bis(η^5-2,4-cyclopentadien-1-yl)dimethylniobium, Nb-00028

$C_{12}H_{16}O_2W$
(Acetato-*O*)[(1,2,3,4,5-η)-methylbenzene](η^3-2-propenyl)-tungsten, W -00077

$C_{12}H_{16}O_2Zr$
Bis(η^5-2,4-cyclopentadien-1-yl)dimethoxyzirconium, Zr-00024

$C_{12}H_{16}Re^{\oplus}$
Bis(η^5-2,4-cyclopentadien-1-yl)dimethylrhenium(1+), Re-00057

$C_{12}H_{16}S_2Ti$
Bis(η^5-2,4-cyclopentadien-1-yl)bis(methanethiolato)-titanium, Ti-00059

$C_{12}H_{16}Ti$
Bis(η^5-2,4-cyclopentadien-1-yl)dimethyltitanium, Ti-00060

$C_{12}H_{16}V$
Bis(2,4-cyclopentadien-1-yl)dimethylvanadium, V -00056

$C_{12}H_{16}Zr$
Bis(η^5-2,4-cyclopentadien-1-yl)dimethylzirconium, Zr-00025

$C_{12}H_{18}Al_2Cl_8Ti$
Tetra-μ-chlorobis(dichloroaluminium)[(1,2,3,4,5,6-η)-hexamethylbenzene)titanium, Ti-00061

$C_{12}H_{18}CrN_2O_4S$
(Di-*tert*-butylsulfurdiimide-*N*,*N'*)tetracarbonylchromium, Cr-00107

$C_{12}H_{18}Mo$
Tris(η^4-1,3-butadiene)molybdenum, Mo-00070

$C_{12}H_{18}O_3W$
Tris[(3,4-η)-3-buten-2-one-*O*]tungsten, W -00078

$C_{12}H_{18}W$
Tris(η^4-1,3-butadiene)tungsten, W -00079

$C_{12}H_{19}Cl_2Ta$
Dichloro(η^2-ethene)(η^5-1,2,3,4,5-pentamethyl-2,4-cyclopentadien-1-yl)tantalum, Ta-00018

$C_{12}H_{19}LuSi$
(η^8-1,3,5,7-Cyclooctatetraene)(trimethylsilylmethyl)-lutetium, Lu-00007

$C_{12}H_{20}Br_2MnNO_4$
Dibromotetracarbonylmanganate(1−); Tetraethylammonium salt, *in* Mn-00003

$C_{12}H_{20}Cl_2MnNO_4$
Tetracarbonyldichloromanganate(1−); Tetraethylammonium salt, *in* Mn-00004

$C_{12}H_{20}Cr_2$
Tetra-2-propenyldichromium, Cr-00108

$C_{12}H_{20}Cr_2I_2$
Di-μ-iodotetrakis(η^3-2-propenyl)dichromium, *in* Cr-00026

$C_{12}H_{20}Hf$
Tetrakis(η^3-2-propenyl)hafnium, Hf-00014

$C_{12}H_{20}MoN_2O_2S_4$
Dicarbonylbis(diethylcarbamodithioato-*S*,*S'*)molybdenum, Mo-00071

$C_{12}H_{20}Mo_2$
Bis[μ-[(1-η:2,3-η)-2-propenyl)]bis(η^3-2-propenyl)-dimolybdenum, Mo-00072

$C_{12}H_{20}Re_2$
Tetrakis(η^3-2-propenyl)dirhenium, Re-00058

$C_{12}H_{20}Th$
Tetrakis(η^3-2-propenyl)thorium, Th-00003

$C_{12}H_{20}U$
Tetrakis(η^3-2-propenyl)uranium, U -00010

$C_{12}H_{20}Zr$
Tetrakis(η^3-2-propenyl)zirconium, Zr-00026

$C_{12}H_{23}MnO_6P_2S$
Carbonothioyl(η^5-2,4-cyclopentadien-1-yl)-bis(trimethylphosphite-*P*)manganese, Mn-00084

$C_{12}H_{27}CrO_3P_3$
Tricarbonyltris(trimethylphosphine)chromium, Cr-00109

$C_{12}H_{28}Cr$
▷Tetraisopropylchromium, Cr-00110

$C_{12}H_{30}LaP_3$
Tris[(dimethylphosphinidenio)bis(methylene)]lanthanum, La-00002

$C_{12}H_{30}LuP_3$
Tris[(dimethylphosphinidenio)bis(methyene)]lutetium, Lu-00008

$C_{12}H_{33}ErSi_3$
Tris[(trimethylsilyl)methyl]erbium, Er-00008

$C_{12}H_{33}OSi_3V$
Oxotris(trimethylsilylmethyl)vanadium, V -00057

$C_{12}H_{33}Tb$
Tris[(trimethylsilyl)methyl]terbium, Tb-00001

$C_{12}H_{33}Yb$
Tris[(trimethylsilyl)methyl]ytterbium, Yb-00010

$C_{12}H_{34}N_2Si_2Ti$
Bis(dimethylamido)bis(trimethylsilylmethyl)titanium, Ti-00062

$C_{12}MnO_{12}Os_2^{\ominus}$
Dodecacarbonylmanganatediosmate(1−), Mn-00085

$C_{12}O_{12}Os_2Re^{\ominus}$
Dodecacarbonylrhenatediosmate(1−), Re-00059

$C_{13}H_3MnO_{13}Os_3$
Tridecacarbonyltrihydromanganesetriosmium, Mn-00086

$C_{13}H_3O_{13}Os_3Re$
Tridecacarbonyltrihydrorheniumtriosmium, Re-00060

$C_{13}H_5ClF_{12}Mo$
Chloro(η^5-2,4-cyclopentadien-1-yl)bis[(2,3-η)-1,1,1,4,4,4-hexafluoro-2-butyne]molybdenum, Mo-00073

$C_{13}H_5ClF_{12}W$
Chloro(η^5-2,4-cyclopentadien-1-yl)bis[(2,3-η)-1,1,1,4,4,4-hexafluoro-2-butyne]tungsten, W -00080

$C_{13}H_5ClO_4W$
Tetracarbonylchloro(3-phenyl-2-propynylidyne)tungsten, W -00081

$C_{13}H_5F_{12}IMo$
(η^5-2,4-Cyclopentadien-1-yl)bis[(2,3-η)-1,1,1,4,4,4-hexafluoro-2-butyne]iodomolybdenum, Mo-00074

$C_{13}H_7FeMnO_6$
Tricarbonyl[μ-[(1,2,3-η:4,5,6,7-η)-2,4,6-cycloheptatrien-1-yl]](tricarbonyliron)manganese, Mn-00087

$C_{13}H_8BF_4MoN_3O_4$
(2,2′-Bipyridine-*N*,*N*′)tricarbonylnitrosylmolybdenum(1+); Tetrafluoroborate, *in* Mo-00075

$C_{13}H_8CrO_3$
Tricarbonyl[(1,2,3,4,4*a*,8*a*-η)naphthalene]chromium, Cr-00111

$C_{13}H_8CrO_6$
Pentacarbonyl(methoxyphenylmethylene)chromium, Cr-00112

$C_{13}H_8MoN_3O_4^{\oplus}$
(2,2′-Bipyridine-*N*,*N*′)tricarbonylnitrosylmolybdenum(1+), Mo-00075

$C_{13}H_8O_5W$
Pentacarbonyl[(η^2-ethenyl)benzene]tungsten, W -00082

$C_{13}H_8O_6W$
Pentacarbonyl(methoxyphenylmethylene)tungsten, W -00083

$C_{13}H_{10}ClF_6NTi$
Chlorodi-π-cyclopentadienyl[2,2,2-trifluoro-1-(trifluoromethyl)ethylideniminato]titanium, Ti-00063

$C_{13}H_{10}CrO_3$
Tricarbonyl(2,3,4,5-η:9,10-η-9-methylenebicyclo[4.2.1]-nona-2,4,7-triene)chromium, Cr-00113

$C_{13}H_{12}CrFeP_2O_9$
Tetracarbonyl[μ-(tetramethyldiphosphine)]-(pentacarbonylchromium)iron, Cr-00114

$C_{13}H_{12}CrO_4$
Tricarbonyl[(4*a*,5,6,7,8,8*a*-η)-1,2,3,4-tetrahydro-1-naphthalenol]chromium, Cr-00115

$C_{13}H_{12}MnNO_4$
Tetracarbonyl[2-[(dimethylamino)methyl]phenyl-*C*,*N*]-manganese, Mn-00088

$C_{13}H_{13}ClZr$
Chloro[η^8-1,3,5,7-cyclooctatetraene](η^5-2,4-cyclopentadien-1-yl)zirconium, Zr-00027

$C_{13}H_{13}F_6PMo$
(η^8-1,3,5,7-Cyclooctatetraene)(η^5-2,4-cyclopentadien-1-yl)molybdenum(1+); Hexafluorophosphate, *in* Mo-00076

$C_{13}H_{13}Ho$
(η^8-1,3,5,7-Cyclooctatetraenyl)(η^5-2,4-cyclopentadien-1-yl)holmium, Ho-00004

$C_{13}H_{13}Mo^{\oplus}$
(η^8-1,3,5,7-Cyclooctatetraene)(η^5-2,4-cyclopentadien-1-yl)molybdenum(1+), Mo-00076

$C_{13}H_{13}Nd$
(η^8-1,3,5,7-Cyclooctatetraene)(η^5-2,4-cyclopentadien-1-yl)neodymium, Nd-00003

$C_{13}H_{13}O_2V$
Acetylcarbonylbis(η^5-2,4-cyclopentadien-1-yl)vanadium, V -00058

$C_{13}H_{13}Sc$
(η^8-1,3,5,7-Cyclooctatetraene)(η^5-2,4-cyclopentadien-1-yl)scandium, Sc-00003

$C_{13}H_{13}Sm$
(η^8-1,3,5,7-Cyclooctatetraene)(η^5-2,4-cyclopentadien-1-yl)samarium, Sm-00007

$C_{13}H_{13}Ti$
(η^8-1,3,5,7-Cyclooctatetraene)(η^5-2,4-cyclopentadien-1-yl)titanium, Ti-00064

$C_{13}H_{13}Y$
(η^8-1,3,5,7-Cyclooctatetraene)(η^5-2,4-cyclopentadien-1-yl)yttrium, Y -00004

$C_{13}H_{14}CeCl$
Chloro[1,3-propanediylbis(η^5-2,4-cyclopentadien-1-ylidene)]cerium, Ce-00002

$C_{13}H_{14}ClLa$
Chloro[1,3-propanediylbis(η^5-2,4-cyclopentadien-1-ylidene)]lanthanum, La-00003

$C_{13}H_{14}Cl_2Hf$
Dichloro[1,3-propanediylbis[(1,2,3,4,5-η)-2,4-cyclopentadien-1-ylidene]]hafnium, Hf-00015

$C_{13}H_{14}Cl_2Ti$
Dichloro[1,3-propanediylbis[(1,2,3,4,5-η)-2,4-cyclopentadien-1-ylidene]]titanium, Ti-00065

$C_{13}H_{14}Cl_2Zr$
Dichloro[1,3-propanediylbis[(1,2,3,4,5-η)-2,4-cyclopentadien-1-ylidene]]zirconium, Zr-00028

$C_{13}H_{15}F_6MoP$
Bis(η^5-2,4-cyclopentadien-1-yl)(η^3-2-propenyl)-molybdenum(1+); Hexafluorophosphate, *in* Mo-00077

$C_{13}H_{15}F_6PW$
Bis(η^5-2,4-cyclopentadien-1-yl)(η^3-2-propenyl)-tungsten(1+); Hexafluorophosphate, *in* W -00084

$C_{13}H_{15}Mo^{\oplus}$
Bis(η^5-2,4-cyclopentadien-1-yl)(η^3-2-propenyl)-molybdenum(1+), Mo-00077

$C_{13}H_{15}Nb$
Bis(η^5-2,4-cyclopentadien-1-yl)(η^3-2-propenyl)niobium, Nb-00029

$C_{13}H_{15}O_3Re$
Tricarbonyl[(1,2,3,4,5-η)-1,2,3,4,5-pentamethyl-2,4-cyclopentadien-1-yl)]rhenium, Re-00061

$C_{13}H_{15}Ta$
Bis(η^5-2,4-cyclopentadien-1-yl)(η^3-2-propenyl)tantalum, Ta-00019

$C_{13}H_{15}V$
Bis(2,4-cyclopentadien-1-yl)(η^1-2-propenyl)vanadium, V -00059

$C_{13}H_{15}W^{\oplus}$
Bis(η^5-2,4-cyclopentadien-1-yl)(η^3-2-propenyl)-tungsten(1+), W -00084

$C_{13}H_{16}CrO_3Si$
Tricarbonyl(η^6-benzyltrimethylsilyl)chromium, Cr-00116

$C_{13}H_{16}CrO_3Sn$
(Benzyltrimethylstannane)tricarbonylchromium, Cr-00117

$C_{13}H_{16}OV^{\oplus}$
Bis(η^5-2,4-cyclopentadien-1-yl)(2-propanone)-vanadium(1+), V -00060

$C_{13}H_{16}OZr$
(η^2-Acetyl)bis(η^5-2,4-cyclopentadien-1-yl)-methylzirconium, Zr-00029

$C_{13}H_{17}Cl_2N_2Ta$
(2,2′-Bipyridine-*N*,*N*′)dichlorotrimethyltantalum, Ta-00020

$C_{13}H_{17}F_6PW$
Bis(η^5-2,4-cyclopentadien-1-yl)(η^2-ethene)-methyltungsten(1+); Hexafluorophosphate, *in* W -00085

$C_{13}H_{17}N_2O_3V$
Tricarbonyl(cyano-*C*)(η^5-2,4-cyclopentadien-1-yl)-vanadate(1−); Tetramethylammonium salt, *in* V -00018

$C_{13}H_{17}Ta$
Bis(cyclopentadienyl)hydropropenetantalum, Ta-00021

$C_{13}H_{17}W^{\oplus}$
Bis(η^5-2,4-cyclopentadien-1-yl)(η^2-ethene)-methyltungsten(1+), W -00085

$C_{13}H_{18}AlClTi$
μ-Chlorobis(η^5-cyclopentadienyl)(dimethylaluminum)-μ-methylenetitanium, Ti-00066

$C_{13}H_{19}As_2O_3V$
Tricarbonyltrihydro[1,2-phenylenebis[dimethylarsine]-*As*,*As*′]vanadium, V -00061

$C_{13}H_{19}ClSiTi$
Chlorobis(η^5-2,4-cyclopentadien-1-yl)(trimethylsilyl)-titanium, Ti-00067

$C_{13}H_{19}ClSiZr$
Chlorobis(η^5-2,4-cyclopentadien-1-yl)(trimethylsilyl)-zirconium, Zr-00030

$C_{13}H_{19}Ta$
Bis(η^5-2,4-cyclopentadien-1-yl)trimethyltantalum, Ta-00022

$C_{13}H_{20}CrClNO_5$
Pentacarbonylchlorochromate(1−); Tetraethylammonium salt, *in* Cr-00006

$C_{13}H_{20}CrINO_5$
Pentacarbonyliodochromate(1−); Tetraethylammonium salt, *in* Cr-00007

$C_{13}H_{20}IMoNO_5$
Pentacarbonyliodomolybdate(1−); Tetraethylammonium salt, *in* Mo-00004

$C_{13}H_{20}N_2O_3S_4W$
Tricarbonylbis(diethylcarbamodithioato-*S,S'*)tungsten, W -00086

$C_{13}H_{21}NaO_4PW$
Tetracarbonyl(triisopropylphosphine)tungstate(2−); Di-Na salt, *in* W -00087

$C_{13}H_{21}O_4PW^{\ominus\ominus}$
Tetracarbonyl(triisopropylphosphine)tungstate(2−), W -00087

$C_{13}H_{24}Cl_3P_2Ta$
Benzylidenetrichlorobis(trimethylphosphine)tantalum, Ta-00023

$C_{13}H_{27}Cl_2CrO_3$
Dichloromethyltris(tetrahydrofuran)chromium, Cr-00118

$C_{13}H_{27}Si_2Zr$
(η^5-Cyclopentadienyl)bis(trimethylsilylmethyl)-zirconium, Zr-00031

$C_{13}H_{29}Cl_4NW$
Tetrachloro(1,1-dimethylpropylidyne)tungsten(1−); Tetraethyl-ammonium salt, *in* W -00004

$C_{13}H_{33}N_3Ti$
Tris(diethylaminato)methyltitanium, Ti-00068

$C_{13}H_{36}BrMnO_{13}P_4$
Bromocarbonyltetrakis(trimethylphosphite-*P*)manganese, Mn-00089

$C_{13}H_{37}ClP_4W$
Chloromethylidynetetrakis(trimethylphosphine)tungsten, W -00088

$C_{13}H_{37}MnO_{13}P_4$
Carbonylhydrotetrakis(trimethylphosphite-*P*)manganese, Mn-00090

$C_{14}FeMnO_{14}Re$
Pentacarbonyl(pentacarbonylmanganese)-(tetracarbonyliron)rhenium, Re-00062

$C_{14}FeMn_2O_{14}$
Tetracarbonylbis(pentacarbonylmanganio)iron, Mn-00092

$C_{14}FeO_{14}Re_2$
Tetracarbonylbis(pentacarbonylrhenio)iron, Re-00063

$C_{14}HO_{14}Re_3$
Tetradecacarbonylhydrotrirhenium, Re-00064

$C_{14}H_5F_{12}IMoO$
Carbonyl(η^5-2,4-cyclopentadien-1-yl)iodo[(1,2,3,4-η)-1,2,3,4-tetrakis(trifluoromethyl)cyclobutadiene]-molybdenum, Mo-00078

$C_{14}H_9CrNO_5$
[1-(Benzylideneamino)ethylidene]pentacarbonylchromium, Cr-00119

$C_{14}H_{10}BCl_4MnO_2$
Dicarbonyl(η^5-2,4-cyclopentadien-1-yl)-(phenylmethylidyne)manganese(1+); Tetrachloroborate, *in* Mn-00093

$C_{14}H_{10}BCl_4O_2Re$
Dicarbonyl(η^5-2,4-cyclopentadien-1-yl)-(phenylmethylidyne)rhenium(1+); Tetrachloroborate, *in* Re-00066

$C_{14}H_{10}Cl_2O_4W_2$
Tetracarbonyldichlorobis(η^5-2,4-cyclopentadien-1-yl)-ditungsten, W -00089

$C_{14}H_{10}Cr_2O_4$
Tetracarbonylbis(η^5-2,4-cyclopentadien-1-yl)dichromium, Cr-00120

$C_{14}H_{10}F_6O_4Ti$
Bis(η^5-2,4-cyclopentadien-1-yl)bis(trifluoroacetato-*O*)-titanium, Ti-00069

$C_{14}H_{10}F_6O_4Zr$
Bis(η^5-2,4-cyclopentadien-1-yl)bis(trifluoroacetato-*O*)-zirconium, Zr-00032

$C_{14}H_{10}MnO_2^{\oplus}$
Dicarbonyl(η^5-2,4-cyclopentadien-1-yl)-(phenylmethylidyne)manganese(1+), Mn-00093

$C_{14}H_{10}MnO_4Re$
(η^5-2,4-Cyclopentadien-1-yl)[μ-(η:η^5-2,4-cyclopentadien-1-ylidene)]-hydrotetracarbonylmanganeserhenium, Re-00065

$C_{14}H_{10}Mo_2O_4$
Tetracarbonylbis(η^5-2,4-cyclopentadien-1-yl)-dimolybdenum, Mo-00079

$C_{14}H_{10}N_2S_2Ti$
Bis(η^5-2,4-cyclopentadien-1-yl)[2,3-dimercapto-2-butenedinitrilato(2−)-*S,S'*]titanium, Ti-00070

$C_{14}H_{10}O_2Re^{\oplus}$
Dicarbonyl(η^5-2,4-cyclopentadien-1-yl)-(phenylmethylidyne)rhenium(1+), Re-00066

$C_{14}H_{10}O_4W_2$
Tetracarbonylbis(η^5-2,4-cyclopentadien-1-yl)ditungsten, W -00090

$C_{14}H_{12}MoO_2$
Benzyldicarbonyl-π-cyclopentadienylmolybdenum, Mo-00080

$C_{14}H_{12}O_2W$
Dicarbonyl(η^5-2,4-cyclopentadien-1-yl)(η^3-6-methylene-2,4-cyclohexadien-1-yl)tungsten, W -00091

$C_{14}H_{14}Cl_3Mo_2$
Trichlorobis(η^7-2,4,6-cycloheptatrien-1-yl)-dimolybdenum, Mo-00081

$C_{14}H_{15}NbO_4$
Tetracarbonyl[(1,2,3,4,5-η)-1,2,3,4,5-pentamethyl-2,4-cyclopentadien-1-yl]niobium, Nb-00030

$C_{14}H_{15}O_4Ta$
Tetracarbonyl[(1,2,3,4,5-η)-1,2,3,4,5-pentamethyl-2,4-cyclopentadien-1-yl]tantalum, Ta-00024

$C_{14}H_{16}Cr$
Bis[(1,2,3,4,5,6-η)methylbenzene]chromium, Cr-00121

$C_{14}H_{16}F_6NPW$
(Cyanomethyl)bis(η^5-2,4-cyclopentadien-1-yl)(η^2-ethene)-tunsten(1+); Hexafluorophosphate, *in* W -00092

$C_{14}H_{16}F_{12}N_2P_2V$
Bis(acetonitrile)bis(η^5-2,4-cyclopentadien-1-yl)-vanadium(2+); Bis(hexafluorophosphate), *in* V -00062

$C_{14}H_{16}Mo$
Bis[(1,2,3,4,5,6-η)methylbenzene]molybdenum, Mo-00082

$C_{14}H_{16}NW^{\oplus}$
(Cyanomethyl)bis(η^5-2,4-cyclopentadien-1-yl)(η^2-ethene)-tunsten(1+), W -00092

$C_{14}H_{16}N_2V^{\oplus\oplus}$
Bis(acetonitrile)bis(η^5-2,4-cyclopentadien-1-yl)-vanadium(2+), V -00062

$C_{14}H_{16}O_2W$
Dicarbonylbis[(1,2,3,4-η)-1,3-cyclohexadiene]tungsten, W -00093

$C_{14}H_{16}O_4Ti$
Bis(acetato-*O*)bis(η^5-2,4-cyclopentadien-1-yl)titanium, Ti-00071

$C_{14}H_{16}Ti$
Bis[(1,2,3,4,5,6-η)-methylbenzene]titanium, Ti-00072

$C_{14}H_{16}V$
Bis[(1,2,3,4,5,6-η)-1,3,5-cycloheptatriene]vanadium, V -00063

$C_{14}H_{16}W$
Bis[(1,2,3,4,5,6-η)methylbenzene]tungsten, W -00094

$C_{14}H_{16}Zr$
(η^4-1,3-Butadiene)bis(η^5-2,4-cyclopentadien-1-yl)-zirconium, Zr-00033

$C_{14}H_{17}BF_4MoO$
Bis[(2,3-η)-2-butyne]carbonyl(η^5-2,4-cyclopentadien-1-yl)molybdenum(1+); Tetrafluoroborate, *in* Mo-00083

$C_{14}H_{17}F_6MoOP$
Bis[(2,3-η)-2-butyne]carbonyl(η^5-2,4-cyclopentadien-1-yl)molybdenum(1+); Hexafluorophosphate, *in* Mo-00083

$C_{14}H_{17}F_6OPW$
Bis[(2,3-η)-2-butyne]carbonyl(η^5-2,4-cyclopentadien-1-yl)tungsten(1+); Hexafluorophosphate, *in* W -00095

$C_{14}H_{17}MoO^{\oplus}$
Bis[(2,3-η)-2-butyne]carbonyl(η^5-2,4-cyclopentadien-1-yl)molybdenum(1+), Mo-00083

$C_{14}H_{17}OW^{\oplus}$
Bis[(2,3-η)-2-butyne]carbonyl(η^5-2,4-cyclopentadien-1-yl)tungsten(1+), W -00095

$C_{14}H_{18}CrF_6NO_3P$
Dicarbonyl[(1,2,3,4,5,6-η)hexamethylbenzene]-nitrosylchromium(1+); Hexafluorophosphate, *in* Cr-00122

$C_{14}H_{18}CrNO_3^{\oplus}$
Dicarbonyl[(1,2,3,4,5,6-η)hexamethylbenzene]-nitrosylchromium(1+), Cr-00122

$C_{14}H_{18}Hf$
(η^8-1,3,5,7-Cyclooctatetraene)bis(η^3-2-propenyl)-hafnium, Hf-00016

$C_{14}H_{18}OSm$
Bis(η^5-2,4-cyclopentadien-1-yl)samarium; THF complex, *in* Sm-00005

$C_{14}H_{18}Ti$
(1,4-Butanediyl)bis(η^5-2,4-cyclopentadien-1-yl)-titanium, Ti-00073

$C_{14}H_{18}Zr$
(η^8-1,3,5,7-Cyclooctatetraene)bis(η^3-2-propenyl)-zirconium, Zr-00034

$C_{14}H_{19}Er$
tert-Butylbis(η^5-cyclopentadienyl)erbium, Er-00009

$C_{14}H_{19}Lu$
Butylbis(η^5-2,4-cyclopentadien-1-yl)lutetium, Lu-00009
tert-Butylbis(η^5-cyclopentadienyl)lutetium, Lu-00010

$C_{14}H_{19}LuO$
Bis(η^5-2,4-cyclopentadien-1-yl)hydro(tetrahydrofuran)-lutetium, *in* Lu-00005

$C_{14}H_{19}Nb$
Bis(η^5-2,4-cyclopentadien-1-yl)(η^2-ethene)ethylniobium, Nb-00031

$C_{14}H_{19}OPTi$
Carbonylbis(η^5-2,4-cyclopentadien-1-yl)-(trimethylphosphine)titanium, Ti-00074

$C_{14}H_{20}AlCl_2Ti$
Di-μ-chlorobis(η^5-cyclopentadien-1-yl)(diethylaluminum)-titanium, Ti-00075

$C_{14}H_{20}MoN_4O_4$
Tetracarbonylbis(1,3-dimethyl-2-imidazolidinylidene)-molybdenum, Mo-00084

$C_{14}H_{20}N_2O_5W$
Pentacarbonyl(cyano-*C*)tungstate(1−); Tetraethylammonium salt, *in* W -00012

$C_{14}H_{20}N_4O_4W$
Tetracarbonylbis(1,3-dimethyl-2-imidazolidinylidene)-tungsten, W -00096

$C_{14}H_{20}V$
Bis(η^5-2,4-cyclopentadien-1-yl)diethylvanadium, V -00064

$C_{14}H_{21}ClHfSi$
Chlorobis(η^5-2,4-cyclopentadien-1-yl)[(trimethylsilyl)-methyl]hafnium, Hf-00017

$C_{14}H_{21}ClSiZr$
Chlorobis(η^5-2,4-cyclopentadien-1-yl)[(trimethylsilyl)-methyl]zirconium, Zr-00035

$C_{14}H_{21}ErSi$
Bis(η^5-2,4-cyclopentadien-1-yl)[(trimethylsilyl)methyl]-erbium, Er-00010

$C_{14}H_{21}LuSi$
Bis(η^5-2,4-cyclopentadien-1-yl)[(trimethylsilyl)methyl]-lutetium, Lu-00011

$C_{14}H_{21}SiSm$
Bis(η^5-2,4-cyclopentadien-1-yl)[(trimethylsilyl)methyl]-samarium, Sm-00008

$C_{14}H_{21}SiYb$
Bis(η^5-2,4-cyclopentadien-1-yl)[(trimethylsilyl)methyl]-ytterbium, Yb-00011

$C_{14}H_{22}AlSc$
Bis(2,4-cyclopentadien-1-yl)(dimethylaluminum)di-μ-methylscandium, Sc-00004

$C_{14}H_{22}AlY$
Bis(η^5-2,4-cyclopentadien-1-yl)(dimethylaluminum)di-μ-methylyttrium, Y -00005

$C_{14}H_{22}AlYb$
Bis(η^5-2,4-cyclopentadien-1-yl)(dimethylaluminum)di-μ-methylytterbium, Yb-00012

$C_{14}H_{22}HfN_2$
Bis(η^5-cyclopentadienyl)bis(dimethylamido)hafnium, Hf-00018

$C_{14}H_{22}IO_4PW$
Carbonyl(η^7-cycloheptatrienylium)iodo(triethyl phosphite)tungsten, W -00097

$C_{14}H_{22}N_2Ti$
Bis(η^5-cyclopentadienyl)bis(dimethylamido)titanium, Ti-00076

$C_{14}H_{22}N_2Zr$
Bis(η-cyclopentadienyl)bis(dimethylamido)zirconium, Zr-00036

$C_{14}H_{22}V$
Bis(η^5-2,4-dimethyl-2,4-pentadien-1-yl)vanadium, V -00065

$C_{14}H_{23}Cl_3O_2U$
Trichloro[(1,2,3,4,5-η)-1-methyl-2,4-cyclopentadien-1-yl]uranium; Bis-THF complex, *in* U -00003

$C_{14}H_{23}NO_5W$
N,*N*,*N*-Triethylethanaminium(1+) pentacarbonylmethyltungstate(1−), *in* W -00009

$C_{14}H_{23}N_3S_6Ti$
(η^5-2,4-Cyclopentadien-1-yl)-tris(dimethylcarbamodithioato-*S*,*S'*)titanium, Ti-00077

$C_{14}H_{23}N_3S_6Zr$
(η^5-2,4-Cyclopentadien-1-yl)-tris(dimethylcarbamodithioato-*S*,*S'*)zirconium, Zr-00037

$C_{14}H_{27}Ta$
[(1,2,3,4,5-η)-1,2,3,4,5-Pentamethyl-2,4-cyclopentadien-1-yl]tetramethyltantalum, Ta-00025

$C_{14}H_{28}P_2Ti$
Bis(η^4-1,3-butadiene)[1,2-ethanediylbis[dimethylphosphine]-*P*,*P'*]titanium, Ti-00078

$C_{14}H_{30}Cl_2O_2P_2Re$
Dicarbonyldichlorobis(triethylphosphine)rhenium, Re-00067

$C_{14}H_{33}TaO_2P_4$
Dicarbonylbis[1,2-ethanediylbis[dimethylphosphine]-*P*,*P'*]hydrotantalum, Ta-00026

$C_{14}H_{35}N_3V$
Tris(diethylamido)ethylvanadium, V -00066

$C_{14}H_{36}CrO_2P_4$
Dicarbonyltetrakis(trimethylphosphine)chromium, Cr-00123

$C_{14}H_{38}MnP_4$
Bis[1,2-ethanediylbis[dimethylphosphine]]-*P*,*P'*]-dimethylmanganese, Mn-00095

$C_{14}H_{42}HfN_2Si_4$
Dimethylbis[bis(trimethylsilyl)amido]hafnium, Hf-00019

$C_{14}H_{42}N_2Si_4Zr$
Dimethylbis[bis(trimethylsilyl)amido]zirconium, Zr-00038

$C_{14}Mn_2O_{14}Os$
Tetracarbonylbis(pentacarbonylmanganio)osmium, Mn-00096

$C_{14}Mn_2O_{14}Ru$
Tetracarbonylbis(pentacarbonylmanganio)ruthenium, Mn-00097

$C_{14}Mn_3O_{14}^{\ominus}$
Tetradecacarbonyltrimanganate(1−), Mn-00098

$C_{14}O_{14}OsRe_2$
Tetracarbonylbis(pentacarbonylrhenio)osmium, Re-00068

$C_{15}H_5BrO_8Re_2$
μ-Bromooctacarbonyl[μ-(phenylmethylidyne)dirhenium, Re-00069

$C_{15}H_8F_6N_3PO_4W$
Tricarbonylnitrosyl(1,10-phenanthroline-N^1,N^{10})-tungsten(1+); Hexafluorophosphate, *in* W -00098

$C_{15}H_8N_3O_4W^{\oplus}$
Tricarbonylnitrosyl(1,10-phenanthroline-N^1,N^{10})-tungsten(1+), W -00098

$C_{15}H_9BrCrFeO_4$
Bromotetracarbonyl(ferrocenylmethylidyne)chromium, Cr-00124

$C_{15}H_9BrFeMoO_4$
Bromotetracarbonyl(ferrocenylmethylidyne)molybdenum, Mo-00085

$C_{15}H_9BrFeO_4W$
Bromotetracarbonyl(ferroceniumylmethylidyne)tungsten, W -00099

$C_{15}H_9BrO_4RuW$
Bromotetracarbonyl(ruthenocenylmethylidyne)tungsten, W -00100

$C_{15}H_9ClO_4RuW$
Tetracarbonylchloro(ruthenocenomethylidyne)tungsten, W -00101

$C_{15}H_9MnMoO_5$
Molybdenumcarbonyl(η^5-2,4-cyclopentadien-1-yl)[μ-(η:η^5-2,4-cyclopentadien-1-ylidene)]-tetracarbonylmanganese, Mn-00099

$C_{15}H_{10}CrFeO_5$
Tricarbonyl(η^5-2,4-cyclopentadien-1-yl)[dicarbonyl(η^5-2,4-cyclopentadien-1-yl)iron]chromium, Cr-00125

$C_{15}H_{10}CrO_6$
Pentacarbonyl(1-methoxy-3-phenyl-2-propenylidene)-chromium, Cr-00126

$C_{15}H_{10}FeMoO_5$
Tricarbonyl(η^5-2,4-cyclopentadien-1-yl)[dicarbonyl(η^5-2,4-cyclopentadien-1-yl)iron]molybdenum, Mo-00087

$C_{15}H_{10}FeO_5W$
Tricarbonyl(η^5-2,4-cyclopentdien-1-yl)[dicarbonyl(η^5-2,4-cyclopentadien-1-yl)iron]tungsten, W -00102

$C_{15}H_{10}O_5V_2$
Di-μ-carbonyltricarbonylbis(η^5-2,4-cyclopentadien-1-yl)-divanadium, V -00067

$C_{15}H_{11}MnO_2$
Dicarbonyl(η^5-2,4-cyclopentadien-1-yl)(2-phenylethenylidene)manganese, Mn-00100

$C_{15}H_{12}BMnO_3$
Tricarbonyl[(1,2,3,4,5,6-η)-4-methyl-1-phenylboratabenzene]manganese, Mn-00101

$C_{15}H_{12}Mn_2O_4$
Tetracarbonylbis(η^5-2,4-cyclopentadien-1-yl)-μ-methylenedimanganese, Mn-00102

$C_{15}H_{13}BF_4MoN_2O_2$
Bis(acetonitrile)dicarbonyl[(1,2,3,3*a*,7*a*-η)-1*H*-inden-1-yl]molybdenum(1+); Tetrafluoroborate, *in* Mo-00088

$C_{15}H_{13}BF_4MoO_2$
(η^4-1,3-Butadiene)dicarbonyl[(1,2,3,3*a*,7*a*-η)-1*H*-inden-1-yl]molybdenum(1+); Tetrafluoroborate, *in* Mo-00089

$C_{15}H_{13}ClN_2O_2W$
(2,2'-Bipyridine-*N*,*N'*)dicarbonylchloro(η^3-2-propenyl)-tungsten, W -00103

$C_{15}H_{13}MnO_3$
Dicarbonyl(η^5-2,4-cyclopentadien-1-yl)(2-methoxy-1-phenylethylidene)manganese, Mn-00103

$C_{15}H_{13}MoN_2O_2^{\oplus}$
Bis(acetonitrile)dicarbonyl[(1,2,3,3*a*,7*a*-η)-1*H*-inden-1-yl]molybdenum(1+), Mo-00088

$C_{15}H_{13}MoO_2^{\oplus}$
(η^4-1,3-Butadiene)dicarbonyl[(1,2,3,3*a*,7*a*-η)-1*H*-inden-1-yl]molybdenum(1+), Mo-00089

$C_{15}H_{15}Am$
Tris(η^5-2,4-cyclopentadien-1-yl)americium, Am-00001

$C_{15}H_{15}Bk$
Tris(η^5-2,4-cyclopentadien-1-yl)berkelium, Bk-00001

$C_{15}H_{15}BrTh$
Bromotris(η^5-2,4-cyclopentadien-1-yl)thorium, Th-00004

$C_{15}H_{15}BrU$
Bromotris(η^5-2,4-cyclopentadien-1-yl)uranium, U -00012

$C_{15}H_{15}Ce$
Tris(η^5-2,4-cyclopentadien-1-yl)cerium, Ce-00003

$C_{15}H_{15}CeCl$
Chlorotris(η^5-2,4-cyclopentadien-1-yl)cerium, Ce-00004

$C_{15}H_{15}Cf$
Tris(η^5-2,4-cyclopentadien-1-yl)californium, Cf-00001

$C_{15}H_{15}ClNp$
Chlorotris(η^5-2,4-cyclopentadien-1-yl)neptunium, Np-00001

$C_{15}H_{15}ClPu$
Chlorotris(η^5-2,4-cyclopentadien-1-yl)plutonium, Pu-00001

$C_{15}H_{15}ClTh$
Chlorotris(η^5-2,4-cyclopentadien-1-yl)thorium, Th-00005

$C_{15}H_{15}ClU$
Chlorotris(η^5-2,4-cyclopentadien-1-yl)uranium, U -00013

$C_{15}H_{15}ClZr$
Chlorotris(η-cyclopentadienyl)zirconium, Zr-00039

$C_{15}H_{15}Cm$
Tris(η^5-2,4-cyclopentadien-1-yl)curium, Cm-00001

$C_{15}H_{15}Dy$
Tris(η^5-2,4-cyclopentadien-1-yl)dysprosium, Dy-00004

$C_{15}H_{15}Er$
Tris(η^5-2,4-cyclopentadien-1-yl)erbium, Er-00011

$C_{15}H_{15}Eu$
Tris(η^5-2,4-cyclopentadien-1-yl)europium, Eu-00004

$C_{15}H_{15}FNp$
Fluorotris(η^5-2,4-cyclopentadien-1-yl)neptunium, Np-00002
$C_{15}H_{15}FTh$
Fluorotris(η^5-2,4-cyclopentadien-1-yl)thorium, Th-00006
$C_{15}H_{15}FU$
Tris(η^5-2,4-cyclopentadien-1-yl)fluorouranium, U-00016
$C_{15}H_{15}Gd$
Tris(η^5-2,4-cyclopentadien-1-yl)gadolinium, Gd-00004
$C_{15}H_{15}Ho$
Tris(η^5-2,4-cyclopentadien-1-yl)holmium, Ho-00005
$C_{15}H_{15}ITh$
Iodotris(η^5-2,4-cyclopentadien-1-yl)thorium, Th-00007
$C_{15}H_{15}IU$
Iodotris(η^5-2,4-cyclopentadien-1-yl)uranium, U -00014
$C_{15}H_{15}La$
Tris(η^5-2,4-cyclopentadien-1-yl)lanthanum, La-00004
$C_{15}H_{15}Lu$
Tris(η^5-2,4-cyclopentadien-1-yl)lutetium, Lu-00012
$C_{15}H_{15}Mn_3N_4O_4$
Tris(η^5-2,4-cyclopentadien-1-yl)-μ_3-nitrosyltri-μ-nitrosyltrimanganese, Mn-00104
$C_{15}H_{15}Nd$
Tris(η^5-2,4-cyclopentadien-1-yl)neodymium, Nd-00004
$C_{15}H_{15}Pm$
Tris(η^5-2,4-cyclopentadien-1-yl)promethium, Pm-00001
$C_{15}H_{15}Pr$
Tris(η-^{5}-2,4-cyclopentadien-1-yl)praseodymium, Pr-00003
$C_{15}H_{15}Pu$
Tris(η^5-2,4-cyclopentadien-1-yl)plutonium, Pu-00002
$C_{15}H_{15}Sc$
Tris(η^5-2,4-cyclopentadien-1-yl)scandium, Sc-00005
$C_{15}H_{15}Sm$
Tris(η^5-2,4-cyclopentadien-1-yl)samarium, Sm-00009
$C_{15}H_{15}Tb$
Tris(η^5-2,4-cyclopentadien-1-yl)terbium, Tb-00002
$C_{15}H_{15}Th$
Tris(η^5-2,4-cyclopentadien-1-yl)thorium, Th-00008
$C_{15}H_{15}Ti$
[(1,2-η)-2,4-cyclopentadien-1-yl]bis(η^5-2,4-cyclopentadien-1-yl)titanium, Ti-00079
$C_{15}H_{15}Tm$
Tris(η^5-2,4-cyclopentadien-1-yl)thulium, Tm-00004
$C_{15}H_{15}U$
Tris(η^5-2,4-cyclopentadien-1-yl)uranium, U -00015
$C_{15}H_{15}Y$
Tris(η^5-2,4-cyclopentadien-1-yl)yttrium, Y -00006
$C_{15}H_{15}Yb$
Tris(η^5-2,4-cyclopentadien-1-yl)ytterbium, Yb-00013
$C_{15}H_{16}CrO_3$
Tricarbonyl[(1,2,3,4,5,6-η)-1,3,5,7-tetramethyl-1,3,5,7-cyclooctatetraene]chromium, Cr-00127
$C_{15}H_{16}MoO_3$
Tricarbonyl(η^6-1,3,5,7-tetramethyl-1,3,5,7-cyclooctatetraene)molybdenum, Mo-00090
$C_{15}H_{17}BMoN_6O_2$
Dicarbonyl[hydrotris(1*H*-pyrazolato-*N*[1])borato(1−)-N^2,N^2,N^2][(1,2,3-η-2-methyl-2-propenyl]-molybdenum, Mo-00091
$C_{15}H_{17}BN_6O_2W$
Dicarbonyl[hydrotris(pyrazolato)borato(1−)](η^3-2-methylpropenyl)tungsten, W -00104
$C_{15}H_{17}ClO_6Ti$
Di-π-cyclopentadienyl(2,4-pentanedionato)titanium(1+); Perchlorate, *in* Ti-00080
$C_{15}H_{17}ClO_6V$
▷Bis(η^5-2,4-cyclopentadien-1-yl)(pentanedionato)-vanadium(1+); Perchlorate, *in* V -00068
$C_{15}H_{17}O_2Ti^{\oplus}$
Di-π-cyclopentadienyl(2,4-pentanedionato)titanium(1+), Ti-00080
$C_{15}H_{17}O_2V^{\oplus}$
Bis(η^5-2,4-cyclopentadien-1-yl)(pentanedionato)-vanadium(1+), V -00068
$C_{15}H_{18}NYb$
Tris(2,4-cyclopentadien-1-yl)ammineytterbium, *in* Yb-00013
$C_{15}H_{18}O_3W$
Tricarbonyl[(1,2,3,4,5,6-η)-hexamethylbenzene]tungsten, W -00105
$C_{15}H_{19}As_2O_4V$
Tetracarbonylmethyl[1,2-phenylenebis(dimethylarsine)-*As*,*As'*]vanadium, V -00069
$C_{15}H_{19}ClHfO_4$
Chloro-π-cyclopentadienylbis(2,4-pentanedionato)-hafnium, Hf-00020
$C_{15}H_{19}ClO_4Zr$
Chloro(η^5-2,4-cyclopentadien-1-yl)bis(2,4-pentanedionato-*O*,*O'*)zirconium, Zr-00040
$C_{15}H_{19}LuO$
Bis(η^5-2,4-cyclopentadien-1-yl)(2,2-dimethyl-1-oxopropyl-*C*,*O*)lutetium, Lu-00013
$C_{15}H_{19}Ti$
Bis(η^5-2,4-cyclopentadien-1-yl)[(1,2,3-η)-2-methyl-2-butenyl]titanium, Ti-00081
$C_{15}H_{20}BF_4NS_2V$
Bis(η^5-2,4-cyclopentadien-1-yl)-(diethylcarbamodithioato-*S*,*S'*)vanadium(1+); Tetrafluoroborate, *in* V -00070
$C_{15}H_{20}I_3NS_2V$
Bis(η^5-2,4-cyclopentadien-1-yl)-(diethylcarbamodithioato-*S*,*S'*)vanadium(1+); Triiodide, *in* V -00070
$C_{15}H_{20}NS_2V^{\oplus}$
Bis(η^5-2,4-cyclopentadien-1-yl)-(diethylcarbamodithioato-*S*,*S'*)vanadium(1+), V -00070
$C_{15}H_{20}NTa$
Bis(η^5-2,4-cyclopentadien-1-yl)(isocyanomethane)-propyltantalum, Ta-00027
$C_{15}H_{21}ClTi$
Chlorobis(η^5-2,4-cyclopentadien-1-yl)(2,2-dimethylpropyl)titanium, Ti-00082
$C_{15}H_{21}Lu$
Bis(η^5-2,4-cyclopentadien-1-yl)(2,2-dimethylpropyl)-lutetium, Lu-00014
$C_{15}H_{21}LuO$
Bis(η^5-2,4-cyclopentadien-1-yl)methyllutetium; THF complex, *in* Lu-00006
$C_{15}H_{22}Zr$
Bis(η^5-cyclopentadienyl)(2,2-dimethylpropyl)-hydrozirconium, Zr-00041
$C_{15}H_{25}Cl_2Ta$
Dichloro(2,2-dimethylpropylidene)[(1,2,3,4,5-η)-1,2,3,4,5-pentamethyl-2,4-cyclopentadien-1-yl]-tantalum, Ta-00028
$C_{15}H_{26}NO_2Re$
Dicarbonyl(η^5-2,4-cyclopentadien-1-yl)hydrorhenate(1−); Tetraethylammonium salt, *in* Re-00027
$C_{15}H_{26}O_3Ti$
Phenyltris(2-propanolato)titanium, Ti-00083
$C_{15}H_{27}Zr$
(η^5-2,4-Cyclopentadien-1-yl)bis(2,2-dimethylpropyl)-zirconium, Zr-00042

$C_{15}H_{35}TaO_2P_4$
Dicarbonylbis[1,2-ethanediylbis[dimethylphosphine]-*P,P'*]methyltantalum, Ta-00029

$C_{16}HMnO_{16}Os_3$
Hexadecacarbonylhydromanganesetriosmium, Mn-00105

$C_{16}HO_{16}Os_3Re$
Hexadecacarbonylhydrorheniumtriosmium, Re-00070

$C_{16}H_8Mo_2O_6$
[*μ*-[(1,2,3,3*a*,8*a*-*η*:4,5,6,7,8-*η*)azulene]]-hexacarbonyldimolybdenum, Mo-00092

$C_{16}H_9MnN_2O_4$
Tetracarbonyl[2-(phenylazo)phenyl]manganese, Mn-00106

$C_{16}H_9N_2O_4Re$
Tetracarbonyl[2-(phenylazo)phenyl]rhenium, Re-00071

$C_{16}H_{10}Cr_2HgO_6$
▷Hexacarbonyldi-*π*-cyclopentadienyl-*μ*-mercuriodichromium, Cr-00128

$C_{16}H_{10}Cr_2O_6$
Hexacarbonylbis(*η*⁵-2,4-cyclopentadien-1-yl)dichromium, Cr-00129

$C_{16}H_{10}F_5Ti$
Bis(η^5-2,4-cyclopentadien-1-yl)-pentafluorophenyltitanium, Ti-00084

$C_{16}H_{10}F_6O_4S_2W_2$
Tetracarbonylbis(η^5-2,4-cyclopentadien-1-yl)bis[*μ*-(trifluoromethanethiolato)]ditungsten, W -00106

$C_{16}H_{10}F_6Ti$
▷Bis-*π*-cyclopentadienylbis(3,3,3-trifluoro-1-propynyl)-titanium, Ti-00085

$C_{16}H_{10}Mo_2O_6$
Hexacarbonylbis(η^5-2,4-cyclopentadien-1-yl)-dimolybdenum, Mo-00093

$C_{16}H_{10}O_6W_2$
Hexacarbonylbis(η^5-2,4-cyclopentadien-1-yl)ditungsten, W -00107

$C_{16}H_{10}Yb$
Bis(phenylethynyl)ytterbium, Yb-00014

$C_{16}H_{11}CrNO_5$
Pentacarbonyl[3-(dimethylamino)-3-phenyl-1,2-propadienylidene]chromium, Cr-00130

$C_{16}H_{11}FeMnO_5$
(2-Acetylferrocenyl-*C,O*)tetracarbonylmanganese, Mn-00107
Pentacarbonyl(ferrocenylmethyl)manganese, Mn-00108

$C_{16}H_{12}CrO_3$
Tricarbonyl[(1,2,3,4,5,6-*η*)-7-phenyl-1,3,5-cycloheptatriene]chromium, Cr-00131

$C_{16}H_{12}MnNO_{12}Os_2$
Dodecacarbonylmanganatediosmate(1−); Tetramethylammonium salt, *in* Mn-00085

$C_{16}H_{12}Mn_4O_{12}S_4$
Dodecacarbonyltetrakis(μ_3-methanethiolato)-tetramanganese, *in* Mn-00005

$C_{16}H_{12}NO_{12}Os_2Re$
Dodecacarbonylrhenatediosmate(1−); Tetramethylammonium salt, *in* Re-00059

$C_{16}H_{12}O_{12}Re_4S_4$
Dodecacarbonyltetrakis(μ_3-methanethiolato)tetrarhenium, *in* Re-00004

$C_{16}H_{14}ClF_{12}MoP$
Chloro(η^5-2,4-cyclopentadien-1-yl)bis[(2,3-*η*)-1,1,1,4,4,4-hexafluoro-2-butyne]-(trimethylphosphine)molybdenum, Mo-00094

$C_{16}H_{14}CrO_7$
Tricarbonyl[(3*a*,4,5,6,7,7*a*-*η*)-2,3-dihydro-1*H*-inden-1-ol]chromium; Monosuccinate, *in* Cr-00099

$C_{16}H_{14}O_2Ti$
[1,2-Benzenediolato(2−)-*O,O'*]bis(η^5-2,4-cyclopentadien-1-yl)titanium, Ti-00086

$C_{16}H_{14}S_2Ti$
[1,2-Benzenedithiolato(2−)-*S,S'*]bis(η^5-2,4-cyclopentadien-1-yl)titanium, Ti-00087

$C_{16}H_{15}ClTi$
Chlorobis(η^5-2,4-cyclopentadien-1-yl)phenyltitanium, Ti-00088

$C_{16}H_{15}Gd$
Bis(η^5-2,4-cyclopentadien-1-yl)phenylgadolinium, Gd-00005

$C_{16}H_{15}NOV^{\oplus}$
Carbonylbis(η^5-2,4-cyclopentadien-1-yl)(pyridine)-vanadium(1+), V -00071

$C_{16}H_{15}NSU$
Tris(η^5-2,4-cyclopentadien-1-yl)(thiocyanato-*N*)uranium, U -00017

$C_{16}H_{15}NU$
(Cyano-*C*)tris(η^5-2,4-cyclopentadien-1-yl)uranium, U -00018

$C_{16}H_{15}SV$
(Benzenethiolato)bis(η^5-2,4-cyclopentadien-1-yl)-vanadium, V -00072

$C_{16}H_{15}Ti$
Bis(η^5-2,4-cyclopentadien-1-yl)phenyltitanium, Ti-00089

$C_{16}H_{15}V$
Bis(η^5-2,4-cyclopentadien-1-yl)phenylvanadium, V -00073

$C_{16}H_{15}Yb$
Bis(η^5-2,4-cyclopentadien-1-yl)phenylytterbium, Yb-00015

$C_{16}H_{16}Ce^{\ominus}$
Bis(η^8-1,3-5,7-cyclooctatetraene)cerate(1−), Ce-00005

$C_{16}H_{16}CeK$
Bis(η^8-1,3-5,7-cyclooctatetraene)cerate(1−); K salt, *in* Ce-00005

$C_{16}H_{16}Cr$
($\eta^{1,2}$-[2,2]Paracyclophane)chromium, Cr-00132

$C_{16}H_{16}KLa$
Bis(η^8-1,3,5,7-cyclooctatetraene)lanthanate(1−); K salt, *in* La-00005

$C_{16}H_{16}KNd$
Bis(η^8-1,3,5,7-cyclooctatetraene)neodymate(1−); K salt, *in* Nd-00005

$C_{16}H_{16}KPr$
Bis(η^8-1,3,5,7-cyclooctatetraene)praesodymate(1−); K salt, *in* Pr-00004

$C_{16}H_{16}KSm$
Bis(η^8-1,3,5,7-cyclooctatetraene)samarate(1−); K salt, *in* Sm-00010

$C_{16}H_{16}KTb$
Bis(η^8-1,3,5,7-cyclooctatetraene)terbate(1−); K salt, *in* Tb-00003

$C_{16}H_{16}KY$
Bis(η^8-1,3,5,7-cyclooctatetraene)yttrate(1−); K salt, *in* Y -00007

$C_{16}H_{16}La^{\ominus}$
Bis(η^8-1,3,5,7-cyclooctatetraene)lanthanate(1−), La-00005

$C_{16}H_{16}Nd^{\ominus}$
Bis(η^8-1,3,5,7-cyclooctatetraene)neodymate(1−), Nd-00005

$C_{16}H_{16}Np^{\ominus}$
Bis(η^8-1,3,5,7-cyclooctatetraene)neptunate(1−), Np-00003
Bis(η^8-1,3,5,7-cyclooctatetraene)neptunium, Np-00004

$C_{16}H_{16}O_4V$
Bis(η^5-2,4-cyclopentadien-1-yl)[(2,3-*η*)-dimethyl 2-butynedioate]vanadium, V -00074

$C_{16}H_{16}Pa$
Bis(η^8-1,3,5,7-cyclooctatetraene)protactinium, Pa-00001

$C_{16}H_{16}Pr^{\ominus}$
Bis(η^8-1,3,5,7-cyclooctatetraene)praesodymate(1−), Pr-00004

$C_{16}H_{16}Pu^{\ominus}$
Bis(η^8-1,3,5,7-cyclooctatetraene)plutonate(1−), Pu-00003

Bis(η^{8}-1,3,5,7-cyclooctatetraene)plutonium, Pu-00004

$C_{16}H_{16}Sm^{\ominus}$
Bis(η^{8}-1,3,5,7-cyclooctatetraene)samarate(1−), Sm-00010

$C_{16}H_{16}Tb^{\ominus}$
Bis(η^{8}-1,3,5,7-cyclooctatetraene)terbate(1−), Tb-00003

$C_{16}H_{16}Th$
▷Bis(η^{8}-1,3,5,7-cyclooctatetraene)thorium, Th-00009

$C_{16}H_{16}Ti$
Bis(1,3,5,7-cyclooctatetraene)titanium, Ti-00090

$C_{16}H_{16}U$
Uranocene, U -00019

$C_{16}H_{16}V$
Bis(η^{8}-1,3,5,7-cyclooctatetraene)vanadium, V -00075

$C_{16}H_{16}W$
Bis(η^{5}-2,4-cyclopentadien-1-yl)hydrophenyltungsten, W -00108

$C_{16}H_{16}Y^{\ominus}$
Bis(η^{8}-1,3,5,7-cyclooctatetraene)yttrate(1−), Y -00007

$C_{16}H_{17}CrNO_{6}$
Benzoylpentacarbonylchromate(1−); Tetramethylammonium salt, *in* Cr-00094

$C_{16}H_{17}Mo_{2}O_{4}P$
Tetracarbonylbis(η^{5}-2,4-cyclopentadien-1-yl)[μ-(dimethylphosphino)]-μ-hydrodimolybdenum, Mo-00095

$C_{16}H_{18}Cl_{2}CrN_{3}$
Dichloromethyltrispyridinechromium, Cr-00133

$C_{16}H_{18}CrO_{4}$
Tetracarbonyl(1,2,3,4,5,6-hexamethylbicyclo[2.2.0]hexa-2,5-diene)chromium, Cr-00134

$C_{16}H_{18}NSc$
Bis(η^{5}-2,4-cyclopentadien-1-yl)methylscandium; Py complex, *in* Sc-00002

$C_{16}H_{18}Th$
Tris(η^{5}-2,4-cyclopentadien-1-yl)methylthorium, Th-00010

$C_{16}H_{18}U$
Tris(η^{5}-2,4-cyclopentadien-1-yl)methyluranium, U -00020

$C_{16}H_{19}Er$
Bis(2,4-cyclopentadien-1-yl)(3,3-dimethylbutynyl)-erbium, Er-00012

$C_{16}H_{20}Cr$
Bis[(1,2,3,4,5,6-η)ethylbenzene]chromium, Cr-00135

$C_{16}H_{20}V$
Bis[(1,2,3,4,5,6-η)ethylbenzene]vanadium, V -00076

$C_{16}H_{20}Zr$
Bis(η^{5}-2,4-cyclopentadien-1-yl)[(1,2,3,4-η)-2,3-dimethyl-1,3-butadiene]zirconium, Zr-00043
Bis(η^{5}-2,4-cyclopentadien-1-yl)-η^{1}-2-propenyl-η^{3}-2-propenylzirconium, Zr-00044

$C_{16}H_{21}BF_{4}MoO_{2}$
Dicarbonyl(η^{4}-2-methylene-1,3-propanediyl)[(1,2,3,4,5-η)-1,2,3,4,5-pentamethyl-2,4-cyclopentadien-1-yl]-molybdenum(1+); Tetrafluoroborate, *in* Mo-00096

$C_{16}H_{21}B_{2}FeMnO_{3}S$
Tricarbonyl[(η^{5}-2,4-cyclopentadien-1-yl)iron][μ-[(1,2,3,4,5-η:1,2,3,4,5-η)-3,4-diethyl-2,5-dihydro-2,5-dimethyl-1,2,5-thiadiborole]]-manganese, Mn-00109

$C_{16}H_{21}MoO_{2}^{\oplus}$
Dicarbonyl(η^{4}-2-methylene-1,3-propanediyl)[(1,2,3,4,5-η)-1,2,3,4,5-pentamethyl-2,4-cyclopentadien-1-yl]-molybdenum(1+), Mo-00096

$C_{16}H_{22}ClOSm$
Chlorobis[(1,2,3,4,5-η)-1-methyl-2,4-cyclopentadien-1-yl]samarium; THF adduct, *in* Sm-00006

$C_{16}H_{22}Cl_{4}O_{2}Ti_{2}$
Tetrachlorobis(η^{5}-2,4-cyclopentadien-1-yl)[μ-[2,3-dimethyl-2,3-butanediolato(2−)-*O*:*O′*]dititanium, Ti-00091

$C_{16}H_{22}OYb$
Bis[(1,2,3,4,5-η)-1-methyl-2,4-cyclopentadien-1-yl]-ytterbium; THF complex, *in* Yb-00009

$C_{16}H_{23}NO_{2}W$
Dicarbonyl(η^{5}-2,4-cyclopentadien-1-yl)(2,2,4,4-tetramethyl-3-pentaniminato)tungsten, W -00109

$C_{16}H_{23}Y$
tert-Butylbis(methylcyclopentadienyl)yttrium, Y -00008

$C_{16}H_{24}ClLaO_{2}$
(η^{8}-1,3,5,7-Cyclooctatetraene)bis(tetrahydrofuran)-lanthanum, *in* La-00001

$C_{16}H_{24}Cl_{2}O_{2}Th$
Dichloro(η^{8}-1,3,5,7-cyclooctatetraene)thorium; Di-THF complex, *in* Th-00001

$C_{16}H_{26}ClP_{2}Ta$
Bis(η^{5}-2,4-cyclopentadien-1-yl)[1,2-ethanediylbis[dimethylphosphine]-*P*,*P′*]-tantalum(1+); Chloride, *in* Ta-00031

$C_{16}H_{26}NO_{3}V$
Tricarbonyl(η^{5}-2,4-cyclopentadien-1-yl)-hydrovanadate(1−); Tetraethylammonium salt, *in* V -00017

$C_{16}H_{26}PTa$
Bis(η^{5}-2,4-cyclopentadien-1-yl)hydro(triethylphosphine)-tantalum, Ta-00030

$C_{16}H_{26}P_{2}Ta^{\oplus}$
Bis(η^{5}-2,4-cyclopentadien-1-yl)[1,2-ethanediylbis[dimethylphosphine]-*P*,*P′*]-tantalum(1+), Ta-00031

$C_{16}H_{26}Si_{2}Yb$
Bis[(1,2,3,4,5-η)-1-(trimethylsilyl)-2,4-cyclopentadien-1-yl]ytterbium, Yb-00016

$C_{16}H_{28}AlClNY$
Chlorobis(cyclopentadienyl)[trihydro(triethylamino)-aluminum]yttrium, Y -00009

$C_{16}H_{28}O_{6}P_{2}Ti$
Bis(η^{5}-2,4-cyclopentadien-1-yl)bis(trimethylphosphite)-titanium, Ti-00092

$C_{16}H_{30}N_{3}OReS_{6}$
Carbonyltris(diethylcarbamodithioato-*S*,*S′*)rhenium, Re-00072

$C_{16}H_{32}Cr_{2}^{\ominus\ominus\ominus\ominus}$
Tetrakis(μ-tetramethylene)dichromate(4−), Cr-00136

$C_{16}H_{32}Cr_{2}Li_{4}$
Tetrakis(μ-tetramethylene)dichromate(4−); Tetra-Li salt, *in* Cr-00136

$C_{16}H_{36}Er^{\ominus}$
Tetra-*tert*-butylerbate(1−), Er-00013

$C_{16}H_{36}O_{5}W$
Tetrakis(2-methyl-2-propanolato)oxotungsten, W -00110

$C_{16}H_{36}Sm^{\ominus}$
Tetra-*tert*-butylsamarate(1−), Sm-00011

$C_{16}H_{36}Yb^{\ominus}$
Tetra-*tert*-butylytterbate(1−), Yb-00017

$C_{16}H_{40}Cr_{2}P_{4}$
Tetrakis[(dimethylphosphonio)bis(methylene)]dichromium, Cr-00137

$C_{16}H_{41}NbP_{4}$
Bis[1,2-ethanediylbis[dimethylphosphine]-*P*,*P′*]bis(η^{2}-ethene)hydroniobium, Nb-00032

$C_{16}H_{44}CrSi_{4}$
Tetrakis[(trimethylsilyl)methyl]chromium, Cr-00138

$C_{16}H_{44}HfSi_{4}$
▷Tetrakis(trimethylsilylmethyl)hafnium, Hf-00021

$C_{16}H_{44}LuSi^{\ominus}$
Tetrakis[(trimethylsilyl)methyl]luteate(1−), Lu-00015

$C_{16}H_{44}Si_{4}Ti$
Tetrakis(trimethylsilylmethyl)titanium, Ti-00093

$C_{16}H_{44}Si_4V$
▷Tetrakis[(trimethylsilyl)methyl]vanadium, V -00077

$C_{16}H_{44}Si_4Y^{\ominus}$
Tetrakis[(trimethylsilyl)methyl]yttrate(1−), Y -00010

$C_{16}H_{44}Si_4Zr$
▷Tetrakis(trimethylsilylmethyl)zirconium, Zr-00045

$C_{16}H_{45}CrO_{16}P_5$
Carbonylpentakis(trimethylphosphite-*P*)chromium, Cr-00139

$C_{16}H_{46}HfN_2Si_4$
Diethylbis[bis(trimethylsilyl)amido]hafnium, Hf-00022

$C_{16}H_{48}Mo_2P_4$
Tetramethyltetrakis(trimethylphosphine)dimolybdenum, Mo-00097

$C_{16}H_{56}B_{20}N_2Ti$
Bis[η^6-decahydro-*C,C'*-dimethyldicarbadodecaborato(2−)]-titanate(2−); Bis(tetramethylammonium) salt, *in* Ti-00019

$C_{16}Na_2O_{16}Re_4$
Hexadecacarbonyltetrarhenate(2−); Di-Na salt, *in* Re-00073

$C_{16}O_{16}Re_4^{\ominus\ominus}$
Hexadecacarbonyltetrarhenate(2−), Re-00073

$C_{17}H_4NO_{15}Os_3Re$
[(Acetonitrile)decacarbonyl-μ-hydrotriosmium]-pentacarbonylrhenium, Re-00074

$C_{17}H_8O_{10}Tc_2$
Nonacarbonyl(methoxyphenylmethylene)ditechnetium, Tc-00009

$C_{17}H_{10}CrO_3$
Anthracenetricarbonylchromium, Cr-00140
Tricarbonyl-η^6-phenanthrenechromium, Cr-00141

$C_{17}H_{10}MnNO_4$
Tetracarbonyl[2-[(phenylimino)methyl]phenyl-*C,N*]-manganese, Mn-00110

$C_{17}H_{10}MnO_5Tl$
▷Pentacarbonyl(diphenylthallium)manganese, Mn-00111

$C_{17}H_{10}NO_4Re$
Tetracarbonyl[2-[(phenylimino)methyl]phenyl-*C,N*]-rhenium, Re-00075

$C_{17}H_{12}CrFeO_6$
Pentacarbonyl(ferrocenylmethoxymethylene)chromium, Cr-00142

$C_{17}H_{14}MoO_4$
Tetracarbonylbis(η^5-2,4-cyclopentadien-1-yl)[μ-[(1,2-η:2,3-η)-1,2-propadiene]]dimolybdenum, Mo-00098

$C_{17}H_{14}OV$
Carbonyl(η^5-2,4-cyclopentadien-1-yl)(η^5-1-phenyl-2,4-cyclopentadien-1-yl)vanadium, V -00078

$C_{17}H_{15}N_2S_2U^{\ominus}$
Tris(η^5-cyclopentadienyl)diisothiocyanatouranate(IV), U -00021

$C_{17}H_{16}CrO_7$
Tricarbonyl[(4*a*,5,6,7,8,8*a*-η)-1,2,3,4-tetrahydro-1-naphthalenol]chromium; Monosuccinate, *in* Cr-00115

$C_{17}H_{16}U$
Tris(η^5-2,4-cyclopentadien-1-yl)ethynyluranium, U -00022

$C_{17}H_{17}Lu$
Benzylbis(η^5-2,4-cyclopentadienyl)lutetium, Lu-00016
Bis(η^5-2,4-cyclopentadien-1-yl)(4-methylphenyl)-lutetium, Lu-00017

$C_{17}H_{17}Ti$
Benzylbis(η^5-cyclopentadienyl)titanium, Ti-00094

$C_{17}H_{17}V$
Benzylbis(cyclopentadienyl)vanadium, V -00079

$C_{17}H_{19}Cl_2DyO_3$
▷Dichloro(η^5-2,4-cyclopentadien-1-yl)dysprosium; Tris-THF adduct, *in* Dy-00001

$C_{17}H_{19}F_6PMo$
(η^6-Benzene)[(1,2,3,4-η)-1,3,5,7-cyclooctatetraene](η^3-2-propenyl)molybdenum(1+); Hexafluorophosphate, *in* Mo-00099

$C_{17}H_{19}Mo^{\oplus}$
(η^6-Benzene)[(1,2,3,4-η)-1,3,5,7-cyclooctatetraene](η^3-2-propenyl)molybdenum(1+), Mo-00099

$C_{17}H_{19}Nb$
Bis(1,3,5,7-cyclooctatetraene)methylniobium, Nb-00033

$C_{17}H_{20}N_2O_4V$
Bis(η^5-2,4-cyclopentadien-1-yl)(η^2-diethyldiazomalonate-*N,N'*)vanadium, V -00080

$C_{17}H_{20}OU$
Ethoxytris(η^5-2,4-cyclopentadien-1-yl)uranium, U -00023

$C_{17}H_{23}Mo_2O_3$
Bis(η^7-2,4,6-cycloheptatrien-1-yl)tri-μ-methoxydimolybdenum, Mo-00100

$C_{17}H_{25}HfP$
Bis[(1,2,3,4,5,6-η)methylbenzene](trimethylphosphine)-hafnium, Hf-00023

$C_{17}H_{25}PZr$
Bis[(1,2,3,4,5,6-η)methylbenzene](trimethylphosphine)-zirconium, Zr-00046

$C_{17}H_{28}NbP$
Bis(cyclopentadienyl)methyl(triethylphosphine)niobium, Nb-00034

$C_{17}H_{29}ClSi_2Zr$
[Bis(trimethylsilyl)methyl]chlorobis(η^5-2,4-cyclopentadien-1-yl)zirconium, Zr-00047

$C_{17}H_{29}Cl_2ErO_3$
Dichloro(η^5-2,4-cyclopentadien-1-yl)erbium; Tris-THF adduct, *in* Er-00001

$C_{17}H_{29}Cl_2EuO_3$
Dichloro(η^5-2,4-cyclopentadien-1-yl)europium; Tris-THF adduct, *in* Eu-00001

$C_{17}H_{29}Cl_2GdO_3$
Dichloro(η^5-2,4-cyclopentadien-1-yl)gadolinium; Tris-THF adduct, *in* Gd-00001

$C_{17}H_{29}Cl_2HoO_3$
Dichloro(η^5-2,4-cyclopentadien-1-yl)holmium; Tris-THF adduct, *in* Ho-00001

$C_{17}H_{29}Cl_2LuO_3$
Dichloro(η^5-2,4-cyclopentadien-1-yl)lutetium; Tris-THF adduct, *in* Lu-00001

$C_{17}H_{29}Cl_2O_3Sm$
Dichloro(η^5-2,4-cyclopentadien-1-yl)samarium; Tris-THF adduct, *in* Sm-00001

$C_{17}H_{29}Cl_2O_3Y$
Dichloro(η^5-2,4-cyclopentadien-1-yl)yttrium; Tris-THF adduct, *in* Y -00001

$C_{17}H_{29}Cl_2O_3Yb$
Dichloro(η^5-2,4-cyclopentadien-1-yl)ytterbium; Tris-THF adduct, *in* Yb-00001

$C_{17}H_{37}NbP_4$
(η^5-2,4-Cyclopentadien-1-yl)bis[1,2-ethanediylbis[dimethylphosphine]-*P,P'*]niobium, Nb-00035

$C_{17}H_{37}O_3Ta$
(2,2-Dimethylpropylidene)tris(2-methyl-2-propanolato)-tantalum, Ta-00032

$C_{17}H_{38}Si_3Ti$
(η^5-2,4-Cyclopentadien-1-yl)tris[(trimethylsilyl)-methyl]titanium, Ti-00095

$C_{18}H_{10}CrO_5$
Pentacarbonyl(diphenylmethylene)chromium, Cr-00143

$C_{18}H_{10}MoO_5$
Pentacarbonyl(diphenylmethylene)molybdenum, Mo-00101

$C_{18}H_{10}O_5W$
Pentacarbonyl(diphenylmethylene)tungsten, W -00111

$C_{18}H_{10}O_8V_2$
Dicarbonylbis(2,4-cyclopentadien-1-yl)vanadium(1+); Hexacarbonylvanadate, *in* V -00050

$C_{18}H_{13}FeMnO_3$
Tricarbonyl[(1,2,3,4,5-η)-ferrocenyl-2,4-cyclopentadien-1-yl]manganese, Mn-00112

$C_{18}H_{14}CeCl_2$
Dichlorobis[(1,2,3,3*a*,7*a*-η)-1*H*-inden-1-yl]cerium, Ce-00006

$C_{18}H_{14}Cl_2Ti$
▷Dichlorobis[(1,2,3,3*a*,7*a*-η)-1*H*-inden-1-yl]titanium, Ti-00096

$C_{18}H_{14}Cl_2Zr$
Dichlorobis[(1,2,3,3*a*,7*a*-η)-1*H*-inden-1-yl]zirconium, Zr-00048

$C_{18}H_{14}CrO_6Ru$
Pentacarbonyl(ethoxyruthenocenylmethylene)chromium, Cr-00144

$C_{18}H_{14}FeMoO_3$
Tricarbonyl(η^5-2,4-cyclopentadien-1-yl)-ferrocenylmolybdenum, Mo-00102

$C_{18}H_{14}FeMoO_6$
Pentacarbonyl(ethoxyferrocenylmethylene)molybdenum, Mo-00103

$C_{18}H_{14}FeO_2W$
Dicarbonyl(η^5-2,4-cyclopentadien-1-yl)-(ferrocenylmethylidyne)tungsten, W -00112

$C_{18}H_{14}FeO_3W$
Tricarbonyl(η^5-2,4-cyclopentadien-1-yl)-ferrocenyltungsten, W -00113

$C_{18}H_{14}FeO_6W$
Pentacarbonyl(ethoxyferrocenylmethylene)tungsten, W -00114

$C_{18}H_{14}MoO_6Ru$
Pentacarbonyl(ethoxyruthenocenylmethylene)molybdenum, Mo-00104

$C_{18}H_{14}O_3W$
Tricarbonyl[(1,2,3,4,5,6-η)-7-(1-phenylethylidene)-1,3,5-cycloheptatriene]tungsten, W -00115

$C_{18}H_{14}O_6RuW$
Pentacarbonyl(ethoxyruthenocenylmethylene)tungsten, W -00116

$C_{18}H_{15}ErO_3W$
[μ-(Carbonyl-*C*,*O*)]bis(η^5-2,4-cyclopentadien-1-yl)-[dicarbonyl(η^5-2,4-cyclopentadien-1-yl)tungsten]-erbium, W -00117

$C_{18}H_{15}Gd$
Bis(η^5-2,4-cyclopentadien-1-yl)(phenylethynyl)-gadolinium, Gd-00006

$C_{18}H_{15}O_2V$
Dicarbonyl(η^5-2,4-cyclopentadien-1-yl)-(phenylcyclopentadiene)vanadium, V -00081

$C_{18}H_{15}O_3WYb$
[μ-(Carbonyl-*C*,*O*)]bis(η^5-2,4-cyclopentadien-1-yl)-[dicarbonyl(η^5-2,4-cyclopentadien-1-yl)tungsten]-ytterbium, W -00118

$C_{18}H_{15}Sc$
Triphenylscandium, Sc-00006

$C_{18}H_{15}Ti$
Bis(η^5-2,4-cyclopentadien-1-yl)(phenylethynyl)titanium, *in* Ti-00145

$C_{18}H_{15}Y$
Triphenylyttrium, Y -00011

$C_{18}H_{15}Yb$
Bis(η^5-2,4-cyclopentadien-1-yl)(phenylethynyl)-ytterbium, Yb-00019

$C_{18}H_{16}Mo_2O_4$
[η-[(2,3-η:2,3-η)-2-Butyne]]tetracarbonylbis(η^5-2,4-cyclopentadien-1-yl)dimolybdenum, Mo-00105

$C_{18}H_{18}BMnO$
Carbonyl(η^5-2,4-cyclopentadien-1-yl)[(1,2,3,6,7-η)-4,5-dihydro-1-phenyl-1*H*-borepin]manganese, Mn-00113

$C_{18}H_{18}N_2SU$
Tris(η^5-2,4-cyclopentadien-1-yl)(thiocyanato-*N*)uranium; MeCN adduct, *in* U -00017

$C_{18}H_{18}N_2Ti$
Bis(η^5-2,4-cyclopentadien-1-yl)di-1*H*-pyrrol-1-yltitanium, Ti-00097

$C_{18}H_{18}N_2Zr$
Bis(η^5-cyclopentadien-1-yl)di-1*H*-pyrrol-1-ylzirconium, Zr-00049

$C_{18}H_{18}Zr$
Bis(η^5-2,4-cyclopentadien-1-yl)[1,2-phenylenebis[methylene]]zirconium, Zr-00050

$C_{18}H_{19}Ti$
Bis(η^5-2,4-cyclopentadien-1-yl)(2,6-dimethylphenyl)-titanium, Ti-00098

$C_{18}H_{20}U$
Tris(η^5-2,4-cyclopentadien-1-yl)-2-propenyluranium, U -00024

$C_{18}H_{21}BrNbP$
Bromobis(η^5-2,4-cyclopentadien-1-yl)-(dimethylphenylphosphine)niobium, Nb-00036

$C_{18}H_{21}ClZr$
Chlorotris(η-methylcyclopentadienyl)zirconium, Zr-00051

$C_{18}H_{21}Cr_2NO_{10}$
Decacarbonyl-μ-hydrodichromate(1−); Tetraethylammonium salt, *in* Cr-00067

$C_{18}H_{21}NO_{10}W_2$
Decacarbonyl-μ-hydroditungstate(1−); Tetraethylammonium salt, *in* W -00040

$C_{18}H_{21}Nd$
Tris[(1,2,3,4,5-η)-1-methyl-2,4-cyclopentadien-1-yl]-neodymium, Nd-00006

$C_{18}H_{21}O_2PW$
Carbonyl(η^5-2,4-cyclopentadien-1-yl)[(1,2-η)(4-methylphenyl)oxoethenyl](trimethylphosphine)-tungsten, W -00119

$C_{18}H_{22}Cl_2Mo_2$
Bis(η^6-benzene)di-μ-chlorobis(η^3-2-propenyl)-dimolybdenum, *in* Mo-00028

$C_{18}H_{22}NbP$
Bis(η^5-2,4-cyclopentadien-1-yl)-hydro(dimethylphenylphosphine)niobium, Nb-00037

$C_{18}H_{22}O_4V$
Bis(η^5-2,4-cyclopentadien-1-yl)[(2,3-η)-diethyl 2-butenedioate]vanadium, V -00082

$C_{18}H_{22}O_8V_2$
Tetrakis[μ-(acetato-*O*,*O′*)]bis(η^5-2,4-cyclopentadien-1-yl)divanadium, V -00083

$C_{18}H_{23}CrNO_5Sn$
[Di-*tert*-butyl(pyridine)tin]pentacarbonylchromium, Cr-00145

$C_{18}H_{24}Mo$
Bis[(1,2,3,4,5,6-η)-1,3,5-trimethylbenzene]molybdenum, Mo-00106

$C_{18}H_{24}N_2Ti$
Bis(dimethylamido)(η^5-methylcyclopentadienyl)-phenylethynyltitanium, Ti-00099

$C_{18}H_{24}Ti$
Bis[(1,2,3,4,5,6-η)-1,3,5-trimethylbenzene]titanium, Ti-00100

$C_{18}H_{25}Nb$
Bis(cyclopentadienyl)hydro(η^2-4-octyne)niobium, Nb-00038

$C_{18}H_{25}Ta$
(η^2-Benzyne)dimethyl[(1,2,3,4,5-η)-1,2,3,4,5-pentamethyl-2,4-cyclopentadien-1-yl]tantalum, Ta-00033

$C_{18}H_{26}AlClZr$
Chlorobis(η^5-2,4-cyclopentadien-1-yl)[1-(dimethylalumino)-2-methyl-1-pentenyl]zirconium, Zr-00052

$C_{18}H_{27}ErO$
tert-Butylbis(η^5-cyclopentadienyl)erbium; THF complex, *in* Er-00009

$C_{18}H_{27}LuO$
Butylbis(η^5-2,4-cyclopentadien-1-yl)lutetium; THF complex, *in* Lu-00009
tert-Butylbis(η^5-cyclopentadienyl)lutetium; THF complex, *in* Lu-00010

$C_{18}H_{28}NaO_{12}V$
Hexacarbonylvanadate(1−); Na salt, bis(diglyme) complex, *in* V -00012

$C_{18}H_{29}ErOSi$
Bis(η^5-2,4-cyclopentadien-1-yl)[(trimethylsilyl)methyl]-erbium; THF complex, *in* Er-00010

$C_{18}H_{29}LuOSi$
Bis(η^5-2,4-cyclopentadien-1-yl)[(trimethylsilyl)methyl]-lutetium; THF complex, *in* Lu-00011

$C_{18}H_{29}N_2O_5V$
Pentacarbonyl(2-isocyano-2-methylpropane)vanadate(1−); Tetraethylammonium salt, *in* V -00025

$C_{18}H_{29}OSiYb$
Bis(η^5-2,4-cyclopentadien-1-yl)[(trimethylsilyl)methyl]-ytterbium; THF complex, *in* Yb-00011

$C_{18}H_{30}Er_2$
Tris(3-hexyne)dierbium, Er-00015

$C_{18}H_{30}HfN_2$
Bis(η^5-cyclopentadienyl)bis(diethylamido)hafnium, Hf-00024

$C_{18}H_{30}N_2U$
Bis(η^5-2,4-cyclopentadien-1-yl)bis(*N*-ethylethanaminato)-uranium, U -00025

$C_{18}H_{30}N_2Zr$
Bis(η^5-cyclopentadienyl)bis(diethylamido)zirconium, Zr-00053

$C_{18}H_{31}Cl_3O_2Th$
[(1,2,3,4,5-η)-1,2,3,4,5-Pentamethyl-2,4-cyclopentadien-1-yl]trichlorothorium; Bis-THF complex, *in* Th-00002

$C_{18}H_{31}Cl_3O_2U$
[(1,2,3,4,5-η)-1,2,3,4,5-Pentamethyl-2,4-cyclopentadienyl]trichlorouranium; Bis-THF complex, *in* U -00008

$C_{18}H_{32}HfSi_2$
Bis(η^5-2,4-cyclopentadien-1-yl)bis[(trimethylsilyl)-methyl]hafnium, Hf-00025

$C_{18}H_{32}Si_2Ti$
Bis(η^5-2,4-cyclopentadien-1-yl)bis[(trimethylsilyl)-methyl]titanium, Ti-00101

$C_{18}H_{32}Si_2Zr$
Bis(η^5-2,4-cyclopentadien-1-yl)bis[(trimethylsilyl)-methyl]zirconium, Zr-00054

$C_{18}H_{35}Cl_2PZr$
Dichloro(2,2-dimethylpropyl)[(1,2,3,4,5-η)-1,2,3,4,5-pentamethyl-2,4-cyclopentadien-1-yl]-(trimethylphosphine)zirconium, Zr-00055

$C_{18}H_{46}N_4W_2$
Dimethyltetrakis(*N*-dimethylamino)ditungsten, W -00120

$C_{19}H_{14}CrFeO_3$
Tricarbonyl(phenylferrocene)chromium, Cr-00146

$C_{19}H_{14}FeO_4W$
Tricarbonyl(η^5-2,4-cyclopentadien-1-yl)-(ferrocenylcarbonyl)tungsten, W -00121

$C_{19}H_{15}FeMnO_3$
Tricarbonyl[(1,2,3,4,5-η)-1-(ferrocenylmethyl)-2,4-cyclopentadien-1-yl]manganese, Mn-00114

$C_{19}H_{15}N_3O_2W$
Dicarbonyl(η^5-2,4-cyclopentadien-1-yl)(1,3-diphenyltriazenato-N^1,N^3)tungsten, W -00122

$C_{19}H_{15}O_4Ti$
Bis(μ-benzoato)(η^5-cyclopentadienyl)titanium, Ti-00102

$C_{19}H_{16}CrO_3$
Tricarbonyl(3-8-η-[2.2]paracyclophane)chromium, Cr-00147

$C_{19}H_{18}Cl_4NbP$
Tetrachloromethyl(triphenylphosphine)niobium, Nb-00039

$C_{19}H_{18}MoO_3Zr$
[μ-(Acetyl-*C*:*O*)][μ-(η^2-carbonyl)][carbonyl(η^5-2,4-cyclopentadien-1-yl)molydenum]bis(η^5-2,4-cyclopentadien-1-yl)zirconium, Mo-00107

$C_{19}H_{19}Mn$
[(1,2,3,4,5-η)-1-methyl-2,4-cyclopentadien-1-yl]-[(1,2,3,4,5,6-η)-7-phenyl-1,3,5-cycloheptatriene]-manganese, Mn-00115

$C_{19}H_{19}Zr$
Bis(benzyl)(η-cyclopentadienyl)zirconium, Zr-00057

$C_{19}H_{22}NSc$
Bis(η^5-2,4-cyclopentadien-1-yl)[2-[(dimethylamino)-methyl]phenyl-*C*,*N*]scandium, Sc-00007

$C_{19}H_{22}U$
Tris(η^5-2,4-cyclopentadien-1-yl)(2-methyl-1-propenyl)-uranium, U -00026

$C_{19}H_{23}Cr_2NO_{10}S$
Decarbonyl(μ-methanethiolato)dichromate(1−); Tetraethylammonium salt, *in* Cr-00083

$C_{19}H_{23}OPu$
Tris(η^5-2,4-cyclopentadienyl)tetrahydrofuranplutonium, *in* Pu-00002

$C_{19}H_{23}OU$
Tris(η^5-2,4-cyclopentadien-1-yl)uranium; THF complex, *in* U -00015

$C_{19}H_{23}OYb$
Tris(η^5-2,4-cyclopentadien-1-yl)-tetrahydrofuranytterbium, *in* Yb-00013

$C_{19}H_{24}OTh$
Butoxytris(η^5-2,4-cyclopentadien-1-yl)thorium, Th-00011

$C_{19}H_{24}OU$
Butoxytris(η^5-2,4-cyclopentadien-1-yl)uranium, U -00027

$C_{19}H_{24}O_2PW^{\oplus}$
Dicarbonyl(dimethylphenylphosphine)hydro[(1,2,3,4,5,6-η)-1,3,5-trimethylbenzene]tunsten(1+), W -00123

$C_{19}H_{24}Th$
Butyltris(η^5-2,4-cyclopentadien-1-yl)thorium, Th-00012

$C_{19}H_{24}U$
tert-Butyltricyclopentadienyluranium, U -00028
Butyltris(η^5-2,4-cyclopentadien-1-yl)uranium, U -00029
Tris(η^5-2,4-cyclopentadien-1-yl)(1-methylpropyl)-uranium, U -00030

$C_{19}H_{28}ClNTi$
(1-*tert*-Butyl-2,2-dimethylpropylideniminato)chlorodi-π-cyclopentadienyltitanium, Ti-00103

$C_{19}H_{30}OW$
Carbonyltris(3-hexyne)tungsten, W -00124

$C_{19}H_{30}U$
[(1,2,3,4,5-η)-1,2,3,4,5-Pentamethyl-2,4-cyclopentadien-1-yl)]tris(2-propenyl)uranium, U -00031

$C_{19}H_{31}BF_4MoO_6P_2$
[(2,3-η)-2-Butyne][(1,2,3,3*a*,7*a*-η)-1*H*-inden-1-yl]-bis(trimethylphosphite)molybdenum(1+); Tetrafluoroborate, *in* Mo-00108

$C_{19}H_{31}BF_4MoP_2$
[(2,3-η)-2-Butyne][(1,2,3,3*a*,7*a*-η)-1*H*-inden-1-yl]-bis(trimethylphosphine)molybdenum(1+); Tetrafluoroborate, *in* Mo-00109

$C_{19}H_{31}MoO_6P_2^{\oplus}$
[(2,3-η)-2-Butyne][(1,2,3,3a,7a-η)-1H-inden-1-yl]-bis(trimethylphosphite)molybdenum(1+), Mo-00108

$C_{19}H_{31}MoP_2^{\oplus}$
[(2,3-η)-2-Butyne][(1,2,3,3a,7a-η)-1H-inden-1-yl]-bis(trimethylphosphine)molybdenum(1+), Mo-00109

$C_{19}H_{32}OSi_2Zr$
Bis(trimethylsilyl)acetyl-C,O)bis(η^5-2,4-cyclopentadien-1-yl)methylzirconium, Zr-00058

$C_{19}H_{38}CrO_5Si_4Sn$
[Bis[bis(trimethylsilyl)methyl]tin]-pentacarbonylchromium, Cr-00148

$C_{19}H_{38}MoO_5PbSi_4$
[Bis[bis(trimethylsilyl)methyl]plumbylene]-pentacarbonylmolybdenum, Mo-00110

$C_{19}H_{38}O_3Zr$
Tris(diethyl ether)methylphenylzirconium, Zr-00059

$C_{19}H_{40}Cl_3N_2O_3Re$
Tricarbonyltrichlororhenate(2−); Bistetraethylammonium salt, *in* Re-00002

$C_{19}H_{57}N_3Si_6Th$
Methyltris[(hexamethyldisilyl)amino]thorium, Th-00013

$C_{19}H_{57}N_3Si_6U$
Methyltris[(hexamethyldisilyl)amido]uranium, U -00032

$C_{20}H_2O_{20}Os_3Re_2$
Decacarbonyl(decacarbonyldi-μ-hydrotriosmium)-diruthenium, Re-00076

$C_{20}H_{10}ClCo_3HfO_{10}$
Chlorobis(η^5-2,4-cyclopentadien-1-yl)[μ_4-[methanolato(4−)-*C:C:C:O*]](nonacarbonyltricobalt)-dihafnium, Hf-00026

$C_{20}H_{10}ClCo_3O_{10}Zr$
Chlorobis(η^5-2,4-cyclopentadien-1-yl)[μ_4-[methanolato(4−)-*C:C:C:O*]](nonacarbonyltricobalt)-zirconium, Zr-00060

$C_{20}H_{10}Co_2MoO_8$
Dicarbonyl(η^5-2,4-cyclopentadien-1-yl)-(hexacarbonyldicobalt)[μ_3-(phenylmethylidyne)]-molybdenum, Mo-00111

$C_{20}H_{10}FeMnO_8P$
Tetracarbonyl[μ-(diphenylphosphino)]-(tetracarbonylmanganio)iron, Mn-00116

$C_{20}H_{10}MoO_5$
Pentacarbonyl(2,3-diphenyl-2-cyclopropen-1-ylidene)-molybdenum, Mo-00112

$C_{20}H_{10}O_8Re_2S_2$
Bis(μ-benzenethiolato)octacarbonyldirhenium, *in* Re-00041

$C_{20}H_{11}Mn_2O_8P$
Octacarbonyl-μ-(diphenylphosphino)-μ-hydrodimanganese, Mn-00117

$C_{20}H_{11}O_5Re$
Pentacarbonyl(1,2-diphenyl-2-cyclopropen-1-yl)rhenium, Re-00077

$C_{20}H_{15}ClMoO$
Carbonylchloro(η^5-2,4-cyclopentadien-1-yl)-(diphenylacetylene)molybdenum, Mo-00113

$C_{20}H_{15}NO_2W$
Dicarbonyl(η^5-2,4-cyclopentadien-1-yl)(α-phenylbenzenemethaniminato)tungsten, W -00125

$C_{20}H_{15}O_3TlW$
▷Tricarbonyl(η^5-2,4-cyclopentadien-1-yl)-(diphenylthallium)tungsten, W -00126

$C_{20}H_{16}Cr$
Bis[(1,2,3,4,4a,8a-η)naphthalene]chromium, Cr-00149

$C_{20}H_{16}Cr_2GeO_6$
(Dimethyldiphenylgermane)bis[tricarbonylchromium], Cr-00150

$C_{20}H_{16}F_{12}P_2V_2$
1,1″:1′,1‴-Bivanadocenium(2+); Bis(hexafluorophosphate), *in* V -00085

$C_{20}H_{16}Mo$
Bis[(1,2,3,4,4a,8a-η)naphthalene]molybdenum, Mo-00114

$C_{20}H_{16}O_8W_2$
Tetracarbonylbis(η^5-2,4-cyclopentadien-1-yl)[μ-[(2,3-η:2,3-η)-dimethyl-2-butynedioate]]ditungsten, W -00127

$C_{20}H_{16}V_2$
▷1,1″;1,1‴-Bivanadocene, V -00084
1,1″:1′,1‴-Bivanadocenium(2+), V -00085

$C_{20}H_{18}Mo_2$
Bis(η^5-2,4-cyclopentadien-1-yl)bis[μ-(η:η^5-2,4-cyclopentadien-1-ylidene)]dimolybdenum, Mo-00115

$C_{20}H_{18}N_2V$
(2,2′-Bipyridine-*N,N′*)bis(η^5-2,4-cyclopentadien-1-yl)-vanadium, V -00086

$C_{20}H_{19}DyO_3W$
[μ-(Carbonyl-*C,O*)]bis[(1,2,3,4,5-η)-1-methyl-2,4-cyclopentadien-1-yl][tricarbonyl(η^5-2,4-cyclopentadien-1-yl)tungsten]dysprosium, W -00128

$C_{20}H_{19}FeMnO_3$
Dicarbonyl[(1,2,3,4,5-η)-1-methyl-2,4-cyclopentadien-1-yl](ferrocenylmethoxymethylene)manganese, Mn-00118

$C_{20}H_{19}HoO_3W$
[μ-(Carbonyl-*C,O*)]bis[(1,2,3,4,5-η)-1-methyl-2,4-cyclopentadien-1-yl][tricarbonyl(η^5-2,4-cyclopentadien-1-yl)tungsten]holmium, W -00129

$C_{20}H_{19}Ti_2$
▷μ(η^1:η^5-Cyclopentadienyl)tris(η-cyclopentadienyl)-dititanium(Ti-Ti), Ti-00104

$C_{20}H_{20}Br_2Ti_2$
Di-μ-bromotetrakis(η^5-2,4-cyclopentadien-1-yl)-dititanium, *in* Ti-00023

$C_{20}H_{20}Ce_2Cl_2$
Dichlorotetrakis(η^5-2,4-cyclopentadien-1-yl)dicerium, Ce-00007

$C_{20}H_{20}Cl_2Dy_2$
Di-μ-chlorotetrakis(η^5-2,4-cyclopentadien-1-yl)-didysprosium, *in* Dy-00002

$C_{20}H_{20}Cl_2Er_2$
Dichlorotetrakis(η^5-2,4-cyclopentadien-1-yl)dierbium, *in* Er-00004

$C_{20}H_{20}Cl_2Gd_2$
Di-μ-chlorotetrakis(η^5-2,4-cyclopentadien-1-yl)-digadolinium, *in* Gd-00002

$C_{20}H_{20}Cl_2Hf_2O$
Dichlorotetrakis(η^5-2,4-cyclopentadien-1-yl)-μ-oxodihafnium, Hf-00027

$C_{20}H_{20}Cl_2Ho_2$
Di-μ-chlorotetrakis(η^5-2,4-cyclopentadien-1-yl)-diholmium, *in* Ho-00002

$C_{20}H_{20}Cl_2Lu_2$
Di-μ-chlorotetrakis(η^5-2,4-cyclopentadien-1-yl)-dilutetium, *in* Lu-00004

$C_{20}H_{20}Cl_2OZr_2$
Dichlorotetrakis(η^5-2,4-cyclopentadien-1-yl)-μ-oxodizirconium, Zr-00061

$C_{20}H_{20}Cl_2Sc_2$
Di-μ-chlorotetrakis(η^5-2,4-cyclopentadien-1-yl)-discandium, *in* Sc-00001

$C_{20}H_{20}Cl_2Sm_2$
Di-μ-chlorotetrakis(η^5-2,4-cyclopentadien-1-yl)-disamarium, *in* Sm-00004

$C_{20}H_{20}Cl_2Ti_2$
Di-μ-chlorotetrakis(η^5-2,4-cyclopentadien-1-yl)-dititanium, *in* Ti-00025

$C_{20}H_{20}Cl_2Tm$
Di-μ-chlorotetrakis(η^5-2,4-cyclopentadien-1-yl)-dithulium, *in* Tm-00002

$C_{20}H_{20}Cl_2Yb_2$
Di-μ-chlorotetrakis(η^5-2,4-cyclopentadien-1-yl)-diytterbium, *in* Yb-00004

$C_{20}H_{20}Cl_4O_4Ti_4$
Tetrachlorotetrakis(η^5-2,4-cyclopentadien-1-yl)tetra-μ-oxotetratitanium, Ti-00105

$C_{20}H_{20}F_2Ti_2$
Tetrakis(η^5-2,4-cyclopentadien-1-yl)di-μ-fluorodititanium, *in* Ti-00028

$C_{20}H_{20}Fe_2MnNO_{12}$
Tetracarbonyl(di-μ-carbonylhexacarbonyldiferrate)-manganate(1−); Tetraethylammonium salt, *in* Mn-00072

$C_{20}H_{20}Hf$
Bis[(1,2,3,3*a*,7*a*-η)-1*H*-inden-1-yl]dimethylhafnium, Hf-00028

$C_{20}H_{20}I_2Ti_2$
Tetrakis(η^5-2,4-cyclopentadien-1-yl)di-μ-iododititanium, *in* Ti-00030

$C_{20}H_{20}Nb_2$
Bis(η^5-2,4-cyclopentadien-1-yl)bis[μ-[(1,η:1,2,3,4,5-η)-2,4-cyclopentadien-1-ylidene]]dihydrodiniobium, Nb-00040

$C_{20}H_{20}Np$
Tetrakis(η^5-2,4-cyclopentadien-1-yl)neptunium, Np-00005

$C_{20}H_{20}O_2Ti_2$
[μ-[(1,2,3,4,5-η:1′,2′,3′,4′,5′-η)[Bi-2,4-cyclopentadien-1-yl]-1,1′-diyl]bis(η^5-2,4-cyclopentadien-1-yl)di-μ-hydroxodititanium, Ti-00106

$C_{20}H_{20}Pa$
Tetrakis(η^5-2,4-cyclopentadien-1-yl)protactinium, Pa-00002

$C_{20}H_{20}Th$
Tetrakis(η^5-2,4-cyclopentadien-1-yl)thorium, Th-00014

$C_{20}H_{20}Ti$
Di-2,4-cyclopentadien-1-ylbis(η^5-2,4-cyclopentadien-1-yl)titanium, Ti-00107

$C_{20}H_{20}Ti_2$
μ(η^5:η^5-Fulvalene)di-μ-hydridobis(cyclopentadienyltitanium), Ti-00108

$C_{20}H_{20}U$
Bis[(1,2,3,4,5,6,7,8-η)-1-ethenyl-1,3,5,7-cyclooctatetraene]uranium, U -00033
Bis[(1,2,3,3*a*,7*a*-η)-1*H*-inden-1-yl]dimethyluranium, U -00034
Tetrakis(η^5-2,4-cyclopentadien-1-yl)uranium, U -00035

$C_{20}H_{20}Zr$
Bis[(1,2,3,3*a*,7*a*-η)-1*H*-inden-1-yl]dimethylzirconium, Zr-00062

$C_{20}H_{21}ClTi$
Chlorobis(η^5-cyclopentadienyl)(2,2-dimethyl-1-phenylethenyl)titanium, Ti-00109

$C_{20}H_{21}ClZr$
Chlorobis(η^5-2,4-cyclopentadien-1-yl)(2-methyl-1-phenyl-1-propenyl)zirconium, Zr-00063

$C_{20}H_{22}Ti_2$
Tetra-π-cyclopentadienyldi-μ-hydrodititanium, *in* Ti-00037

$C_{20}H_{23}ClO_4W_2$
Tetrakis(η^5-2,4-cyclopentadien-1-yl)-μ-hydrodihydroditungsten(1+); Perchlorate, *in* W -00130

$C_{20}H_{23}W_2^{\oplus}$
Tetrakis(η^5-2,4-cyclopentadien-1-yl)-μ-hydrodihydroditungsten(1+), W -00130

$C_{20}H_{24}F_3O_5PSW$
Dicarbonyl(dimethylphenylphosphine)hydro[(1,2,3,4,5,6-η)-1,3,5-trimethylbenzene]tunsten(1+); Trifluoromethanesulfonate, *in* W -00123

$C_{20}H_{24}HfSn_2$
Bis[(1,2,3,4,5,6-η)methylbenzene]bis(trimethylstannyl)-hafnium, Hf-00029

$C_{20}H_{24}Si_2Ti_2$
Tetrakis(η^5-2,4-cyclopentadien-1-yl)di-μ-silylenedititanium, Ti-00110

$C_{20}H_{26}Th$
Tris(η^5-2,4-cyclopentadien-1-yl)(2,2-dimethylpropyl)-thorium, Th-00015

$C_{20}H_{26}U$
Tris(η^5-2,4-cyclopentadien-1-yl)(2,2-dimethylpropyl)-uranium, U -00036

$C_{20}H_{28}MoON_2O_4$
Dicarbonylbis(2-methyl-2-propanolato)bis(pyridine)-molybdenum, Mo-00116

$C_{20}H_{30}AlCl_4Yb$
Di-μ-chloro(dichloroaluminum)bis[(1,2,3,4,5-η)-1,2,3,4,5-pentamethyl-2,4-cyclopentadienyl]-ytterbium, Yb-00020

$C_{20}H_{30}ClU$
Chlorobis(pentamethylcyclopentadienyl)uranium, U -00037

$C_{20}H_{30}ClYb$
Chlorobis[(1,2,3,4,5-η)-1,2,3,4,5-pentamethyl-2,4-cyclopentadien-1-yl]ytterbium, Yb-00021

$C_{20}H_{30}Cl_2Nd^{\ominus}$
Dichlorobis[(1,2,3,4,5-η)-pentamethyl-2,4-cyclopentadien-1-yl]neodymate(1−), Nd-00008

$C_{20}H_{30}Cl_2Th$
Dichlorobis[(1,2,3,4,5-η)-1,2,3,4,5-pentamethyl-2,4-cyclopentadien-1-yl]thorium, Th-00016

$C_{20}H_{30}Cl_2Ti$
Dichlorobis[(1,2,3,4,5-η)-1,2,3,4,5-pentamethyl-2,4-cyclopentadien-1-yl]titanium, Ti-00111

$C_{20}H_{30}Cl_2U$
Dichlorobis[(1,2,3,4,5-η)-1,2,3,4,5-pentamethyl-2,4-cyclopentadien-1-yl]uranium, U -00038

$C_{20}H_{30}Cl_2Yb^{\ominus}$
Dichlorobis[(1,2,3,4,5-η)-1,2,3,4,5-pentamethyl-2,4-cyclopentadien-1-yl]ytterbate(1−), Yb-00022

$C_{20}H_{30}Eu$
Bis(pentamethylcyclopentadienyl)europium, Eu-00005

$C_{20}H_{30}F_6MnP$
Decamethylmanganocenium; Hexafluorophosphate, *in* Mn-00120

$C_{20}H_{30}I_2Yb^{\ominus}$
Diiodobis[(1,2,3,4,5-η)-1,2,3,4,5-pentamethyl-2,4-cyclopentadien-1-yl]ytterbate(1−), Yb-00023

$C_{20}H_{30}Mn$
Decamethylmanganocene, Mn-00119
Decamethylmanganocenium, Mn-00120

$C_{20}H_{30}Mn_2O_8P_2$
Octacarbonylbis(triethylphosphine)dimanganese, Mn-00121

$C_{20}H_{30}Sm$
Bis[(1,2,3,4,5-η)-1,2,3,4,5-pentamethyl-2,4-cyclopentadien-1-yl]samarium, Sm-00012

$C_{20}H_{30}Ti$
Bis[(1,2,3,4,5-η)-1,2,3,4,5-pentamethyl-2,4-cyclopentadien-1-yl]titanium, Ti-00112

$C_{20}H_{30}V$
Bis[(1,2,3,4,5-η)-1,2,3,4,5-pentamethyl-2,4-cyclopentadien-1-yl]vanadium, V -00087

$C_{20}H_{30}Yb$
Bis[(1,2,3,4,5-η)-1,2,3,4,5-pentamethyl-2,4-cyclopentadien-1-yl]ytterbium, Yb-00024

$C_{20}H_{31}Lu$
Hydridobis[(1,2,3,4,5-η)-1,2,3,4,5-pentamethyl-2,4-cyclopentadien-1-yl]lutetium, Lu-00018

$C_{20}H_{31}OY$
tert-Butylbis(methylcyclopentadienyl)yttrium; THF complex, *in* Y -00008

$C_{20}H_{31}Sm$
Hydrobis(pentamethylcyclopentadienyl)samarium, Sm-00013

$C_{20}H_{32}Th$
Dihydrobis(pentamethylcyclopentadienyl)thorium, Th-00017

$C_{20}H_{32}U$
Dihydrobis(pentamethylcyclopentadienyl)uranium, U -00039

$C_{20}H_{32}Zr$
Bis(η^5-2,4-cyclopentadien-1-yl)bis(2,2-dimethylpropyl)-zirconium, Zr-00064
Dihydrobis[(1,2,3,4,5-η)-1,2,3,4,5-pentamethyl-2,4-cyclopentadien-1-yl]zirconium, Zr-00065

$C_{20}H_{33}Al_2ClZr$
μ-Chlorobis(η^5-2,4-cyclopentadien-1-yl)-bis(diethylaluminum)-μ-1-ethanyl-2-ylidenezirconium, Zr-00066

$C_{20}H_{34}Sn_2Zr$
Bis[(1,2,3,4,5,6-η)-methylbenzene]bis(trimethylstannyl)-zirconium, Zr-00067

$C_{20}H_{35}LuO_2Si$
(η^8-1,3,5,7-Cyclooctatetraene)(trimethylsilylmethyl)-bis(tetrahydrofuran)lutetium, *in* Lu-00007

$C_{20}H_{38}PTa$
(2,2-Dimethylpropylidene)-η^2-ethene(η^5-pentamethyl-2,4-cyclopentadien-1-yl)(trimethylphosphine)tantalum, Ta-00034

$C_{20}H_{43}Ta$
Tris(2,2-dimethylpropyl)(2,2-dimethylpropylidene)-tantalum, Ta-00035

$C_{20}H_{44}Cr$
Tetrakis(2,2-dimethylpropyl)chromium, Cr-00151

$C_{20}H_{44}Hf$
Tetrakis(2,2-dimethylpropyl)hafnium, Hf-00030

$C_{20}H_{44}Ti$
Tetrakis[2,2-dimethylpropyl)titanium, Ti-00113

$C_{20}H_{44}Zr$
Tetrakis(2,2-dimethylpropyl)zirconium, Zr-00068

$C_{20}H_{48}Mo_2N_4$
Bis[μ-[2-methyl-2-propanaminato(2−)]]bis[2-methyl-2-propanaminato(2−)]tetramethyldimolybdenum, Mo-00117

$C_{20}H_{48}N_4W_2$
Bis[μ-[2-methyl-2-propanaminato(2−)]]bis[2-methyl-2-propanaminato(2−)]tetramethylditungsten, W -00131

$C_{20}H_{49}ErO_2Si_3$
Bis(tetrahydrofuran)tris[(trimethylsilyl)methyl]erbium, *in* Er-00008

$C_{20}H_{49}O_2Si_3Tb$
Bis(tetrahydrofuran)tris[(trimethylsilyl)methyl]-terbium, *in* Tb-00001

$C_{20}H_{49}O_2Si_3Yb$
Bis(tetrahydrofuran)tris[(trimethylsilyl)methyl]-ytterbium, *in* Yb-00010

$C_{20}Mn_4O_{20}U$
Tetrakis(pentacarbonylmanganio)uranium, Mn-00122

$C_{21}H_{11}O_6Re$
Pentacarbonyl[(2,3-diphenyl-2-cyclopropen-1-yl)]-carbonylrhenium, Re-00078

$C_{21}H_{14}CrO_3$
Tricarbonyl(diphenylmethylenecyclopentadiene)chromium, Cr-00152

$C_{21}H_{14}O_3W$
Tricarbonyl(diphenylmethylidenecyclopentadiene)-tungsten, W -00132

$C_{21}H_{15}ClNO_4PW$
Tricarbonylchloronitrosyl(triphenylphosphine)tungsten, W -00133

$C_{21}H_{15}CrNO_6$
Pentacarbonyl[[(diphenylmethylene)amino]-ethoxymethylene]chromium, Cr-00153

$C_{21}H_{15}NbO_2$
Dicarbonyl(η^5-2,4-cyclopentadien-1-yl)-(diphenylacetylene)niobium, Nb-00041

$C_{21}H_{15}O_2V$
▷Dicarbonyl(η^5-2,4-cyclopentadien-1-yl)-(diphenylacetylene)vanadium, V -00088

$C_{21}H_{20}Ce$
Tris(η^5-2,4-cyclopentadien-1-yl)phenylcerium, Ce-00008

$C_{21}H_{24}CrO_3$
Tricarbonyl(η^6-octamethylnaphthalene)chromium, Cr-00154

$C_{21}H_{24}HfO_5$
η^5-2,4-Cyclopentadien-1-ylbis(2,4-pentanedionato-*O,O′*)-phenoxyhafnium, Hf-00031

$C_{21}H_{24}MoO_3$
Tricarbonyl(η^6-octamethylnaphthalene)molybdenum, Mo-00118

$C_{21}H_{24}O_3W$
Tricarbonyl(η^6-octamethylnaphthalene)tungsten, W -00134

$C_{21}H_{24}O_5Zr$
π-Cyclopentadienylbis(2,4-pentanedionato)-phenoxyzirconium, Zr-00069

$C_{21}H_{25}LuO$
Benzylbis(η^5-2,4-cyclopentadienyl)lutetium; THF complex, *in* Lu-00016
Bis(η^5-2,4-cyclopentadien-1-yl)(4-methylphenyl)-lutetium; THF complex, *in* Lu-00017

$C_{21}H_{27}U$
▷Tris(2,4-dimethylpentadienyl)uranium, U -00041

$C_{21}H_{28}O_2P_2W$
Carbonyl(η^5-2,4-cyclopentadien-1-yl)[1,2-ethanediylbis(dimethylphosphine)]-*P,P′*][(4-methylphenyl)oxoethenyl]tungsten, W -00135

$C_{21}H_{33}ClTh$
Chloro(methyl)bis[(1,2,3,4,5-η)-1,2,3,4,5-pentamethyl-2,4-cyclopentadien-1-yl]thorium, Th-00018

$C_{21}H_{33}ClU$
Chloro(methyl)bis[(1,2,3,4,5-η)-1,2,3,4,5-pentamethyl-2,4-cyclopentadien-1-yl]uranium, U -00042

$C_{21}H_{33}Lu$
Methylbis[(1,2,3,4,5-η)-1,2,3,4,5-pentamethyl-2,4-cyclopentadien-1-yl]lutetium, Lu-00019

$C_{21}H_{33}Nd$
Tris[(1,2,3,4,5-η)-2,4-dimethyl-2,4-pentadienyl]-neodymium, Nd-00009

$C_{21}H_{33}Y$
Methylbis[(1,2,3,4,5-η)-1,2,3,4,5-pentamethyl-2,4-cyclopentadien-1-yl]yttrium, Y -00012

$C_{21}H_{42}F_3MnO_{11}P_2S$
Dicarbonyl(trifluoromethanesulfonate-*O,O′*)-bis(triisopropylphosphite-*P*)manganese, Mn-00123

$C_{21}H_{57}ClHfSi_6$
Tris[bis(trimethylsilyl)methyl]chlorohafnium, Hf-00032

$C_{21}H_{57}ClSi_6Yb^{\ominus}$
Tris[bis(triethylsilyl)methyl]chloroytterbate(1−), Yb-00025

$C_{21}H_{57}ClSi_6Zr$
Tris[bis(trimethylsilyl)methyl]chlorozirconium, Zr-00070

$C_{21}H_{57}ScSi_6$
Tris[bis(trimethylsilyl)methyl]scandium, Sc-00008

$C_{21}H_{57}Si_6Ti$
Tris[bis(trimethylsilyl)methyl]titanium, Ti-00114
$C_{21}H_{57}Si_6V$
Tris[bis(trimethylsilyl)methyl]vanadium, V -00089
$C_{21}H_{57}Si_6Y$
Tris[bis(trimethylsilyl)methyl]yttrium, Y -00013
$C_{22}H_{10}F_{10}Hf$
Bis(η^5-2,4-cyclopentadien-1-yl)bis(pentafluorophenyl)-hafnium, Hf-00033
$C_{22}H_{10}F_{10}Ti$
Bis(η^5-2,4-cyclopentadien-1-yl)bis(pentafluorophenyl)-titanium, Ti-00115
$C_{22}H_{10}F_{10}Zr$
▷Bis(η^5-cyclopentadienyl)bis(pentafluorophenyl)-zirconium, Zr-00071
$C_{22}H_{10}KO_{10}PW$
Decacarbonyl(μ-diphenylphosphido)ditungstate(1−); K salt, *in* W -00136
$C_{22}H_{10}O_{10}PW^{\ominus}$
Decacarbonyl(μ-diphenylphosphido)ditungstate(1−), W -00136
$C_{22}H_{14}O_4PRe$
Tetracarbonyl[2-(diphenylphosphinophenyl-*C,P*)]rhenium, Re-00079
$C_{22}H_{15}KMnO_4P$
Tetracarbonyl(triphenylphosphine)manganate(1−); K salt, *in* Mn-00124
$C_{22}H_{15}MnNaO_4P$
Tetracarbonyl(triphenylphosphine)manganate(1−); Na salt, *in* Mn-00124
$C_{22}H_{15}MnO_4P^{\ominus}$
Tetracarbonyl(triphenylphosphine)manganate(1−), Mn-00124
$C_{22}H_{15}NO_5PV$
Tetracarbonylnitrosyl(triphenylphosphine)vanadium, V -00090
$C_{22}H_{15}Nb_3O_7$
Heptacarbonyltris(cyclopentadienyl)triniobium, Nb-00042
$C_{22}H_{16}MnO_4P$
Tetracarbonylhydro(triphenylphosphine)manganese, Mn-00125
$C_{22}H_{16}Mn_2O_4$
Tetracarbonylbis(η^5-2,4-cyclopentadien-1-yl)[μ-(phenylethenylidene)dimanganese, Mn-00126
$C_{22}H_{20}Hf$
Bis(η^5-2,4-cyclopentadien-1-yl)diphenylhafnium, Hf-00034
$C_{22}H_{20}HfSe_2$
Bis(benzeneselenolato)bis(η^5-2,4-cyclopentadien-1-yl)-hafnium, Hf-00035
$C_{22}H_{20}N_2V$
(Azobenzene)bis(cyclopentadienyl)vanadium, V -00091
$C_{22}H_{20}NbS_2$
Bis(benzenethiolato)bis(η^5-2,4-cyclopentadien-1-yl)-niobium, Nb-00043
$C_{22}H_{20}O_2Ti$
Bis(η^5-2,4-cyclopentadien-1-yl)bis(phenolato-*O*)-titanium, Ti-00116
$C_{22}H_{20}S_2Ti$
Bis(benzenethiolato)bis(η^5-2,4-cyclopentadien-1-yl)-titanium, Ti-00117
$C_{22}H_{20}S_2V$
Bis(benzenethiolato)bis(η^5-2,4-cyclopentadien-1-yl)-vanadium, V -00092
$C_{22}H_{20}S_2Zr$
Bis(benzenethiolato)bis(η^5-2,4-cyclopentadien-1-yl)-zirconium, Zr-00072
$C_{22}H_{20}Ti$
Bis(η^5-cyclopentadien-1-yl)diphenyltitanium, Ti-00118
$C_{22}H_{20}Zr$
Bis(η^5-2,4-cyclopentadien-1-yl)diphenylzirconium, Zr-00073
$C_{22}H_{21}CoO_4Sm$
[μ-(Carbonyl-*C,O*)]tris[(1,2,3,4,5-η)-1-methyl-2,4-cyclopentadien-1-yl](tetracarbonylcobalt)-samarium, Sm-00014
$C_{22}H_{21}Nb$
Bis(1,3,5,7-cyclooctatetraene)phenylniobium, Nb-00044
$C_{22}H_{22}U$
Benzyltricyclopentadienyluranium, U -00043
$C_{22}H_{24}U$
Dicyclopentenouranocene, U -00044
$C_{22}H_{26}BScSi_4$
Bis[(1,2,3,4,5-η)-1,3-bis(trimethylsilyl)-2,4-cyclopentadien-1-yl]tetrahydroboratoscandium, Sc-00009
$C_{22}H_{26}DyN$
Tris(2,4-cyclopentadien-1-yl)-isocyanocyclohexanedysprosium, *in* Dy-00004
$C_{22}H_{26}Dy_2$
Tetrakis(η^5-2,4-cyclopentadienyl)-di-μ-methyldidysprosium, *in* Dy-00003
$C_{22}H_{26}ErN$
Tris(2,4-cyclopentadien-1-yl)isocyanocyclohexaneerbium, *in* Er-00011
$C_{22}H_{26}Er_2$
Tetrakis(η^5-2,4-cyclopentadien-1-yl)di-μ-methyldierbium, *in* Er-00005
$C_{22}H_{26}HoN$
Tris(η^5-2,4-cyclopentadien-1-yl)-isocyanocyclohexaneholmium, *in* Ho-00005
$C_{22}H_{26}Ho_2$
Tetrakis(η^5-2,4-cyclopentadien-1-yl)di-μ-methyldiholmium, *in* Ho-00003
$C_{22}H_{26}LuN$
Tris(η^5-2,4-cyclopentadien-1-yl)-isocyanocyclohexanelutetium, *in* Lu-00012
$C_{22}H_{26}NTb$
Tris(η^5-2,4-cyclopentadien-1-yl)-isocyanocyclohexaneterbium, *in* Tb-00002
$C_{22}H_{26}NTm$
Tris(2,4-cyclopentadien-1-yl)-isocyanocyclohexanethulium, *in* Tm-00004
$C_{22}H_{26}NU$
Tris(η^5-2,4-cyclopentadien-1-yl)-isocyanocyclohexaneuranium, *in* U -00015
$C_{22}H_{26}NYb$
Tris(η^5-2,4-cyclopentadien-1-yl)-isocyanocyclohexaneytterbium, *in* Yb-00013
$C_{22}H_{26}Tm_2$
Tetrakis(η^5-2,4-cyclopentadien-1-yl)di-μ-methyldithulium, *in* Tm-00003
$C_{22}H_{26}Y_2$
Tetrakis(η^5-2,4-cyclopentadien-1-yl)di-μ-methyldiyttrium, *in* Y -00003
$C_{22}H_{26}Yb_2$
Tetrakis(η^5-2,4-cyclopentadien-1-yl)di-μ-methyldiytterbium, *in* Yb-00006
$C_{22}H_{30}CeKO_3$
Bis(η^8-1,3-5,7-cyclooctatetraene)cerate(1−); K salt, diglyme adduct, *in* Ce-00005
$C_{22}H_{30}F_6O_2PV$
Dicarbonylbis[(1,2,3,4,5-η)-1,2,3,4,5-pentamethyl-2,4-cyclopentadien-1-yl]vanadium(1+); Hexafluorophosphate, *in* V -00093

$C_{22}H_{30}HfO_2$
Dicarbonylbis[(1,2,3,4,5-η)-1,2,3,4,5-pentamethyl-2,4-cyclopentadien-1-yl]hafnium, Hf-00036
$C_{22}H_{30}O_2V^{\oplus}$
Dicarbonylbis[(1,2,3,4,5-η)-1,2,3,4,5-pentamethyl-2,4-cyclopentadien-1-yl]vanadium(1+), V -00093
$C_{22}H_{30}O_2Zr$
Dicarbonylbis[(1,2,3,4,5-η)-1,2,3,4,5-pentamethyl-2,4-cyclopentadien-1-yl)zirconium, Zr-00074
$C_{22}H_{32}B_4FeMn_2O_6S_2$
Hexacarbonylbis[μ-[(1,2,3,4,5-η:1,2,3,4,5-η)-3,4-diethyl-2,5-dihydro-2,5-dimethyl-1,2,5-thiadiborole]](iron)dimanganese, Mn-00127
$C_{22}H_{35}NZr$
Hydro[(methylimino)methyl]bis[(1,2,3,4,5-η)-1,2,3,4,5-pentamethyl-2,4-cyclopentadien-1-yl]zirconium, Zr-00075
$C_{22}H_{36}ClNTh$
Chloro(dimethylamino)bis(pentamethylcyclopentadienyl)-thorium, Th-00019
$C_{22}H_{36}INZr$
Dimethylamidoiodobis(η^5-pentamethylcyclopentadienyl)-zirconium, Zr-00076
$C_{22}H_{36}Th$
Dimethylbis[(1,2,3,4,5-η)-1,2,3,4,5-pentamethyl-2,4-cyclopentadien-1-yl]thorium, Th-00020
$C_{22}H_{36}Ti$
Dimethylbis[(1,2,3,4,5-η)-1,2,3,4,5-pentamethyl-2,4-cyclopentadien-1-yl]titanium, Ti-00119
$C_{22}H_{36}U$
Dimethylbis[(1,2,3,4,5-η)-1,2,3,4,5-pentamethyl-2,4-cyclopentadien-1-yl]uranium, U -00045
Tris(2-methyl-1-propenyl)[(1,2,3,4,5-η)-1,2,3,4,5-pentamethyl-2,4-cyclopentadien-1-yl]uranium, U -00046
$C_{22}H_{36}Zr$
Dimethylbis[(1,2,3,4,5-η)-1-(2-methylbutyl)-2,4-cyclopentadien-1-yl]zirconium, Zr-00077
$C_{22}H_{37}NZr$
Dimethylamidohydro(pentamethylcyclopentadienyl)-zirconium, Zr-00078
$C_{22}H_{39}FeN_3Ti$
Tris(diethylaminato)ferrocenyltitanium, Ti-00120
$C_{22}H_{40}ClP_4Ta$
Chlorobis[1,2-ethanediylbis[dimethylphosphine]-*P*,*P'*]-(η^4-naphthalene)tantalum, Ta-00036
$C_{22}H_{42}ClPrSi_4$
Bis[1,3-bis(trimethylsilyl)cyclopentadienyl)-chloropraesodymium, Pr-00006
$C_{22}H_{42}ClScSi_4$
Bis[1,3-bis(trimethylsilyl)cyclopentadienyl]-chloroscandium, Sc-00010
$C_{22}H_{42}Cl_2NdSi_4^{\ominus}$
Dichlorobis[(1,2,3,4,5-η)-1,3-bis(trimethylsilyl)-2,4-cyclopentadien-1-yl]neodymate(1−), Nd-00010
$C_{22}H_{62}Cr_2P_2Si_4$
Bis(trimethylphosphine)bis[μ-[(trimethylsilyl)methyl]]-bis[(trimethylsilyl)methyl]dichromium, Cr-00155
$C_{23}H_{15}BrO_4SiW$
Bromotetracarbonyl[(triphenylsilyl)methylidyne]-tungsten, W -00137
$C_{23}H_{15}F_6MnO_5P_2$
Pentacarbonyl(triphenylphosphine)manganese(1+); Hexafluorophosphate, *in* Mn-00129
$C_{23}H_{15}FeMnN_2O_7P$
Dinitrosyl(pentacarbonylmanganio)(triphenylphosphine)-iron, Mn-00128
$C_{23}H_{15}MnO_5P^{\oplus}$
Pentacarbonyl(triphenylphosphine)manganese(1+), Mn-00129
$C_{23}H_{15}MoO_5P$
Pentacarbonyl(triphenylphosphine)molybdenum, Mo-00119
$C_{23}H_{15}O_5PV^{\ominus}$
Pentacarbonyl(triphenylphosphine)vanadate(1−), V -00094
Pentacarbonyl(triphenylphosphine)vanadium, V -00095
$C_{23}H_{16}MoO_3$
Tricarbonyl[(1,2,3,4,5,6-η)-7-(diphenylmethylene)-1,3,5-cycloheptatriene]molybdenum, Mo-00120
$C_{23}H_{17}ClHfN_2O_2$
Chloro-π-cyclopentadienylbis(8-quinolinolato)hafnium, Hf-00037
$C_{23}H_{17}ClN_2O_2Zr$
Chloro(η^5-cyclopentadienyl)bis(8-quinolinolato)-zirconium, Zr-00079
$C_{23}H_{20}ClNTi$
Chlorodi-π-cyclopentadienyl(1,1-diphenylmethyleniminato)titanium, Ti-00121
$C_{23}H_{20}HfO$
(η^2-Benzoyl)bis(η^5-2,4-cyclopentadien-1-yl)-phenylhafnium, Hf-00038
$C_{23}H_{20}IMnNOP$
(η^5-2,4-Cyclopentadien-1-yl)-iodonitrosyl(triphenylphosphine)manganese, Mn-00130
$C_{23}H_{20}OZr$
(η^2-Benzoyl)bis(η^5-2,4-cyclopentadien-1-yl)-phenylzirconium, Zr-00080
$C_{23}H_{20}O_3Ti$
(η^5-2,4-Cyclopentadien-1-yl)triphenoxytitanium, Ti-00122
$C_{23}H_{20}U$
Tris(η^5-2,4-cyclopentadien-1-yl)(phenylethynyl)uranium, U -00047
$C_{23}H_{21}ClMo$
Chlorobis(1-phenyl-1-propyne)(η^5-cyclopentadienyl)-molybdenum, Mo-00121
$C_{23}H_{22}ClPZr$
Chlorobis(η^5-2,4-cyclopentadien-1-yl)-[(diphenylphosphino)methyl]zirconium, Zr-00081
$C_{23}H_{22}CrGeO_5$
[Bis(2,4,6-trimethylphenyl)germylene]-pentacarbonylchromium, Cr-00156
$C_{23}H_{24}U$
[Tris(η^5-2,4-cyclopentadien-1-yl)][(4-methylphenyl)-methyl]uranium, U -00048
$C_{23}H_{33}ClN_2U$
Chlorobis[(1,2,3,4,5-η)-1,2,3,4,5-pentamethyl-2,4-cyclopentadien-1-yl)(1*H*-pyrazolato-N^1,N^2)uranium, U -00049
$C_{23}H_{34}Cl_2N_2U$
Dichlorobis[(1,2,3,4,5-η)-1,2,3,4,5-pentamethyl-2,4-cyclopentadien-1-yl](1*H*-pyrazole-N^2)uranium, U -00050
$C_{23}H_{34}Si_2Zr$
[Bis(trimethylsilyl)methyl]bis(η^5-2,4-cyclopentadien-1-yl)phenylzirconium, Zr-00082
$C_{23}H_{39}NTh$
Methyl(*N*-methylmethanaminato)bis[(1,2,3,4,5-η)-1,2,3,4,5-pentamethyl-2,4-cyclopentadien-1-yl]-thorium, Th-00021
$C_{23}H_{45}ClSi_4Zr$
[Bis(trimethylsilyl)methyl]chlorobis[(1,2,3,4,5-η)-1-(trimethylsilyl)-2,4-cyclopentadien-1-yl]-zirconium, Zr-00083
$C_{23}H_{47}MnO_8P_2$
Dicarbonyl(η^3-2-propenyl)bis(triisopropylphosphite-*P*)-manganese, Mn-00131
$C_{24}H_{14}O_6W_2$
Hexacarbonylbis[(1,2,3,3*a*,7*a*-η)-1*H*-inden-1-yl]-ditungsten, W -00138

$C_{24}H_{15}NbO_6Sn$
Hexacarbonyl(triphenylstannyl)niobium, Nb-00045
$C_{24}H_{15}O_6SnV$
Hexacarbonyl(triphenylstannyl)vanadium, V -00096
$C_{24}H_{15}Sc$
Tris(phenylethynyl)scandium, Sc-00011
$C_{24}H_{17}MoO_4P$
Tetracarbonyl[[2-(η^2-ethenyl)phenyl]diphenylphosphine-*P*]molybdenum, Mo-00122
$C_{24}H_{17}PO_4W$
Tetracarbonyl[2-(η^2-ethenyl)phenyldiphenylphosphine-*P*]-tungsten, W -00139
$C_{24}H_{18}AlMn_3O_{18}$
Hexakis[μ-(acetyl-*C:O*)](aluminum)-dodecacarbonyltrimanganese, *in* Mn-00037
$C_{24}H_{18}CrIO_5P$
Pentacarbonyliodochromate(1−); Methyltriphenylphosphonium salt, *in* Cr-00007
$C_{24}H_{18}N_2O_8Ti$
Bis(η^5-2,4-cyclopentadien-1-yl)bis(4-nitrobenzoato-*O'*)-titanium, Ti-00123
$C_{24}H_{20}CrO_4Tl_2$
▷μ-(Chromato-*O,O'*)tetraphenyldithallium, Cr-00157
$C_{24}H_{20}Cr_2O_6Sn$
(Dibenzyldimethylstannane)bis[tricarbonylchromium], Cr-00158
$C_{24}H_{20}F_6MnNO_2P_2$
Carbonyl(η^5-2,4-cyclopentadien-1-yl)-nitrosyl(triphenylphosphine)manganese; Hexafluorophosphate, *in* Mn-00132
$C_{24}H_{20}MnNO_2P^{\oplus}$
Carbonyl(η^5-2,4-cyclopentadien-1-yl)-nitrosyl(triphenylphosphine)manganese, Mn-00132
$C_{24}H_{20}NO_2PW$
Carbonyl(η^5-2,4-cyclopentadien-1-yl)-nitrosyl(triphenylphosphine)tungsten, W -00140
$C_{24}H_{20}OV$
Bis(2,4-cyclopentadien-1-yl)(η^2-*C,O*-diphenylketene)-vanadium, V -00097
$C_{24}H_{20}O_4V_4$
Tetracarbonyltetrakis(cyclopentadienyl)tetravanadium, V -00098
$C_{24}H_{20}Ti$
Tetraphenyltitanium, Ti-00124
$C_{24}H_{22}BF_4NOPRe$
(η^5-2,4-Cyclopentadien-1-yl)-methylenenitrosyl(triphenylphosphine)rhenium(1+); Tetrafluoroborate, *in* Re-00080
$C_{24}H_{22}Cl_2Zr$
Bis(η^5-benzylcyclopentadienyl)dichlorozirconium, Zr-00084
$C_{24}H_{22}NOPRe^{\oplus}$
(η^5-2,4-Cyclopentadien-1-yl)-methylenenitrosyl(triphenylphosphine)rhenium(1+), Re-00080
$C_{24}H_{24}Ce_2$
Tris(1,3,5,7-cyclooctatetraene)dicerium, Ce-00009
$C_{24}H_{24}Cr_2$
[μ-[(1,2,3,4-η:5,6,7,8-η)-1,3,5,7-Cyclooctatetraene]-[bis(1,2,3,4-η)-1,3,5,7-cyclooctatetraene]-dichromium, Cr-00159
$C_{24}H_{24}Hf$
Bis(benzyl)bis(η^5-cyclopentadienyl)hafnium, Hf-00039
$C_{24}H_{24}Mo_2$
[μ-[(1,2,3,4-η:5,6,7,8-η)-1,3,5,7-cyclooctatetraene]]-bis[(1,2,3,4-η)-1,3,5,7-cyclooctatetraene]-dimolybdenum, Mo-00123
$C_{24}H_{24}PSc$
Bis(η^5-2,4-cyclopentadien-1-yl)-[(diphenylphosphinidenio)bis(methylene)]scandium, Sc-00012
$C_{24}H_{24}Ti$
Dibenzylbis(η^5-2,4-cyclopentadien-1-yl)titanium, Ti-00125
$C_{24}H_{24}Ti_2$
Tris(1,3,5,7-cyclooctatetraene)dititanium, Ti-00126
$C_{24}H_{24}V$
Dibenzylbis(cyclopentadienyl)vanadium, V -00099
$C_{24}H_{24}W$
Dibenzylbis(η^5-2,4-cyclopentadien-1-yl)tungsten, W -00141
$C_{24}H_{24}W_2$
[μ-[(1,2,3,4-η:5,6,7,8-η)-1,3,5,7-cyclooctatetraene]]-bis[(1,2,3,4-η)-1,3,5,7-cyclooctatetraene]-ditungsten, W -00142
$C_{24}H_{24}Zr$
Bis(benzyl)bis(η^5-cyclopentadienyl)zirconium, Zr-00085
$C_{24}H_{25}Cl_2Ta$
Dichloro(diphenylacetylene)[(1,2,3,4,5-η)-1,2,3,4,5-pentamethyl-2,4-cyclopentadien-1-yl]tantalum, Ta-00037
$C_{24}H_{27}PU$
Tris(η^5-2,4-cyclopentadien-1-yl)-[(dimethylphenylphosphoranylidene)methyl)]-uranium, U -00051
$C_{24}H_{28}Cl_2Er_2$
Di-μ-chlorotetrakis[(1,2,3,4,5-η)-1-methyl-2,4-cyclopentadien-1-yl]dierbium, *in* Er-00007
$C_{24}H_{28}Cl_2Yb_2$
Di-μ-chlorotetrakis[(1,2,3,4,5-η)-1-methyl-2,4-cyclopentadien-1-yl]diytterbium, *in* Yb-00008
$C_{24}H_{30}CoO_4Yb$
[μ-(Carbonyl-*C,O*)]bis[(1,2,3,4,5-η)-1,2,3,4,5-pentamethyl-2,4-cyclopentadien-1-yl]-(tricarbonylcobalt)ytterbium, Yb-00026
$C_{24}H_{30}Cr_2O_4$
Tetracarbonylbis[(1,2,3,4,5-η)-1,2,3,4,5-pentamethyl-2,4-cyclopentadien-1-yl]dichromium, Cr-00160
$C_{24}H_{30}Mo_2O_4$
Tetracarbonylbis[(1,2,3,4,5-η)-1,2,3,4,5-pentamethyl-2,4-cyclopentadien-1-yl]dimolybdenum, Mo-00124
$C_{24}H_{30}O_4W_2$
Tetracarbonylbis[(1,2,3,4,5-η)-1,2,3,4,5-pentamethyl-2,4-cyclopentadien-1-yl]ditungsten, W -00143
$C_{24}H_{32}ClN_2V$
Bis(η^5-2,4-cyclopentadien-1-yl)bis(isocyanocyclohexane)-vanadium(1+); Chloride, *in* V -00100
$C_{24}H_{32}KO_2Np$
Bis(η^8-1,3,5,7-cyclooctatetraene)neptunate(1−); K salt, bis-THF complex, *in* Np-00003
$C_{24}H_{32}KO_2Pu$
Bis(η^8-1,3,5,7-cyclooctatetraene)plutonate(1−); K salt, bis-THF complex, *in* Pu-00003
$C_{24}H_{32}N_2Ti_2$
Tetra-π-cyclopentadienylbis[μ-(dimethylaminato)]-dititanium, *in* Ti-00058
$C_{24}H_{32}N_2V^{\oplus}$
Bis(η^5-2,4-cyclopentadien-1-yl)bis(isocyanocyclohexane)-vanadium(1+), V -00100
$C_{24}H_{32}N_4U$
[(1,2,3,4,5-η)-2,5-Dimethyl-1*H*-pyrrol-1-yl]tris(2,5-dimethyl-1*H*-pyrrol-1-yl)uranium, U -00052
$C_{24}H_{32}U$
Bis[(1,2,3,4,5,6,7,8-η)-1,3,5,7-tetramethyl-1,3,5,7-cyclooctatetraene]uranium, U -00053
$C_{24}H_{33}AlCl_2OY_2$
μ-Chloro-μ_3-chlorotetrakis(η^5-2,4-cyclopentadien-1-yl)-trihydro[1,1'-oxybis[ethane]]aluminumdiyttrium, Y -00014

$C_{24}H_{34}HfSi_2$
Bis(η^5-2,4-cyclopentadien-1-yl)[1,2-phenylenebis[(trimethylsilyl)methylene]]hafnium, Hf-00040

$C_{24}H_{34}Si_2Zr$
Bis(η^5-2,4-cyclopentadien-1-yl)[1,2-phenylenebis[(trimethylsilyl)methylene]]-zirconium, Zr-00086

$C_{24}H_{36}Er^{\ominus}$
Tetrakis(3,3-dimethyl-1-butynyl)erbate(1−), Er-00016

$C_{24}H_{36}F_6PV$
Bis(acetonitrile)bis(η^5-pentamethyl-2,4-cyclopentadien-1-yl)vanadium(1+); Hexafluorophosphate, *in* V -00101

$C_{24}H_{36}Lu^{\ominus}$
Tetrakis(3,3-dimethyl-1-butynyl)lutetate(1−), Lu-00020

$C_{24}H_{36}N_2V^{\oplus}$
Bis(acetonitrile)bis(η^5-pentamethyl-2,4-cyclopentadien-1-yl)vanadium(1+), V -00101

$C_{24}H_{36}Sm^{\ominus}$
Tetrakis(3,3-dimethyl-1-butynyl)samarate(1−), Sm-00015

$C_{24}H_{38}ClOYb$
Chlorobis[(1,2,3,4,5-η)-1,2,3,4,5-pentamethyl-2,4-cyclopentadien-1-yl]ytterbium; THF complex, *in* Yb-00021

$C_{24}H_{38}EuO$
Bis(pentamethylcyclopentadienyl)europium; THF complex, *in* Eu-00005

$C_{24}H_{38}OYb$
Bis[(1,2,3,4,5-η)-1,2,3,4,5-pentamethyl-2,4-cyclopentadien-1-yl]ytterbium; THF complex, *in* Yb-00024

$C_{24}H_{40}ClNTh$
Chloro(diethylamino)bis(pentamethylcyclopentadienyl)-thorium, Th-00022

$C_{24}H_{40}ClNU$
Chloro(diethylamino)bis(pentamethylcyclopentadienyl)-uranium, U -00054

$C_{24}H_{40}Zr$
Hydro(2-methylpropyl)bis[(1,2,3,4,5-η)-1,2,3,4,5-pentamethyl-2,4-cyclopentadien-1-yl)zirconium, Zr-00087

$C_{24}H_{41}MoNO_3$
Tricarbonyl(η^5-2,4-cyclopentadien-1-yl)molybdate(1−); Tetrabutylammonium salt, *in* Mo-00014

$C_{24}H_{41}O_2NSW$
(Carbonothioyl)dicarbonyl(η^5-2,4-cyclopentadien-1-yl)-tungstate(1−); Tetrabutylammonium salt, *in* W -00022

$C_{24}H_{42}Cr_2N_2O_8$
Octacarbonyl-μ-dihydrodichromate(2−); Bis(tetraethylammonium) salt, *in* Cr-00037

$C_{24}H_{42}Mo_2N_2O_8$
Octacarbonyl-μ-dihydrodimolybdate(2−); Tetraethylammonium salt, *in* Mo-00009

$C_{24}H_{42}N_2O_8W_2$
Octacarbonyldi-μ-hydroditungstate(2−); Bis(tetraethylammonium) salt, *in* W -00017

$C_{24}H_{42}N_2Th$
Bis(dimethylamino)bis(pentamethylcyclopentadienyl)-thorium, Th-00023

$C_{24}H_{42}N_2U$
Bis(dimethylamino)bis(pentamethylcyclopentadienyl)-uranium, U -00055

$C_{24}H_{42}O_2Si_2Yb$
Bis[(1,2,3,4,5-η)-1-(trimethylsilyl)-2,4-cyclopentadien-1-yl]ytterbium; Bis-THF complex, *in* Yb-00016

$C_{24}H_{44}Ti$
▷Tetracyclohexyltitanium, Ti-00127

$C_{24}H_{48}O_4U_2$
Bis[μ-(2-propanolato)]bis(2-propanolato)tetrakis(η^3-2-propenyl)diuranium, U -00056

$C_{24}H_{54}Mo_2O_6$
Hexakis(2-methyl-2-propanolato)dimolybdenum, Mo-00125

$C_{24}H_{56}Li_4Mo_2O_4$
Octamethyldimolybdenum(4−); Tetra-Li salt, tetrakis-THF complex, *in* Mo-00020

$C_{24}H_{57}ErO_3Si_3$
Tris(tetrahydrofuran)tris[(trimethylsilyl)methyl]-erbium, *in* Er-00008

$C_{24}H_{62}Nb_2Si_6$
Tetrakis[(trimethylsilyl)methyl]bis[μ-(trimethylsilyl)-methylidyne]diniobium, Nb-00046

$C_{24}H_{62}Si_6W_2$
Bis[μ-[(trimethylsilyl)methylidyne]-tetrakis[(trimethylsilyl)methyl]ditungsten, W -00144

$C_{24}H_{66}ErLi_3N_6$
Hexamethylerbate(3−); Tris[(tetramethylethylenediamine)lithium] salt, *in* Er-00002

$C_{24}H_{66}Li_3LuN_6$
Hexamethyllutetate(3−); Tris[(tetramethylethylenediamine)-lithium] salt, *in* Lu-00002

$C_{24}H_{66}Li_3N_6Nd$
Hexamethylneodymate(3−); Tris[(tetramethylethylenediamine)-lithium] salt, *in* Nd-00001

$C_{24}H_{66}Li_3N_6Pr$
Hexamethylpraesodymate(3−); Tris[(tetramethylethylenediamine)lithium] salt, *in* Pr-00001

$C_{24}H_{66}Li_3N_6Sm$
Hexamethylsamarate(3−); Tris[(tetramethylethylenediamine)-lithium] salt, *in* Sm-00002

$C_{24}H_{66}Li_3N_6Tm$
Hexamethylthulate(3−); Tris[(tetramethylethylenediamine)lithium] salt, *in* Tm-00001

$C_{24}H_{66}Li_3N_6Yb$
Hexamethylytterbate(3−); Tris[(tetramethylethylenediamine)-lithium] salt, *in* Yb-00002

$C_{24}H_{66}Mo_2Si_6$
Hexakis[(trimethylsilyl)methyl]dimolybdenum, Mo-00126

$C_{24}H_{66}Si_6W_2$
Hexakis[(trimethylsilyl)methyl]ditungsten, W -00145

$C_{25}H_{20}ClMoO_2P$
Dicarbonylchloro(η^5-2,4-cyclopentadien-1-yl)-(triphenylphosphine)molybdenum, Mo-00127

$C_{25}H_{20}O_4V$
Tetracarbonyl(η^3-propenyl)(triphenylphosphine)vanadium, V -00102

$C_{25}H_{21}MoO_5P$
Tetracarbonyl(1-methoxyethylidene)(triphenylphosphine)-molybdenum, Mo-00128

$C_{25}H_{23}MnNO_3P$
η^5-2,4-Cyclopentadien-1-yl(methoxycarbonyl)-nitrosyl(triphenylphosphine)manganese, Mn-00133

$C_{25}H_{24}FeU$
Tris(η^5-2,4-cyclopentadien-1-yl)ferrocenyluranium, U -00057

$C_{25}H_{26}FeO_2Sm$
[μ-(Carbonyl-*C,O*)]tris[(1,2,3,4,5-η)-1-methyl-2,4-cyclopentadien-1-yl][dicarbonyl(η^5-2,4-cyclopentadien-1-yl)iron]samarium, Sm-00016

$C_{25}H_{29}N_2Pu$
Tris(η^5-2,4-cyclopentadien-1-yl)(nicotine)plutonium, *in* Pu-00002

$C_{25}H_{29}N_2U$
Tris(η^5-2,4-cyclopentadienyl)(nicotine)uranium, *in* U -00015

$C_{25}H_{33}Cl_3OP_3Tc$
Carbonyltrichlorotris(dimethylphenylphosphine)-technetium, Tc-00010

$C_{25}H_{33}OTi$
Bis(η^5-cyclopentadienyl)(2,6-di-*tert*-butyl-4-methylphenoxy)titanium, Ti-00128

$C_{25}H_{34}MoO_6P_2$
(η^5-2,4-Cyclopentadien-1-yl)[(1,2-η)-1,2-diphenylethenyl]bis(trimethylphosphite-*P*)-molybdenum, Mo-00129

$C_{25}H_{35}LuN$
[Bis(1,2,3,4,5-η)-1,2,3,4,5-pentamethyl-2,4-cyclopentadien-1-yl]-2-pyridinyllutetium, Lu-00021

$C_{25}H_{38}Al_2Zr$
[2,2-Bis(diethylalumino)ethyl]tris(η^5-2,4-cyclopentadien-1-yl)zirconium, Zr-00088

$C_{25}H_{40}ClNOTh$
Chloro[(diethylamino)carbonyl-*C*,*O*]bis[(1,2,3,4,5-η)-1,2,3,4,5-pentamethyl-2,4-cyclopentadien-1-yl]-thorium, Th-00024

$C_{25}H_{40}NS_2Yb$
Bis[(1,2,3,4,5-η)-1,2,3,4,5-pentamethyl-2,4-cyclopentadien-1-yl]diethylcarbamodithioato-*S*,*S'*-ytterbium, Yb-00027

$C_{25}H_{41}ClOSiTh$
Chlorobis[(1,2,3,4,5-η)-1,2,3,4,5-pentamethyl-2,4-cyclopentadien-1-yl][1-(trimethylsilyl)-ethenolato]thorium, Th-00025

$C_{25}H_{41}ClTh$
Chloro(2,2-dimethylpropyl)bis[(1,2,3,4,5-η)-1,2,3,4,5-pentamethyl-2,4-cyclopentadien-1-yl]thorium, Th-00026

$C_{25}H_{43}LuO$
Methylbis[(1,2,3,4,5-η)-1,2,3,4,5-pentamethyl-2,4-cyclopentadien-1-yl]lutetium; Et_2O complex, *in* Lu-00019

$C_{25}H_{44}ClP_2Yb$
[Bis(1,2,3,4,5-η)-1,2,3,4,5-pentamethyl-2,4-cyclopentadien-1-yl][bis(dimethylphosphino)-methane]chloroytterbium, Yb-00028

$C_{25}H_{45}ClSi_2Zr$
Bis(η^5-*tert*-butylcyclopentadienyl)[bis(trimethylsilyl)-methyl]chlorozirconium, Zr-00089

$C_{25}H_{48}Zr$
Tris(2,2-dimethylpropyl)[(1,2,3,4,5-η)-1,2,3,4,5-pentamethyl-2,4-cyclopentadien-1-yl]zirconium, Zr-00090

$C_{25}H_{49}P_2Ta$
Bis(2,2-dimethylpropylidene)(2,4,6-trimethylphenyl)-bis(trimethylphosphine)tantalum, Ta-00038

$C_{25}H_{54}Mo_2O_7$
Hexa-*tert*-butoxy-μ-carbonyldimolybdenum, Mo-00130

$C_{26}H_{15}O_5V$
Pentacarbonyl[(1,2,3-η)-1,2,3-triphenyl-2-cyclopropen-1-yl]vanadium, V -00103

$C_{26}H_{17}CrO_3P$
Tricarbonyl[(1,2,3,4,5,6-η)-2,4,6-triphenylphosphorin]-chromium, Cr-00161

$C_{26}H_{20}BF_4O_2ReSi$
Dicarbonyl(η^5-2,4-cyclopentadien-1-yl)[(triphenylsilyl)-methylidyne]rhenium(1+); Tetrafluoroborate, *in* Re-00081

$C_{26}H_{20}CrO_3Pb$
Tricarbonyl(η^5-2,4-cyclopentadien-1-yl)-(triphenylplumbyl)chromium, Cr-00162

$C_{26}H_{20}MoO_6Si$
Pentacarbonyl[ethoxy(triphenylsilyl)methylene]-molybdenum, Mo-00131

$C_{26}H_{20}NbO_3P$
Tricarbonyl(η^5-2,4-cyclopentadien-1-yl)-(triphenylphosphine)niobium, Nb-00047

$C_{26}H_{20}O_2ReSi^{\oplus}$
Dicarbonyl(η^5-2,4-cyclopentadien-1-yl)[(triphenylsilyl)-methylidyne]rhenium(1+), Re-00081

$C_{26}H_{20}O_3PV$
Tricarbonyl(η^5-2,4-cyclopentadien-1-yl)-(triphenylphosphine)vanadium, V -00104

$C_{26}H_{20}O_6SiW$
Pentacarbonyl[ethoxy(triphenylsilyl)methylene]tungsten, W -00146

$C_{26}H_{20}O_6Zr$
(η^5-2,4-Cyclopentadien-1-yl)tris(2-hydroxy-2,4,6-cycloheptatrien-1-onato-*O*,*O'*)zirconium, Zr-00091

$C_{26}H_{20}Ti$
Bis(η^5-2,4-cyclopentadien-1-yl)bis(phenylethynyl)-titanium, Ti-00129

$C_{26}H_{20}Zr$
Bis(η^5-2,4-cyclopentadien-1-yl)bis(phenylethynyl)-zirconium, Zr-00092

$C_{26}H_{21}CrNO_5Si$
Pentacarbonyl[(dimethylamino)(triphenylsilyl)methylene]-chromium, Cr-00163

$C_{26}H_{21}CrOPS$
(η^6-Benzene)(carbonothioyl)carbonyl(triphenylphosphine)-chromium, Cr-00164

$C_{26}H_{24}Zr$
Bis(η-cyclopentadienyl)bis(2-phenylethenyl)zirconium, Zr-00093

$C_{26}H_{26}Zr$
Tribenzyl(η-cyclopentadienyl)zirconium, Zr-00094

$C_{26}H_{27}MnNO_4P$
Tetracarbonyl(triphenylphosphine)manganate(1−); Tetramethylammonium salt, *in* Mn-00124

$C_{26}H_{30}Cr_2O_6$
Hexacarbonylbis[(1,2,3,4,5-η)-1,2,3,4,5-pentamethyl-2,4-cyclopentadien-1-yl]dichromium, Cr-00165

$C_{26}H_{31}NTi$
Tribenzyl(piperidinyl)titanium, Ti-00130

$C_{26}H_{32}U$
Di-*tert*-butyldiindenyluranium, U -00058

$C_{26}H_{33}ClO_2P_3Re$
Dicarbonylchlorotris(dimethylphenylphosphine)rhenium, Re-00082

$C_{26}H_{36}N_4U$
Bis[(1,2,3,4,5-η)-1,2,3,4,5-pentamethyl-2,4-cyclopentadien-1-yl]bis(1*H*-pyrazolato-N^1,N^2)-uranium, U -00059

$C_{26}H_{38}Zr$
[1,3-Bis(methylene)-1,4-butanediyl]bis(1,2,3,4,5-η)-1,2,3,4,5-pentamethyl-2,4-cyclopentadien-1-yl]-zirconium, Zr-00095

$C_{26}H_{40}Cr_2N_2O_{10}$
Decacarbonyldichromate(2−); Bis(tetraethylammonium) salt, *in* Cr-00066

$C_{26}H_{41}ClOTh$
Chloro(3,3-dimethyl-1-oxobutyl-*C*,*O*)bis[(1,2,3,4,5-η)-1,2,3,4,5-pentamethyl-2,4-cyclopentadien-1-yl]-thorium, Th-00027

$C_{26}H_{42}N_2O_2U$
Bis[(dimethylamino)carbonyl-*C*,*O*]bis[(1,2,3,4,5-η)-1,2,3,4,5-pentamethyl-2,4-cyclopentadien-1-yl]-uranium, U -00060

$C_{26}H_{46}Cl_2LiN_2Yb$
Dichlorobis[(1,2,3,4,5-η)-1,2,3,4,5-pentamethyl-2,4-cyclopentadien-1-yl]ytterbate(1−); Li salt, tetramethylethylenediamine complex, *in* Yb-00022

$C_{26}H_{47}P_2U$
[1,2-Ethanediylbis(dimethylphosphine)-*P*,*P'*]-hydrobis[(1,2,3,4,5-η)-1,2,3,4,5-pentamethyl-2,4-cyclopentadien-1-yl]uranium, U -00061

$C_{26}H_{50}ClSi_4Yb$
Bis[1,3-bis(trimethylsilylmethyl)cyclopentadienyl]-chloroytterbium, Yb-00029

$C_{26}H_{50}Cl_2Nd^{\ominus}$
Bis[(1,2,3,4,5-η)-1,3-bis(trimethylsilylmethyl)-2,4-cyclopentadien-1-yl]dichloroneodymium(1−), Nd-00011

$C_{27}H_{15}Mn_2O_9P$
Nonacarbonyl(triphenylphosphine)dimanganese, Mn-00134

$C_{27}H_{21}ClTh$
Chlorotris[(1,2,3,3*a*,7*a*-η)-1*H*-inden-1-yl)]thorium, Th-00028

$C_{27}H_{21}ClU$
Chlorotris[(1,2,3,3*a*,7*a*-η)-1*H*-inden-1-yl]uranium, U -00062

$C_{27}H_{21}Dy$
Triindenyldysprosium, Dy-00006

$C_{27}H_{21}Gd$
Triindenylgadolinium, Gd-00008

$C_{27}H_{21}La$
Triindenyllanthanum, La-00006

$C_{27}H_{21}Sm$
Triindenylsamarium, Sm-00017

$C_{27}H_{21}Tb$
Triindenylterbium, Tb-00005

$C_{27}H_{21}Yb$
Triindenylytterbium, Yb-00030

$C_{27}H_{22}Br_3OPU$
Tribromo[(1,2,3,3*a*,7*a*-η-1*H*-inden-1-yl]-(triphenylphosphine oxide-*O*)uranium, U -00063

$C_{27}H_{26}MoO_2PTl$
Dicarbonyl(η[5]-2,4-cyclopentadien-1-yl)-(dimethylthallium)(triphenylphosphine)molybdenum, Mo-00132

$C_{27}H_{26}O_2PTlW$
▷Dicarbonyl(η[5]-2,4-cyclopentadien-1-yl)-(dimethylthallium)(triphenylphosphine)tungsten, W -00147

$C_{27}H_{29}BMnN_3O_3$
Triamminetricarbonylmanganese(1+); Tetraphenylborate, *in* Mn-00002

$C_{27}H_{33}V$
Tris(2,4,6-trimethylphenyl)vanadium, V -00105

$C_{27}H_{35}ClOU$
(Benzoyl-*C*,*O*)chlorobis[(1,2,3,4,5-η)-1,2,3,4,5-pentamethyl-2,4-cyclopentadien-1-yl]uranium, U -00064

$C_{27}H_{36}CrN_3$
▷Tris[[2-(dimethylamino)phenyl]methyl-*C*,*N*]chromium, Cr-00166

$C_{27}H_{36}N_3Sc$
Tris[2-[(dimethylamino)methyl]phenyl-*C*,*N*]scandium, Sc-00013
[Tris(2-(dimethylamino)phenyl]methyl-*C*,*N*)scandium, Sc-00014

$C_{27}H_{57}MnNO_3P$
Bis(η[4]-1,3-butadiene)(trimethyl phosphite)manganate(1−) ; Tetrabutylammonium salt, *in* Mn-00069

$C_{28}H_{20}Mo_2O_4$
μ-Carbonyltricarbonylbis(η[5]-2,4-cyclopentadien-1-yl)[μ-[1,1'-(η[2]:η[2]-1,2-ethynediylbisbenzene]]-dimolybdenum, Mo-00133

$C_{28}H_{22}MnP$
(η[5]-2,4-Cyclopentadien-1-yl)[(1,2,3,4,5,6-η)-2,4,6-triphenylphosphorin]manganese, Mn-00135

$C_{28}H_{24}Th$
Tris[(1,2,3,3*a*,7*a*-η)-1*H*-indenyl]methylthorium, Th-00029

$C_{28}H_{24}U$
Bis[(1,2,3,4,5,6,7,8-η)-1-phenyl-1,3,5,7-cyclooctatetraene]uranium, U -00065
Tris[(1,2,3,3*a*,7*a*-η)-1*H*-inden-1-yl]methyluranium, U -00066

$C_{28}H_{24}Zr$
Bis[(4*a*,4*b*,8*a*,9*a*-η)-9*H*-fluoren-9-yl]dimethylzirconium, Zr-00096

$C_{28}H_{25}ClGeHf$
Chlorobis(η[5]-2,4-cyclopentadien-1-yl)(triphenylgermyl)-hafnium, Hf-00041

$C_{28}H_{25}ClGeTi$
Chlorobis(η[5]-cyclopentadienyl)(triphenylgermyl)-titanium, Ti-00131

$C_{28}H_{25}ClGeZr$
Chlorobis(η[5]-2,4-cyclopentadien-1-yl)(triphenylgermyl)-zirconium, Zr-00097

$C_{28}H_{25}ClHfSi$
Chlorobis(η[5]-2,4-cyclopentadien-1-yl)(triphenylsilyl)-hafnium, Hf-00042

$C_{28}H_{25}ClHfSn$
Chlorobis(η[5]-2,4-cyclopentadien-1-yl)(triphenylstannyl)-hafnium, Hf-00043

$C_{28}H_{25}ClSiZr$
Chlorobis(η[5]-2,4-cyclopentadien-1-yl)(triphenylsilyl)-zirconium, Zr-00098

$C_{28}H_{25}ClSnTi$
Chlorobis(η[5]-cyclopentadienyl)(triphenylstannyl)-titanium, Ti-00132

$C_{28}H_{25}ClSnZr$
Chlorobis(η[5]-2,4-cyclopentadien-1-yl)(triphenylstannyl)-zirconium, Zr-00099

$C_{28}H_{25}GeTi$
Bis(η[5]-cyclopentadienyl)triphenylgermyltitanium, Ti-00133

$C_{28}H_{25}P_3Ti$
Bis(η[5]-2,4-cyclopentadien-1-yl)[1,2,3-triphenyltriphosphinato(2−)-*P*[1],*P*[3]]titanium, Ti-00134

$C_{28}H_{25}P_3Zr$
Bis(η[5]-2,4-cyclopentadien-1-yl)[1,2,3-triphenyltriphosphinato(2−)-*P*[1],*P*[3]]zirconium, Zr-00100

$C_{28}H_{25}SnTi$
Bis(η[5]-cyclopentadienyl)triphenylstannyltitanium, Ti-00135

$C_{28}H_{26}GeW$
Bis(η[5]-2,4-cyclopentadien-1-yl)hydro(triphenylgermyl)-tungsten, W -00148

$C_{28}H_{28}ClTa$
Tetrabenzylchlorotantalum, Ta-00039

$C_{28}H_{28}Hf$
Tetrabenzylhafnium, Hf-00044

$C_{28}H_{28}Th$
Tetrabenzylthorium, Th-00030

$C_{28}H_{28}Ti$
Tetrabenzyltitanium, Ti-00136

$C_{28}H_{28}V$
Tetrabenzylvanadium, V -00106

$C_{28}H_{28}W$
Tetrabenzyltungsten, W -00149

$C_{28}H_{28}Zr$
Tetrabenzylzirconium, Zr-00101

$C_{28}H_{30}O_3Ti$
[(1,2,3,4,5-η)-1,2,3,4,5-Pentamethyl-2,4-cyclopentadien-1-yl]triphenoxytitanium, Ti-00137

$C_{28}H_{32}Co_3ErO_{16}$
Tris(tetracarbonylcobalto)erbium; Tetrakis-THF complex, *in* Er-00006

$C_{28}H_{32}F_{10}O_4Yb$
Bis(pentafluorophenyl)ytterbium; Tetrakis-THF complex, *in* Yb-00007

$C_{30}H_{24}O_4P_2V^{\ominus}$
Tetracarbonyl[1,2-ethanediylbis[diphenylphosphine]-*P*,*P*′]vanadate(1–), V -00108
Tetracarbonyl[1,2-ethanediylbis[diphenylphosphine]-*P*,*P*′]vanadium, V -00109

$C_{30}H_{24}Ti$
Bis[(1,2,3,3*a*,7*a*-η)-1*H*-inden-1-yl]diphenyltitanium, Ti-00141

$C_{30}H_{25}O_4P_2V$
Tetracarbonyl[1,2-ethanediylbis[diphenylphosphine]-*P*,*P*′]hydrovanadium, V -00110

$C_{30}H_{28}Fe_2Ti$
Bis(η⁵-2,4-cyclopentadien-1-yl)diferrocenyltitanium, Ti-00142

$C_{30}H_{28}Th_2$
Tetrakis(η⁵-2,4-cyclopentadien-1-yl)bis[μ-(1-η:1,2,3,4,5-η)-2,4-cyclopentadien-1-ylidene]-dithorium, Th-00033

$C_{30}H_{28}Zr_2$
Tetrakis(η⁵-2,4-cyclopentadien-1-yl)-μ-hydro[μ-[(1,2-η:2-η)-2-naphthalenyl]]dizirconium, Zr-00103

$C_{30}H_{29}OPU$
Tris(η⁵-2,4-cyclopentadien-1-yl)-[(methyldiphenylphosphoranylidene)acetyl]uranium, U -00070

$C_{30}H_{33}ClEr_3^{\ominus}$
μ-Chlorohexakis(η⁵-2,4-cyclopentadien-1-yl)di-μ-hydro-μ₃-hydrotrierbate(1–), Er-00017

$C_{30}H_{36}Cl_5LiO_2U_2$
Di-μ-chlorodi-μ₃-chloro(μ-chlorobis[methylenebis(η⁵-2,4-cyclopentadien-1-ylidene))]-diuraniumbis(tetrahydrofuran)lithium, U -00071

$C_{30}H_{36}HfS_2$
Bis(η-*tert*-butylcyclopentadienyl)bis(phenylthiolato)-hafnium, Hf-00046

$C_{30}H_{38}Hf_2O_9$
Bis(η⁵-2,4-cyclopentadien-1-yl)-μ-oxotetrakis(2,4-pentanedionato-*O*,*O*′)dihafnium, Hf-00047

$C_{30}H_{39}CrO_3$
Triphenyltris(tetrahydrofuran)chromium, Cr-00171

$C_{30}H_{40}N_2Yb$
Bis[(1,2,3,4,5-η)-1,2,3,4,5-pentamethyl-2,4-cyclopentadienyl]bis(pyridine)ytterbium, Yb-00031

$C_{30}H_{44}Zr_2$
Tetrakis(η⁵-2,4-cyclopentadien-1-yl)bis(2,2-dimethylpropyl)di-μ-hydrodizirconium, *in* Zr-00041

$C_{30}H_{50}ClOPTh$
Chloro[1-hydroxy-4,4-dimethyl-1-(trimethylphosphoranylidene)-2-pentanonato-*O*,*O*′]-bis[(1,2,3,4,5-η)-1,2,3,4,5-pentamethyl-2,4-cyclopentadien-1-yl]thorium, Th-00034

$C_{30}H_{58}Cl_2LiNdO_2Si_4$
Dichlorobis[(1,2,3,4,5-η)-1,3-bis(trimethylsilyl)-2,4-cyclopentadien-1-yl]neodymate(1–); Li salt, Bis-THF complex, *in* Nd-00010

$C_{30}H_{66}Mo_2$
Hexakis(2,2-dimethylpropyl)dimolybdenum, Mo-00136

$C_{31}H_{24}O_5P_2V^{\ominus}$
Pentacarbonyl[η¹-1,2-ethanediylbis[diphenylphosphine]-*P*]vanadate(1–), V -00111

$C_{31}H_{29}ClMoO_2P_2$
Dicarbonylchloro[1,2-ethanediylbis[diphenylphosphine]-*P*,*P*′](η³-2-propenyl)molybdenum, Mo-00137

$C_{31}H_{29}DyO$
Triindenyldysprosium; THF adduct, *in* Dy-00006

$C_{31}H_{29}GdO$
Triindenylgadolinium; THF complex, *in* Gd-00008

$C_{31}H_{29}LaO$
Triindenyllanthanum; THF adduct, *in* La-00006

$C_{31}H_{29}OSm$
Triindenylsamarium; THF adduct, *in* Sm-00017

$C_{31}H_{29}OTb$
Triindenylterbium; THF adduct, *in* Tb-00005

$C_{31}H_{29}OYb$
Triindenylytterbium; THF complex, *in* Yb-00030

$C_{31}H_{30}Br_3O_2PU$
Tribromo[(1,2,3,3*a*,7*a*-η-1*H*-inden-1-yl]-(triphenylphosphine oxide-*O*)uranium; THF adduct, *in* U -00063

$C_{31}H_{30}Th$
Butyltris[(1,2,3,3*a*,7*a*-η)-1*H*-inden-1-yl]thorium, Th-00035

$C_{31}H_{33}O_7P_3Re_2$
Heptacarbonyltris(dimethylphenylphosphine)dirhenium, Re-00083

$C_{31}H_{35}NO_5PV$
Pentacarbonyl(triphenylphosphine)vanadate(1–); Tetraethylammonium salt, *in* V -00094

$C_{31}H_{36}LiO_4PSnW$
Tetracarbonyl(triphenylstannane)-(triisopropylphosphine)tungstate(1–); Li salt, *in* W -00152

$C_{31}H_{36}O_4PSnW^{\ominus}$
Tetracarbonyl(triphenylstannane)-(triisopropylphosphine)tungstate(1–), W -00152

$C_{31}H_{36}Th$
Tribenzyl(pentamethylcyclopentadienyl)thorium, Th-00036

$C_{31}H_{36}U$
Tribenzyl(pentamethylcyclopentadienyl)uranium, U -00072

$C_{31}H_{41}OV$
Tris(2,4,6-trimethylphenyl)vanadium; THF complex, *in* V -00105

$C_{31}H_{42}OWZr$
[Bis(η⁵-2,4-cyclopentadien-1-yl)hydro[μ-[methanolato(3–)-*C*:*O*]]bis[(1,2,3,4,5-η)-1,2,3,4,5-pentamethyl-2,4-cyclopentadien-1-yl]-zirconium, W -00153

$C_{32}H_{23}HfN_3O_3$
π-Cyclopentadienyltris(8-quinolinolato)hafnium, Hf-00048

$C_{32}H_{23}N_3O_3Zr$
π-Cyclopentadienyltris(8-quinolinolato)zirconium, Zr-00105

$C_{32}H_{26}CrIN_4$
Bis(2,2′-bipyridine-*N*,*N*′)diphenylchromium(1+); Iodide, *in* Cr-00172

$C_{32}H_{26}CrN_4^{\oplus}$
Bis(2,2′-bipyridine-*N*,*N*′)diphenylchromium(1+), Cr-00172

$C_{32}H_{28}CrFe_2$
Bis[(η⁶-phenyl)ferrocene]chromium, Cr-00173
Bis[(η⁶-phenyl)ferrocene]chromium(1+), Cr-00174

$C_{32}H_{28}CrFe_2I$
Bis[(η⁶-phenyl)ferrocene]chromium(1+); Iodide, *in* Cr-00174

$C_{32}H_{30}OS_2Zr_2$
Bis(benzenethiolato)tetrakis(η⁵-2,4-cyclopentadien-1-yl)-μ-oxodizirconium, Zr-00106

$C_{32}H_{33}GeOTi$
Bis(η⁵-cyclopentadienyl)triphenylgermyltitanium; THF complex, *in* Ti-00133

$C_{32}H_{33}SnOTi$
Bis(η⁵-cyclopentadienyl)triphenylstannyltitanium; THF complex, *in* Ti-00135

$C_{32}H_{36}Cr_2O_8$
Tetrakis[μ-(2,6-dimethoxyphenyl-*C*:*O*)]dichromium, Cr-00175

$C_{32}H_{36}Lu^{\ominus}$
Tetrakis(2,6-dimethylphenyl)lutetate(1–), Lu-00023

$C_{32}H_{36}Mo_2O_8$
Tetrakis[μ-(2,6-dimethoxyphenyl)-*C*:*O*)]dimolybdenum, Mo-00138

$C_{32}H_{36}O_8V_2$
Tetrakis[μ-(2,6-dimethoxyphenyl-*C:O*)]divanadium, V -00112

$C_{32}H_{36}Yb^{\ominus}$
Tetrakis(2,6-dimethylphenyl)ytterbate(1−), Yb-00032

$C_{32}H_{38}Er_2$
Tetrakis(η[5]-2,4-cyclopentadien-1-yl)bis[μ-(3,3-dimethylbutynyl)]dierbium, *in* Er-00012

$C_{32}H_{38}Lu_2O_4$
Tetrakis(η[5]-2,4-cyclopentadien-1-yl)[μ-(2,2,7,7-tetramethyl-3,4,5,6-octanetetrone-*O*[3],*O*[5]:*O*[4],*O*[6])]-dilutetium, Lu-00024

$C_{32}H_{40}Er_2O_2$
Tris(1,3,5,7-cyclooctatetraene)bis(tetrahydrofuran)-dierbium, Er-00018

$C_{32}H_{40}La_2O_2$
Tris(1,3,5,7-cyclooctatetraene)bis(tetrahydrofuran)-dilanthanum, La-00007

$C_{32}H_{40}N_2O_{16}Re_4$
Hexadecacarbonyltetrarhenate(2−); Bis(tetraethylammonium) salt, *in* Re-00073

$C_{32}H_{40}Nd_2O_2$
[μ-[(1,2-η:1,2,3,4,5,6,7,8-η)-1,3,5,7-Cyclooctatetraene][bis(η[8]-1,3,5,7-cyclooctatetraene)bis(tetrahydrofuran)-dineodymium, Nd-00012

$C_{32}H_{40}Zr$
Bis[(1,2,3,4,5-η)-1,2,3,4,5-pentamethyl-2,4-cyclopentadien-1-yl]diphenylzirconium, Zr-00107

$C_{32}H_{46}O_2Y_2$
Di-μ-hydrotetrakis[(1,2,3,4,5-η)-1-methyl-2,4-cyclopentadien-1-yl]bis(tetrahydrofuran)-diyttrium, Y -00015

$C_{32}H_{48}Ce_2Cl_2O_4$
Chloro(cyclooctatetraene)cerium; Bis-THF adduct, *in* Ce-00001

$C_{32}H_{48}Cl_2Nd_2O_4$
Chloro(cyclooctatetraene)neodymium; Bis-THF adduct, *in* Nd-00002

$C_{32}H_{48}Cl_2O_4Pr_2$
Chloro(cyclooctatetraene)praesodymium; Bis-THF complex, *in* Pr-00002

$C_{32}H_{48}Cl_2O_4Sm_2$
Chloro(cyclooctatetraene)samarium; THF complex, *in* Sm-00003

$C_{32}H_{56}Al_2Cl_2N_2Y_2$
Di-μ-chlorotetrakis(η[5]-2,4-cyclopentadien-1-yl)-bis[trihydro(triethylalumino)aluminum]diyttrium, *in* Y -00009

$C_{32}H_{58}EuO_3$
Bis(pentamethylcyclopentadienyl)europium; Complex with $2Et_2O$ + 1THF, *in* Eu-00005

$C_{32}H_{68}ErLiO_4$
Tetra-*tert*-butylerbate(1−); Li salt, tetra-THF complex, *in* Er-00013

$C_{32}H_{68}LiO_4Sm$
Tetra-*tert*-butylsamarate(1−); Li salt, Tetra-THF complex, *in* Sm-00011

$C_{32}H_{76}LiO_4Si_4Y$
Tetrakis[(trimethylsilyl)methyl]yttrate(1−); Li salt, tetrakis-THF complex, *in* Y -00010

$C_{32}H_{84}LiLuO_4Si$
Tetrakis[(trimethylsilyl)methyl]luteate(1−); Li salt, tetrakis(tetrahydrofuranate), *in* Lu-00015

$C_{33}H_{30}PYb$
Tris(η[5]-2,4-cyclopentadien-1-yl)-triphenylphosphineytterbium, *in* Yb-00013

$C_{33}H_{36}LuP$
Bis(η[5]-2,4-cyclopentadien-1-yl)(*tert*-butyl)-(triphenylphosphonium η-methylide)lutetium, Lu-00025

$C_{33}H_{38}LuPSi$
Bis(η[5]-2,4-cyclopentadien-1-yl)[(trimethylsilyl)methyl]-(triphenylphosphonium η-methylide)lutetium, Lu-00026

$C_{33}H_{44}ClOP_4Re$
Carbonylchlorotetrakis(dimethylphenylphosphine)rhenium, Re-00084

$C_{33}H_{62}MoO_2P_2$
[(2,3,5,6-η)-Bicyclo[2.2.1]hepta-2,5-diene]-dicarbonylbis(tributylphosphine)molybdenum, Mo-00139

$C_{34}H_{25}BF_4MoO$
Carbonyl(η[5]-cyclopentadienyl)bis(diphenylacetylene)-molybdenum(1+); Tetrafluoroborate, *in* Mo-00140

$C_{34}H_{25}MoO^{\oplus}$
Carbonyl(η[5]-cyclopentadienyl)bis(diphenylacetylene)-molybdenum(1+), Mo-00140

$C_{34}H_{25}NbO$
Carbonyl-η[5]-2,4-cyclopentadien-1-ylbis(diphenylacetylene)niobium, Nb-00048

$C_{34}H_{25}OV$
Carbonyl-η[5]-2,4-cyclopentadien-1-ylbis(diphenylacetylene)vanadium, V -00113

$C_{34}H_{30}N_2U$
Bis(η[5]-2,4-cyclopentadien-1-yl)bis(*N*-phenylbenzenaminato)uranium, U -00073

$C_{34}H_{30}S_4V_2$
Tetrakis[μ-(benzenethiolato)]bis(η[5]-2,4-cyclopentadien-1-yl)divanadium, V -00114

$C_{34}H_{34}N_2Yb_2$
Hexakis(η[5]-2,4-cyclopentadien-1-yl)[μ-(pyrazine-*N*[1]:*N*[4])]-diytterbium, Yb-00033

$C_{34}H_{34}OZr_2$
Tetrakis(η[5]-2,4-cyclopentadien-1-yl)bis(2-methylphenyl)-μ-oxodizirconium, Zr-00108

$C_{34}H_{40}Co_2Mo_2O_4$
Di-μ-carbonyl[μ[4]-(η[2]-carbonyl-*C:C:C*)][μ-carbonylbis[(1,2,3,4,5-η)-1,2,3,4,5-pentamethyl-2,4-cyclopentadien-1-yl]dicobalt]bis(η[5]-2,4-cyclopentadien-1-yl)dimolybdenum, Mo-00141

$C_{34}H_{54}Al_2Cl_2Zr_2$
Di-μ-chlorotetrakis(η[5]-2,4-cyclopentadien-1-yl)-μ-1,2-ethanediylbis(triethylaluminum)dizirconium, Zr-00109

$C_{34}H_{68}FeN_6Ti_2$
Hexakis(diethylaminato)-μ-1,1′-ferrocenediyldititanium, Ti-00143

$C_{34}H_{68}N_6RuTi_2$
Hexakis(diethylaminato)-μ-1,1′-ruthenocenediyldititanium, Ti-00144

$C_{35}H_{31}BMoN_2O_2$
Bis(acetonitrile)dicarbonyl(η[5]-2,4-cyclopentadien-1-yl)-molybdenum(1+); Tetraphenylborate, *in* Mo-00054

$C_{35}H_{35}Ta$
Pentabenzyltantalum, Ta-00040

$C_{36}H_{25}O_3V$
Dicarbonyl-η[5]-2,4-cyclopentadien-1-yl(η[4]-tetraphenylcyclopentadienone)vanadium, V -00115

$C_{36}H_{30}BO_2V$
Dicarbonylbis(2,4-cyclopentadien-1-yl)vanadium(1+); Tetraphenylborate, *in* V -00050

$C_{36}H_{30}HfN_2$
Bis(η[5]-cyclopentadienyl)bis(diphenylketimido)hafnium, Hf-00049

$C_{36}H_{30}Li_2U$
Hexaphenyluranate(2−); Di-Li salt, *in* U -00074

$C_{36}H_{30}N_2Zr$
Bis(η^5-cyclopentadienyl)bis(diphenylketimido)zirconium, Zr-00110

$C_{36}H_{30}Ti_2$
Tetra-π-cyclopentadienylbis(phenylethynyl)dititanium, Ti-00145

$C_{36}H_{30}U^{\ominus\ominus}$
Hexaphenyluranate(2−), U -00074

$C_{36}H_{32}Hf$
Bis(η^5-2,4-cyclopentadien-1-yl)bis(diphenylmethyl)-hafnium, Hf-00050

$C_{36}H_{32}Zr$
Bis(η^5-2,4-cyclopentadien-1-yl)bis(diphenylmethyl)-zirconium, Zr-00111

$C_{36}H_{33}ClU$
Chlorotris(benzylcyclopentadienyl)uranium, U -00075

$C_{36}H_{34}Cl_2Si_2Yb$
Dichlorobis[(1,2,3,4,5-η)-1-(methyldiphenylsilyl)-2,4-cyclopentadien-1-yl]ytterbate(1−), Yb-00034

$C_{36}H_{34}P_2V^{\oplus}$
Bis(η^5-2,4-cyclopentadien-1-yl)[1,2-ethanediylbis[diphenylphosphine]-*P*,*P'*]-vanadium(1+), V -00116

$C_{36}H_{39}ClTh$
Chlorotris[(1,2,3,3*a*,7*a*-η)-1,4,7-trimethyl-1*H*-inden-1-yl]thorium, Th-00037

$C_{36}H_{39}ClU$
Chlorotris[(1,2,3,3*a*,7*a*-η)-1,4,7-trimethyl-1*H*-inden-1-yl]uranium, U -00076

$C_{36}H_{42}Yb_2$
Tetrakis(1,2,3,4,5-η-1-methyl-2,4-cyclopentadien-1-yl)-bis[μ-3,3-dimethyl-1-butyn-1-yl)diytterbium, Yb-00035

$C_{36}H_{44}V^{\ominus}$
Tetrakis(2,4,6-trimethylphenyl)vanadate(1−), V -00117
Tetrakis(2,4,6-trimethylphenyl)vanadium, V -00118

$C_{36}H_{46}Sm_2$
Bis[μ-(3,3-dimethyl-1-butyn-1-yl)]tetrakis(1,2,3,4,5-η-1-methyl-2,4-cyclopentadien-1-yl)disamarium, Sm-00018

$C_{36}H_{48}Mn_2N_4$
Bis[μ-[[2-(dimethylamino)phenyl]methyl-*C*:*C*,*N*]][[2-(dimethylamino)phenyl]methyl][[2-(dimethylamino)-phenyl]methyl-*C*,*N*]dimanganese, Mn-00137

$C_{36}H_{54}Cl_6Nb_3^{\oplus}$
Hexa-μ-chlorotris(hexamethylbenzene)triniobium(1+), Nb-00049

$C_{36}H_{54}Cl_7Nb_3$
Hexa-μ-chlorotris(hexamethylbenzene)triniobium(1+); Chloride, *in* Nb-00049

$C_{36}H_{86}B_{18}Cl_2Li_2O_8U$
Dichlorobis[(7,8,9,10,11-η)undecahydro-7,8-dicarbaundecaborato]uranate(2−); Di-Li salt, octa-THF complex, *in* U -00001

$C_{37}H_{30}MoO_2PTl$
Dicarbonyl(η^5-2,4-cyclopentadien-1-yl)-(diphenylthallium)(triphenylphosphine)molybdenum, Mo-00142

$C_{37}H_{30}O_2PTlW$
Dicarbonyl(η^5-2,4-cyclopentadien-1-yl)-(diphenylthallium)(triphenylphosphine)tungsten, W -00154

$C_{37}H_{36}BOV$
Bis(η^5-2,4-cyclopentadien-1-yl)(2-propanone)-vanadium(1+); Tetraphenylborate, *in* V -00060

$C_{37}H_{89}ClLiO_4Si_6Yb$
Tris[bis(triethylsilyl)methyl]chloroytterbate(1−); Li salt, tetrakis-THF complex, *in* Yb-00025

$C_{38}H_{20}AsMn_3O_{14}$
Tetradecacarbonyltrimanganate(1−); Tetraphenylarsonium salt, *in* Mn-00098

$C_{38}H_{30}CrINO_3P_2$
Dicarbonyliodonitrosylbis(triphenylphosphine)chromium, Cr-00177

$C_{38}H_{30}Hf$
Bis(η^5-2,4-cyclopentadien-1-yl)(1,2,3,4-tetraphenyl-1,3-butadiene-1,4-diyl)hafnium, Hf-00051

$C_{38}H_{30}MnNO_3P_2$
Dicarbonylnitrosylbis(triphenylphosphine)manganese, Mn-00138

$C_{38}H_{30}NO_3P_2Re$
Dicarbonylnitrosylbis(triphenylphosphine)rhenium, Re-00085

$C_{38}H_{30}O_8Ti_2$
Tetrakis[μ-benzoato-*O*:*O'*]bis(η^5-2,4-cyclopentadien-1-yl)dititanium, *in* Ti-00102

$C_{38}H_{30}Ti$
Bis(η^5-2,4-cyclopentadien-1-yl)(1,2,3,4-tetraphenyl-1,3-butadiene-1,4-diyl)titanium, Ti-00146

$C_{38}H_{30}V$
Bis(η^5-2,4-cyclopentadien-1-yl)(1,2,3,4-tetraphenyl-1,3-butadiene-1,4-diyl)vanadium, V -00119

$C_{38}H_{30}Zr$
Bis(η^5-2,4-cyclopentadien-1-yl)(1,2,3,4-tetraphenyl-1,3-butadiene-1,4-diyl)zirconium, Zr-00112

$C_{38}H_{42}Cl_4N_4O_2U_3$
Bis[bis(acetonitrile)tris(η^5-cyclopentadienyl)-uranium(1+)]tetrachlorodioxouranate(2−), U -00077

$C_{38}H_{44}NO_4P_2V$
Tetracarbonyl[1,2-ethanediylbis[diphenylphosphine]-*P*,*P'*]vanadate(1−); Tetraethylammonium salt, *in* V -00108

$C_{38}H_{45}P_2Re$
Bis(diethylphenylphosphine)triphenylrhenium, Re-00086

$C_{39}H_{30}BClF_4O_3P_2Re$
Tricarbonylchlorobis(triphenylphosphine)rhenium(1+); Tetrafluoroborate, *in* Re-00087

$C_{39}H_{30}ClO_3P_2Re^{\oplus}$
Tricarbonylchlorobis(triphenylphosphine)rhenium(1+), Re-00087

$C_{39}H_{30}ClO_3P_2Tc$
Tricarbonylchlorobis(triphenylphosphine)technetium, Tc-00012

$C_{39}H_{31}MnO_3P_2$
Tricarbonylhydrobis(triphenylphosphine)manganese, Mn-00139

$C_{39}H_{44}NO_5P_2V$
Pentacarbonyl[η^1-1,2-ethanediylbis[diphenylphosphine]-*P*]vanadate(1−); Tetraethylammonium salt, *in* V -00111

$C_{39}H_{56}NO_4PSnW$
Tetracarbonyl(triphenylstannane)-(triisopropylphosphine)tungstate(1−); Tetraethylammonium salt, *in* W -00152

$C_{40}H_{20}Co_6Hf_2O_{21}$
Tetrakis(η^5-2,4-cyclopentadien-1-yl)bis[μ_4-[methanolato(4−)-*C*:*C*:*C*:*O*]-bis(nonacarbonyltricobalt)-μ-oxodihafnium, Hf-00052

$C_{40}H_{20}Co_6O_{21}Zr_2$
Tetrakis(η^5-2,4-cyclopentadien-1-yl)bis[μ_4-[methanolato(4−)-*C*:*C*:*C*:*O*]-bis(nonacarbonyltricobalt)-μ-oxodizirconium, Zr-00113

$C_{40}H_{30}F_6MnO_4P_3$
Tetracarbonylbis(triphenylphosphine)manganese(1+); Hexafluorophosphate, *in* Mn-00140

$C_{40}H_{30}F_6O_4P_3Re$
Tetracarbonylbis(triphenylphosphine)rhenium(1+); Hexafluorophosphate, *in* Re-00088

$C_{40}H_{30}MnO_4P_2^{\oplus}$
Tetracarbonylbis(triphenylphosphine)manganese(1+), Mn-00140

$C_{40}H_{30}Nb_2O_2$
Dicarbonylbis(η^5-cyclopentadien-1-yl)bis[1,1′-(1,2-ethynediyl)bis[benzene]]diniobium, Nb-00050

$C_{40}H_{30}O_4P_2Re^{\oplus}$
Tetracarbonylbis(triphenylphosphine)rhenium(1+), Re-00088

$C_{40}H_{35}BNOV$
Carbonylbis(η^5-2,4-cyclopentadien-1-yl)(pyridine)vanadium(1+); Tetraphenylborate, *in* V -00071

$C_{40}H_{38}Ti_2$
[μ-(1,4-Diphenyl-1,3-butadiene-1,2,3,4-tetrayl-*C*^{1},*C*^{3}:*C*^{2},*C*^{4})]tetrakis[(1,2,3,4,5-η)-1-methyl-2,4-cyclopentadien-1-yl]dititanium, Ti-00147

$C_{40}H_{44}Hf$
Tetrakis(2-methyl-1-phenyl-1-propenyl)hafnium, Hf-00053

$C_{40}H_{44}Li_4Mo_4$
Tetrakis[bis(η^5-2,4-cyclopentadien-1-yl)-hydromolybdenum]tetralithium, *in* Mo-00040

$C_{40}H_{44}Li_4W_4$
Bis(η^5-2,4-cyclopentadien-1-yl)hydro(tungsten)lithium; Tetramer, *in* W -00052

$C_{40}H_{44}Ti$
Tetrakis(2-methyl-1-phenyl-1-propenyl)titanium, Ti-00148

$C_{40}H_{44}Zr$
Tetrakis(2-methyl-1-phenyl-1-propenyl)zirconium, Zr-00114

$C_{40}H_{52}Cr$
Tetrakis(2-methyl-2-phenylpropyl)chromium, Cr-00178

$C_{40}H_{52}Mn_2$
Bis[μ-(2-methyl-2-phenylpropyl)]bis(2-methyl-2-phenylpropyl)dimanganese, Mn-00141

$C_{40}H_{60}N_2Ti_2$
[η-(Dinitrogen-*N*,*N*′)]tetrakis[(1,2,3,4,5-η)-1,2,3,4,5-pentamethyl-2,4-cyclopentadien-1-yl]dititanium, Ti-00149

$C_{40}H_{62}Sm_2$
Di-μ-hydrotetrakis[(1,2,3,4,5-η)-1,2,3,4,5-pentamethyl-2,4-cyclopentadien-1-yl]disamarium, *in* Sm-00013

$C_{40}H_{64}Th_2$
Di-μ-hydrodihydrotetrakis[(1,2,3,4,5-η)-1,2,3,4,5-pentamethyl-2,4-cyclopentadien-1-yl]dithorium, *in* Th-00017

$C_{40}H_{64}U_2$
Di-μ-hydrodihydrotetrakis[(1,2,3,4,5-η)-1,2,3,4,5-pentamethyl-2,4-cyclopentadien-1-yl]diuranium, *in* U -00039

$C_{40}H_{88}Li_2O_8Re_2$
Octamethyldirhenate(2−); Bis[tetrakis(tetrahydrofuran)lithium(1+)]salt, *in* Re-00036

$C_{41}H_{30}ClCrNO_5P_2$
Pentacarbonylchlorochromate(1−); Bis(triphenylphosphine)iminium salt, *in* Cr-00006

$C_{41}H_{30}IMoNO_5P_2$
Pentacarbonyliodomolybdate(1−); Bis(triphenylphosphine)iminium salt, *in* Mo-00004

$C_{41}H_{30}MnNO_5P_2$
Pentacarbonylmanganate(1−); Bis(triphenylphosphine)iminium salt, *in* Mn-00014

$C_{41}H_{33}N_2O_5P_2V$
Amminepentacarbonylvanadate(1−); Bis(triphenylphosphine)-iminium salt, *in* V -00001

$C_{41}H_{35}AsN_2S_2U$
Tris(η^5-cyclopentadienyl)diisothiocyanatouranate(IV); Tetraphenylarsonium salt, *in* U -00021

$C_{41}H_{35}Cl_3O_2P_2U$
Trichloro(η^5-2,4-cyclopentadien-1-yl)-bis(triphenylphosphine oxide-*O*)uranium, U -00078

$C_{41}H_{35}F_6MnNO_7P_3$
(η^5-2,4-Cyclopentadien-1-yl)nitrosylbis(triphenyl phosphite-*P*)manganese(1+); Hexafluorophosphate, *in* Mn-00142

$C_{41}H_{35}MnNO_7P_2^{\oplus}$
(η^5-2,4-Cyclopentadien-1-yl)nitrosylbis(triphenyl phosphite-*P*)manganese(1+), Mn-00142

$C_{42}H_{30}CrN_2O_5P_2$
Pentacarbonyl(cyano-*C*)chromate(1−); Bis(triphenylphosphine)-iminium salt, *in* Cr-00016

$C_{42}H_{30}CrN_3O_4P_2$
Tetracarbonyldicyanochromate(2−); Bis(triphenylphosphine)-iminium salt, *in* Cr-00017

$C_{42}H_{30}MoN_2P_2O_5$
Pentacarbonyl(cyano-*C*)molybdate(1−); Bis(triphenylphosphine)iminium salt, *in* Mo-00006

$C_{42}H_{30}N_2O_5P_2W$
Pentacarbonyl(cyano-*C*)tungstate(1−); Bis(triphenylphosphine)-iminium salt, *in* W -00012

$C_{42}H_{30}Nb_2O_4$
Tetracarbonylbis(cyclopentadienyl)-bis(diphenylacetylene)diniobium, *in* Nb-00041

$C_{42}H_{33}CrNO_5P_2$
(*T*-4)Triphenyl(*P*,*P*,*P*-triphenylphosphineimidato-*N*)-phosphorus(1+) pentacarbonylmethylchromate(1−), *in* Cr-00024

$C_{42}H_{33}NO_5P_2W$
(T-4)Triphenyl(*P*,*P*,*P*-triphenylphosphineimidato-*N*)-phosphorus(1+) pentacarbonylmethyltungstate(1−), *in* W -00009

$C_{42}H_{42}Cl_3Re_3$
Hexabenzyltri-μ-chlorotrirhenium, Re-00089

$C_{42}H_{42}O_6Re_2$
Tetrakis[μ-(2-methoxyphenyl-*C*:*O*)]bis(2-methoxyphenyl-*C*,*O*)dirhenium, Re-00090

$C_{42}H_{62}I_2O_2Zr_2$
[μ-[1,2-Ethenediolato(2−)-*O*:*O*′]]-diodotetrakis[(1,2,3,4,5-η)-1,2,3,4,5-pentamethyl-2,4-cyclopentadien-1-yl]dizirconium, Zr-00115

$C_{42}H_{64}O_2Zr_2$
[μ-[1,2-Ethenediolato(2−)-*O*:*O*′]]-dihydrotetrakis[(1,2,3,4,5-η)-1,2,3,4,5-pentamethyl-2,4-cyclopentadien-1-yl]dizirconium, Zr-00116

$C_{42}H_{66}Lu_2$
Methylbis[(1,2,3,4,5-η)-1,2,3,4,5-pentamethyl-2,4-cyclopentadien-1-yl]lutetium; Dimer, *in* Lu-00019

$C_{42}H_{66}Y_2$
Methylbis[(1,2,3,4,5-η)-1,2,3,4,5-pentamethyl-2,4-cyclopentadien-1-yl]yttrium; Dimer, *in* Y -00012

$C_{43}H_{30}OW$
Carbonyltris(diphenylacetylene)tungsten, W -00155

$C_{43}H_{35}BF_4MoO_2P_2$
Dicarbonyl(η^5-2,4-cyclopentadien-1-yl)-bis(triphenylphosphine)molybdenum(1+); Tetrafluoroborate, *in* Mo-00143

$C_{43}H_{35}MoO_2P_2^{\oplus}$
Dicarbonyl(η^5-2,4-cyclopentadien-1-yl)-bis(triphenylphosphine)molybdenum(1+), Mo-00143

$C_{43}H_{48}Cl_4N_4O_2U_3$
Bis[bis(acetonitrile)tris(η^5-cyclopentadienyl)-uranium(1+)]tetrachlorodioxouranate(2−); Cyclopentadiene adduct, *in* U -00077

$C_{44}H_{30}O_8P_2Re_2$
Octacarbonylbis(triphenylphosphine)dirhenium, Re-00091

$C_{44}H_{30}O_8P_2Tc_2$
Octacarbonylbis(triphenylphosphine)ditechnetium, Tc-00013

$C_{44}H_{36}NO_3P_2V$
Tricarbonyl(η^5-2,4-cyclopentadien-1-yl)-hydrovanadate(1−); Bis(triphenylphosphine)iminium salt, *in* V -00017

$C_{44}H_{54}Cl_2O_2Si_2Yb$
Dichlorobis[(1,2,3,4,5-η)-1-(methyldiphenylsilyl)-2,4-cyclopentadien-1-yl]ytterbate(1−); Li salt, bis(diethyl ether) complex, *in* Yb-00034

$C_{44}H_{84}Cl_2Pr_2Si_8$
Tetrakis[(1,2,3,4,5-η)-1,3-bis(trimethylsilyl)-2,4-cyclopentadien-1-yl]di-μ-chlorodipraesodymium, *in* Pr-00006

$C_{44}H_{84}Cl_2Sc_2Si_8$
Bis[1,3-bis(trimethylsilyl)cyclopentadienyl]-chloroscandium; Dimer, *in* Sc-00010

$C_{45}H_{30}FeMnNO_9P_2$
Nonacarbonylferratemanganate(1−); Bis(triphenylphosphine)-iminium salt, *in* Mn-00041

$C_{45}H_{38}NO_3P_2V$
Tricarbonyl(η^5-2,4-cyclopentadien-1-yl)-methylvanadate(1−); Bis(triphenylphosphine)iminium salt, *in* V -00021

$C_{45}H_{45}MoP_3$
(η^6-Benzene)tris(methyldiphenylphosphine)molybdenum, Mo-00144

$C_{46}H_{30}CrINO_{10}P_2$
Decacarbonyl-μ-iododichromate(1−); Bis(triphenylphosphine)-iminium salt, *in* Cr-00065

$C_{46}H_{31}CrNO_{10}P_2$
Decacarbonyl-μ-hydrodichromate(1−); Bis(triphenylphosphine)-iminium salt, *in* Cr-00067

$C_{46}H_{31}NO_{10}P_2W_2$
Decacarbonyl-μ-hydroditungstate(1−); Bis(triphenylphosphine)-iminium salt, *in* W -00040

$C_{46}H_{40}Ge_2Ti$
Bis(η^5-2,4-cyclopentadien-1-yl)bis(triphenylgermyl)-titanium, Ti-00150

$C_{46}H_{40}O_2Si_2Ti$
Bis(η^5-2,4-cyclopentadien-1-yl)bis(triphenylsilyloxy)-titanium, Ti-00151

$C_{46}H_{40}Sn_2Ti$
Bis(η-cyclopentadienyl)bis(triphenylstannyl)titanium, Ti-00152

$C_{46}H_{54}Cl_4N_4O_2U_3$
Bis[bis(acetonitrile)tris(η^5-cyclopentadienyl)-uranium(1+)]tetrachlorodioxouranate(2−); Bis(1,3-butadiene) adduct, *in* U -00077

$C_{46}H_{65}ClEr_3LiO_4$
μ-Chlorohexakis(η^5-2,4-cyclopentadien-1-yl)di-μ-hydro-μ_3-hydrotrierbate(1−); Li salt, tetra-THF complex, *in* Er-00017

$C_{47}H_{33}Cr_2NO_{10}P_2S$
Decarbonyl(μ-methanethiolato)dichromate(1−); Bis(triphenylphosphine)iminium salt, *in* Cr-00083

$C_{47}H_{35}O_5P_2V$
Pentacarbonyl(triphenylphosphine)vanadate(1−); Tetraphenylphosphonium salt, *in* V -00094

$C_{47}H_{47}P_3U$
(η^5-2,4-Cyclopentadien-1-yl)-tris[(diphenylphosphinidenio)bis(methylene)]-uranium, U -00079

$C_{48}H_{30}Fe_2MnNO_{12}P_2$
Tetracarbonyl(di-μ-carbonylhexacarbonyldiferrate)-manganate(1−); Bis(triphenylphosphine)iminium salt, *in* Mn-00072

$C_{48}H_{35}AsMnO_2P$
Dicarbonyl[(1,2,3,4,5-η)-2,3,4,5-tetraphenyl-1*H*-arsol-1-yl](triphenylphosphine)manganese, Mn-00143

$C_{48}H_{35}NbO$
Cyclopentadienylcarbonyl(diphenylacetylene)-(tetraphenylcyclobutadiene)niobium, Nb-00051

$C_{48}H_{46}P_2U_2$
Tetrakis(η^5-2,4-cyclopentadien-1-yl)bis[μ-[methylene(diphenylphosphinidenio)methylidyne]]-diuranium, U -00080

$C_{48}H_{68}LiLuO_4$
Tetrakis(2,6-dimethylphenyl)lutetate(1−); Li salt, tetra-THF complex, *in* Lu-00023

$C_{48}H_{68}LiO_4Yb$
Tetrakis(2,6-dimethylphenyl)ytterbate(1−); Li salt, tetra-THF complex, *in* Yb-00032

$C_{48}H_{72}Th_2O_4$
Bis[μ-(2-butene-2,3-diolato(2-)-*O*,*O*′]-tetrakis[(1,2,3,4,5-η)-1,2,3,4,5-pentamethyl-2,4-cyclopentadien-1-yl]dithorium, Th-00038

$C_{50}H_{70}AsCl_2NdSi_4$
Bis[(1,2,3,4,5-η)-1,3-bis(trimethylsilylmethyl)-2,4-cyclopentadien-1-yl]dichloroneodymium(1−); Tetraphenylarsonium salt, *in* Nd-00011

$C_{51}H_{60}Fe_3Yb_2$
Tetrakis[μ_3-(carbonyl-*C*,*C*,*O*)](heptacarbonyltriiron)-tetrakis[(1,2,3,4,5-η)-1,2,3,4,5-pentamethyl-2,4-cyclopentadien-1-yl]diytterbium, Yb-00036

$C_{51}H_{72}Cl_2NdP$
Bis[(1,2,3,4,5-η)-1,3-bis(trimethylsilylmethyl)-2,4-cyclopentadien-1-yl]dichloroneodymium(1−); Benzyltriphenylphosphonium salt, *in* Nd-00011

$C_{52}H_{48}MoN_4P_4$
Bis(dinitrogen)bis[1,2-bis(diphenylphosphino)ethane]-molybdenum, Mo-00145

$C_{52}H_{48}N_4P_4W$
Bis(dinitrogen)bis[1,2-bis(diphenylphosphino)ethane]-tungsten, W -00156

$C_{52}H_{76}LiO_4V$
Tetrakis(2,4,6-trimethylphenyl)vanadate(1−); Li salt, tetrakis(THF) complex, *in* V -00117

$C_{52}H_{100}Cl_2Si_8Yb_2$
Tetrakis[(1,2,3,4,5-η)-1,3-bis(trimethylsilylmethyl)-2,4-cyclopentadien-1-yl]di-μ-chlorodiytterbium, *in* Yb-00029

$C_{53}H_{48}Br_3ClOP_4Re$
Carbonylchlorobis[1,2-ethanediylbis(diphenylphosphine-*P*,*P*′)]rhenium(1+); Tribromide, *in* Re-00092

$C_{53}H_{48}ClF_6OP_5Re$
Carbonylchlorobis[1,2-ethanediylbis(diphenylphosphine-*P*,*P*′)]rhenium(1+); Hexafluorophosphate, *in* Re-00092

$C_{53}H_{48}ClOP_4Re^{\oplus}$
Carbonylchlorobis[1,2-ethanediylbis(diphenylphosphine-*P*,*P*′)]rhenium(1+), Re-00092

$C_{54}H_{40}NO_5P_2V$
▷Pentacarbonyl(cyano-*C*)vanadate(2−); Bis(tetraphenylphosphonium) salt, *in* V -00011

$C_{54}H_{47}NOP_3Re$
Dihydronitrosyltris(triphenylphosphine)rhenium, Re-00093

$C_{54}H_{48}ClMnO_2P_4$
Dicarbonylbis[1,2-ethanediylbis[diphenylphosphine]-*P*,*P*′]manganese(1+); Chloride, *in* Mn-00144

$C_{54}H_{48}ClMnO_6P_4$
Dicarbonylbis[1,2-ethanediylbis[diphenylphosphine]-*P*,*P*′]manganese(1+); Perchlorate, *in* Mn-00144

$C_{54}H_{48}MnO_2P_4^{\oplus}$
Dicarbonylbis[1,2-ethanediylbis[diphenylphosphine]-*P*,*P*′]manganese(1+), Mn-00144

$C_{54}H_{82}Cl_2O_4Th_2$
Dichlorotetrakis[(1,2,3,4,5-η)-1,2,3,4,5-pentamethyl-2,4-cyclopentadien-1-yl][μ-(2,2,9,9-tetramethyl-4,5,6,7-decanetetrone-*O*[4],*O*[5],*O*[6],*O*[7])]dithorium, Th-00039

$C_{55}H_{44}Mn_2O_5P_4$
μ-Carbonyltetracarbonylbis[μ-[methylenebis[diphenylphosphine]-*P*,*P*′]]-dimanganese, Mn-00145

$C_{56}H_{48}BCrFe_2$
Bis[(η[6]-phenyl)ferrocene]chromium(1+); Tetraphenylborate, *in* Cr-00174

$C_{56}H_{56}MoP_4$
Bis[1,2-ethanediylbis[diphenylphosphine]-*P*,*P*′]bis(η[2]-ethene)molybdenum, Mo-00146

$C_{58}H_{45}MnNO_4P_3$
Tetracarbonyl(triphenylphosphine)manganate(1−); Bis(triphenylphosphine)iminium salt, *in* Mn-00124

$C_{58}H_{50}Si_4Ti$
Bis(η[5]-2,4-cyclopentadien-1-yl)(1,1,2,2,3,3,4,4-octaphenyl-1,4-tetrasilanediyl)titanium, Ti-00153

$C_{60}H_{54}BP_2V$
Bis(η[5]-2,4-cyclopentadien-1-yl)[1,2-ethanediylbis[diphenylphosphine]-*P*,*P*′]-vanadium(1+); Tetraphenylborate, *in* V -00116

$C_{60}H_{90}Cl_3U_3$
Tri-μ-chlorohexakis[(1,2,3,4,5-η)-1,2,3,4,5-pentamethyl-2,4-cyclopentadien-1-yl]triuranium, *in* U -00037

$C_{64}H_{48}U$
Bis[(1,2,3,4,5,6,7,8-η)-1,3,5,7-tetraphenyl-1,3,5,7-cyclooctatetraene]uranium, U -00081

$C_{82}H_{60}Cr_2N_2O_{10}P_4$
Decacarbonyldichromate(2−); Bis(bis(triphenylphosphine)iminium)salt, *in* Cr-00066

$C_{82}H_{60}Mo_2N_2O_{10}P_4$
Decarbonyldimolybdate(2−); Bis[bis(triphenylphosphine)iminium]salt, *in* Mo-00045

CAS Registry Number Index

309-89-7 Bis(η^5-2,4-cyclopentadien-1-yl)difluorotitanium, Ti-00029

1270-98-0 Trichloro(η^5-2,4-cyclopentadien-1-yl)titanium, Ti-00009

1271-18-7 Di-μ-chlorotetrakis(η^5-2,4-cyclopentadien-1-yl)dititanium, *in* Ti-00025

1271-19-8 ▷Dichlorobis(η^5-2,4-cyclopentadien-1-yl)titanium, Ti-00026

1271-24-5 Chromocene, Cr-00075

1271-27-8 ▷Manganocene, Mn-00058

1271-29-0 ▷Titanocene, Ti-00035

1271-31-4 Bis(η^5-2,4-cyclopentadien-1-yl)ytterbium, Yb-00005

1271-32-5 Bis(η^5-2,4-cyclopentadien-1-yl)hydrorhenium, Re-00042

1271-33-6 Bis(η^5-2,4-cyclopentadien-1-yl)dihydrotungsten, W -00055

1271-43-8 (η^6-Benzene)(η^5-2,4-cyclopentadien-1-yl)manganese, Mn-00067

1271-54-1 ▷Bis(η^6-benzene)chromium, Cr-00101

1271-66-5 Bis(η^5-2,4-cyclopentadien-1-yl)dimethyltitanium, Ti-00060

1272-21-5 Tris(η^5-2,4-cyclopentadien-1-yl)gadolinium, Gd-00004

1272-22-6 Tris(η^5-2,4-cyclopentadien-1-yl)holmium, Ho-00005

1272-23-7 Tris(η^5-2,4-cyclopentadien-1-yl)lanthanum, La-00004

1272-24-8 Tris(η^5-2,4-cyclopentadien-1-yl)lutetium, Lu-00012

1272-25-9 Tris(η^5-2,4-cyclopentadien-1-yl)terbium, Tb-00002

1272-26-0 Tris(η^5-2,4-cyclopentadien-1-yl)thulium, Tm-00004

1273-09-2 Bis(η^5-cyclopentadien-1-yl)diphenyltitanium, Ti-00118

1273-98-9 Tris(η^5-2,4-cyclopentadien-1-yl)neodymium, Nd-00004

1277-47-0 Bis(η^5-2,4-cyclopentadien-1-yl)vanadium, V -00033

1278-83-7 Chlorobis(η^5-2,4-cyclopentadien-1-yl)methyltitanium, Ti-00045

1282-41-3 (π-Cyclopentadienyl)triethoxytitanium, Ti-00048

1282-42-4 Tris(acetato-*O*)(η^5-2,4-cyclopentadien-1-yl)titanium, Ti-00047

1282-45-7 Bis(η^5-2,4-cyclopentadien-1-yl)-bis(trifluoroacetato-*O*)titanium, Ti-00069

1282-51-5 Bis(acetato-*O*)bis(η^5-2,4-cyclopentadien-1-yl)titanium, Ti-00071

1284-81-7 Chlorotris(η^5-2,4-cyclopentadien-1-yl)uranium, U -00013

1284-82-8 Chlorotris(η^5-2,4-cyclopentadien-1-yl)thorium, Th-00005

1284-95-3 Butoxytris(η^5-2,4-cyclopentadien-1-yl)uranium, U -00027

1291-32-3 ▷Dichlorobis(η^5-2,4-cyclopentadien-1-yl)zirconium, Zr-00007

1291-40-3 Bis(η^5-2,4-cyclopentadien-1-yl)dihydromolybdenum, Mo-00043

1292-47-3 Bis(benzenethiolato)bis(η^5-2,4-cyclopentadien-1-yl)titanium, Ti-00117

1293-64-7 Dichloro(η^5-2,4-cyclopentadien-1-yl)oxovanadium, V -00004

1293-73-8 Dibromobis(η^5-2,4-cyclopentadien-1-yl)titanium, Ti-00024

1294-07-1 Tris(η^5-2,4-cyclopentadien-1-yl)yttrium, Y -00006

1294-67-3 Dibromobis(η^5-2,4-cyclopentadien-1-yl)zirconium, Zr-00006

1295-18-7 Tris(η^5-2,4-cyclopentadien-1-yl)americium, Am-00001

1295-20-1 Tris(η^5-2,4-cyclopentadien-1-yl)ytterbium, Yb-00013

1298-37-9 Diazidobis(η^5-2,4-cyclopentadien-1-yl)titanium, Ti-00032

1298-41-5 Bis(η^5-2,4-cyclopentadien-1-yl)diiodozirconium, Zr-00009

1298-53-9 Tris(η^5-2,4-cyclopentadien-1-yl)cerium, Ce-00003

1298-54-0 Tris(η^5-2,4-cyclopentadien-1-yl)scandium, Sc-00005

1298-55-1 Tris(η^5-2,4-cyclopentadien-1-yl)samarium, Sm-00009

1298-75-5 Tetrakis(η^5-2,4-cyclopentadien-1-yl)thorium, Th-00014

1298-76-6 Tetrakis(η^5-2,4-cyclopentadien-1-yl)uranium, U -00035

1317-00-6 Chlorotris(η^5-2,4-cyclopentadien-1-yl)neptunium, Np-00001

1317-21-1 Bis(η^5-2,4-cyclopentadien-1-yl)(1,2,3,4-tetraphenyl-1,3-butadiene-1,4-diyl)titanium, Ti-00146

2371-70-2 ▷Tetramethyltitanium, Ti-00006

2371-72-4 Tetraphenyltitanium, Ti-00124

2747-38-8 Trichloromethyltitanium, Ti-00002

2948-60-9 Dichlorotrimethylniobium, Nb-00004

3020-02-8 Dichlorotrimethyltantalum, Ta-00003

4984-20-7 Trichloroethyltitanium, Ti-00003

10170-69-1 Decacarbonyldimanganese, Mn-00063

11057-26-4 Tris(η-^{5}-2,4-cyclopentadien-1-yl)praseodymium; Cyclohexylisocyanide adduct, *in* Pr-00003

11057-67-3 Tetra-π-cyclopentadienylbis[μ-(dimethylaminato)]-dititanium, *in* Ti-00058

11058-18-7 Bis(η-cyclopentadienyl)bis(2-phenylethenyl)zirconium, Zr-00093

11065-40-0 (η^8-1,3,5,7-Cyclooctatetraene)(η^5-2,4-cyclopentadien-1-yl)titanium, Ti-00064

11067-83-7 Di-π-cyclopentadienylbis[μ(dimethylaminato)]-bis(dimethylaminato)dititanium, *in* Ti-00022

11077-28-4 Bis(η^5-2,4-cyclopentadienyl)pentathiovanadium, V -00032

11077-50-2 Bis(η^6-benzene)rhenium(1+), Re-00055

11077-51-3 Bis(η^6-benzene)technetium(1+), Tc-00008

11077-55-7 Chlorodi-π-cyclopentadienyl[2,2,2-trifluoro-1-(trifluoromethyl)ethylideniminato]titanium, Ti-00063

11077-58-0 Tris(η^5-2,4-cyclopentadien-1-yl)curium, Cm-00001

11077-59-1 Tris(η-^{5}-2,4-cyclopentadien-1-yl)praseodymium, Pr-00003

11077-71-7 Bis(η^5-2,4-cyclopentadien-1-yl)hydrophenyltungsten, W -00108

11077-90-0 Bis(η^5-2,4-cyclopentadien-1-yl)di-1*H*-pyrrol-1-yltitanium, Ti-00097

11077-95-5 Bis(η^5-2,4-cyclopentadien-1-yl)-bis[(trimethylsilyl)methyl]hafnium, Hf-00025

11077-96-6 Bis(η^5-2,4-cyclopentadien-1-yl)-bis[(trimethylsilyl)methyl]titanium, Ti-00101

11077-97-7 Bis(η^5-2,4-cyclopentadien-1-yl)-bis[(trimethylsilyl)methyl]zirconium, Zr-00054

11078-10-7 Tetra-π-cyclopentadienyldi-μ-hydrodititanium, *in* Ti-00037

11079-26-8 Uranocene, U -00019

11082-39-6 Chlorobis(η^5-2,4-cyclopentadien-1-yl)-[(trimethylsilyl)methyl]zirconium, Zr-00035

11082-41-0 Chlorobis(η^5-2,4-cyclopentadien-1-yl)(2,2-dimethylpropyl)titanium, Ti-00082

11082-60-3 Bis(η^5-2,4-cyclopentadien-1-yl)bis(4-nitrobenzoato-*O'*)titanium, Ti-00123

11082-70-5 Chlorotris[(1,2,3,3*a*,7*a*-η)-1*H*-inden-1-yl]uranium, U -00062

11087-14-2 Di-μ-chlorotetrakis(η^5-2,4-cyclopentadien-1-yl)digadolinium, *in* Gd-00002

11087-26-6 Chlorobis(η-cyclopentadienyl)methoxyzirconium, Zr-00016

11087-29-9 Bis(η^5-2,4-cyclopentadien-1-yl)dimethoxyzirconium, Zr-00024

11088-05-4 Tetrakis[μ-benzoato-*O*:*O′*]bis(η^5-2,4-cyclopentadien-1-yl)dititanium, *in* Ti-00102
11090-85-0 Bis(η^5-2,4-cyclopentadien-1-yl)difluorozirconium, Zr-00008
11105-67-2 Bis(η^5-2,4-cyclopentadien-1-yl)trihydroniobium, Nb-00014
11105-68-3 Carbonylbis(η^5-2,4-cyclopentadien-1-yl)hydroniobium, Nb-00018
11105-69-4 Carbonylbis(η^5-2,4-cyclopentadien-1-yl)hydrotantalum, Ta-00014
11105-70-7 Bis(η^5-2,4-cyclopentadien-1-yl)(η^2-ethene)hydroniobium, Nb-00027
11105-81-0 Bis(η^5-2,4-cyclopentadien-1-yl)-hydro(triethylphosphine)tantalum, Ta-00030
11105-88-7 (1-*tert*-Butyl-2,2-dimethylpropylideniminato)-chlorodi-π-cyclopentadienyltitanium, Ti-00103
11105-93-4 Bis(η^5-2,4-cyclopentadien-1-yl)bis[μ-[(1,η:1,2,3,4,5-η)-2,4-cyclopentadien-1-ylidene]]dihydrodiniobium, Nb-00040
11106-00-6 Chlorodi-π-cyclopentadienyl(1,1-diphenylmethyleniminato)titanium, Ti-00121
11106-17-5 Bis(η^5-cyclopentadienyl)bis(diphenylketimido)hafnium, Hf-00049
11106-18-6 Bis(η^5-cyclopentadienyl)bis(diphenylketimido)zirconium, Zr-00110
11108-17-1 Bis(η^5-2,4-cyclopentadien-1-yl)trihydrotungsten(1+), W -00056
11108-17-1 Bis(η^5-2,4-cyclopentadien-1-yl)trihydrotungsten(1+); Hexafluorophosphate, *in* W -00056
11108-30-8 Bis(η-cyclopentadienyl)bis(dimethylamido)zirconium, Zr-00036
11133-05-4 Chlorotris[(1,2,3,3*a*,7*a*-η)-1*H*-inden-1-yl)]thorium, Th-00028
11136-26-8 Bis(η^5-2,4-cyclopentadien-1-yl)phenyltitanium, Ti-00089
11136-36-0 Dichlorobis[(1,2,3,4,5-η)-1,2,3,4,5-pentamethyl-2,4-cyclopentadien-1-yl]titanium, Ti-00111
11136-37-1 Bis[(1,2,3,4,5-η)-1,2,3,4,5-pentamethyl-2,4-cyclopentadien-1-yl]titanium, Ti-00112
11136-41-7 Dimethylbis[(1,2,3,4,5-η)-1,2,3,4,5-pentamethyl-2,4-cyclopentadien-1-yl]titanium, Ti-00119
11136-46-2 [η-(Dinitrogen-*N*,*N′*)]tetrakis[(1,2,3,4,5-η)-1,2,3,4,5-pentamethyl-2,4-cyclopentadien-1-yl]dititanium, Ti-00149
12071-51-1 Chloro(η^5-2,4-cyclopentadien-1-yl)dinitrosylchromium, Cr-00015
12078-15-8 Tricarbonylthiophenechromium, Cr-00030
12079-65-1 ▷Cyclopentadienyltricarbonylmanganese, Mn-00033
12079-73-1 Tricarbonyl(η^5-2,4-cyclopentadien-1-yl)rhenium, Re-00033
12079-79-7 Bromotricarbonyl(η^5-2,4-cyclopentadien-1-yl)molybdenum, Mo-00011
12082-03-0 Tricarbonyl(η^6-chlorobenzene)chromium, Cr-00052
12082-05-2 (η^6-Fluorobenzene)tricarbonylchromium, Cr-00053
12082-07-4 Tricarbonyl(carboxy-π-cyclopentadienyl)manganese, Mn-00044
12082-08-5 ▷(η^6-Benzene)tricarbonylchromium, Cr-00057
12082-10-9 Dicarbonyl(η^7-cycloheptatrienylium)iodotunsten, W -00033
12082-25-6 Tricarbonyl(η^5-2,4-cyclopentadien-1-yl)methylmolybdenum, Mo-00025
12082-27-8 Tricarbonyl(η^5-2,4-cyclopentadien-1-yl)methyltungsten, W -00034
12082-46-1 Tris(η^3-2-propenyl)chromium, Cr-00064
12083-01-1 Decacarbonyl-μ-hydroditungstate(1−); Tetraethylammonium salt, *in* W -00040
12083-16-8 Tricarbonyl(η^7-cycloheptatrienylium)vanadium, V -00023
12083-17-9 Tricarbonyl(η^7-cycloheptatrienylium)tunsten(1+); Tetrafluoroborate, *in* W -00041
12083-24-8 Tricarbonyl[(1,2,3,4,5,6-η)methylbenzene]chromium, Cr-00072
12083-34-0 Tricarbonyl[(1,2,3,4,5,6-η)methylbenzene]molybdenum, Mo-00034
12083-48-6 Dichlorobis(η^5-2,4-cyclopentadien-1-yl)vanadium, V -00030
12083-77-1 Bis(η^5-2,4-cyclopentadien-1-yl)-bis[tetrahydroborato(1−)-*H*,*H′*]zirconium, Zr-00015
12086-52-1 Bis(η^5-cyclopentadienyl)bis(dimethylamido)titanium, Ti-00076
12086-80-5 Tetradecacarbonylhydrotrirhenium, Re-00064
12087-16-0 Bis(η^5-2,4-cyclopentadien-1-yl)[2,3-dimercapto-2-butenedinitrilato(2−)-*S*,*S′*]titanium, Ti-00070
12087-58-0 Bis[(1,2,3,4,5,6-η)methylbenzene]chromium, Cr-00121
12088-04-9 Tris(η^5-2,4-cyclopentadien-1-yl)dysprosium, Dy-00004
12088-73-2 Pentacarbonyl[dicarbonyl(η^5-2,4-cyclopentadien-1-yl)iron]manganese, Mn-00075
12088-83-4 Tricarbonyl[(3*a*,4,5,6,7,7*a*-η)-2,3-dihydro-1*H*-inden-1-ol]chromium, Cr-00099
12089-15-5 (η^6-1,3,5-Trimethylbenzene)tricarbonylmolybdenum, Mo-00064
12089-23-5 Bis(η^6-benzene)tungsten, W -00073
12089-78-0 Bis(η^5-2,4-cyclopentadien-1-yl)-bis(methanethiolato)titanium, Ti-00059
12090-34-5 Tetrakis(η^3-2-propenyl)zirconium, Zr-00026
12091-64-4 Hexacarbonylbis(η^5-2,4-cyclopentadien-1-yl)dimolybdenum, Mo-00093
12092-01-2 Tetracarbonylbis(η^5-2,4-cyclopentadien-1-yl)[μ-(dimethylphosphino)]-μ-hydrodimolybdenum, Mo-00095
12093-03-7 Tricarbonyl[(1,2,3,4,5,6-η)-1,3,5,7-cyclooctatetraene]chromium, Cr-00085
12093-14-0 Tricarbonyl[(1,2,3,4,5,6-η)-1,3,5-cyclooctatriene]molybdenum, Mo-00051
12093-29-7 Tricarbonyl(η^5-2,4-cyclopentadien-1-yl)-(trimethylstannyl)tungsten, W -00065
12093-81-1 (η^7-Cycloheptatrienylium)(η^5-2,4-cyclopentadien-1-yl)chromium, Cr-00102
12094-55-2 Tricarbonyl-η^6-phenanthrenechromium, Cr-00141
12097-04-0 Dichlorotetrakis(η^5-2,4-cyclopentadien-1-yl)-μ-oxo-dizirconium, Zr-00061
12098-32-7 Tris(η^5-2,4-cyclopentadien-1-yl)isocyanocyclohexaneholmium, *in* Ho-00005
12098-33-8 Tris(η^5-2,4-cyclopentadien-1-yl)neodymium; Cyclohexylisocyanide adduct, *in* Nd-00004
12098-34-9 Tris(η^5-2,4-cyclopentadien-1-yl)isocyanocyclohexaneterbium, *in* Tb-00002
12098-35-0 Tris(η^5-2,4-cyclopentadien-1-yl)yttrium; Cyclohexylisocyanide adduct, *in* Y -00006
12098-36-1 Tris(η^5-2,4-cyclopentadien-1-yl)isocyanocyclohexaneytterbium, *in* Yb-00013
12103-45-6 Hexa-μ-chlorotris(hexamethylbenzene)triniobium(1+); Chloride, *in* Nb-00049
12107-08-3 Dicarbonyltrichloro(η^5-2,4-cyclopentadien-1-yl)tungsten, W -00015
12107-36-7 Tricarbonyl(η^5-2,4-cyclopentadien-1-yl)tungstate(1−); Na salt, *in* W -00023
12108-03-1 Tetracarbonyl(η^5-2,4-cyclopentadien-1-yl)niobium, Nb-00009
12108-04-2 Tetracarbonyl(η^5-2,4-cyclopentadien-1-yl)vanadium, V -00020
12108-11-1 (η^6-Aniline)tricarbonylchromium, Cr-00059
12108-13-3 ▷(Methylcyclopentadienyl)tricarbonylmanganese, Mn-00046
12108-14-4 Tricarbonyl[(1,2,3,4,5-η)-2,4-cyclohexadien-1-yl]manganese, Mn-00047
12108-93-9 Tricarbonyl[(1,2,3,4,5,6-η)-1,3,5,7-cyclooctatetraene]molybdenum, Mo-00047
12108-95-1 Tricarbonyl[(1,2,3,4,5,6-η)-1,3,5,7-cyclooctatetraene]tungsten, W -00060
12109-14-7 Bis(acetonitrile)dicarbonyl(η^5-2,4-cyclopentadien-1-yl)molybdenum(1+), Mo-00054
12109-61-4 Dicyanatobis(η^5-cyclopentadienyl)titanium, Ti-00050
12109-63-6 Bis(cyanato-*N*)bis(η^5-2,4-cyclopentadien-1-yl)zirconium, Zr-00019
12109-67-0 Bis(η^5-2,4-cyclopentadien-1-yl)bis(thiocyanato-*N*)zirconium, Zr-00020
12109-74-9 Tetracarbonyl[(1,2,5,6-η)-1,5-cyclooctadiene]molybdenum, Mo-00065

26901-33-7 Heptacarbonyltris(dimethylphenylphosphine)dirhenium, Re-00083
26901-38-2 Tetracarbonyl(dimethylphenylphosphine)iodorhenium; *trans-form*, *in* Re-00053
27323-98-4 Hexacarbonyl(triphenylstannyl)vanadium, V -00096
27436-93-7 Pentacarbonyl(methoxyphenylmethylene)chromium, Cr-00112
27674-37-9 (Acetonitrile)pentacarbonylmanganese(1+), Mn-00020
28300-66-5 Pentacarbonyl(ethoxycarbonyl)manganese, Mn-00035
28480-04-8 Pentacarbonyl(hexamethyldistannatellurane)chromium, Cr-00093
28597-01-5 ▷Dichlorooxophenylvanadium, V -00009
28939-17-5 Dinitrosyl(pentacarbonylmanganio)-(triphenylphosphine)iron, Mn-00128
29421-17-8 (Acetonitrile)pentacarbonylrhenium(1+), Re-00018
29454-58-8 (Arsine)pentacarbonylchromium, Cr-00010
29890-04-8 Tetracarbonyl[1,2-ethanediylbis[diphenylphosphine]-*P*,*P′*]chromium, Cr-00169
30043-33-5 Tribromomethyltitanium, Ti-00001
30092-08-1 Pentacarbonyl(hydrazine-*N*)chromium, Cr-00014
30169-70-1 ▷Tetracyclohexyltitanium, Ti-00127
30476-93-8 Tricarbonyltris(trimethylphosphine)chromium, Cr-00109
30476-94-9 Dicarbonyltetrakis(trimethylphosphine)chromium, Cr-00123
31011-82-2 Oxotris(trimethylsilylmethyl)vanadium, V -00057
31135-22-5 Tetrakis(pentacarbonylmanganio)uranium, Mn-00122
31172-83-5 Pentacarbonyl(dimethyl sulfide)chromium, Cr-00034
31246-02-3 [1-(Benzylideneamino)ethylidene]pentacarbonylchromium, Cr-00119
31406-67-4 Tetrabenzylhafnium, Hf-00044
31424-66-5 Tetrakis(μ-tetramethylene)dichromate(4−); TetraLi salt, *in* Cr-00136
31724-95-5 (Benzyltrimethylstannane)tricarbonylchromium, Cr-00117
31725-05-0 Pentacarbonyl(ferrocenylmethoxymethylene)chromium, Cr-00142
31741-76-1 (Carbonothioyl)dicarbonyl(η^5-2,4-cyclopentadien-1-yl)manganese, Mn-00031
31760-74-4 Tris(acetato)(η-cyclopentadienyl)zirconium, Zr-00018
31781-62-1 Dichloro(η^5-2,4-cyclopentadien-1-yl)titanium, Ti-00008
31811-41-3 Dicarbonyl(η^5-2,4-cyclopentadien-1-yl)(η^3-2-propenyl)tungsten, W -00050
31811-51-5 Carbonyl(η^5-2,4-cyclopentadien-1-yl)dinitrosylvanadium, V -00010
31833-59-7 Dicarbonyl(η^5-2,4-cyclopentadien-1-yl)(α-phenylbenzenemethaniminato)tungsten, W -00125
31854-87-2 Tricarbonyl-π-cyclopentadienyl(trimethylstannyl)chromium, Cr-00092
31868-92-5 Tricarbonyl(trimethylphenylstannane)chromium, Cr-00106
31868-99-2 Tricarbonylhydro(η^6-1,3,5-trimethylbenzene)vanadium, V -00053
31869-16-6 Dicarbonyl(cyclopentadiene)-π-cyclopentadienylniobium, Nb-00023
31870-69-6 Tricarbonyl(η^5-2,4-cyclopentadien-1-yl)iodotungsten, W -00021
31870-76-5 Tricarbonyl(η^6-trimethylstannylbenzene)molybdenum, Mo-00067
31884-35-2 Tricarbonyl-π-cyclopentadienyl(trimethylgermyl-)chromium, Cr-00091
31921-90-1 Dicarbonyl(η^5-2,4-cyclopentadien-1-yl)nitrosylmanganese(1+), Mn-00023
31921-90-1 Dicarbonyl(η^5-2,4-cyclopentadien-1-yl)nitrosylmanganese(1+); Hexafluorophosphate, *in* Mn-00023
31921-91-2 Dicarbonyl(η^5-2,4-cyclopentadien-1-yl)-(hydrazine-*N*)manganese, Mn-00026
31922-27-7 Pentacarbonyl(ethene)rhenium(1+); Hexafluorophosphate, *in* Re-00021
31960-40-4 Dicarbonyl(η^5-2,4-cyclopentadien-1-yl)nitrosylrhenium(1+); Tetrafluoroborate, *in* Re-00023
31960-41-5 Carbonyl(η^5-2,4-cyclopentadien-1-yl)-(hydroxymethyl)nitrosylrhenium, Re-00030
32356-12-0 Pentacarbonyl(stibine)chromium, Cr-00013
32370-44-8 (Aminophenylmethylene)pentacarbonylchromium, Cr-00096
32518-07-3 Tetrakis(cyclohexylisocyanide)tetraiodouranium, U -00067
32610-90-5 Dicarbonyl(η^5-2,4-cyclopentadien-1-yl)(2,2,4,4-tetramethyl-3-pentaniminato)tungsten, W -00109
32628-95-8 Tetracarbonyl(η^5-2,4-cyclopentadien-1-yl)tantalum, Ta-00009
32629-44-0 π-Cyclopentadienyl[dimercaptomaleonitrilato(2−)]-nitrosylmanganate(1−), Mn-00043
32648-85-4 Tricarbonyl[(2,3,4,5,6,7-η)-2,4,6-cycloheptatrien-1-one]chromium, Cr-00069
32649-18-6 Tetracarbonyl[μ-(diphenylphosphino)]-(tetracarbonylmanganio)iron, Mn-00116
32665-18-2 ▷Tetrakis(trimethylsilylmethyl)zirconium, Zr-00045
32680-07-2 π-Cyclopentadienyl(diethyldithiocarbamato)nitrosylmanganese, Mn-00061
32761-36-7 Tricarbonyl-π-pyrrolylmanganese, Mn-00022
32798-86-0 Tricarbonyl[(1,2,3,4,5-η)-2,4-cycloheptadien-1-yl]manganese, Mn-00056
32802-03-2 Tricarbonyl(η^6-phenol)chromium, Cr-00058
32825-28-8 Tricarbonyl(ethoxy-π-cycloheptatrienylium)chromium(1+); Tetrafluoroborate, *in* Cr-00100
32880-58-3 Carbonyl(η^5-2,4-cyclopentadien-1-yl)-nitrosyl(triphenylphosphine)manganese; Hexafluorophosphate, *in* Mn-00132
32966-00-0 Dibromodicarbonyl(η^5-2,4-cyclopentadien-1-yl)rhenium, Re-00022
33085-81-3 Bis[(1,2,3,4,4*a*,8*a*-η)naphthalene]chromium, Cr-00149
33113-68-7 Carbonyl(η^5-2,4-cyclopentadien-1-yl)-nitrosyl(triphenylphosphine)tungsten, W -00140
33114-15-7 Tetrachloro(η^5-2,4-cyclopentadien-1-yl)niobium, Nb-00006
33135-95-4 Dicarbonylchloro[1,2-ethanediylbis[diphenylphosphine]-*P*,*P′*](η^3-2-propenyl)molybdenum, Mo-00137
33152-40-8 Tetracarbonylbis(pentacarbonylmanganio)ruthenium, Mn-00097
33153-70-7 Tetracarbonylbis(pentacarbonylrhenio)osmium, Re-00068
33221-75-9 Bis(acetonitrile)dicarbonylchloro(η^3-2-propenyl)molybdenum, Mo-00029
33221-77-1 Bis(acetonitrile)dicarbonyliodo(η^3-2-propenyl)molybdenum, Mo-00030
33221-79-3 Tetrakis[μ-(acetato-*O*,*O′*)]bis(η^5-2,4-cyclopentadien-1-yl)divanadium, V -00083
33248-13-4 [Trimethyl(η^6-phenyl)silane]tricarbonylchromium, Cr-00105
33292-90-9 Tetracarbonylbis(pentacarbonylmanganio)osmium, Mn-00096
33306-91-1 Tricarbonyl(η^5-2,4-cyclopentadien-1-yl)-(trimethylgermyl)molybdenum, Mo-00056
33306-93-3 Tricarbonyl(η^5-2,4-cyclopentadien-1-yl)-(trimethylgermyl)tungsten, W -00063
33307-28-7 Tetracarbonyl(η^3-2-propenyl)manganese, Mn-00025
33307-29-8 Tetracarbonyl(η^3-2-propenyl)rhenium, Re-00024
33505-53-2 Tricarbonyl[(1,2,3,4,5,6-η)-hexamethylbenzene]tungsten, W -00105
33506-43-3 Tricarbonyl(1-methylpyrrole)chromium, Cr-00043
33634-75-2 Pentacarbonylrhenate(1−); Na salt, *in* Re-00012
33678-85-2 Pentacarbonyl(1,2-dioxopropyl)manganese, Mn-00029
33948-28-6 Tetrakis(trimethylsilylmethyl)titanium, Ti-00093
33958-72-4 Pentacarbonyl(pentacarbonylmanganese)-(tetracarbonyliron)rhenium, Re-00062
34089-19-5 Pentacarbonyl[η^1-1,2-ethanediylbis[diphenylphosphine]-*P*]vanadate(1−); Tetraethylammonium salt, *in* V -00111
34216-63-2 Tris(tetracarbonylcobalto)erbium; Tetrakis-THF complex, *in* Er-00006

34406-98-9 Tetracarbonyl[μ-(tetramethyldiphosphine)]-(pentacarbonylchromium)iron, Cr-00114
34439-17-3 Hexakis[(trimethylsilyl)methyl]dimolybdenum, Mo-00126
34691-61-7 Carbonyl(η^5-2,4-cyclopentadien-1-yl)-nitrosyl(triphenylphosphine)manganese, Mn-00132
34691-66-2 Pentacarbonyl(ethoxyferrocenylmethylene)tungsten, W -00114
34738-51-7 (Benzene)hexa-μ-chlorohexachlorotrialuminumuranium, U -00002
34767-30-1 Trichloro(η^5-2,4-cyclopentadien-1-yl)vanadium, V -00005
34767-44-7 Trichloro(η^5-2,4-cyclopentadien-1-yl)zirconium, Zr-00002
34788-74-4 Pentaaquabenzylchromium(2+), Cr-00036
34844-61-6 Tetrakis[(trimethylsilyl)methyl]bis[μ-(trimethylsilyl)methylidyne]diniobium, Nb-00046
35004-32-1 (η^6-Benzene)tris(methyldiphenylphosphine)molybdenum, Mo-00144
35084-05-0 Tricarbonyltrichlororhenate(2−); Di-Cs salt, *in* Re-00002
35226-88-1 Dicarbonyl[(1,2,3,4,5-η)-1-methyl-2,4-cyclopentadien-1-yl]-(ferrocenylmethoxymethylene)manganese, Mn-00118
35244-74-7 Pentacarbonyl(trimethylphosphine sulfide-*S*)chromium, Cr-00048
35394-18-4 Tetrakis[(trimethylsilyl)methyl]chromium, Cr-00138
35625-70-8 (η^6-Benzene)bis(η^3-2-propenyl)molybdenum, Mo-00069
35796-61-3 Bis(η^8-1,3,5,7-cyclooctatetraene)vanadium, V -00075
35886-41-0 η^5-2,4-Cyclopentadien-1-yl(methoxycarbonyl)-nitrosyl(triphenylphosphine)manganese, Mn-00133
36133-73-0 ▷Hexamethyltungsten, W -00011
36153-02-3 Tricarbonyl[(1,2,3,4,5,6-η)-2,4,6-triphenylphosphorin]chromium, Cr-00161
36153-92-1 Dichloromethyltris(tetrahydrofuran)chromium, Cr-00118
36153-96-5 Dichloromethyltrispyridinechromium, Cr-00133
36223-58-2 (η^5-2,4-Cyclopentadien-1-yl)-nitrosylbis(triphenyl phosphite-*P*)manganese(1+); Hexafluorophosphate, *in* Mn-00142
36312-04-6 Dicarbonyl(η^5-2,4-cyclopentadien-1-yl)nitrosylchromium, Cr-00032
36333-76-3 Tetrakis(bicyclo[2.2.1]-hept-1-yl)titanium, Ti-00139
36333-77-4 Tetrakis(bicyclo[2.2.1]hept-1-yl)vanadium, V -00107
36354-35-5 Dichlorobis[(1,2,3,3*a*,7*a*-η)-1*H*-inden-1-yl]cerium, Ce-00006
36495-37-1 Tricarbonyl(η^5-2,4-cyclopentadien-1-yl)hydrochromium, Cr-00041
36523-12-3 [(1,2,3,4,5-η)-1,2,3,4,5-Pentamethyl-2,4-cyclopentadien-1-yl]triphenoxytitanium, Ti-00137
36527-71-6 Pentacarbonyl(trimethylplumbyl)manganese, Mn-00038
36571-13-8 Hexacarbonyl(ethylmercury)vanadium, V -00014
36580-33-3 Carbonyl(η^7-cycloheptatrienylium)iodo(triethyl phosphite)tungsten, W -00097
36580-51-5 Amminepentacarbonylvanadate(1−); Tetraphenylphosphonium salt, *in* V -00001
36580-53-7 ▷Pentacarbonyl(cyano-*C*)vanadate(2−); Bis(tetraphenylphosphonium) salt, *in* V -00011
36594-09-9 Tetracarbonyl[2-[(phenylimino)methyl]phenyl-*C,N*]manganese, Mn-00110
36607-01-9 Bis(η^5-2,4-cyclopentadien-1-yl)di-μ-nitrosyldinitrosyldichromium, Cr-00078
36643-37-5 Hexakis[(trimethylsilyl)methyl]ditungsten, W -00145
36669-56-4 Hexacarbonyl(triphenylstannyl)niobium, Nb-00045
36681-92-2 Tetracarbonyl[1,2-ethanediylbis[diphenylphosphine]-*P,P'*]hydrovanadium, V -00110
36955-48-3 Bis[(1,2,3,4,5,6-η)ethylbenzene]vanadium, V -00076
37007-84-4 Tetrakis(2,2-dimethylpropyl)chromium, Cr-00151
37007-87-7 Tetrakis(2-methyl-2-phenylpropyl)chromium, Cr-00178
37048-11-6 Tricarbonyl[(1,2,3,4,5-η)-ferrocenyl-2,4-cyclopentadien-1-yl]manganese, Mn-00112
37131-50-3 Bromotricarbonyl(η^5-2,4-cyclopentadien-1-yl)tungsten, W -00018
37176-05-9 Tetracarbonyldiiodomanganate(1−), Mn-00006
37176-07-1 Dibromotetracarbonylmanganate(1−), Mn-00003
37205-20-2 Bis(η^5-2,4-cyclopentadien-1-yl)-bis(trifluoroacetato-*O*)zirconium, Zr-00032
37205-22-4 Bis(acetonitrile)bis(η^5-2,4-cyclopentadien-1-yl)vanadium(2+); Bis(hexafluorophosphate), *in* V -00062
37205-28-0 Tris(η^5-2,4-cyclopentadien-1-yl)methyluranium, U -00020
37205-32-6 Di-μ-chlorotetrakis(η^5-2,4-cyclopentadien-1-yl)discandium, *in* Sc-00001
37206-34-1 Bis(benzenethiolato)bis(η^5-2,4-cyclopentadien-1-yl)zirconium, Zr-00072
37206-36-3 Benzyltricyclopentadienyluranium, U -00043
37206-41-0 Bis(benzyl)bis(η^5-cyclopentadienyl)zirconium, Zr-00085
37215-22-8 Bis(η^5-2,4-cyclopentadien-1-yl)-(diethylcarbamodithioato-*S,S'*)vanadium(1+); Tetrafluoroborate, *in* V -00070
37216-56-1 Tetrakis(η^5-2,4-cyclopentadien-1-yl)neptunium, Np-00005
37260-83-6 Dibromobis(η^5-2,4-cyclopentadien-1-yl)hafnium, Hf-00002
37260-84-7 Bis(η^5-2,4-cyclopentadien-1-yl)difluorohafnium, Hf-00004
37260-85-8 Bis(η^5-2,4-cyclopentadien-1-yl)diiodohafnium, Hf-00005
37260-88-1 Bis(η^5-2,4-cyclopentadien-1-yl)dimethylhafnium, Hf-00013
37274-09-2 Bis[(1,2,3,4,5,6,7,8-η)-1-ethenyl-1,3,5,7-cyclooctatetraene]uranium, U -00033
37274-13-8 Bis[(1,2,3,4,5,6,7,8-η)-1-phenyl-1,3,5,7-cyclooctatetraene]uranium, U -00065
37281-22-4 Bis(η^8-1,3,5,7-cyclooctatetraene)neptunium, Np-00004
37281-23-5 Bis(η^8-1,3,5,7-cyclooctatetraene)plutonium, Pu-00004
37298-41-2 ▷Bis(η^5-2,4-cyclopentadien-1-yl)-[tetrahydroborato(1−)-*H,H'*]niobium, Nb-00015
37298-76-3 Tris(η^5-2,4-cyclopentadien-1-yl)-2-propenyluranium, U -00024
37298-77-4 Bromobis(η^5-2,4-cyclopentadien-1-yl)-(dimethylphenylphosphine)niobium, Nb-00036
37298-84-3 Butyltris(η^5-2,4-cyclopentadien-1-yl)uranium, U -00029
37298-90-1 Tris(η^5-2,4-cyclopentadien-1-yl)(2,2-dimethylpropyl)uranium, U -00036
37298-97-8 Tris(η^5-2,4-cyclopentadien-1-yl)cerium; Cyclohexylisocyanide adduct, *in* Ce-00003
37298-98-9 Tris(2,4-cyclopentadien-1-yl)isocyanocyclohexanedysprosium, *in* Dy-00004
37298-99-0 Tris(2,4-cyclopentadien-1-yl)isocyanocyclohexaneerbium, *in* Er-00011
37299-00-6 Tris(η^5-2,4-cyclopentadien-1-yl)europium; Cyclohexylisocyanide adduct, *in* Eu-00004
37299-01-7 Tris(η^5-2,4-cyclopentadien-1-yl)gadolinium; Cyclohexylisocyanide adduct, *in* Gd-00004
37299-02-8 Tris(η^5-2,4-cyclopentadien-1-yl)lanthanum; Cyclohexylisocyanide adduct, *in* La-00004
37299-03-9 Tris(η^5-2,4-cyclopentadien-1-yl)isocyanocyclohexanelutetium, *in* Lu-00012
37299-04-0 Tris(η^5-2,4-cyclopentadien-1-yl)samarium; Cyclohexylisocyanide adduct, *in* Sm-00009
37299-05-1 Tris(2,4-cyclopentadien-1-yl)isocyanocyclohexanethulium, *in* Tm-00004
37299-11-9 Dibenzylbis(η^5-2,4-cyclopentadien-1-yl)titanium, Ti-00125
37299-12-0 Tetracarbonylbis[(1,2,3,4,5-η)-1,2,3,4,5-pentamethyl-2,4-cyclopentadien-1-yl]dichromium, Cr-00160

39912-34-0 Benzoylpentacarbonylchromate(1−); Li salt, *in* Cr-00094
40270-21-1 Nonacarbonyl-μ-hydronitrosylditungsten, W -00029
40334-04-1 ▷Tetrakis(trimethylsilylmethyl)hafnium, Hf-00021
40844-50-6 Tetrakis(2,6-dimethylphenyl)lutetate(1−); Li salt, tetra-THF complex, *in* Lu-00023
40901-26-6 Pentacarbonyl-1,3-dioxalan-2-ylidenemanganese(1+), Mn-00030
41311-89-1 Cyclopentadienyltricarbonylmethylchromium, Cr-00060
41328-40-9 Tetrabenzylvanadium, V -00106
41354-56-7 [(2,3,4,5,6,7-η)-Bicyclo[6.1.0]nona-2,4,6-triene]tricarbonylmolybdenum, Mo-00062
41354-64-7 Tricarbonyl(3-8-η-[2.2]paracyclophane)chromium, Cr-00147
41395-41-9 Di-μ-chlorodichlorobis(η^5-2,4-cyclopentadien-1-yl)dinitrosyldimolybdenum, *in* Mo-00003
41453-01-4 Tetrachloromethylniobium, Nb-00001
41453-02-5 Trichlorodimethylniobium, Nb-00002
41453-04-7 Tetrachloromethyltantalum, Ta-00001
41453-05-8 Trichlorodimethyltantalum, Ta-00002
41556-44-9 Chlorotris(η^3-2-propenyl)uranium, U -00006
41558-36-5 (η^8-1,3,5,7-Cyclooctatetraene)bis(η^3-2-propenyl)hafnium, Hf-00016
41572-08-1 ▷Tetrakis[(trimethylsilyl)methyl]vanadium, V -00077
41576-43-6 Tricarbonyl(chloromercury)[μ-[(1-η:1,2,3,4,5,6-η)phenyl]]chromium, Cr-00051
41618-11-5 Tetracarbonyl(η^5-2,4-cyclopentadien-1-yl)molybdenum(1+); Hexafluorophosphate, *in* Mo-00022
41654-98-2 ▷Pentacarbonyl(diphenylthallium)manganese, Mn-00111
41678-48-2 Bis(2,2′-bipyridine-*N,N′*)bis[(trimethylsilyl)-methyl]chromium(1+); Iodide, *in* Cr-00167
41699-43-8 Di-μ-carbonyltricarbonylbis(η^5-2,4-cyclopentadien-1-yl)divanadium, V -00067
41700-58-7 Bis(dinitrogen)bis[1,2-bis(diphenylphosphino)-ethane]molybdenum, Mo-00145
41772-92-3 [Bis[bis(trimethylsilyl)methyl]tin]pentacarbonyl-chromium, Cr-00148
41944-00-7 Hexacarbonylrhenium(1+), Re-00017
42077-05-4 Hexakis(2,2-dimethylpropyl)dimolybdenum, Mo-00136
42077-14-5 (η^8-1,3,5,7-cyclooctatetraene)dihydrozirconium, Zr-00005
42087-89-8 Bis(η^6-chlorobenzene)chromium, Cr-00097
42087-90-1 Bis(η^6-fluorobenzene)chromium, Cr-00098
42167-74-8 Bis(pentacarbonylmanganese)dimethyllead, Mn-00078
42178-05-2 Tetrachlorobis(η^5-2,4-cyclopentadien-1-yl)[μ-[2,3-dimethyl-2,3-butanediolato(2−)-*O*:*O′*]-dititanium, Ti-00091
42423-66-5 Dichloro(η^8-1,3,5,7-cyclooctatetraene)zirconium, Zr-00004
42536-19-6 Tricarbonylchloronitrosyl(triphenylphosphine)tungsten, W -00133
42584-07-6 Tricarbonylchlorobis(triphenylphosphine)rhenium(1+); *mer,trans-form*, Tetrafluoroborate, *in* Re-00087
42612-73-7 Di-μ-chlorotetrakis(η^5-2,4-cyclopentadien-1-yl)diytterbium, *in* Yb-00004
42801-92-3 Iodotris(η^3-2-propenyl)uranium, U -00007
42801-93-4 Bromotris(η^3-2-propenyl)uranium, U -00005
42920-73-0 Hexakis(diethylaminato)-μ-1,1′-ferrocenediyldititanium, Ti-00143
42993-54-4 Tris(diethylaminato)ferrocenyltitanium, Ti-00120
43105-73-3 Dicarbonyl[(1,2,3,4,5,6-η)hexamethylbenzene]nitrosylchromium(1+); Hexafluorophosphate, *in* Cr-00122
43184-81-2 Dibromotetracarbonylmanganate(1−); Tetraethyl-ammonium salt, *in* Mn-00003
45113-87-9 Pentacarbonyl(isocyanomethane)manganese(1+), Mn-00021
45167-07-5 Tricarbonyltris(isocyanomethane)manganese(1+), Mn-00049
45211-82-3 Carbonylpentakis(isocyanomethane)manganese(1+), Mn-00068
45264-01-5 Decacarbonyldichromate(2−), Cr-00066
45978-17-4 Dicarbonyl(η^5-2,4-cyclopentadien-1-yl)nitrosylrhenium(1+), Re-00023
46238-41-9 Tricarbonyl(η^5-2,4-cyclopentadien-1-yl)(η^2-ethene)molybdenum(1+), Mo-00035
46238-43-1 Tricarbonyl(η^7-cycloheptatrienylium)chromium(1+), Cr-00070
46238-45-3 Tricarbonyl(η^7-cycloheptatrienylium)molybdenum(1+), Mo-00032
46238-46-4 Tricarbonyl(η^7-cycloheptatrienylium)tunsten(1+), W -00041
46238-54-4 Tricarbonyl(cyano-*C*)(η^5-2,4-cyclopentadien-1-yl)vanadate(1−); Na salt, *in* V -00018
48121-47-7 Tricarbonyl(η^5-2,4-cyclopentadien-1-yl)chromate(1−), Cr-00038
49547-63-9 Nonacarbonyl(1-methoxyethylidene)dirhenium, Re-00052
49547-64-0 Pentacarbonyl[tetracarbonyl(1-methoxyethylidene)]manganeserhenium, Re-00050
49596-04-5 Bis[(1,2,3,3*a*,7*a*-η)-1*H*-inden-1-yl]dimethylzirconium, Zr-00062
49596-06-7 Bis[(1,2,3,3*a*,7*a*-η)-1*H*-inden-1-yl]dimethylhafnium, Hf-00028
49596-07-8 Bis[(1,2,3,3*a*,7*a*-η)-1*H*-inden-1-yl]diphenyltitanium, Ti-00141
49598-13-2 Chloro(η^5-2,4-cyclopentadien-1-yl)trimethyltantalum, Ta-00007
49736-18-7 Decacarbonyldichromate(2−); Bis(bis(triphenyl-phosphine)iminium)salt, *in* Cr-00066
49736-19-8 Decarbonyldimolybdate(2−); Bis[bis(triphenylphosphine)iminium]salt, *in* Mo-00045
49742-48-5 Bromodicarbonyltris(trimethylphosphine)rhenium, Re-00047
50276-12-5 Pentacarbonyl(diphenylmethylene)tungsten, W -00111
50358-90-2 (Carbonothioyl)pentacarbonylchromium, Cr-00018
50358-92-4 (Carbonothioyl)pentacarbonyltungsten, W -00013
50589-81-6 (2,2′-Bipyridine-*N,N′*)dichlorotrimethyltantalum, Ta-00020
50643-51-1 Tris(η^5-2,4-cyclopentadien-1-yl)(2-methyl-1-propenyl)uranium, U -00026
50647-46-6 *tert*-Butyltricyclopentadienyluranium, U -00028
50654-35-8 Tetrakis(2,2-dimethylpropyl)hafnium, Hf-00030
50701-13-8 Bromotetracarbonyl(phenylmethylidyne)chromium, Cr-00084
50701-14-9 Tetracarbonyliodoethylidynechromium, Cr-00023
50726-31-3 Bromotetracarbonylethylidynetungsten, W -00005
50726-32-4 Tetracarbonylethylidyneiodotungsten, W -00008
50831-23-7 (2-Acetylphenyl-*C,O*)tetracarbonylmanganese, Mn-00080
50923-29-0 Tetrakis(η^5-2,4-cyclopentadien-1-yl)di-μ-silylenedititanium, Ti-00110
50982-07-5 [Di-*tert*-butyl(pyridine)tin]pentacarbonylchromium, Cr-00145
51056-18-9 Bis(η^8-1,3,5,7-cyclooctatetraene)protactinium, Pa-00001
51159-66-1 (Azobenzene)bis(cyclopentadienyl)vanadium, V -00091
51177-48-1 Chloro(cyclooctatetraene)praesodymium; Bis-THF complex, *in* Pr-00002
51177-49-2 Chloro(cyclooctatetraene)neodymium; Bis-THF adduct, *in* Nd-00002
51177-53-8 Bis(η^8-1,3,5,7-cyclooctatetraene)praesodymate(1−); K salt, *in* Pr-00004
51177-54-9 Bis(η^8-1,3,5,7-cyclooctatetraene)neodymate(1−); K salt, *in* Nd-00005
51177-55-0 Bis(η^8-1,3,5,7-cyclooctatetraene)samarate(1−); K salt, *in* Sm-00010
51177-57-2 Bis(η^8-1,3,5,7-cyclooctatetraene)terbate(1−); K salt, *in* Tb-00003
51177-89-0 Bis(η^5-2,4-cyclopentadien-1-yl)diphenylzirconium, Zr-00073
51185-44-5 Tetracarbonyldichlorobis(η^5-2,4-cyclopentadien-1-yl)ditungsten, W -00089
51187-41-8 Bis(η^8-1,3,5,7-cyclooctatetraene)yttrate(1−); K salt, *in* Y -00007
51187-42-9 Bis(η^8-1,3,5,7-cyclooctatetraene)lanthanate(1−); K salt, *in* La-00005

51187-43-0 Bis(η^8-1,3-5,7-cyclooctatetraene)cerate(1−); K salt, *in* Ce-00005
51203-49-7 (η^7-Cycloheptatrienylium)(η^5-2,4-cyclopentadien-1-yl)titanium, Ti-00055
51231-81-3 Bis(η^5-2,4-cyclopentadien-1-yl)diphenylhafnium, Hf-00034
51233-19-3 Pentacarbonylchromate(2−); Di-Na salt, *in* Cr-00008
51733-16-5 Tris(η^4-1,3-butadiene)tungsten, W -00079
51733-17-6 Tris(η^4-1,3-butadiene)molybdenum, Mo-00070
51809-70-2 Trichlorobis(η^7-2,4,6-cycloheptatrien-1-yl)dimolybdenum, Mo-00081
51831-93-7 Carbonothioyl(η^5-2,4-cyclopentadien-1-yl)-bis(trimethylphosphite-*P*)manganese, Mn-00084
51837-59-3 Tetrakis[bis(η^5-2,4-cyclopentadien-1-yl)-hydromolybdenum]tetralithium, *in* Mo-00040
51877-88-4 Tetrachloromethyl(triphenylphosphine)niobium, Nb-00039
51886-84-1 Carbonylchlorobis[1,2-ethanediylbis(diphenylphosphine-*P*,*P′*)]rhenium(1+); *trans-form*, Hexafluorophosphate, *in* Re-00092
52124-67-1 (1,4-Butanediyl)bis(η^5-2,4-cyclopentadien-1-yl)titanium, Ti-00073
52138-78-0 Pentacarbonyl[(cyano-*C*)trihydroborato(1−)-*N*]-chromate(1−); Tetramethylammonium salt, *in* Cr-00021
52193-18-7 Carbonyltris(diethylcarbamodithioato-*S*,*S′*)rhenium, Re-00072
52308-70-0 [(1,2,3,4,5,6-η)-1,3,5-Cycloheptatriene](η^5-2,4-cyclopentadien-1-yl)manganese, Mn-00083
52308-73-3 (η^7-Cycloheptatrienylium)(η^5-2,4-cyclopentadien-1-yl)manganese(1+); Hexafluorophosphate, *in* Mn-00082
52340-90-6 Tetradecacarbonyltrimanganate(1−); Tetraphenylarsonium salt, *in* Mn-00098
52346-34-6 ▷Bis(η^6-chlorobenzene)molybdenum, Mo-00061
52346-44-8 Bis[(1,2,3,4,5,6-η)methylbenzene]tungsten, W -00094
52409-63-9 Tricarbonyl(η^5-2,4-cyclopentadien-1-yl)methanesulfonyltungsten, W -00035
52445-38-2 Tricarbonyl[(1,2,3,4,5,6-η)-7-(diphenylmethylene)-1,3,5-cycloheptatriene]molybdenum, Mo-00120
52456-36-7 Tricarbonyl(η^5-2,4-cyclopentadien-1-yl)-(dimethylarsino)tungsten, W -00051
52462-43-8 Bis(η^6-benzene)titanium, Ti-00054
52471-95-1 Tricarbonyl[(1,2,3,4,5,6-η)-7-(1-phenylethylidene)-1,3,5-cycloheptatriene]tungsten, W -00115
52472-03-4 Tricarbonyl[(1,2,3,4,5,6-η)-7-methylene-1,3,5-cycloheptatriene]chromium, Cr-00086
52472-04-5 Tricarbonyl(η^5-2,4-cyclopentadien-1-yl)ferrocenyltungsten, W -00113
52542-59-3 Pentacarbonylmanganate(1−); Bis(triphenylphosphine)iminium salt, *in* Mn-00014
52550-20-6 Tris(η^5-2,4-cyclopentadien-1-yl)thorium, Th-00008
52571-36-5 Tricarbonyl(η^5-2,4-cyclopentadien-1-yl)-(dimethylstilbino)tungsten, W -00053
52589-03-4 Decacarbonyl[μ-(diphosphine-*P*:*P′*)]dichromium, Cr-00068
52646-26-1 Bromocarbonyltetrakis(trimethylphosphite-*P*)manganese, Mn-00089
52668-25-4 (η^8-1,3,5,7-Cyclooctatetraene)(η^5-2,4-cyclopentadien-1-yl)yttrium, Y -00004
52668-26-5 (η^8-1,3,5,7-Cyclooctatetraene)(η^5-2,4-cyclopentadien-1-yl)neodymium, Nd-00003
52668-27-6 (η^8-1,3,5,7-Cyclooctatetraene)(η^5-2,4-cyclopentadien-1-yl)samarium, Sm-00007
52668-28-7 (η^8-1,3,5,7-Cyclooctatetraenyl)(η^5-2,4-cyclopentadien-1-yl)holmium, Ho-00004
52676-23-0 μ(η^5:η^5-Fulvalene)di-μ-hydridobis(cyclopentadienyltitanium), Ti-00108
52679-41-1 Dichloro(η^5-2,4-cyclopentadien-1-yl)samarium; Tris-THF adduct, *in* Sm-00001
52679-42-2 Dichloro(η^5-2,4-cyclopentadien-1-yl)holmium; Tris-THF adduct, *in* Ho-00001
52679-44-4 Chloro(cyclooctatetraene)samarium; THF complex, *in* Sm-00003
52680-31-6 Tricarbonyl(η^5-2,4-cyclopentadien-1-yl)-(ferrocenylcarbonyl)tungsten, W -00121
52700-41-1 [(1,2-η)-2,4-cyclopentadien-1-yl]bis(η^5-2,4-cyclopentadien-1-yl)titanium, Ti-00079
52720-93-1 Tricarbonyl(η^5-2,4-cyclopentadien-1-yl)thalliummolybdenum, Mo-00015
52810-84-1 Dichloro(η^5-2,4-cyclopentadien-1-yl)yttrium; Tris-THF adduct, *in* Y -00001
52827-35-7 Tris(η^5-2,4-cyclopentadien-1-yl)ethynyluranium, U -00022
52827-36-8 Tris(η^5-2,4-cyclopentadien-1-yl)ferrocenyluranium, U -00057
52841-89-1 Bromotricarbonylbis(trimethylphosphite-*P*)manganese, Mn-00053
52901-72-1 Tetracarbonyldichloromanganate(1−); Tetraethylammonium salt, *in* Mn-00004
53011-14-6 Tricarbonyl[(1,2,3,4,5-η)-2,4,6-cycloheptatrien-1-yl]manganese, Mn-00054
53022-70-1 Tetramethyloxorhenium, Re-00005
53041-79-5 Tricarbonyl[(1,2,3,4,5,6-η)ethenylbenzene]molybdenum, Mo-00048
53079-53-1 Dichloro(η^5-2,4-cyclopentadien-1-yl)-(diethylcarbamodithioato-*S*,*S′*)titanium, Ti-00039
53127-56-3 (η^6-Benzene)dicarbonyl(methoxycarbonyl)manganese, Mn-00057
53183-18-9 Dicarbonylnitrosylbis(triphenylphosphine)rhenium, Re-00085
53224-33-2 Bis(η^5-2,4-cyclopentadien-1-yl)(phenylethynyl)gadolinium, Gd-00006
53224-34-3 Bis(η^5-2,4-cyclopentadien-1-yl)(phenylethynyl)ytterbium, Yb-00019
53224-35-4 Chlorobis(η^5-2,4-cyclopentadien-1-yl)erbium, Er-00004
53235-29-3 Tetrabenzyltungsten, W -00149
53237-33-5 Tetrakis[(dimethylphosphonio)bis(methylene)]dichromium, Cr-00137
53291-02-4 Bis(η^5-2,4-cyclopentadien-1-yl)iodovanadium, V -00031
53291-19-3 Carbonylbis(η^5-2,4-cyclopentadien-1-yl)iodovanadium, V -00039
53307-60-1 Octamethyldimolybdenum(4−); Tetra-Li salt, tetrakis-THF complex, *in* Mo-00020
53321-77-0 Bis(η^7-2,4,6-cycloheptatrien-1-yl)tri-μ-methoxydimolybdenum, Mo-00100
53322-18-2 Bis(η^5-2,4-cyclopentadien-1-yl)hydro(tungsten)lithium; Tetramer, *in* W -00052
53339-41-6 Carbonylbis(η^5-2,4-cyclopentadien-1-yl)vanadium, V -00040
53378-72-6 ▷Pentamethyltantalum, Ta-00005
53418-18-1 Tetracarbonyl(triphenylphosphine)manganate(1−), Mn-00124
53419-02-6 Carbonyl(η^5-2,4-cyclopentadien-1-yl)dinitrosyltungsten(1+), W -00010
53419-03-7 Carbonyl(η^5-2,4-cyclopentadien-1-yl)dinitrosyltungsten(1+); Hexafluorophosphate, *in* W -00010
53419-14-0 Chloro(η^5-cyclopentadien-1-yl)dinitrosyltungsten, W -00003
53433-49-1 Tetracarbonyl(di-μ-carbonylhexacarbonyldiferrate)manganate(1−); Tetraethylammonium salt, *in* Mn-00072
53433-58-2 Bis(η^5-2,4-cyclopentadien-1-yl)(1,2,3,4-tetraphenyl-1,3-butadiene-1,4-diyl)zirconium, Zr-00112
53433-59-3 Bis(η^5-2,4-cyclopentadien-1-yl)(1,2,3,4-tetraphenyl-1,3-butadiene-1,4-diyl)hafnium, Hf-00051
53437-03-9 [1,2-Ethanediylbis[dimethylphosphine]-*P*,*P′*]pentamethylniobium, Nb-00021
53449-94-8 Bis(η^5-2,4-cyclopentadien-1-yl)(η^3-2-propenyl)molybdenum(1+), Mo-00077
53449-95-9 Bis(η^5-2,4-cyclopentadien-1-yl)(η^3-2-propenyl)molybdenum(1+); Hexafluorophosphate, *in* Mo-00077
53470-46-5 Trichlorotrimethylniobate(1−); Tetraethylammonium salt, *in* Nb-00005
53470-78-3 Tricarbonyl(η^6-benzyltrimethylsilyl)chromium, Cr-00116
53504-75-9 Bis(η^5-2,4-cyclopentadien-1-yl)(η^2-ethene)methyltungsten(1+), W -00085

55987-17-2 (Carbonoselenoyl)dicarbonyl(η^5-2,4-cyclopentadien-1-yl)manganese, Mn-00032

56009-67-7 Bis(benzenethiolato)bis(η^5-2,4-cyclopentadien-1-yl)vanadium, V -00092

56028-53-6 Bis(η^5-2,4-cyclopentadien-1-yl)diiodoniobium, Nb-00013

56090-02-9 ▷Hexamethylrhenium, Re-00016

56172-01-1 Decacarbonyl-μ-hydroditungstate(1−); Bis(triphenylphosphine)iminium salt, *in* W -00040

56200-25-0 Di-μ-chlorotetrakis(η^5-2,4-cyclopentadien-1-yl)disamarium, *in* Sm-00004

56200-26-1 Chlorobis(η^5-2,4-cyclopentadien-1-yl)holmium, Ho-00002

56200-27-2 Tetracarbonylbis(η^5-2,4-cyclopentadien-1-yl)dimolybdenum, Mo-00079

56307-56-3 Bis[1,2-ethanediylbis[diphenylphosphine]-*P*,*P′*]-bis(η^2-ethene)molybdenum, Mo-00146

56328-27-9 Pentacarbonyl(triphenylphosphine)vanadate(1−); Tetraethylammonium salt, *in* V -00094

56379-20-5 Di-μ-chlorotetrakis(η^5-2,4-cyclopentadien-1-yl)-μ-1,2-ethanediylbis(triethylaluminum)dizirconium, Zr-00109

56408-47-0 Tetracarbonyl[1,2-ethanediylbis[dimethylphosphine]-*P*,*P′*]hydrovanadium, V -00035

56408-48-1 Tetracarbonylmethyl[1,2-phenylenebis(dimethylarsine)-*As*,*As′*]vanadium, V -00069

56420-26-9 Bis(η^5-2,4-cyclopentadien-1-yl)-bis[tetrahydroborato(1−)-*H*,*H′*]hafnium, Hf-00009

56483-30-8 Tetracarbonyl[2-(diphenylphosphinophenyl-*C*,*P*)]-rhenium, Re-00079

56665-73-7 μ-Carbonyltetracarbonylbis[μ-[methylenebis[diphenylphosphine]-*P*,*P′*]]dimanganese, Mn-00145

56708-99-7 (2-Acetylferrocenyl-*C*,*O*)tetracarbonylmanganese, Mn-00107

56765-58-3 Tris(diethylamido)ethylvanadium, V -00066

56770-59-3 Bis(benzyl)bis(η^5-cyclopentadienyl)hafnium, Hf-00039

56770-61-7 Carbonylbis(η^5-2,4-cyclopentadien-1-yl)-(trimethylphosphine)titanium, Ti-00074

57088-86-5 (η^5-2,4-Cyclopentadien-1-yl)dioxovanadium, V -00007

57088-89-8 Bis(η^5-2,4-cyclopentadien-1-yl)[1,2-ethanediylbis[diphenylphosphine]-*P*,*P′*]vanadium(1+); Tetraphenylborate, *in* V -00116

57088-91-2 Dicarbonylbis(2,4-cyclopentadien-1-yl)vanadium(1+); Tetraphenylborate, *in* V -00050

57088-95-6 Carbonylbis(η^5-2,4-cyclopentadien-1-yl)-(pyridine)vanadium(1+); Tetraphenylborate, *in* V -00071

57091-06-2 Bromotetracarbonyl[(diethylamino)methylidyne]-chromium, Cr-00062

57127-95-4 Pentacarbonylchromate(2−); Di-Cs salt, *in* Cr-00008

57196-01-7 Pentacarbonyl(methyl thiocyanate-*N*)chromium, Cr-00028

57288-89-8 Hexacarbonyltantalate(1−); Tetraphenylarsonium salt, *in* Ta-00006

57347-27-0 Bis[(1,2,3,4,5,6-η)-1,3,5-trimethylbenzene]titanium, Ti-00100

57348-61-5 Decacarbonyldichromate(2−); Di-K salt, *in* Cr-00066

57348-62-6 Decarbonyldimolybdate(2−), Mo-00045

57348-62-6 Decarbonyldimolybdate(2−); Di-K salt, *in* Mo-00045

57398-65-9 Bis(η^5-2,4-cyclopentadien-1-yl)-(dimethylaluminum)di-μ-methylyttrium, Y -00005

57560-35-7 Pentacarbonyl(1,3-dimethyl-2-imidazolidinylidene)molybdenum, Mo-00038

57574-51-3 *N*,*N*,*N*-Triethylethanaminium(1+) pentacarbonylmethyltungstate(1−), *in* W -00009

57595-19-4 Pentachloro[(trimethylsilyl)methyl]tungsten, W -00001

57603-41-5 Tetracarbonylbis(η^5-2,4-cyclopentadien-1-yl)-μ-methylenedimanganese, Mn-00102

57812-91-6 (Chlorobenzene)tricarbonylmanganese(I), Mn-00042

57891-59-5 Diacetyltetracarbonylmanganate(1−); Li salt, *in* Mn-00037

57913-16-3 Bis(η^5-2,4-cyclopentadien-1-yl)-methyl(methylidene)tantalum, Ta-00017

57956-59-9 Nonacarbonyl(isocyanomethane)dimanganese, Mn-00065

57987-16-3 Bromotetracarbonyl[(4-methylphenyl)methylidyne]-chromium, Cr-00095

58034-11-0 Hexakis[μ-(acetyl-*C*:*O*)](aluminum)dodecacarbonyltrimanganese, *in* Mn-00037

58034-14-3 Octacarbonylbis(isocyanomethane)dimanganese, Mn-00077

58057-99-1 (η^5-Cyclopentadienyl)tris(dimethylamido)titanium, Ti-00049

58355-39-8 [(1,2,3,4,5-η)-1-methyl-2,4-cyclopentadien-1-yl][(1,2,3,4,5,6-η)-7-phenyl-1,3,5-cycloheptatriene]manganese, Mn-00115

58384-14-8 μ-Bromooctacarbonyl[μ-(phenylmethylidyne)di-rhenium, Re-00069

58689-69-3 Bis(η^5-benzylcyclopentadienyl)dichlorozirconium, Zr-00084

58694-74-9 Dihydronitrosyltris(triphenylphosphine)rhenium; *mer,cis-form*, *in* Re-00093

58832-09-0 Tetrakis(2-methyl-2-propanolato)oxotungsten, W -00110

58920-14-2 Tris(η^5-2,4-cyclopentadien-1-yl)(2,2-dimethylpropyl)thorium, Th-00015

58938-56-0 Pentacarbonyl[chloro(dimethylamino)methylene]-manganese(1+), Mn-00036

58938-57-1 Pentacarbonyl[chloro(dimethylamino)methylene]-manganese(1+); Perchlorate, *in* Mn-00036

59123-04-5 Tetracarbonylmanganate(3−); Tri-Na salt, *in* Mn-00008

59172-20-2 (η^5-2,4-Cyclopentadien-1-yl)bis[(2,3-η)-1,1,1,4,4,4-hexafluoro-2-butyne]iodomolybdenum, Mo-00074

59218-83-6 Pentacarbonyl[chloro(diethylamino)methylene]chromium, Cr-00074

59218-84-7 Tetracarbonylchloro[(diethylamino)methylidyne]-chromium, Cr-00063

59296-77-4 Bis(1,3,5,7-cyclooctatetraene)phenylniobium, Nb-00044

59296-78-5 Bis(1,3,5,7-cyclooctatetraene)methylniobium, Nb-00033

59412-84-9 Chlorobis(η^5-2,4-cyclopentadien-1-yl)oxoniobium, Nb-00011

59423-86-8 Dicarbonyl(η^5-2,4-cyclopentadien-1-yl)-(tetrahydrofuran)rhenium, Re-00046

59424-00-9 Bis(η^5-2,4-cyclopentadien-1-yl)ethylvanadium, V -00055

59458-55-8 [μ-[(1,2-η:1,2,3,4,5,6,7,8-η)-1,3,5,7-Cyclooctatetraene][bis(η^8-1,3,5,7-cyclooctatetraene)bis(tetrahydrofuran)dineodymium, Nd-00012

59464-56-1 (η^6-Benzene)[(1,2,3,4-η)-1,3,5,7-cyclooctatetraene](η^3-2-propenyl)molybdenum(1+); Hexafluorophosphate, *in* Mo-00099

59464-63-0 Carbonyl(η^5-2,4-cyclopentadien-1-yl)[(2,3-η)-1,1,1,4,4,4-hexafluoro-2-butyne]-(trifluoromethanethiolato)molybdenum, Mo-00046

59487-44-4 Tetrakis[μ-(benzenethiolato)]bis(η^5-2,4-cyclopentadien-1-yl)divanadium, V -00114

59487-85-3 Dicarbonylbis(η^5-2,4-cyclopentadien-1-yl)zirconium, Zr-00021

59487-86-4 Dicarbonylbis(η^5-2,4-cyclopentadien-1-yl)hafnium, Hf-00012

59512-63-9 Tribenzyl(piperidinyl)titanium, Ti-00130

59512-73-1 Bis(dimethylamido)bis(trimethylsilylmethyl)titanium, Ti-00062

59512-75-3 Tris(dimethylamido)(trimethylsilylmethyl)titanium, Ti-00043

59512-77-5 Bis(dimethylamido)dimethyltitanium, Ti-00013

59547-67-0 Bis(η^5-2,4-cyclopentadien-1-yl)(1,2,3,4-tetraphenyl-1,3-butadiene-1,4-diyl)vanadium, V -00119

59589-12-7 Dicarbonylchloro(η^7-cycloheptatrienylium)tungsten, W -00032

59645-02-2 [μ-(1,4-Diphenyl-1,3-butadiene-1,2,3,4-tetrayl-C^1,C^3:C^2,C^4)]tetrakis[(1,2,3,4,5-η)-1-methyl-2,4-cyclopentadien-1-yl]dititanium, Ti-00147

61663-90-9 Octacarbonylbis(triphenylphosphine)ditechnetium, Tc-00013
61732-19-2 Di-μ-chlorotetrakis(η^5-2,4-cyclopentadien-1-yl)diterbium, Tb-00004
61769-24-2 Tetracarbonylmanganate(3−), Mn-00008
61770-99-8 Pentacarbonyl[(dimethylamino)methylidyne]chromium(1+); Tetrachloroborate, *in* Cr-00039
61771-01-5 Pentacarbonyl[(dimethylamino)methylidyne]chromium(1+); Hexafluorophosphate, *in* Cr-00039
61771-03-7 Pentacarbonyl[cyano(dimethylamino)methylene]-chromium, Cr-00055
61771-06-0 [Bis(dimethylamino)methylene]pentacarbonylchromium, Cr-00079
61818-15-3 Dicarbonyldichlorobis(triethylphosphine)rhenium, Re-00067
61906-04-5 Trichloro(η^5-2,4-cyclopentadien-1-yl)hafnium, Hf-00001
61916-37-8 Dicarbonylbis[1,2-ethanediylbis[dimethylphosphine]-*P*,*P'*]methyltantalum, Ta-00029
61917-87-1 Tricarbonyltrihydro[1,2-phenylenebis[dimethylarsine]-*As*,*As'*]vanadium, V -00061
61930-16-3 (η^8-1,3,5,7-Cyclooctatetraene)(η^5-2,4-cyclopentadien-1-yl)molybdenum(1+); Hexafluorophosphate, *in* Mo-00076
61993-63-3 Dicarbonyl(η^5-2,4-cyclopentadien-1-yl)-(phenylmethylidyne)rhenium(1+), Re-00066
61993-64-4 Dicarbonyl(η^5-2,4-cyclopentadien-1-yl)-(phenylmethylidyne)rhenium(1+); Tetrachloroborate, *in* Re-00066
62015-65-0 Tricarbonyl(η^5-2,4-cyclopentadien-1-yl)-(tetracarbonylcobalt)molybdenum, Mo-00060
62044-68-2 Pentabenzyltantalum, Ta-00040
62050-66-2 Pentacarbonyl[(1,2,3-η)-1,2,3-triphenyl-2-cyclopropen-1-yl]vanadium, V -00103
62050-78-6 Dicarbonylbis[1,2-ethanediylbis[dimethylphosphine]-*P*,*P'*]hydrotantalum, Ta-00026
62171-36-2 Octamethyldirhenate(2−); Bis[tetrakis(tetrahydrofuran)lithium(1+)]salt, *in* Re-00036
62197-79-9 (T-4)Triphenyl(*P*,*P*,*P*-triphenylphosphineimidato-*N*)phosphorus(1+) pentacarbonylmethyltungstate(1−), *in* W -00009
62197-85-7 (*T*-4)Triphenyl(*P*,*P*,*P*-triphenylphosphineimidato-*N*)phosphorus(1+) pentacarbonylmethylchromate(1−), *in* Cr-00024
62303-78-0 Tetracarbonylbis(η^5-2,4-cyclopentadien-1-yl)[μ-[(2,3-η:2,3-η)-dimethyl-2-butynedioate]]ditungsten, W -00127
62319-82-8 Bromotetracarbonyl[(triphenylsilyl)methylidyne]-tungsten, W -00137
62337-31-9 Octamethyltungstate(2−); Di-Li salt, *in* W -00027
62341-83-7 Decacarbonyl-μ-hydrodichromate(1−); Bis(triphenylphosphine)iminium salt, *in* Cr-00067
62363-03-5 Bis(2,4-cyclopentadien-1-yl)dimethylvanadium, V -00056
62450-90-2 Bis[μ-(2-methyl-2-phenylpropyl)]bis(2-methyl-2-phenylpropyl)dimanganese, Mn-00141
62559-95-9 Dichlorobis[(7,8,9,10,11-η)undecahydro-7,8-dicarbaundecaborato]uranate(2−); Di-Li salt, octa-THF complex, *in* U -00001
62573-31-3 Tetracarbonylbis(η^5-2,4-cyclopentadien-1-yl)[μ-[(1,2-η:2,3-η)-1,2-propadiene]]dimolybdenum, Mo-00098
62589-11-1 Pentacarbonyl(diphenylmethylene)chromium, Cr-00143
62613-18-7 Tetrakis[μ-(2,6-dimethoxyphenyl-*C*:*O*)]divanadium, V -00112
62651-63-2 Bis[1,2-ethanediylbis[dimethylphosphine]-*P*,*P'*]-bis(η^2-ethene)hydroniobium, Nb-00032
62680-77-7 Bis(acetonitrile)dicarbonylchloro(η^3-2-propenyl)tungsten, W -00039
62789-86-0 [Bis[bis(trimethylsilyl)methyl]plumbylene]pentacarbonylmolybdenum, Mo-00110
62798-73-6 Carbonyl(η^5-2,4-cyclopentadien-1-yl)(η^2-ethyne)-methyltungsten, W -00038
62853-03-6 Tetracarbonylbis(η^5-2,4-cyclopentadien-1-yl)ditungsten, W -00090
62866-03-9 Tricarbonyl(η^5-2,4-cyclopentadien-1-yl)tungstate(1−); K salt, *in* W -00023
62866-16-4 Tricarbonyl(η^5-2,4-cyclopentadien-1-yl)(η^2-ethene)molybdenum(1+); Tetrafluoroborate, *in* Mo-00035
62927-98-4 Tetrachloro(η^5-2,4-cyclopentadien-1-yl)tantalum, Ta-00004
62938-49-2 Tetracarbonylchloroethylidynechromium, Cr-00022
62980-54-5 Tricarbonyl[(1,2,3,4,5,6-η)-4-methyl-1-phenylboratabenzene]manganese, Mn-00101
62997-61-9 Pentacarbonyl(cyano-*C*)chromate(1−); $(THF)_3Cr(3+)$ salt, *in* Cr-00016
63079-59-4 Tricarbonyl(η^5-2,4-cyclopentadien-1-yl)(η^2-ethene)tungsten(1+); Tetrafluoroborate, *in* W -00045
63118-93-4 Bis(η^5-2,4-cyclopentadien-1-yl)diethylvanadium, V -00064
63118-94-5 Dibenzylbis(cyclopentadienyl)vanadium, V -00099
63139-42-4 Pentacarbonylrhenate(1−); K salt, *in* Re-00012
63166-76-7 (η^5-2,4-Cyclopentadien-1-yl)trifluorotitanium, Ti-00010
63243-40-3 Di-*tert*-butoxydimethylvanadium, V -00036
63339-27-5 Tetracarbonyl(η^5-methyl-2,4-cyclopentadien-1-yl)vanadium, V -00024
63356-86-5 (η^6-Benzene)(carbonothioyl)dicarbonylchromium, Cr-00056
63356-87-6 (Carbonoselenoyl)pentacarbonylchromium, Cr-00019
63415-86-1 Ethoxytris(η^5-2,4-cyclopentadien-1-yl)uranium, U -00023
63422-29-7 Tetrakis(2-methyl-1-phenyl-1-propenyl)titanium, Ti-00148
63422-30-0 Tetrakis(2-methyl-1-phenyl-1-propenyl)zirconium, Zr-00114
63422-31-1 Tetrakis(2-methyl-1-phenyl-1-propenyl)hafnium, Hf-00053
63422-41-3 Chlorobis(η^5-2,4-cyclopentadien-1-yl)(2-methyl-1-phenyl-1-propenyl)zirconium, Zr-00063
63428-34-2 [Bis(2,4,6-trimethylphenyl)germylene]pentacarbonylchromium, Cr-00156
63590-28-3 (η^6 Benzene)(carbonothioyl)-carbonyl(triphenylphosphine)chromium, Cr-00164
63641-12-3 Dicarbonyl(η^5-2,4-cyclopentadien-1-yl)(1,3-diphenyltriazenato-N^1,N^3)tungsten, W -00122
63643-43-6 Butyltris[(1,2,3,3*a*,7*a*-η)-1*H*-inden-1-yl]thorium, Th-00035
63643-54-9 Tris[(1,2,3,3*a*,7*a*-η)-1*H*-indenyl]methylthorium, Th-00029
63643-55-0 Tris[(1,2,3,3*a*,7*a*-η)-1*H*-inden-1-yl]methyluranium, U -00066
63720-03-6 Bis[(1,2,3,3*a*,7*a*-η)-1*H*-inden-1-yl]dimethyluranium, U -00034
63720-04-7 Di-*tert*-butyldiindenyluranium, U -00058
63726-15-8 Di-2,4-cyclopentadien-1-ylbis(η^5-2,4-cyclopentadien-1-yl)titanium, Ti-00107
63782-46-7 Carbonylchlorobis(η^5-2,4-cyclopentadien-1-yl)niobium, Nb-00017
63816-92-2 Hexaphenyluranate(2−); Di-Li salt, *in* U -00074
63816-93-3 Hexamethyluranate(3−); Tri-Li salt, *in* U -00004
63835-71-2 Tricarbonyl[η^6-5-(1-methylethylidene)-1,3-cyclopentadiene]molybdenum, Mo-00052
63882-22-4 Tricarbonyl(diphenylmethylidenecyclopentadiene)-tungsten, W -00132
63890-69-7 (2,2'-Bipyridine-*N*,*N'*)tricarbonylnitrosylmolybdenum(1+), Mo-00075
63890-70-0 (2,2'-Bipyridine-*N*,*N'*)tricarbonylnitrosylmolybdenum(1+); *fac-form*, Tetrafluoroborate, *in* Mo-00075
63890-73-3 Tricarbonylnitrosyl(1,10-phenanthroline-N^1,N^{10})-tungsten(1+), W -00098
63890-74-4 Tricarbonylnitrosyl(1,10-phenanthroline-N^1,N^{10})-tungsten(1+); *mer-form*, Hexafluorophosphate, *in* W -00098
63947-89-7 (η^4-1,3-Butadiene)dicarbonyl(η^5-2,4-cyclopentadien-1-yl)molybdenum(1+); Tetrafluoroborate, *in* Mo-00055
63950-86-7 Hexakis(η^5-2,4-cyclopentadien-1-yl)[μ-(pyrazine-N^1:N^4)]diytterbium, Yb-00033
64041-01-6 Tricarbonyl(η^5-2,4-cyclopentdien-1-yl)-[dicarbonyl(η^5-2,4-cyclopentadien-1-yl)-iron]tungsten, W -00102

66615-33-6 Bis(acetonitrile)dicarbonyl[(1,2,3,3*a*,7*a*-η)-1*H*-inden-1-yl]molybdenum(1+); Tetrafluoroborate, *in* Mo-00088

66632-54-0 Bis(η^5-2,4-cyclopentadien-1-yl)[1,2-ethanediylbis[dimethylphosphine]-*P*,*P'*]tantalum(1+); Chloride, *in* Ta-00031

66705-66-6 Bis(η^5-2,4-cyclopentadien-1-yl)bis[μ-(η:η^5-2,4-cyclopentadien-1-ylidene)]dimolybdenum, Mo-00115

66719-22-0 [μ-[(1,2,3,4-η:5,6,7,8-η)-1,3,5,7-cyclooctatetraene]]bis[(1,2,3,4-η)-1,3,5,7-cyclooctatetraene]dimolybdenum, Mo-00123

66719-23-1 [μ-[(1,2,3,4-η:5,6,7,8-η)-1,3,5,7-cyclooctatetraene]]bis[(1,2,3,4-η)-1,3,5,7-cyclooctatetraene]ditungsten, W -00142

66862-11-1 Hexamethylerbate(3−); Tris[(tetramethylethylenediamine)lithium] salt, *in* Er-00002

66862-13-3 Hexamethyllutetate(3−); Tris[(tetramethylethylenediamine)lithium] salt, *in* Lu-00002

66901-71-1 ($\eta^{1,2}$-[2,2]Paracyclophane)chromium, Cr-00132

66908-84-7 Chlorobis(η^5-2,4-cyclopentadien-1-yl)methylvanadium, V -00046

66915-54-6 Pentacarbonyl(diethylcyanamide-*N'*)chromium, Cr-00076

66921-70-8 Hexacarbonylbis[μ-[(1,2,3,4,5-η:1,2,3,4,5-η)-3,4-diethyl-2,5-dihydro-2,5-dimethyl-1,2,5-thiadiborole]](iron)dimanganese, Mn-00127

66921-71-9 Tricarbonyl[(η^5-2,4-cyclopentadien-1-yl)iron]-[μ-[(1,2,3,4,5-η:1,2,3,4,5-η)-3,4-diethyl-2,5-dihydro-2,5-dimethyl-1,2,5-thiadiborole]]manganese, Mn-00109

66940-26-9 μ-Carbonyloctacarbonyldirhenium(2−); Di-K salt, *in* Re-00040

66964-10-1 Diacetyltetracarbonylmanganate(1−); Bis(triphenylphosphine)iminium salt, *in* Mn-00037

66979-94-0 (η^5-2,4-Cyclopentadien-1-yl)iodonitrosyl(η^3-2-propenyl)molybdenum, Mo-00019

67030-82-4 Dimethyltetrakis(*N*-dimethylamino)dimolybdenum, Mo-00044

67047-38-5 Tetracarbonyl(triphenylphosphine)manganate(1−); Bis(triphenylphosphine)iminium salt, *in* Mn-00124

67063-43-8 Tetrakis(η^5-2,4-cyclopentadien-1-yl)bis(2,2-dimethylpropyl)di-μ-hydrodizirconium, *in* Zr-00041

67063-52-9 Hexa-*tert*-butoxy-μ-carbonyldimolybdenum, Mo-00130

67108-86-5 Hydro(2-methylpropyl)bis[(1,2,3,4,5-η)-1,2,3,4,5-pentamethyl-2,4-cyclopentadien-1-yl)zirconium, Zr-00087

67158-14-9 Carbonyl(η^5-2,4-cyclopentadien-1-yl)-[(1,2,3,6,7-η)-4,5-dihydro-1-phenyl-1*H*-borepin]manganese, Mn-00113

67178-24-9 Tetrakis(η^3-2-propenyl)dirhenium, Re-00058

67202-46-4 ▷Tetracarbonyltungstate(4−); Tetra-Na salt, *in* W -00002

67202-62-4 ▷Tetracarbonylchromate(4−); Tetra-Na salt, *in* Cr-00002

67202-63-5 ▷Tetracarbonylmolybdate(4−); Tetra-Na salt, *in* Mo-00001

67204-23-3 Tetracarbonyl[1,2-ethanediylbis[diphenylphosphine]-*P*,*P'*]vanadate(1−); Tetraethylammonium salt, *in* V -00108

67204-49-3 Tetracarbonyl(triphenylphosphine)manganate(1−); K salt, *in* Mn-00124

67251-59-6 Tetracarbonyl(η^5-2,4-cyclopentadien-1-yl)molybdenum(1+); Tetrafluoroborate, *in* Mo-00022

67407-28-7 Tricarbonyl(η^5-2,4-cyclopentadien-1-yl)-(dimethylaminomethylene)vanadium, V -00044

67421-16-3 [μ(η:η^6-Phenyl)](tricarbonylchromium)lithium, Cr-00054

67421-95-8 (η^5-2,4-Cyclopentadien-1-yl)[μ_4-[methanolato(4−)-*C*:*C*:*C*:*O*]]-(nonacarbonyltricobalt)-(tetracarbonylcobalt)titanium, Ti-00140

67422-50-8 Tetrakis(η^5-2,4-cyclopentadien-1-yl)bis[μ_4-[methanolato(4−)-*C*:*C*:*C*:*O*]-bis(nonacarbonyltricobalt)-μ-oxodihafnium, Hf-00052

67422-52-0 Tetrakis(η^5-2,4-cyclopentadien-1-yl)bis[μ_4-[methanolato(4−)-*C*:*C*:*C*:*O*]-bis(nonacarbonyltricobalt)-μ-oxodizirconium, Zr-00113

67422-53-1 Bis(η^5-2,4-cyclopentadien-1-yl)bis[μ_4-[methanolato(4−)-*C*:*C*:*C*:*O*]]-bis(nonacarbonyltricobalt)dizirconium, Zr-00102

67422-55-3 Bis(η^5-2,4-cyclopentadien-1-yl)bis[μ_4-[methanolato(4−)-*C*:*C*:*C*:*O*]]-bis(nonacarbonyltricobalt)dihafnium, Hf-00045

67422-56-4 Chlorobis(η^5-2,4-cyclopentadien-1-yl)[μ_4-[methanolato(4−)-*C*:*C*:*C*:*O*]]-(nonacarbonyltricobalt)dihafnium, Hf-00026

67422-57-5 Chlorobis(η^5-2,4-cyclopentadien-1-yl)[μ_4-[methanolato(4−)-*C*:*C*:*C*:*O*]]-(nonacarbonyltricobalt)zirconium, Zr-00060

67422-84-8 Bis[(1,2,3,4,4*a*,8*a*-η)naphthalene]molybdenum, Mo-00114

67483-82-3 Bis(tetrahydrofuran)tris[(trimethylsilyl)-methyl]terbium, *in* Tb-00001

67483-83-4 Bis(tetrahydrofuran)tris[(trimethylsilyl)-methyl]erbium, *in* Er-00008

67483-85-6 Tetrakis[(trimethylsilyl)methyl]yttrate(1−); Li salt, tetrakis-THF complex, *in* Y -00010

67483-92-5 Tris[bis(triethylsilyl)methyl]chloroytterbate(1−); Li salt, tetrakis-THF complex, *in* Yb-00025

67483-97-0 [Bis(ethylthio)methylene]pentacarbonylchromium, Cr-00077

67506-86-9 Decamethylmanganocene, Mn-00119

67506-88-1 Dichlorobis[(1,2,3,4,5-η)-1,2,3,4,5-pentamethyl-2,4-cyclopentadien-1-yl]thorium, Th-00016

67506-89-2 Dichlorobis[(1,2,3,4,5-η)-1,2,3,4,5-pentamethyl-2,4-cyclopentadien-1-yl]uranium, U -00038

67506-90-5 Dimethylbis[(1,2,3,4,5-η)-1,2,3,4,5-pentamethyl-2,4-cyclopentadien-1-yl]thorium, Th-00020

67506-91-6 Chloro(methyl)bis[(1,2,3,4,5-η)-1,2,3,4,5-pentamethyl-2,4-cyclopentadien-1-yl]uranium, U -00042

67506-92-7 Di-μ-hydrodihydrotetrakis[(1,2,3,4,5-η)-1,2,3,4,5-pentamethyl-2,4-cyclopentadien-1-yl]dithorium, *in* Th-00017

67507-08-8 Bromotris(η^5-2,4-cyclopentadien-1-yl)uranium, U -00012

67507-18-0 Trichloro[(1,2,3,4,5-η)-1-methyl-2,4-cyclopentadien-1-yl]uranium; Bis-THF complex, *in* U -00003

67507-20-4 Trichloro(η^5-2,4-cyclopentadien-1-yl)-bis(triphenylphosphine oxide-*O*)uranium, U -00078

67507-24-8 μ-Carbonyltricarbonylbis(η^5-2,4-cyclopentadien-1-yl)[μ-[1,1'-(η^2:η^2-1,2-ethynediyl)bisbenzene]]dimolybdenum, Mo-00133

67552-61-8 Bis(cyclopentadienyl)methyloxoniobium, Nb-00020

67574-52-1 Bis[(η^6-phenyl)ferrocene]chromium, Cr-00173

67588-76-5 Di-μ-hydrodihydrotetrakis[(1,2,3,4,5-η)-1,2,3,4,5-pentamethyl-2,4-cyclopentadien-1-yl]diuranium, *in* U -00039

67605-92-9 Dimethylbis[(1,2,3,4,5-η)-1,2,3,4,5-pentamethyl-2,4-cyclopentadien-1-yl]uranium, U -00045

67624-22-0 Bis(tetrahydrofuran)tris[(trimethylsilyl)-methyl]ytterbium, *in* Yb-00010

67719-69-1 μ-Chlorobis(η^5-cyclopentadienyl)-(dimethylaluminum)-μ-methylenetitanium, Ti-00066

67771-71-5 Tetrakis(η^5-2,4-cyclopentadien-1-yl)bis[μ-[methylene(diphenylphosphinidenio)-methylidyne]]diuranium, U -00080

67891-23-0 (η^5-2,4-Cyclopentadien-1-yl)-tris(dimethylcarbamodithioato-*S*,*S'*)titanium, Ti-00077

68033-40-9 Dicarbonyl(η^4-1,3-cyclobutadiene)(η^5-2,4-cyclopentadien-1-yl)vanadium, V -00038
68079-61-8 [η-[(2,3-η:2,3-η)-2-Butyne]]-tetracarbonylbis(η^5-2,4-cyclopentadien-1-yl)dimolybdenum, Mo-00105
68088-94-8 Bis(η^6-benzene)niobium, Nb-00024
68182-14-9 Dichloro(η^5-2,4-cyclopentadien-1-yl)(phenylazo)titanium, Ti-00044
68185-45-5 Dicarbonyl(η^5-2,4-cyclopentadien-1-yl)-(hexacarbonyldicobalt)[μ_3-(phenylmethylidyne)]molybdenum, Mo-00111
68349-04-2 Dicarbonyl(η^5-2,4-cyclopentadien-1-yl)(η^3-6-methylene-2,4-cyclohexadien-1-yl)tungsten, W -00091
68520-37-6 Tetracarbonyl(di-μ-carbonylhexacarbonyldiferrate)manganate(1−); Bis(triphenylphosphine)iminium salt, *in* Mn-00072
68550-41-4 Tricarbonyl(η^5-2,4-cyclopentadien-1-yl)molybdate(1−); Li salt, *in* Mo-00014
68586-58-3 Hexacarbonylbis[(1,2,3,3*a*,7*a*-η)-1*H*-inden-1-yl]ditungsten, W -00138
68586-68-5 Bis(cyclopentadienyl)hydropropenetantalum, Ta-00021
68643-99-2 Tricarbonyl(η^5-2,4-cyclopentadien-1-yl)-(hydrazine-*N*)molybdenum(1+); Chloride, *in* Mo-00018
68688-11-9 Tricarbonyl(η^5-2,4-cyclopentadien-1-yl)vanadate(2−); Di-Na salt, *in* V -00015
68688-15-3 Tricarbonyl(η^5-2,4-cyclopentadien-1-yl)methylvanadate(1−); Bis(triphenylphosphine)iminium salt, *in* V -00021
68738-02-3 Tricarbonyl(η^5-2,4-cyclopentadien-1-yl)hydrovanadate(1−); Bis(triphenylphosphine)iminium salt, *in* V -00017
68795-84-6 Tetracarbonyl(dimethylcarbamodithioato-*S*,*S'*)rhenium, Re-00026
68830-04-6 Bromotetracarbonyl(ferroceniumylmethylidyne)tungsten, W -00099
68830-11-5 Pentacarbonyl(ethoxyferrocenylmethylene)molybdenum, Mo-00103
68830-12-6 Bromotetracarbonyl(ferrocenylmethylidyne)molybdenum; *trans-form*, *in* Mo-00085
68830-17-1 Bromotetracarbonyl(ferrocenylmethylidyne)chromium; *trans-form*, *in* Cr-00124
68866-90-0 *N*,*N*,*N*-Trimethylmethanaminium(1+) pentacarbonylmethyltungstate(1−), *in* W -00009
68868-08-6 Bis(acetonitrile)dicarbonyl(η^5-2,4-cyclopentadien-1-yl)molybdenum(1+); Tetraphenylborate, *in* Mo-00054
68868-91-7 Tetra-*tert*-butylytterbate(1−); Li salt, tri-THF complex, *in* Yb-00017
68868-92-8 Tetra-*tert*-butylsamarate(1−); Li salt, Tetra-THF complex, *in* Sm-00011
68878-72-8 Bis(η^5-2,4-cyclopentadien-1-yl)[2-[(dimethylamino)methyl]phenyl-*C*,*N*]scandium, Sc-00007
68893-45-8 Tetra-*tert*-butylerbate(1−); Li salt, tetra-THF complex, *in* Er-00013
68927-82-2 Pentacarbonyl(cyano-*C*)tungstate(1−), W -00012
68927-89-9 (η^6-Benzene)tricarbonylrhenium(1+), Re-00037
68927-90-2 (η^6-Benzene)tricarbonylrhenium(1+); Hexafluorophosphate, *in* Re-00037
68963-85-9 Bis[(1,2,3,4,5-η)-1,2,3,4,5-pentamethyl-2,4-cyclopentadien-1-yl]bis[(trimethylsilyl)-methyl]uranium, U -00068
68963-89-3 Chloro[(1,2,3,4,5-η)-1,2,3,4,5-pentamethyl-2,4-cyclopentadien-1-yl][1-(trimethylsilyl)-ethenolato]thorium, Th-00025
69005-92-1 Chlorotris(η-methylcyclopentadienyl)zirconium, Zr-00051
69005-93-2 Chlorotris(η-cyclopentadienyl)zirconium, Zr-00039
69021-32-5 Bis[μ-(2-butene-2,3-diolato(2-)-*O*,*O'*]-tetrakis[(1,2,3,4,5-η)-1,2,3,4,5-pentamethyl-2,4-cyclopentadien-1-yl]dithorium, Th-00038
69030-40-6 Iodotris(η^5-2,4-cyclopentadien-1-yl)uranium, U -00014
69040-88-6 Bis[(1,2,3,4,5-η)-1,2,3,4,5-pentamethyl-2,4-cyclopentadien-1-yl]bis[(trimethylsilyl)-methyl]thorium, Th-00031
69109-78-0 Dibromobis(η^5-2,4-cyclopentadien-1-yl)vanadium, V -00027
69120-57-6 (η^5-2,4-Cyclopentadien-1-yl)-iodonitrosyl(triphenylphosphine)manganese, Mn-00130
69276-81-9 Tetracarbonyltetrakis(cyclopentadienyl)tetravanadium, V -00098
69289-49-2 [Bis(η^5-2,4-cyclopentadien-1-yl)hydro[μ-[methanolato(3−)-*C*:*O*]]bis[(1,2,3,4,5-η)-1,2,3,4,5-pentamethyl-2,4-cyclopentadien-1-yl]zirconium, W -00153
69302-77-8 (η^2-Benzyne)dimethyl[(1,2,3,4,5-η)-1,2,3,4,5-pentamethyl-2,4-cyclopentadien-1-yl]tantalum, Ta-00033
69377-30-6 Tris(η^5-2,4-cyclopentadien-1-yl)(thiocyanato-*N*)-uranium; MeCN adduct, *in* U -00017
69439-84-5 Cyclopentadienyltrinitrosylvanadium(1+); Hexafluorophosphate, *in* V -00006
69517-43-7 Methyltris[(hexamethyldisilyl)amino]thorium, Th-00013
69517-44-8 Methyltris[(hexamethyldisilyl)amido]uranium, U -00032
69526-48-3 Tris(η^5-2,4-cyclopentadien-1-yl)(thiocyanato-*N*)-uranium, U -00017
69552-43-8 Bis(2,2-dimethylpropylidene)(2,4,6-trimethylphenyl)bis(trimethylphosphine)tantalum, Ta-00038
69593-26-6 Carbonylchlorobis(η^5-2,4-cyclopentadien-1-yl)tantalum, Ta-00013
69593-37-9 Tris(diethyl ether)methylphenylzirconium, Zr-00059
69621-06-3 Carbonyl(η^5-2,4-cyclopentadien-1-yl)-formyl(nitrosyl)rhenium, Re-00025
69662-01-7 Bis(benzenethiolato)tetrakis(η^5-2,4-cyclopentadien-1-yl)-μ-oxodizirconium, Zr-00106
69793-83-5 Bis(η^5-2,4-cyclopentadien-1-yl)bis(2,2-dimethylpropyl)zirconium, Zr-00064
70083-62-4 Trichloro(2,2-dimethylpropylidene)-bis(trimethylphosphine)tantalum, Ta-00015
70083-76-0 Tetracarbonylbis(triphenylphosphine)manganese(1+); Hexafluorophosphate, *in* Mn-00140
70197-13-6 Methyltrioxorhenium, Re-00001
70388-23-7 Tetrakis[(trimethylsilyl)methyl]luteate(1−); Li salt, tetrakis(tetrahydrofuranate), *in* Lu-00015
70445-59-9 Fluorotris(η^5-2,4-cyclopentadien-1-yl)neptunium, Np-00002
70585-10-3 Tricarbonyl(2,3,4,5-η:9,10-η-9-methylenebicyclo[4.2.1]nona-2,4,7-triene)chromium, Cr-00113
70605-18-4 Hexacarbonylbis[(1,2,3,4,5-η)-1,2,3,4,5-pentamethyl-2,4-cyclopentadien-1-yl]dichromium, Cr-00165
70616-56-7 Chlorobis(η^5-2,4-cyclopentadien-1-yl)methylniobium, Nb-00019
70616-59-0 Bis(cyclopentadienyl)methyl(triethylphosphine)niobium, Nb-00034
70634-77-4 Tetracarbonylbis[(1,2,3,4,5-η)-1,2,3,4,5-pentamethyl-2,4-cyclopentadien-1-yl]ditungsten, W -00143
70675-84-2 Bis(η^5-2,4-cyclopentadien-1-yl)-hydro(triphenylgermyl)tungsten, W -00148
70756-51-3 Trichloro[(1,2,3,4,5-η)-1,2,3,4,5-pentamethyl-2,4-cyclopentadien-1-yl]titanium, Ti-00040
70844-68-7 Tetrakis(η^5-2,4-cyclopentadien-1-yl)di-μ-methyldithulium, *in* Tm-00003
70969-30-1 Dimethylbis[bis(trimethylsilyl)amido]zirconium, Zr-00038
70969-31-2 Diethylbis[bis(trimethylsilyl)amido]hafnium, Hf-00022
71000-84-5 Dimethylbis[bis(trimethylsilyl)amido]hafnium, Hf-00019
71110-90-2 Tetrakis[μ-(2-methoxyphenyl-*C*:*O*)]bis(2-methoxyphenyl-*C*,*O*)dirhenium, Re-00090
71163-80-9 Decamethylmanganocenium, Mn-00120
71163-81-0 Decamethylmanganocenium; Hexafluorophosphate, *in* Mn-00120
71191-42-9 Chlorobis(η^5-2,4-cyclopentadien-1-yl)oxovanadium, V -00028

79194-27-7 [Bis(trimethylsilyl)methyl]bis(η^5-2,4-cyclopentadien-1-yl)phenylzirconium, Zr-00082

79255-11-1 [(Acetonitrile)decacarbonyl-μ-hydrotriosmium]pentacarbonylrhenium, Re-00074

79292-30-1 Bis(η^5-2,4-cyclopentadien-1-yl)-hydro(tetrahydrofuran)lutetium, *in* Lu-00005

79301-20-5 Chloro(methyl)bis[(1,2,3,4,5-η)-1,2,3,4,5-pentamethyl-2,4-cyclopentadien-1-yl]thorium, Th-00018

79372-14-8 Bis[(1,2,3,4,5-η)-1,2,3,4,5-pentamethyl-2,4-cyclopentadien-1-yl]samarium; Bis-THF complex, *in* Sm-00012

79391-56-3 Tricarbonyl(η^6-octamethylnaphthalene)chromium, Cr-00154

79391-57-4 Tricarbonyl(η^6-octamethylnaphthalene)molybdenum, Mo-00118

79391-58-5 Tricarbonyl(η^6-octamethylnaphthalene)tungsten, W -00134

79503-47-2 Bis(η^5-2,4-cyclopentadien-1-yl)(2,2-dimethyl-1-oxopropyl-*C,O*)lutetium, Lu-00013

79533-62-3 Tris(η^5-2,4-cyclopentadien-1-yl)lanthanum; THF adduct, *in* La-00004

79643-31-5 Pentacarbonyl[(2,3-diphenyl-2-cyclopropen-1-yl)]-carbonylrhenium, Re-00078

79643-33-7 Pentacarbonyl(1,2-diphenyl-2-cyclopropen-1-yl)rhenium, Re-00077

79716-32-8 Tetrakis(η^5-2,4-cyclopentadien-1-yl)[μ-(2,2,7,7-tetramethyl-3,4,5,6-octanetetrone-O^3,O^5:O^4,O^6)]dilutetium, Lu-00024

79739-15-4 Bis[μ-[(1-η:2,3-η)-2-propenyl)]bis(η^3-2-propenyl)dimolybdenum, Mo-00072

79847-76-0 Bis[(1,2,3,4,5-η)-1,2,3,4,5-pentamethyl-2,4-cyclopentadien-1-yl]diphenylzirconium, Zr-00107

79932-08-4 Hydrobis[(1,2,3,4,5-η)-1,2,3,4,5-pentamethyl-2,4-cyclopentadien-1-yl](2,2,4,4-tetramethyl-3-pentanolato)thorium, Th-00032

80049-78-1 Decacarbonyl(μ-diphenylphosphido)ditungstate(1−), W -00136

80145-92-2 Methylbis[(1,2,3,4,5-η)-1,2,3,4,5-pentamethyl-2,4-cyclopentadien-1-yl]lutetium; Et_2O complex, *in* Lu-00019

80287-59-8 Dicyclopentenouranocene, U -00044

80410-05-5 Tris(η^5-2,4-cyclopentadien-1-yl)methylthorium, Th-00010

80410-09-9 Tris(η^5-2,4-cyclopentadien-1-yl)(1-methylpropyl)uranium, U -00030

80410-12-4 Iodotris(η^5-2,4-cyclopentadien-1-yl)thorium, Th-00007

80432-28-6 Tetracarbonyl[(1,2,3,4,5-η)-1,2,3,4,5-pentamethyl-2,4-cyclopentadien-1-yl]niobium, Nb-00030

80432-35-5 Tetrachloro[(1,2,3,4,5-η)-1,2,3,4,5-pentamethyl-2,4-cyclopentadien-1-yl]niobium, Nb-00016

80502-52-9 Tris[(1,2,3,4,5-η)-2,4-dimethyl-2,4-pentadienyl]neodymium, Nd-00009

80602-96-6 [1,2-Ethanediylbis(dimethylphosphine)-*P,P′*]-hydrobis[(1,2,3,4,5-η)-1,2,3,4,5-pentamethyl-2,4-cyclopentadien-1-yl]uranium, U -00061

80642-67-7 *tert*-Butylbis(methylcyclopentadienyl)yttrium; THF complex, *in* Y -00008

80658-44-2 Di-μ-hydrotetrakis[(1,2,3,4,5-η)-1-methyl-2,4-cyclopentadien-1-yl]bis(tetrahydrofuran)diyttrium, Y -00015

80679-14-7 Bis(η^5-2,4-cyclopentadien-1-yl)(η^2-diethyldiazomalonate-*N,N′*)vanadium, V -00080

80679-48-7 Bis(acetonitrile)bis(η^5-pentamethyl-2,4-cyclopentadien-1-yl)vanadium(1+); Hexafluorophosphate, *in* V -00101

80679-49-8 Dicarbonylbis[(1,2,3,4,5-η)-1,2,3,4,5-pentamethyl-2,4-cyclopentadien-1-yl]vanadium(1+); Hexafluorophosphate, *in* V -00093

80737-39-9 Bis(η^5-2,4-cyclopentadien-1-yl)(η^2-*C,O*-formaldehyde)vanadium, V -00045

80758-23-2 [μ-(Carbonyl-*C,O*)]bis[(1,2,3,4,5-η)-1,2,3,4,5-pentamethyl-2,4-cyclopentadien-1-yl]-(tricarbonylcobalt)ytterbium; THF adduct, *in* Yb-00026

80795-37-5 μ-Chlorohexakis(η^5-2,4-cyclopentadien-1-yl)di-μ-hydro-μ_3-hydrotrierbate(1−); Li salt, tetra-THF complex, *in* Er-00017

80878-91-7 Tetrakis[μ_3-(carbonyl-*C,C,O*)]-(heptacarbonyltriiron)tetrakis[(1,2,3,4,5-η)-1,2,3,4,5-pentamethyl-2,4-cyclopentadien-1-yl]diytterbium, Yb-00036

80952-45-0 Dicarbonyl(η^5-2,4-cyclopentadien-1-yl)dihydrorhenium, Re-00028

80952-46-1 Dicarbonyl(η^5-2,4-cyclopentadien-1-yl)dimethylrhenium, Re-00039

81033-04-7 Chlorotris(η^5-2,4-cyclopentadien-1-yl)plutonium, Pu-00001

81276-69-9 Bis[(1,2,3,4,5-η)-1,2,3,4,5-pentamethyl-2,4-cyclopentadien-1-yl]-diethylcarbamodithioato-*S,S′*-ytterbium, Yb-00027

81276-72-4 Bis[(1,2,3,4,5-η)-1,2,3,4,5-pentamethyl-2,4-cyclopentadienyl]bis(pyridine)ytterbium, Yb-00031

81277-17-0 Dichlorobis[(1,2,3,4,5-η)-1,2,3,4,5-pentamethyl-2,4-cyclopentadien-1-yl](1*H*-pyrazole-N^2)uranium, U -00050

81277-18-1 Chlorobis[(1,2,3,4,5-η)-1,2,3,4,5-pentamethyl-2,4-cyclopentadien-1-yl)(1*H*-pyrazolato-N^1,N^2)uranium, U -00049

81293-74-5 Bis[(1,2,3,4,5-η)-1,2,3,4,5-pentamethyl-2,4-cyclopentadien-1-yl]bis(1*H*-pyrazolato-N^1,N^2)uranium, U -00059

81388-97-8 Tris(η^5-2,4-cyclopentadien-1-yl)-[(methyldiphenylphosphoranylidene)acetyl]-uranium, U -00070

81496-90-4 (Dibenzyldimethylstannane)bis[tricarbonylchromium], Cr-00158

81507-33-7 Dichlorobis[(1,2,3,4,5-η)-1,3-bis(trimethylsilyl)-2,4-cyclopentadien-1-yl]neodymate(1−); Li salt, Bis-THF complex, *in* Nd-00010

81507-53-1 Bis[1,3-bis(trimethylsilyl)cyclopentadienyl]chloroscandium; Dimer, *in* Sc-00010

81507-56-4 Bis[1,3-bis(trimethylsilyl)cyclopentadienyl)chloropraesodymium, Pr-00006

81536-99-4 Tetrakis[(1,2,3,4,5-η)-1,3-bis(trimethylsilylmethyl)-2,4-cyclopentadien-1-yl]di-μ-chlorodiytterbium, *in* Yb-00029

81703-54-0 Dichloro(η^5-2,4-cyclopentadien-1-yl)erbium; Tris-THF adduct, *in* Er-00001

81831-31-4 Amminepentacarbonylvanadate(1−); Na salt, *in* V -00001

81831-32-5 Amminepentacarbonylvanadate(1−); Bis(triphenylphosphine)iminium salt, *in* V -00001

82293-70-7 Butylbis(η^5-2,4-cyclopentadien-1-yl)lutetium; THF complex, *in* Lu-00009

82293-72-9 Bis(η^5-2,4-cyclopentadien-1-yl)-[(trimethylsilyl)methyl]ytterbium; THF complex, *in* Yb-00011

82293-73-0 Bis(η^5-2,4-cyclopentadien-1-yl)-[(trimethylsilyl)methyl]samarium, Sm-00008

82311-89-5 Bis(η^5-2,4-cyclopentadien-1-yl)-[(trimethylsilyl)methyl]erbium; THF complex, *in* Er-00010

82511-71-5 [(1,2,3,4,5-η)-1,2,3,4,5-Pentamethyl-2,4-cyclopentadien-1-yl]trichlorothorium; Bis-THF complex, *in* Th-00002

82511-72-6 [(1,2,3,4,5-η)-1,2,3,4,5-Pentamethyl-2,4-cyclopentadienyl]trichlorouranium; Bis-THF complex, *in* U -00008

82511-73-7 Tribenzyl(pentamethylcyclopentadienyl)thorium, Th-00036

82511-74-8 Tribenzyl(pentamethylcyclopentadienyl)uranium, U -00072

82963-77-7 Bis(η^4-1,3-butadiene)(trimethyl phosphite)manganese, Mn-00070

82963-81-3 Bis(η^4-1,3-butadiene)(trimethyl phosphite)manganate(1−); Tetrabutylammonium salt, *in* Mn-00069

83243-22-5 μ-Chloro-μ_3-chlorotetrakis(η^5-2,4-cyclopentadien-1-yl)trihydro[1,1′-oxybis[ethane]]aluminumdiyttrium, Y -00014

83350-15-6 μ-Carbonylpentacarbonyl(η^5-2,4-cyclopentadienyl)iron-μ-methylenemanganese, Mn-00079

83378-77-2 Dicarbonylnitrosyl(η^4-tetraphenylcyclobutadiene)manganese, Mn-00136

83463-59-6 Hydridobis[(1,2,3,4,5-η)-1,2,3,4,5-pentamethyl-2,4-cyclopentadien-1-yl]lutetium, Lu-00018

84254-54-6 [Bis(1,2,3,4,5-η)-1,2,3,4,5-pentamethyl-2,4-cyclopentadien-1-yl]-[bis(dimethylphosphino)methane]chloroytterbium, Yb-00028

84474-41-9 Di-μ-carbonyl[μ_4-(η^2-carbonyl-*C:C:C*)][μ-carbonylbis[(1,2,3,4,5-η)-1,2,3,4,5-pentamethyl-2,4-cyclopentadien-1-yl]-dicobalt]bis(η^5-2,4-cyclopentadien-1-yl)dimolybdenum, Mo-00141

84582-80-9 (η^8-1,3,5,7-Cyclooctatetraene)-(trimethylsilylmethyl)bis(tetrahydrofuran)lutetium, *in* Lu-00007

84623-27-8 Chlorobis[(1,2,3,4,5-η)-1-methyl-2,4-cyclopentadien-1-yl]samarium; THF adduct, *in* Sm-00006

84642-15-9 Bis[μ-(3,3-dimethyl-1-butyn-1-yl)]-tetrakis(1,2,3,4,5-η-1-methyl-2,4-cyclopentadien-1-yl)disamarium, Sm-00018

84850-68-0 Octacarbonyl-μ-dihydrodimolybdate(2−); Tetraethylammonium salt, *in* Mo-00009

84850-72-6 Octacarbonyl-μ-dihydrodichromate(2−); Bis-(tetraethylammonium) salt, *in* Cr-00037

84895-68-1 Tris(2-methyl-1-propenyl)[(1,2,3,4,5-η)-1,2,3,4,5-pentamethyl-2,4-cyclopentadien-1-yl]uranium, U -00046

84895-69-2 [(1,2,3,4,5-η)-1,2,3,4,5-Pentamethyl-2,4-cyclopentadien-1-yl)]tris(2-propenyl)uranium, U -00031

85962-87-4 Methylbis[(1,2,3,4,5-η)-1,2,3,4,5-pentamethyl-2,4-cyclopentadien-1-yl]lutetium, Lu-00019

86528-31-6 Bis-μ-*tert*-butylformimidoyltetrakis[(1,2,3,4,5-η)-1-methyl-2,4-cyclopentadien-1-yl]diyttrium, Y -00016

86543-76-2 Chloro[1-hydroxy-4,4-dimethyl-1-(trimethylphosphoranylidene)-2-pentanonato-*O,O′*]bis[(1,2,3,4,5-η)-1,2,3,4,5-pentamethyl-2,4-cyclopentadien-1-yl]thorium, Th-00034

87136-56-9 Methylbis[(1,2,3,4,5-η)-1,2,3,4,5-pentamethyl-2,4-cyclopentadien-1-yl]yttrium, Y -00012

87145-50-4 Dicarbonyl(η^5-2,4-cyclopentadien-1-yl)tetramethylenerhenium, Re-00045

87145-51-5 Dicarbonyl(η^5-2,4-cyclopentadien-1-yl)hydrorhenate(1−); K salt, *in* Re-00027

87145-52-6 Dicarbonyl(η^5-2,4-cyclopentadien-1-yl)hydrorhenate(1−); Tetraethylammonium salt, *in* Re-00027

87450-50-8 Bis[1,2-ethanediylbis[dimethylphosphine]]-*P,P′*]-dimethylmanganese, Mn-00095